PEOPLING THE LANDSCAPE OF ÇATALHÖYÜK

REPORTS FROM THE 2009–2017 SEASONS

Edited by
Ian Hodder

BRITISH INSTITUTE AT ANKARA
Monograph 53
Çatalhöyük Research Project Series Volume 13
2021

Published by
British Institute at Ankara
10 Carlton House Terrace, London SW1Y 5AH
www.biaa.ac.uk

This book is available from
Oxbow Books
10 Hythe Bridge Street, Oxford, OX1 2EW
www.oxbowbooks.com

ISBN 978 1 912090 78 5

Typeset by Abby Robinson
Printed by Short Run Press Ltd, Exeter

Contents

Online supplementary material

Supplementary material available online (https://doi.org/10.18866/BIAA/e-13) comprises additional elements for chapters 5, 8, 12 and 13, and the entirety of chapter 19. Colour versions of selected figures are available for chapters 2, 4, 11, 14, 15 and 16.

Contributors

N. Ezgi Altınışık	Human_G Group, Department of Anthropology, Hacettepe University
Evilena Anastasiou	Department of Archaeology, University of Cambridge
Eleni Asouti	Department of Archaeology, Classics and Egyptology, University of Liverpool
Çiğdem Atakuman	Graduate School of Social Sciences, Middle East Technical University (METU)
Gianna Ayala	Department of Archaeology, University of Sheffield
Julia Best	School of History, Archaeology and Religion, Cardiff University
Barbara Betz	Department of Anthropology, Ohio State University
Amy Bogaard	School of Archaeology, University of Oxford
Michael Charles	School of Archaeology, University of Oxford
Maciej Chyleński	Institute of Human Biology and Evolution, Faculty of Biology, Adam Mickiewicz University in Poznań
Nihan Dilşad Dağtaş	Department of Biological Sciences, Middle East Technical University (METU)
Beatrice Demarchi	Department of Life Sciences and Systems Biology, University of Turin
G. Arzu Demirergi	Department of Anthropology, SUNY Stony Brook
Ayça Doğu	Department of Biological Sciences, Middle East Technical University (METU)
Chris Doherty	School of Archaeology, Oxford University
Irene Dori	UMR-5199 De la Préhistoire à l'Actuel: Culture, Environnement, et Anthropologie (PACEA), University of Bordeaux
Claudia Engel	Department of Anthropology and Stanford Libraries, Stanford University
Yılmaz Selim Erdal	Human_G Group, Department of Anthropology, Hacettepe University
Jane Evans	British Geological Survey, Nottingham
Michelle Feider	Department of Archaeology and Anthropology, Bournemouth University
Evrim Fer	Department of Genetics, University of Arizona
Dragana Filipović	Institute for Balkan Studies, Serbia
Dorian Q. Fuller	Institute of Archaeology, University College London
Michelle Gamble	Austrian Archaeological Institute, Vienna
Aroa García-Suárez	Department of Archaeology, University of Reading
Evan Garofalo	College of Medicine, University of Arizona
Hasan Can Gemici	Graduate School of Social Sciences, Middle East Technical University (METU)
Ayshin Ghalichi	Department of Biological Sciences, Middle East Technical University (METU)
Bonnie Glencross	Department of Archaeology and Heritage Studies, Wilfrid Laurier University
Anders Götherström	Department of Archaeology and Classical Studies, Stockholm University
Lara González Carretero	Institute of Archaeology, University College London
Laura Green	School of Archaeology, University of Oxford
Scott D. Haddow	Department of Cross-Cultural and Regional Studies, University of Copenhagen
Sheila Hamilton-Dyer	Department of Archaeology and Anthropology, Bournemouth University
Ian Hodder	Department of Anthropology, Stanford University
Ulaş Işıldak	Department of Biological Sciences, Middle East Technical University (METU)
Emma Jenkins	Institute for the Modelling of Socio-Environmental Transitions (IMSET), Bournemouth University
Anna Juras	Institute of Human Biology and Evolution, Faculty of Biology, Adam Mickiewicz University in Poznań
Ceren Kabukcu	Department of Archaeology, Classics and Egyptology, University of Liverpool
Damla Kaptan	Department of Biological Sciences, Middle East Technical University (METU)
Cansu Karamurat	Graduate School of Social Sciences, Middle East Technical University (METU)

Gülşah Merve Kılınç	Department of Bioinformatics, Graduate School of Health Sciences, Hacettepe University
Christopher J. Knüsel	UMR-5199 De la Préhistoire à l'Actuel: Culture, Environnement, et Anthropologie (PACEA), University of Bordeaux
Dilek Koptekin	Department of Health Informatics, Middle East Technical University (METU)
Vendela Kempe Lagerholm	Department of Archaeology and Classical Studies, Stockholm University
Angela Lamb	British Geological Survey, Nottingham
Clark S. Larsen	Department of Anthropology, Ohio State University
Marissa Ledger	Department of Archaeology, University of Cambridge
Melanie Leng	British Geological Survey of Environmental Science Centre, Nottingham
Jerry M. Lloyd	Department of Geography, Durham University
Joseph Losh	Department of Archaeology, Classics and Egyptology, University of Liverpool
Marco Madella	Department of Humanities, Universitat Pompeu Fabra Barcelona
Igor Mapelli	Department of Biological Sciences, Middle East Technical University (METU)
Arkadiusz Marciniak	Faculty of Archaeology, Adam Mickiewicz University
Wendy Matthews	Department of Archaeology, University of Reading
Camilla Mazzucato	Department of Anthropology, Stanford University
Marco Milella	Department of Physical Anthropology, Institute of Forensic Medicine, University of Bern
Piers Mitchell	Department of Archaeology, University of Cambridge
Rachel Hodara Nelson	Haleakalā National Park
Carla Lancelotti	Department of Humanities, Universitat Pompeu Fabra Barcelona
Javier Lara-Recuero	Department of Humanities, Universitat Pompeu Fabra Barcelona
Jacqui A. Mulville	School of History, Archaeology and Religion, Cardiff University
Füsun Özer	Human_G Group, Department of Anthropology, Hacettepe University
Jessica Pearson	Department of Archaeology, Classics and Egyptology, University of Liverpool
Marin A. Pilloud	Department of Anthropology, University of Nevada, Reno
Marta Portillo	Department of Archaeology and Anthropology, Institució Milà i Fontanals, Spanish National Research Council (CSIC) Barcelona
Samantha Presslee	Department of Archaeology, York University
Tina Roushannafas	School of Archaeology, University of Oxford
Christopher B. Ruff	Center for Functional Anatomy and Evolution, Johns Hopkins University
Joshua W. Sadvari	University Libraries, Ohio State University
Carlos G. Santiago-Marrero	Department of Humanities, Universitat Pompeu Fabra Barcelona
Eline M.J. Schotsmans	UMR-5199 De la Préhistoire à l'Actuel: Culture, Environnement, et Anthropologie (PACEA), University of Bordeaux
Mehmet Somel	Department of Biological Sciences, Middle East Technical University (METU)
Elizabeth Stroud	School of Archaeology, University of Oxford
Elif Sürer	Department of Modeling and Simulation, Graduate School of Informatics, Middle East Technical University (METU)
Belinda Tibbetts	Department of Archaeology, Exeter University
İnci Togan	Department of Biological Sciences, Middle East Technical University (METU)
Katheryn C. Twiss	Department of Anthropology, SUNY Stony Brook
Petra Vaiglova	Department of Anthropology, Washington University in St. Louis
Wim Van Neer	Laboratory of Biodiversity and Evolutionary Genomics, Katholieke Universiteit Leuven
Milena Vasić	Independent researcher, Berlin

Kıvılcım Başak Vural	Department of Biological Sciences, Middle East Technical University (METU)
John Wainwright	Department of Geography, Durham University
Joanna R. Walker	Department of Archaeology, University of Cambridge
Jesse Wolfhagen	Department of Anthropology, SUNY Stony Brook
Wim Wouters	Department of Zooarchaeology, Royal Belgian Institute of Natural Sciences
Rena Veropoulidou	Wiener Laboratory for Archaeological Science, American School of Classical Studies at Athens, Hellenic Ministry of Culture and Sports
Reyhan Yaka	Department of Biological Sciences, Middle East Technical University (METU)
Sevgi Yorulmaz	Department of Biological Sciences, Middle East Technical University (METU)

List of figures

List of tables

Chapter 17

Acknowledgements

This publication of four volumes (Volumes 12, 13, 14 and 15 in the British Institute at Ankara Monograph Series) would not have been possible without the help of a large number of individuals, institutions and sponsors. We have attempted to list everyone who assisted the Çatalhöyük Research Project during the period of excavation and post-excavation work (2009–2018) covered in these volumes and apologies are extended to anyone who has been inadvertently overlooked.

Funding for the field research was provided by a wide variety of corporate and academic bodies. The main corporate sponsors were Boeing, Yapı Kredi Bankası, Shell and Koçtaş. I am particularly grateful to them for providing support over most of the 25-year period of the project. They sustained the project through crises and celebrations and provided incalculable scientific, cultural, social and economic benefits to many. The John Templeton Foundation also provided key support for the period covered by these volumes, and I am particularly grateful for the advice and help provided by Paul Wason. During this period of research funding was also provided by the following: British Institute at Ankara, Global Heritage Fund, Foundation for Polish Science, Free University Berlin, Hedef Alliance, Humboldt Foundation, Imitatio (Thiel Foundation), Kaplan Foundation, Konya Çimento, Konya Şeker, National Geographic Society, National Science Foundation, Polish Heritage Council, Polish Ministry of Science and Higher Education, Polish National Science Center, Stanford Archaeology Center, Stanford University, SUNY Buffalo, TAV, Turkish Cultural Foundation, University College London, University of Gdansk, University of Poznan, US Embassy in Ankara.

The project worked in Turkey with a permit from the Ministry of Culture and Tourism, General-Directorate of Cultural Heritage and Museums. Over the period covered by these publications much support and advice was given by the department and Director Generals, through their representatives on site (bakanlık temsilcileri). The project worked under the auspices of the British Institute at Ankara and I would like to thank in particular Lutgarde Vandeput, Gülgün Girdivan, Tamar Hodos, Stephen Mitchell and Shahina Farid. Additionally I am grateful for the support of the BIAA committee members and for the assistance of the Ambassadors and staff at the Turkish Embassy and consulates in London, Washington and Los Angeles and the British Ambassadors in Ankara. The patrons of the project are Professor Lord Renfrew of Kaimsthorn and Sir David Attenborough.

In the region, help and support were provided at many levels, in particular by the Konya Vali, the Konya Koruma Kurulu, the Cultural Director, the Konya Museums Director Yusuf Benli and his staff, officers at the Emniyet, our bank managers at Konya Yapı Kredi, the Çumra Kaymakam and Belediye Başkan. We would also like to extend our thanks to the Jandarma commitants, and the managers of the Dedeman Hotel in Konya, and Asim Kaplan from Karavan. Since 2009 the project managers have been Shahina Farid, Banu Aydınoğlugil, Yıldız Dirmit and Bilge Küçükdoğan. The latter in particular played a major role in extracting us from the site in 2016 and in overseeing the winding up of the project at Stanford and in Turkey. I am also forever grateful to the Assistant Director Serap Özdöl for her guidance and contributions to many aspects of the project over many years.

We owe an enormous debt to those who managed the project on site, particularly the camp manager Levent Özer whose wise advice steered the project through many trials and successes. Our guards at the site were our close companions over the years: Ibrahim Eken and Mustafa and Hasan Tokyağsun. These volumes are dedicated to the memory of Ibrahim whose long-term devotion to the project and gentle manner warmed the hearts of all who interacted with him.

From the local village of Küçükköy we would like to thank the people and their mukhtar. Those who worked at the site and contributed directly to the project in various ways are included in the list of team members that follows.

Doğu Furkan ACARER, Donovan ADAMS, Sabrina AGARWAL, Rifat AHSAN, Sam AINSWORTH, Atiye AKBULUT, Bünyamin AKBULUT, Hanafi AKCAN, Hasan AKÇAY, Enver AKGÜN, Kiraz AKOĞLU, Mustafa AKYURT, Moussab ALBESSO, Sophie ALCOCK, Richard ALLEN, Thomas ALLEN, Mehmet ALTINAY, Emma ANDERSON, Jennie ANDERSON, Veysel APAYDIN, Renata ARAUJO, Theodore ARNOLD-FORSTER, Mehmet ARSLAN, Numan ARSLAN, Monique ARNTZ, Althea ASARO, Gemma ASHBURY, Eleni ASOUTI, Mert ATALAR, Sonya ATALAY, Soner ATEŞOĞULLARI, Christopher ATKINSON, Quentin ATKINSON, Deanna AUBERT, Jeffrey AVISS, Gianna AYALA, Fahri AYÇİN, Banu AYDINOĞLUGİL, İnan AYDOĞAN, Melike AYHAN, Sema BAĞCI, Mustafa BAHÇECİ, Jack BAIGENT, Roseleen BAINS, Daniella BAR-YOSEF MAYER, Marek Zbigniew BARAŃSKI,

Judit BARASTEGUI, Alexandra BARMETTLER, Daniela A. BARRANTES, Marta BARTKOWIAK, Isabel BARTLEY, Célia BASSET, Rachel BASSINGER, Purnur Ece BAŞ, Emmeline BATCHELOR, Erin BAXTER, Alexandra BAYLISS, Umut BAYRAM, Tolga BAYRAM, Carlos BAZUA, Joel BEATH, Menna BELL, Cristina BELMONTE SANTISTEBAN, Brenda BENAVIDES, Lise BENDER JØRGENSEN, Lucy BENNISON CHAPMAN, Åsa BERGGREN, Johanna M. BERGKVIST, Mary BERMAN, Julia BEST, Barbara BETZ, Elisa BIANCI-FIORI, Peter BIEHL, Patrycja BIELSKA, Rachel BINGHAM, Tom BIRCH, İsa BİLGİÇ, Serdar BİLİŞ, Stephanie BLACK, Emmy BOCAEGE, Amy BOGAARD, Nikita BOGDANOV, Patrick BOLL, Jennie BORGSTROM, Sezgin BOŞLAMAZ, Hannah J. BOWDEN, Garrett BOYD, Mathew BOYD, Başak BOZ, Ahmet BOZGEYIK, Malwina Ewa BRACH-MANSKA, Henry BRADFORD, Jacob BRADY, Maxime BRAMI, Matthew BRITTEN, Kelly BROWN, Nicholas BROWN, Hallvard BRUVOLL, Laura BUCCIERI, Eniko BUDAK, Mikolaj BUDNER, Bayram BULUT, Narcis BURGUES, Oliver BURTON, Gesualdo BUSACCA, Emine BÜLÜÇ, İsmail BÜLÜÇ, Numan BÜLÜÇ, Jennifer BYRNS, Agnieszka BYSTRON, Katarzyna BZDUCH, Tiffany CAIN, Kelly CALDWELL, Stefano CAMPANA, Erica CAMURRI, Gözde CAN, Frank CARPENTIER, Christopher CARLTON, Robert Bergman CARTER, Tristan CARTER, Julie CASSIDY, Gianluca CATANZARITI, Rebecca CESSFORD, Merve CEYLAN, Benjamin CHAN, Ian CHANNELL, Michael CHARLES, Jessica CHATBORN, Claire CHRISTENSEN, Kimberly CHRISTENSON, Angeliki CHRYSANTHI, Maciej CHYLENSKI, Piotr CIESIELSKI, Marguerite CLARKE, Christopher CLEERE, Julia CLINE, Andrew COCHRANE, Jon COGDALE, Alana COLBERT, Tara COPPLESTONE, Grant COX, Michelle CREPEAU, Kyle CROSSET, Caitlin L. CURTIS, Cassy CUTULLE, Lech CZERNIAK, Agata CZESZEWSKA, Duygu ÇAMURCUOĞLU, Hatice ÇELİK, Lokman ÇELIK, Mahmut ÇELİK, Mustafa ÇEŞŞUR, Elif S. ÇIPLAK, Leyla E. ÇIPLAK, Mehmet ÇIRAK, Davide D'ERRICO, Nihan Dilşad DAĞTAŞ, Nevio DANELON, Julie DAUJAT, Anna DAVENPORT, Antonia DAVIDOVIC WALTHER, Neil DAVIES, Tudur DAVIES, Danielle DE CARLE, Alysha DE SOUZA, Funda DEĞER, Nicolo DELL'UNTO, Mateusz DEMBOWIAK, Marvin DEMICOLI, Burcu DEMİR, Arzu DEMİRERGİ, Işıl DEMİRTAŞ, Meghan DENNIS, Lindsay DER, Emma DEVEREUX, Paola DI GIUSEPPANTONIO, Charlotte DIFFEY, Bela DIMOVA, Emilie DINGLER, Sermin DİNÇ, Ayşe Ş. DİNÇER, Filiz DİRİ, Yıldız DİRMİT, Triantafyllia Eirini DOGIAMA, Chris DOHERTY, Irene DORI, Sean DOYLE, Kelly DU RAND, Güneş DURU, Graeme EARL, David EBNER, Daniel EDDISFORD, Leslie EDMONDS, Selma EFELER, Erol EKEN, Fadimana EKEN, Fatma EKEN, Selda EKEN, Saliha EKEN, Turgut EKEN, Ümmügülsüm EKEN, Sophie EKSTRAND, İzettin ELALMIŞ, Hermione ELDERTON, Nada ELIAS, Mustafa Özgür ELMACIOGLU, Erica EMOND, Claudia ENGEL, Ahmet ERDOĞAN, Burçin ERDOĞU, Kerim E. ERGEN, Duygu ERGENÇ, Müge ERGÜN, Gunhild ERIKSDOTTER, Rebecka ERNTELL, Tuğçe ERTABAK, Duygu ERTEMİN, Osman ERTÜRK, Cumhur ERTÜZÜN, Üğür EYİLİK, Catherine FAIRLESS, Chris FARIA, Shahina FARID, Sayeh FATTAHI, Michelle FEIDER, Haşim FERAHKAYA, Lauren FIELD-FIDLER, Rose FIGURA, Clara FILET, Dragana FILIPOVIC, Patrycja FILIPOWICZ, Ashley FISHER, MAX FORREST, Maurizio FORTE, Hayley FOSTER, Katrina FOXTON, Jenna FOWLER, Sheelagh FRAM, Tom FRANKLAND, Ingmar FRANZ, Dorian FULLER, Fabrizio GALEAZZI, Michelle GAMBLE, Eleonora GANDOLFI, Mary GANIS, Virginia GARCIA-DÍAZ, Aroa GARCIA-SUAREZ, Katrina GARGETT, Evan GAROFALO, Gary GIBBONS, Bonnie GLENCROSS, Andrew GOLDMAN, Sarah GONZAGA, Lara GONZÁLEZ CARRETERO, Donna Rae GOULD, Nuriye GÖKÇE, Juan Jose GARCIA GRANERO FOS, Sarah GRANT, Laura GREEN, Rachel GREENBERG, Haskell GREENFIELD, Janet GRIFFITHS, Daniel GRISWOLD, Lisa M. GUERRE, Hilal GÜLTEKIN, Burçin GÜMÜŞ, Ramazan GÜNDÜZ, Gülgün GÜRCAN, Erkan GÜRÇAL, Sevgi GÜRDAL, Simge GÜREŞ, Nergis GURSES, Küpra GÜVEN, Anna HABERLAND, Piraye HACIGÜZELLER, Remi HADAD, Scott D. HADDOW, Christoffer HAGBERG, Lori HAGER, Cordelia HALL, Julie HAMILTON, Anette HANSEN, Katarzyna Weronika HARABASZ, Karen HARDY, Menekşe HAREMKAHYA, Beth HARLEY, Karl HARRISON, Laura HARRISON, Caroline HEBRON, Juliette HEMELAAR, Andrew HENDERSON, Liz HENTON, Xose HERMOSO-BUXAN, Lucia HERRERO, Simon HILLSON, Rachel HODARA, Claire HODSON, Kerrie HOFFMAN, Milicent HOLMAN, John HOLSTON, Phillip HOLT, Braxton M. HOOD, Jedrez HORDECKI, Rosemary HOSHINO, Michael HOUSE, Helen HUMAN, Susan HYDEN, Trevor ILIFF, Rachel IRESON, Justine ISSAVI, Graham ISTED, Resul İBİŞ, Tunç İLEDA, Bianca JACKSON, Mark JACKSON, Antonia JAMES, Rosemary JEFFREYS, Emma JENKINS, Erik

JOHANSSON, Emily JOHNSON, Karolina JOKA, Jennifer JONES, Kimberly JONES, Sarah JONES, Sian JONES, Kristina JONSSON, Emma JORDAN, Rosemary JOYCE, Friederike JÜRCKE, Ceren KABUKCU, Tuukka KAIKKONEN, George KAMBOUROGLOU, Till S. KAPPUS, Aydan KARADEMIR, Akrivi KATIFORI, Ali KAVAS, Kevin KAY, Ramazan KAYA, Vahap KAYA, Nurcan KAYACAN, Nuray KAYGAZ, Courtney KEMNITZ, Sheena KETCHUM, Kübra KILIÇ, Katy KILLACKEY, Hyunyoung KIM, Laurie KING, Ian KIRKPATRICK, Galip KİRAZ, Arkadiusz KLIMOWICZ, Adam KLUPS, Christopher KNÜSEL, Georgia KOROMILA, Vasileios KOURTIS, Vasiliki KOUTRAFOURI, Tomasz KOZLOWSI, Marcin KRZEWICKI, Milena KUBIACZYK, Gülbin KULBAY, Cansu KURT, Nejla KURT, Sevim KURTULDU, Sıla KURTULUS, Aldona KURZAWSKA, Melek KUŞ, Orhan KUŞÇUOĞLU, Bilge KÜÇÜKDOĞAN, Ahmet KÜRKMEN, Ditte Kannegaard KVIST, Jacquelyn KYLE, Florence LAINO, Carla LANCELOTTI, Clark LARSEN, Mikael LARSSON, Jinok LEE, Christina LEMORINI, Amanda LEON, Nicola LERCARI, Andrzej LESZCZEWICZ, Xuelei LI, Amanda LINDSEY, Arzu LINGA, Ashley Morgan LINGLE, Mikolaj LISOWSKI, Yan LIU, Alexandra LIVARDA, Rafael LIZERRALDE, Catherine LONGFORD, Serena LOVE, Jackie LOW, Leilani LUCAS, Dominik LUKAS, Julius LUNDIN, Stella MACHERIDIS, Helen MACKAY, David MACKIE, Marco MADELLA, Richard D. W. MADGWICH, Wiebke MAINUSH, Anna MARCH-LEWSKA, Arkadiusz MARCINIAK, Elizabeth MARGOLIN, Darko MARICEVIC, Gemma MARTIN, Louise MARTIN, Jack MARTINEZ, Michele MASSA, Wendy MATTHEWS, Richard MAY, Camilla MAZZUCATO, Graeme MCARTHUR, Romy MCINTOSH, Claudia MCKENZIE, Katherine MCKUSTER, Sanaz MEHRAN, Teddy MENDOZA, Mehmet MERTEK, Lynn MESKELL, Gamze MEŞE, Alison MICKELL, Danica MIHAILOVIC, Eva Maria MIHAN, Marco MILELLA, Marina MILIĆ, Slobodan MITROVIĆ, Olja MLADJENOVIĆ, Lauren MONKS, Lucie MONO, Sophie MOORE, Gianfranco MORELLI, Colleen MORGAN, Jacob MORIS, Stephanie MOSER, Chiara MOTTOLESE, Elmas MOTUK, Mehmet Ali MOTUK, Jacqui MULVILLE, Charlene MURPHY, Daniel MURPHY, Inbal NACHMAN, Carolyn NAKAMURA ALDRICH, Goce NAUMOV, Adam NAZAROFF, Kate NELSON, Alexandra NEUMANN, Kelly NGUYEN, Bjorn NILSSON, Dorthe NISTAD, Antoni NOWAK, Selin E. NUGENT, Katie O'CONNELL, Aslı OFLAZ, Jessica OGDEN, Sinan OMACAN, Llonel ONSUREZ, David ORTON, Sonia OSTAPTCHOUK, Sara OUENES, Lütfi ÖNEL, Yasemin ÖZARSLAN, Mihriban ÖZBAŞARAN, Onur ÖZBEK, Özlem ÖZBEN, Ozan ÖZBUDAK, Serap ÖZDÖL KUTLU, Füsun ÖZER, Hakan ÖZER, Levent ÖZER, Özlem ÖZHABES, Özge ÖZKAN, Hembo PAGI, Francesca PAJNO, Philip PARKES, Kamilla PAWLOWSKA, Aruna PAWSON, Halle PAYNE, Chloe PEARCE, Jessica PEARSON, Daniela PEDROZA, Marta PERLINSKA, Sara PERRY, Anda PETROVIC, Paul PETTERSSON, Matteo PILATI, Camile PILLIOUGINE, Marin PILLOUD, Sharmini PITTER, Charles PIVER, Marek POLCYN, Marta PORTILLO RAMIREZ, Adrienne POWELL, Tera PRUITT, Laia PUJOL-TOST, Liz PYE, Lyla PYNCH-BROCK, Joanna PYZEL, Ling QIN, Jason QUINLAN, Antoinette RAST-EICHER, Flavia RAVAIOLI, Roddy REGAN, Katarzyna REGULSKA, Emily RICHARDSON, Megan RIDSDALE, Jana ROGASCH, Nolwen M. ROL, Kate ROSE, Elizabeth ROSEN, Eva ROSENSTOCK, Jamie ROWE, Abel RUIZ-GIRALT, Nerissa RUSSELL, Eugen RUZI, Philippa RYAN, Anna RYBARCZYK, Freya SADARANGANI, Josh SADVARI, Hannah SAINSBURY, İsmail SALMANCI, Nicole SAM, Carlos G. SANTIAGO MARRERO, Judre SAPRANAUSKAITE, Marta SAJ, Jill SAUNDERS, Billy SAWOYO SANKEI, Melania SAVINO, Heeli C. SCHECHTER, Sophie SCHMIDT, Eline SCHOTSMANS, Mesa SCHUMACHER, Jessica SCORRER, Mitchell SCOTT, Jerrod SEIFERT, Uğurcan O. SELÇUK, Recep Yunus SERİN, Gülay SERT, Kent SEVERSON, Harish SHARMA, Daniel SHAW, Russell SHEPTAK, Hannah SHILLING, Lisa-Marie SHILLITO, Anna SHOEMAKER, Ruth SIDDALL, Matilda SIEBRECHT, Maroles SIJSTERMANS, Ahmet SİVAZ, Ebru SİVAZ, Fadimana SİVAZ, Havva SİVAZ, Keziban SİVAZ, Mevlüt SİVAZ, Saliha SİVAZ, Zekeriya SİVAZ, Arne SJÖSTROM, Cassie SKIPPER, Dean SMITH, Kierstyn SMITH, Mehmet SOMEL, Tiffany SOULE, Muhammet SÖKEN, Abdurrahman SÖNMEZ, Charlotte SPIERING, Mira STEVANOVIC, Shannon STEWART, Ivana STOJANOVIĆ, Helen STOKES, Weronika STOSIK, Marketa STOVÍCKOVÁ, Elizabeth A. STROUD, Kristian STRUTT, Amy STYRING, Thomas SUTCLIFFE, Lauren SWEET, Martyna SZYMCZAK, Melike ŞAHİN, Neriman ŞAHİN GÜÇHAN, Elmas ŞENER, Esra ŞENER, Nevriye ŞENER, Muhsin ŞENOL, Ayşegül TABAKOĞLU, Wang TAO, Duygu TARKAN, Dena TASSE-WINTER, James S. TAYLOR, Beliz TERCELI, Kilian TEUWSEN, Gregory THOMA, Johanna THUNBERG, Belinda TIBBETTS, Jenna TINNING, Hatice TOKYAĞSUN, Mavili

TOKYAĞSUN, Yusuf TOKYAĞSUN, Margaret TOMASZCZUK, Angela TORNEY, Jovana TRIPKOVIC, Hoang Anh N. (Elizabeth) TRINH, Christina TSORAKI, Gemma TULLY, Mevriziye TUNCAY, Burcu TUNG, Didem TURAN, Özge TUTAR, Mustafa TUTUMLULAR, Talu TÜNTAŞ, Ali TÜRKCAN, Ülcan TÜRKKAN, Katheryn TWISS, Hakkı UNCU, İdris USLU, Özgür Can USLU, Bilgehan USTA, Oktay UZUN, Ekin ÜNAL, Petra VAIGLOVA, Anne VAN GIJN, Mirjam VAN SAANE, Milena VASIĆ, Maria VAYANOU, Renee VD LOCHT, Rena VEROPOULIDOU, Owen VINCE, Alice VINET, Sophie VULLINGS, John WAINWRIGHT, Sam WAKEFORD, Gillian WALKER, Jiajing WANG, Johnathan WANG, Marcin WAS, Amanda WATTS, Sadie WEBER, Willemina WENDRICH, Elizabeth WESSELLS, Joanne WESTBROOK, Lucy WHEELER, Harvey WHITEHOUSE, Jade WHITLAM, Patrick WILLETT, Alice WILLIAMS, Chelsea WISEMAN, Jessie WOLFHAGEN, Karen WRIGHT, Nurcan YALMAN, Thaer YARTAH, Mustafa YAŞ, Hasan YAŞLI, Hatice YAŞLI, Hulusi YAŞLI, Hüseyin YAŞLI, İsmail YAŞLI, Lokman YAŞLI, Metin YAŞLI, Mustafa YAŞLI, Osman YAŞLI, Rabia YAŞLI, Senay YAŞLI, Tulin YAŞLI, Lisa YEOMANS, Gökhan YEŞIL, Gülay YILANKAYA- ERDOĞU, Nejat YÜCEL, Onur YÜKSEL, Eren YÜNCÜ, Mustafa ZEYTIN and Bright ZHOU.

The volumes 12–15 in this series would not have been produced without the work of Scott Haddow, Jason Quinlan, Kathryn Killackey, Dominik Lukas and Camilla Mazzucato. I am very grateful to them for their commitment.

1. Changing Çatalhöyük worlds

Ian Hodder

Çatalhöyük is a 9,000-year-old tell site in central Turkey. First excavated by James Mellaart (e.g., Mellaart 1967) in the 1960s, a new project began in 1993 (Hodder 1996; 2000a). The site is of international significance because it is large (13.5ha) at an early date, had dense population (probably over 1,000 people), has rich symbolism and sub-floor burial, and was occupied for 1,500 years; the Neolithic Çatalhöyük East mound dates from 7100 BCE to 5900 BCE, with the Chalcolithic Çatalhöyük West Mound overlapping in time in the last quarter of the seventh millennium BC and continuing on until 5600 BCE (Orton et al. 2018). The well-preserved buildings and rich art in the Neolithic mound give a unique insight into early village life. The site allows study of many of the main questions dealing with the early formation of settled villages/towns and the early intensification of agriculture. The site was inscribed on the UNESCO World Heritage list in 2012.

This volume is one of four which discuss interpretation of the material obtained during excavation of the site from 2009 to 2017. Early work on the surface of the mound in 1993–1995 is published in Volume 1 (Hodder 1996), and the methodology is described in full in its own volume, Volume 2 (Hodder 2000a; for updates see Hodder, Marciniak 2015 and Berggren et al. 2015). The excavation and the interpretation of contexts and buildings excavated in 1995–1999 are described in Volumes 3–6 (Hodder 2005; 2006a; 2006b; 2007), and Volumes 7–10 describe the excavation and interpretation of material obtained in the 2000–2008 seasons (Hodder 2013a; 2013b; 2014a; 2014b). Volume 11 is an account of the excavation of Building 3 by a team from the University of California at Berkeley (Tringham, Stevanović 2012).

The new series of four volumes (Volumes 12 to 15) presents reports on the results of excavations in 2009–2017, although many contributors also take the chance to summarise and compare results over the 25-year span of the Çatalhöyük Research Project. Volume 12 describes the excavation results regarding architecture and the deposits uncovered in the 2009–2017 seasons. The current Volume 13 reports on the ways in which humans engaged in their material and biotic environments, using a wide range of archaeological evidence. This volume will also summarise work on the skeletal remains recovered from the site, as well as analytical research on isotopes and aDNA. Volume 14 turns from 'ecofacts' and biology to material artefacts recovered from the site in these seasons, including a range of clay-based objects (ceramics, clay balls, tokens, figurines) as well as those made of stone, shell and textile. Volume 15 is also based on the 2009–2017 excavations, but its aim is synthetic, drawing on material from Volumes 12 to 14 to deal with broad themes. Data from architecture and excavation contexts are linked into broader discussion of topics such as social networks, seasonality, curation, colour, community engagement and outreach.

Future volumes will describe other excavations and analyses undertaken over recent years under the permit provided to the Çatalhöyük Research Project (ÇRP). Volume 16 will describe the excavations on the West Mound (Anvari, Biehl forthcoming), and future volumes will describe the work in the TP and TPC Areas (Marciniak et al. forthcoming) and on the historic materials found on and off the mound (Jackson, Moore forthcoming).

Research questions

The long (25-year) aim of the Çatalhöyük Research Project has been to situate the elaborate symbolic production at the site within its full environmental, economic and social context. Why did this rich outpouring of symbolism and ritual occur at this time and this place? Can we understand why the art and symbolism were produced and can they help us understand why people formed large agglomerations and intensified resource procurement?

In the earliest phase of the current project (1993–1995), minimal excavation took place. The work concentrated on regional survey (Baird 2002; 2005), on planning and studying the surface of the mounds, conducting surface pickup, drawing eroded profiles of the earlier excavation trenches and using geophysical prospection. The project also undertook a re-evaluation of the material in museums that had been excavated by Mellaart. All this work was published in 1996 (Hodder 1996).

During the second phase of excavation (1996–1999), analysis and publication (2000–2002), the research aim focused on individual buildings, asking questions about site formation processes and trying to understand whether the mudbrick buildings on the site were sometimes shrines (as Mellaart had suggested) or whether they were all houses (even if at times with much symbolic and ritual content). For the results of this work see Hodder (2005; 2006a; 2006b; 2007).

In the third phase of excavation (2000–2008), followed by analysis and publication (2009–2012), the research aims turned from individual houses to the social geography of the settlement as a whole and larger community structure. The focus on social geography had a number of more specific components. One component concerned social and economic differentiation. Mellaart argued that he had discovered an area of the site inhabited by priests and their shrines and houses. But right from the first phase of the current project we had found uniformity across the site. Surface survey, geophysical work, surface scraping and excavation all showed that wherever one dug on Çatalhöyük, one tended to find the same thing: houses containing evidence of domestic activity varying only slightly in size and elaboration, and areas of refuse or midden. There seemed little on which to base any argument for social and economic differentiation.

So the fourth phase of excavation (2009–2017), followed by analysis and publication (2018–2019), focused on trying to understand how social and economic organisation varied through time. The excavation in 2000–2008 had concentrated on earlier and later levels of occupation on the East Mound, but it became increasingly clear that there was evidence for major shifts in many aspects of the evidence, especially around 6500 BCE and also in the latest levels as occupation became more dispersed, including spread to the West Mound. It became critical to examine the middle levels of occupation, especially South M-O (Mellaart's Levels VII and VI), and to open more contiguous buildings of the same time period in the North Area (North G). At the same time, Arek Marciniak undertook excavations on the latest levels of occupation on the East Mound (Marciniak et al. 2015), followed by related work by Marek Barański (Barański et al. 2015; Barański 2016) in the adjacent GDN Area.

The overall trajectory of the project has thus been from the study of individual houses to communities and larger-scale social geography. But I have so far given the impression of a limited set of research questions that animated the project as a whole. In fact, the size of the project meant that many different sources of funding had to be sought and a range of different research questions asked. For example, funding for detailed dating was obtained by Hodder and Bayliss from NSF and AHRC, while NSF funding was also obtained by Twiss and Bogaard in order to study the integration of faunal and botanical data. A large number of graduate students have focused on particular research questions. While all such projects worked within the general overarching aims just described, the project has benefited from the diversity of perspectives brought by these individuals with their different funding sources (see Hodder, Marciniak 2015).

This is particularly true of a series of grants obtained from the John Templeton Foundation that all dealt with various aspects of the rich symbolism from the site in asking questions about religion and cognition. These grants were also important in adding to the multivocality and interdisciplinarity which have always been at the heart of the project. The Templeton projects involved bringing specialists in religion, philosophy and anthropology to spend a week each year engaging with the archaeologists at the site and ultimately producing a series of edited volumes.

In 2006, research funding for a three-year period from the John Templeton Foundation was directed towards the following four questions: (1) How can archaeologists recognise the spiritual, religious and transcendent in early time periods? (2) Are changes in spiritual life and religious ritual a necessary prelude to the social and economic changes that lead to civilisation? (3) Do human forms take on a central role in the spirit world in the early Holocene, and does this centrality lead to new conceptions of human agency that themselves provide the possibility for the domestication of plants and animals? (4) Do violence and death act as the foci of transcendent religious experience during the transitions of the early Holocene in the Near East, and are such themes central to the creation of social life in the first large agglomerations of people?

Answering this first set of Templeton questions proved very productive and resulted in a publication entitled *Religion in the Emergence of Civilization: The Example of Çatalhöyük* (Hodder 2010). One of the results of this work was the identification of different general house types at Çatalhöyük based on burial and architectural and symbolic elaboration. The most notable were 'history houses' (Hodder, Pels 2010), which are architecturally elaborate buildings with multiple burials that endure for generations and undergo numerous rebuilding phases. Houses can also have multiple burials whether we have evidence they were part of a long sequence or not. There are also elaborate houses without many burials and re-buildings, and there are non-elaborate houses. These four different house types and their associated burials at Çatalhöyük allowed us to explore social differentiation in relation to the symbolic and religious features in the houses.

In a second Templeton project (2009–2012), entitled 'Religion as the basis for power and property in the first civilizations. The analysis and publication of Çatalhöyük', another set of four questions explored the relationship between symbolism and differentiation in relation to the already excavated data from 2000 to 2008. The four research questions were as follows, the first two relating to a focus on social geography and the second two relating to change through time: (1) At Çatalhöyük what is the relationship between religious and other symbolism and the control of production in the different house types? For example, are there differences in the productive activities associated with history houses, elaborate houses and other houses? Is there evidence for differences in health and in skeletal markers of mobility and workload between history houses, elaborate houses and other houses? (2) Is there secure evidence that important symbolic and religious objects were handed down in houses, and preferentially in 'history houses', at Çatalhöyük? For example, what are the sequential processes by which some buildings became transformed into history houses while others declined? Is there evidence that objects were handed down from building to building through time creating histories, and does such transfer occur preferentially in the history houses? (3) Through time how does the 'history house' system change, and is there a concomitant decline in the role of religious symbolism? This question relates directly to the question identified above about changes in the social geography of Çatalhöyük through time. For example, what are the economic changes evident at the site in the upper levels? Were cattle domesticated, for example? What are the social changes associated with the upper levels? Do households become larger, more integrated and with less focus on continuity with the past? (4) Is religion related to power and property elsewhere in the emergence of civilisation in the Neolithic of the Middle East? For example, how does the changing role of ritual and differentiation at Çatalhöyük play a part in larger and longer-term changes across the Middle East and Anatolia during the Neolithic? This second Templeton grant allowed us to pursue the overarching aim of opening extensive areas in the middle phases of occupation at the site. The results were published as *Religion at Work in a Neolithic Society: Vital Matters* (Hodder 2014c).

A third Templeton project (2012–2015) asked a further four questions that followed on from the results of the second project, and explored the hypothesis that religion had a primary role in the origin of settled life because it allowed the production of the two main struts of that life – historical depth and attachment to place. The first question considered whether Çatalhöyük could be considered a 'history town'? If groups of houses were tied together by history houses, was it possible that the settlement as a whole was constructed so that the inhabitants could 'read' the history of their relationships with each other in the layout of their settlement? A second question extended the same point to the landscape as a whole. Was that too, including the organisation of fields, linked into the arrangement of houses in the settlement? Third, how did the 'history house', 'history town' and 'history landscape' at Çatalhöyük emerge? Fourth, what is the relationship between religion and history making at other sites in the Neolithic of the Middle East? The results of this project were published as *Religion, History and Place in the Origin of Settled Life* (Hodder 2018a).

A fourth Templeton project (2015–2017) shifted focus from religion to aspects of cognition, in particular consciousness, creativity and notions of self. Again four questions were asked. The first was whether levels of consciousness have increased over time, leading to greater abstraction and objectification of thought. The second question was whether rates of innovation and creativity increased during the Neolithic as seen through the sequence at Çatalhöyük. The third question was whether different forms of cognitive change in the Neolithic might have been related to a greater awareness of an integrated self. And finally, the aim was to explore whether any observed cognitive changes might occur suddenly or were stretched out over time. The answers to these questions were published in *Consciousness, Creativity and Self at the Dawn of Settled Life* (Hodder 2020).

Overlapping with these projects on religion and cognition at Çatalhöyük, between 2013 and 2016 funding was obtained from Imitatio, a group of scholars following the mimetic theory of René Girard. The general hypothesis explored was that there is a link between violence and the sacred. Specifically at Çatalhöyük it was argued that the process of mimesis whereby humans desire the same things as each other leads to violence that can easily break out into violence of all against all. This type of rampant violence can be resolved by the participants turning against one – the scapegoat. The latter may then take on a sanctified position as the one that brought peace to society, and through time the sacrifice of the scapegoat is repeated and re-enacted in ritual in order to sustain peaceful order. This hypothesis was seen to be relevant to the ritual treatment of the bull at Çatalhöyük. This work was published as *Violence and the Sacred in the Ancient Near East: Girardian Conversations at Çatalhöyük* (Hodder 2019).

Overall, then, the Templeton and Imitatio projects, as well as funding from research foundations and commercial sponsors, allowed excavation and post-excavation work to pursue the main questions of the third and fourth phases of excavation at Çatalhöyük. As the main research concerns shifted from individual buildings to overall social geography and change through time, the funding allowed large areas of excavation to take place in the North Area, and it allowed deep excavations to take place in the South Area. And it did so by involving the issue of variation in symbolism and ritual. In this way the larger aim of the project as a whole since its inception could be effectively pursued: 'to situate the elaborate symbolic production at the site within its full environmental, economic and social context' (see above).

Regional trends

The above research questions can be situated within wider understanding of the Neolithic in Turkey and the Middle East. A full account of the regional trends within which Çatalhöyük is situated is provided in Volume 15, Chapter 2; only a brief summary is provided here.

In Central Anatolia, by the second half of the ninth millennium there are both small settlements as at Boncuklu and Pınarbaşı (Baird 2007b) and the large highly agglomerated village at Aşıklı Höyük (8400–7400 BC) (Esin, Harmankaya 1999; Özbaşaran 2011). At the latter site there is very little evidence of storage facilities in domestic houses. As a result of large-scale area excavation, a clear social geography has been identified, with public ritual buildings (frequently renewed) separated from other dwellings by a monumental street. The dwellings are organised into sectors by narrow spaces or narrow streets and there are collective common middens. The area with public ritual buildings and perhaps storage is distinctive in having larger numbers of pressure-retouched projectiles and higher percentages of cattle bones. There is evidence of butchering and sharing of meat. Overall, the evidence is interpreted in terms of the collective and communal rather than in terms of centralised elites.

In cultural and social terms, Çatalhöyük emerges from a strong local tradition on the Konya Plain, as seen at Boncuklu (Baird 2007a). The parallels with this site of the late ninth and early eighth millennia include below-floor burial, painting and wild animal installations, chipped stone point typologies, and most tellingly, a separation in houses between 'clean' floors (with burial and symbolism) and 'dirty' floors (with hearths and activity traces). There are of course many differences between the sites, in terms of layout and economy (for example, the shift to a heavy dependence on sheep at Çatalhöyük), but the socio-cultural continuities are strong, especially when contrasted with Cappadocian sites such as Aşıklı Höyük. It is fascinating to note that the north/south, clean/dirty division within houses at Boncuklu is inverted through most of the sequence at Çatalhöyük (Hodder, Cessford 2004). But the earliest levels see ovens and hearths in the northern part of buildings (in Buildings 161 and 17).

It is possible to conceive of the early population at Çatalhöyük being largely derived from small sites such as Boncuklu and Sancak and others identified by Baird in the Konya Plain survey (Baird 2005). Can Hasan III is another possible source (French 1972) and there are undoubted similarities in settlement organisation and use of a microblade lithic tradition. Although there are also similarities in lithic tradition with Cappadocia, the Konya Plain sites differ markedly from Aşıklı Höyük, where there is less symbolism in houses.

However, new genomic evidence (summarised in Chapter 18) indicates closer links between Çatalhöyük and Aşıklı Höyük than between Çatalhöyük and Boncuklu. On the other hand, the aDNA data indicate that, beyond a general derivation from the east, the Cappadocian-Konya Plain grouping is relatively distinct from the Levant, indicating a somewhat separate trajectory. Epipalaeolithic and aceramic Neolithic Central Anatolians (as seen at Pınarbaşı and Boncuklu) shared traits with groups in the Caucasus, Zagros and Europe. But the genomic profiles in Central Anatolia in the early Neolithic are closely related to Epipalaeolithic populations in the same region (Lazaridis et al. 2016; Feldman et al. 2019; Chyleński et al. 2019), suggesting the spread of technology, ritual and domestic species rather than predominantly people.

From the aceramic period to the ceramic Neolithic period, human movement between Anatolia and neighbouring regions increases. Populations of the PPNB Levant (for example, 'Ain Ghazal) show higher genetic affinity to Neolithic Anatolians than did the Natufians who lived in the same region 2,000 years earlier (Kılınç et al. 2017). Similarly the population of Tepecik-Çiftlik had more links to Levantine and Zagros populations than did people at aceramic Boncuklu, people who lived in the same broad region 2,000 years earlier. These changing genetic affinities can be explained by some gene flow from the Levant into Anatolia, alongside the cultural influences (Cauvin 2000).

There is also evidence of genetic affinities between Central Anatolia and the early Neolithic in northwest Anatolia and Europe. Indeed it is tempting to argue that Çatalhöyük was one source of the spread of early farming into these areas. As we will see below, there is evidence of population increase and higher densities of occupation, as well as stress and disease on human bodies, in the Middle phases of Çatalhöyük from about 6600 to 6500 BCE. After 6500 BCE there is evidence at the site for greater mobility and dispersal. Regional survey has found a scarcity of sites contemporary with the main ceramic layers at Çatalhöyük (Baird 2005). It is almost as if the site sucked in population and then spat it out again in the later levels of occupation, culminating in an increase of Chalcolithic sites on the plain.

Seen at a larger scale it is possible that such forces stimulated larger-scale dispersal. In Volume 15, Chapter 2 Marciniak makes the claim that the spread of farming into western and northwestern Anatolia, as well as new developments in Cappadocia, can be seen as the result of migration from Çatalhöyük. The examples he gives include Gökhöyük, the houses of which are indeed remarkably similar to Çatalhöyük, with the same internal division of clean and dirty space. Other examples are Suberde, Erbaba, Bademağacı, Höyücek, Hacılar IX–VI and Kuruçay in the Lake District and adjacent areas. In northwestern Anatolia cultural influence is seen at Uğurlu, Pendik, Fikirtepe, Aktopraklık, Menteşe and Barcın (Özdoğan 2010; Karul, Avcı 2011; Roodenberg et al. 2003; Gerritsen et al. 2013; Erdoğu 2013) and suggested by the dark burnished ware of the Fikirtepe culture, which is parallel to the dark burnished wares of Central Anatolia. In western Cappadocia cultural influence is seen at Tepecik-Çiftlik and Köşk Höyük (Bıçakçı et al. 2012; Öztan 2012).

Central Anatolia is located at a cusp between east and west in terms of the spread of farming (Özdoğan 2010). In many aspects the cultural assemblage from Çatalhöyük has similarities with the Marmara and Balkan region (bone spoons, bone belt hooks) rather than with the southeast of Turkey. This difference also includes the social organisation of settlements and ritual. As Özdoğan has argued, to the east special buildings are more commonly found and a general focus on ancestors and history making. But to the west of Çatalhöyük special buildings do not dominate and there are extramural cemeteries (for example, in the Fikirtepe culture sites with rectangular architecture in the Marmara region) (Özdoğan 2011a), although many of these regional differences are cross-cut by change through time.

Overall, then, Çatalhöyük is part of a distinct local tradition on the Konya Plain, certainly linked culturally and by the exchange of obsidian with Cappadocia but with its own distinctive focus on symbolic elaboration. Through time there is increasing human dispersal to and from the Konya Plain. There is also much evidence of a wider cultural koine. There are similarities in myth and symbolism (Hodder, Meskell 2011) over enormous areas and time periods, and overall general trends from northern Mesopotamia and the Levant into Anatolia. Throughout the region houses often increase in size from the Epipalaeolithic onwards into the large settled villages of the MPPNB. Çatalhöyük is one of a number of 'megasites' (Verhoeven 2011) that, as at Çatalhöyük itself, often break up into smaller more dispersed units in the ceramic Neolithic. There are intriguing similarities in this process over extensive areas. In Chapter 15 Haddow et al. describe a wider process of increased secondary burial and separation of burial from houses as seen at both Çatalhöyük and Tell Sabi Abyad (Akkermans 1987), and there are indeed marked similarities in the multi-mound structure of the latter and of later Çatalhöyük. The emergence of elaborately painted pottery in northern Syria is reminiscent of the painted pottery at Çatalhöyük West. Çatalhöyük can thus provide an insight into important regional trends in social and economic processes, as well as a potential impetus for the spread of farming and a source for population dispersal.

Background and method

Çatalhöyük was first excavated by James Mellaart in 1961–1965 (Mellaart 1962; 1963; 1964; 1965; 1966; 1967). His work focused on the southwest area of the East Mound where over 160 buildings were excavated in 14 levels numbered from 0, I, II to XII. Level VI was subdivided into VIA and VIB. These excavations established the importance of the site as a large and densely packed Neolithic 'town'. No further work was conducted at the site until the present project began in 1993. As noted above, the first years of the project were spent in surface survey, and excavation began in earnest in 1995. This volume is concerned with presenting the results of specialist analysis of material from two areas of excavation at Çatalhöyük (fig. 1.1) known as South and North.

The excavations in the South Area between 1995 and 1999 were initially focused upon a 20 x 20m area in the southwest part of the mound, which was investigated by James Mellaart in the 1960s. The main aim of our work in this area was to investigate the earlier phases at the site, and the lowermost deposits above the lake marl were excavated and designated Level Pre XII.A-E. This work was reported on in Volume 3. In the 2000s, work continued in the South Area under a shelter. The Foundation Trenches for this shelter were first excavated (in 2002), and then excavation concentrated on later levels within the shelter in the upstanding deposits around the edges of the original 20m x 20m area. The aim was to reduce these surrounding deposits so that we could safely return to excavation of the lowest levels in the 20m x 20m area.

In the 1995–1999 period our excavations in the North Area aimed to contribute to the research aim of understanding the formation processes in individual buildings. In the northern area, buildings were excavated in great detail (Buildings 1 and 5 discussed in Volume 3). In 1997, a team from the University of California, Berkeley began excavating Building 3 in the BACH Area and continued until 2002 (published in Tringham, Stevanović 2012). The KOPAL excavations consisted of a long trench across the northern flank of the mound to investigate site formation processes (1996–1997) and an offsite area to the north of this to determine what natural and cultural deposits were present (1997 and 1999). In the 2000–2008 period of work

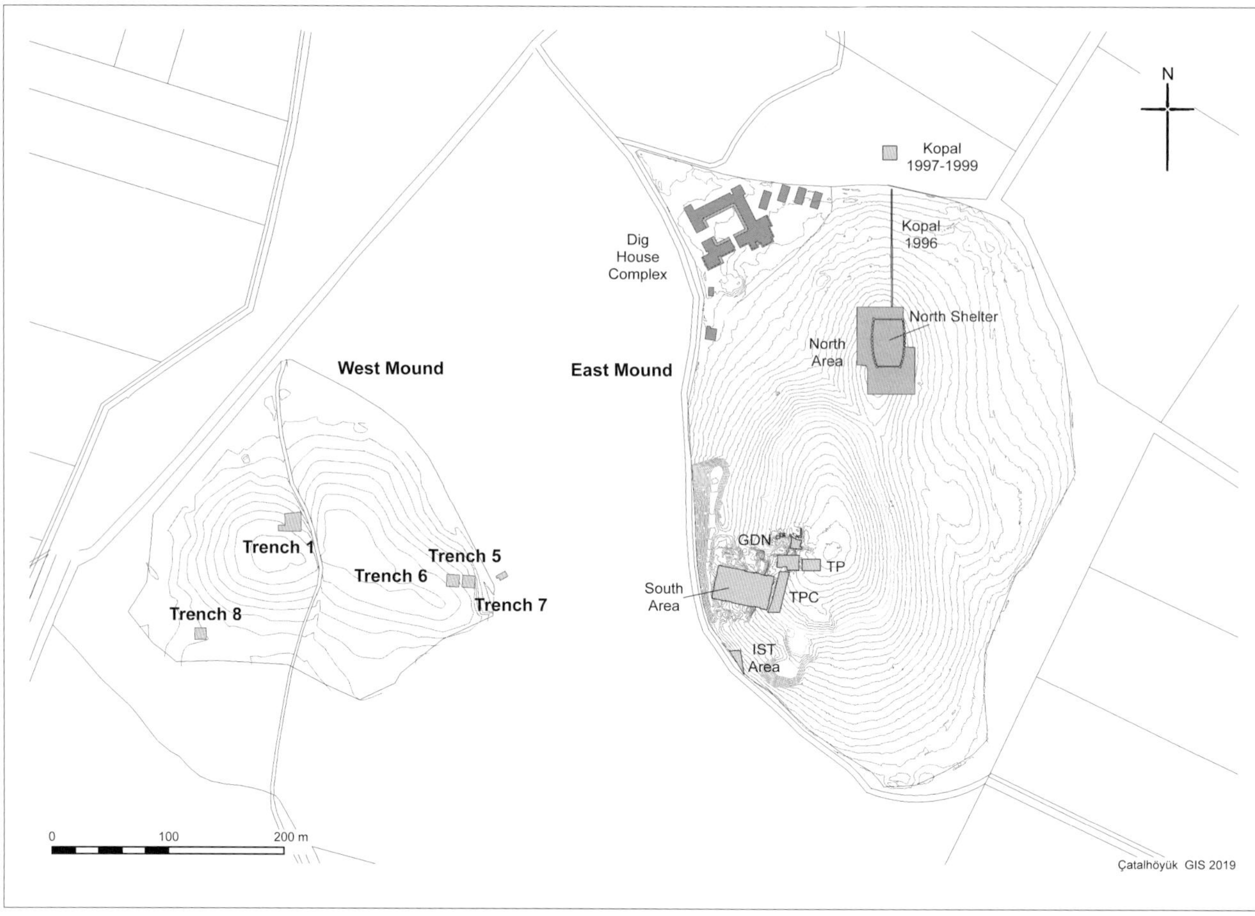

Figure 1.1. Map of excavation areas on the East and West Mounds at Çatalhöyük (map: Camilla Mazzucato).

the smaller excavations in the northern part of the mound were extended in order to understand the overall social geography of settlement in this area. A 40m x 40m trench was identified as being relatively undisturbed by later historic activity on the basis of surface research conducted in 1993–1995 (Hodder 1996). In 2004 to 2008 the 4040 Area was investigated in order to understand the social geography and change through time in this part of the site. In 2007 the foundations for a shelter covering the 4040 Area, Buildings 3 and 5 were excavated. The shelter was then built, allowing excavations in 2008 to take place in a more controlled setting.

Excavations by other teams, especially the TP (Team Poznan) team, led by Arek Marciniak of Poznan University and Lech Czerniak of the University of Gdansk in Poland, and the IST team, led by Mihriban Özbaşaran of Istanbul University, allowed further exploration of the upper levels of the East Mound. On the Chalcolithic West Mound, excavation by three teams (University of Thrace at Edirne led by Burçin Erdoğu, Selçuk University at Konya led by Ahmet Tırpan and Asuman Baldıran, and Berlin University and SUNY Buffalo led by Peter Biehl and Eva Rosenstock) allowed an increased understanding of the developments in the sixth millennium BC. Apart from the inclusion of the IST results in Volumes 7–10, all the excavations by other teams (TP, West Mound) are being published in separate volumes.

In the period 2009–2017 excavation has continued under the direction of Arek Marciniak on the late levels found between TP and the South shelter. The area is designated TPC (Team Poznan Connect), and targeted excavation in Mellaart's old trenches (GDN) from levels 0 to III has been undertaken by a team led by Marek Barański (the latter reported on in Volume 12). The main focus of excavation has been in the South and North Areas. In the South Area excavations concentrated on middle level buildings in order to fill gaps in the sequence. It also proved possible to return to the earliest levels in this part of the site. In 1999 the deep sounding had discovered open areas and animal penning deposits above the Pleistocene marl, and we had not found the earliest houses at the site. In the period since 1999 it was hoped to return to these early deposits beneath Buildings 17 and 43 in the hope of finding houses. We did manage to achieve this and dig below these two buildings in 2017, but again open areas were uncovered. In the North Area the focus was on understanding the overall plan of buildings and exploring the variation between buildings and their inter-relations.

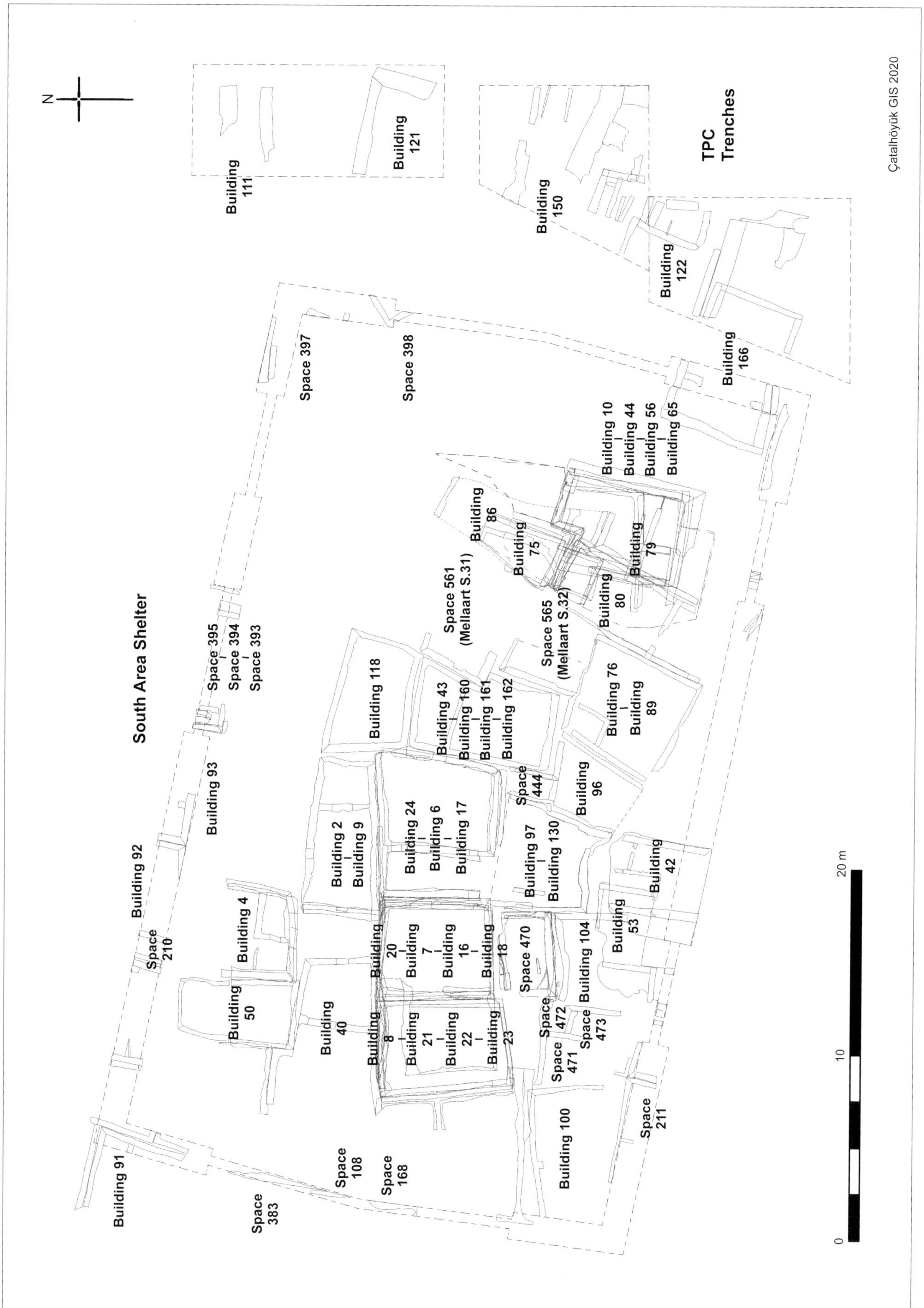

Figure 1.2. Map of buildings in the South Area (map: Camilla Mazzucato).

Figure 1.3. Map of buildings in the North Area (map: Camilla Mazzucato).

A modified form of single-context ('unit' in the Çatalhöyük Research Project terminology) excavation and recording was employed, combined with an intensive and extensive sampling regime, including both wet and dry sieving. The unit forms the basic element of a nested hierarchical system that includes features (groups of related units), spaces (spatially bounded entities generally defined by the walls of buildings), buildings (groups of spaces forming a structural entity), areas (spatially discrete locations where excavation has occurred) and mounds. Chronological grouping is provided by phases/subphases (temporal divisions within a particular space or building, usually reflecting identified changes in the spatial organisation or arrangement of a space of building) and levels (broadly contemporary groups of spaces and buildings – see below).

Units are divided into nine primary data categories of fill, floor, construction, midden, activity, arbitrary, cut, cluster and skeleton. For the numbers of units excavated in these categories see table 1.1. A cut is any recognisable event that led to the removal of another deposit, and an arbitrary layer is a subdivision of a layer or a grouping of layers carried out for practical reasons during excavation. Clusters are any deposits defined primarily by their material inclusions such as obsidian, bone, botanical material etc. Skeletons are effectively a particular type of cluster, one composed of human remains. The other data categories represent types of layer. Fills are layers that have been deposited in any form of negative space, be that a cut, the limits of an upstanding feature or a whole space or building. Construction/make-up/packing layers are deposits related to the construction of buildings and/or features, while a floor is any deposit inside or outside a structure that forms a surface upon which activities of any sustained duration occur. Middens are deposits primarily related to the deliberate disposal of cultural material, again either outside or inside a structure. All these deposits seem to have been unroofed and more recent research has shown they were used for a variety of activities including disposal of waste. The research conducted by Justine Issavi summarised in Volume 15, Chapter 16 indicates that the term 'midden' should be replaced with 'open area' (although the 'midden' term is found in the database at www.catalhoyuk.com and is retained in table 1.1 in order to retain consistency). Activity deposits relate to a heterogeneous range of activities that produce distinctive remains including fire spots. These primary data categories are then subdivided according to whether they are in situ, their location, their basic description, the material they are composed of, the nature of their deposition and whether they are a basal deposit.

During excavation, selected units were prioritised for study based on discussions between the excavators and laboratory specialists. Every two three days during the excavations members of the laboratory teams (faunal, botanical, chipped stone, phytolith, ceramic etc.) went on 'priority tours' of the excavation areas in order to (a) identify units and features for quick processing in the laboratories and (b) discuss with the excavators the information obtained from the quick processing of earlier material. These discussions form the basis for the detailed discussion of individual units, features and buildings in this volume. The aim was to identify a list of units that would be studied by all labs during post-excavation so that data could be compared. A list of 456 priority units was selected for analysis including 71 units from the TPC Area. During the first stage of the selection process each laboratory team provided a 'wish-list' of preferred units for study. Depending on how many labs chose the same unit, units were given different weights. The final list, which contained both priority units (that is, units identified as priority during excavation) and other units prioritised in the post-excavation process included: a) units selected by two or more labs, b) units from certain buildings and spaces that were prioritised (based on Level attribution and type of context), such as B. 77, 79, 80, 160, 131, 132 and Sp. 489, 490, and c) for the final publication cycle it was deemed necessary to provide a more contextualised and integrated study of burials, and therefore a list of units from burial contexts was selected by the Human Remains team and these were analysed by all labs. To ensure comparability of results among different materials all labs were requested to provide the same consistent information for all units selected as part of this process. This includes basic taphonomic information, count, weight, and basic identification of species/artefact type (or size class for faunal material) for each priority unit.

Bayesian analysis of all reliable radiocarbon determinations from Çatalhöyük using BCal indicates that the main occupational sequence at Çatalhöyük East revealed so far probably lasted from 7100 to 6000 cal BC with considerable overlap with the West Mound that continued into the early sixth millennium. In Volume 7, Chapter 3 Bayliss et al. describe the long history of radiometric dating at Çatalhöyük East culminating in their renewed large-scale project integrating Bayesian statistical analysis. This renewed radiometric programme is ongoing and until it is complete, we do not have accurate links between the northern and southern areas of excavation. A summary of the current state of the programme is found in Volume 12, Chapter 3 (see also Bayliss et al. 2015; Marciniak et al. 2015), but for the moment we do not have precise correlation between the deposits in the north and south, and to some degree we remain reliant on comparisons of the cultural material between levels in the North and South Areas (see below).

	2000–2008		2009–2017		Total (1993–2017)	
North Area	*Percentage*	*No. of units*	*Percentage*	*No. of units*	*Percentage*	*No. of units*
Activity	1%	31	1%	40	1%	80
Arbitrary	5%	141	2%	66	5%	391
Cluster	4%	93	1%	31	2%	162
Construction/make-up/packing	28%	709	39%	1,088	32%	2,525
Cuts	10%	250	13%	374	11%	910
Fills	24%	624	21%	589	25%	1,979
Floors	15%	371	14%	400	15%	1,189
Middens	6%	142	2%	55	3%	236
Skeletons	7%	177	6%	178	7%	530
Total	**100%**	**2,538**	**100%**	**2,821**	**100%**	**8,002**
South Area						
Activity	4%	86	2%	43	2%	171
Arbitrary	4%	79	1%	26	4%	295
Cluster	3%	74	3%	68	3%	231
Construction/make-up/packing	37%	795	37%	934	36%	2,752
Cuts	10%	211	11%	281	10%	725
Fills	17%	364	23%	577	20%	1,480
Floors	15%	321	17%	441	17%	1,269
Middens	7%	149	3%	89	6%	451
Skeletons	3%	68	4%	90	3%	204
Total	**100%**	**2,147**	**100%**	**2,549**	**100%**	**7,578**
TOTAL		**4,685**		**5,370**		**15,580**

Table 1.1. Excavated units by data category.

However, we have since the 2000–2008 period come to realise that the grouping of buildings and deposits into levels numbered from 0 to XII by Mellaart no longer captures the complexity of the stratigraphy of Çatalhöyük (as discussed by Farid in Volume 7, Chapter 2). The main problems identified are as follows. Mellaart himself vacillated as he tried to fit buildings and open areas into a clear sequence of overarching levels. In particular, he had to divide Level VI into two, VIA and VIB, and at various times he re-allocated buildings between Levels VII and VI. The problem that he was struggling against is the same that we have faced – that in fact there are few large-scale horizons of activity that stretch across large areas of the site. In fact, the stratigraphy at Çatalhöyük is mainly house-based. Each house is rebuilt at its own pace and on its own footings. There is very little inter-cutting from house to adjacent house. Although there are some entrance holes between houses, the columns of individual houses rarely inter-connect; each follows its own sequence. Houses are built onto open areas at times, but there are few large-scale events that link adjacent houses together.

In the earlier phases of our work in 1993–1999 we had managed to fit newly excavated buildings into the Mellaart system of levels in the South Area. This is partly because we were mainly working in the lower levels where stratigraphical relationships seem more regular. But as we worked in the South Area in 2000–2008, particularly in the upper levels, the system of Mellaart's levels began to break down. For example, counting upwards from buildings excavated by Mellaart, B.65 in the 65-56-44-10 sequence should have been Level V. This would have placed Buildings 56–10 in Levels IV to II. But still higher levels have been excavated in TP where the Polish team had stratigraphical links to Mellaart's Levels 0-III. So, clearly something was wrong. In response to these difficulties, in 2008 we introduced a new system of levels based on the stratigraphical relations in our own overall Harris Matrix. Our matrices for the separate South, IST, TP and North Areas were not linked stratigraphically, so we developed separate sequences of levels for each area. In each separate sequence, levels are given letters from lower (nearer A) to higher (nearer Z). Thus South G is lower than South H. It is important to note that the letters given to levels in each sequence are not linked to levels in other sequences. Thus it is not the case that South G is equivalent in time to North G. The current understanding of the relationships between levels in the different sequences, based on ceramic and chipped stone data, is shown in table 1.2.

Table 1.3 gives the numbers of units excavated in each level in the North and South Areas. This shows a much more even spread of units than was achieved in the earlier campaigns (the differences between the unit totals in tables 1.1 and 1.3 result from the ways in which Levels are attributed via Space). While 100 building numbers have been assigned by the project, it is important to note that not all buildings were fully excavated. Many buildings could not be excavated for structural reasons or could only be partially excavated as they had to be left on display. Table 1.4 provides the amount excavated per building for 81 at least partially excavated buildings. For the more fully excavated buildings, table 1.4 also shows the numbers of burials and an elaboration index. The latter measure of architectural complexity was devised by Hodder and Pels (2010) and it simply sums, for any one phase of a building, the numbers of floor segments, basins, benches, installations (protuberances on walls, including bucrania and other animal fixtures), pillars and paintings in the main room of the building. It could be objected that this measure mixes mundane architectural features such as floor segments and wall pillars on the one hand, and symbolic features such as paintings and bucrania. However, it is difficult to discern which features are more or less symbolic – for example, floor segments seem to have played an important symbolic role in differentiating areas of activity. The simple sum of features does seem to provide a measure that correlates with other variables such as numbers of burials (see Hodder 2016).

Temporal groupings of levels	*South*	*North*	*Cal BC*
Final	TP.O-R and TPC Trenches 1 and 2 (B109 and 115)		6300–5950 BC
Late	GDN	North.H,I,J and IST	6500–6300 BC
	South.T. TP.N. TPC B110 and B150		
	South.S. TP.M. TPC B150 and B122		
	South.R		
	South.Q		
	South.P		
Middle	South.O	North.F, G	6700–6500 BC
	South.N		
	South.M		
Early	South.L		7100–6700 BC
	South.K		
	South.J		
	South.I		
	South.H		
	South.G		

Table 1.2. Current understanding of the relationships between levels in the South and North Areas.

The project worked through changing and complex political changes in Turkey that impacted methods in various ways. For example, the attempted coup in July 2016 resulted in the season's work being curtailed and shifted to following years. Towards the end of the project it became increasingly difficult to export material from Turkey. For example, we were not able to export phytolith and starch samples until too late for full analysis that could be included in this volume (see Chapter 7). At times, as in the case of macrobotanical remains, samples could be sent within Turkey and studied, for example, at the British Institute in Ankara. In other cases, all analysis had to be carried out in the labs at the site.

It should also be noted that most of the laboratory teams changed through the lifetime of the project. Many of the chapters in this volume mention such changes. Some of the changes were the result of individual decisions by lab members and heads, but I also made the decision to change lab heads from time to time in order to promote critical reflection and diversity of debate. I am extremely grateful to all the lab heads who through time contributed to constructive and productive research. It is the sum total of this work that is reported on here.

Synopsis and integration

As noted above, our main research aims in the 2009–2017 seasons were to understand how social and economic organisation varied through time, especially in the middle levels. There was also a continued focus on the social geography of the settlement and the role of the 'history houses' that were rebuilt more times, were more elaborate (see elaboration index above) and had more burials beneath the floors. During this period there was also an emphasis on cognitive questions and the organisation and social efficacy of ritual and religion. These questions led us to examine the entanglements between humans and their material worlds at various scales, from the overall use of the landscape around the site, to the arrangement of buildings on the site, and to the social lives of the inhabitants of the mounds. These themes will be the subject of the summary of our findings as described in this volume.

2000–2008		*2009–2017*			*Total (1993–2017)*		
Levels - North Area	*Number*	*Units*	*Spaces*	*Buildings*	*Units*	*Spaces*	*Buildings*
North ?F		17	3	1	50	7	2
North F	3	554	21	8	814	23	9
SCRAPE ?G	355						
North ?G	86	290	16	6	1,038	37	7
North G	936	1,744	44	12	4,584	63	14
SCRAPE ?H	2						
North ?H	11	13	1	0	73	2	0
North H	471	208	23	11	854	37	13
North I	224	15	6	2	223	12	3
North J	25				29	1	1
Total	**2,113**	**2,841**	**114**	**40**	**7,665**	**182**	**49**
Levels - South Area	*Number*						
South G	1				102	1	0
South ?H		18	1	0	18	1	0
South H		58	1	1	75	2	1
South ?I		41	1	0	41	1	0
South I					11	1	0
South ?J		14	2	1	15	3	2
South J		3	1	1	342	5	2
South ?K	1	59	3	2	510	4	2
South K		332	6	2	685	8	4
South ?L	25	2	2	1	432	12	2
South L	29	260	11	4	611	12	5
South ?M	160	21	5	3	785	28	10
South M		90	7	2	90	7	2
South ?N		30	1	1	48	2	1
South N		360	6	3	360	6	3
South O	1	1,428	10	7	1,440	10	7
South ?P	10				10	1	0
South P	334	221	9	2	570	12	2
South Q	429	9	4	2	470	11	3
South R	377				390	7	3
South S	493	1		1	572	5	1
South ?T	33				39	1	0
South T	27	9	1	0	73	4	1
Total	**1,920**	**2,956**	**71**	**33**	**7,689**	**144**	**51**
TOTAL	**4,033**	**5,797**	**185**	**73**	**15,354**	**326**	**100**

Table 1.3. The numbers of units excavated in each level.

The landscape – changing perspectives

The Konya Plain is part of the Central Anatolian Plateau, and is made up of Pleistocene lake marls. Located at about 1,000m altitude, the central plateau has a continental climate with a yearly precipitation (winter snow and spring rain) of <400mm (only <350mm in the Konya Plain itself). The marls on the Konya Plain are not suitable for early agriculture. They are impermeable, nutrient-poor and become salty with heavy evaporation. However, bordering the lacustrine plains, the alluvial fans are very suitable. Çatalhöyük is located in the centre of the largest of the fans on the south edge of the plain – the Çarşamba Fan, fed by the Çarşamba and May Rivers from the southwest. The Çarşamba Fan was created in two phases (Roberts et al. 1999), from 7600–7100 cal BC to 5100 cal BC and around 3700 cal BC.

The data collected from the site in the 1995–1999 excavations provided support for the view based on the KOPAL sedimentological studies (Roberts et al. 1999) that the site was located in a wetland environment. For example, a herbaceous riparian flora was well represented in the seed assemblages (Volume 4, Chapter 8). The wetland species in the flora recovered include club-rush and *Alopecurus* sp. The assemblage of bird bones indicated a clear predominance of wetland species (Volume 4, Chapter 3 and this volume, Chapter 13). *Unio* and other freshwater shells found in large numbers at Çatalhöyük confirm that plentiful fresh-water was close to the site (Volume 4, Chapters 5 and 6 and this volume, Chapter 3). Most of the small fish on the site are Cyprinidae and the rest are loaches. These species would have thrived in small pools across the local landscape (Volume 4, Chapter 15 and this volume, Chapter 12).

Previous work had emphasised the scale of spring flooding. Geomorphological analysis (Boyer et al. 2006) had suggested that the site was located on a natural marl rise where it was protected from severe spring flooding that deposited 'backswamp clay' around the site. Phytolith evidence, however, suggested that cereals were grown in dryland habitats (Rosen 2005) and thus must have been located in fields on the nearest dry terraces, about 12km from the site (Roberts, Rosen 2009).

More recent research (Volume 9, Chapter 3) started to provide a rather different view and in Chapter 2 Ayala et al. attempt to disentangle the different views that have been presented regarding the palaeoenvironment of Çatalhöyük (see also Doherty 2017). Rather than the previous perspective of the alluvial plain as a very marshy environment subject to significant seasonal flooding (Boyer et al. 2006; Roberts, Rosen 2009; Roberts et al. 1999), new coring on a more detailed scale near the site has led to an alternative model that during the occupation of Çatalhöyük the alluvial plain was dominated by small low-energy channels that were interconnected to create a mosaic of dry and wet patches of some diversity and with a variety of rich habitats.

Ayala et al. in Chapter 2 summarise the landscape sequence in their figure 2.10. After the retreat of the Pleistocene Konya lake in Phase 1 there was a single channel leading through a rolling landscape of marl, with height variations of up to 7m. An interconnected branching system of water channels emerged in Phase 2. In Phase 3, at the time of Çatalhöyük East, the rolling landscape is more filled-in with alluvium but there is also expansion of the branching system of channels. In Phase 4, at the time of the Chalcolithic West Mound or later, there is a shift to a reduced range of channels, leading ultimately to the formation of the Çarşamba, which has more recently been channelised as part of an intensive irrigation system on the Konya Plain.

Other evidence supports this scenario of the landscape around the site. In Chapter 3, Veropoulidou discusses the archaeomalacological results that indicate a landscape with localised areas of dry conditions alongside river channels and smaller or larger pools. The molluscan remains indicate the presence of rich and variable freshwater vegetation throughout the sequence of habitation, and similar results are derived from bioarchaeological, phytolith and micromorphological studies (Chapters 5–7 and 14).

Changing climate

In general terms (Kuzucuoğlu, Roberts 1997; Roberts et al. 1999; Kuzucuoğlu 2002; Roberts et al. 2008), Anatolia experienced a rise in humidity at the end of the Pleistocene. Early Holocene average annual precipitation was higher than at present (Roberts et al. 2001). Many of the early sites in the region are located near springs and river mouths or in wetlands, such as Suberde, Pınarbaşı, Canhasan III and Boncuklu. The occupation of Çatalhöyük coincided with a period of relatively warm, wet conditions in comparison to the cold, dry Younger Dryas period. The earliest occupation at Çatalhöyük seems to be after a relatively dry period towards the end of the eighth millennium BCE (Ayala et al., this volume, Chapter 2). The evidence, based on isotope sequences including that from Nar Lake (Roberts et al. 2016), indicates that at the end of the eighth millennium and in the first 400–500 years of the seventh millennium BCE there was an intermediate period of moisture followed by a relatively dry phase leading up to ca 6200 cal BCE.

There is mounting evidence to support this picture of increasing dryness through the Çatalhöyük occupation. In Volume 9, Chapter 11, Pitter et al. had shown change through time in stable isotopes obtained from fatty acids in ceramic residues from the different levels at Çatalhöyük (see also Roffet-Salque et al. 2018). The isotopes suggested a trend towards less depleted values, perhaps best interpreted in terms of a move towards a more arid climate. Charcoal data suggest that the vegetation of the region prior to occupation is characterised by oak-pistacia woodland, typical of fairly dry conditions. The early occupation of Çatalhöyük in contrast coincides with the development of oak steppe-parkland characteristic of wetter conditions.

Building	*Area*	*Level*	*% excavated*	*Complete in plan*	*Internal msq*	*Side room msq*	*Plat-form msq*	*Burnt Y/N*	*Elabration index (Eı)*	*Total no. of individuals*	*No. of known rebuilds*
1	North	North G	100	Y	37.96	9.83	11.91	Y (partially)	33	61	2 (B1-B.5)
2	South	South K	75	Y	22.13	5.14	6.27	N	16	0	no known rebuilds
3	North/ BACH	North G	100	Y	27.54	6.26	10.11	N	17	8	no known rebuilds
4	South	South L	50	N	NA	NA	NA	N	NA	0	no known rebuilds
5	North	North F	50	Y	39.04	10.37	2.59	N	10	4	2 (B1-B.5)
6	South	South L	25	Y	38.89	10.06	6.85	N	9	10	4 (S.VI.10-B.24-B.6-B.17)
7	South	South L	0	Y	NA	NA	NA	N	NA	0	5 (S.VI.8-B.20-B.7-B.16-B.18)
8	South	South M	0	Y	NA	NA	NA	N	NA	0	5 (S.VI.1-B.8-B.21- B.22- B.23)
10	South	South T	75	N	NA	NA	NA	N	NA	0	4 (B.10-B.44-B.56-B.65)
12	North	North ?H	15	N	NA	NA	NA	N	NA	1	2(B.12-B.77)
13	North	North H	15	N	NA	NA	NA	N	NA	0	no known rebuilds
16	South	South K	25	Y	23.73	NA	NA	N	12	0	5 (S.VI.8-B.20-B.7-B.16-B.18)
17	South	South K	75	Y	35.68	9.05	6.11	N	22	22	4 (S.VI.10-B.24-B.6-B.17)
18	South	South J	25	Y	24.15	4.16	NA	N	3	1	5 (S.VI.8-B.20-B.7-B.16-B.18)
20	South	South M	0	Y	NA	NA	NA	N	NA	0	5 (S.VI.8-B.20-B.7-B.16-B.18)
21	South	South L	0	Y	NA	NA	NA	N	NA	0	5 (S.VI.1-B.8-B.21-B.22-B.23)
22	South	South K	0	Y	26.32	NA	NA	N	6	0	5 (S.VI.1-B.8-B.21-B.22-B.23)
23	South	South J	50	Y	21.88	4.38	NA	N	6	2	5 (S.VI.1-B.8-B.21-B.22-B.23)
24	South	South M	25	Y	NA	NA	NA	N	12	0	4 (S.VI.10-B.24-B.6-B.17)
40	South	South M	<25	N	NA	NA	NA	N	NA	0	no known rebuilds
42	South	South R	50	N	18.75	NA	NA	N	NA	8	no known rebuilds
43	South	South L	75	Y	24.13	4.92	5.7	N	NA (very eroded + excavated by Mellaart)	9	4 (B.43-B.160-B.161-B.162)
44	South	South S	100	Y	25.72	NA	13.51	N	10	16	4 (B.10-B.44-B.56-B.65)

Table 1.4. Summary information about buildings excavated.

Infilling material	*Infill/ closure*	*Part occupation sequence only*	*Complete occupation sequence*	*Construction phase excavated*	*Walls excavated*	*Notes*
clean infill	Y	N	Y	Y	Y	
abandoned	Y	Y	N	partially	N	
clean infill	Y	N	Y	Y	Y	
clean infill	Y	N	Y	Y	Y	
clean infill	Y	Y	N	N	N	
unknown	NA	Y	N	Y	N	Mostly excavated in 1960s. Finished by Hodder Team in 1999
unknown	NA	N	NA	NA	N	Outline only. Excavated by Mellaart Team as Shrine E. VIII.8. Excavated in 1960s leaving extant walls only
unknown	NA	N	NA	NA	N	Outline only. Mellaart Shrine 1.VII. Complete in plan. Excavated in 1960s leaving extant walls only. Walls recorded by Hodder team.
NA	N	N	Y	Y	Y	Excavated by team from Thessaloniki (96 – 97)
N	N	N	N	N	Y	
N	N	N	N	N	N	
clean infill	N	Y	N	N	Y	Mellaart shrine E.IX.8
clean infill	Y	Y	N	Y	partially	Sp.182 has not been fully excavated
unknown	Y	Y	N	N	N	Mellaart shrine E.X.8
unknown	NA	NA	NA	NA	partially	Mellaart E.VII.1. Outline only
N	N	N	N	N	N ?	Mellaart shrine VIII.1
N	N	N	N	N	recorded	Mellaart shrine 1. E.IX.I Hodder Team recorded walls
uknown	N	Y	N	Y	N	Mellaart shrine E.X.1
clean infill	Y	Y	N	N	N	Mellaart S.10.VII
unknown	N	partially	N	partially	partially	Mostly excavated in 1960s. Mellaart house E.VII.2 and E.Vii.12
clean infill	Y	N	Y	Y	Y	Incomplete plan. Heavily eroded and collapsed area on S edge
unknown	N	N	Y	Y	Y	Mellaart E.VIII.27. Complete in plan. Fully defined and excavated by Mellaart as Shrine. E.VIII.27 down to occupation sequence. Re-opened by Hodder Team in 2004 with partial excavation of some features only. Reopened in 2012
clean infill	Y	N	Y	Y	Y	

Building	*Area*	*Level*	*% excavated*	*Complete in plan*	*Internal msq*	*Side room msq*	*Plat-form msq*	*Burnt Y/N*	*Elabration index (Eı)*	*Total no. of individuals*	*No. of known rebuilds*
45	North	North H	25	Y	46.84	11.57	18.49	Y	10	3	no known rebuilds
46	North	North H	0	N	NA	NA	NA	N	NA	0	2 (B.46-B.48)
47	North	North J	25	N	16.87	NA	3.83	N	NA	0	no known rebuilds
48	North	North G	25	Y	12.31	3.37	5.51	N	7	1	2 (B.46-B.48)
49	North	North G	100	Y	15.08	2.25	5.55	N	12	15	no known rebuilds
50	South	South M	75	Y	15.95	3.06	6.59	N	NA	15	no known rebuilds
51	North	North G	100	Y	11.06	NA	3.86	N	8	0	no known rebuilds
52	North	North G	85	Y	44.66	8.45	9.32	Y	42	18	no known rebuilds
53	South	South Q	50	N	NA	NA	NA	N	NA	8	no known rebuilds
54	North	North I	25	Y	28.69	7.53	NA	N	NA	12	no known rebuilds
55	North	North H	25	Y	42.79	NA	7.77	N	NA	0	no known rebuilds
56	South	South R	100	Y	30.98	4.67	13.01	N	7	4	4 (B.10-B.44-B.56-B.65)
57	North	North H	25	Y	26.83	5.8	7.21	N	7	3	no known rebuilds
58		North H	25	Y	26.2	6.79	8.39	N	10	9	no known rebuilds
59	North	North G	100	Y	50.05	29.54	14.53	N	13	1	3 (B.83-B.59-B.60)
60	North	North H	25	N	20.16	NA	NA	N	NA	10	3 (B.83-B.59-B.60)
64	North	North H	25	Y	31.68	NA	NA	N	2	0	no known rebuilds
65	South	South Q	100	Y	24.01	5.27	8.04	N	10	16	4 (B.10-B.44-B.56-B.65)
66	North	North G	25	Y	19.37	NA	NA	N	NA	0	no known rebuilds
67	North	North H	25	N	24.86	7.95	1.49	N	NA	0	no known rebuilds
70	North	North I	0	N	NA	NA	NA	N	NA	0	no known rebuilds
71	North	North I	25	N	NA	NA	NA	N	NA	0	no known rebuilds
75	South	South P	25	N	NA	NA	NA	N	NA	3	no known rebuilds
76	South	South O	50	N	27.49	NA	10.34	Y	NA	5	no known rebuilds
77	North	North G	100	Y	28.19	7.93	8.44	Y	20	37	2 (B.12-B.77)
79	South	South O	25	N	NA	NA	NA	Y	NA	0	no known rebuilds
80	South	South O	75	N	18.7	NA	9.37	Y	25	17	no known rebuilds
82	North	North G	50	Y	22.07	NA	NA	N	NA	0	no known rebuilds
88	North	North H	25	Y	13.77	NA	NA	N	NA	7	no known rebuilds
89	South	South N	75	N	35.11	NA	7.98	N	11	14	no known rebuilds
91	South	South M	25	N	NA	NA	NA	N	NA	4	no known rebuilds
92	South	South M	25	N	NA	NA	NA	N	NA	5	no known rebuilds
96	South	South O	25	N	NA	2.58	NA	N	NA	9	no known rebuilds

Table 1.4 (continued). Summary information about buildings excavated.

Infilling material	*Infill/ closure*	*Part occupation sequence only*	*Complete occupation sequence*	*Construction phase excavated*	*Walls excavated*	*Notes*
clean infill	Y	Y	N	N	N	
unknown	N	N	N	N	Y	Almost completely eroded
unknown	Y	Y	N	Y	Y	
clean infill	Y	Y	N	N	N	
clean infill	Y	N	Y	Y	N	
NA	N	N	N	Y	Y	Excavated by Mellaart as E.VII.9
clean infill	Y	N	Y	Y	Y	
rubble	Y	N	Y	N	N	
clean infill	Y	Y	N	Y	Y	Highly eroded building
unknown	N	Y	N	N	Y	
clean infill	Y	N	N	N	N	Excavated to last occupation sequence only. Truncated ın the NW corner
clean infill	Y	N	Y	Y	Y	
clean infill	Y	N	N	N	N	
clean infill	Y	Y	N	N	N	
clean infill	Y	N	Y	Y	Y	
unknown	N	Y	N	Y	Y	Very eroded, only the eastern part of the building survived (25% of building survived)
clean infill	Y	N	N	N	N	
clean infill	Y	N	Y	Y	Y	
abandoned	Y	N	N	N	N	
unknown	N	Y	N	N	N	
NA	N	N	N	Y	Y	Heavily eroded
clean infill	Y	Y	N	Y	Y	Truncated
unknown	Y (Mellaart)	N	Y	Y	partially	Excavated to the floor level by Mellaart. Re-opened in 2008, fully excavated ın 2009. Very eroded and damaged by Neolithic fire. Truncated to the north
rubble	Y	N	Y	Y	Y	
rubble	Y	Y	N	N	N	
rubble	Y	Y	N	N	partially	Side room truncated
clean infill	Y	Y	N	N	N	
clean infill	Y	Y	N	N	N	Eroded
clean infill	Y	Y	N	N	N	Underneath 76
clean infill	partially	Y	N	N	N	FT South
	partially	Y	N	N	N	FT South. Correlated to Mellarts's house E.VII.18
abandoned	Y	Y	N	N	N	

Building	*Area*	*Level*	*% excavated*	*Complete in plan*	*Internal msq*	*Side room msq*	*Plat-form msq*	*Burnt Y/N*	*Elabration index (Eı)*	*Total no. of individuals*	*No. of known rebuilds*
97	South	South O	100	N	21.84	3.89	8.28	Y	9	14	no known rebuilds
100	South	South M	25	N	NA	NA	NA	N	NA	1	no known rebuilds
101	North	North G	<25	N	NA	NA	NA	N	NA	2	no known rebuilds
102	North	North G	50%	N	NA	NA	NA	N	NA	36	no known rebuilds
108	North	North H	25	N		NA	NA	N	NA	9	no known rebuilds
112	North	North G	50	N	28	NA	NA	N	NA	0	2(B.112-B.119)
113	North	North G	0	Y	31.86	10.05	NA	Y	NA	0	no known rebuilds
114	North	North G	75	N	18.75	NA	1.65	N	11	16	no known rebuilds
118	South	South H	25	N	NA	NA	NA	N	NA	0	no known rebuilds
119	North	North F	75	Y	27.52	5.97	11.33	N	12	0	2 (B.112-B.119)
122	TPC	South S – TP.M								12	
128	North	North G	50	Y	16.76	1.62	NA	N	NA	12	no known rebuilds
129	North	North H	25	N	45.8	NA	NA	N	NA	23	2 (B.129-B.131)
130	South	South L	25	Y	NA	NA	NA	N	NA	0	no known rebuilds
131	North	North G	100	Y	46.6	15.7	10.13	Y	18	42	3 (B.129-B.131 - B.139)
132	North	North F	75	N	72	NA	8.75	N	15	11	no known rebuilds
139	North	North F		N						0	3 (B.129-B.131-B.139)
150	TPC	South T – TP.N								26	
160	South	South K	100	Y	22.8	6.7	3.07	N	5	4	4 (B.43-B.160-B.161-B.162)
161	South	South J	50	N	17	NA	NA	N	NA	8	4 (B.43-B.160-B.161-B.162)
162	South	South J	50	N	16.84	NA	NA	N	NA	0	4 (B.43-B.160-B.161-B.162)
163	North	to be assigned	0	N	NA	NA	NA	N	NA	0	no known rebuilds
165	North	to be assigned	0	N	NA	NA	NA	N	NA	0	no known rebuilds
166	TPC	to be assigned								6	
167	North	to be assigned	<25	N	NA	NA	NA	N	NA	4	no known rebuilds

Table 1.4 (continued). Summary information about buildings excavated.

and only rare consumption of venison. Fallow and roe deer were uncommon. People ate pork very rarely. Bear remains arrived at Çatalhöyük only occasionally, in the form of pelts. As already noted, goats were a minor resource in relation to sheep, but Chapter 9 finds evidence of a symbolic interest in goat crania and horn cores. The bird bone data discussed in Chapter 13 also indicate a symbolic or social use of the landscape, since many birds seem to have been taken for use as feathers and Russell and McGowan (2003) had earlier indicated their use in costume and dance.

The inhabitants of Çatalhöyük followed a strong set of socio-symbolic codes in the arrangement of life and death in the house (see Chapters 15 and 17, and Hodder, Cessford 2004). North and south parts of the house, and the different platforms, were used differently for burial and activities. Despite variability that will be discussed further below, each house seems to have followed much the same rules. It might have been expected that there would be a high degree of conformity in the ways in which the landscape was used. In fact, the opposite is found. All the chapters in this volume point to variability in the ways in which the landscape was exploited. For example, in Chapter 5 Bogaard et al. describe a broad spectrum of domesticated and wild plants made use of at the site. Stable carbon and nitrogen isotope analysis of crop remains suggest that growing conditions at Çatalhöyük were highly variable, at least partly because of the complex mosaic of localised conditions created by the branching and interconnected water systems (see Chapter 2). Building 52 is associated with the broadest range of growing conditions, probably as a result of using near and distant plots in the landscape.

Much the same variability characterises the exploitation of animals. In Chapter 11 Pearson et al. note that animal carbon and nitrogen isotope values are surprisingly diverse, reflecting natural and anthropogenic feeding regimes that result in large ranges (up to 9.5‰ variation) in isotope values even amongst animals with narrow, herbivorous diets. This variability is especially true of sheep that often consumed substantial amounts of C_4 plants. Much the same variability is found in several other aspects of the evidence. For example, in Chapter 4 Kabukcu describes the variety of mainly oak, juniper and elm used in house construction. There does not seem to be a clear pattern of different types of wood used for different functions. Clays were extracted from local sources for floor and wall plasters, and in Chapter 14 García-Suárez et al. point to the enormous range of different floor plaster types used through time.

But can we be more specific in identifying how the diverse mosaic of landscape conditions was used by social groups on site? For example, was the landscape managed differently, or were its products consumed differently, by different sex and age groups? The answer very broadly seems to be no (regarding sex) and yes (regarding age). In Chapter 11 Pearson et al. note nearly identical carbon and nitrogen values for males and females, and this result accords with burial evidence for a lack of gendered difference in the orientation and placement of graves and in most aspects of burial goods (Chapters 15 and 17). For example, both men and women had skulls removed after death. Instead the population differentiated between social groups through age (Pearson, Meskell 2015). Isotope data show clear differences in diet according to age. It is not that clear what the isotopic differences in diet between age groups represented, but in Chapter 11 Pearson et al. suggest that the young may have eaten more wild meat and more plants with more nitrogen, but it is also possible that older groups may have been consuming more C_4 plants. Age differences are echoed in the figurine corpus and grave goods (Nakamura, Meskell 2009). In Chapter 15 clear age differences are noted in the placement of burials in houses. As Haddow et al. (ibid.) note

> adult (including adolescents) primary burials occur more frequently than subadult primary burials in the east, northeast and west platforms, as well as under central floors, whereas subadults predominate within side rooms, north and northwest platforms, or the south area of the central room. Secondary deposits of adult individuals occur more often than subadults in north and east platforms, and exclusively in the south and west areas of the central room. Secondary deposits of subadult individuals occur most often in side rooms and northwest platforms. Secondary deposits of human skeletal remains are never found in the southern area of the central room or under central room floors.

A number of indicators suggest that age was the main social determinant at Çatalhöyük rather than sex or status. This is clear in the burial data in Chapter 17. For example, Vasić notes that ‘the older the buried individual was, the more likely he or she was to be associated with pigments and decorated with beads’.

Variation by house and house group

While sex- and age-based differences in landscape use and diet may have been relatively subtle, there is clearer evidence for variation by house or house grouping. An important new discovery in this new phase of research is that the group of people buried together in a house also ate together and/or obtained food resources from the same part of the landscape mosaic. It is important to note that some buildings have many burials (up to 62 in Building 1) while others have few or none (see Chapter 15). It seems very likely, therefore, that at least some of the people buried in buildings with many burials had lived in other buildings. In Chapter 11 Pearson et al. find statistically significant evidence for dietary differences between those buried in different buildings. This is significant for the Middle and Late period levels. That such differences concern both diet and production is suggested by the functional ecology approach described in Chapter 5 that finds building-specific signatures. The crops and weed flora in individual buildings suggest a distinct set of growing conditions, and perhaps particular fields in the arable landscape. This notion that each house or house-group used the landscape differently is also suggested in some of the faunal data. For example, in Chapter 8 a distinctive cluster of pathological caprine metapodia is reported in the South Area's Building 104, and in Volume 8, Chapter 11, a distinct set of sheep pathologies was associated with the 65-56-44-10 sequence of buildings. The residents of B.65-56-44 also had unusual access to, or atypical consumption of, wolf remains (Chapter 8).

As noted above, the buildings with multiple burials are often those with more symbolic and other architectural elaboration, and they are frequently rebuilt to produce 'stacks' of houses on the same footprint over time. These 'history houses' often seem, based on the evidence above, to have managed the landscape and its resources in distinctive ways. The evidence suggests that the rich landscape mosaic was exploited separately by the different history house groupings (although we shall see other larger and cross-cutting groupings below). The diverse landscape was thus matched by a diverse and fragmented social use.

There is much evidence in this volume for the continuity of history houses and the construction of memories, especially in the Early and Middle periods but also to some degree in the Late period. For example, in the Middle period there is a concentration of secondary burials in both Building 131 and the ensuing Building 129. These two buildings are linked in other ways, including the very rare occurrence of obsidian mirrors in both buildings in association with human remains. In Chapter 17 Vasić et al. note the enormous variability in burial practices and in the deposition of burial goods but they also note some very distinctive continuities, such as the placement of imitation red deer canine beads in two burials in B.6 and the following B.17, occurrence of pigments and associated items in the stack of buildings B.22, B.21 and B.8, belt hooks and eyes in E.VIB.20 and E.VIA.20, rings in E.VII.14 and E.VI.14 and wooden bowls in B.1 and B.5.

Throughout this volume examples are given of elaborate foundation or abandonment rituals that occurred between buildings that were rebuilt on the same footprint. There are cases of gaps between rebuilds, as when latrines, refuse or animal scat are found between rebuilds in the 162-161-43 building sequence (García-Suárez et al. Chapter 14 and Vasić et al. Chapter 17). In Chapter 15 Haddow et al. describe the burials that took place in the time gap between B.132 and B.77. But in many cases there is a focus on rituals that found or end houses as part of the rebuilding process. For example, in Chapter 8 there is discussion of animal deposits, often indicating feasting, in the abandonment and/or founding of Buildings 77, 80, 97 and 160.

It might have been expected that the more continuous history houses that dominated the co-burying and co-eating groups would be inhabited by people with distinct and better diets. Since we cannot identify between those who lived in a history house and those who were buried in a history house, it is difficult to test this question. But we can at least see if those buried in the history houses had different diets from or were better protected from disease than those not buried in history houses. In Chapter 16 correlations are sought between indicators of health and well-being and elaboration in history houses. The authors find that the extent of building elaboration and its inclusion in a history house sequence is unrelated to the living conditions of the buried individuals. In Chapter 11 Pearson et al. also note no difference in diet related to building elaboration or to numbers of burials in a building. Even those individuals whose heads had been removed after burial had no isotopic evidence of dietary distinction.

As was discussed in Volume 10 and will be discussed further in Volumes 12, 14 and 15, the current project has been unable to find much in the way of evidence linking history houses to high-status individuals. In Chapter 17, Vasić et al. confirm earlier evidence that building size and elaboration do not correlate with the presence of different aspects of burial treatment, and the percentage of buried individuals with pigment associations, adornment or objects did not depend on the size of building nor its elaboration. Similarly, no exceptional treatment has been observed in the burial assemblage in buildings defined as 'history houses'. It seems necessary to conclude, therefore, that the history houses acted as foci for a group of houses, the inhabitants of which worked in the landscape and ate together, buried their dead together, and probably conducted rituals together (in the history house buildings), but within that collective there were not marked differences in the status of individuals. This conclusion is supported by the evidence that the more elaborate houses did not have more evidence of storage capacity or productive capacity (see Volume 8, Chapter 1 and Hodder 2016).

It is for these reasons that I have preferred to call Çatalhöyük society egalitarian. Clearly there were marked differences in the elaboration of houses, and the more elaborate houses had more burials and were probably rebuilt more frequently on the same spot. These houses managed memory making for a group of houses, but there is no evidence that this socio-ritual role led to the emergence of houses or individuals with better diets or better health and no evidence that the more elaborate houses had better access to storage and production. There is no evidence that the inhabitants of the less elaborate houses 'worked for' the central and more elaborate houses within the history house group. Rather, the evidence suggests collaborative working of the landscape by history house groups, with a relative lack of differentiation within those groupings. The fact that the more elaborate buildings are not larger than the non-elaborate buildings also suggests that those living in the more elaborate houses were not able to control the labour of those in other buildings. It also suggests that it is not the case that large groups of people congregated in the elaborate buildings for eating and ritual. Evidence from reconstructions of the size of feasting groups (see Chapter 8 and see Demirergi et al. 2014; see also Bogaard et al. 2009) also suggests relatively small-scale groupings for the most part. Overall, then, the landscape mosaic of varying and rich resources seems to have been used by a mosaic of small social groupings.

This conclusion is especially true of the Middle period. The focus on history making and continuity is a way of establishing horizontal linkages and dependencies. All those that buried in the same history houses could call upon others in the same grouping during hardship. But we have seen pressures build up in this system. The numbers buried in history houses increase, as does the scale and intensity of ritual in houses. By Level North G in Building 1, 62 people were buried and in Building 52 there is an impressive pile of 13 wild bull horns stacked above a bucranium. By the end of the Middle period, it is as if the history houses are competing with each other in order to attract adherents. This may have been simply an accretionary process, or it may have been impacted by climatic and environmental change as described above. But the pressures were real, with decreases in health and increases in disease. As we have already seen, in the Late levels there is a gradual shift. Rather than building alliances through history making, there is greater focus on building horizontal links between increasingly independent and self-sufficient house units. This is particularly seen in the emergence of large houses with attached yards and work areas, a decreased focus on continuity, and the shift of symbolism from installations in walls to mobile stamp seals and decorated ceramics.

Other supra-house groupings: the neighbourhood

Already in the Early and Middle periods there is evidence of other types of grouping which may or may not correlate with the history house groupings. We do not know where the people lived who were buried in the elaborate houses. It is possible that they lived immediately adjacent to them in local neighbourhoods, and this has been suggested by Düring and Marciniak (2006), who have described the settlement pattern at Çatalhöyük and elsewhere as made up of neighbourhood clusters. But it remains unclear whether the houses in the history house groups were all adjacent to each other or not. Groupings of history houses occur in both the South (Mellaart's Shrines 10, 8, 1) and the North (Buildings 1, 129, 77) Areas, which seems to suggest that the houses affiliated to individual history houses were more widely dispersed. It is possible, then, that history house groupings of houses cross-cut neighbourhoods.

There is now much evidence for local clusters of houses with distinct architectural or other features. The evidence for groupings of adjacent houses based on architectural and symbolic features are discussed by Mazzucato (2019), and in this volume there is much evidence based on burial practices. The highest numbers of burials we have found at the site occur in three adjacent buildings in the North, Buildings 1, 77 and 131. In Chapter 17 Vasić et al. note that this group, as well as B.129 which was built on top of B.131, also has the largest concentration of secondary burials. Buildings 1 and 131 also had burials of young people with very high numbers of beads (ibid.). Another possible example related to this grouping and adjacent B.114 is the deposition of stone maceheads in burials. As Vasić et al. note, these artefacts are not common on site, and they are largely absent from burials. Only four examples have been recovered from burial features in 25 years of excavation; complete maceheads were found in burials in B.5 (below B.1) and B.114, whilst two fragments were recovered from the infill of two burial features in B.131.

In the South Area two groups of adjacent houses are suggested by the burial data (Chapter 17). Collars of boar tusks are extremely rare on site, but two individuals with sets of two and five collars were buried in the neighbouring buildings (E.VII.12 and B.50) in the South Area. The early houses B.160-B.161-B.43-B.17-B.6 also form a distinctive group. These buildings see the repeated placement of wooden planks, carnivore scat and carnivore pellets, which are distinctive practices rarely seen elsewhere on site. The fact that they occur in four burials located in three neighbouring buildings makes them even more exceptional. The fact that these buildings are not all exactly contemporary is of interest: it suggests that these local neighbourhood groups existed through time, with practices passed down between buildings in the group. The same was noted for the North B1, B.5, B.114, B.129, B.131 group, since B.5 is below B.1.

Other evidence for these neighbourhood groupings includes portholes between buildings (for example, B.5 is connected to B.131 by an access hole). There is further discussion of these and other examples of connections between adjacent buildings in Volume 7, Chapter 1. But is there evidence that neighbourhood groupings of adjacent houses produced and consumed together? In Chapter 5, the archaeobotanical evidence for weed assemblages associated with cropping suggests that buildings that were directly adjacent (B.79, B.80), located in the same neighbourhood (B.1, B.52) and/or in the same sector of the settlement (B.63, B.122), shared particular similarities. The authors suggest this indicates some form of supra-household organisation of land use. Neighbouring or nearby houses were linked with cultivation areas that were also proximate in the surrounding arable landscape, and so shared detailed similarities in weed composition. The authors again note that these groupings seem to continue through time since nearby buildings with similar weed compositions may not be precisely contemporary.

As regards consumption, in Chapter 11 Pearson et al. assess the carbon and nitrogen isotopes of humans buried in adjacent groups of houses to see whether there were distinct neighbourhood diets. When six house clusters were compared overall, regardless of phase, the groups' isotopes were found to be significantly different (p=0.001). However, when the data were separated into Middle and Late periods, significant differences were not found within a particular phase.

Open spaces and neighbourhoods. Despite the claims made by Kuijt (2018), there is no evidence that houses at Çatalhöyük were grouped around public spaces or 'plazas'. Mellaart identified 'courtyards' amongst houses, and the current project has referred to these areas as middens. More recent excavation and research by Issavi (forthcoming) has indicated that these spaces were used for a variety of functions and that the term 'open space' may be more appropriate. At times these are spaces dedicated to open-air activities over a long period of time, and in other cases they are shorter term, often occurring in abandoned buildings. They are scattered across the site, and there is no good evidence that they were assigned to particular groups of buildings. Indeed, comparison of the carbon and nitrogen isotopes of humans buried in houses and the expected values for animal bone discarded in adjacent open areas found a mismatch, suggesting that the occupants of houses had not consumed the animal bone found in adjacent open areas (Pearson in Volume 8, Chapter 13).

Nevertheless, it might be assumed that to some degree adjacent houses were more likely to make use of nearby open spaces, even if this usage was more generalised and collective than specific to a tight group of houses. In Chapter 14 García-Suárez et al. describe the diversity of uses of these open spaces. There is much evidence of omnivore faecal material (largely human, given the paucity of boar on site), animal dung and small-scale sweepings from hearths and ovens, dumped construction material, but also fire spots, walls and pits. But in many cases, especially in the upper levels, the open areas are covered in extensive layers of ash and clay, creating relatively clean surfaces on which a range of activities, fire spots and ovens are at times found. There is much evidence that the open areas were carefully managed, and 'cleaned' to reduce negative impacts from organic refuse and to create activity areas. The burning of middens would have helped reduce disease, and also the calcitic clay layers would have had anti-microbial and insect-repellent properties. The build-up of 'midden' refuse was actually very minimal and well controlled. The amount of activity in these areas seems to have been relatively limited – there is limited evidence of trampling and repeated movement.

Rarely, burial took place in open areas. Individual cases occur but there are also 'cemeteries' of burials that occur prior to the construction of columns of houses. After the abandonment of B.132 the open Space 602 was used for a series of burials before the construction of B.77. Other examples of burials in external spaces at Çatalhöyük include a series of neonate and infant burials in Sp.144, an external area that preceded the B.65-B.56-B.44 house sequence in the South Area. Burials also occurred prior to the construction of B.17, the start of Mellaart's 'Shrine 10' sequence.

North v South

Given that there are many fewer open spaces than individual houses, their use for burial and a range of activities suggests some form of collectivity, however loosely defined. There is also evidence of larger-scale groupings. In the North in particular there appear to be 'alleyways' that separate zones of clustered housing. These 'alleyways' were used as open areas in the ways described above and there is no evidence of intensive trampling suggestive of 'streets'. Rather, these linear open areas seem to define zones of houses, although we have found no evidence of cultural practices that distinguish these groupings. There are also, in both the South and North Areas, wall alignments suggestive of a radial arrangement. Chapter 5 discusses the possibility that the resulting 'wedges' may have extended out into the landscape and may have been the basis for the division of the landscape into zones of shared exploitation.

2. Disentangling the palaeoenvironmental reconstructions of Çatalhöyük

Gianna Ayala, John Wainwright, Jerry M. Lloyd, Joanna R. Walker, Rachel Hodara Nelson, Melanie Leng, Chris Doherty and Michael Charles

Introduction

This chapter focuses on the ongoing landscape investigations in the immediate vicinity of the two tell sites at Çatalhöyük, concentrating on how the natural world shaped peoples' engagement with their environment. The chapter builds upon a long history of environmental research in the region, not just at Çatalhöyük (Roberts 1991; 1995; Roberts et al. 1996; 1999; Rosen, Roberts 2005; Roberts, Rosen 2009) but more widely in the Konya Plain (Boyer 1999: 63; Boyer et al. 2006: 684; Boyer et al. 2007; Roberts et al. 2016).

There are two main aims of this chapter. Our first aim is to evaluate previous palaeoenvironmental models and refine them with additional and enhanced datasets from the current programme of landscape research (with field sampling from 2007–2015). The focus of our study has therefore been the high-resolution palaeoenvironmental reconstruction of the alluvial landscape in the vicinity of the site. The modern Çarşamba River flows close to the edge of the site and extends southwards until the termination of the Konya Plain at the limestone hills that border the Taurus Mountains (fig. 2.1). However, the river historically had a branch which flowed between the East and West Mounds. Previous geoarchaeological research has characterised the alluvial plain as a very marshy environment subject to significant seasonal flooding (Roberts et al. 1999; Rosen, Roberts 2005; Boyer et al. 2006; Roberts, Rosen 2009), which was based on sedimentological data derived from nine coring locations and trench sections near the tells as well as the investigation of 16 archaeological sites (four of which date from the Palaeolithic to Bronze Age) further away in the area of the Konya palaeolake (Boyer 1999: 63; Boyer et al. 2006: 684; Boyer et al. 2007).

In this chapter we present the recent programme of research (Ayala et al. 2017) and describe how the current palaeohydrological model we propose has been constructed. The evidence comes from a coring programme undertaken between 2007 and 2015 that targeted 29 coring locations within a radius of up to 1.6km of Çatalhöyük using both more intensive and more extensive sampling protocols than previously utilised in this region (fig. 2.1). A combination of sediment and isotope analysis, and 3D modelling of the stratigraphic sequence has enabled us to construct a more refined understanding of the hydrology and resulting dynamic topography of the low-lying alluvial plain (Ayala et al. 2017). After summarising the results from Ayala et al. 2017, we show how an additional 34 cores taken in 2015 allow us to refine the palaeoenvironmental model further. Pollen and isotope data are used to evaluate the broader landscape and climate context of the site. We also reinterpret the existing hydrological data used as a basis for the spring-flooding hypothesis and demonstrate that its use is highly problematic.

The second aim of the chapter is to highlight how the new geoarchaeological data not only enhance and refine the previous landscape models of the region, but importantly support new ways of thinking about the archaeological record. Our research presents a narrative of landscape formation that is fundamental for any interpretation of how people at Çatalhöyük were entwined with their world, both constrained by the very physicality of the environment and transforming the matter and substances within which they were entangled. Hodder has highlighted the experiences and processes which became part of daily life, from construction of structures to the domestication of foodstuffs, as a complex and dynamic relationship interlacing the materiality of human existence with the environment (Hodder 2011: 160–62; Hodder 2012). Through this understanding of human-thing relationships, people's entanglements with environmental processes have been regarded as a driving force for change and adaption (Hodder 2011).

Our understanding of these processes and behavioural reactions is dependent on accurate and high-resolution reconstruction of environmental parameters. Excavation and off-site investigations at Çatalhöyük have provided intriguing evidence that has been used to suggest lines of this entangled relationship (Roberts et al. 2007; Roberts, Rosen 2009; Love 2012; Charles et al. 2014; Doherty 2017). The classic example of an entangled environment-thing relationship based on one

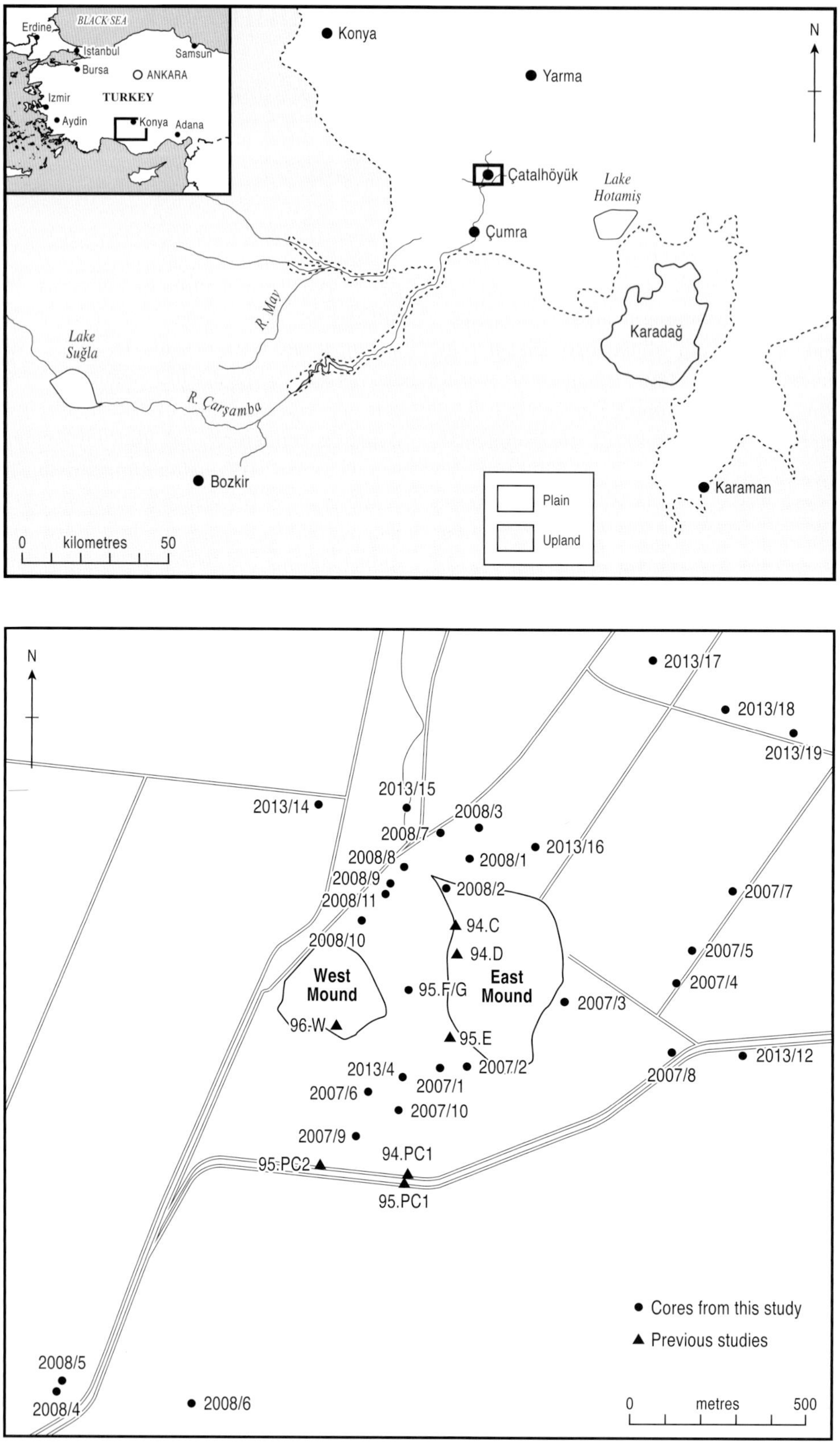

Figure 2.1. Location of site and coring locations (KOPAL and 2007–2013) (Ayala et al. 2017: fig. 1). See fig. 2.4 for additional cores taken in 2015.

of these paleoenvironmental reconstructions is the identification of depressions near the East Mound which have been interpreted as clay extraction pits for the production of clay bricks (Roberts et al. 2007; Hodder 2011: 161–62). These depressions reach the marl sediment underlying the alluvium. Archaeometric analysis of plaster and mudbrick from the site has identified marl as a key constituent in the earliest occupation levels (Love 2012; Doherty 2017). The creation of the pits has been attributed to the occupants' search for appropriate building material in the earliest habitation phases. The pits have also been attributed with providing the perfect damp conditions that would have encouraged the spread of invasive wetland vegetation. It has been suggested that the extraction of clay for building material and the invasion of wetland plant species affected the local topography and hydrology surrounding the site (Charles et al. 2014: 86; Philippa Ryan cited in Hodder 2011). These interpretations are clearly interdependent on the reconstruction of the type of river system near the site and its flood regime, as well as the interpretation of features such as these depressions in the underlying marl near the tell (Charles et al. 2014; Doherty 2017).

The importance of a high-resolution reconstruction of the palaeoenvironment in the vicinity of the settlement becomes increasingly greater as we attempt to interpret behavioural patterns and choices that were open to and taken by the occupants of Çatalhöyük. Previous attempts to reconstruct land-use patterns or 'taskscapes' of Çatalhöyük have been based on environmental proxies (Roberts, Rosen 2009; Bogaard et al. 2014a; Charles et al. 2014; Bogaard et al. 2017). To investigate more closely one of the most fundamental environmental parameters key to the raft of human-thing relationships, namely water, a programme of intensive coring was undertaken in the vicinity of the tell (Ayala et al. 2017) to define the morphology and development of the palaeo-Çarşamba River. This research is designed to feed into the multitude of ways in which the materiality of the archaeological record can be interpreted in relation to the environmental record. It is hoped that with an accurate model of the development of the alluvial landscape surrounding Çatalhöyük we will be able to enhance the interpretations of issues including but not limited to resource provision, the mapping of seasonal tasks, and the development and evolution of agriculture and animal husbandry, as well as the construction and maintenance of the tell itself (Doherty 2017) and the lifeways within.

In the first section of this chapter we review the previous palaeoenvironmental work undertaken at the site and within the Konya Plain. From this basis we assess the reconstruction of the palaeolandscape from this evidence and the implications it had for previous land-use models. We then present the findings to date from the recent programme of coring, followed by a discussion of the evidence and reinterpretation of the palaeohydrology on the alluvial plain upon which the site is located. Finally, we present future goals and directions for the project.

Palaeoenvironmental context

Previous palaeoenvironmental research in the Konya Basin

Çatalhöyük is located in the southern part of the Konya Basin, which is a closed pluvial basin. The present course of the Çarşamba River is to the west of Çatalhöyük, but it has been diverted over the last 50 years and is currently constrained by a concrete channel. Previously it would have had much more ability to move across the low-angled fan surface, after its debouchment from the constrained upland section south of Çumra, and through the sand ridge (a palaeolake shoreline) north of Çumra. Evidence from the past century suggests that in recent historical times, the Çarşamba flowed in a single-thread channel between the East and West Mounds (Rosen, Roberts 2005; Roberts, Rosen 2009).

Early studies of the palaeoenvironment were carried out by the pedologist de Meester (1970: 86), who considered that that the Neolithic soils were formed under 'semi-lacustrine marsh' conditions in a deltaic context. The KOPAL (KOnya basin PALaeoenvironmental research) project concurred with de Meester's (1970) assessment of soil formation. Roberts et al. (1999) defined three broad sets of deposit: (i) a dark, organic clay deposit interpreted as a marsh or backswamp deposit; (ii) the 'Lower Alluvium' made up of dark-grey-brown silt-clay representing a seasonally flooding environment; and (iii) the 'Upper Alluvium', which has a slightly coarser grain size and increased carbonate content. The first of these deposits was interpreted (ibid.: 624) as forming just before settlement of Neolithic Çatalhöyük (ca 7400 cal BCE: Bayliss et al. 2015), the second as being contemporaneous with the occupation of Çatalhöyük (from ca 7000 cal BCE) (Roberts et al. 1999: 625), and the third at some point in the mid-Holocene (ibid.: 627). Roberts and Rosen (2009) asserted that the shift to the Upper Alluvium represented a reduction in alluvial flooding that correlated with the 8.2kya event that is, ca 6250 cal BCE), which is seen in Greenland ice cores (Gasse 2000; Alley, Ágústsdóttir 2005), even though this shift is very poorly dated locally. We will return to the interpretation of the 8.2kya event in the section on palaeoenvironmental proxies.

Despite the previous, extensive work carried out in the Konya Basin, the current palaeoenvironmental reassessment was driven forward by the perspective that previous approaches have tended not to recognise the inherent heterogeneity of dryland environments (Parsons, Abrahams 2009; Müller et al. 2013). Spatial heterogeneity is reinforced by temporal variability in Mediterranean climates (Wainwright, Thornes 2004). Small samples in space and time are therefore problematic when reconstructing dynamic landscape change, which seems to have been the case for Neolithic Çatalhöyük. Thus, the present study has started to focus on more intensive, detailed approaches to sampling and reconstruction.

The 2007–2013 field seasons

Recovery and laboratory methods. A total of 29 sediment cores from different coring locations were taken in 2007–2013 (fig. 2.1). These were focused on the immediate environment surrounding the two tells. In the first instance, the coring programme of 2007–2008 concentrated on an area within 1km of the site, with coring locations spread out in order to ensure representation of potentially varied microenvironments. All cores were extracted with a percussion corer. The cores in 2007 were taken in discontinuous 0.5m sections, while in 2008 a system of coring parallel sets of overlapping cores 1–2m apart was employed to ensure that a continuous sequence was recovered. Cores were taken from 21 different locations roughly following east/west transects. The cores went to a depth of 3 to 5m and were extracted in 0.5m sections which were then described and photographed in the field, wrapped in cellophane and placed in plastic guttering for transportation back to the UK, where they were refrigerated prior to analysis. Subsampling for sediment was carried out at 0.05m intervals on the 2007 cores, while sampling was focused on the identified lithological units on 2008 and 2013 cores. The second set of cores in this set were taken in the summer of 2013, a further eight coring locations were sampled from an area ca 2km^2 centred around the Çatalhöyük settlement mounds, using transects that concentrated on areas that had not previously been sampled. At each location a parallel set of overlapping cores was taken 2–3m apart to a depth of 5m (8 × 4.50m from each borehole; the top 0.5m was discarded due to considerable modern reworking of sediments by agriculture since the Hellenistic-Byzantine period) (Boyer et al. 2006). Following transportation, all cores were then refrigerated to prevent degradation before analysis (Tirlea et al. 2014). The lithology of the cores was studied, and colour, sediment type and grain size were recorded (Melville, Atkinson 1985; Munsell Color Company 1994). Sediment analyses included magnetic susceptibility, loss on ignition, particle size and isotope analysis (Ayala et al. 2017). These observations were then mapped and plotted using RockWorks™ v16 software.

Results. The analysis of the cores allowed us to identify the stratigraphy shown in figure 2.2. Based on this evidence, we determined sediment complexes and plotted this stratigraphical information as a fence diagram, revealing the distribution of the subsurface sequence in the environs of the two tells (fig. 2.3).

The Basal Complex. The lowest part of the sequence is made up of marl, and sands with gravel in units of 0.1–0.5m thickness. The sands and gravels tend to be moderately to well sorted, all subrounded to rounded. Poorly sorted layers are present locally and tend to be constituted by mixed granules of different lithologies derived from the local limestone bedrock and surrounding sand ridges, as well as from igneous and other bedrocks from further upstream in the Çarşamba catchment (up to small pebbles of 5mm). These sands and gravels are typically light brown in colour, although locally they are darker brown. However, lateral variation in texture at equivalent elevations is apparent across the landscape. All locations sampled are capped by a marl layer that varies in thickness from 0.01m (core 2007/7) to 1.04m (2013/12). The marl has a clay texture in the lower parts of the complex and a silty clay texture towards the top of the complex and is predominantly light grey to white in colour.

Because of its ubiquity, the upper part of this complex was taken as the uppermost appearance of marl in the core, and its elevation varies between locations. At its deepest (core 2007/6), the upper boundary is at 6.33m below the modern ground surface, and at its shallowest (core 2007/4) it is at 1.65m. The upper surface tends to be lower between and immediately to the south of the mounds, but it also undulates in a north–south and east–west direction between cores. In the shorter cores, this complex is absent from locations 2008/8 and 9 and 2013/4.

It is of significance that within the Basal Complex there is also the presence of a laminated Dark Clay layer 1.1m below the marl, and another thin Dark Clay layer in between two marl units (see further discussion of the Dark Clay below). Two dates were obtained from the Basal Complex. One from a level of laminated Dark Clay at a depth of 3.865–3.88m produced a date of 27,617–27,011 cal BCE (2σ) on bulk organics. At a depth of 3.82–3.83m, a date of 44,666–42,555 cal BCE (2σ) on large (up to 20mm) angular shell fragments was obtained (Ayala et al. 2017: table 1).

The Lower Complex. The Lower Complex consists of silts, silty clays and clays with some reworked marl in places (fig. 2.2). The boundary with the Basal Complex is locally expressed in a variety of ways, but tends to be

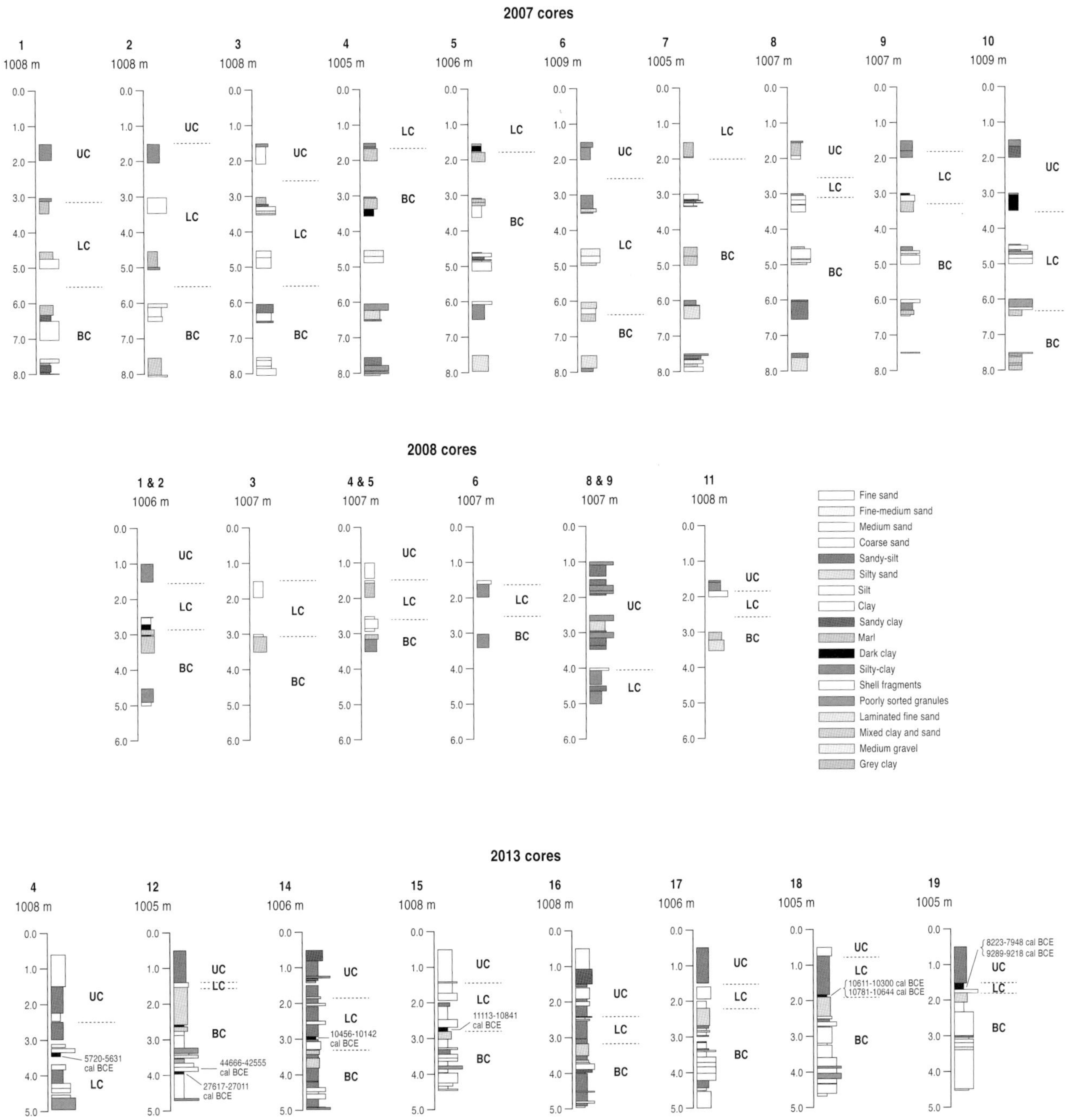

Figure 2.2. Stratigraphic sequence (2007–2013 cores) (Ayala et al. 2017: fig. 2a–c).

abrupt and smooth or occasionally wavy, suggesting erosional contacts. In a number of coring locations there is a dark grey or black clay (subsequently called Dark Clay) directly overlying the marl at the top of the Basal Complex. In some places, the boundary between the Dark Clay and the marl is irregular, suggesting intermixing due to vertic processes, as also noted by de Meester (1970: 86). Elsewhere, where the Dark Clay is absent, the Lower Complex starts with lighter coloured silts and clays, or in the case of core 2007/10, a gravel with silty matrix. In yet another core (2007/4), there is a transitionary boundary of 0.04m with the Basal Complex characterised by a mix of marl and the silt. The upper contact of the marl at the top of the Basal Complex was not observed in the other 11 cores.

The dark grey or black clay layer is also found at higher points stratigraphically in the Lower Complex in a number of cores, while being absent elsewhere. The Dark Clay varies from 1mm thick (2008/3) to between 5 and 15mm thick and is made up of coarse clay to fine silt. The dark and grey clays make up 15% (by number) of the described units in the Lower Complex from the

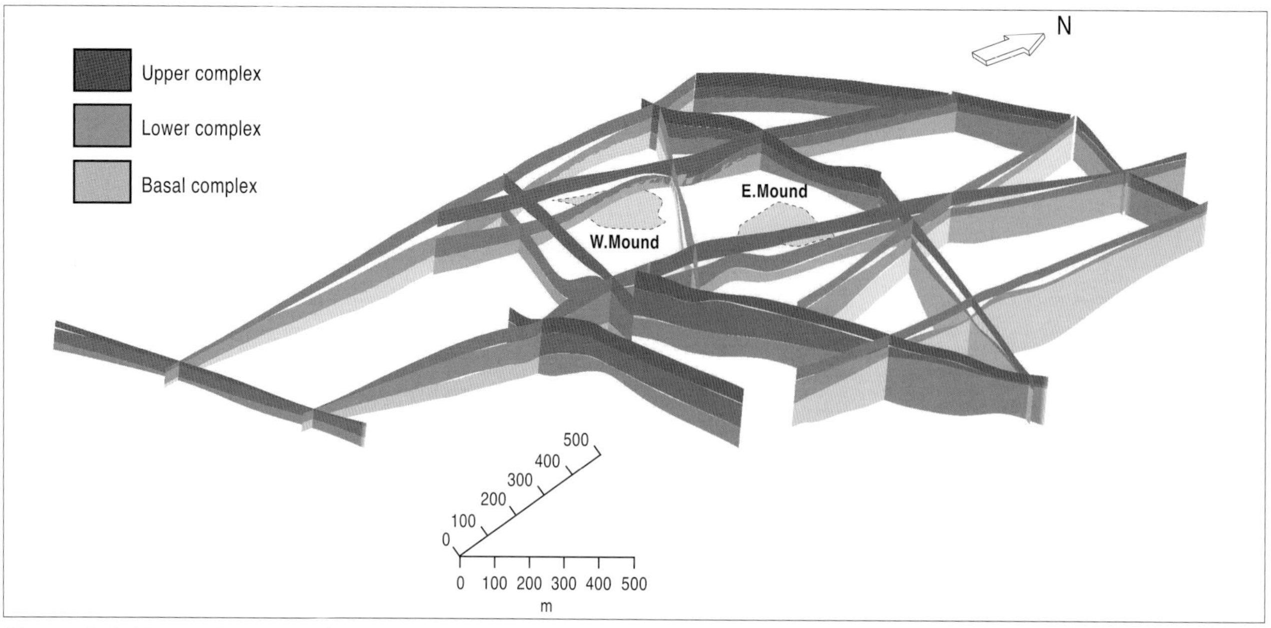

Figure 2.3. Fence diagram of the 2007–2013 cores (Ayala et al. 2017: fig. 5). The sequence at Çatalhöyük is here summarised: full sedimentological description and supporting evidence is available in Ayala et al. 2017.

2007–2013 cores. Of the remaining units in the Lower Complex, 43% are made up of silty-clays or silts, and a further 11% of clays. However, there is also a range of sands, granules and gravels, occasionally with silt matrices. Colours are predominantly from dark grey/grey to light yellowish brown.

Dates were obtained on bulk organic carbon from sediments from seven samples of the Dark Clay layer. The dates (all 2σ) range from 11,113–10,841 cal BCE to 5720–5631 cal BCE (Ayala et al. 2017: table 1).

The Upper Complex. The units of the Upper Complex are predominantly (51%) silty-clays or silts, followed by 11% of clays. Coarse sands are less frequent than in the Lower Complex, but there are still relatively frequently recorded poorly sorted granules (10%) or sandy silts (15%). There is a slight tendency for the Upper Complex sediments to be lighter than Lower Complex sediments (more dark grey/grey to light yellowish browns and fewer black to very dark greyish browns).

The transition from the Lower to Upper Complex occurs at a wide range of depths. In the majority of locations the transition sits at 1.5–2.5m below the modern surface; it can vary from 0.74 to 4.07m. Near the mounds, it is possible that colluvial deposits from the tells lead to some of this variability through the introduction of colluvial material into the alluvial sediments (e.g., Boyer 1999). This topic is currently under further investigation. However, in all locations, the Upper Complex grades up into the modern plough-soil in the upper 0.5m or so.

The Upper Complex is defined by the combination of the colour change from the grey to brown expressions of hue and the lower frequency of coarser material (sand and granule fractions). None of the measured variables (specifically, mass-specific magnetic susceptibility or organic content) were found to be significantly different from the values recorded in the Lower Complex. Further to this, it was unfortunately impossible to identify any unit with sufficiently concentrated bulk organics to provide a radiocarbon date.

Isotopic analyses comparing $\delta^{13}C$ and C/N ratios suggested that the Dark Clay layers were suggestive of standing water, riverbank soils or cutoff deposits, in slightly wetter conditions than riverbank deposits above them, or (where present) below them. Ayala et al. (2017) used these data as well as the lithostratigraphic interpretation to suggest increasingly dry conditions during the early part of the Holocene and thus during the Neolithic occupation.

Additional results from the 2015 cores

In July 2015, a further 34 cores were drilled, using the same truck-based percussion coring system as in 2013. Each core is made up of a series of three offsets in 0.5m intervals from 0.5m below the surface (as the top 0.5m is typically dry, crumbly, disturbed topsoil or track sediment which proved impossible to retain in the sample tubes) to 5.0m below the surface. The first set of offset samples is at 0.5–1.0m, 2.0–2.5m and 3.5–4.0m; the second at 1.0m–1.5m, 2.5–3.0m and 4.0–4.5m, and the third at 1.5–2.0m, 3.0–3.5m and 4.5–5.0m. This offsetting of samples is

a more dryland anabranching system. Lewis et al. (2017) studied *Unio* fossil shells from the site (spanning the occupation period ca 7200–6050 cal BCE) and suggested that the bivalves likely lived in small lakes/wetlands around the site before being gathered and taken to Çatalhöyük. Wet-dry seasonal cycles occur in the shell oxygen isotope ($\delta^{18}O$) profiles with low winter values reflecting winter precipitation and high $\delta^{18}O$ in the summer resulting from evaporation. The most striking trend in the $\delta^{18}O$ data is the drop in maximum summer $\delta^{18}O$ ca 6350 years BCE, interpreted as lower summer evaporation and hence a reduction in seasonality. The shift in seasonal climate could be due to solar-forced climate change beginning ca 6650 cal BCE (see also fig. 2.11). Although these data have previously been used to support the fission-fusion farming hypothesis in that they show wet winter/early spring conditions during the Early Holocene which could have caused flooding, they also support the present reconstruction of seasonally flooding anastomosing channels. Again, localised pockets of wetness persist into the Chalcolithic in this setting, with a single date on a Dark Clay at 5720–5631 cal BCE (2σ). Therefore, our fourth phase probably starts during or just after the Chalcolithic, with a shift to a larger, probably single-thread channel (with a significantly higher flow: see below), which probably persisted to the present period until the Çarşamba was channelised.

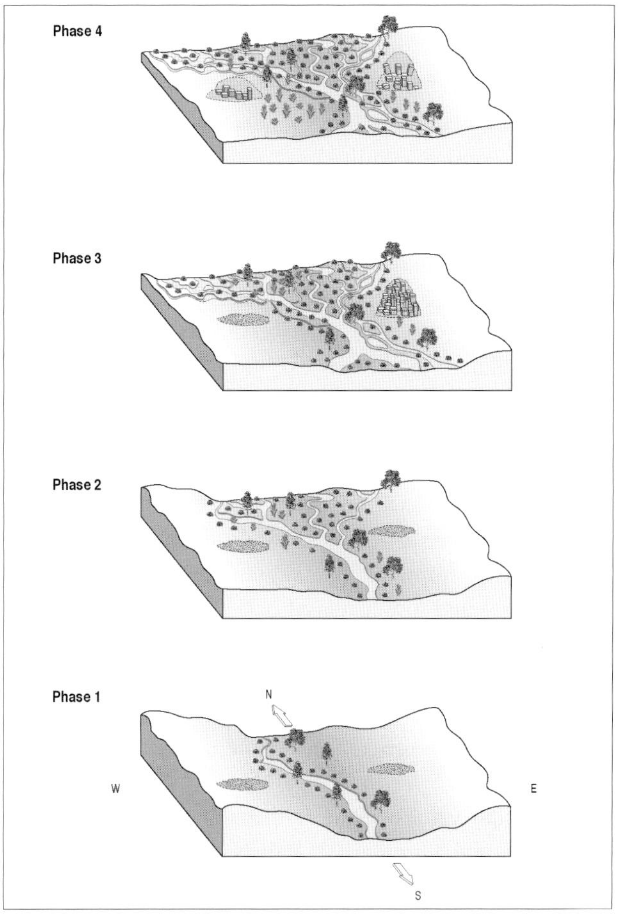

Figure 2.10. Palaeohydrological reconstruction of Ayala et al. (2017: fig. 9) with superimposed interpretation of vegetation patterns based on the discussion in the text.

Pollen and palaeoenvironmental proxies

Vegetation of the Konya Plain during the occupation of Çatalhöyük

Since the palaeoenvironmental review of Çatalhöyük (Charles et al. 2014), more information has emerged regarding the climate and vegetation of the Central Anatolian Plateau (Roberts et al. 2016; Eastwood 2018; Woodbridge et al. 2019). This information helps to refine our previous model for the nature of vegetation and land use around Çatalhöyük during its occupation.

Charles et al. (2014) suggest that site occupation of Çatalhöyük (7100–6050 cal BCE = 9100–8000 cal BP) (Bayliss et al. 2015; Marciniak et al. 2015a; Orton et al. 2018) coincided with a period of relatively warm, wet conditions in comparison to the cold, dry Younger Dryas period. The vegetation reconstructed from pollen cores (Lake Nar, Eski Acıgöl and Akgöl) was relatively moist forest/grass steppe characterised by high levels of Oak (*Quercus* sp.), and grass (Poaceae) pollen. In their subsequent work, Roberts et al. (2016) compare the results from two lakes (Lake Nar and Eski Acıgöl). They include oxygen-isotope data that provide the basis for subsequent discussion. In addition, data from the cores allows inference of vegetation (pollen), lake levels (diatoms) and river influxes (calcite). These additions provide more evidence for broad vegetation trends for our reconstruction (table 2.1), though it must be noted that the period of site occupation is still only represented by fewer than ten samples.

Based on a set of hydroclimate proxies from Nar Lake, Roberts et al. (2016) infer wet and dry phases for the period of occupation at Çatalhöyük (table 2.1). These indicate that Çatalhöyük was initially settled towards the end of a relatively dry period (7550–7150 BCE). The subsequent 500 years are shown as an intermediate period of moisture before a ca 500-year period of relatively dry conditions terminating ca 6250 cal BCE. The vegetation of the region prior to occupation (8750–7250 cal BCE) is characterised as that of oak-pistacia woodland, typical of fairly dry conditions. The early occupation of Çatalhöyük in contrast coincides with the development of oak steppe-parkland characteristic of wetter conditions.

Woodbridge et al. (2015) use the pollen records from a series of sites across Southern Anatolia to create a chronological sequence of the predominant inferred vegetation present in the region. Eski Acıgöl, the only site on the Anatolian Plateau for which there is a continuous record for the period of Çatalhöyük's occupation, shows a remarkable level of vegetation continuity through the Early to Late periods, a period made up of parkland/grassland. This 'vegetation cluster' comprises a relatively open landscape characteristic of relatively moist conditions, dominated by grasses, members of the Asteraceae family and open oak woodland (Woodbridge et al. 2018). It changes to a more arid vegetation made up of sclerophyllous parkland in the Final period. Similar changes to more arid vegetation are also seen at Akgöl (on the plateau) and Beyşehir to the southwest, though nearby Gölhisar shows no change.

Moving our attention to a more local picture of vegetation on the Konya Plain, Kabukcu (2017) provides a detailed view based on wood charcoal evidence from Çatalhöyük and two other sites (Boncuklu, Can Hasan III). Evidence from the latter shows high levels of willow (Salicaceae) and low levels of oak in the period prior to occupation at Çatalhöyük. The Early period of Çatalhöyük continues this picture before there is a substantial increase in oak values in Levels South G-H that coincides with the rise in oak pollen values in the Eski Acıgöl and (to a lesser extent) the Nar cores. The high oak charcoal values continue through to Level South P when juniper (*Juniperus*) values become dominant, along with almond (*Amygdalus*) and pistachio (*Pistacia*), potentially indicating increasing aridity. Evidence for river conditions based on riparian woodland taxa shows an abundance of willow and the elm family (Ulmaceae) in the Early period that is replaced by ash (*Fraxinus*) in the later period that could indicate a change in river hydrology consistent with the other evidence presented in this chapter.

Isotopic evidence

Roffet-Salque et al. (2018) used δ^2H analysis from lipids on pottery in well-dated contexts at Çatalhöyük (fig. 2.11a) to suggest that there was a climate change that correlated with the 8.2kya event recorded in Greenland ice cores (fig. 2.11b) and elsewhere. Following our critique (Wainwright, Ayala 2019), Roffet et al. (2019) agreed that the fluctuation in their data is not statistically significant and that the sequence shows a high degree of variability. This observation is consistent with the idea that the local climate was dominated then as now by high degrees of interannual variability. Figure 2.11c shows that the ~8.2kya excursion at Nar that Roffet-Salque et al. (2018) used to suggest there was regional evidence for was actually part of a longer series of oscillations that started much earlier and thus cannot be correlated with the Greenland event. Wainwright and Ayala (2019) also used the proxy data from Jones et al. (2005) to suggest that the maximum summer temperature was a reasonable proxy for the Nar isotope data, and that it varied little over the period in question (fig. 2.11d). Roffet-Salque et al. (2018) followed Dean et al. (2016) in stating that the proxy is more related to water availability (precipitation/evaporation), but if there is no evidence that precipitation varied other than on an interannual basis (and the proxy estimates are smoothed to minimise this effect), then these two interpretations are not incompatible, as Dean et al. (2016) calculated evaporation from temperature data. In figure 2.11e, we address the effectiveness of the Nar proxy (in our case, for temperature) for Çatalhöyük, using the relation between the CRU TS4.01 reconstructed JAS maximum temperature 1901–2016 (Harris et al. 2014) for the 0.5° grid cells containing Nar and Çatalhöyük. This relation is highly significant (r^2=0.943, $p < 2.2 \times 10^{-16}$), but because these data are both interpolations from other sites, the strength of the relation is likely to be overestimated. Thus, figure 2.11f has instrumental data for JAS mean temperature from Niğde and Konya for the period 1951–2001, which is also highly significant (r^2=0.428 and p=1.919 $\times 10^{-7}$).

Reconstructing the palaeohydrology of the Çatalhöyük landscape

In Ayala et al. (2017), we challenged the interpretation of Roberts and Rosen (2009: 397) that the Neolithic environment was of 'backswamp basins gradually infilled' by spring floodwaters for two reasons: first, the more waterlogged conditions suggested by the Dark Clay layers are earlier than occupation (except for a localised outlier that dates to the Chalcolithic); and second, the anastomosing channel system is not compatible with the interpretation of a backswamp (see further discussions on the latter point by Charles et al. 2014 and Doherty 2017). Here, we address further issues with the spring floodwater model of Roberts and Rosen (2009) through a discussion of the data used to build their model and the modelling method that was subsequently used to estimate flooding regime at the time. The data upon which they build their hydrological interpretation come from Bozkir because the extensive channelisation schemes in the late 19th and early 20th centuries (especially the Beyşehir-Çarşamba Canal) and the dam at Apa significantly altered the flows in the lower part of the Çarşamba, so that modern instrumental records are not useful for interpreting the possible Neolithic flow regime. It should be noted that Bozkir is a long way upstream (32° 12' 34" N, 37° 10' 24" E), at an elevation of 1156m with a catchment area of only 206km^2, compared to an estimated 2,079km^2 for the Neolithic Çarşamba just downstream of Çumra (estimated using ArcMap hydrological analysis on the SRTM 30m DTM by adding an artificial blockage to the catchment just upstream of

<table>
<tr><th>Çatal Occupation [a]</th><th>Dates cal BCE [b]</th><th>climate [c]</th><th>Lake Nar [d]</th><th>Eski Acıgöl [e]</th><th>Akgöl [e]</th><th>Beyşehir [e]</th><th>Gölhisar [e]</th></tr>
<tr><td>POST-</td><td></td><td></td><td></td><td></td><td></td><td></td><td></td></tr>
<tr><td rowspan="3">Final 6300-5950 BCE</td><td rowspan="2">6200-5300</td><td rowspan="2">wetter</td><td rowspan="8">oak steppe-parkland [grasses & artemisia] (NG-4)</td><td rowspan="3">dry park</td><td rowspan="2">dry park</td><td rowspan="3">dry park</td><td>park/grass</td></tr>
<tr><td rowspan="7">pine wood</td></tr>
<tr><td rowspan="3">6700-6200</td><td rowspan="3">drier</td><td>steppe park</td></tr>
<tr><td>Late 6500-6300 BCE</td><td rowspan="8">park/grass</td><td rowspan="3">pine steppe</td><td rowspan="4">pine wood</td></tr>
<tr><td>Middle 6700-6500 BCE</td></tr>
<tr><td rowspan="3">Early 7100-6700 BCE</td><td rowspan="3">7200-6700</td><td rowspan="3">intermediate</td></tr>
<tr><td rowspan="2">[no data]</td></tr>
<tr><td>dry park</td></tr>
<tr><td rowspan="3">PRE-</td><td>7550-7200</td><td>drier</td><td rowspan="3">oak-pistacia (NG-3)</td><td rowspan="3">steppe park</td><td rowspan="2"></td><td rowspan="2"></td></tr>
<tr><td rowspan="2"><7550</td><td rowspan="2">wetter</td></tr>
<tr><td>pine wood</td><td>deciduous oak parkland</td></tr>
</table>

Table 2.1. Palaeoenvironmental reconstruction of conditions on the Anatolian Plateau and surrounding region during the occupation of Çatalhöyük: (a) occupation periods and their chronology; (b) dates – BCE calibrated for palaeoenvironmental indicators; (c) climate as defined by Roberts et al. (2016), shading showing their interpretation of relative dryness (but cf. figure 2.11); (d) Lake Nar data summarised from Roberts et al. (2016); (e) pollen records summarised by Woodbridge et al. (2019). Park/grass = parkland/grassland; dry park = sclerophyllous parkland; steppe park = steppe parkland.

where the Beyşehir-Çarşamba Canal joins the main stem of the Çarşamba River). Thus, the catchment area feeding the channel system at Çatalhöyük is an order of magnitude larger than that producing the flow at Bozkir. As noted by Roberts and Rosen (2009: 396), 'the Taurus Mountain region above Bozkir receives high precipitation, much of it as snowfall, whose melting gives rise to the main annual river flood peak in March and April'. However, much of the rest of the catchment is at lower elevation and receives less precipitation as snowfall. Precipitation is also more evenly spread through the year. For example, at Çumra, the 1961–1990 monthly means are very similar from October to May, between 31 and 45mm, dropping to 19mm in June and 0mm between July and September. Although some of this precipitation is in the form of snowfall, more of it is as rain, so that runoff and river flow may well have been more evenly spread through the year in the lower Çarşamba.

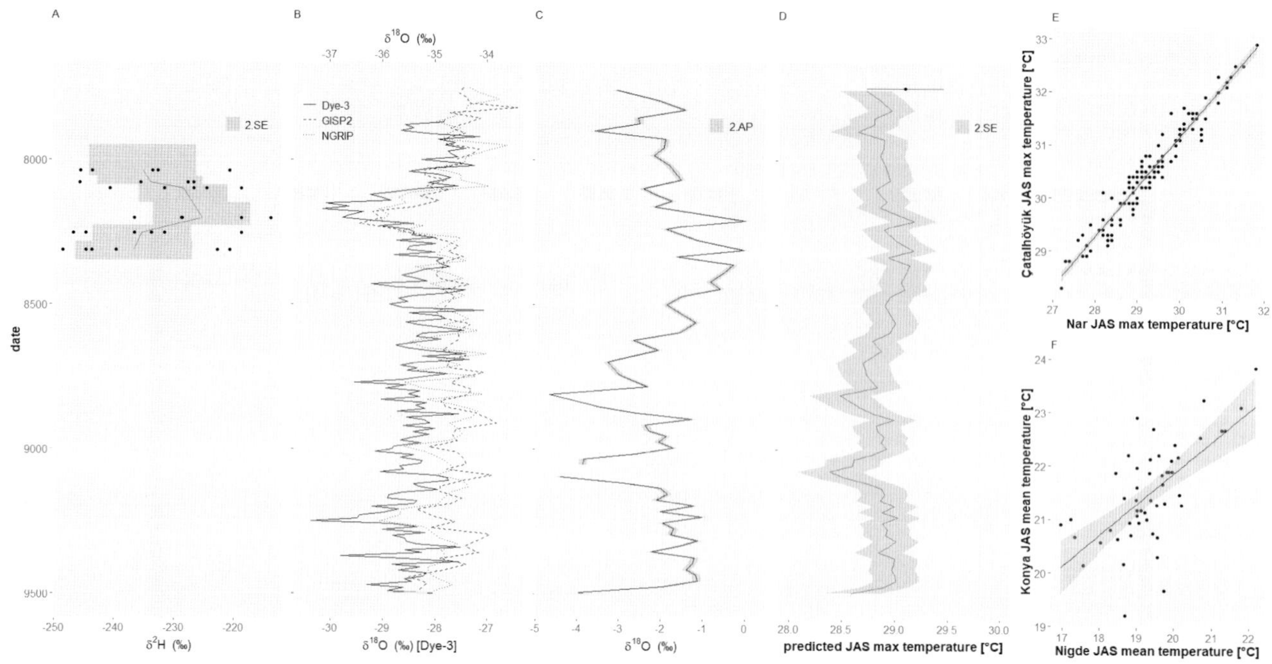

Figure 2.11. Comparison of the $\delta^2H_{18:0}$ proxy of Roffet-Salque et al. (2018) with global and regional climate records. This plot uses a window from 9000 to 7800 calibrated B.P. to place the proxy data in a longer-term context consistent with the discussion in the text here: (a) replots the data of figure 1D in Roffet-Salque et al. (2018) using shading to show the 2-SE variation of data around the mean (solid line); original data points are plotted at the midpoint of the archaeological phase, but shading is shown vertically to demonstrate the full uncertainty of the Bayesian estimates of these phases. There is clear overlap between the phases, and variability is more important than mean values; (b) water $\delta^{18}O$ values of three Greenland ice cores (Stuiver et al. 1995; NGRIP Dating Group 2006; Vinther et al. 2006; Rasmussen et al. 2007) used to demonstrate the onset and duration of the 8.2kya event; (c) $\delta^{18}O$ record from the sediments in Nar Lake (Jones et al. 2005), ~150 km from Çatalhöyük, which provides the best available regional information for climate proxies. Shaded area shows 2× analytical precision (AP) of the measurements; (d) estimate of summer maximum temperature for Nar Lake, based on the proxy derived by Jones et al. (2005) from modern climate measurements. The shaded area is 2 SEs, using a RMS combination of the uncertainties from the measurement analytical precision and the SE from the proxy model. The point and error bar at the top of the plot show the mean and 2-SE range for the 1961–1990 climatic observations; (e): relation between CRU TS4.01 (Harris et al. 2014) reconstructed JAS maximum temperature 1901–2016, $r^2=0.943$, $p < 2.2 \times 10^{-16}$; (f): instrumental data for JAS mean temperature from Niğde and Konya 1951–2001, $r^2=0.428$, $p=1.919 \times 10^{-7}$ (based on Wainwright, Ayala 2018).

A common, simple method for estimating flows progressively downstream in an ungauged catchment is to fit a power-law model of the form $Q = a A^b$, where Q is the discharge (in $m^3 s^{-1}$), A is the catchment area (in km^2) and a and b are empirical coefficients. Based on monthly average data from the Turkish Hydrological Service (DSI: http://www.dsi.gov.tr), an attempt was made to extrapolate flows using the available sites upstream of the modern diversions in order to estimate the pre-modern flow accumulation. Measurements are available from the stations at Sorkun (32° 8' 58" N, 37° 9' 56" E), Bozkir and Pinarcik (32° 16' 49" N, 37° 15' 47" E), which have catchment areas of 84.0km², 205.6km² and 343km², respectively. However, for all months except February, the mean monthly flow at Pinarcik is lower than Bozkir, and it is not possible to fit a significant model using the catchment area. The implication of this observation is that the runoff coefficient (the proportion of precipitation that is converted into flow) decreases in the lower part of the catchment and/or there are significant transmission losses (infiltration from the channel into the channel bed). Both are common characteristics of dryland channels, as a result of spatial variability in vegetation and soils, temporal variability of rainfall intensity during thunderstorms, and unsaturated subsurface conditions between storm events (Parsons et al. 1999; Wainwright, Parsons 2002; Yair, Kossovsky 2002; Wainwright, Bracken 2011), and some parts of the Çarşamba catchment are also karstic.

Given these limitations of the data, a simple way to estimate likely mean flows for the lower Çarşamba is to use estimates from the flow yield (discharge per unit area of catchment) at each of the known stations, and then multiply by the catchment area at the point of interest (2,079km^2 in this case). The estimated values are as follows for the 'spring flood' period (table 2.2).

Because of the downstream losses, all of these estimates are likely to be overestimates, with those based on the most downstream station (Pinarcik) integrating more of these effects and thus likely to be more reasonable. Based on modern data, the highest mean flows are about 2 to 2.5× these values, although it should be remembered that these values would still be for monthly averages, not peak flood flows. In comparison, it is possible to estimate the bankfull conveyance (discharge that can be carried before flooding occurs) of the two palaeochannels observed near to Çatalhöyük using the Manning equation. Assuming there is no truncation of the channel profiles in Boyer (1999), Chalcolithic palaeochannel 95/PC2 has an area of 38.2m^2, an estimated wetted perimeter of 34.9m, and thus a hydraulic radius of 1.1m. If the former channel slope was subparallel to the modern surface (probably an underestimate), and using Manning's n values of 0.025 to 0.033, characteristic of a channel that is 'clean, straight, full stage, [with] no rifts or deep pools' (Chow 1959), the bankfull conveyance would have been between 62 and 82m^3 s^{-1}. Palaeochannel 94/PC1 is much larger, with an area of 126.2m^2, wetted perimeter of 54.4m and hydraulic radius of 2.3m, so that with the same assumptions, its bankfull conveyance would be between 338 and 446m^3 s^{-1}. Bankfull discharges of channels tend to adjust to the mean annual flood (Dury 1973). If both 94/PC1 and 95/PC2 were single-thread channels (as the Çarşamba was known to be historically), then these estimates would imply that 94/PC1 was formed in a much drier period than 95/PC2, which is consistent with other climatic observations from the late Pleistocene and Chalcolithic, respectively, when these channels were formed. Palaeochannel 95/PC2 would easily have been able to convey the peak mean flows estimated above without flooding. Although it must be remembered that the above estimates do not account for extreme flood flows, the data suggest that the Chalcolithic channel was adjusted to higher flows than at present.

Estimate using station	*Estimated mean monthly flow m^3 s^{-1}*			
	Feb	*Mar*	*Apr*	*May*
Sorkun	61.1	106.5	159.5	116.2
Bozkır	4.3	85.7	108.8	63.9
Pınarcık	23.7	45.1	53.7	2.8

Table 2.2. Estimated mean monthly flows for the lower River Çarşamba using three measuring stations in the upper reaches of the River Çarşamba during the months of February–May based on data from Devlet Su Işleri (http://www.dsi.gov.tr/).

However, Ayala et al. (2017) demonstrated that the earlier Holocene channels of the Çarşamba were most likely to be multi-thread, anastomosing channels. The problem with these channels is that it is impossible to estimate the past channel dimensions directly from the stratigraphic record (Nanson, Knighton 1996), even if there were available exposures locally. Furthermore, there have been very few studies of the flooding characteristics of this style of channel, not least because of the difficulties of making measurements. Entwistle et al. (2018) used a modelling approach to look at contrasting single-thread and anastomosing reaches of the River Wear to show that the styles of flooding were very different in the different reaches. Although care must be taken in extrapolating from an upland, gravel-bed channel in the UK, their results suggest that the anastomosing system will typically flood more frequently, but it also has the effect of spreading out the flow more broadly. As a consequence, the anastomosing system has lower shear stresses, leading to more stable surfaces, that is, less likely to erode and more likely to have sediment deposition. Sediment deposition would have led to there being progressively higher capacity for subsurface water storage, and thus progressively less extreme floods through time, and potentially the increasingly 'dry' anastomosing channel types in the record. Flooding may have been extensive but restricted to the locally lower points in the landscape, which suggests it could have been only a kilometre or so across, thus leaving extensive, unflooded areas close to the site that would have been available for cultivation. As noted above, the inherent heterogeneity of dryland environments is key to understanding the complexity and ephemerality of the environment, which would have influenced the choices made by the occupants of Çatalhöyük over time.

Conclusions

The palaeoenvironmental record of the Çarşamba River plain provides key information to further our understanding of the entangled nature of the relationship that occupants of Çatalhöyük had with their landscape. Fundamental questions surrounding agricultural production and organisation at the site require high-resolution palaeoenvironmental evidence which can support robust environmental models and interpretations of behaviour. The intensive coring programme has enhanced the previous landscape models, derived from a limited number of data points (Boyer 1999; Boyer et al. 2006). We have redefined the stratigraphical record and hence re-propose the alluvial development of the area into four phases (Ayala et al. 2017). During the early Holocene, when the site was occupied, the immediate environment was dominated by a multi-thread anastomosing channel instead of a single meandering one. The analysis of the most recent cores (2015) and modelling of the palaeohydrological evidence suggests that flooding in the area would have not created the backswamp once imagined (Roberts, Rosen 2009) but would instead have left areas of high ground which would have been suitable for agriculture. This more nuanced reconstruction of the Çarşamba River plain will enable us to write more accurate and meaningful interpretations of the relationships, and long-term interdependencies, that the Neolithic inhabitants had with their environment.

Returning to previous interpretations of entangled behaviour, whilst there is undoubtedly evidence of human manipulation of the floodplain, for example, the pits near the tell, the interpretation of their impact on the alluvial environment and the watertable requires a higher resolution model of the alluvial landscape. Our work has demonstrated how powerful this level of accuracy can be to test traditional models of climate change (Wainwright, Ayala 2019) and to reinterpret past land use (Roberts, Rosen 2009). The reinterpretation of the sedimentary sequence coupled with the modelling of the channel flow has supported our reconstruction of the development of the dryland anastomosing channel system and hence the increasing potential for dryland agriculture in the immediate vicinity of the tells throughout the Neolithic and Chalcolithic, which supports evidence for demographic growth of the settlement and the development of garden-type agriculture throughout the Neolithic (Bogaard 2005).

Although the localised flooding in lower-lying areas occurred throughout the period of occupation, the anastomosing system created a range of ecological niches that could have supported a diverse range of cropping strategies in the areas immediately surrounding the site. Archaeobotanical evidence suggesting a preference for more drought-tolerant barley species being selected at the tells (Bogaard et al. 2017) is consistent with this observation. The entangled nature of life in the alluvial environment thus requires us to consider how environmental niches that support resilience in the Neolithic settlement of Çatalhöyük are created through the agency of the alluvial landscape, bringing water and sediment to the local landscape. This agency is modified by the use of sediment and water for cultivation, creating a landscape that is accreting vertically and helping to enhance the dryness of the landscape, buffering the fluctuations in climate. As the tell built up, it provided a further sediment source, and the coupling of colluvial and alluvial processes created a yet further, more complex landscape. The agency of the mound produces conditions where the channel belt tends to be diverted away from the settlement, and thus locally more conducive conditions for habitation and sustainable agriculture.

Future work is needed to refine these entangled models of human-landscape interactions, because the complexity of dryland environments requires evidence at as high a resolution as possible. As well as further sedimentological analyses immediately surrounding the site (for example, looking at colluvial-alluvial interactions) and in the immediate vicinity (for example, to look at changing patterns of the channel belt through time), a larger landscape context is needed to look at the source and sinks of sediment, and thus patterns of land available for cultivation and other uses. Future palaeohydrological modelling will evaluate patterns of seasonality in more detail, as well as the impacts of longer-term climate variability. Scales of variability – in both space and time – are fundamental to the understanding of this landscape.

Acknowledgements
The authors wish to thank Ian Hodder for supporting sampling in the field; Hannah Russ and Harriet White, Bradley Brandt and Sophia Lapidaru, who all ran samples at the University of Sheffield; and Neil Tunstall, Chris Longley, Alison George, Frank Davies and Katheryn Melvin, who assisted with samples at Durham University. Thanks also to Chris Orton and Michele Allan for their assistance with the figures in this chapter.

Supplementary material
For supplementary material related to this chapter, please visit https://doi.org/10.18866/BIAA/e-13. It comprises colour versions of figures 2.2, 2.3, 2.5, 2.6, 2.7, 2.8, 2.9 and 2.11.

3. The archaeomalacology of the 2009–2017 excavations at Neolithic Çatalhöyük

Rena Veropoulidou

Shells of freshwater molluscs are one of the most ubiquitous categories of finds at Çatalhöyük, being present in every level and type of deposit throughout the sequence of habitation (Mitrović, Vasić 2013). Previous work on shell remains has allowed insights into a variety of freshwater mollusc and shell uses, from intentional gathering for food consumption and artefact manufacture to unintentional collection along with materials from the local freshwater bodies. The discussions also included questions related to environmental reconstruction, seasonality and climate changes (Reese 2005; Bar-Yosef Mayer et al. 2012; Bar-Yosef Mayer 2013; Gümüş, Bar-Yosef Mayer 2013; Lewis et al. 2017).

The archaeomalacological investigation of nearly 300 units from the 2009–2017 excavations in the South and North Areas is reported here. Building on earlier work, the study focuses on gathering detailed information to explore a number of mollusc/shell-related issues that demand a full spatio-temporal analysis and attention to the contexts of deposition. In detail, the aims of this phase of research can be summarised as follows: (a) to address taxa representation and relative proportions through time to contribute to the reconstruction of local freshwater environments and discuss human-environmental interactions; (b) to analyse mussel frequencies through time and categories of deposits to obtain a detailed understanding of molluscan food consumption and secondary treatment of food remains; (c) to provide a better understanding of the exploitation of aquatic materials and the seasonality of this activity; (d) to expand the temporal range of shell material with samples from the mid/late levels in order to monitor shifts that may have resulted from changes in the local environments or human practices; (e) to increase the sample of *Unio* shells for isotope analysis in order to examine the seasonality of molluscan gathering; (f) to shed light into the relationship of the farming community with a 'wild' food resource and to understand the social context of molluscan food consumption and of shell deposition; (g) to integrate shell data with information from excavation and the study of other materials to contribute to the understanding of the depositional history of excavated deposits.

After the presentation of materials and methods, the chapter is organised into three sections: the first gives a general overview of the shell assemblage and, on the basis of the ecological requirements of molluscs, provides a description of exploited freshwater habitats around the site; the second section presents the freshwater mussels, discussing gathering methods, seasonality of activity, food consumption and secondary use of remains with regard to temporal changes and context-related variation; the third section presents all freshwater shells that were brought on site unintentionally to open up insights into how they were incorporated into the archaeological deposits.

Materials and methods

Sample size and recovery

The studied sample includes 265 units (58.2%) that were analysed in full detail and 33 units (7.2%) that received a first-level analysis out of the 455 units from South, North and TPC Areas prioritised for publication by the excavation team and specialists (table 3.1). In cases where more than one flotation sample existed for one excavation unit, all samples were studied. Moreover, there were 88 units (19.3%) that were void of shells, while 69 units (15.2%) were not studied. Therefore, there is detailed shell data from all occupation levels and different types of deposits that enable statistically valid comparisons.

As stated above, the goal of the current phase of archaeomalacological research was to recover as much data as possible. In this vein, full study included all hand-picked and dry-sieved material (4mm mesh), as well as all material recovered from the sorting of the 1mm, 2mm and 4mm sieves of the heavy residue (fig. 3.1). The study of all fractions of the heavy residue and of all specimens, regardless of size and state of preservation, was performed for the first time at Çatalhöyük. Despite being very time-consuming, it proved particularly useful, as it allowed a detailed understanding of the exploitation of wetland materials and the context of secondary shell use, but most importantly because it shed light on the archaeological context of shell deposition.

Level	*Units for publication*	*Full analysis*	*Full analysis %*	*First-level analysis*	*No shells*	*Not studied*
North F/G	2	1	50.0			1
North F	52	33	63.5	5	12	2
North G	86	60	69.8	1	13	12
North H	8	2	25.0		4	2
North Unassigned	1	1	100.0			
North area: Sub-total	149	97	65.1	6	29	17
South H	6	6	100.0			
South ?J	2	2	100.0			
South K	32	25	78.1		4	3
South L	27	19	70.4		6	2
South M (O?)	2	2	100.0			
South N	56	22	39.3	20	12	2
South O	103	70	68.0	6	15	12
South P	5	3	60.0		1	1
South R	2				2	
South T	1	1	100.0			
South area: Sub-total	236	150	63.6	26	40	20
North-South areas: Total	385	247	64.2	32	69	37
TPC area: Total	70	18	25.7	1	19	32
Grand total	**455**	**265**	**58.2**	**33**	**88**	**69**

Table 3.1. Summary of units by level and area that received full and first-level archaeomalacological analysis. Units void of shells and not studied are also noted.

Methods of analysis

All shells were examined under a low magnification stereoscope following a specific archaeomalacological methodology developed by Veropoulidou (2011) and adjusted to meet the specificities of this material (Veropoulidou 2017). In detail, species identification was based on comparisons with web databases (AnimalBase Project Group 2005–2019) and publications on freshwater molluscs of Turkey (Yildirim 1999; 2004; Yildirim et al. 2006; Kalyoncu et al. 2008; Şereflişan et al. 2009; Arslan et al. 2013; Girod 2013; Gürelli, Özbek 2012; Koşal Şahin, Zeybek 2016).

Recording of specimens included: intact, partially intact and which part is missing, diagnostic body parts (apices and apertures for gastropods; right and left umbos for bivalves) and all non-diagnostic fragmented material. Apart from the Number of Identified Specimens (NISP), the Number of Intact (NI) and the Number of Fragments (NF) that are directly extracted, the Minimum Number of Individuals (MNI) was calculated on the basis of the largest number of apices or apertures for gastropods and the largest number of right or left umbos for bivalves per distinct cultural deposit.

All quantifications are expressed in absolute numbers and as density per 1 litre of soil in order to allow valid comparisons between deposits of different size. Note that the amount of heavy residue sorted for the 2mm and 1mm fractions varies from 6.25 to 100% depending on the amount of heavy residue produced. In cases where the heavy residue was not 100% sorted, the volume of the 1mm and 2mm fractions was divided according to the proportion sorted to obtain the exact density (NISP/MNI/NI/FR) per litre, instead of multiplying up to represent the amount one would expect if the 100% of the sample had been available.

In addition, measurements of the vast majority of intact and partially intact specimens (height, length/width) were taken using a Vernier calliper (in mm) to shed light onto gathering practices and to estimate the age of gastropods. In addition, a large random sample of *Unio* fragments from each unit was measured (length/width, thickness) to infer whether valves came from adult or juvenile individuals and to discuss post-depositional processes. The study of taphonomic attributes included the separation of fresh and worn collected shells, and recording of the state of surface preservation and the type of wear, as well as of any natural modifications – such as abrasion, corrosion, bio-erosion, or encrustation – and human or post-depositional modifications – such as degrees of burning, shape and edges of fragments.

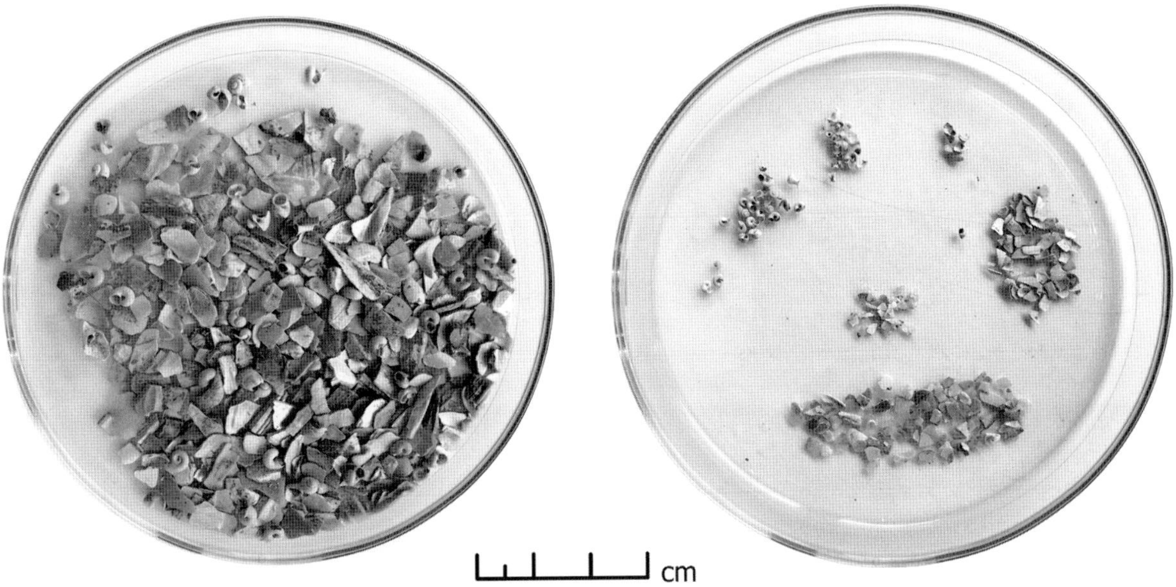

Figure 3.1. Unsorted and sorted shells from the heavy residue: left: >2mm fraction, flot (12220), unit (22512), right: >1mm fraction, flot (12436), unit (22351) (photograph: R. Veropoulidou).

First-level analysis included a semi-quantitative and qualitative description of each sample, that is, species identification, taxa representation, description of human and natural modifications. For archival purposes and quick documentation, all units and sieved fractions were photographed by the author.

Overview of the shell assemblage

In total, there are 29,395 specimens distributed unequally among 20 different freshwater species. There are also 566 specimens (ca 1mm) preserved in extremely bad condition that could only be attributed to the class of bivalves or gastropods, as well as three terrestrial snails, which due to their small number are not further discussed. Therefore, in total 29,961 shells were recorded (table 3.2). Raw data are available online in the Çatalhöyük database (http://catalhoyuk.com/research/database).

Identified species are grouped into three categories on the basis of size (macro- and micro-molluscs) and age (adult and juveniles) of shells: (a) adult bivalves (*U. mancus*) (fig. 3.2) and gastropods (*Viviparus* sp., *L. stagnalis*, *R. auricularia*, *St. palustris*), (b) juvenile bivalves (*Dreissena* sp.) and gastropods (*Pl. corneus*, *M. praemorsa*), usually below 10–15 mm in height/length, (c) juvenile and adult micro-molluscs, that is, shells under 10–15 mm in height (*Bithynia* sp., *Bythinella* sp., *G. truncatula*, *G. albus*, *Pisidium* sp., *Pl. carinatus*, *Pl. planorbis*, *Th. heldreichi*, *Th. anatolicus*, *V. piscinalis*). A broadly similar taxa variety and representation was observed in the 2000–2008 shell material (table 3.2; Gümüs, Bar-Yosef Mayer 2013: fig. 5.1), which allows inferences on continuities in the exploitation of freshwater bodies and related wetland materials throughout the occupation of the settlement.

Based on MNI, the vast majority are the remains of *V. piscinalis* (fig. 3.3), followed by *Unio* sp., *Pl. corneus*, *Bithynia* sp. and *Viviparus* sp. The remaining 15 species are represented by small amounts (table 3.2). The assemblage is characterised by very low diversity (Shannon-Weaver formula, H'=0.063), due to the dominance of *V. piscinalis* individuals and very low equitability (Sheldon formula, V'=0.022) as the number of individuals is not distributed evenly across identified species.

Previous interpretations of shells

On the basis of size, omnipresence and concentrations in deposits, especially the 'shell middens' in the earliest level of habitation (South G), the edible *Unio* specimens have been interpreted as food consumption refuse (Reese 2005; Bar-Yosef Mayer 2013). Reese (2005: 123) also indicated the presence of burnt *Unio* and attributed them to lime production. The use of shells for plaster production was also one of the hypotheses put forward by Bar-Yosef Mayer (2013: 330) to interpret the decrease of the amount of *Unio* in the late levels of habitation. The potential of the remaining freshwater species for the reconstruction of the local environmental conditions was recognised (Gümüş, Bar-Yosef Mayer 2013), but their presence was seen as puzzling. Their varying amounts and miniscule size, excluding some worked shells, led to the suggestion that they were brought on site accidentally with water (Reese 2005: 123) or muds utilised in construction (Bar-Yosef Mayer 2013) or plants (Bar-Yosef Mayer 2013: 337; Gümüs, Bar-Yosef Mayer 2013: 85).

Species	Gümüs, Bar-Yosef Mayer 2013	Habitat	NISP	NISP/lt	NISP %	MNI	MNI/lt	MNI %	Fragments	Fragments %	Intact	Intact %	Ratio NF:NI
Unio mancus	**x**	**Mixed [P]**	**18,642**	**0.137**	**61.6**	**644**	**0.005**	**11.6**	**17,230**	**74.3**	**290**	**7.3**	**59.41**
Dreissena sp.	x	Lac [P]	268	0.002	0.9	76	0.001	1.4	65	0.3	9	0.2	7.22
Pisidium sp.	x	Riv [P]	87	0.001	0.3	20		0.4	42	0.2	41	1.0	1.02
Amphimelania sp.		Riv [P]	12		0.0	1		0.0	11	0.0		0.0	
Bithynia sp.	x	Mixed [P]	1,159	0.009	3.9	280	0.002	5.0	797	3.4	4	0.1	
Bythinella sp.	x	Riv	2		0.0	2		0.0		0.0	1	0.0	0.00
Gyraulus albus	x	Lac [P]	48		0.2	22		0.4	17	0.1	10	0.3	1.70
Galba truncatula		Lac [P]	26		0.1	6		0.1	20	0.1	1	0.0	20.00
Lymnaea stagnalis	x	Lac [P]	147	0.001	0.5	15		0.3	127	0.5	3	0.1	42.33
Melanospsis praemorsa		Riv [P]	5		0.0			0.0	5	0.0		0.0	
Fagotia sp.	x	Mixed [P]	671	0.005	2.2	34		0.6	617	2.7	26	0.7	23.73
Planorbis carinatus	x	Lac [P]	2		0.0	1		0.0		0.0	2	0.1	0.00
Planorbarius corneus	x	Lac [P]	448	0.003	1.5	366	0.003	6.6	29	0.1	379	9.5	0.08
Planorbis planobris		Lac [P]	14		0.0	11		0.2	3	0.0	8	0.2	0.38
Radix auricularia	x	Mixed [P]	6		0.0	2		0.0	3	0.0	3	0.1	1.00
Stagnicola palustris	x	Lac [P]	213	0.002	0.7	13		0.2	200	0.9	3	0.1	66.67
Theodoxus anatolicus		Riv	4		0.0			0.0		0.0	3	0.1	0.00
Theodoxus heldreichi	x	Lac	388	0.003	1.3	56		1.0	168	0.7	132	3.3	1.27
Valvata piscinalis	x	Mixed [P]	4,812	0.036	16.1	3,798	0.028	68.3	884	3.8	3,001	75.3	0.29
Viviparus sp.	x	Lac	2,621	0.020	8.7	204	0.002	3.7	2,406	10.4	65	1.6	37.02
Micro-molluscs			**10,933**	**0.082**	**38.4**	**4,907**	**0.036**	**88.3**	**5,394**	**23.3**	**3,691**	**92.6**	
Unidentified biv.			19		0.1	1		0.0	17	0.1	2	0.1	8.50
Unidentified gastr.			547	0.004	1.8	6		0.1	540	2.3	5	0.1	
Grand total			**29,961**	**0.223**	**100.0**	**5,558**	**0.041**	**100.0**	**23,181**	**100.0**	**3,988**	**100.0**	

Table 3.2. Representation of freshwater mollusc remains: NISP, MNI, NF, NI in absolute numbers, Density per litre of soil and frequencies, as well as Ratio of Fragments and Intact specimens. Information on habitat is provided: 'Lac' stands for 'Lacustrine' and 'Riv' for 'Riverine', while P in brackets indicates association with plants.

F.8015, F.8017, F.8204, F.8205, F.8206 (B.17, Level South K). Other contexts worth mentioning with high densities of *Unio* include the infant burials in baskets F.8320 (B.132, Level North F) and F.7859 (B.77, Level North G) and the child burial F.7860 (B.77, Level North G). Interestingly, the general decline of *Unio* from the Early to the Late periods is not so evident in the analysis of burial contexts. Maybe this indicates the intentional deposition of burnt shells into burials to increase the pH of the soil and help the preservation of the bones and maybe also to reduce the smell of decomposition (Okumura, Eggers 2014).

Moving to middens (fig. 3.12), the higher densities are observed in the Early period, in particular Sp.620 (Level South I?; collection in spring), in Sp.583 (Level South K) and in Sp.559 (Level South L; collection in winter). On the other hand, low/moderate densities are observed in the Middle period and even lower in the Late period. Therefore, the temporal distribution of *Unio* in middens also points towards a general decline in its exploitation through time.

Last, the distribution of *Unio* (fig. 3.13) within buildings is not uniform and there are great differences even among buildings that belong to the same level and occupation period. In addition, the distribution does not follow the general decline from the Early to the Late periods. Though it seems that densities tend to increase across time, this is a result of very high densities at particular buildings (for example, B.110, B.150 and Sp.159) that skew the results. In any case, the significant differences observed among buildings is not only related to the types of deposits studied from each building, but may be taken to suggest that each household expressed different attitudes towards food ingredients and the secondary treatment of food remains, which implies more autonomous households in the Late and Final periods.

The analysis of freshwater fish from the site (Van Neer et al. 2013; Hamilton-Dyer this volume, Chapter 12) is of particular interest here, as the reproduction and population dynamics of *Unio* populations have positive correlation with the abundance of cyprinids and loaches that host the *Unio* glochidia (Aldridge 1999; Österling, Schneider 2012; Soylu 2014). As with *Unio*, there is also a decrease in the quantity of fish across time. The lack of suitable host fishes may have affected *Unio* populations, thus suggesting the mutual decrease of interdependent species. The decline of fish bones in the later levels of habitation has been interpreted as evidence for smaller water bodies around the site and thus has been related with the proposed changes in the suitable environmental niches (Van Neer et al. 2013).

Whether the decline of *Unio* exploitation is due to environmental changes that affected the availability of mussel populations, or to economic and social changes, or to a combination of both factors can be addressed on the basis of the analysis of the size of *Unio* valves across time (fig. 3.14). The largest valves are found in the Early period.

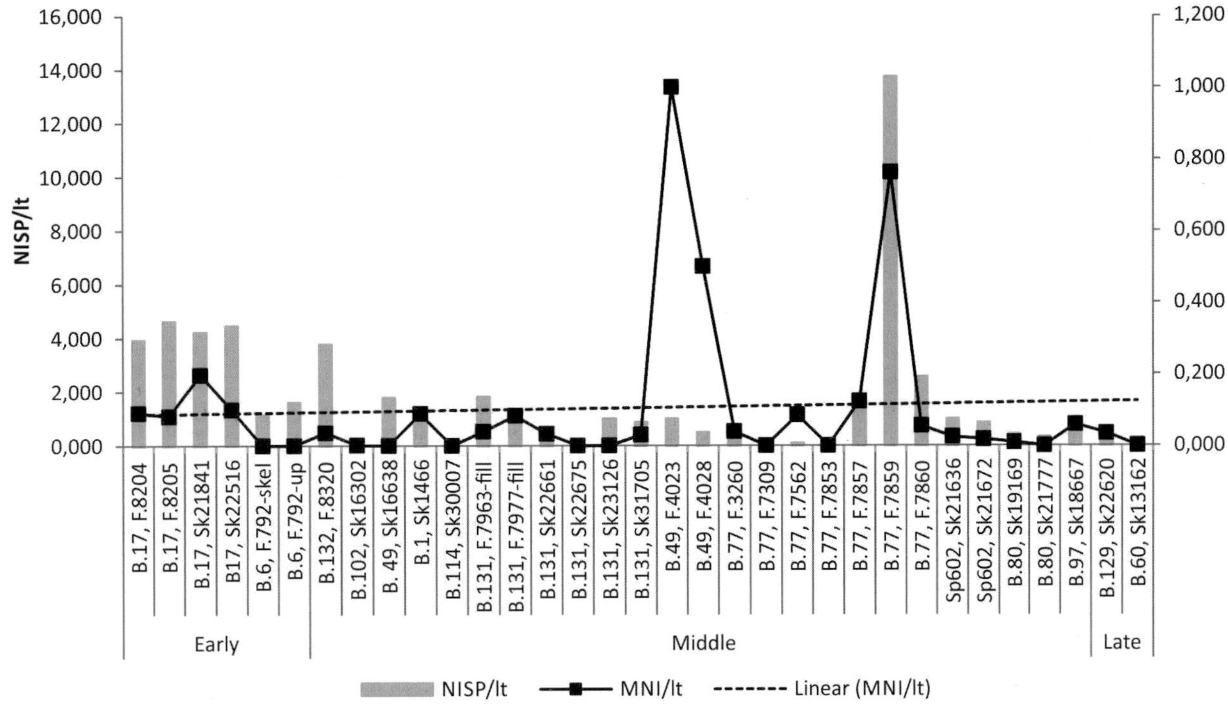

Figure 3.11. Temporal distribution of Unio *in burials (NISP=6,349; MNI=202).*

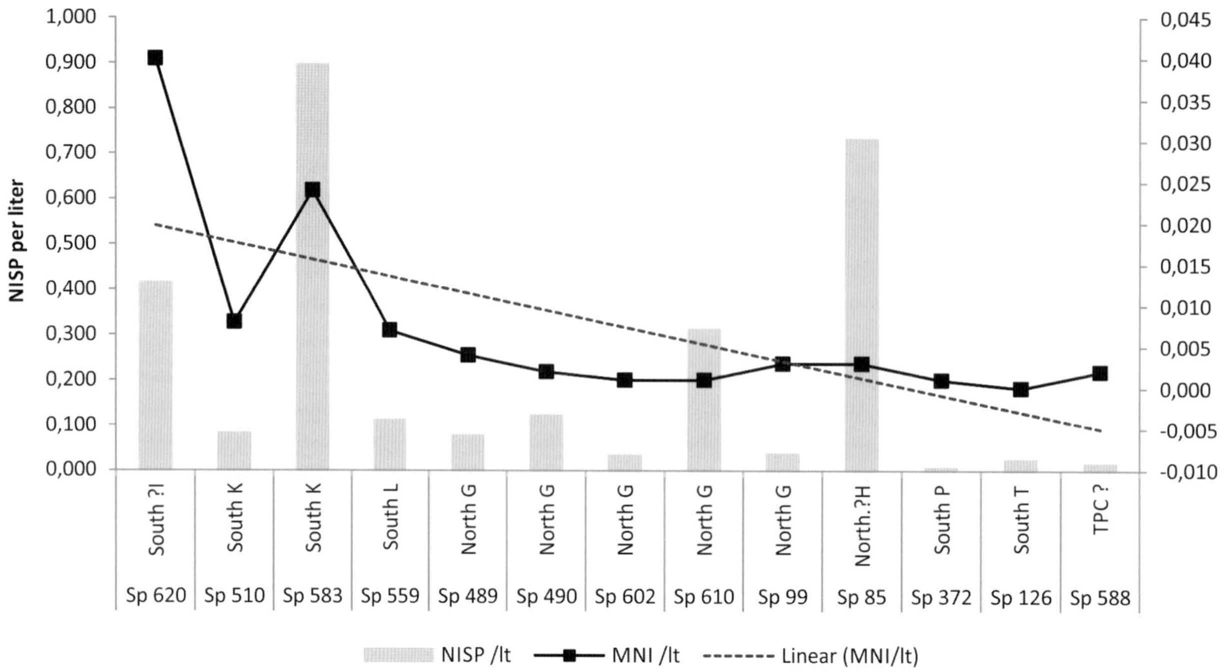

Figure 3.12. Temporal distribution of Unio *in middens (NISP=4,070; MNI=199).*

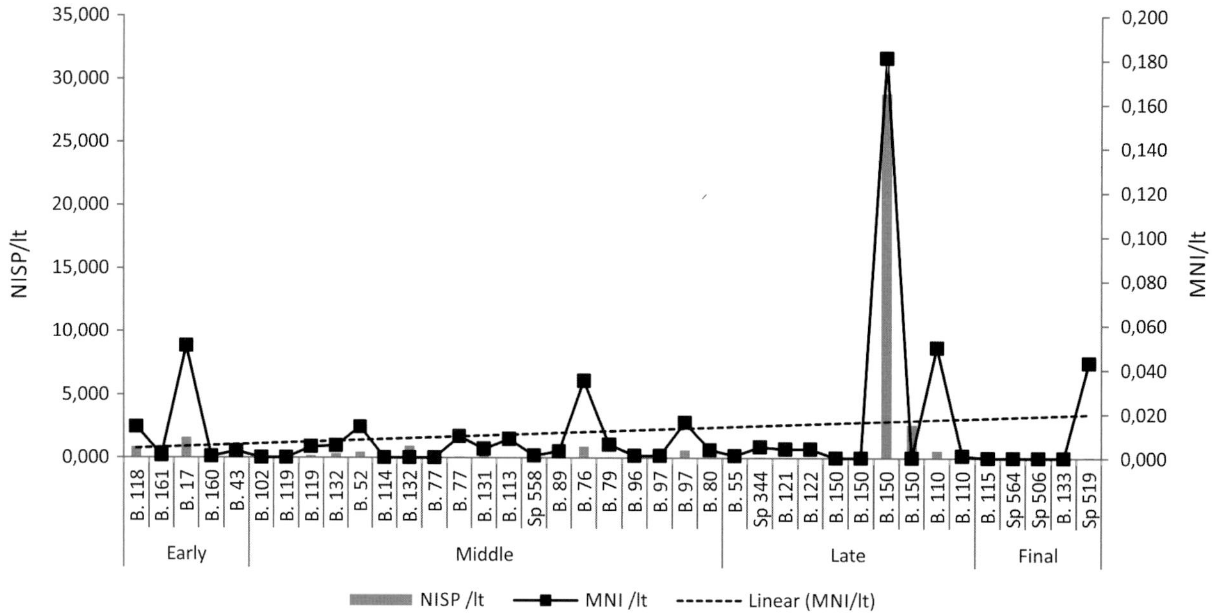

Figure 3.13. Temporal distribution of Unio *in buildings (NISP= 8,027; MNI=273).*

This suggests good environmental conditions for mussels, but also that gatherers were attracted by large individuals that offered more meat for consumption. This observation coupled with the largest quantities of remains corroborates the suggestion of a more intensive scale of exploitation in the Early period. The range of valve heights declines considerably during the Middle period. The low median implies that the available natural populations were smaller in size. In several archaeomalacological studies, this has been related to the depletion of natural populations due to previous over-exploitation (e.g., Swadling 1976). If indeed this was also the case at Çatalhöyük, then the large range of sizes coupled with the presence of outliers (very large-in-size molluscs would have existed in some niches, and would have been targeted) in the Middle period means that people were forced to gather what was available in nature. This explanation is corroborated by the expansion of harvesting in several seasons discussed above. This picture is inverted in the Late period. On the one hand, a higher median indicates that larger mussels were again available, maybe because natural populations were replenished. However, the large spread of values coupled with the evidence for low amounts of *Unio* on-site (figs 3.8b, 3.9, 3.10) and harvesting in various seasons of the year (table 3.4) are the factors that point towards the circumstantial exploitation of freshwater mussels in this period.

A gradual decline of the scale of exploitation of *Unio* through time is observed. Apart from environmental factors, several lines of evidence suggest significant economic and social changes at the settlement (Twiss 2012; Asouti 2013; Russell et al. 2013; Marciniak et al. 2015b). In this vein, the results from *Unio* analysis can be interpreted as indicating a shift from a larger scale of consumption of mussels, especially at a communal level, in the Early period to a lower (domestic) scale of consumption and more diverse activities of households in the Middle and especially the Late/Final periods.

Freshwater molluscs and wetland materials

This section discusses the juvenile bivalves, juvenile and adult gastropods and micro-molluscs that were incorporated into the anthropogenic deposits accidentally. The most frequently encountered species is *V. piscinalis* followed by *Pl. corneus*, *Bithynia* sp. and *Viviparus* sp. (table 3.2). *V. piscinalis* (fig. 3.3) is of particular importance due to its high frequency, but also due to the existence of a large corpus of literature on the ecology and biology of modern populations that opens up insights into the ancient populations.

State of preservation

Freshwater molluscs are represented by a large quantity of intact specimens (NI=92.55% of the total NI, n=3,691) and diagnostic body parts (n=1,848) that usually bear

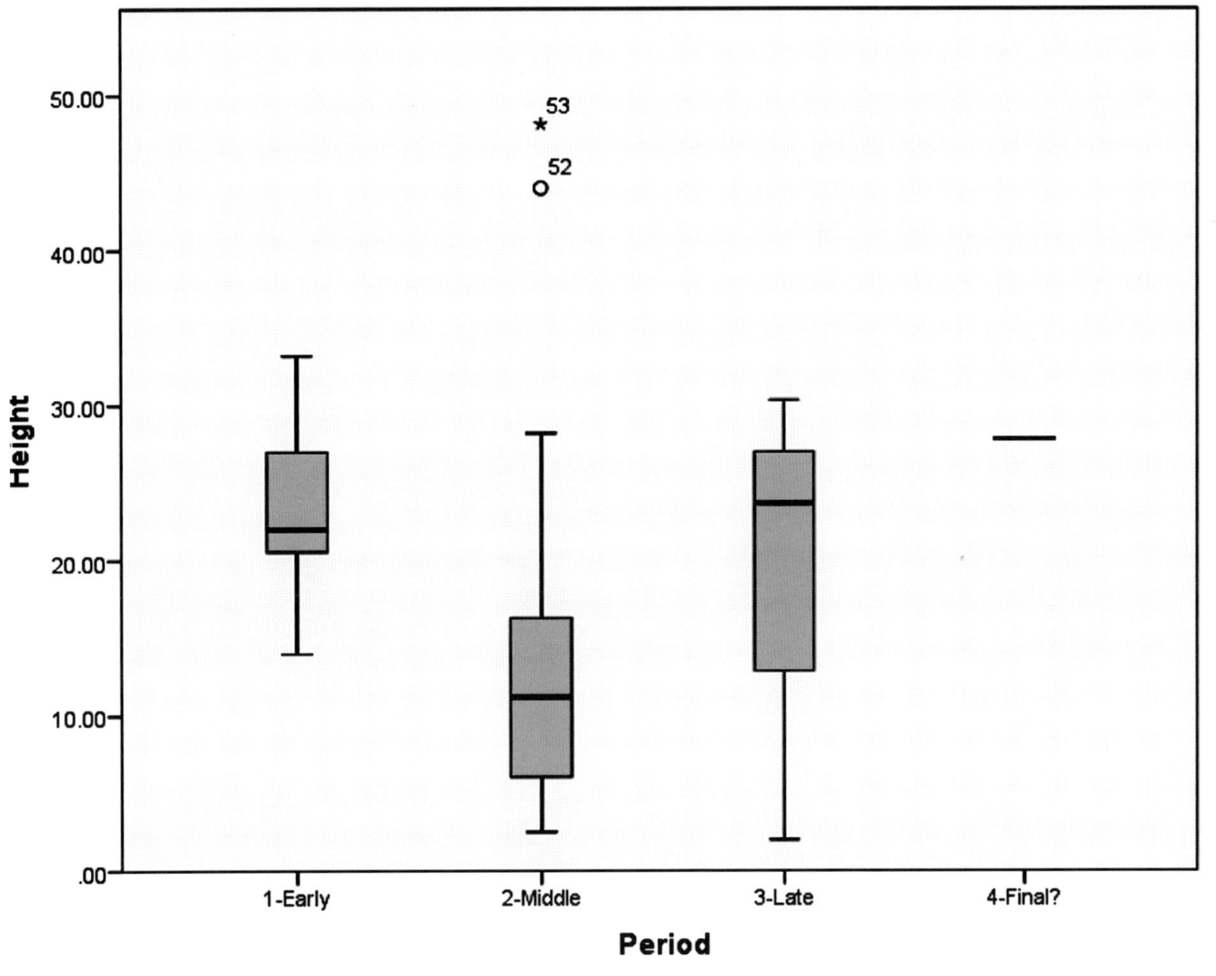

Figure 3.14. The height (in mm) distribution of Unio *valves across major occupation periods.*

slight evidence of burning or none (table 3.5). Therefore, these specimens are characterised by a completely different state of preservation compared to *Unio* specimens. The amount of fragmented material is still large (NF=23.27% of the total NF), but these fragments belong to larger-sized molluscs with thin shells (*Dreissena* sp., *Lymnaea* sp., *Stagnicola* sp., *Viviparus* sp.), while there are only a few fragments of more robust shells (*Fagotia* sp., *Th. heldreichi*). The vast majority of specimens exhibits a moderate to high degree of wear that indicates delayed burial after deposition.

Measurements

The height of the majority of intact shells was measured to shed light on the issue of their accidental introduction into the anthropogenic deposits. The focus is placed on the *V. piscinalis* individuals, with the aim of gathering information on the represented age groups, which can serve as a proxy for the season of mollusc death and subsequently allow inferences on the season of collection of wetland materials.

The majority of *V. piscinalis* intact specimens are below 1.9mm in height (45.8%), while there are significant numbers measuring 2 to 2.9mm (25.6%) and 3 to 3.9mm in height (26%) (fig. 3.15). Only few specimens (n=3) have a height greater than 5mm. In other words, the vast majority of *Valvata* specimens are juveniles and young-in-age individuals. Similar are the age groups represented in the *Pl. corneus* assemblage (92.1%, n=280), as well as in some other species, like *G. albus*, *G. truncatela* and *Fagotia* sp.

Studies of the life-cycle of *V. piscinalis* (Hunter 1961; Lodge, Kelly 1985; Myzyk 2007; Mouthon, Daufresne 2008) have shown that a size of 1.5–3mm (2–3 months old) is expected at the end of summer to mid-autumn, while a size of 3–5mm (young adults and adults) is usually expected in spring. Mortality rates are high in juveniles 1.5 to 3 months after hatching, usually occurring in the end of summer to mid-autumn. Therefore, the age groups represented in the anthropogenic assemblages suggest that the majority of *V. piscinalis* snails were deposited in the natural environment sometime in late summer/autumn, while a smaller group was deposited sometime in spring.

	V. piscinalis		*Remaining 18 species*	
Quantification	*Absolute*	*Density*	*Absolute*	*Density*
NISP [total]	4,812	0.036	6,121	0.046
NISP [no traces]	3,326	0.025	3,724	0.026
NISP [worn]	87	0.001	610	0.003
NISP [burnt]	1,252	0.01	1,586	0.01
MNI	3,798	0.028	1,109	0.008
NF	884	0.007	4,510	0.033
NI	3,001	0.022	690	0.004
Ratio NF:NI	0.29		13.49	

Table 3.5. Summary of absolute numbers, density per litre of soil and taphonomic attributes of V. piscinalis *and the remaining 18 freshwater species (table 3.2).*

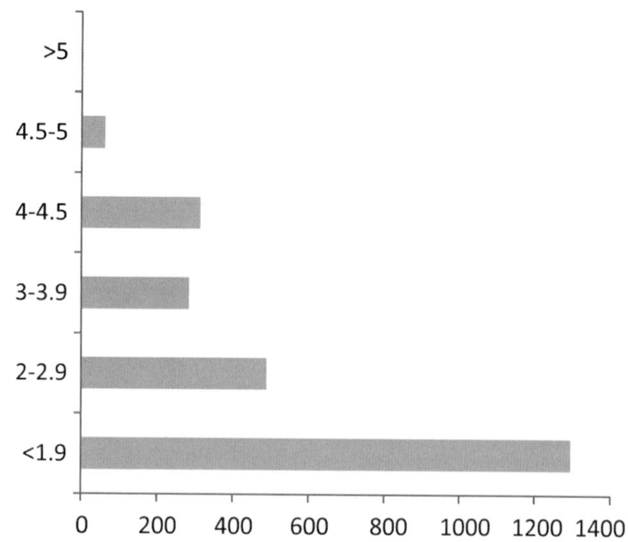

Figure 3.15. The height (in mm) distribution of V. piscinalis *individuals (n=2,448).*

In the natural environment, *V. piscinalis* co-exists and has a similar life-cycle to other species, such as *Pl. corneus* (Lodge, Kelly 1985). These two species are usually found together in the Çatalhöyük assemblage. Therefore, it is suggested that these molluscs were collected from specific loci in the wetlands that favour the occurrence of juvenile individuals of co-existent species (life community) and form particular thanatocoenoses (death assemblages). In other words, these snails were brought on-site together, among similar materials. The question that arises is where these life communities and subsequent death assemblages are to be found in a freshwater ecosystem. Further information on this issue can be gained from the identification of fresh and worn shells and the fragmentation rates of each species.

Taphonomy

The state of preservation of the vast majority of intact juvenile specimens of *V. piscinalis* and *Pl. corneus* (72.9% and 92.1% of their total MNIs respectively) is pristine, bearing unworn surfaces and sometimes preserving slight colour and gloss. Worn surfaces, traces of abrasion and erosion, chalky and whitish shells occurred very rarely in the assemblage. In addition, the fragmentation rates of these two species are low (Ratio NF:NI is 0.29 for *V. piscinalis* and 0.08 for *Pl. corneus*). This preservation cannot be related to their small and globular shape, which makes them less prone to fragmentation. Rather, the evidence for very little, if any, post-mortem transportation leads to the plausible suggestion that the majority were collected either alive or very recently dead from the natural environment.

The remaining taxa, especially adult bivalves and gastropods (for example, *Dreissena* sp., *Viviparus* sp., *Bithynia* sp.), are represented by worn fragments with abraded or chalky surfaces and a few intact or partially intact shells with evidence of post-mortem wear (worn apices, traces of bio-erosion). In other words, the moderate state of their surfaces points towards worn shells deposited in calm conditions with low current velocity and little/moderate post-mortem transportation. It must be noted here that shells bearing different conditions of preservation and wear co-existed in the majority of anthropogenic deposits.

To allow further insights into the taphonomic processes that shaped the anthropogenic assemblage, the temporal distribution of intact and fragmented shells is explored (fig. 3.16). The density of NI is rather similar across periods followed by an increase in the density of the NF towards the late levels. Considering that taxa representation is broadly similar among different periods (fig. 3.4), then the increase of the frequency of fragments must be a result of the re-use and recycling of the anthropogenic deposits by the inhabitants, as has been suggested by Doherty (2017).

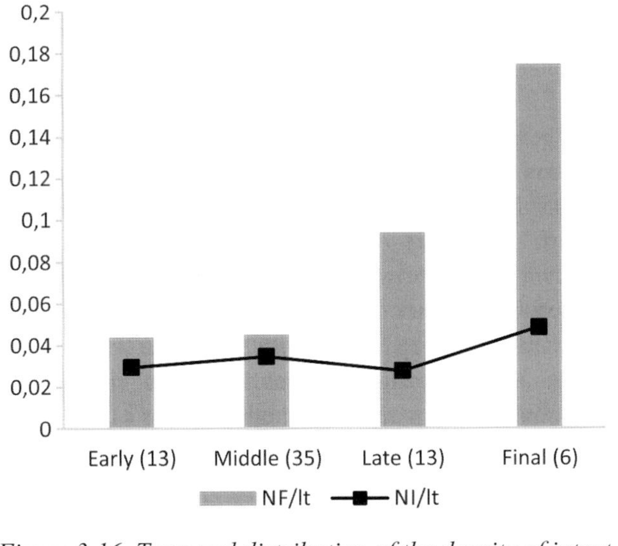

Figure 3.16. Temporal distribution of the density of intact and fragmented micro-molluscs. The number of studied deposits is noted in parentheses.

As regards traces of human modification, a moderate amount of remains (NISP=25.96%, n=2,838) shows evidence of fire/heat alteration: the majority (17.89%) bears light brown and light grey hues all over the shell surface, indicating burning in varying temperatures for a short period of time, while a small amount (8.07%) bears a dark grey to black colour indicating prolonged burning in high temperatures. Greyish spots that indicate the exact point of direct contact with fire are present in several of these specimens. The moderate amount of these burnt remains is in sharp contrast with what has been observed in the *Unio* assemblage, thus indicating that each group of shells passed through different depositional and post-depositional processes.

How were these shells introduced on-site?

As mentioned above, Reese (2005) suggested that these shells were introduced with water. On the other hand, Gümüs and Bar-Yosef Mayer (2013: 85) proposed that they were introduced in muds used for construction, noting that they were mainly discovered in the 'debris of mudbrick and mortar used to build the houses of Çatalhöyük'. This proposition was corroborated by the inspection of bricks and mortar in the field and by micromorphological studies that identified shell fragments inside the building material (Shillito et al. 2011a; Love 2012). They also assumed that 'bricks and mortars were constructed from newly collected materials in each building phase' (Gümüs, Bar-Yosef Mayer 2013: 86). However, they also raised the possibility that these shells were introduced with plants, such as reeds and sedges, used in construction and for matting (Gümüs, Bar-Yosef Mayer 2013: 85; Bar-Yosef Mayer 2013: 337).

In this framework, a central aim of the present study was to investigate if the above interpretations are plausible or not, employing detailed spatial and contextual analysis of these remains. Moreover, the present study included the analysis of different data categories (dirty floors, room fills, hearths, ovens, construction material, clusters, dumps and bin infills) to obtain a more holistic understanding of the presence and frequency of shells in each one.

The analysis of the spatial distribution of freshwater molluscan remains (fig. 3.17) shows that the lowest densities are associated with building/construction material and room fills (for example, B.160, Level South K; B.97, Level South O; B.77, Level North G; B.132, Level North F), while the highest densities are observed in ash rake-outs, hearths and ovens (for example, B.160, Level South K; B.80 and B.97, Level South O; Sp.85, Level North H?). Interestingly, in all types of data categories micro-molluscs exhibit various taphonomic characteristics, as unburnt, slightly burnt and heavily burnt shells co-exist. Even in the cases of hearths, ash rake-outs and ovens the low frequencies of burnt shells indicate that they were included among materials that protected at least some of them from extensive burning.

Turning to other lines of evidence, research on construction materials (Matthews 2005a; Tung 2013; Love 2012; Doherty 2017) has concluded (notwithstanding different opinions on specific issues) that there is a shift in the use of clay sources from dark clay in early levels to red-brown colours from Level South M onwards coupled with extensive recycling of building materials in the Middle/Late periods. Moreover, mudbricks contained leaves and stems from sedges (Cyperaceae) and reeds (*Phragmites* sp.), as well as very tiny shell fragments (<0.5 mm), but rarely larger fragments or intact shells. Therefore, if these shells had been introduced in muds used for construction, one would expect a different state of preservation and spatial distribution of micro-molluscs, as well as shifts in taxa representation between the early and late levels of the sequence. In other words, research on mudbricks does not corroborate the suggestion that micro-molluscs were mainly introduced with muds used for construction.

Complementary information on these molluscs was sought from ecological, biological and geological research in various geographic areas (Young 1975; Lodge, Kelly 1985; Kumari et al. 2007; Myzyk 2007; Mouthon, Daufresne 2008; Şereflişan et al. 2009; Girod 2013; Altınsaçlı et al. 2014; Qazar 2016). These studies provide compelling proof for accumulations of *V. piscinalis* and other co-existent molluscan species (*Pl. corneus*, *Lymnaea* sp. *Bithynia* sp.) in relation to aquatic plants, including submersed weeds, sedges, rushes, reeds and roots of reeds, as well as in rotting and dying vegetation along lake and river banks during summer droughts. Snails of various ages and conditions of preservation are found in such accumulations, while juvenile snails that have suffered very little post-mortem transportation are particularly common. Interestingly, the description of snails and taxa representation in these modern studies have many similarities with what has been observed in the anthropogenic deposits of Çatalhöyük.

On the basis of all lines of evidence, it seems plausible to suggest that micro-molluscs were mainly brought on site along with aquatic plants, which were used extensively to answer several human needs, including food, fuel, temper in mudbricks, roofing, matting, cording and basketry (Ryan 2013; Matthews 2005a; Matthews et al. 2014; Shillito, Matthews 2013; Shillito, Ryan 2013; Asouti 2013; Bogaard et al. 2013; Charles et al. 2014). More specifically, the low fragmentation rates, the pristine condition of preservation and the very large numbers of juvenile individuals in the anthropogenic assemblage are factors that point towards the introduction of these micro-molluscs in the roots of reeds and drying vegetation from the banks of river channels and pools. The higher densities observed in the ashy remains of fuels are in tandem with the high amounts of phytoliths of

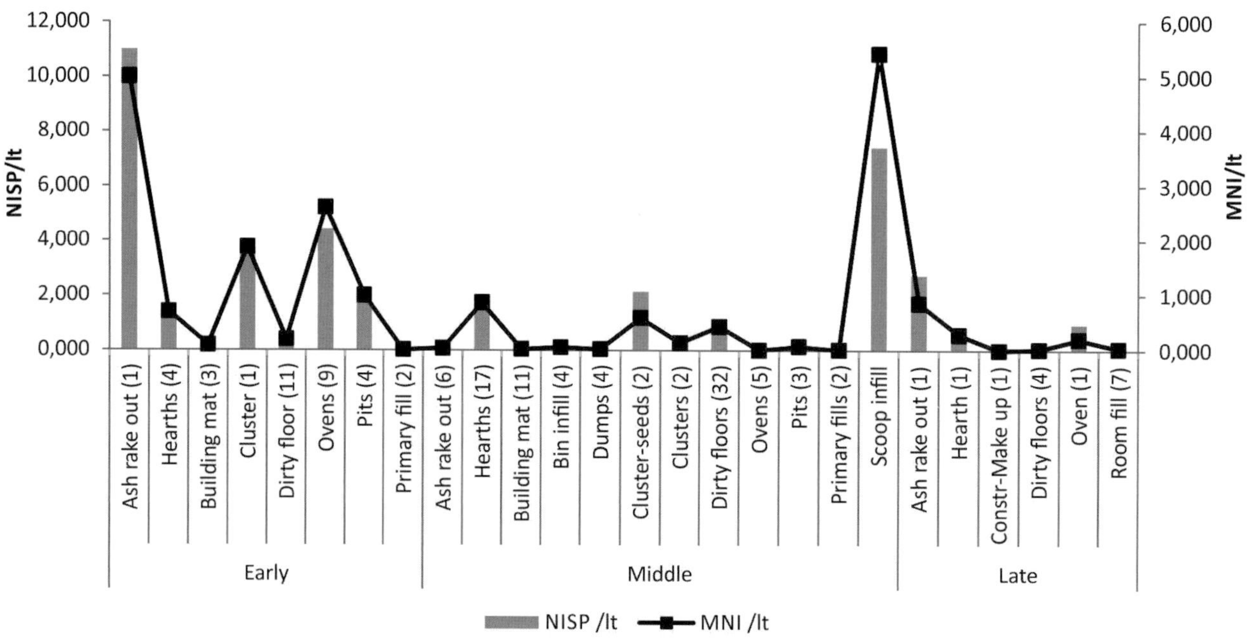

Figure 3.17. Temporal distribution of micro-molluscs according to data categories. The number of studied deposits is noted in parentheses.

sedges and *Phragmites* reeds that were used as fuel in the same types of deposits (Ryan 2013; Matthews et al. 2014). A compelling proof in this scenario was given by the fuel experiments with dry roots of reeds at Konya Plain conducted by C. Kabukcu (2013), as a similar taxa representation and a similar state of preservation of snails were observed in the remains of these fuels.

Temporal and spatial distribution

As stated above, these freshwater molluscs are present in the vast majority of deposits throughout the sequence, albeit with some differences in taxa variety and representation across periods (fig. 3.4). The density of freshwater molluscs is particularly low (<1 specimens per litre), which makes sense considering the low/moderate amounts observed in the natural assemblages in the modern studies mentioned above. On the other hand, the fact that some species are present in all periods underlines the continuous exploitation of particular wetland niches and materials (reeds, sedges) across time.

The analysis of the temporal distribution of these remains shows similar frequencies of MNI densities (fig. 3.18), thus pointing towards a broadly similar scale of exploitation of these wetland niches and materials across the sequence. This similarity in MNI values seems to be in contrast with the increasing densities of *Phragmites* in the phytolith assemblage in the later levels (Asouti 2013; Ryan 2013). However, archaeomalacological analysis cannot differentiate between the wetland plants exploited and it cannot accurately detect possible differences in ecological micro-niches exploited, which are assumed on the basis of less taxa diversity and higher fragmentation rates of shells in the Late/Final periods (fig. 3.16, 3.18b). The latter can be also justified by the proposed recycling of anthropogenic deposits in later levels (Doherty 2017).

The analysis of temporal distribution of these shells in different types of deposits reveals some specific patterns (fig. 3.19). The higher frequencies throughout the sequence are observed in burials, low to moderate densities in buildings (burials excluded) and the lowest densities in middens. In addition, the high abundance of these shells in burials declines across time, followed by a small increase in buildings and middens. Considering that many temporal differences may in fact be explained as spatial differences (Hodder 2013c: 11), individual contexts and data categories were analysed.

Starting with burials (fig. 3.20), fills rich in freshwater shells are present in several levels, despite the general decrease in the passage of time. The higher densities were recovered in the infant burial in basket F.8320 (B.132, Level North F), as was also found with *Unio* remains. Other interesting concentrations include the child burial F.4013 (B.49, Level North G) and the burial fill F.492 (B.6, Level South L). It must be underlined here that *Unio* and other freshwater mollusc distributions are usually different, with only a few exceptions (for example, F.8320 in B.132; F.8204 and F.8206 in B.17). On the other hand, there are significant differences in burials in the same building, as, for instance, the different burials in B.17 (Level South K) and B.131 (Level North G). These can be taken as evidence for different activities performed during mortuary practices. The comparative analysis of other inclusions is needed to provide further insights into the formation of burial fills.

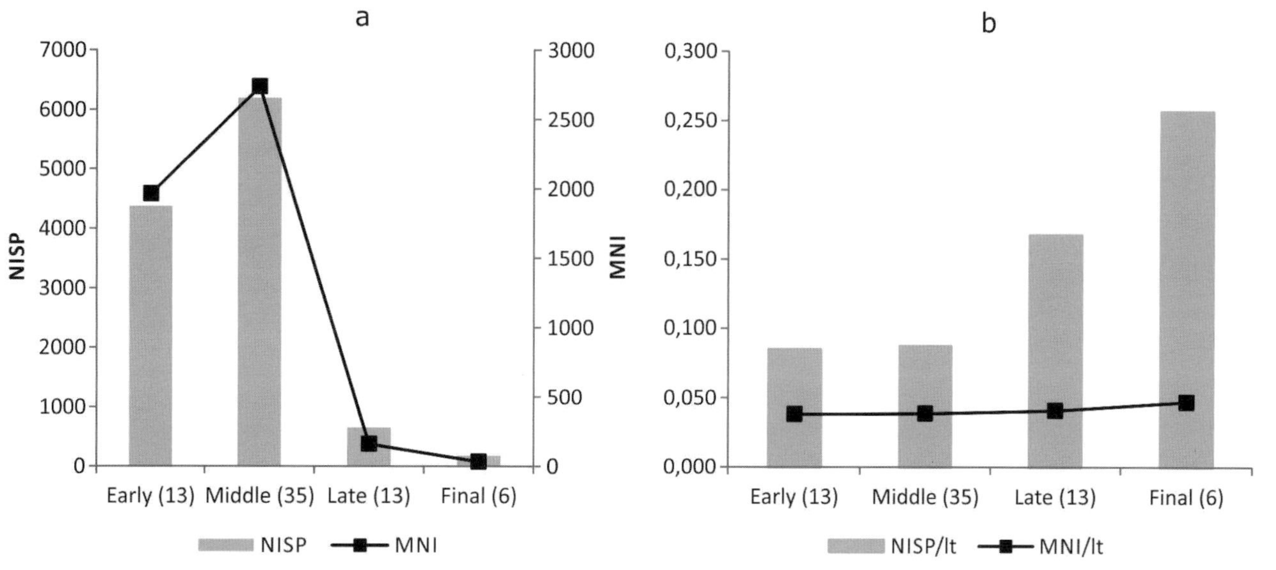

Figure 3.18. Distribution of freshwater mollusc remains (NISP, MNI) in absolute numbers and density across grouped levels. The number of studied deposits is noted in parentheses.

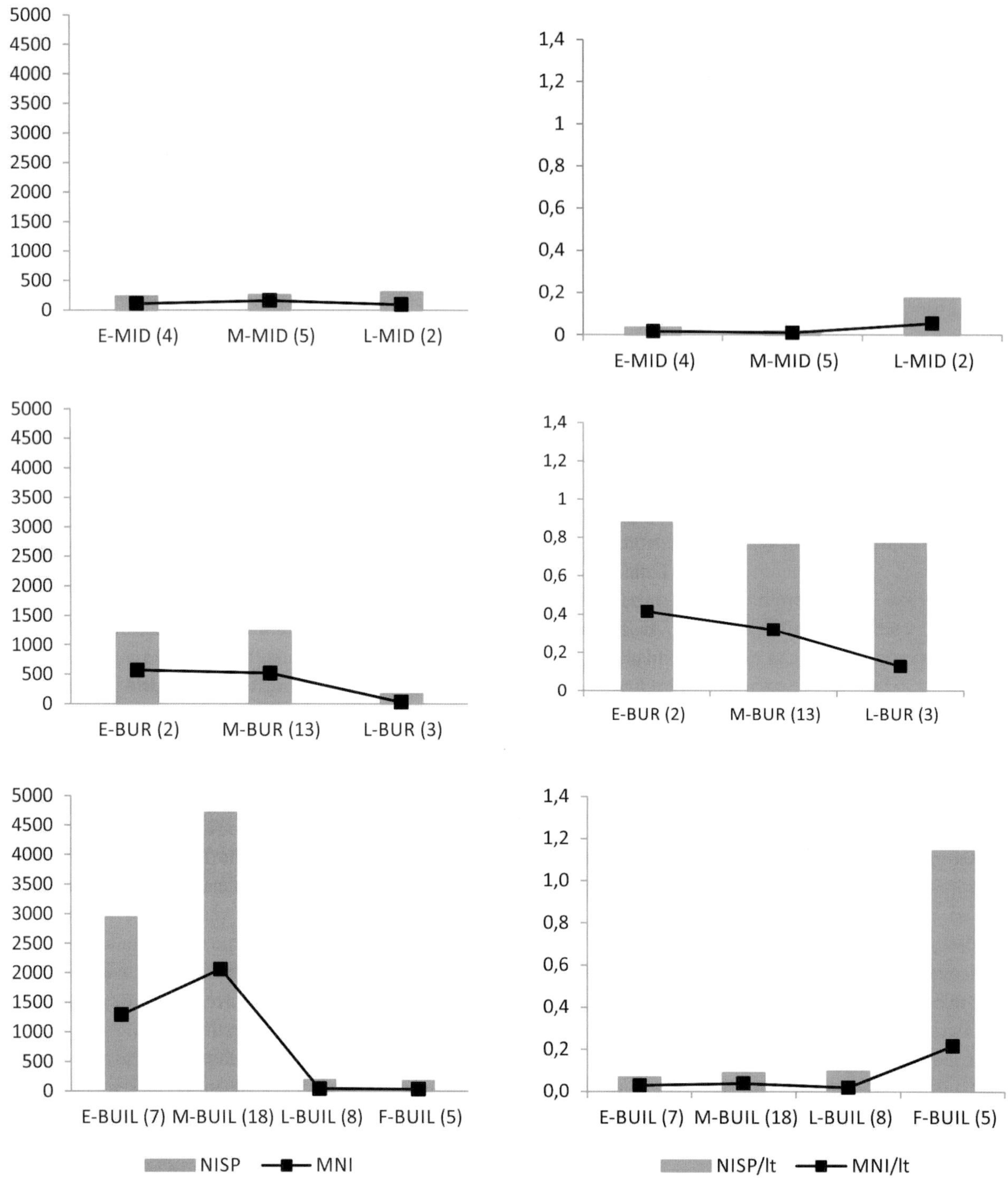

Figure 3.19. Temporal distribution of freshwater molluscs in middens, burials and buildings. The number of studied deposits is noted in parentheses.

Moving to middens (fig. 3.21), the higher taxa richness and the most significant densities are observed in Sp.583 (Level South K) and Sp.85 (Level North H?) coupled with high densities of *Unio* remains (fig. 3.12). In both spaces, species variety and abundance, the large amounts of *Bithynia* opercula, the presence of two age groups of *V. piscinalis* (juvenile and young in age) and juvenile *Pl. corneus* individuals point towards *death assemblages* in calm conditions of deposition among rich aquatic vegetation in a stagnant or slow-moving freshwater body. Therefore, it is tempting to interpret these assemblages as directly representing drying roots of reeds and sedges in the banks of a pool or a river channel that were used as fuel.

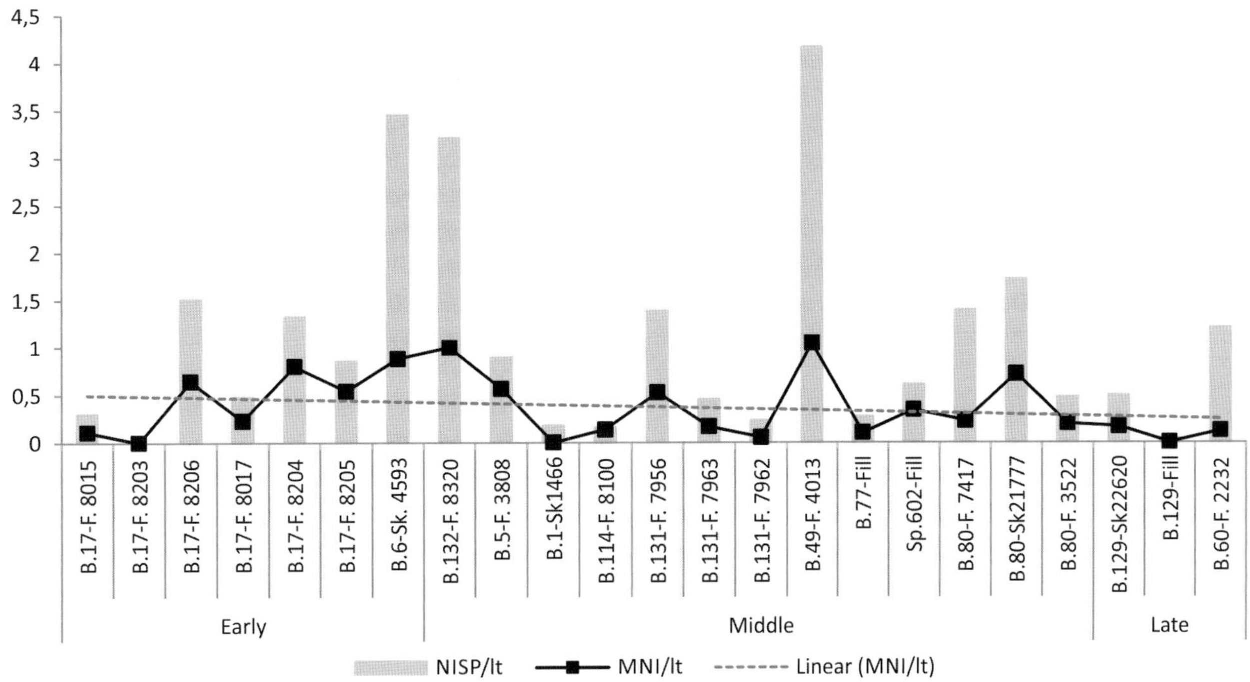

Figure 3.20. Temporal distribution of freshwater molluscs in different burials (NISP=2,595; MNI=1,108).

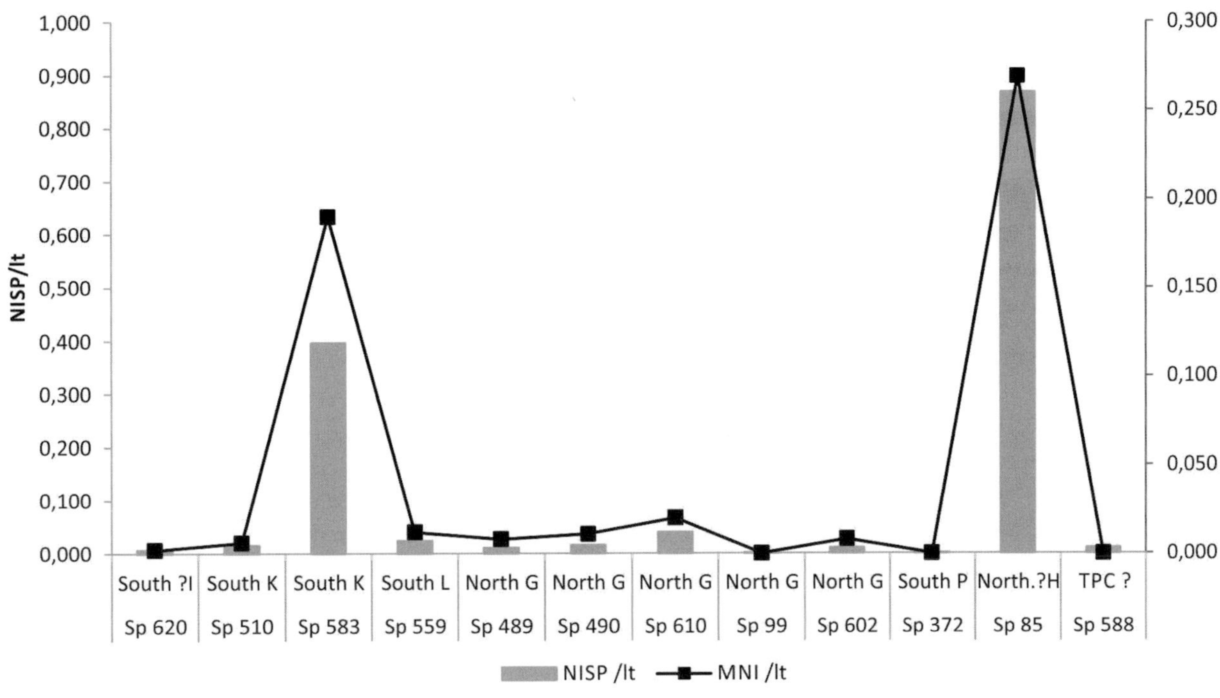

Figure 3.21. Temporal distribution of freshwater molluscs in different middens (NISP=874; MNI=423).

Last, the taxa representation in buildings is not uniform (fig. 3.22). One of the most prominent peaks in the density of remains is observed in B.17 (Level South K) due to an oven floor and oven fill (F.579), followed by B.76 (Level South O) due to an ash rake-out (18704). There are also high densities in B.132 (Level North F) due to a scoop infill (F.7731), a fire spot (F.7741) and a special hearth closure (F.8322), as well as in Sp.506 due to a hearth (F.3930). Therefore, the frequencies of shell remains directly depend on the types of deposits included in analysis, in particular, whether there were deposits related to the use of fire or not.

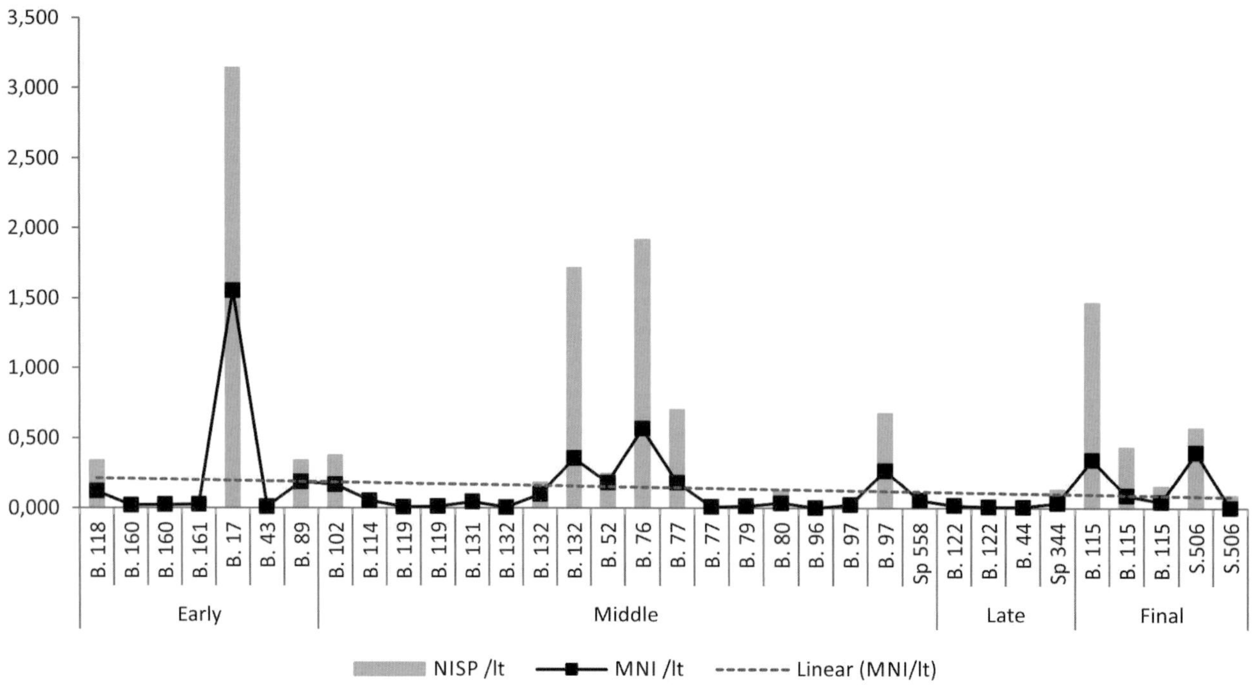

Figure 3.22. Temporal distribution of freshwater molluscs (NISP/lt, MNI/lt) in different buildings (NISP=7,821; MNI=3,402).

Conclusions

The work reported in this chapter allows a series of insights into the social context of mollusc- and shell-related activities in this early farming community. The ubiquitous presence of freshwater molluscs, both *Unio* and other macro- and micro-molluscs, contributes to our knowledge of the ecology and biology of fresh-water bodies around the site and complements other palaeoenvironmental proxies. The ecological requirements and the temporal distribution of shells point towards the presence of small channels and pools, where flowing, slow moving and stagnant water co-existed, creating a mosaic of wetlands with rich vegetation. It has also been argued that juvenile and minuscule molluscs were introduced accidently into the anthropogenic deposits mainly through plants, in particular reeds and sedges, used as fuel and as temper in mudbricks.

Building on the observations of Reese (2005) and Bar-Yosef Mayer (2013), the Çatalhöyük shell assemblage indicates constant use of freshwater mussels for food consumption that transcends levels and buildings. Despite the overall continuity, certain chronological changes are apparent. Notwithstanding the potential skewing of results due to the destructive treatment of food refuse and the inclusion of secondary and tertiary deposits in analysis, the integration of different proxies indicates that the decline in late levels can be better explained on economic and social grounds rather than by environmental changes. In fact, the evidence for more intensive exploitation of mussels in early levels may reflect communal and shared activities, labour and food consumption (gathering of mussels is considered as a *per se* communal activity, Meehan 1982). Meanwhile, the decline in late levels can be taken to indicate more autonomous households and consumption of food at the household level.

Leaving aside archaeological biases and preconceptions about molluscan food (Moss 1993; Claassen 1998: 176), there is compelling proof of the consumption of molluscs as food in several early farming communities, usually depending on local environmental conditions (Veropoulidou 2011; Çakırlar 2013a). In this vein, it is argued that mussel gathering was embedded into the farming way of life; harvesting of freshwater mussels was probably combined with other off-site activities in the same environmental niches, for instance, plant and wood collection and animal herding. From a dietary perspective, mussels added proteins and minerals (calcium, phosphorus and iron) to diets and variety to the menu either as a seasonal food or as a year-round supplement.

Moreover, the secondary treatment of *Unio* food refuse for the production of mortar and maybe also lime is also observed in all buildings and levels. The inclusion of shell temper in clay, mudbricks and plasters has been observed on-site, but shell lime could also have been used in various ways, such as in the processing of plants, in cooking, as food for birds and animals, as calcium in fields, that do not leave archaeologically recognisable traces. In any case, what is important to underline is the management of this particular food refuse with its unique properties, thus maximising the potential of materials and demonstrating an economising behaviour towards available resources.

The above are key archaeomalacological inferences based on several analytical methods not previously applied, at least in this scale, in Çatalhöyük and other eastern Mediterranean archaeomalacological assemblages. The study of all material, regardless of size and state of preservation, recovered from all fractions of the heavy residue, and the systematic recording of taphonomic attributes has allowed discussion of food consumption practices, opened up insights into the secondary treatment of food refuse, provided further evidence for the exploitation of various wetland materials and complemented work on the environmental conditions around the site. It is hoped that the implementation of various analytical methods and the qualitative and quantitative analysis of all shell remains in different scales demonstrated the potential of archaeomalacological analysis in disentangling human-environmental relations.

Acknowledgements

I am grateful to Ian Hodder for inviting me to participate in the Çatalhöyük Research Project and Christina Tsoraki for enabling the study. This chapter would not have been possible without the work of all project members and the stimulating discussions during the Catania meeting. Particular thanks go to Marek Z. Barański, Amy Bogaard, Lisa Guerre, Scott D. Haddow, Justin Issavi, Ceren Kabukcu, Christopher J. Knüsel, Dominik Lukas and Arkadiusz Marciniak for sharing information, data and advice. Many thanks also to Jason Quinlan for the photographs and to Nikos Valasiadis for editing the figures.

4. Woodland vegetation, fuelwood and timber use at Çatalhöyük: the anthracological remains from the 1996 to 2017 excavations

Ceren Kabukcu, Eleni Asouti and Joseph Losh

Aims and scope

To date, the anthracological assemblage of Çatalhöyük East represents one of the largest and most intensively studied prehistoric wood charcoal assemblages in Southwest Asia. As indicated in previous publications, the foremost condition that has enabled this intensity and depth of analysis is the exceptional (by comparison to other prehistoric sites in the Eastern Mediterranean region) preservation of charred wood and non-wood archaeobotanical macro-remains at Çatalhöyük. The principal aim of this chapter is to provide a comprehensive account of the anthracological research carried out at the site to date, incorporating the results obtained from laboratory and field analyses between 2015 and 2017 with those published previously by Asouti (2001; 2005; 2013) and Kabukcu (2015; 2017; 2018a). The new work reported in this chapter includes analyses of context-related variation in charcoal sample composition and its interpretation in terms of charcoal taphonomy and prehistoric fuelwood selection (both in terms of species and log size), alongside the results of ongoing work on the identification of woodland management practices and tree growth conditions in south-central Anatolia. Finally, new data are also presented on the use of timber in Çatalhöyük East.

Materials and methods

Routine analytical methods (charcoal identification and basic quantitative analysis) follow those described in detail in earlier publications (Asouti 2005; 2013). More recent work has also established protocols for the systematic collection and analysis of dendroanthracological data (Kabukcu 2015; 2017; 2018a; 2018b). Dendroanthracological features recorded at Çatalhöyük are classified into three main groups: (i) features relating to the general condition of unburnt wood (evidence of fungal decay, collapsed fibres/vessels/tracheids, insect/woodworm boreholes), (ii) signs of physiological stress (narrow growth rings, discontinuous/false rings, traumatic canals/ducts, the formation of scar/callus tissue) and (iii) indicators of log sizes used as fuel and their positioning on the tree as branch, twig and/or stem wood (estimated *inter alia* through the recording of the presence of tyloses, pith and/or bark, alongside qualitative and quantitative estimations of growth ring curvature) (Kabukcu 2018a; 2018b).

Following on from previous anthracological work at the site, sample selection focused on deposits containing accumulations of fuelwood waste debris, including midden and other secondary charcoal deposition contexts (for example, building infills, 'dirty floor' deposits, etc.), which permit determining diachronic patterns of routine fuelwood use (Asouti 2005; Asouti, Austin 2005; Kabukcu 2018b). Contextual variation in fuelwood use was explored via multivariate analyses (Correspondence Analysis, CA) of charcoal sample composition involving both internal and external fire features (for example, wood charcoal from hearths, ovens, fire spots). Spatial patterning in charcoal sample composition was also explored by applying CA on anthracological data originating from midden, fire feature and 'dirty floor' deposits.

Results

Midden and midden-like deposits

Previous analyses of fuelwood use in the North and South Areas focused predominantly on Levels North G and H/I and South G, G-H, I, J, K, L, M, O, P, Q, R, S and T (Asouti 2013; Kabukcu 2015; 2017). New results presented in this chapter include midden samples from the previously unreported South N and midden units from North G, H and I, plus 'dirty floor' deposits from North F (from which no midden deposits were available for analysis). The combined results of previous and new analyses are presented in table 4.1 and in the form of an anthracological diagram in figure 4.1.

Charcoal sample composition during the earliest levels of occupation at Çatalhöyük (South G) is dominated by a diverse range of riparian woodland taxa including elm (*Ulmus*), hackberry (*Celtis*), elm/hackberry (Ulmaceae), willow/poplar (Salicaceae) and tamarisk (*Tamarix*).

Level	*South G (N=18)*		*South G-H (N=13)*		*South I (N=6)*		*South J (N=2)*		*South K (N=9)*		*South L (N=11)*		*South M (N=4)*		*South N (N=7)*		*South O (N=2)*	
	C	% C	C	% C	C	% C	C	% C	C	% C	C	% C	C	% C	C	% C	C	% C
Juniperus	9	0.80	7	0.56	1	0.16	2	0.94	23	2.51	28	2.40	20	5.00	27	7.96	36	20.81
Quercus	31	2.75	651	52.46	384	61.44	130	61.03	371	40.46	636	54.55	253	63.25	160	47.20	78	45.09
Amygdalus	21	1.87	4	0.32	1	0.16	1	0.47	7	0.76	8	0.69	1	0.25	10	2.95	10	5.78
Pistacia	39	3.46	19	1.53	1	0.16	1	0.47	26	2.84	19	1.63	14	3.50	9	2.65	2	1.16
Prunus	1	0.09	2	0.16						0.00	8	0.69	2	0.50	1	0.29		
Maloideae	16	1.42	19	1.53	1	0.16	7	3.29	68	7.42	19	1.63	4	1.00			1	0.58
Anacardiaceae			1	0.08					1	0.11	1	0.09					4	2.31
Rhamnus			3	0.24														
Ulmaceae	473	42.01	119	9.59	88	14.08	26	12.21	110	12.00	61	5.23	5	1.25	56	16.52	7	4.05
Ulmus	86	7.64	28	2.26	7	1.12	2	0.94	37	4.03	70	6.00	38	9.50	2	0.59	16	9.25
Celtis	132	11.72	126	10.15	12	1.92	3	1.41	67	7.31	69	5.92	11	2.75	9	2.65	1	0.58
Salicaceae	281	24.96	226	18.21	126	20.16	28	13.15	127	13.85	146	12.52	31	7.75	47	13.86	15	8.67
Fraxinus			1	0.08	1	0.16	12	5.63	7	0.76	24	2.06	2	0.50	6	1.77		
Acer									2	0.22	5	0.43	2	0.50				
Chenopodiaceae	7	0.62	9	0.73	2	0.32			10	1.09	18	1.54	3	0.75	5	1.47		
Artemisia	14	1.24	15	1.21					11	1.20	8	0.69	2	0.50	5	1.47		
Leguminoseae	1	0.09	8	0.64					38	4.14	31	2.66	2	0.50			2	1.16
Capparis			1	0.08					4	0.44	1	0.09	1	0.25				
Tamarix	3	0.27							1	0.11	1	0.09	3	0.75	2	0.59	1	0.58
Caprifoliaceae	1	0.09											1	0.25				
Alnus											1	0.09	2	0.50				
Vitex	2	0.18	2	0.16							4	0.34	1	0.25				
Platanus											1	0.09						
Ficus carica											1	0.09						
Ephedra	1	0.09																
Labiateae	5	0.44			1	0.16	1	0.47	6	0.65	5	0.43	1	0.25				
Rosaceae	3	0.27							1	0.11	1	0.09	1	0.25				
Total Identified	1,126		1,241		625		213		917		1,166		400		339		173	

Table 4.1. Per level summary of fragment counts (C) and percentage fragment counts (C%) from midden and midden-like contexts at Çatalhöyük East, incorporating datasets produced by Asouti (2003; 2005; 2013) and Kabukcu (2015; 2017); N: number of analysed samples.

Level	*South P* (*N*=7)		*South Q* (*N*=6)		*South R* (*N*=6)		*South S* (*N*=7)		*South T* (*N*=5)		*North F* (*N*=7)		*North G* (*N*=12)		*North H* (*N*=2)		*North I* (*N*=9)	
	C	% C	C	% C	C	% C	C	% C	C	% C	C	% C	C	% C	C	% C	C	% C
Juniperus	42	7.14	104	34.90	116	41.43	294	67.28	91	52.91	14	2.87	116	10.14	75	28.63	324	41.38
Quercus	270	45.91	83	27.85	53	18.93	48	10.98	28	16.28	147	30.18	540	47.20	102	38.93	239	30.52
Amygdalus	8	1.36	11	3.69	23	8.21	35	8.01	12	6.98	4	0.82	20	1.75	5	1.91	36	4.60
Pistacia	6	1.02	10	3.36	7	2.5	15	3.43	13	7.56	11	2.26	31	2.71	5	1.91	21	2.68
Prunus	2	0.34									3	0.62	1	0.09				
Maloideae	6	1.02									26	5.34	2	0.17				
Anacardiaceae																		
Rhamnus											1	0.21					2	0.26
Ulmaceae	96	16.32	27	9.06	25	8.93	7	1.60	13	7.56	143	29.36	184	16.08	22	8.40	74	9.45
Ulmus	22	3.74	11	3.69	4	1.43	4	0.92	1	0.58	15	3.08	79	6.91	11	4.20	31	3.96
Celtis	17	2.89	5	1.68	7	2.5	1	0.23			54	11.09	29	2.53	7	2.67	7	0.89
Salicaceae	102	17.35	21	7.05	3	1.07	2	0.46	5	2.91	56	11.50	69	6.03	10	3.82	23	2.94
Fraxinus	3	0.51	23	7.72	41	14.64	28	6.41	9	5.23			50	4.37	24	9.16	22	2.81
Acer					1	0.36	2	0.46			2	0.41	1	0.09				
Chenopodiaceae											2	0.41	6	0.52				
Artemisia	6	1.02	2	0.67			1	0.23			6	1.23	11	0.96	1	0.38	1	0.13
Leguminoseae	1	0.17																
Capparis	2	0.34									1	0.21	2	0.17				
Tamarix	4	0.68	1	0.34									2	0.17			3	0.38
Caprifoliaceae																		
Alnus																		
Vitex																		
Platanus																		
Ficus carica	1	0.17																
Ephedra																		
Labiateae											2	0.41						
Rosaceae																		
Total Identified	588		298		280		437		172		487		1,144		262		783	

Table 4.1 (continued). Per level summary of fragment counts (C) and percentage fragment counts (C%) from midden and midden-like contexts at Çatalhöyük East, incorporating datasets produced by Asouti (2003; 2005; 2013) and Kabukcu (2015; 2017); N: number of analysed samples.

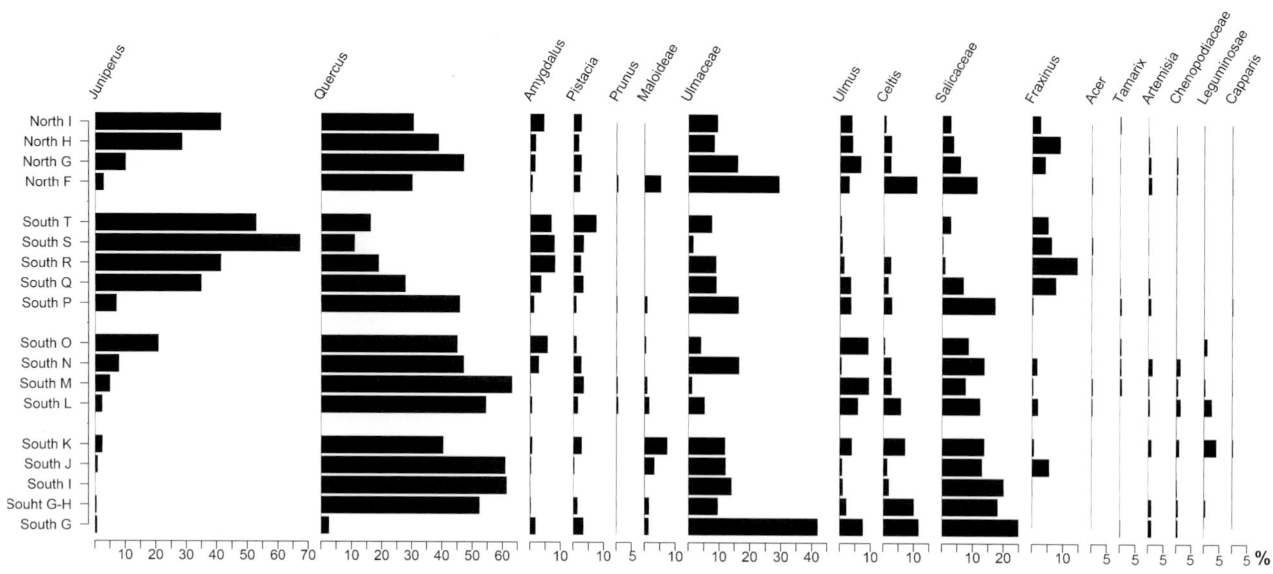

Figure 4.1. Anthracological diagram from all sampled levels at Çatalhöyük East (detailed dataset, cf. table 4.1).

Deciduous oak (*Quercus*) charcoal values represent 3% of the total number of identified fragments in Level South G; a rapid increase is observed during the final phase of Levels South G and in South H up to a maximum of 53% of charcoal composition (percentage fragment counts) (fig. 4.1). This abrupt change in *Quercus* values composition was initially detected by Asouti (2005; 2013) and was subsequently verified by further analyses conducted by Kabukcu (2015; 2017). From Levels South G-H onwards, deciduous oak dominates charcoal sample composition until the end of South P, accounting for ~50% of sample composition. In the transition between South P-Q, *Quercus* values drop from 43% to 27% and continue to decrease gradually through South R, S, T. Juniper (*Juniperus*) charcoal is attested in the Çatalhöyük charcoal stratigraphy from the earliest levels of occupation, albeit in low frequencies (<5% of sample composition). Juniper frequencies increase very gradually to 8% in Level South N and more markedly to 20% in South O (see fig. 4.1). This rapid increase may, at least in part, be an artefact of the low number of midden samples analysed from South O, in light also of the fact that juniper quantities drop again to 7% in subsequent South P. From South Q, juniper values increase to 34% and reach their peak in South S (67%), followed by a small drop to 52% in South T.

The sequence from the North Area mirrors to a certain degree the trends summarised above for the South Area. Similar to the earliest levels excavated in the South Area, Level North F 'dirty floor' deposits reflect a high degree of reliance on riparian taxa (Ulmaceae, *Ulmus* and *Celtis*). However, there is a considerable proportion of deciduous oak charcoal during this level, which is only marginally lower than the frequencies of oak charcoal obtained from South K and L midden samples. At the same time, the proportions of riparian taxa in North F are higher compared to post-South G levels in the South Area. In Levels North H and I there is a pronounced increase in juniper values (from 2.87% in North F through to 10% in North G, 28.6% in North H and 41.38% in North I). This pattern largely mirrors the upward trend in juniper charcoal values and the concomitant reduction in oak charcoal frequencies observed in South P-Q/R (fig. 4.1). Furthermore, a gradual increase in almond (*Amygdalus*) and terebinth (*Pistacia*) charcoal values is observed during the later levels of Neolithic occupation (that is, North I, South R, S and T) in both the North and the South Areas.

In order to explore further diachronic trends in charcoal sample composition and context-related variation across the North and South Areas, a Correspondence Analysis (CA) was carried out on the sample-by-sample charcoal taxon counts of 128 midden and midden-like contexts derived from all sampled levels of occupation (fig. 4.2). The CA biplot demonstrates the affiliation of the Level North F 'dirty floors' with the South K, L and M midden samples (high proportion of oak and similar proportions of riparian taxa). It also shows that the Level North G and South N, O and P middens are similarly positioned with regard to the proportions of *Celtis*, *Ulmus* and *Quercus*. Levels North I and H are clustering together with South R, S and T reflecting similar proportions of *Juniperus*, *Tamarix*,

Fraxinus, *Amygdalus* and *Pistacia* in their samples. Finally, it is possible to observe a greater variability in sample composition within Levels South Q, R, S, T and North I, H, whereby some samples have lower proportions of juniper and others higher proportions of *Fraxinus*, *Amygdalus* and, to a certain extent, *Tamarix* and *Pistacia*. Such variability is not observed within earlier South Area levels (for example, South K, L, N) or in the North G and F samples.

Primary fuel waste deposits (hearths/ovens and fire spots)
Previous anthracological analyses at Çatalhöyük have provided accumulating data on the taxonomic composition of samples derived from primary fuel waste deposits (Kabukcu 2015; see also Asouti 2013). For the purpose of the present study we performed a multivariate analysis of all the available samples in order to (i) investigate variation in charcoal sample composition between different types of fire features and (ii) explore any latent diachronic and/or spatial patterning in taxon representation. CA was applied on the sample-by-sample charcoal taxon counts from 56 internal (hearths/ovens) and external (fire spots) fire features across multiple occupation levels, which preserved at least 25 identified charcoal fragments (fig. 4.3). The CA biplot demonstrates that the majority of internal fire features from all levels in the North and South Areas cluster around deciduous oak (*Quercus*). This pattern holds, regardless of diachronic patterns in taxon representation observed in midden and midden-like samples. For example, two internal fire features from South Q contain predominantly *Quercus*, while juniper predominates in the six midden samples studied from this level. By contrast, as the CA biplot demonstrates, external fire features from all levels contain a much greater diversity of taxa including combinations of oak and/or juniper and riparian taxa. In some cases, sample composition replicates the patterns identified in midden deposits from the same level (for example, South G external fire features contain high quantities of Ulmaceae just like the midden samples from the same level). On the other hand, while Level South P middens are overall dominated by *Quercus*, external and internal fire features in South P are very variable in their composition: some external fire features from South P contain more riparian taxa or *Juniperus* (see fig. 4.3, upper right-hand corner of the CA biplot) while other external and internal South P fire features are dominated by deciduous oak (see fig. 4.3, lower right-hand corner of the CA biplot).

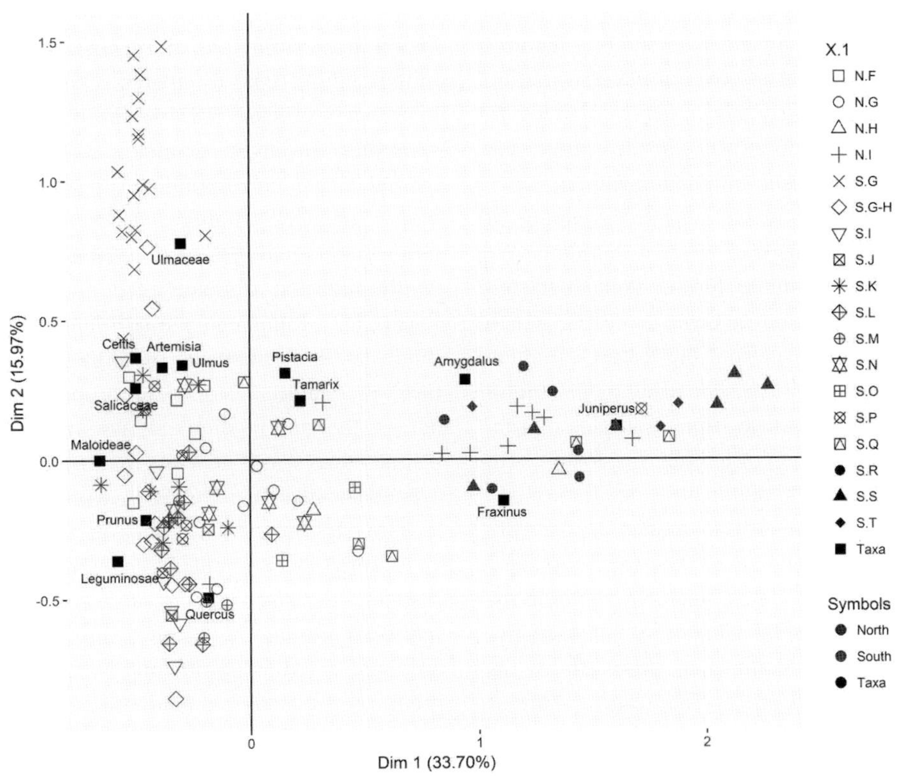

Figure 4.2. Plot of Dimensions 1 and 2, CA run on per sample wood charcoal taxon counts from midden and midden-like contexts.

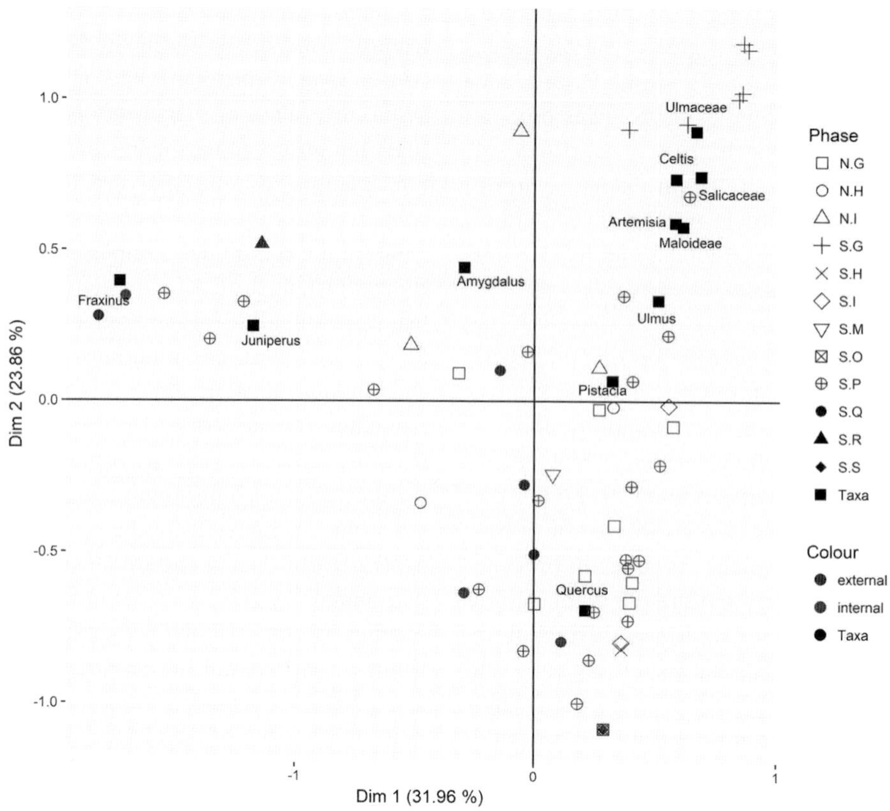

Figure 4.3. Plot of Dimensions 1 and 2, CA run on per sample wood charcoal taxon counts from primary fuel waste deposits (internal and external).

The variability observed in the charcoal-sample composition between internal and external fire features warranted further exploration of context-related variation in taxon representation. 'Dirty floor' deposits at Çatalhöyük represent accumulation of waste inside buildings at variable lengths of time. This accumulation, sometimes at high densities, occurs despite the repeated cleaning of domestic floors and the discard of waste in middens. We therefore decided to investigate further context-related variation in taxon composition in Level North G, which comprises 31 charcoal samples from fire features, midden deposits and 'dirty floor' contexts. The resulting CA biplot (fig. 4.4) demonstrates that 'dirty floor' samples closely follow the sample composition of internal fire features (in this case both containing predominantly deciduous oak charcoal). In addition, while North G midden deposits are also dominated by oak, at the same time they contain a greater diversity of charcoal taxa including *Juniperus*, *Amygdalus*, *Fraxinus*, *Pistacia*, Ulmaceae, etc. (see fig. 4.4, upper right-hand side of the plot). Interestingly, as identified in the CA carried out on fire feature samples (fig. 4.3), there is also variability in the composition of external fire feature charcoal samples, with some being similar in composition to those of internal fire features and others more closely resembling midden samples.

Results of dendroanthracological analyses

Qualitative estimations of log sizes were applied to 3,762 individual charcoal fragments separating them into three main size classes: CD1=weakly curved rings, corresponding to larger diameter wood; CD2=moderately curved rings, corresponding to smaller diameter trunk wood; and CD3=highly curved rings, corresponding to twigs and small branches (see Kabukcu 2018a). The results revealed a greater predominance of medium-size logs (CD2), followed by small-diameter wood (CD3) and much lower frequencies of large-diameter logs (CD 1) (see also table 4.2). These proportions are replicated across the anthracological assemblage including both primary fuel waste deposits and midden and midden-like secondary deposits. Previous work has highlighted the lack of significant variability in the distribution of log sizes across all sampled levels (Kabukcu 2015; 2018a). Instead, log sizes appear to vary in a taxon-specific manner. *Quercus*, *Juniperus*, *Ulmus* and *Fraxinus* and ash are represented predominantly by fragments derived from medium-size logs in both fire features and midden deposits (see figs 4.5 and 4.6). By contrast, charcoals of fruit/nut bearing taxa such as *Amygdalus*, *Pistacia*, and Maloideae (apples, pears and hawthorns) comprise predominantly fragments originating from small diameter

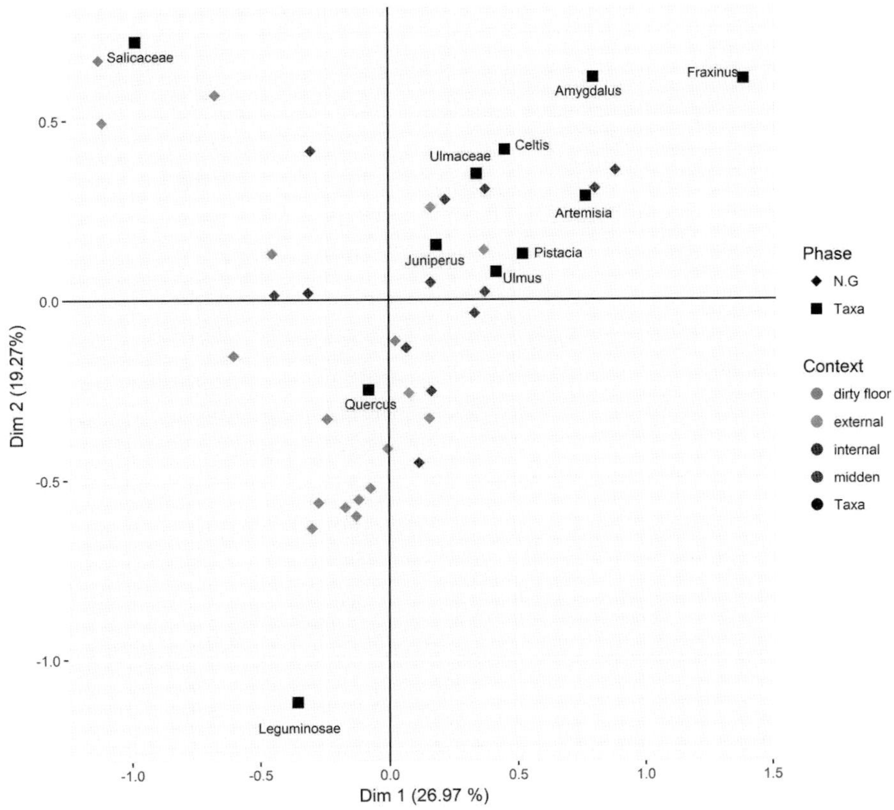

Figure 4.4. Plot of Dimensions 1 and 2, CA run on per sample wood charcoal taxon counts from Level North G midden, fire feature and 'dirty floor' deposits.

wood (fig. 4.5). It is necessary to note here that *Juniperus*, Ulmaceae, Salicaceae and *Fraxinus* included numerous fragments that were not assigned a CD class due to the overall small size of their charcoal fragments, poor preservation or the specific qualities of their wood anatomy. For example, juniper is characterised by naturally wavy growth ring boundaries often including false and/or discontinuous growth rings, hence ring curvature may be very difficult to ascertain for small *Juniperus* charcoal fragments, while Salicaceae may have indistinct growth ring boundaries, particularly in dead/decayed wood.

Quantitative estimations of log diameter were carried out using the trigonometric method, involving the triangulation of the angle between wood rays and growth ring boundaries (see detailed exposition in Kabukcu 2018a). As this method relies on the clear observation of rays, growth-ring boundaries and multiple growth rings, its application on the assemblage was limited to large charcoal fragments (>10mm) of ring-porous taxa characterised by large and clearly visible rays and large earlywood vessels; in the present study these were limited to *Quercus*, *Fraxinus*, *Ulmus* and *Celtis* (150 fragments across all levels in the South Area). Quantitative log diameter estimations yielded similar results to the qualitative growth ring classification (CD classes discussed above) (see fig. 4.7). Firstly, the majority of measured specimens were identified as derived from medium to small diameter logs (<15cm in minimum diameter). Secondly, the comparison between CD classes and log diameter estimations performed on the same specimen indicated that overall the qualitative technique provides a sufficient blueprint for separating between small (<5cm min. diameter), medium (~8-6cm min. diameter) and large (>10cm min. diameter) log sizes across the measured specimens (fig. 4.8).

Further indicators of log size selection and use have derived from recording the frequency of occurrence of pith, bark and tyloses. In some taxa (for example, *Quercus*, *Ulmus*, *Celtis*, *Fraxinus*) the presence of tyloses in >90% of the observable early wood vessels in consecutive growth rings may indicate derivation from the heartwood of mature stems/trunks. Across the studied dendroanthracological assemblage from Çatalhöyük tyloses were commonly observed (>60% of all analysed fragments; see table 4.2). Tyloses were most commonly observed in oak, alongside a higher incidence of CD2 and CD1 diameter size classes. While there is a considerable proportion of CD3 fragments (totalling 958 observations; see table 4.2), observations of pith and/or bark

Dendro-anthracological features		*Midden contexts*		*Fire features*	
		Count	*%*	*Count*	*%*
Curvature degree	1	137	8.0	13	2.3
(Marguerie, Hunot 2007)	2	850	49.4	319	57.2
	3	733	42.6	225	40.4
	N/A	1,048	N/A	437	N/A
Pith		195	7.0	125	12.6
Bark		41	1.5	59	5.9
Tyloses		1,802	65.1	620	62.4
Traumatic resin canals/gum ducts		599	21.6	144	14.5
Fungal hyphae		1,335	48.2	610	61.4
Narrow growth rings		424	15.3	177	17.8
Radial cracks		145	5.2	22	2.2
Collapsed vessels		180	6.5	37	3.7
Boreholes		44	1.6	2	0.2
Scar/callus tissue		30	1.1	1	0.1
Mineral deposits		166	6.0	75	7.5
Reaction wood		44	1.6	9	0.9
False rings		45	1.6	10	1.0
Knots		193	7.0	60	6.0
Total number of analysed wood charcoal fragments		**2,768**		**994**	

Table 4.2. Summary of dendro-anthracological observations (numbers of fragments exhibiting individual features) obtained from wood charcoal fragments from midden/midden-like contexts and fire features at Çatalhöyük.

were considerably low by comparison. Thus, the high numbers of small and medium diameter specimens in the assemblage are likely to have derived from twigs, branches and small diameter heartwood. The under-representation of the sapwood regions (that is, larger diameter log portions) could reflect log burning without prior splitting, thus leading to the complete combustion of the larger diameter log portions.

With regard to the general condition of fuelwood used, there is ample evidence for the use of deadwood and/or wood stored for sufficient periods to provide time for the development of fungal decay: nearly half of all the examined specimens displayed signs of fungal hyphae (see table 4.2). Less frequently, more advanced signs of decay such as insect boreholes and collapsed vessels/tracheids were also recorded.

Traumatic resin canals or gum ducts were relatively frequent (~20% across the studied assemblage). In some cases, other forms of growth irregularities were also recorded under this category, such as tyloses formed in isolated regions within the woody tissue (see fig. 4.9), which could suggest the reaction of riparian taxa to sudden fluctuations in water levels. Narrow (<0.2mm) and false growth rings were also observed in similar frequencies as traumatic growth (~20%). Both types of features relate to physiological stress and were particularly common among *Juniperus* specimens. Narrow and false growth rings have also been recorded in *Quercus*, *Ulmus*, *Celtis* and *Fraxinus* (see fig. 4.9). Stress indicators are not limited to large diameter wood; they have been observed frequently in smaller diameter specimens, possibly indicating prolonged periods of stress resulting in dwarf growth forms.

In tandem with quantitative log diameter calculations, growth-ring width measurements were also carried out, recording annual growth-ring width sequentially. The aim was to explore in a more systematic manner tree growth conditions and characterise possible anthropogenic impacts on woodland habitats. In temperate environments, annual growth increments are distinct and are largely influenced by climatic conditions, tree physiology and other external factors (for example, human and animal impacts on individual trees). Thus, the study of annual growth-ring width provides a method for understanding woodland growth conditions and habitats. As the tree ages, annual increments in growth-ring width (observed along the radius of the tree) narrow with increasing trunk diameter. This situation necessitates plotting ring width against log diameter in order to produce realistic reconstructions of changing tree growth rates. For this reason, ring width measurements were carried out only on specimens on which trigonometric diameter calculations were performed. In order to investigate differences in rates

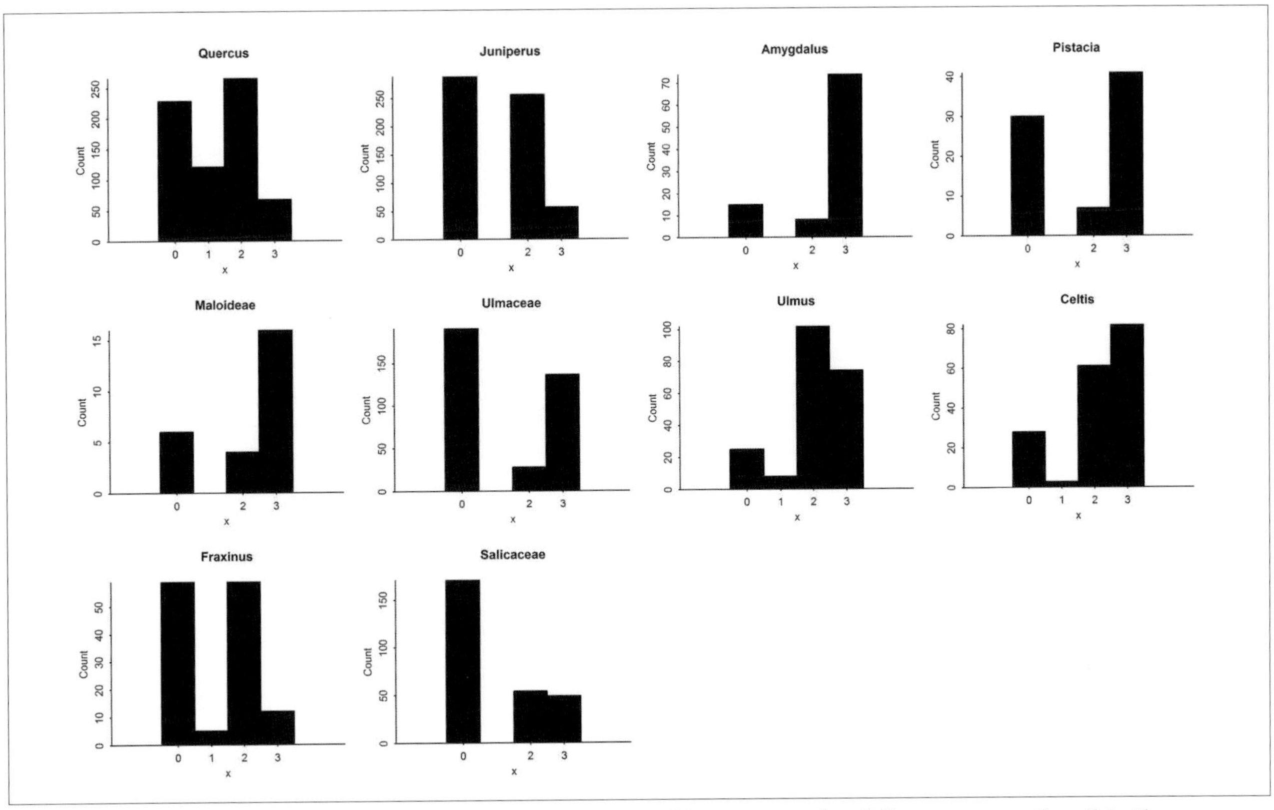

Figure 4.5. Distribution of Curve Degree (CD) classes in midden contexts for different taxa at Çatalhöyük.

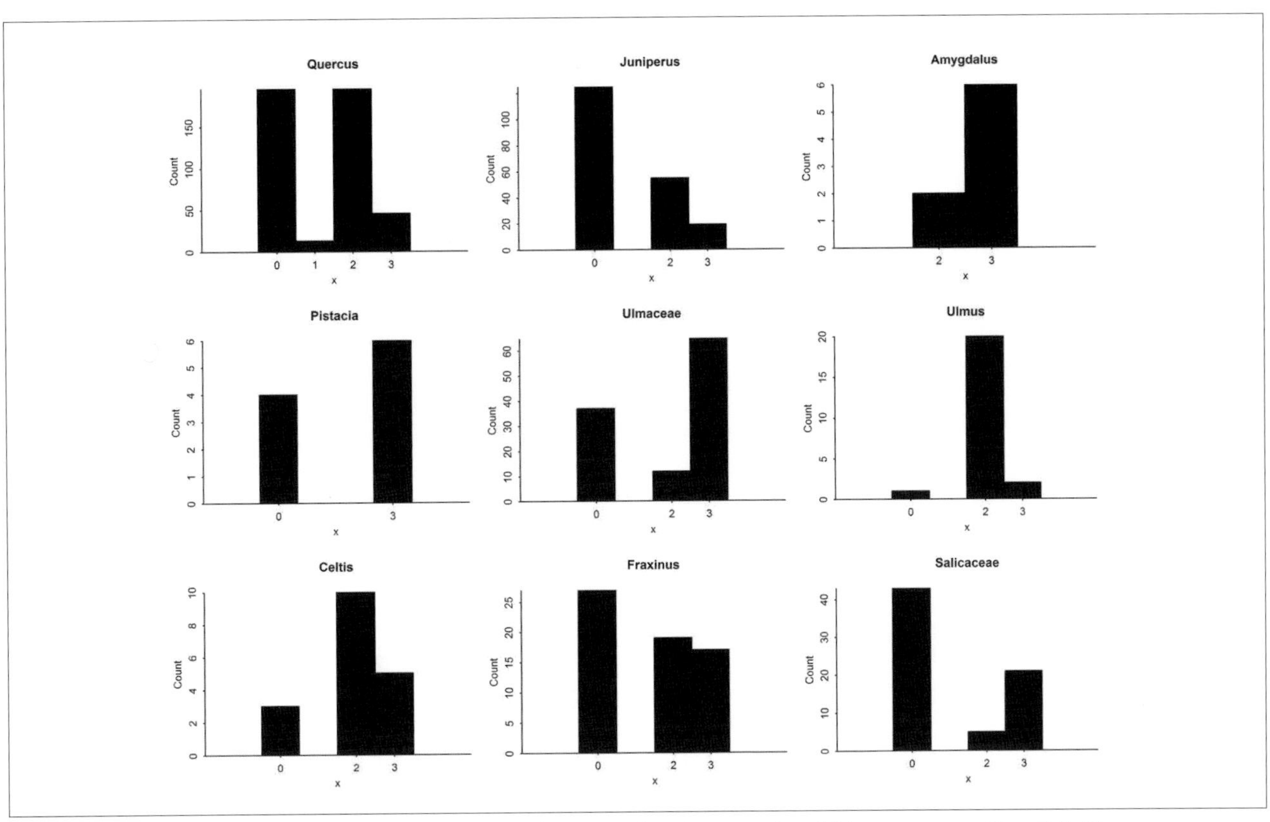

Figure 4.6. Distribution of Curve Degree (CD) classes in fire features for different taxa at Çatalhöyük.

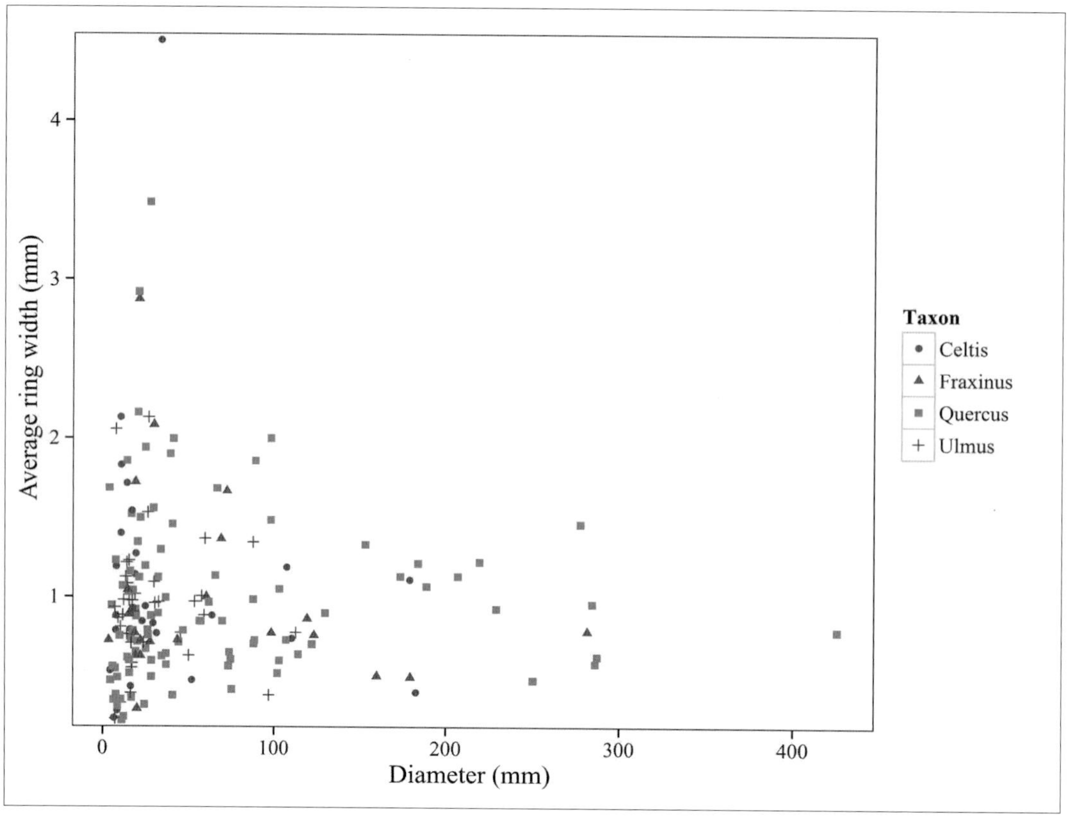

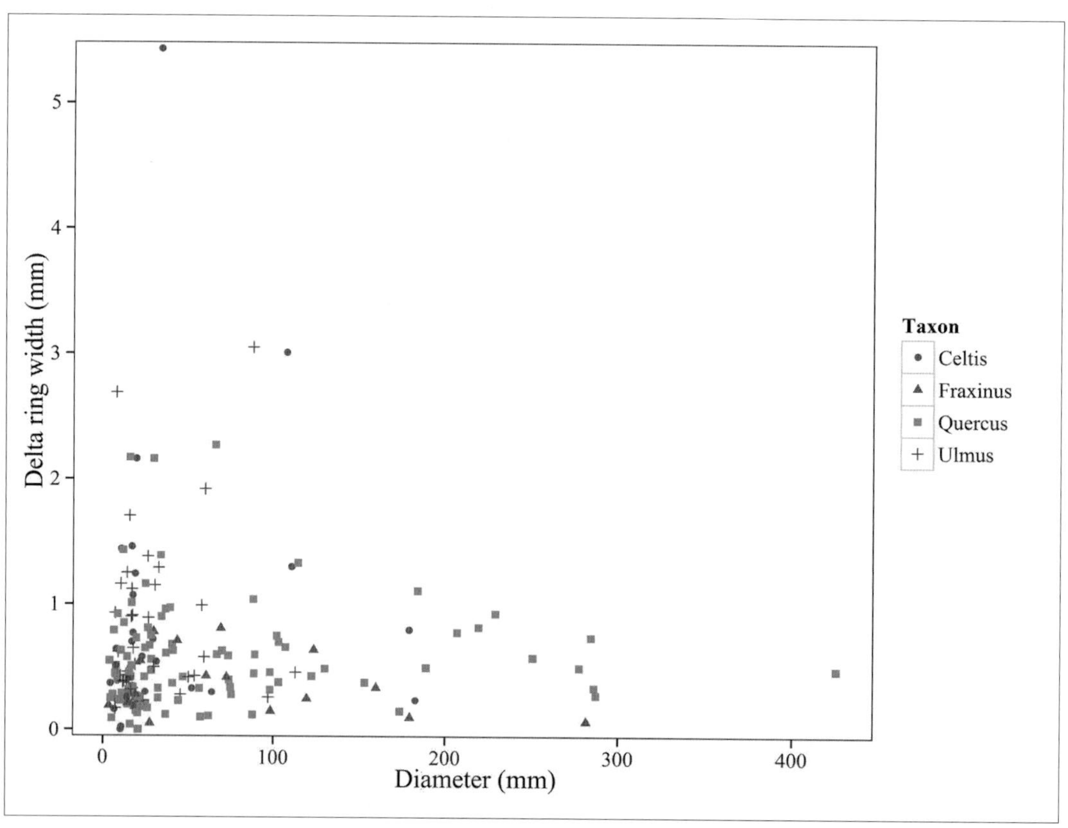

Figure 4.7. Scatter plot of average ring width and diameter measurements for each specimen from Çatalhöyük (top); scatter plot of delta ring width (maximum-minimum) and diameter measurements for each specimen from Çatalhöyük (bottom).

of tree growth between riparian and semi-arid woodlands, average ring width measurements for each specimen were plotted against respective diameter estimations (see fig. 4.7). The plot of average ring width shows that there is a great degree of variability in rates of growth in smaller-diameter size classes (particularly in stems <10cm). A significant number of specimens in this group are characterised by markedly reduced growth (<0.5mm). At the same time, a large number of small-diameter specimens display average rates of growth above the median (between 1 and 2mm). A further group within the <10cm diameter category display average ring width of >3mm. Such variability in average growth rates does not appear to relate to the micro-ecologies of different ecotones (for example, riparian vs semi-arid). All three groups include specimens of both riparian and semi-arid taxa. Furthermore, the larger diameter specimens also demonstrate stable and reduced annual growth rates (0.8–1.5mm). As expected, the effect of increasing trunk diameter and age result in relatively narrower growth-ring width.

Additionally, we investigated variation in growth-ring width observed within the same specimen: delta=Max (ring width) − Min (ring width).

The aim was to capture abrupt shifts in annual growth rates (see fig. 4.7). Studies on present-day managed woodlands indicate that cyclical cutting practices impact on wood anatomy in two main ways: an abrupt increase (two–threefold increase) or a significant decrease in annual growth which are then sustained for a number of years. For example, coppice woodlands comprise species which regenerate vegetatively and rely on the rapid responses of trees to the cutting of their trunks at the base

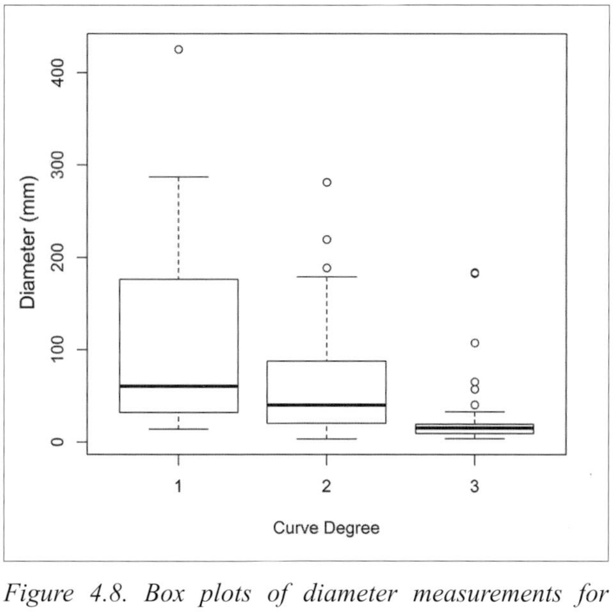

Figure 4.8. Box plots of diameter measurements for specimens of different Curve Degree classes.

by releasing shoots in the following growth season. After a cycle of cutting, the new shoots growing from the cut trees and the remaining trees experience a period of improved growth conditions characterised by an abrupt increase in ring width (referred to as growth release period) (Schweingruber et al. 1990; Corcuera et al. 2006; Schweingruber 2007; Altman et al. 2013). Furthermore, shoots growing from coppice stools tend to have wider growth rings compared to seedlings and saplings growing from seed. This improved period of growth is sustained for 5–10 years, followed by a period of reduced growth rates, as competition for light and nutrients intensifies, brought about by increased canopy density (referred to as growth suppression period) (Schweingruber et al. 1990; Rozas 2004; Bleicher 2014). On the other hand, silvicultural practices such as pollarding, crown lopping and leafy fodder harvesting result in sustained periods of abrupt growth reduction (growth suppression). In pollarded trees, wide-scale loss of foliage leads to significant reduction of annual growth, with recovery to 'normal' rates occurring as the crown regenerates by releasing shoots from the trunk. Similarly, defoliation (for example, by herbivore browsing) can also result in growth reduction (narrow growth rings, often accompanied by false growth rings) (fig. 4.9; see also Thiébault 2006; Schweingruber 2007: 139).

In the Çatalhöyük dendroanthracological assemblage, observations on both average and delta ring-width values display considerable variability. A significant number of specimens display very minor shifts in annual growth; these correspond to the group with low average ring-width values as well and may represent saplings growing under sustained competition (for example, closed canopy, repeated cutting and/or browsing). This group includes both riparian and semi-arid taxa, thus re-affirming that predominant woodland growth patterns are not controlled primarily by the micro-ecologies of the various woodland catchments. Furthermore, taxa displaying higher variability in growth rates (that is, greater delta ring-width values) are in the smaller-diameter group, while more stable growth rates are observed in the larger-diameter classes. We argue that this situation reflects the occurrence of three distinct groups of growth forms in the Çatalhöyük assemblage: twigs/branches, sapling stems and regenerating shoot stems, which are observed in all the taxa from which ring-width measurements were obtained. At the same time, there is ample evidence for severe growth suppression in several specimens (fig. 4.9) derived from both riparian and semi-arid woodland taxa. These are likely to represent the impacts of repeated episodes of defoliation (for example, from leafy fodder collection, animal browsing, etc.).

Evidence for timber use

Several burnt buildings have yielded in situ preserved carbonised timbers and other structural/construction wood remains. In addition to the analyses of timber remains previously reported by Asouti (2013), we report here new evidence obtained from the 2012–2017 excavation seasons (see table 4.3). Burnt buildings are found in Levels South O and North G and North F; for this reason, it is not possible to provide evidence of diachronic trends in timber selection and use. Two structures in the South Area, B.80 and B.97 (South O) contained charred timber elements. Juniper and elm specimens retrieved from B.97 derived from an in situ preserved partition wall. Partition wall timber elements were not recovered in B.80, but the collapse fill of this building contained significant quantities of juniper timber elements, some of which may belong to the collapsed roof structure. Four in situ timber elements preserved inside the pillars were also recovered in B.80 derived from oak (20082), juniper (18961, 18959) and elm (18960). The excavation of (20082) inside pillar F.3428, established that the oak timber sat at the base of the pillar, was ~30cm in width and had been sourced from a halved trunk with a maximum height of ~45cm. The remainder of the pillar above (20082) was filled with coarse and heterogeneous construction/packing material. This building also provided the in situ charred remains of an oak ladder (18963), produced from a halved oak trunk of ~20cm in preserved width.

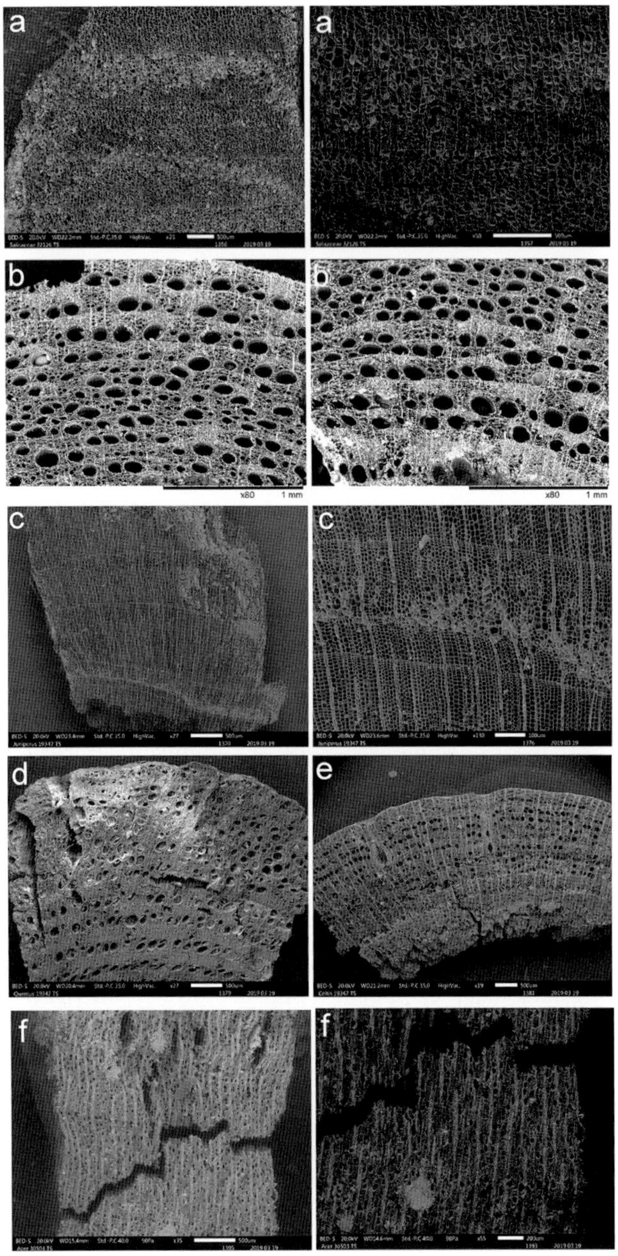

Figure 4.9. Scanning Electron Microscopy on wood charcoal specimens from Çatalhöyük: (a) traumatic tyloses in Salicaceae indicating fluctuations in the water table; (b) narrow and false/discontinuous growth rings in Quercus*; (c) scar/callus tissue and traumatic resin canals associated with frost damage in* Juniperus*; (d) growth suppression in* Quercus *(associated with pollarding/defoliation); (e) growth suppression in* Celtis *(associated with pollarding/defoliation); (f)* Acer*, fragment of wooden bowl (30503) (transverse and tangential longitudinal sections).*

Three burnt buildings in the North Area, B.52, B.77 and B.131, also yielded carbonised timber remains. In all three buildings several small posts (<10cm in diameter) and various timber elements from roof collapse were identified as elm, willow/poplar and oak. While roof collapse remains of juniper timbers were recovered in B.77, juniper was absent from B.52 and B.131. The timbers used in plastered pillars in B.77 and B.131 comprised elm and oak. The low height of the timber used was observed clearly in the east wall pillar of B.131, where an elm timber (23018) of ~42cm width was inserted into the base of the pillar; the preserved height of (23018) did not exceed 30cm. Similar to the pillar features excavated in B.80, this pillar was filled with heterogeneous packing materials. The same method of pillar construction is also attested in buildings from which there is no evidence of burning. For example, the mineralised remains and impression of a timber (21804) were discovered in the eastern wall pillar of B.43 in the South Area. Here again, the timber appears to have been inserted at the base of the pillar and the space above it filled with coarse construction material; the entire pillar was subsequently plastered.

A series of in situ carbonised timber remains discovered in the side room (Sp.504) of B.131 including a large oak timber (~20–25cm in diameter) do not appear to be related to specific features. Three further small posts were also recovered in this space (all identified as elm, ~12–15cm in diameter).

occupation, particularly in relation to the proportions of taxa from semi-arid (*Juniperus*) and riparian (*Fraxinus*) woodland catchments. This may reflect the use of increasingly diverse landscape units and associated woodland catchments (evidenced *inter alia* in carbon and nitrogen increasing variability in isotope ratios of caprines during the later levels of occupation, signalling the possibility that either herds were divided and ranged in different locations or they collectively encountered more diverse environments through their lifespan, see Pearson 2013) and/or heightened seasonal variations in fuel use and discard practices.

Across all occupation levels, the study of charcoals from internal and external fire features points to a distinctive preference for *Quercus* wood as the fuel of choice for domestic fireplaces, whereas external fire spots are taxonomically more diverse, including evidence for the use of lower quality fuels (such as dung, which is largely avoided in indoor contexts) (Bogaard et al. 2013; Filipović 2014; Bogaard et al this volume, Chapter 5). Whether these identified trends in fuel use relate to pyro-technological requirements or seasonal fluctuations in the availability of fuel resources could be investigated further through future multi-disciplinary research. The most important quality of fuelwood regarding 'clean' burning is moisture content. Wood that has not been properly seasoned prior to burning, or that is burned in a green state (that is, shortly after cutting) will invariably produce high amounts of smoke and be a poor heat emitter. Thus, it is not surprising that the use of dry deadwood and seasoned wood is well evidenced across the sampled sequence. In addition to the evidence from Çatalhöyük, other Neolithic sites in the Konya Plain (Pınarbaşı, Boncuklu and Can Hasan III) also contain ample evidence for the use of dry deadwood as fuel (Kabukcu 2017). This pattern points to high levels of deadwood productivity in the Konya Plain semi-arid and riparian woodlands, which was probably enhanced through seasonal woodcutting allowing logs to dry off-site, thus further facilitating the collection and transportation back to the settlement of lighter fuelwood loads. Such practices may explain the lack of evidence to date for on-site fuelwood storage, although stored dry wood could have also been located on rooftops, in open areas (courtyards, middens) and on the settlement edges.

Conclusions

The analysis of the anthracological assemblage from Çatalhöyük has provided important insights into Neolithic woodland composition, ecology and use during the early-mid Holocene in prehistoric south-central Anatolia. This very well-preserved assemblage has provided the earliest documented case for prehistoric woodland management practices in the Eastern Mediterranean region.

This chapter represents the culmination of 20 years of research on the anthracology of prehistoric occupations in the Konya Plain of south-central Anatolia, supported by extensive woodland vegetation survey and documentation that supplemented routine archaeological fieldwork and laboratory analyses. Çatalhöyük formed the centrepiece of this research due to the unprecedented combination of excellent preservation conditions coupled with long-term multidisciplinary contextual analyses, involving archaeobiological, environmental, architectural and material culture studies. It thus provided the springboard for testing and developing novel methodologies in the field of anthracology and their first systematic application outside the geographical limits of the western European temperate environments. Importantly, for the Eastern Mediterranean and Southwest Asia regions, our research on prehistoric habitations in south-central Anatolia has firmly established anthracology as an invaluable source of high-resolution palaeoecological archives on the regional vegetation histories and the evolution of anthropogenic landscapes. Future research at Çatalhöyük and other prehistoric sites in the Konya Plain of south-central Anatolia will undoubtedly continue to provide novel insights and avenues for methodological developments (especially in fields such as wood ecoanatomy, the analysis of artefactual wood and stable isotope applications).

Acknowledgements

Financial support was received from the Staff Research Fund of the Department of Archaeology, Classics and Egyptology of the University of Liverpool. Ceren Kabukcu's doctoral research was supported by an Overseas Research Scholarship awarded by the Faculty of Humanities and Social Sciences of the University of Liverpool (2011–2015). Additional support has been received from the British Institute in Ankara for fieldwork undertaken in 2014 (Strategic Research Initiative Funding).

Supplementary material

For supplementary material related to this chapter, please visit https://doi.org/10.18866/BIAA/e-13. It comprises colour versions of figures 4.2, 4.3, 4.4 and 4.7.

5. The archaeobotany of Çatalhöyük: results from 2009–2017 excavations and final synthesis

Amy Bogaard, Mike Charles, Dragana Filipović, Dorian Q. Fuller, Lara González Carretero, Laura Green, Ceren Kabukcu, Elizabeth Stroud and Petra Vaiglova

Introduction

This chapter brings together new archaeobotanical results from the 2009–2017 excavations with previous work on the archaeobotany of Çatalhöyük. By synthesising the results from 25 years of intensive archaeobotanical investigation, we present an unusually high-resolution picture of plant use and management in an early farming community in western Asia.

Methods

Methods of sampling, recovery, flotation, sorting, scanning and full analysis are detailed in previous reports in the Çatalhöyük monograph series (Hastorf 2005; Fairbairn et al. 2005a; Bogaard et al. 2013), supplemented by annual archive reports available through the project website. The standard onsite sampling strategy for all excavation units (that is, sampling of ca 30 litres of sediment per unit, where available, or all sediment from smaller units) continued through the 2009–2017 excavations, and methods of recovery, flotation, sorting and full analysis also continued without essential changes. Over 13,000 flotation samples were logged by the time excavations finished in 2017.

The protocol for 'level 1' scanning of non-priority units was simplified in the final excavation years to speed up the process (Bogaard et al. 2015). At this stage of the project, the botanical team was very familiar with the full range of deposit types and could predict overall botanical composition without detailed scanning. The protocol was therefore adjusted to identify unusual samples reflecting 'single-activity residues' only.

Sample selection: 2009–2017 excavations

Selection of samples for full analysis was based on a combination of archaeological and archaeobotanical criteria, as in previous work. First, we filtered out samples from 'primary' contexts of in situ burning, including burned buildings, fills of fire installations, dirty floors adjacent to fire installations in buildings and (outdoor and indoor) fire spots. Second, we targeted the botanically richest sample from each unit, analysing additional samples from the same unit only in the case of burned building deposits or gridded floors that were sampled multiple times to capture spatial variation. Thirdly, we analysed all seed-rich primary deposits in burned buildings (for example, Buildings 79 and 80 in Level South O; Buildings 131 and 77 in Level North G; Space 493 of Building 122 and Space 637 of Building 150 in Level TPC M). Fourth, we aimed to analyse ca 20–30 samples minimum per level, and to target those levels under- or unrepresented in previous excavation cycles (for example, mid-lower levels in the South Area, upper levels in TPC). We also tried to ensure that the samples selected for a given level derived from at least two different buildings. We note that this process left some samples from primary contexts in intensively sampled buildings or spaces (for example, B.132, Spaces 85/610/631) and/or Levels (North G, South O, South P) subjected only to level 1 or level 2/priority assessment. We include the results of these less detailed levels of assessment in the archaeobotanical unit-by-unit report on selected buildings and spaces (see online Appendix S5.1).

Additionally, a small set of six samples from the TP excavations of Final period levels (TP O–R) was analysed in full. We selected TP samples that were relatively 'rich' in botanical items from multiple buildings per level and/or primary deposits (for example, hearths and dirty floors). This procedure identified a larger number of samples, but only this set was fully analysed, due to uncertainty about how they had been sorted within the TP team (Marciniak and Polcyn pers. comm.).

All botanical remains were identified and quantified as detailed in Bogaard et al. (2013). Where samples from the same unit were analysed, these were not amalgamated if they differed appreciably in composition or density. Among the 2009–2017 samples selected for analysis, none of those analysed from the same units were amalgamated, though there are multiple samples from intensively sampled floors and burned buildings, and we monitor variation within and among these samples in our analyses.

Collation and standardisation of data from all excavation years

For this final synthesis, all samples fully analysed – by Fairbairn et al. (2005a), Filipović (2012; 2014), Bogaard et al. (2013), Ergun et al. (2013) and Green (2017) – were assembled and taxon categories standardised and updated, along with the new data from the 2009–2017 excavations. Where the same sample had been analysed more than once (Green's thesis targeted new wild/weed identifications for samples from crop 'storage' deposits, including those previously analysed), the more up-to-date results were used. Similar quantification procedures for grains, chaff and fruit/nut items were used in all studies except Fairbairn et al. (2005a); here we have incorporated MNI estimates of cereal grains and fruits/nuts from Fairbairn et al. (2005a) based on charred weight. Additionally, >2mm volumes of wood charcoal, dung, tuber and 'bread/food' remains per sample were measured by Filipović (2012; 2014), Bogaard et al. (2013), Ergun et al. (2013) and Green (2017). Volumes of 'bread/food' remains down to 1 mm per sample were adjusted based on the results of SEM analysis by González Carretero et al. (2017; in preparation).

The results of recent work on wild plant seed identification and related ecological analysis of the arable weed assemblage in plant 'storage' contexts in burned buildings are incorporated here (Green 2017; Green et al. 2018). A review of crop domestication traits is presented in a separate chapter by Charles et al. (Chapter 6). Work on plant 'products' – SEM analysis of the remains of 'food' fragments (González Carretero et al. 2017; in preparation) – is ongoing; selected results are included in a thematic chapter on food by González Carretero et al. (Volume 15, Chapter 19).

Results

We fully analysed a total of 237 samples from the 2009–2017 excavations. Table 5.1 summarises the samples selected, by level and area, alongside fully analysed sample sets from all previous archaeobotanical studies under the aegis of the Çatalhöyük Research Project. The total botanical dataset of 757 fully analysed samples is presented in online table S5.1, incorporating any original amalgamations of similar samples from the same unit to approximate 'behavioural events' (cf. Jones 1991), plus standardisation and updating of taxonomic categories, which have inevitably evolved over the years as we learned more about the archaeobotany of Çatalhöyük. A catalogue with photos and descriptions of all major botanical categories and taxa will be made available online in due course. The final dataset of 757 samples represents only ca 6% of the total assemblage of flotation samples (>13,000), but this small percentage nevertheless offers the most direct reflection of in situ activity residues, given the archaeological selection criteria employed. Online table S5.1 reports some 130,000,000 identifiable plant remains from these 757 samples. In both numbers of fully analysed samples and of plant remains, Çatalhöyük is easily the largest dataset now available for investigating plant-related activity in an early farming community in western Asia.

Previous work has largely established the taphonomy and taxonomy of the abundant plant assemblage from Çatalhöyük. In brief, the charred assemblage can be broadly separated into routinely generated residues of domestic activity, on the one hand, and exceptional preservation of 'storage' concentrations of food plants in burned buildings, on the other.

The routine activity residues range from more or less pure by-products of glume wheat dehusking (fig. 5.1) to deposits of wild plant seeds derived from the burning of sheep/goat dung as fuel. Dehusking of glume wheats was routine because these crops were demonstrably stored as spikelets (grains enclosed by tough outer glumes) following initial threshing (Bogaard et al. 2013). The chaff of the other, 'free-threshing' cereals – bread wheat and (primarily naked) barley – is also encountered frequently but at lower levels, presumably because initial crop processing stages that largely separated grain from chaff (threshing, winnowing) mostly took place off site (fig. 5.1) (Fairbairn et al. 2005a; Bogaard et al. 2013). Pulses are frequently encountered alongside cereal remains, but are primarily represented by their seeds, and only exceptionally by pod and stem remains in certain well-preserved and undisturbed lenses or concentrations (Bogaard et al. 2013; Filipović 2014). Dung fuel residues are indicated by a repeated suite of small-seeded grasses, rushes and sedges (especially the nutlets of *Bolboschoenus glaucus*) that track grazing of sheep/goat herds in different habitats (Bogaard et al. 2013), a topic we revisit with the updated dataset below. Routine activity residues – ranging from dehusking to fuel residues, including abundant wood charcoal as well as dung (see Chapter 4) – were variously generated in indoor and outdoor spaces, and we review this spatial distribution of activity below. Abandoned houses were often relatively 'clean' of charred organic remains in comparison with outdoor 'midden' sequences, and there is a general sense in which middens preserve sequences of activities swept up from house floors. At the same time, excavations have increasingly revealed the complexity and diversity of outdoor space use and the emergence of 'yards', as well as external 'fire spots' and other instal-

	*Levels**	*2009–2017*	*Green 2017*	*Bogaard et al. 2013*	*Ergun et al. 2013*	*Filipović 2012; 2014*	*Fairbairn et al. 2005*†	*Total*
TP[C]+ GDN	TP.R	4						4
	TP.Q	5						5
	TP.P	2						2
	TP.O	1						1
	TP.N	10						10
	TP.M	50	4					54
	unassigned	2						2
South	South.T			6				6
	South.S			28				28
	South.R			22				22
	South.Q			43				43
	South.P	7		54				61
	South.O	44	5					49
	South.N	14				1		15
	South.M	8				4	3	15
	South.L	5				17	6	28
	South.K	15				14	11	40
	South.J					10	4	14
	South.I	7					1	8
	South.H	4				1	4	9
	South.G					7	20	27
	unassigned	5						5
IST			4		28			32
North	North.I			11				11
	North.H	2		11		2		15
	North.G	28	11	135		33	10	217
	North.F	24				2	4	30
	unassigned			1				1
Kopal							3	3
Total		237	24	311	28	91	66	757

Table 5.1. Summary of samples selected, by level and area, alongside fully analysed sample sets from all previous archaeobotanical studies under the aegis of the Çatalhöyük Research Project.
** ? and definite attributions are merged per Level; † includes acorn cluster in B.1 reported by Hastorf 1996*

lations that attest to in situ outdoor activity (Bogaard et al. 2014b; Issavi forthcoming). Again, we review indoor-outdoor activity evidence below in light of the latest botanical results.

Concentrations of selected plants in burned buildings form an obvious contrast to the routine activity plant residues at Çatalhöyük. These deposits preserve direct evidence for the collection and storage of key food plants, including cereal and pulse crops, as well as 'wild' gathered resources. We define 'food plants' as edible species for which we have evidence for storage and in most cases also extensive evidence for processing. These taxa clearly differ from the suite of wild plant species associated with dung burning and hence sheep grazing. As detailed further below, 'storage deposits' in burned buildings through the entire sequence have yielded more or less pure concentrations of more than a dozen major food plants, including cereals (emmer wheat, 'new type' glume wheat, einkorn wheat, bread wheat, naked barley) and pulses (pea, lentil, bitter vetch, grass pea) with the exception of chickpea, plus collected plants that were likely also subject to varying forms of management: annual small-seeded wild mustard (mostly *Descurainia sophia*), nuts including acorn and almond (pistachio

Figure 5.1. Schematic representation of crop processing sequences for the major cereal types at Çatalhöyük: glume wheat, free-threshing wheat and naked barley (illustration: Katy Killackey).

nutshell is also very frequent but this taxon has not been found as a 'store' of whole nuts) and fruits such as hackberry (widely preserved without charring due to its naturally carbonate-rich shell). We describe concentrations of plant material in burned buildings as 'storage' evidence – with inverted commas – to allow for the possibility that they reflect deliberate placement for intentional destruction, rather than haphazardly preserved stores in accidentally burned buildings (for example, see discussion in Twiss et al. 2007; Bogaard et al. 2013). Whatever the circumstances of burning, however, clear spatial and taxonomic patterning in plant concentrations, discussed further below, strongly suggests that these deposits offer direct insight into food plant practices and choices.

Our presentation of results for this final synthesis is framed around three research questions:

1. How did use of food plants change over time?
2. What spatial patterns of plant-related activity are discernible?
3. How was arable land managed?

These questions relate to key themes of innovation, social geography and political ecology. Here we reconsider each of them based on the total assemblage summarised in table 5.1 and reported on a sample-by-sample basis in online table S5.1.

1. Innovation in use of food plants through time

We can refine the chronological overview by Bogaard et al. (2017). In particular, we now have a sharper picture of plant use through the Late–Final period on the East Mound and of its comparison with contemporary occupation on the West Mound (cf. Orton et al. 2018). Figures 5.2, 5.4–5.5 and 5.8–5.10 summarise key aspects of food plant use and change through time. In all cases (un)certain level assignments have been merged, since separate versions based only on certain assignments (not shown) are very similar. As in Bogaard et al. (2017), we include data from 125 fully analysed samples from the West Mound (Stroud 2016 submitted; Charles and Bogaard in preparation).

A shift from six-row naked barley in the Early period to two-row naked barley in the Middle period levels remains clear (fig. 5.2) based on the morphology of barley chaff (rachis) (Charles et al. in preparation), illustrated in figure 5.3(a–b). Our hypothesis is that this was the result of selection for barley with greater drought tolerance (Bogaard et al. 2017). Modern two-row barley has greater water-use efficiency than six-row barley, and stable carbon isotope analysis of grain suggests that barley tended to be grown under drier conditions than wheat and pulses (Wallace et al. 2015; Stroud 2016; Stroud et al. submitted; Vaiglova 2016; Vaiglova et al. in preparation). The ongoing dating programme will be crucial for helping us to refine how this and other crop shifts over time relate to Çatalhöyük's resilience leading up to and through the 8.2kya dry/cold event (cf. Flohr et al. 2016). As far as current correlations between the South and North Area levels allow, this shift appears similar in both areas. The Late period levels in the South Area show a slight increase in six-row naked barley that carries over onto the West Mound.

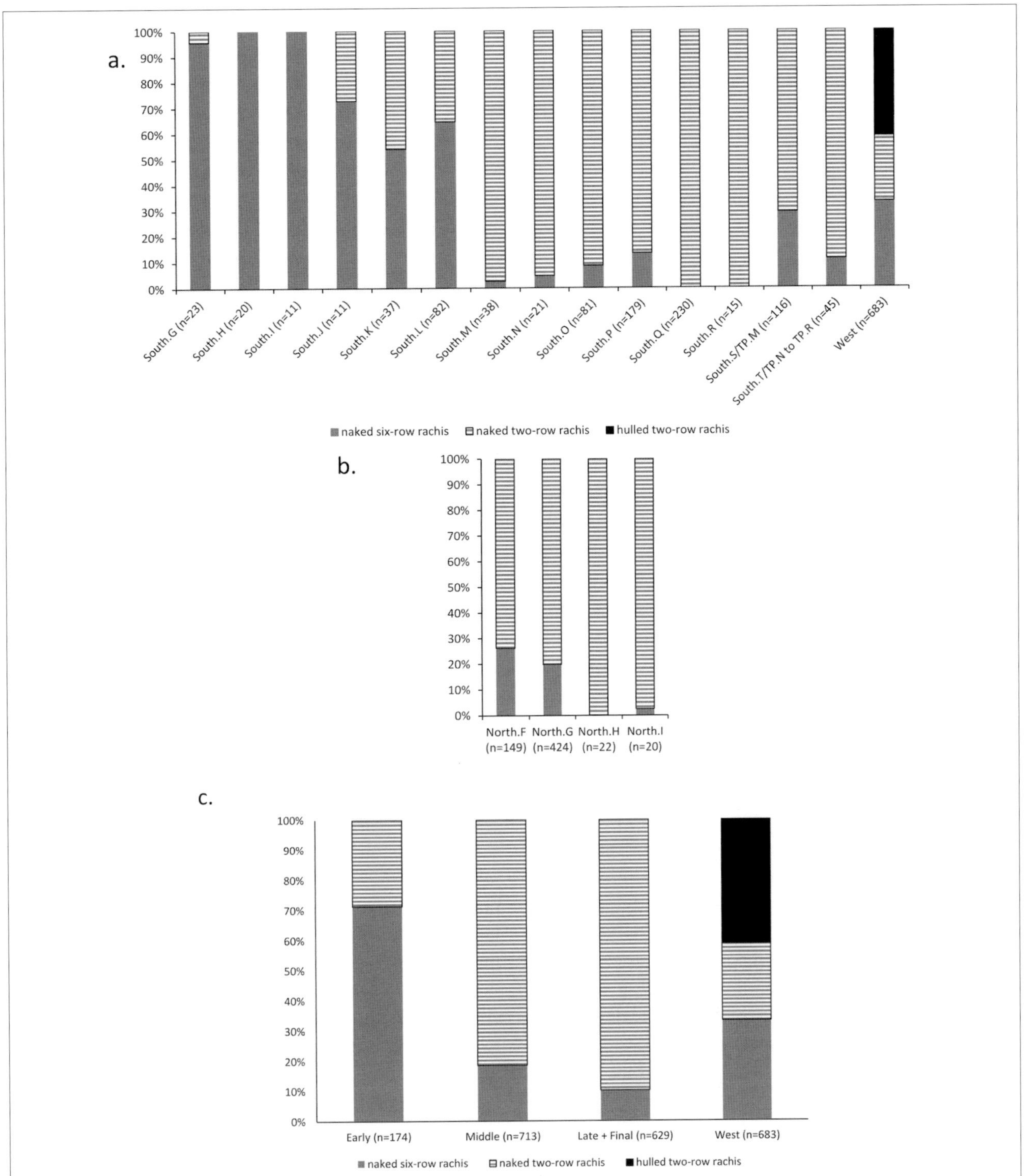

Figure 5.2. Bar charts summarising proportions of barley rachis segments identified as six-row naked, two-row naked and two-row hulled barley through time: (a) the South Area of the East Mound, and West Mound; (b) North Area of the East Mound, aligned to show approximate contemporaneity with the South sequence; (c) combined data for Early, Middle, Late and Final periods.

A distinctive similarity between the crop spectra of the Final period sequence on the East Mound and the West Mound is the occurrence of hulled as well as naked barley, visible in cereal grain proportions (fig. 5.4a). This similarity relates to a period of demonstrable chronological overlap between the two mounds (Orton et al. 2018). The morphological difference between naked and hulled barley grain is shown in figure 5.3(d–e); naked versus hulled

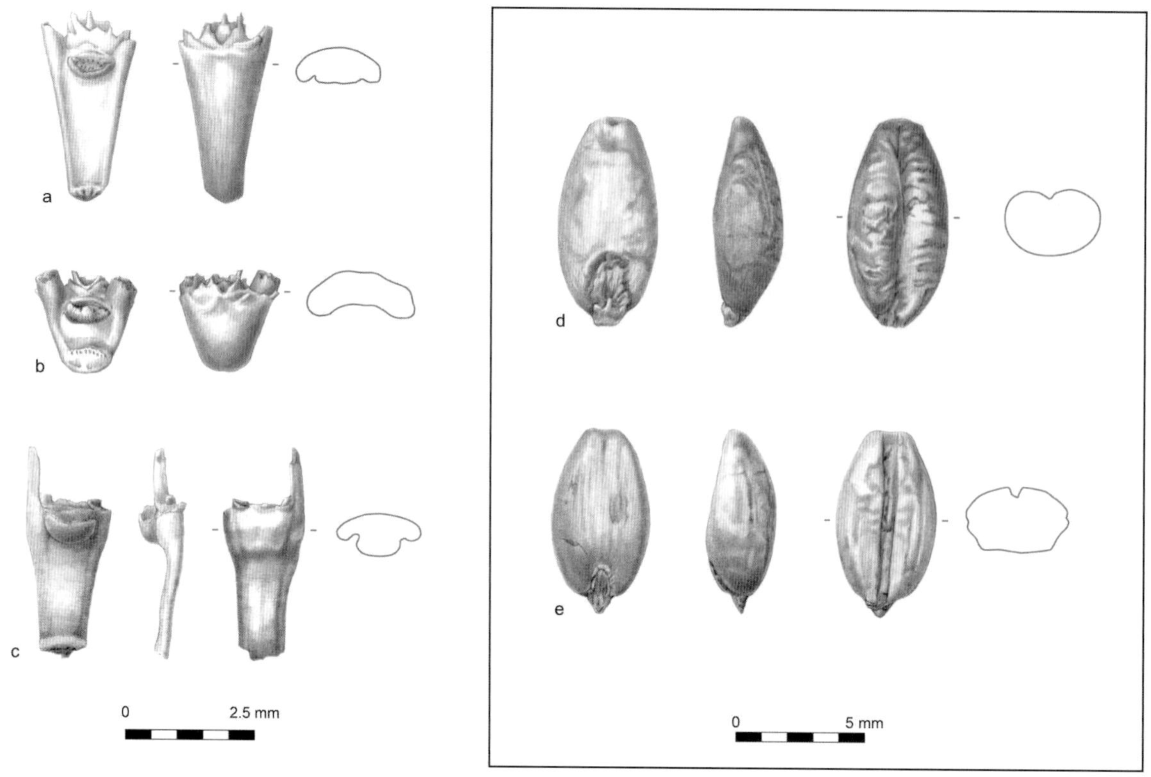

Figure 5.3. Illustration of barley rachis (a–c) and grain (d–e) morphologies at Çatalhöyük (by Katy Killackey): (a) two-row naked barley rachis; (b) six-row naked barley rachis; (c) two-row hulled barley rachis; (d) naked barley grain; (e) hulled barley grain (illustration: Katy Killackey).

barley rachis is illustrated in figure 5.3(a–c). (Very low numbers of barley rachis items (n=8) in the Final period samples meant that these were merged with the Late period in figure 5.2c, obscuring comparison of the Final period with the West Mound on the basis of barley chaff).

Figure 5.4 illustrates cereal proportions through time on the basis of grain counts. The full chronological sequence in the South Area suggests a variable trend from dominance of glume (hulled) wheat grain in the Early period to free-threshing wheat and barley in the Middle to Late periods. This tendency partly reflects 'storage' deposits in Middle period burned buildings (B.1, B.52, B.77, B.131), though these include glume wheats as well. There appears to be a broad tendency towards a more even balance between glume wheats and free-threshing cereals in the Late-Final period (and West Mound). There is a particular abundance of barley in the amalgamated Late period (fig. 5.4c). This reflects the occurrence of barley 'storage' deposits in the three Late period burned buildings: two Level TPC M buildings (B.122, B.150) and B.63 in the IST Area. A concentration of 'new type' glume wheat spikelets was also found in the storeroom (Sp.493) of B.122 but was mostly reduced to ash, and so is underrepresented here (Bogaard et al. in preparation TPC volume).

Figure 5.5 summarises change in cereal proportions through time on the basis of chaff rather than grain. This allows differentiation of glume wheat taxa (einkorn, emmer, 'new type') across the whole assemblage. These taxa are most readily distinguished on their chaff morphology (fig. 5.6a–c) (Jones et al. 2000), though pure deposits of 'new type' spikelets in burned buildings (B.77, B.131) have also confirmed their distinctive grain morphology (fig. 5.6d) compared with einkorn and emmer (Kohler-Schneider 2001). Glume wheat glume bases are overrepresented relative to the rachis of barley (fig. 5.3a–c) and free-threshing wheat (fig. 5.6e) due to storage of glume wheats in spikelet form and piecemeal dehusking, producing chaff-rich residues (for example, glume wheat spikelet 'storage' deposits in B.77, B.131, B.79, B.122). Initial threshing of cereals, breaking glume wheat ears into spikelets and releasing free-threshing grain from chaff (fig. 5.1), likely took place off site (Fairbairn et al. 2005a; Bogaard et al. 2013).

A shift through time from emmer to 'new type' as the major glume wheat remains clear (fig. 5.5). Figure 5.5 (a, c) also underlines further the similarity in crop spectra between the Final period of the East Mound and the West Mound.

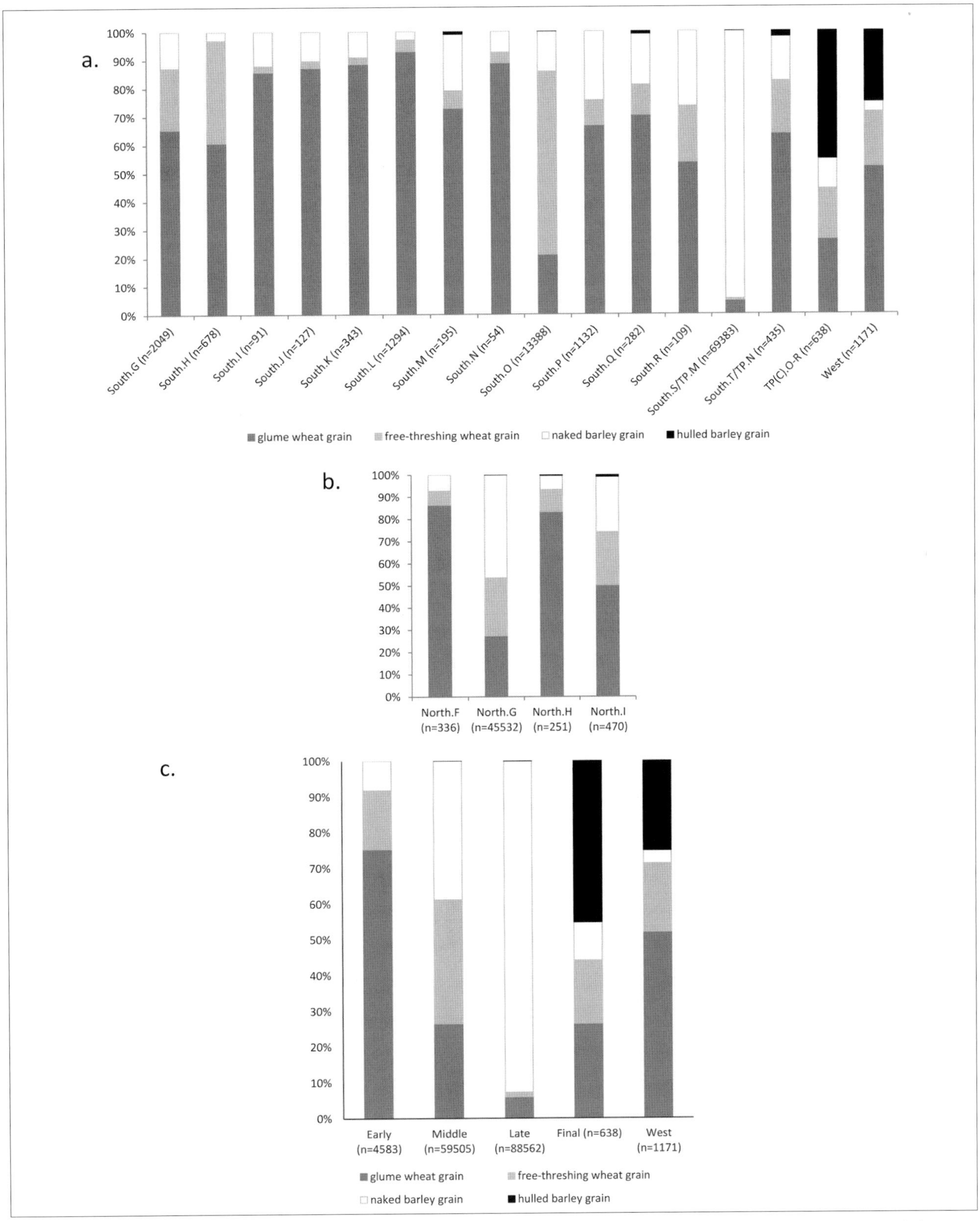

Figure 5.4. Bar charts summarising proportions of cereal grain types through time: (a) the South Area of the East Mound, and West Mound; (b) North Area of the East Mound, aligned to show approximate contemporaneity with the South sequence; (c) combined data for Early, Middle, Late and Final periods.

Despite re-assignment of B.131 with its 'new type' spikelet concentrations from Levels North F to North G since Bogaard et al. (2017) was published, it is clear that the shift away from emmer was *earlier* in the North than in the South Area of the settlement (fig. 5.5). This contrast, the occurrence of 'new type' spikelet concentrations in two

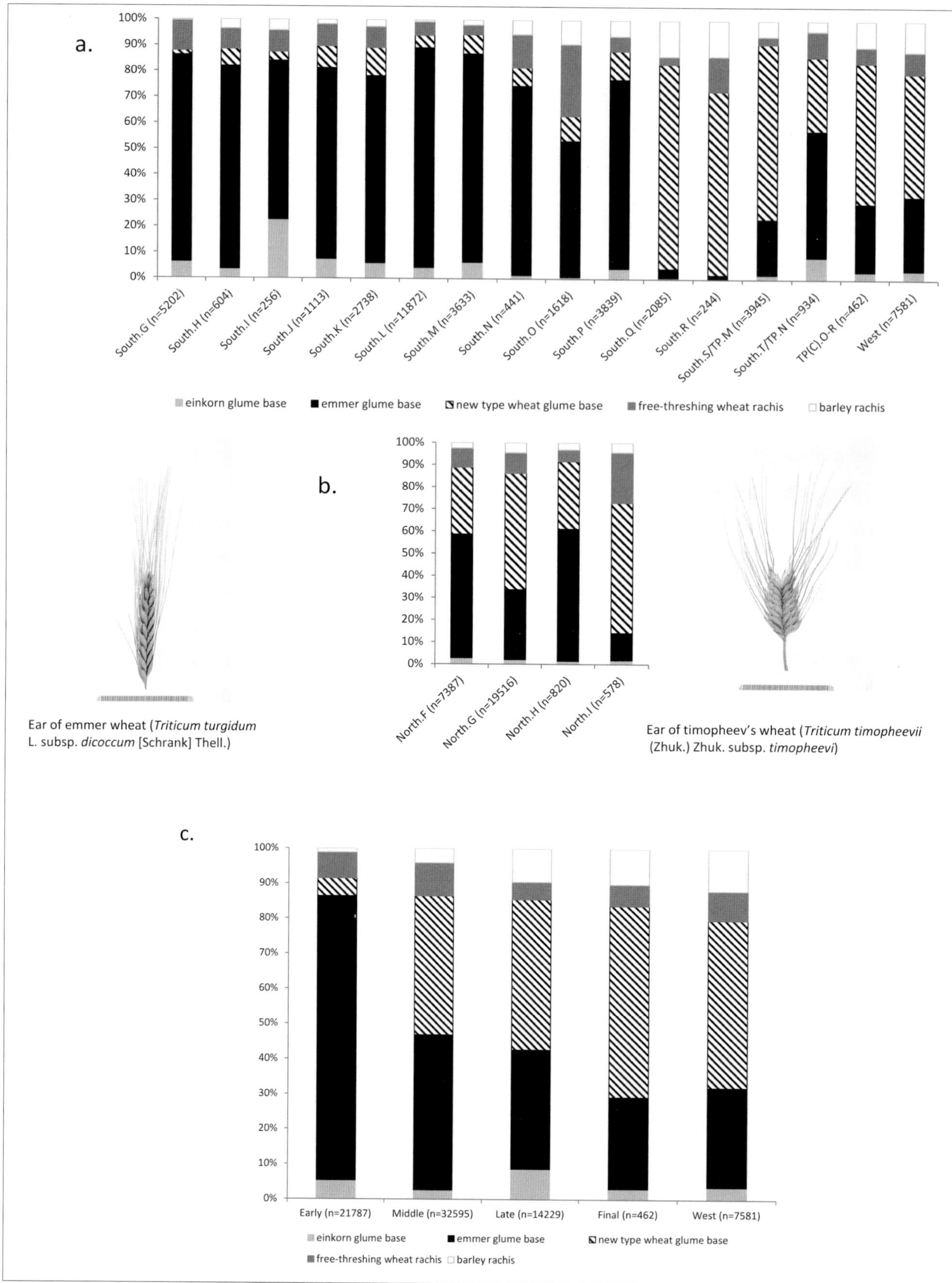

Figure 5.5. Bar charts summarising proportions of cereal chaff remains through time: (a) the South Area of the East Mound, and West Mound; (b) North Area of the East Mound, aligned to show approximate contemporaneity with the South sequence; (c) combined data for Early, Middle, Late and Final periods. Also shown are photographs of modern ears of emmer (left) and timopheev's wheat (right).

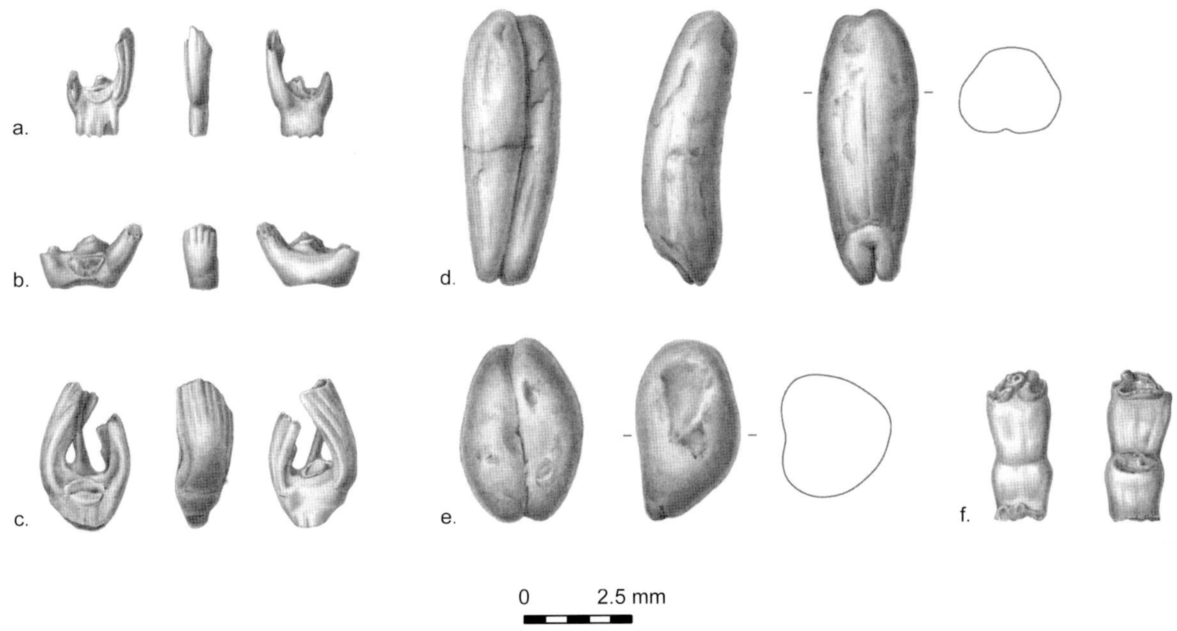

Figure 5.6. Illustration of wheat chaff and grain morphologies at Çatalhöyük (by Katy Killackey): (a) einkorn spikelet fork; (b) emmer spikelet fork; (c) 'new type' spikelet fork; (d) 'new type' grain; (e) free-threshing wheat grain; (f) bread wheat rachis (illustration: Katy Killackey).

neighbouring burned buildings of North G (B.77, B.131), and their absence in other Level North G and South O burned buildings (fig. 5.7), sheds light on the agency of innovation in cropping strategies through time, discussed further below.

Given that both emmer and 'new type' glume wheat would require dehusking of spikelets to release the grain from the chaff, and isotopic evidence that they were grown under similarly variable conditions (Stroud 2016; Stroud et al. submitted; Vaiglova 2016; Vaiglova et al. in preparation), the reason why certain households in the North Area opted for 'new type' in the Middle period over emmer must be sought elsewhere. As figure 5.5 shows, whole ears of emmer and timopheev's wheat (the *Triticum timopheevii* group being the likeliest candidate for the 'new type' – see Czajkowska et al. 2020) are clearly different in appearance. There is also emerging evidence that they differed in domestic status (Chapter 6). Given the observation that Late period 'new type' had become largely non-shattering, it seems that there was continued selection pressure for non-shattering forms of 'new type' through sickle harvesting. The implication is that Çatalhöyük farmers who opted to cultivate 'new type' tended to harvest it by cutting (as they would the other domestic cereals), and that avoidance of shattering by harvesting before full ripeness was not systematic enough to remove selection pressure for non-shattering forms, since these were dominant by the Late period.

A plausible scenario is that the choice of 'new type' versus emmer was connected with distinctions in household and neighbourhood identity, paralleling other differences in practice between the North and South Areas through the Middle period (cf. Larsen et al. 2015; see this volume, Chapters 1, 8, 11). These distinctions may have extended to culinary properties and recipes. Subtle culinary differences between glume wheat species are apparent in the Kastamonu region of northern Turkey today, for example, where einkorn and emmer are variously grown to produce distinctive dishes (Ertuğ 2004).

Figure 5.8 summarises proportions of pulse (legume) crops through the sequence. A shift from lentil to pea remains clear but appears *earlier* in the South Area than in the North. Given isotopic evidence that these two pulse crops were grown under similar conditions (Stroud 2016; Stroud et al. submitted; Vaiglova 2016; Vaiglova et al. in preparation), the most plausible scenario is that the choice of lentil versus pea was connected with household culinary preference and identity. Reduction in bitter vetch from the earlier to later levels is still apparent, and generally signals a shift away from pulses containing toxins that must be a.

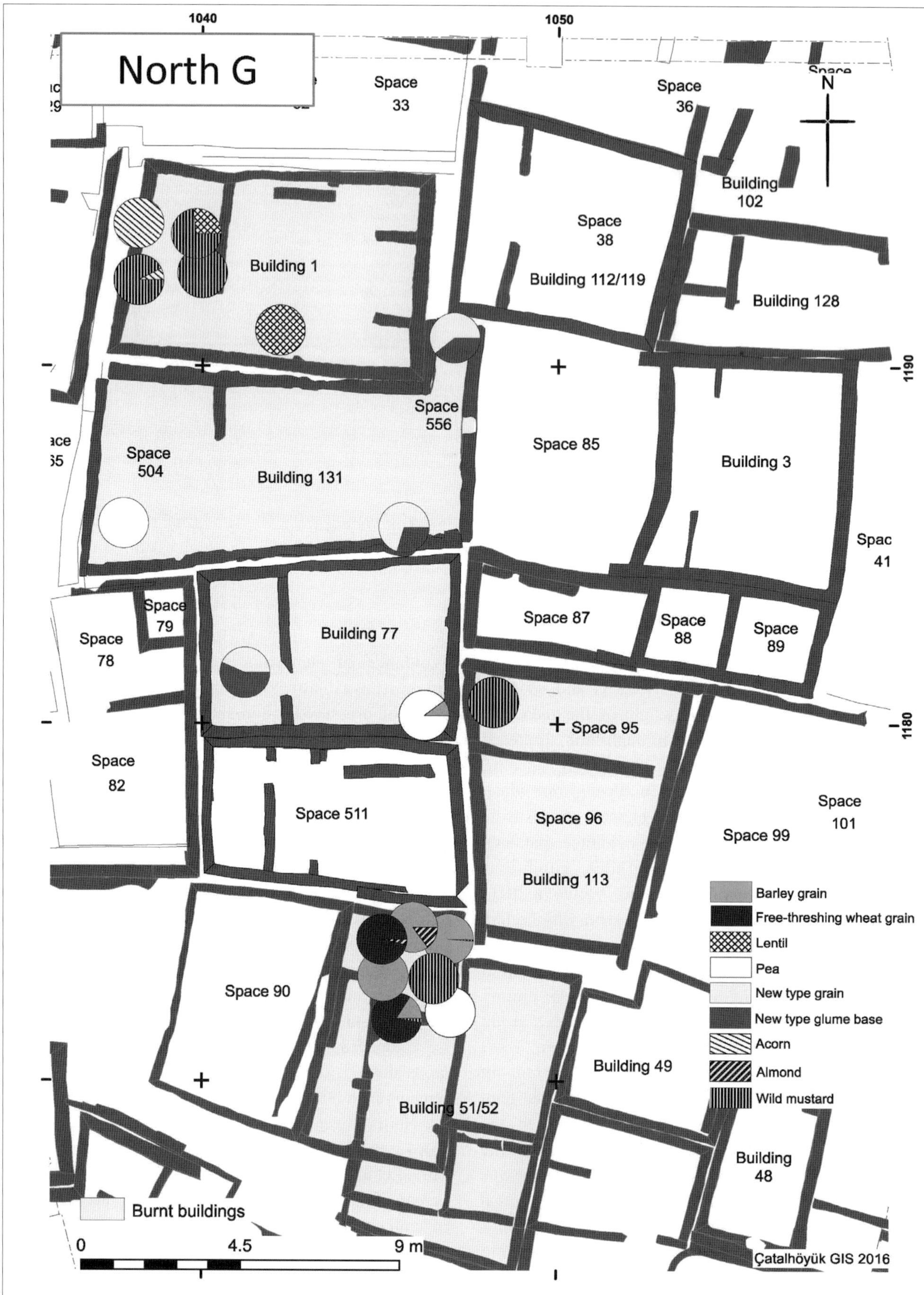
North G
1040
1050
Space 33
Space 36
N
Building 102
Space 38
Building 1
Building 112/119
Building 128
1190
Space 556
Space 85
Space 504
Building 131
Building 3
Space 41
Space 79
Space 78
Building 77
Space 87
Space 88
Space 89
Space 95
1180
Space 82
Space 511
Space 96
Space 101
Space 99
Building 113
Barley grain
Free-threshing wheat grain
Lentil
Pea
New type grain
New type glume base
Acorn
Almond
Wild mustard
Space 90
Building 49
Building 51/52
Building 48
Burnt buildings
0
4.5
9 m
Çatalhöyük GIS 2016

b.

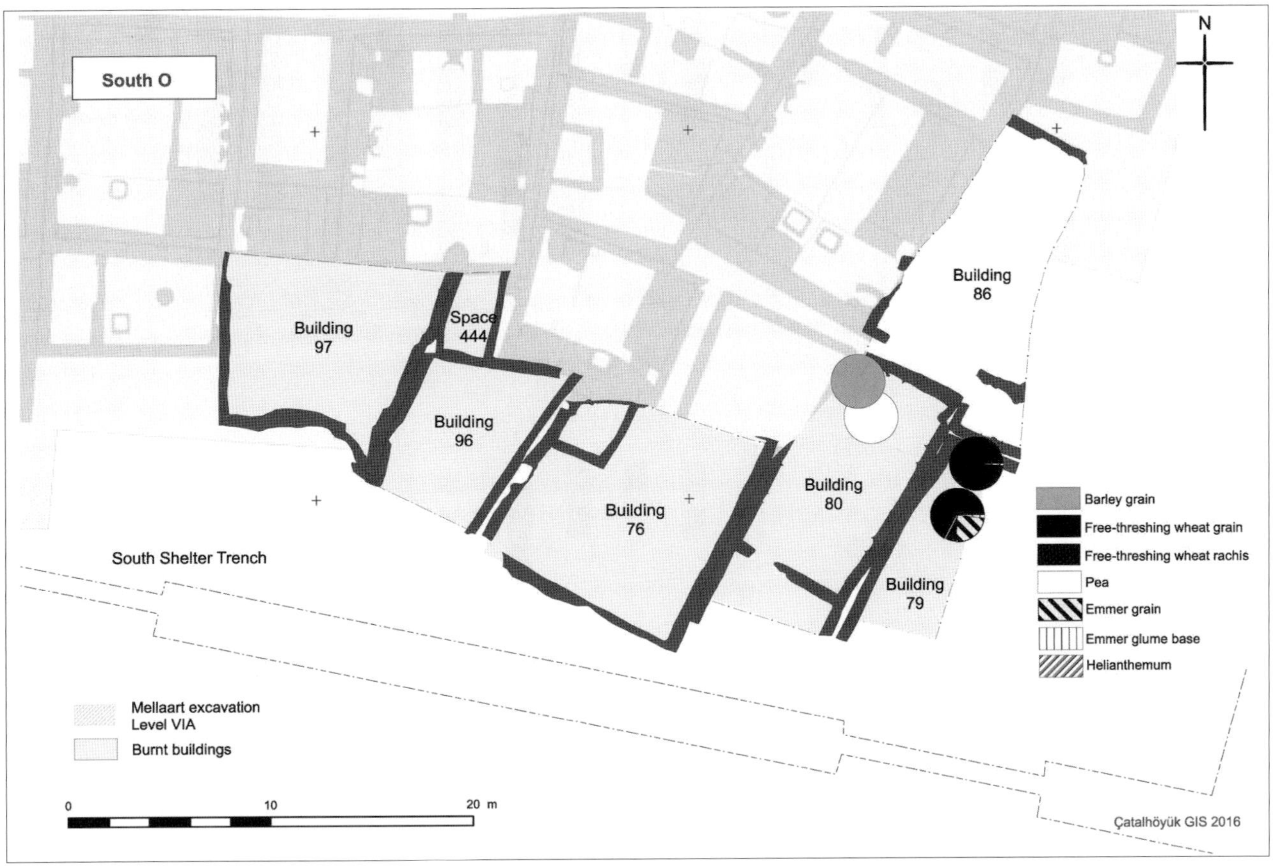

Figure 5.7. Architectural plans of (a) North Area and (b) South Area neighbourhoods with burned buildings (light grey shading) from Middle period Levels North G and South O, respectively. Pie-charts provide a simplified overview of plant concentrations in burned buildings; a single pie-chart is shown where there are multiple similar adjacent concentrations, minor components have been left out and counts of the tiny seeds of wild mustard (for example, Descurainia sophia*) and rock rose (*Helianthemum*) have been divided by 1,000 to improve the visibility of other components for this overview.*

removed before (human) consumption (Bogaard et al. 2017), but other more minor pulse species (chickpea and grass pea) seem to re-emerge at different points in the Late period. The most pronounced of these is the representation of lentil and grass pea in burned Level TPC M Sp.637 in B.150, which included deposits of pulses and barley in baskets (Bogaard et al. in preparation TPC volume). The re-appearance of 'archaic' pulses in the Late period suggests that minor/trace crops were not 'lost' but archived/monitored for potential selection, and again sheds light on how innovation in cropping strategies was achieved.

Figure 5.9 summarises the occurrence of major plant categories through time. Aside from anomalies caused by the concentration of burned buildings in Level South O, North G and TPC M/South S, there is remarkable continuity in the steady occurrence of all categories through time. Chaff clearly overrepresents the cereals relative to the pulses. Unusually, pulse pod fragments have in fact been identified at Çatalhöyük (Bogaard et al. 2013; Filipović 2014) and have implications for the domestication status of pulses (see Chapter 6), but pod material is only preserved in exceptional, undisturbed deposits (for example, B.1 lentil deposit; Sp.637 of B.150; certain fire spots). Cereal grain and pulse seeds are subject to similar taphonomic biases; the fact that cereal grains normally outnumber pulses is therefore a plausible reflection of more frequent use and deposition. From this perspective pulses and cereals are more evenly balanced in Level South G and South H than later, and pulses are especially rare on the West Mound. Fruit/nut proportions are dominated by hackberry seeds (see below), which are preserved without charring due to their carbonate-rich shell. The tiny seeds of mustard are also inherently over-represented relative to cereals and pulses.

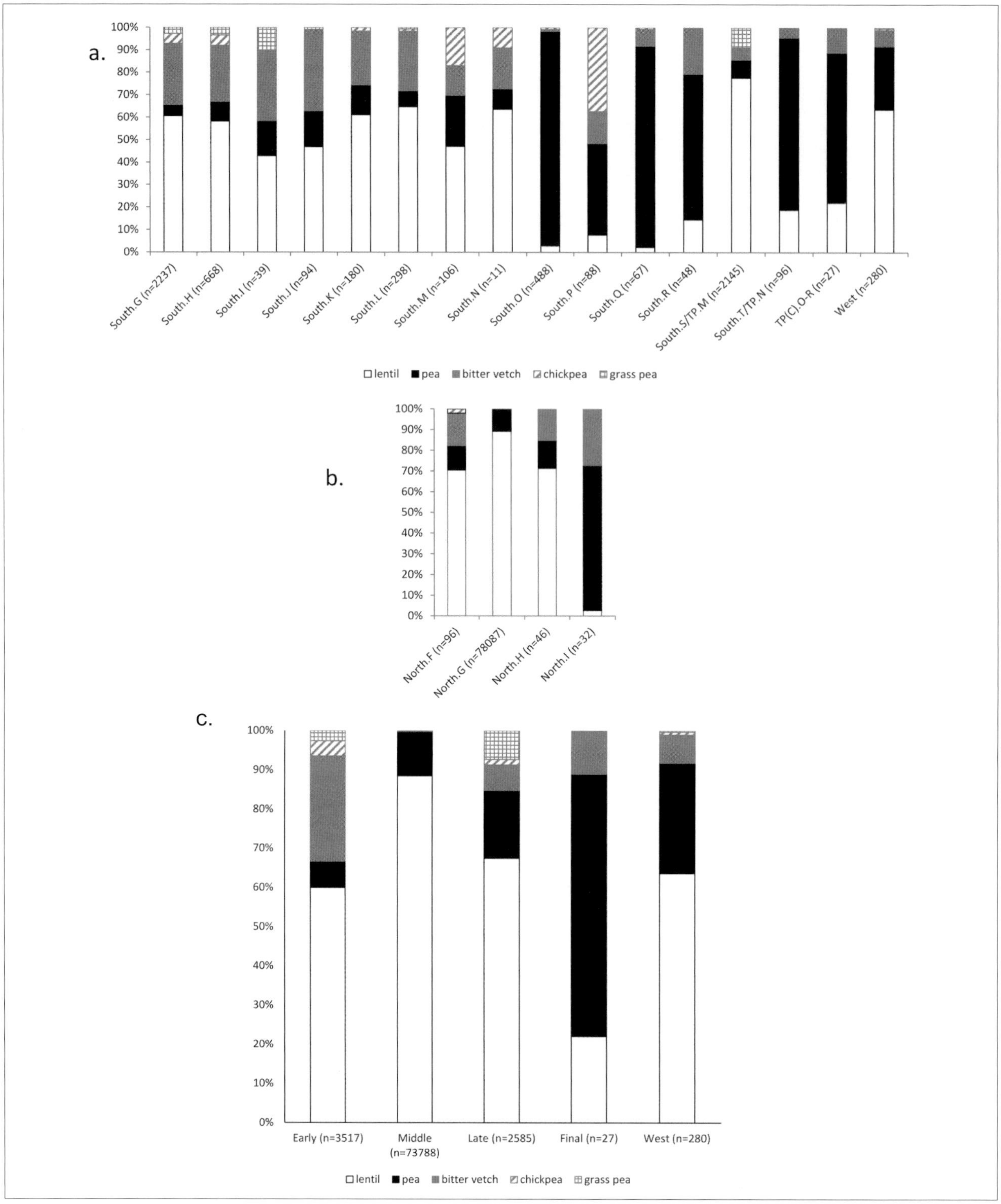

Figure 5.8. Bar charts summarising proportions of pulse taxa through time: (a) the South Area of the East Mound, and West Mound; (b) North Area of the East Mound, aligned to show approximate contemporaneity with the South sequence; (c) combined data for Early, Middle, Late and Final periods.

Finally, figure 5.10 breaks down the 'fruit/nut' category to explore change and continuity through time in the use of collected plants. Hackberry is generally dominant throughout due to its preservation in the absence of charring. After Level South P, acorn is virtually absent (with only single acorn items in some subsequent levels), echoing the shift from oak to juniper in the charcoal record (Asouti 2013; Kabukcu et al. this volume, Chapter 4).

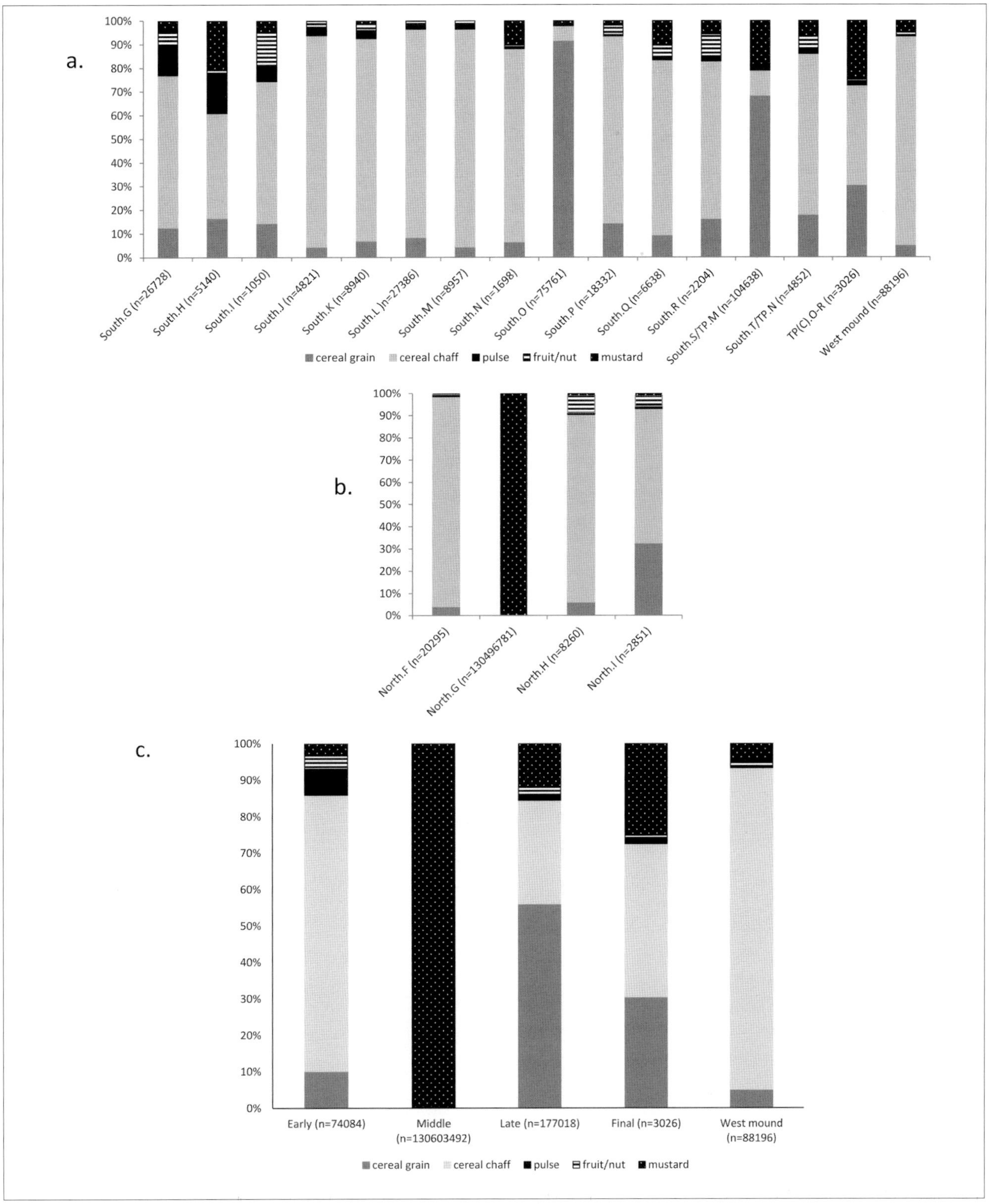

Figure 5.9. Bar charts summarising proportions of major plant food categories through time: (a) the South Area of the East Mound, and West Mound; (b) North Area of the East Mound, aligned to show approximate contemporaneity with the South sequence; (c) combined data for Early, Middle, Late and Final periods.

A potential food plant not included above is club-rush (*Bolboschoenus glaucus*) (Wollstonecroft et al. 2011). The taphonomy of this plant in the assemblage is complex; it offers an object lesson in the perils of blanket assumptions about how/whether plants were used and deposited. The seeds of *B. glaucus* are edible and among the most ubiquitous botanical items across the assemblage as a whole, occurring in 75% of

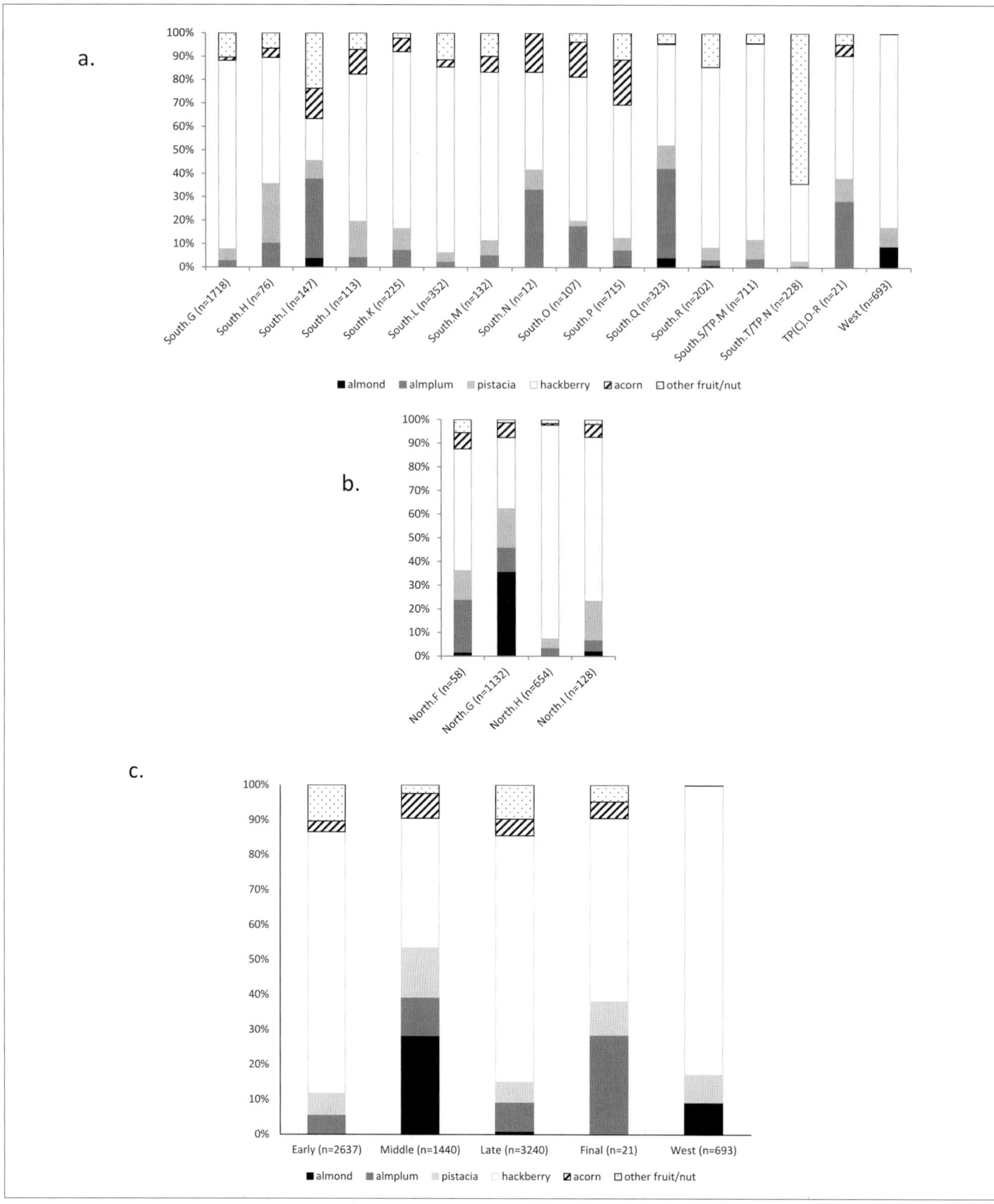

Figure 5.10. Bar charts summarising proportions of fruit and nut taxa through time: (a) the South Area of the East Mound, and West Mound; (b) North Area of the East Mound, aligned to show approximate contemporaneity with the South sequence; (c) combined data for Early, Middle, Late and Final periods.

samples (568 out of 757, online table S5.1). They have often been assumed to represent collected food on Epipalaeolithic-Neolithic sites in western Asia (e.g., Hillman 2000; Savard et al. 2006) but on later sites, such as Bronze Age Abu Salabikh in southern Iraq, are associated with animal dung fuel residues (Charles 1998). At Çatalhöyük they have been found embedded in sheep/goat dung pellets (Bogaard et al. 2013: fig. 7.21c) and are often associated with other wild plants also likely derived from animal dung burned as fuel (discussed further below). The tubers of *B. glaucus* are also edible and very frequent across the assemblage. Though not included in figures 5.9–5.10 above since they were not recorded in the Fairbairn et al. (2005a) dataset, online table S5.1 shows that across the other datasets, tuber material occurs in nearly half of the samples (300 out of 691). Tuber tissues, presumably of *B. glaucus*, have been found embedded in some of the 'food' fragments studied under SEM (González Carretero et al. 2017). The frequent discard of these potentially edible plant components suggests that they were not highly valued foods, but it of course does not exclude consumption. The tubers require intensive processing and are best eaten (after pulverising) 'fresh' in season (Wollstonecroft et al. 2008), which would help to explain why they were routinely discarded. Neither the seeds nor the tubers of club-rush have been found in 'storage' deposits in burned buildings, unlike crops, nuts (almonds in B.52, acorns in B.1), hackberry (B.55), 'wild' mustard (B.1, B.52) and rock rose seeds (B.79) (see table 5.2 below, and online Appendix S5.1).

To summarise this reconsideration of plant use through time at Çatalhöyük, we can demonstrate near continual innovation in cropping strategies (cf. Bogaard et al. 2017). The North and South Areas of the settlement innovated in different ways through the mid-Neolithic levels (for example, a shift from emmer to 'new type' wheat in the North Area, and a shift from lentil to pea in the South Area), but these changes were generally adopted by the Late period. We have previously used a panarchy (resilience) theory perspective to understand how agricultural innovation worked at Çatalhöyük (Bogaard et al. 2017): experimentation by particular households/groups, reflected for example by contrasting burned building 'storage' deposits, was scaled up to the neighbourhood and wider community if successful. Just as crucial as this nested social structure were the diverse spectrum of crops and conservation of minor species by cultivators. Social morphology and agroecology were thus intertwined in Çatalhöyük's resilience and sustained occupation over 1,500 years, notably through the 8.2kya episode of colder/drier climate (cf. Flohr et al. 2016).

2. Spatial (and temporal) patterning in plant-related activity

In this section we explore spatial distribution in two parts: (1) among (burned and unburned) buildings, to assess consistency/divergence in the diversity of plant use; and (2) excluding burned building deposits, among contexts relating to in situ burning in internal and external spaces and through time.

Comparing buildings. We have already touched on variability in representation of crops and other food plants in burned-building 'storage' deposits (fig. 5.7; see also Bogaard et al. 2013). Neighbouring buildings burned in the same level can have identical 'stored' plants (for example, B.77, B.131) or different 'storage' spectra (for example, B.79, B.80) (fig. 5.7). Further mapping of the distribution of plant concentrations is in preparation for the Level TPC M burned buildings, B.122 and B.150 (Bogaard et al. in preparation).

Here the aim is to update previous work on inter-building comparison (Bogaard et al. 2013: table 7.4) by summarising the occurrence of major food plants across all buildings excavated up to the end of the Çatalhöyük Research Project. Table 5.2 presents this building-by-building summary, based only on excavation units linked with in situ burning (n=419), thus excluding building fills or midden deposits that accumulated in buildings after their abandonment. Buildings that have yielded 'use deposits' include those partly or totally destroyed by fire ('burned buildings', n=14) plus a series of unburned buildings (n=33) in which fills and rakeouts from fire installations and adjacent dirty floor layers have been excavated.

Table 5.2 summarises the occurrence of 15 food plant taxa in 'use deposits' across 47 excavated buildings. Rather than simply reporting the raw numbers of seed, chaff, nutshell, etc. items per sample, occurrence is graded as follows: 'x' = trace amounts (<30) within any individual sample; 'xx' = at least 30 items within any individual sample and 'S' = visible, normally pure 'storage' concentration (associated almost entirely with burned buildings). In the context of recent discussion of the quantitative criteria suitable for identifying deliberate collection of a plant species (Wallace et al. 2018), it is apparent quantitatively as well as stratigraphically/contextually that the 'storage' concentrations ('S' in table 5.2) provide the clearest evidence. The distinction between low numbers per sample ('x' occurrences) and higher numbers ('xx') is arbitrary, like other thresholds that have similarly been used to distinguish chance occurrences from deliberate behaviour (for example, Halstead 1994 and Wallace et al. 2018, who use 50 items) and is intended simply to illustrate stark patterns of absolute abundance among plant categories.

Several initial observations relevant to the wider interpretation of plant occurrence and diversity among buildings can be made. First, the only category that is represented by relatively high numbers per sample ('xx') across many unburned buildings is that of glume wheat glume bases. This chaff material is generated by dehusking of glume wheats (emmer, 'new type', einkorn) that were stored as spikelets (fig. 5.1). In relation to quantitative criteria for 'deliberate use', only glume wheats (due

SOUTH

Level	H	J	J	J	K	K	K	K	L	L	N	M	M	O	O	O	O	O	P	Q	Q	Q	R	S	IST	TP(C). M	TP(C). M/GD N	TP(C). M	TP(C). M	TP(C). N/GDN	TP(C). O-R	TP(C). O-R
Building	118	18	23	160	2	16	17	160	6	43	89	50	100	76	79	80	96	97	75	53	65	68	56	44	63	122	142	150	166	140	73	115
Type														**B**	**B**	**B**	**B**	**B**							**B**	**B**		**B**				
No. samples	4	8	3	1	18	2	10	9	3	5	12	1	1	4	16	18	3	10	3	1	11	1	7	11	7	13	5	24	2	1	1	1
glume bases	**xx**	**xx**	**xx**	**xx**	**xx**	**xx**	**xx**	**xx**	**xx**	**xx**	**xx**	**xx**	**xx**	**xx**	**S**	**xx**	**xx**	**xx**	**xx**	**xx**	x	x	x	x	x	**S**	x	**xx**	**xx**	**xx**		**xx**
glume wheat gr.	x	x	x	x	x	x	**xx**	x	x	x	x	x	x	**xx**	**S**	**xx**	x	x	x	x	x	x	x	x	x	**S**	x	**xx**	x			x
barley rachis	x	x	x	x	x		x	x	x	x	x	**xx**	x	x	x	**xx**	x	x	x	x	x			x	x	**xx**	x	x	x			
naked barley gr.	x	x			x	x	x	x	x	x	x	**xx**	x	x	x	**S**	x	x	**xx**	x	x		x	x	**S**	**S**	x	**S**	**xx**			x
f-t wheat rachis		x	x	x	x	x	x	x	x	x	x	x	x	x	**S**	**xx**	x	**xx**	x	x	x			x		x	x	x	x			
f-t wheat grain	x				x	x	x				x		x	x	**S**	**xx**	x	x			x		x	x	**xx**	**xx**	**xx**	**xx**	x			x
pea	x	x	x		x	x		x			x	x		x	x	**S**	x	x		x	x		x	x	x	x	x	**S**	x			x
lentil	x	x		x	x	x	x	x	x		x	x	x	x	x		x	x	x					x		x		**S**	x			
bitter vetch	x	x	x		x	x	x	x	x	x	x	x					x							x	x	x		**S**				
chickpea					x		x				x		x		x																	
grasspea	x																											**S**				
wild mustard	x	x					x	**xx**		x					**xx**	**xx**	x	x			**xx**			x	**xx**	**S**		**xx**				x
almond/plum	x	x	x		x		x	x	x	x	x		x	x	x	x	x	x	x	x	**xx**		x	x	x	x	x				x	
acorn	x	x	x		x	x	x	x	x		x		x			x	x	x							x							
hackberry	x	x	x	x	x	x	x	x	x	x	x	x		x	x	x	x	x	x	x	x	x	x	x	x	x	x	x		x	x	x
rock rose		x				x	x		x	x					**S**	x					x			x				x				
pistachio	x	x	x		x		x	x	x	x			x				x	x			x		x	x	x	x	x	x		x	x	
fig									x															x					x			

Table 5.2. Summary of the occurrence of major food plants across 47 excavated buildings, based on units linked to space use through in situ burning (n=419); 'x'= trace amounts (<30) within any individual sample; 'xx'= at least 30 items within any individual sample and 'S'= visible, normally pure 'storage' concentration (associated almost entirely with burned buildings, indicated with B).

to their chaff-rich dehusking residues) can generally be linked to unburned buildings. Other cereal categories – glume wheat grain, and rachis or grain of naked barley and free-threshing wheat – occasionally occur in relatively high numbers per sample ('xx') in unburned buildings. It is notable that pulse seeds never turn up in such high numbers in unburned buildings. Tiny-seeded wild mustard and almond/plum nutshell/fruitstone occasionally occur at these higher levels in unburned buildings.

Secondly, if all forms of occurrence are considered meaningful – that is, assuming that 'trace' amounts per 'use' deposit reflect likely use of a plant taxon – 40/47 buildings contain some evidence for cereals, pulses *and* fruit/nut taxa. Moreover, buildings represented by at least five samples all contain evidence for the three major cereal types (glume wheat, barley, free-threshing wheat), one to four pulses and two to seven fruit/nut taxa (fig. 5.11). In other words, all reasonably well-sampled buildings have yielded at least trace evidence for this considerable diversity of cereals, pulses and fruits/nuts. Cereal diversity is particularly well represented, due at least in part to the routine preservation of cereal remains as both grain *and* chaff.

Thirdly, the distribution of plant remains in *burned* buildings corroborates the inference that 'trace' amounts of plant taxa in *unburned* buildings reflect genuine diversity in plant use. Some burned buildings are relatively 'clean' and resemble unburned buildings in their plant content (for example, B.76, B.96, B.97), while others contain high-density concentrations of plant material ('S' storage deposits in table 5.2). Those in the latter category (n=10) include some buildings that were not fully exposed, such as B.79 (fig. 5.7), but most (8/10) nevertheless yielded multiple concentrations representing at least two of the major food plant categories (cereal/pulse/fruit). Three of the best preserved (B.52, B.77, B.131) contained at least one cereal, one pulse and one fruit/nut concentration. Moreover, if all plant occurrences of at least 30 items per sample are taken into account (so 'xx' and 'S' in table 5.2), 6/10 of the burned buildings with 'storage' concentrations yielded at least one cereal, one pulse and one fruit/nut taxon concentration; the four exceptions (B.63, B.79, B.122, B.113) were notably eroded and/or not fully exposed. Table 5.3 shows that these six well-preserved/fully exposed burned buildings with 'storage' concentrations corroborate use of (2–)3 cereals, 1(–4) pulses and 1(–4) other taxa. This overall picture closely resembles that observed above across unburned buildings represented by at least five samples (fig. 5.11), indicating widespread use within buildings of all three cereals, at least one pulse crop and at least one other taxon (often wild mustard, plus other fruit/nut taxa).

	NORTH														
Level	F	F	F	G	G	G	G	G	G	G	G	G	H	H	H
Building	119	132	167	1	3	49	52	59	77	113	114	131	45	55	60
Type				**B**			**B**		**B**	**B**		**B**	**B**		
No. samples	4	18	2	28	5	44	48	4	36	3	1	12	5	2	2
glume bases	**xx**	**xx**	**xx**	**xx**	**xx**	**xx**	**xx**	**xx**	**S**	**xx**	**xx**	**S**	**xx**	x	x
glume wheat gr.	x	x	x	**xx**	x	x	**xx**	x	**S**	**xx**	x	**S**	**xx**	x	x
barley rachis	x	**xx**	x	x	x	x	**xx**	x	x	x	x	x	x	x	x
naked barley gr.	x	x		**xx**	x	x	**S**	x	**xx**	**xx**	x	**xx**	x	x	x
f-t wheat rachis	**xx**	x	x	**xx**	**xx**	x	**xx**	x	**xx**	x	x	x	x	x	x
f-t wheat grain	x	x		x	x	x	**S**		x	x		x	x	x	x
pea	x	x		**S**	x	x	**S**		**S**			**S**	x		x
lentil	x	x		**S**	x	x	x		x	x		x	x		
bitter vetch	x	x		x		x	x		x	x		x	x	x	
chickpea		x		x			x					x			
grasspea															
wild mustard	x	x	x	**S**	x	**xx**	**S**	x	**S**	**S**		**S**	**xx**		
almond/plum		x	x	x	x	x	**S**		x			x	x	x	
acorn		x		**S**	x	x	x		x	x		x			
hackberry		x	x	x	x	x	**xx**	x	x	x	x	x	x	**S**	
rock rose	x	x		**xx**		x	**xx**		x				x		
pistachio	x	x		x	x	x	x	x	x	x		x	x	x	
fig						x						x	x		

Table 5.2 (continued). Summary of the occurrence of major food plants across 47 excavated buildings, based on units linked to space use through in situ burning (n=419); 'x' = trace amounts (<30) within any individual sample; 'xx' = at least 30 items within any individual sample and 'S' = visible, normally pure 'storage' concentration (associated almost entirely with burned buildings, indicated with B).

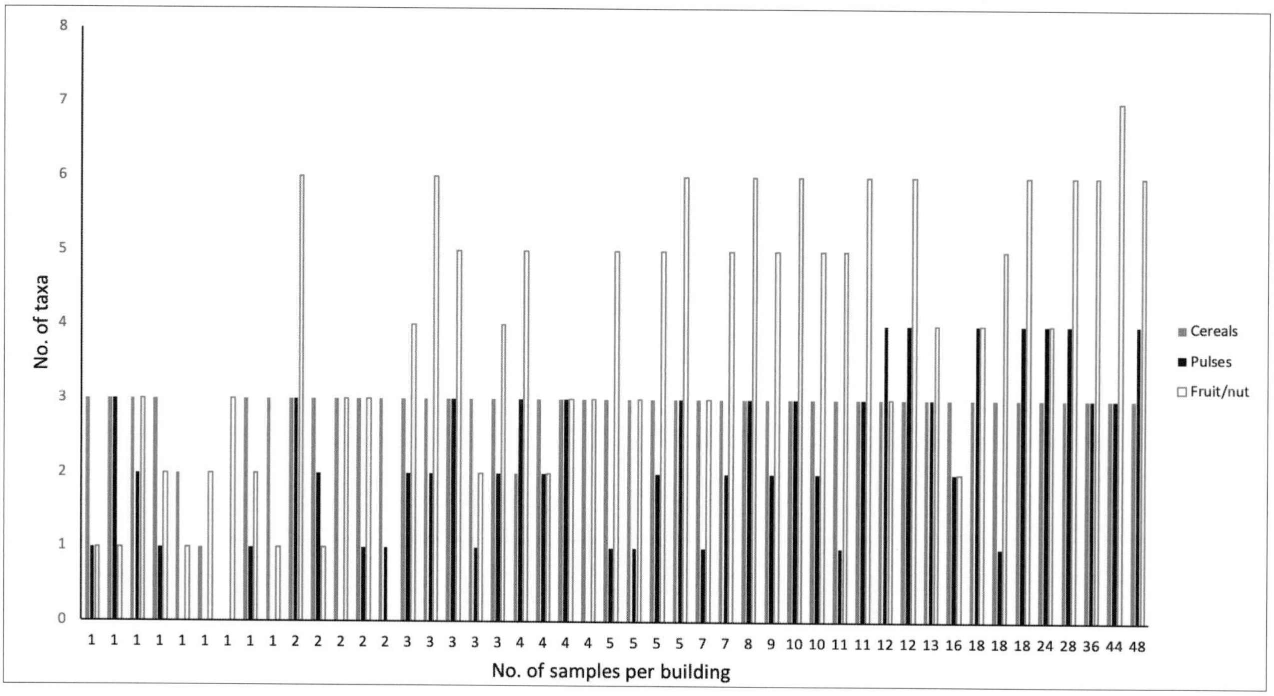

Figure 5.11. The relationship between the number of archaeobotanical samples from in situ burning units in buildings to the counts of cereal, pulse and other plant food taxa recovered.

It appears plausible, therefore, that the occupants of most if not all buildings consumed cereals, pulses and other food plants. Moreover, the desirability of such diverse use is represented in the way well-preserved burned buildings were provisioned. Against this background of similarly diverse plant use, however, there *are* clear differences in the selection of particular taxa among buildings, particularly among the glume wheats, pulses and fruits/nuts. Some of these differences appear broadly to distinguish plant use between the North and South Areas through the mid-Neolithic sequence (for example, 'new type' in the North, emmer in the South; lentil in the North, pea in the South), though this was a question of emphasis or degree, not of exclusive access. For example, 'new type' glume wheat was present at low levels relative to emmer through the Early-Middle periods in the South Area (fig. 5.5a); it was not absent altogether. The broad difference between North and South Area neighbourhoods appears instead to have been a result of conscious decisions to experiment differently with minor crops, such as 'new type' glume wheat in the North and pea in the South. The subsequent permeation of such innovations across buildings and areas of the site speaks to general equality of access to food plant resources.

Comparing 'routine' activity deposits contextually and through time. Here we reconsider broad patterns in the deposition of different kinds of charred plant remains and their implications for routine activity areas through

	SOUTH		NORTH			
Level	O	TP(C).M	G	G	G	G
Building	80	150	1	52	77	131
Type	**B**	**B**	**B**	**B**	**B**	**B**
No. Samples	18	24	28	48	36	12
glume bases	**xx**	**xx**	**xx**	**xx**	**S**	**S**
glume wheat grain	**xx**	**xx**	**xx**	**xx**	**S**	**S**
barley rachis	**xx**	x	x	**xx**	x	x
naked barley grain	**S**	**S**	**xx**	**S**	**xx**	**xx**
f-t wheat rachis	**xx**	x	**xx**	**xx**	**xx**	x
f-t wheat grain	**xx**	**xx**	x	**S**	x	x
pea	**S**	**S**	**S**	**S**	**S**	**S**
lentil		**S**	**S**	x	x	x
bitter vetch		**S**	x	x	x	x
chickpea			x	x		x
grasspea		**S**				
wild mustard	**xx**	**xx**	**S**	**S**	**S**	**S**
almond/plum	x		x	**S**	x	x
acorn	x		**S**	x	x	x
hackberry	x	x	x	**xx**	x	x
rock rose	x	x			x	
pistachio		x	x	x	x	x
fig						x

Table 5.3. Summary of the occurrence of major food plants in six well-preserved/fully exposed burned buildings with 'storage' concentrations; for legend, see table 5.2.

space and over time. We are building on previous work by Filipović (2012: fig. 6.18; 2014) and Bogaard et al. (2013: figs 7.14–7.20; 2014b). As previously, we explore variation in composition among samples using correspondence analysis. We focus only on samples that relate more or less directly to burning in situ – that is, fire installations, adjacent dirty floors, fire spots – but exclude burned buildings, so that we can look at routine activity patterns rather than structured deposition or 'storage'.

Figures 5.12–5.14 and 5.18–5.19 show the results of a correspondence analysis of 292 samples and 113 taxa, from all levels and studies. Figures 5.12 and 5.13 show, respectively, the distribution of samples coded by context type and the plant categories that drive these trends. Figure 5.14 shows the species plot coded by the same plant categories as shown in pie-chart form in Figure 5.13. The dominant trend is one from samples dominated by crops (especially chaff (glume bases) from glume wheat dehusking) and collected plants in the lower-left part of the diagram to samples dominated by dung-derived material in the upper-right part of the plot. In terms of context types, this trend relates to a broad contrast between samples from indoor 'use' deposits (especially floors), concentrated lower left, and external fire spots, upper right (fig. 5.12). There is considerable overlap, with some internal fire installations having dung-rich fuel residues, and some fire spots containing crops and collected plants, but the general contrast remains clear. As in previous analysis of the mid-later levels excavated pre-2009 (Bogaard et al. 2013: fig. 7.15), external fire installations (concentrated in Level South P) are intermediate within this broad 'indoor versus outdoor' contrast (fig. 5.12).

Figure 5.15 shows a contemporary outdoor oven in the nearby village of Küçükköy, where dried dung cakes are used as a major source of (outdoor) cooking fuel. Various attempts to use dung cakes as fuel in the reconstructed 'experimental house' at Çatalhöyük over the years suggest that they tend to generate too much smoke for routine use indoors.

A secondary trend, clearer now than in previous analyses, is that there are two 'kinds' of dung-rich sample: one characterised by a group of taxa including *Aeluropus*, small-seeded legumes and club-rush (*Bolboschoenus glaucus*) seeds (upper left), and one characterised by wormwood (*Artemisia*) and rushes (*Juncus*) (upper right) (figs 5.13–5.14). These groupings

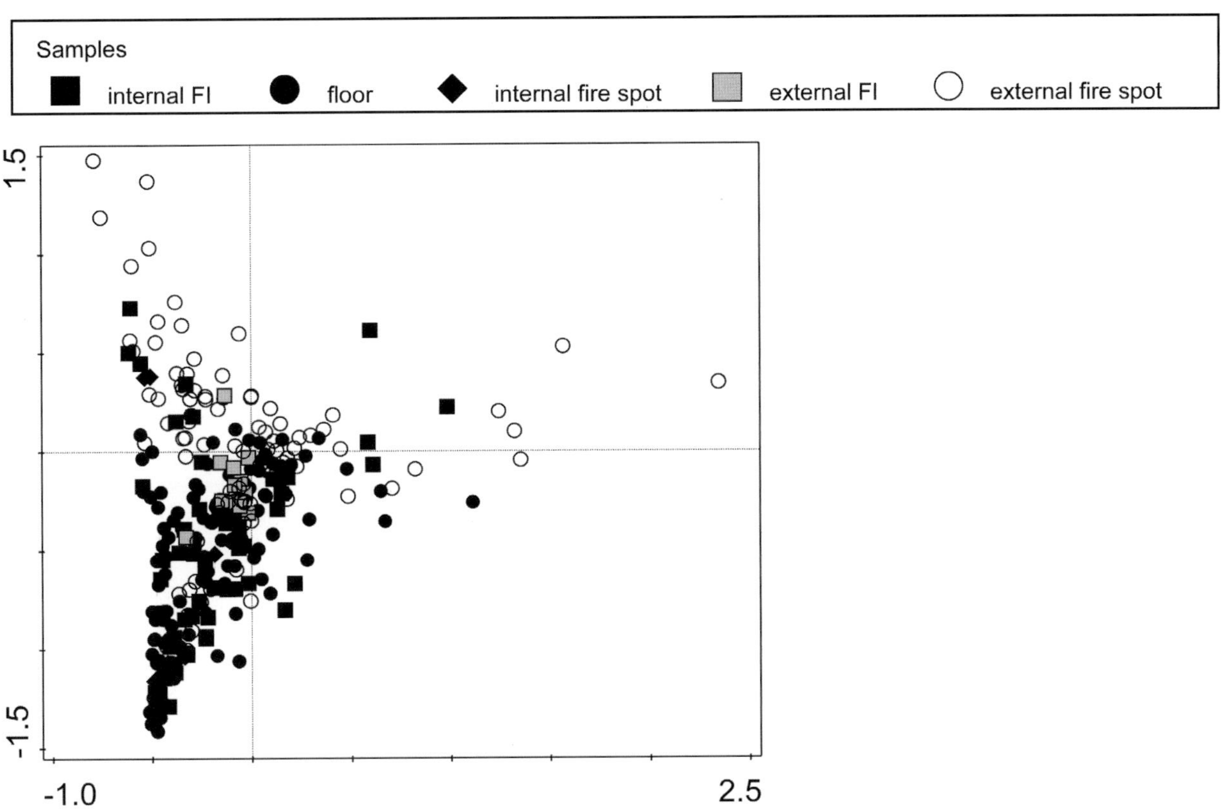

Figure 5.12. Correspondence analysis plot of 292 samples from in situ burning units on the basis of 113 taxa, with sample points coded by context type; FI = fire installation.

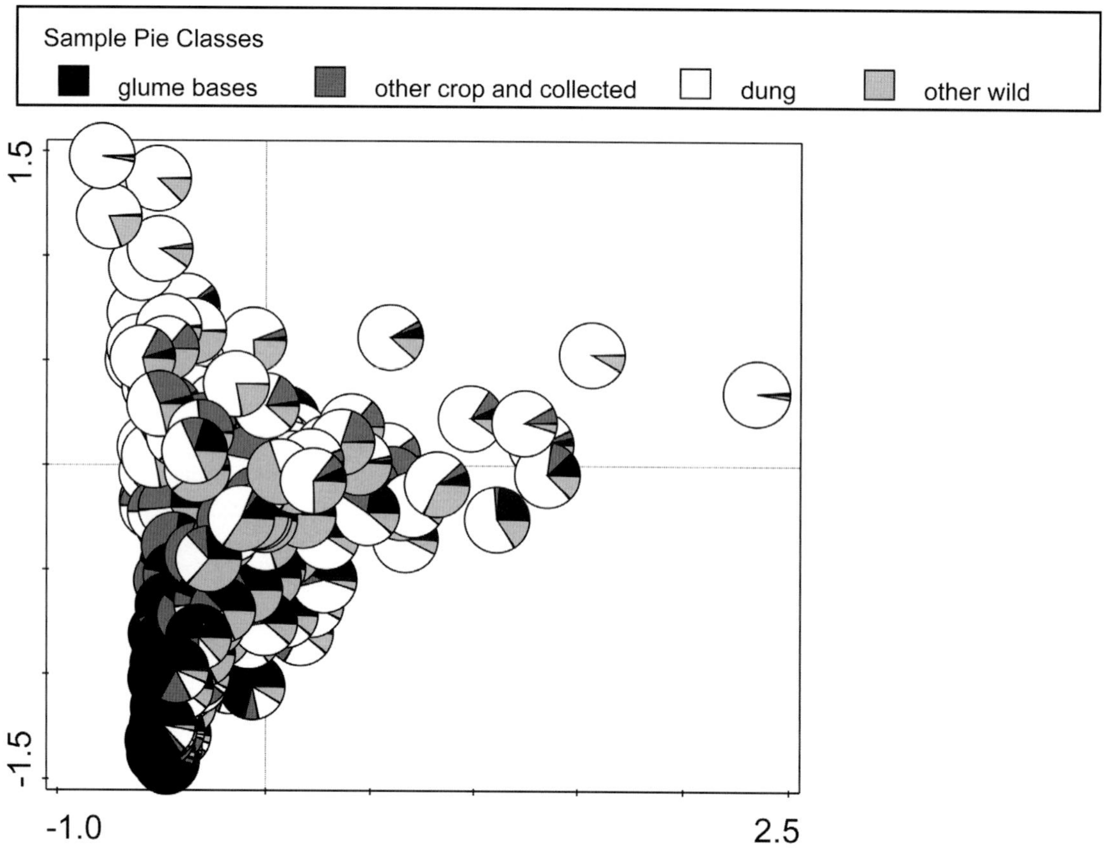

Figure 5.13. Correspondence analysis plot of 292 samples from in situ burning units on the basis of 113 taxa, with sample points represented as pie-charts showing proportions of major plant categories.

broadly suggest grazing in marshy to riverine habitats versus open, dry/seasonally wet steppe, respectively. Many fire spots are clustered between these two extremes and present a mixture of the two.

Figure 5.16 shows photos taken in or around protected wetland in the Seydeşehir area to the west of the southern Konya Plain, including vast stretches of *Bolboschoenus glaucus*, and riverine vegetation and grazing. Figure 5.17 shows a shepherd grazing sheep and goats on marl-steppe near Dedemoğlu, a few kilometres west of Çatalhöyük; here clumps of rushes (*Juncus*) attest to periodic/seasonal wetness (fig. 5.17, bottom).

Figure 5.18a–d combines context type with broad temporal periods (Early, Middle, Late, Final) and North Area versus South Area attributions (especially relevant for the Middle period) to explore the indoor-outdoor continuum further. Several observations can be made. First, the majority of the fire spots – and most of the very dung-rich ones – are Late period, but the indoor-outdoor spectrum can be traced from Early through Middle to Late periods (there are too few Final period samples to judge). Secondly, the Early indoor-outdoor trend falls toward the extreme left (negative) end of axis 1 (the horizontal axis), and most of the early external fire spots (South G) fall within the 'riverine dung' (upper left) region (fig. 5.18b). In the Middle and Late Periods, the 'indoor' end of the continuum remains anchored in glume wheat glume bases (lower left) but the array pivots clockwise, reflecting a more balanced input of dung residues from riverine and steppic landscapes (fig. 5.18c–d). In the Late period, with the most fire spots, there is evidence for the widest variety of 'riverine' through 'steppic' dung composition. Importantly, this Late period diversity of grazing aligns with faunal evidence for a greater focus on sheep/goat herding (Russell et al. 2013; Twiss et al. this volume, Chapter 8). A case can be made that more diverse grazing signatures in the Late period reflect a greater scale/range of herding, with wider variation in grazing areas, despite a reduced *human* population compared with the Middle period.

Figure 5.19a–b shows the same correspondence analysis sample plot, coded by the distribution of charred dung and 'bread/food' fragments. While identifiable dung is found in all but the 'purest' glume base-dominated samples (lower left), potential 'food' remains (including those studied under SEM by González Carretero) are mostly found in samples dominated by

Discussion and conclusions

Çatalhöyük fits into a regional Central Anatolian tradition of diverse crop cultivation combined with caprine herding that can be traced back to the mid-later ninth millennium BC at Aşıklı Höyük in Cappadocia (van Zeist, de Roller 1995; 2003; Stiner et al. 2014; Bogaard et al. 2017: table 1; Ergun 2018; Ergun et al. 2018). More locally, at Boncuklu Höyük the emerging picture from the later ninth–eighth millennia BC is of 'low-level' cultivation and caprine herding in a distinctly wetter landscape than at Çatalhöyük (Baird et al. 2018). It appears that Çatalhöyük inherited both the integrated cultivation and herding approach that had evolved at Aşıklı and the rich wetland foraging tradition of the Boncuklu community.

Systematic sampling combined with rigorous protocols for routine scanning of samples and selection of 'primary' contexts (associated with in situ burning) for full analysis has enabled us to establish new insights into early farming practice and ecology at Çatalhöyük. In particular, we have identified remarkable change and innovation in food plant spectra through time, clear spatial patterning in plant-related activity and the nature and variability of Çatalhöyük's agroecology. Here we discuss the significance of each of these findings in turn.

First, the observation of near continual change and innovation in food plant use through time corrects the pervasive view that Neolithic farming was based on a package of 'founder crops' that, once established, paved the way to unchanging/conservative reliance on this spectrum. The reality at Çatalhöyük – and, we suspect, many other sites lacking detailed archaeobotanical investigation – is that resilience and sustainability depended on creative experimentation with a diverse spectrum of crops and other food plants, and not on rigid adherence to a particular set of species. We see much of Çatalhöyük's changing food plant use through time as *innovation* because (with the exception of hulled barley) all of the species concerned were present from the earliest levels. Thus, experimentation with minor crops, we think by particular households and neighbourhoods, was an important mechanism for 'auditioning' potential staples. This kind of 'contained' experimentation is usefully framed by a panarchical resilience theory perspective (Bogaard et al. 2017; cf. Gunderson, Holling 2002). The mid-Neolithic sequence presents powerful evidence that the North and South Areas were agriculturally distinct 'communities of practice' that innovated in different ways, echoing other independent lines of evidence for contrasts between these sectors of the settlement (e.g., Larsen et al. 2015; Twiss et al. this volume, Chapter 8). A related point is that, throughout the sequence, Çatalhöyük cultivators were also dedicated foragers, perpetuating broad spectrum plant use (cf. Wallace et al. 2018). The recent discovery of a major concentration of rock rose (*Helianthemum*) seeds in burned B.79 (table 5.2; see also online Appendix S5.1) further underscores the evidence that emerged already during Mellaart's excavations (Fairbairn et al. 2007) for collection and storage of a diverse repertoire of wild plant foods at Çatalhöyük.

Secondly, comparison of occupation deposits across dozens of excavated and sampled buildings directly attests to the diversity of plant use – spanning cereals, pulses and wild plant foods – at this 'household' level. Notwithstanding the likelihood that the occupants of multiple buildings regularly cooperated and shared resources (cf. Bogaard 2017; see also below), the archaeobotanical evidence strongly suggests a kind of redundancy that is echoed also in the built storage and cooking features of modular Çatalhöyük houses. Moreover, and building on previous evidence (for example, B.49 – Bogaard et al. 2013), the extended occupation sequences of recently excavated buildings (for example, B.132, B.89 – see online Appendix S5.1) provides fine-grained corroboration of the inference that occupation of these structures entailed regular deposition of plentiful debris from the dehusking of glume wheats (stored as spikelets and processed piecemeal) plus trace amounts of other crops and food plants. Alongside this 'redundancy' in use of cereals, pulses and 'wild' food plants, however, households and neighbourhoods experimented and innovated in different ways with particular species.

Furthermore, there is clear spatial patterning in the use of indoor versus outdoor space at Çatalhöyük. The inference of 'private' storage of food plants and other consumables from storage features (bins) is corroborated by concentrations of food plants in burned buildings, though these sometimes occur outside of storage features, including in main room spaces. A predominant emphasis on *internal* spaces for plant food handling and processing through the use-lives of buildings is demonstrated through analysis of botanical composition from internal and external deposits (for example, figs 5.12–5.13). Distinctively outdoor activities are characterised by deposits rich in dung-derived plants, reflecting intensive use of dung fuel in external spaces accompanied by various activities including craft production (Bogaard et al. 2013; 2014b).

Another dimension of spatial variation explored here is that of grazed plants derived from the burning of caprine dung. Our exploratory multivariate analysis attests to clear shifts through time, from a predominance of riverside grazing in the Early period through a more diverse riverine and steppic grazing pattern through the Middle period to a continued diversity of practices in the Late period (fig. 5.18). The proliferation of fire spots in the Late period in fact exhibits the widest variety of 'riverine' to 'steppic' dung signatures. This observation coincides with faunal evidence for a greater focus on sheep/goat herding (Russell et al. 2013; Twiss et al. this volume, Chapter 8) and supports the case for expansive herding despite a shrinking human population.

Our third major finding, stemming from analysis of arable weed assemblages (Green 2017; Green et al. 2018), is that Çatalhöyük's agroecology was diverse, extending from highly fertile and disturbed ('garden-like') conditions to relatively unproductive and poorly disturbed arable habitats. We think that this spectrum relates at least in part to the distance of cultivation plots from the settlement, determining accessibility for labour-intensive management (cf. Jones et al. 1999) and also nutrient-rich inputs from the tell itself, including manure accumulating in penning areas (Farid 2007a). The observation that the Level North G ecological spectrum is particularly wide supports this interpretation, as does the tendency for Late period weed assemblages in Levels South Q-T and TPC M to shift towards high fertility/disturbance as the scale of the community and its arable catchment shrank, becoming more accessible for intensive management. The weed data thus directly demonstrate how Çatalhöyük's maximum scale, like that of other Neolithic 'megasites', restricted agricultural productivity, meaning that household production was subordinated to collective security (Bogaard, Isaakidou 2010; Bogaard 2017).

Finally, subtle patterning in weed composition among crop 'storage' deposits in burned buildings suggests that social proximity within the built landscape was mirrored in the wider arable one. A hypothetical model of how this could have worked is suggested by the 'radial lines' noted in Çatalhöyük's layout within the North and South Areas (cf. Charles et al. 2014; Hodder 2014; Bogaard 2017). More speculatively, corresponding radial field boundaries in the surrounding landscape may be reflected in some of the radial features recently detected through map regression of the local landscape (Jackson, Moore 2018) (fig. 5.30).

Figure 5.30. Plan of the East and West Mounds showing Middle period architecture on the East Mound, including radial features ('midden' areas or wall alignments) in the South and North areas (solid to dotted black lines), and radial field boundaries in the surrounding landscape detected through map regression (figure: Camilla Mazzucato, after Hodder 2014: fig. 9.3 and Jackson, Moore 2018: fig. 4).

All of these findings provide a rich context for the assessment of domestication traits in Çatalhöyük crops (Chapter 6). It is now clear that the Çatalhöyük community was engaged in the domestication process of the 'new type' glume wheat, and that this domestication episode began and ended much later than those of the other cereals. The domestication of this distinctive Anatolian wheat was a product not only of continued selection pressure from sickle harvesting, but also of the social dynamics and landscape of the community. The decision by Middle period households in the North Area to cultivate (semi-shattering) 'new type' glume wheat served in part to distinguish this area of the mound from the South Area, which continued to favour emmer. The eventual adoption of 'new type' cultivation by South Area households during the Late period coincided with fixation of the non-shattering, domesticated form.

Moreover, reconstruction of Çatalhöyük's agroecology informs understanding of changes in crop seed size through time (Chapter 6). Increases in seed size are observed in most crops, particularly from the Early to Middle periods. Larger seeds confer greater competitiveness from the seedling to established phases of crop plants, but such competitive ability is only advantageous if growing conditions are highly productive (e.g., Preece et al. 2017). It is now clear that arable growing condi stions at Çatalhöyük were variable but tended to be relatively fertile, favouring the development of larger-seeded morphologies and thus contributing to this aspect of the domestication syndrome.

Acknowledgements

Our thanks go to the many students and colleagues who joined the archaeobotanical team from 2009 to 2017. We also thank Dr Osman Tugay of Selçuk University, Konya for many botanical expeditions over the years and for access to the herbarium. Funding was provided by the European Research Council ('AGRICURB' grant no. 312785, PI Bogaard; 'ComPAg' grant no. 323842, PI Fuller); the John Templeton Foundation (grant no. 13463, 22893, 52003; PI Hodder); the Polish National Science Centre (Harmonia research grant, decision DEC-2012/06/M/H3/00286; PI Marciniak); and the Polish Ministry of Science and Higher Education (grant no. 3085/Kultura/2014/2).

Supplementary material

For supplementary material related to this chapter, please visit https://doi.org/10.18866/BIAA/e-13. It comprises appendix S5.1 (unit-by-unit discussion, with 15 figures: S5.1–S5.15), and two tables (S5.1 and S5.2).

6. An assessment of crop plant domestication traits at Çatalhöyük

Mike Charles, Dorian Q. Fuller, Tina Roushannafas and Amy Bogaard

Introduction

Since excavations were renewed at Çatalhöyük in the mid-1990s, there have been key shifts in understanding of plant domestication in Western Asia. Publication of the experiments of Hillman and Davies in the early 1990s had raised the prospect of rapid cereal domestication, potentially even within the span of a human generation, though the authors also cautioned that this unconscious process could have taken much longer (Hillman, Davies 1990; 1992). Furthermore, excavations at Abu Hureyra and the identification of domestic-type rye grains in the late Epipalaeolithic assemblage suggested that initial cereal domestication occurred during the relatively cold/dry conditions of the Younger Dryas (Moore et al. 2000; Hillman et al. 2001).

Doubt has since been cast on both the rapid domestication scenario and the prospect of morphological cereal domestication during the Younger Dryas. The archaeobotanical evidence is now widely understood to align with the co-evolutionary perspective of Rindos (1984), which predicted that the selection process for domestication traits was both unconscious and gradual. The key domestication trait of a tougher rachis in cereals – preventing the 'shattering' of cereal ears when grains ripen – now appears to have taken place over several thousand years (Tanno, Willcox 2006; Willcox et al. 2008; Fuller et al. 2014a; Allaby et al. 2017). Furthermore, thanks in particular to the archaeobotanical investigation of PPNA-ePPNB sites on the Middle Euphrates like Jerf al Ahmar and Dja'de, corresponding to a hiatus between Late Epipalaeolithic Abu Hureyra I and Middle-Late PPNB Abu Hureyra II, it is now clear that there was a lengthy period of intensive cereal use prior to morphological domestication (Willcox et al. 2008; Douché, Willcox 2018). This so-called pre-domestication cultivation phase is now recognised as a multi-regional phenomenon, identified in both the Southern and the Northern Levant as well as in lowland southwest Iran (Willcox 2013; Riehl et al. 2013). The focus on wild cereal progenitors such as wild barley, einkorn, rye and emmer by hunter-gatherers in these disparate regions was accompanied or replaced in others by a focus on other wild foods, such as smaller seeded grasses including perennials, other annual plants with edible nutlets, various tree nuts, etc. (e.g., Savard et al. 2006; Riehl et al. 2012; Fuller et al. 2018; Rössner et al. 2018; Whitlam et al. 2018). Wild cereal (as well as pulse) exploitation was thus a widespread, multi-regional phenomenon pursued by hunter-gatherer groups in conducive settings, but also only one of a number of plant exploitation strategies by early Holocene foragers.

Synthetic accounts of the gradual loss of the shattering habit in cereals such as barley and einkorn have incorporated chaff data from Çatalhöyük (Fairbairn et al. 2005), indicating that wild forms were virtually absent (Asouti, Fuller 2013; Fuller et al. 2014a). The predominant form of barley at Çatalhöyük is in fact naked barley, which by definition confirms its domestic status and distinctiveness from the wild (hulled) form; moreover, it appears that naked forms evolved from domesticated hulled forms (Zohary et al. 2012; Fuller, Weisskopf 2014). More recently published data from Çatalhöyük also document the rarity of wild barley and einkorn chaff (Filipović 2012, 2014; Bogaard et al. 2013). Based on the barley and einkorn data, Çatalhöyük would appear to represent an 'endpoint' in the domestication process (e.g., Shennan 2018: 35).

Recent work by Allaby et al. (2017) on loss of seed shattering in West Asian barley, emmer and einkorn, and in East Asian rice, further confirms low selection pressure and protracted domestication, but also identifies 'inflection points' when selection was more rapid. Relatively rapid selection for non-shattering barley, einkorn and emmer is indicated in the timeframe of 8500–7500 BC (that is, Early-Middle PPNB), well before the Çatalhöyük occupation.

From the earliest Neolithic levels at Çatalhöyük the cereal spectrum included two unquestionably domesticated taxa that lack immediate wild counterparts – naked barley and bread wheat – alongside einkorn and two other 'glume' (or hulled) wheats: emmer and 'new type' glume wheat (NGW) (Fairbairn et al. 2005; Bogaard et al. 2013; 2017; this volume Chapter 5; Filipović 2014). NGW was only recognised relatively recently as a morphologically and taxonomically distinct cereal crop and has been identified in Neolithic Western Asia, Europe and western Central Asia (e.g., Jones et al. 2000; Charles, Bogaard

2010; Bogaard 2011). Çatalhöyük offers a significant Neolithic assemblage of this taxon, which is also represented earlier, from the mid–later ninth millennium BC at Aşıklı Höyük in Cappadocia (Ergun et al. 2018), Cafer Höyük, Eastern Anatolia (De Moulins 1997) and also, in small numbers and more locally, at Boncuklu Höyük (Baird et al. 2018). Amongst the glume wheats at Çatalhöyük, emmer and NGW are far more common than einkorn, and it is now clear that emmer is dominant through the Early period and NGW through to the Late, with a transition phase in the Middle period, when NGW is already dominant in the North Area but emmer remains dominant in the South Area of the settlement (Bogaard et al. 2017; this volume, Chapter 5).

The key domestication trait of a tough (non-shattering) rachis in cereals is diagnosed by distinguishing between 'wild/smooth' and 'domestic/rough' abscission scars, a distinction that is not straightforward in glume wheat chaff (Tanno, Willcox 2012). Weide (2015) recently presented a method for distinguishing wild and domestic type emmer wheat that takes account of variation in scar morphology within the cereal ear, by focusing on the diagnostic profile/section through the scar. In this study we have followed Weide's approach for assessing wild/domestic status in emmer, in order to confirm whether or not the Çatalhöyük material is indeed domestic. We also assessed the applicability of this approach to modern *Triticum timopheevii* (Zhuk.) Zhuk. subsp. *araraticum* Jakubz. and to *Triticum timopheevii* (Zhuk.) Zhuk. subsp. *timopheevi,* which are, respectively, the hypothesised wild and domestic counterparts of NGW (Jones et al. 2000). Henceforth we refer to wild and domestic members of the *Triticum timopheevii* group as araraticum and timopheevii, for the sake of simplicity. It has been noted previously that apparently 'wild type' smooth abscission scars are regularly observed in the Çatalhöyük NGW material (Bogaard et al. 2013; 2017), but this is the first systematic quantitative study of their occurrence. We report below on the results of this investigation of modern araraticum and timopheevii accessions, and on the application of the resulting criteria for distinguishing wild and domestic NGW chaff at Çatalhöyük.

The counterpart of cereal chaff in pulses – the pod – is not usually preserved in charred assemblages, but Çatalhöyük is a remarkable exception. Well-preserved 'fire spots' and pulse 'storage' deposits repeatedly preserve not only seeds but also intact pulse pod and stem fragments (that were not redeposited and so did not disintegrate) (Filipović 2012; 2014; Bogaard et al. 2013). As previously noted, the pod fragments associated with both lentil and pea deposits in such contexts are clearly 'flat' rather than of the 'curled' type associated with wild/dehiscent forms (Filipović et al. 2012; 2014; Bogaard et al. 2013). From this evidence, which dates predominantly to the Middle-Late periods, it is apparent that these major Çatalhöyük pulse crops were indehiscent. To the best of our knowledge this is the earliest direct evidence for pod indehiscence in Neolithic West Asian pulses.

The loss of shattering/dehiscence as the natural seed dispersal mechanism in cereals and pulses, however, was only one part of the domestication syndrome (Hammer 1984). Another dimension that can be monitored archaeobotanically is seed size, which has an important but complex relationship with seedling competitiveness, established plant size/competitiveness, and yield (Harlan et al. 1973; Sadras 2007; Fuller, Stevens 2017; Kluyver et al. 2017; Preece et al. 2017). Recent work on seed size by Fuller et al. (2014a) has demonstrated rates of seed size increase in the major West Asian cereals (barley, emmer, einkorn), as well as in East Asian rice, that are similarly protracted to the emergence of non-shattering forms. In the present study we conducted seed size measurements of all major cereal and pulse crop taxa in order to chart seed/grain size through the thousand-year sequence of the East Mound and to assess how these site-specific data fit within supra-regional trends.

Materials and methods

Assessing scar morphology

Modern material. Modern accessions of wild emmer (*T. turgidum* L. subsp. *dicoccoides* [Korn. ex Asch. & Graebn.] Thell, n=3), domesticated emmer (*Triticum turgidum* L. subsp. *dicoccum* [Schrank] Thell, n=3), araraticum (n=6) and timopheevii (n=4) were prepared for morphological analysis using criteria based on Tanno and Willcox (2012) and Weide (2015). The modern wheat accessions were principally drawn from the archaeobotanical reference collection at the University of Oxford. Additional accessions were sourced from the International Centre for Agriculture Research in the Dry Areas (ICARDA).

Selection of archaeobotanical material. Including non-diagnostic types, a total of 1,201 spikelet forks from 18 archaeological contexts were categorised as part of the chaff analysis. Selected samples represented a minimum of three archaeological units per period ('Early', 'Middle' or 'Late' period) and contained the highest recorded numbers of emmer or NGW chaff remains within their phases. Samples in which analysts' notes indicated the presence of particularly well-preserved specimens were also prioritised. In the case of emmer, chaff remains were primarily in the form of glume bases, while spikelet forks were more common in the apparently more robust NGW, leading overall to a significantly lower number of categorisations for emmer.

Methodology. Z-stacked images of the scars of wild emmer and domesticated emmer were taken using a Leica Z6 APO macroscope with automated focus drive at 3.60x magnification. Analysis of these images using the Leica Application Suite (LAS V4.12) confirmed the usefulness of a vertical profile in distinguishing wild and domestic emmer as described by Weide (2015). Weide's study identified that wild emmer either presents a 'smooth' scar or a 'rough' scar that in profile shows an abrupt transition ('knee') where the upper rachis segment attaches to the lower segment along the bottom part of the abscission area (see fig. 6.1a). In domestic emmer, by contrast, no matter how much of the upper rachis segment remains attached to the scar, making it 'rough', the transition from the lower to upper rachis segment is gradual in profile (fig. 6.1b).

When araraticum and timopheevii were compared in the same way, it was observed that both wild and domestic types exhibit an abrupt transition ('knee') where the upper rachis segment attaches to the lower segment along the bottom part of the abscission area (fig. 6.2a-b). This means that for the NGW counterparts, as for barley and einkorn (Weide pers. comm.), a profile-oriented approach is not suitable for distinguishing wild and domestic types. Thus, a modified version of the Tanno and Willcox (2012) approach was devised distinguishing smooth scars from slightly rough, rough, clearly protruding (figs 6.2-6.3) and 'torn' scar areas potentially damaged by dehusking. Archaeobotanical specimens that were poorly preserved or otherwise damaged were classified as 'non-diagnostic'. The scar type categories were created through observation and classification of both the archaeobotanical material and modern accessions of araraticum and timopheevii. All available reference material of araraticum and timopheevii was analysed as part of the scar-type categorisation process (n=258 spikelets).

Descriptions of the scar categories are as follows:

Smooth: Two types of 'smooth' scar were identified based on the transition from the rachis to interior scar surface, being (1) virtually seamless or (2) a slightly raised 'lip'. The 'smooth' category included those with a more defined 'ridge' as long as this was either consistent all the way round or grading to smooth (this 'ridge' type was observed in modern araraticum and is, therefore, not

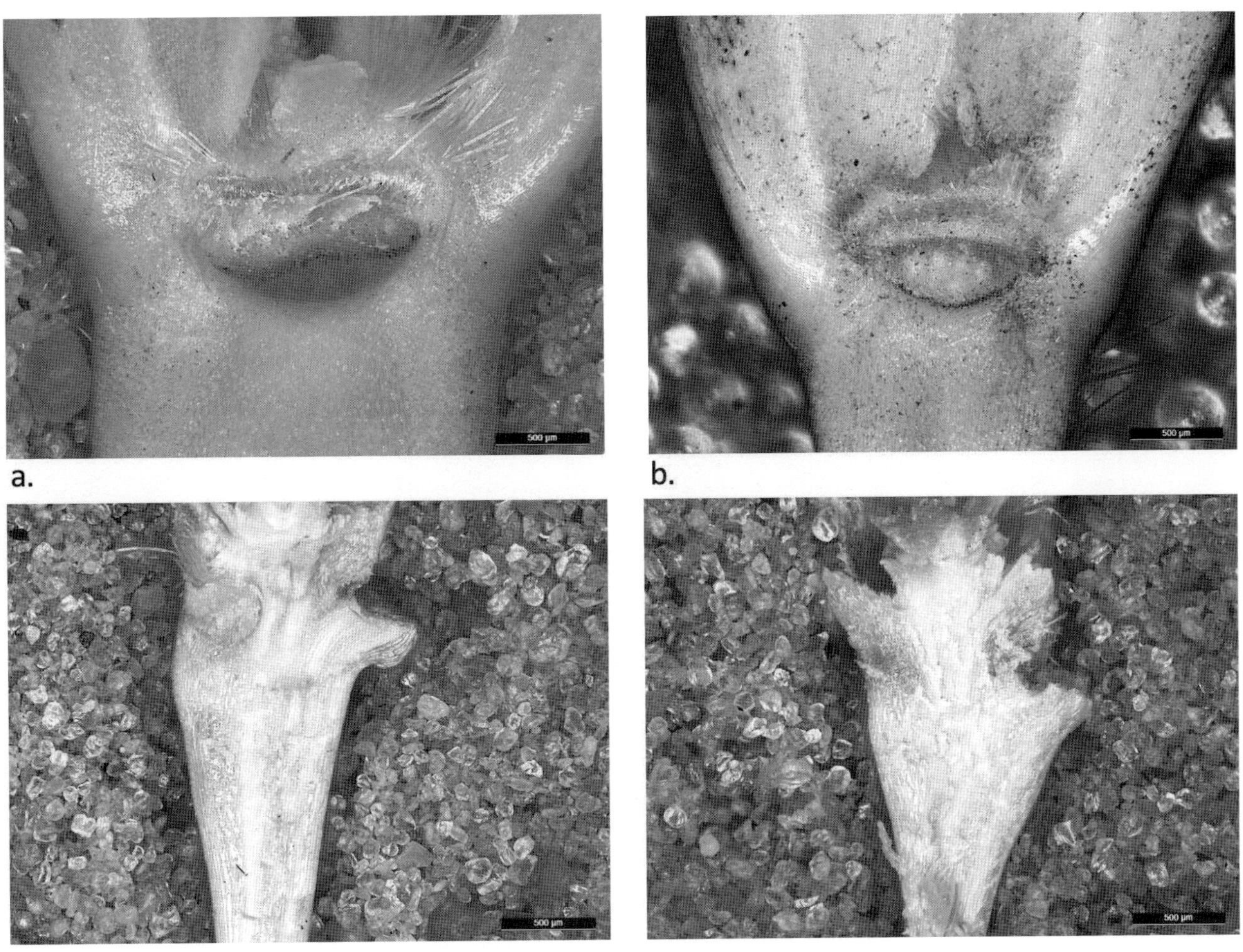

Figure 6.1. Examples of 'protruding' scars in (a) wild emmer and (b) domesticated emmer.

evidence of domestication). Roughness visible only at the transition to the upper perimeter was disregarded, as this appears to be fairly typical. The scar outline is concave, with visible vascular bundles.

Slightly rough: As for 'smooth', but with slight breakage, irregularity or roughness along the lower 'lip' or surface of the scar itself or a very small protrusion from scar. In some cases, the perimeter of the scar may appear very smooth, but the internal surface of the scar is rough.

Rough: These specimens exhibit notable roughness around the perimeter or surface of the scar, in terms of either missing edges or protrusions. The roughness occurs over the majority of surface/perimeter.

The identification of slightly rough/rough NGW scars is subtler than in emmer, with the latter frequently exhibiting a much more abrupt transition between the scar perimeter and internal surface.

Protruding: These specimens have a strong protrusion with clear curvature around at least 50% of the scar: that is, a portion of next rachis segment remains attached. In both the modern accessions and the archaeological NGW these have a distinctive 'knee-shaped' appearance as used to describe wild emmer scars in Weide (2015). In domestic emmer, protruding scars appear 'lifted' (fig. 6.1b).

Torn: These specimens are distinguished by a usually deep tear which exposes underlying tissue, extending in a single direction (often vertically) from the lower perimeter of the upper scar. This type was only found in archaeological material, which suggests that it results from taphonomic factors, most likely processing, and is analogous to the type identified by Tanno and Wilcox (2012: 109) as 'damaged upper portion of spikelet with tear off scar'. Slight lower 'tear' type damage on the

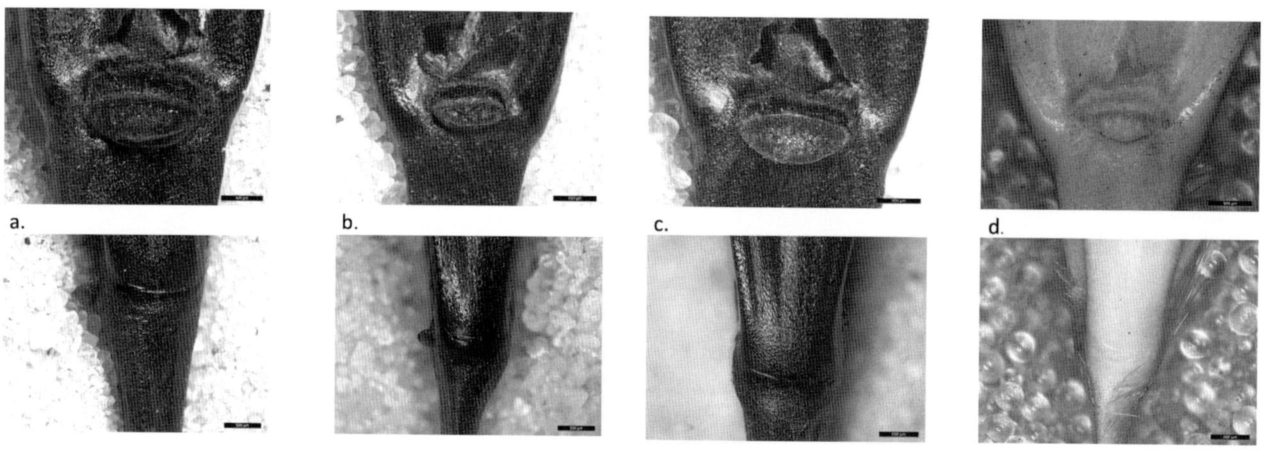

Figure 6.2. Examples of 'protruding' scars in (a) araraticum (wild) and (b) timopheevii (domesticated); examples of 'smooth' scars in (c) araraticum (wild) and (d) timopheevii (domesticated).

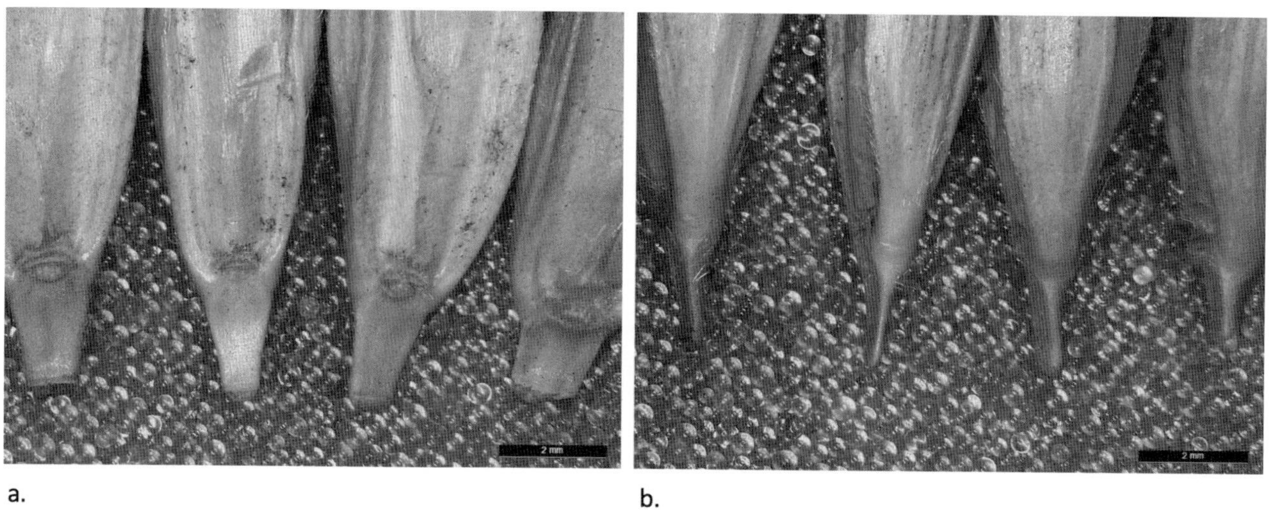

Figure 6.3. Examples of (from left to right) smooth, slightly rough, rough and protruding scar specimens of timopheevii, showing: (a) ventral view and (b) side view.

lower rim of the scar does occur in modern 'rough' types but is less severe and usually appears in conjunction with other irregularities.

Measuring seed size

Material. As in the scar-type analysis, samples representing a minimum of three archaeological units per phase were analysed for each of the eight major crop types found at Çatalhöyük (einkorn, emmer, NGW, free-threshing wheat, barley, bitter vetch, lentil and pea). Wherever possible, three specimens were measured per archaeological unit. The largest, most complete and best-preserved grains or seeds were chosen, with larger specimens excluded where they showed clear signs of distortion resulting from charring at high temperatures. While a more random and less judgement-based approach may have been preferable, this methodology allowed for rapid selection alongside reasonable consistency between samples.

Methodology. All archaeobotanical specimens were measured in three dimensions with a total of 594 measurements taken in total. The widths and heights (excluding protruding embryo if present) of cereal grains were taken in dorsal view, while depth was taken from the thickest part of the grain in lateral view. The height and width of pulses was measured in the 'front' view directly facing the radicle/hilum, with depth obtained by turning the pulse on its side for the lateral view. The height of *Vicia ervilia* was taken from the lateral view, due to the pronounced indentation of the hilum visible from the 'front'.

Each seed was photographed using a Leica Z6 APO macroscope with automated focus drive under a 1.0x objective and 1–1.25x magnification. Measurements were then taken from these images using the point-to-point measuring tool in LAS (V4.12). In a minority of cases, dimensions were obtained from pre-existing images taken in the same views at the University of Sheffield, which were calibrated and measured within the image capture and processing software Infinity Analyze.

Results

Scar types observed in modern araraticum and timopheevii

All of the scar type categories (with the exception of the 'torn' category) described above occur in *both* wild araraticum and domesticated timopheevii, but in different proportions (table 6.1, fig. 6.4). As might be expected, smooth scars dominate amongst spikelet forks of araraticum, but these are accompanied by lesser quantities of both 'rough' and 'slightly rough' types. 'Protruding' scar types were rare, occurring in <5% of modern accessions of araraticum. The distribution of scar-type categories was broadly inverse in timopheevii. Of 100 spikelet forks examined, 'knee-shaped', 'protruding' types are the largest group (37%), closely followed by 'rough' (34%) and a smaller number of 'slightly rough' (21%). Two examples of completely smooth spikelet forks were also identified, indicating that, even in modern domesticated populations, spikelet forks which appear morphologically 'wild' occur.

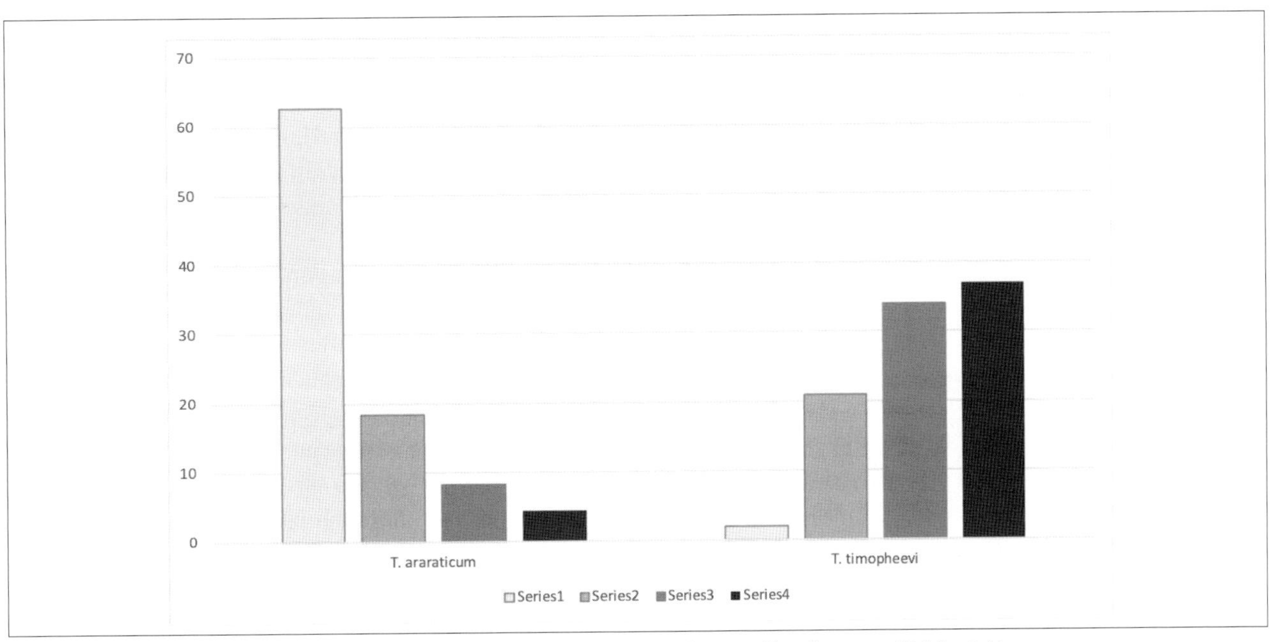

Figure 6.4. Percentages of scar types in araraticum and timopheevii (for data see Table 6.1).

Accession/ Sample no.	*Species/Type*	*Smooth*	*Slightly rough*	*Rough*	*Protruding*	*Torn*	*Basal*	*Terminal*	*Non-diagnostic*	*Total*
RMN 1285	araraticum	17	6	1	1	0	2	0	0	27
RMN 1544	araraticum	30	8	3	1	0	2	0	1	45
RMN 2448	araraticum	11	1	1	1	0	0	1	1	16
ICARDA IG 116169	araraticum	9	6	2	3	0	0	0	2	22
ICARDA IG 113295	araraticum	15	5	3	1	0	0	0	0	24
ICARDA IG 113296	araraticum	17	3	3	0	0	0	1	0	24
Araraticum combined	araraticum	**99**	**29**	**13**	**7**	**0**	**4**	**2**	**4**	**158**
Araraticum combined %	**araraticum**	**62.7**	**18.4**	**8.2**	**4.4**	**0.0**	**2.5**	**1.3**	**2.5**	
RMN 7212	timopheevii	1	4	7	8	0	0	1	0	21
ICARDA IG45220	timopheevii	1	5	8	10	0	0	0	0	24
RMN1282	timopheevii	0	10	11	11	0	0	3	0	35
MPC 1978 (whole ear)	timopheevii	0	2	8	8	0	1	1	0	20
Timopheevii combined	timopheevii	**2**	**21**	**34**	**37**	**0**	**1**	**5**	**0**	**100**
Timopheevii combined %	**timopheevii**	**2**	**21**	**34**	**37**	**0**	**1**	**5**	**0**	

Table 6.1. Summary of scar types in the modern araraticum and timopheevii accessions.

While the link between the *Triticum timopheevii* group and the NGW at Çatalhöyük has recently been strenghthened by aDNA (Czajkowska et al. 2020), this study has confirmed further morphological common ground between the two in the prevalence of 'knee-shaped' scars and a more 'subtle' or gradual transition between scar perimeter and scar surface than is found in emmer. On this basis, the results of the modern spikelet-fork analysis have useful implications in interpretation of archaeological NGW chaff assemblages. While a high proportion of 'smooth' scars would point to a wild/shattering population, the mere presence of 'smooth' scar types would not necessarily demonstrate this, as these occur occasionally within domesticated populations. Similarly, as previously observed in regard to emmer (Weide 2015), the presence of 'protruding' or 'rough' scar types within an NGW assemblage is not necessarily evidence of domestication, but this scar type should be far more common than smooth scars in a domesticated population.

Applying shattering criteria to archaeobotanical emmer and NGW spikelet forks

Archaeobotanical examples of the scar types defined above are shown in figure 6.5. Terminal spikelet forks (that is, forks lacking a scar, from the top of the ear) were also quantified, to assess their occurrence; in shattering forms these would be the first spikelets to fall from the ear and so might tend to be under-represented archaeobotanically, whereas we expect that they would be more prevalent in non-shattering domestic forms. We distinguished NGW terminal spikelet forms with pronounced veins from those of emmer lacking pronounced veins (fig. 6.5e).

Application of Weide's (2015) method of distinguishing wild and domestic emmer to Çatalhöyük samples from rich contexts from Early, Middle and Late periods shows that all of the spikelet forks with protruding scars were of the domestic type (table 6.2 and fig. 6.6). Aside from a few apparently smooth scar specimens (n=4) recovered from Early levels, of the 355 specimens recorded the majority had a more or less rough scar. Whether or not the presence of smooth types in the Early period indicates a co-occurrence of shattering and non-shattering forms remains to be examined in a larger study. In the Early period samples examined, most spikelet forks had a 'torn' scar area and so could not be classified.

The results of the classification of NGW spikelet forms indicate a lower number of NGW spikelet forks from the Early period, reflecting the low frequency of this morphological type through the early sequence (table 6.3 and fig. 6.7). While specimens with some form of rough scar are predominant throughout, there is a decrease in smooth scars through time, together with increasing abundance and proportions of protruding scars and terminal spikelet forks. The implication is that there was a significant minority of shattering forms in the NGW at Early and Middle period Çatalhöyük, with continuing selective pressure for non-shattering forms

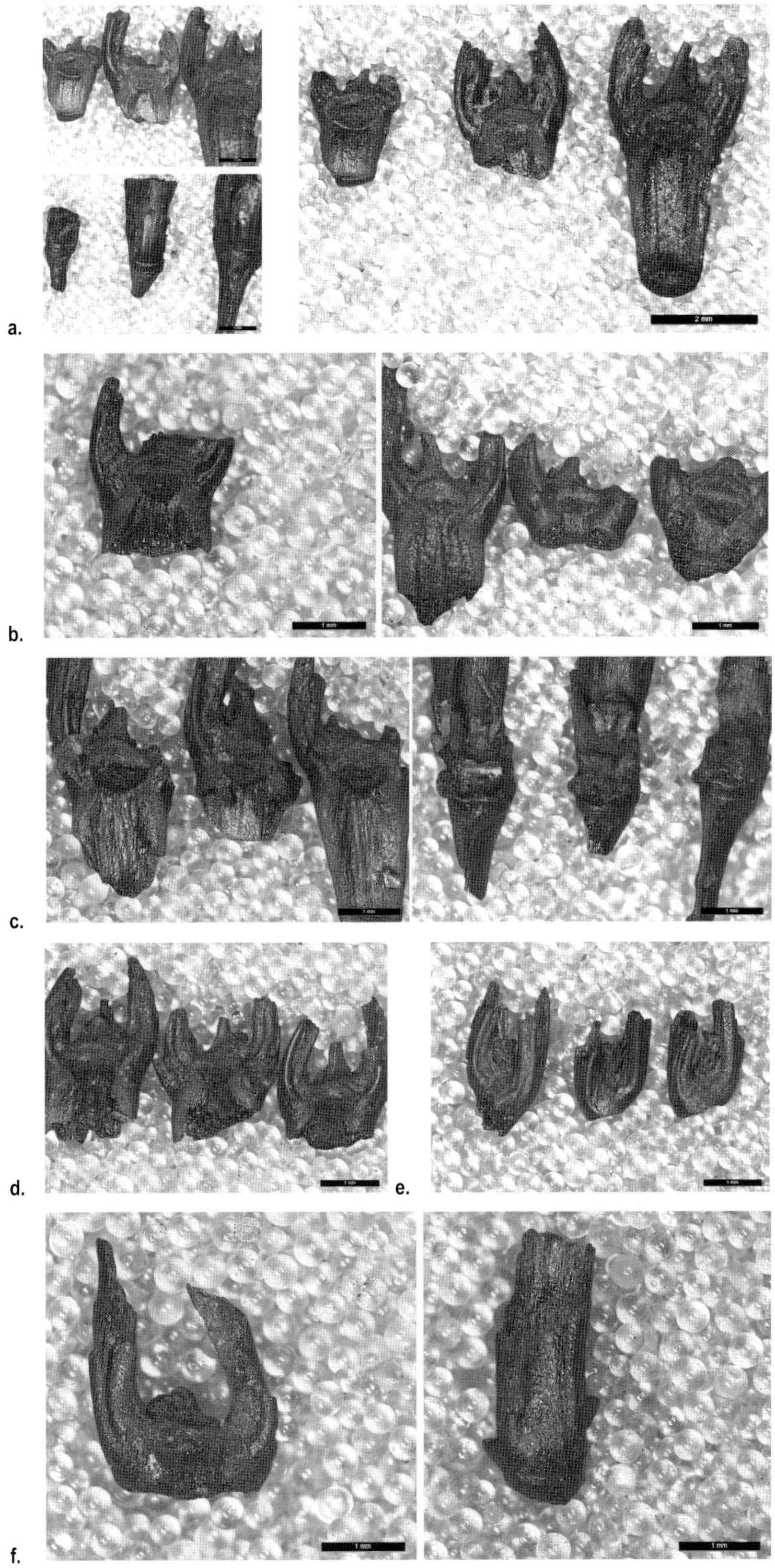

Figure 6.5. Examples of NGW and emmer spikelet forks with different scar types at Çatalhöyük: (a) NGW smooth (ventral view top, side view bottom); (b) NGW slightly rough (left, ventral view) and rough (right, ventral view); (c) timopheevii protruding (ventral view left, side view right); (d) NGW torn (ventral view); (e) NGW terminal (no scar); (f) emmer with protruding-lifted scar type (ventral view left, side view right).

	Smooth	*Smooth rough*	*Rough*	*Protruding-lifted*	*Terminal*	*Torn*	*Total*
Early	4	1	16	8	15	70	114
Middle	0	0	62	55	13	26	156
Late	0	2	25	16	12	30	85
							355

Table 6.2. Summary of scar types in Çatalhöyük emmer spikelet forks.

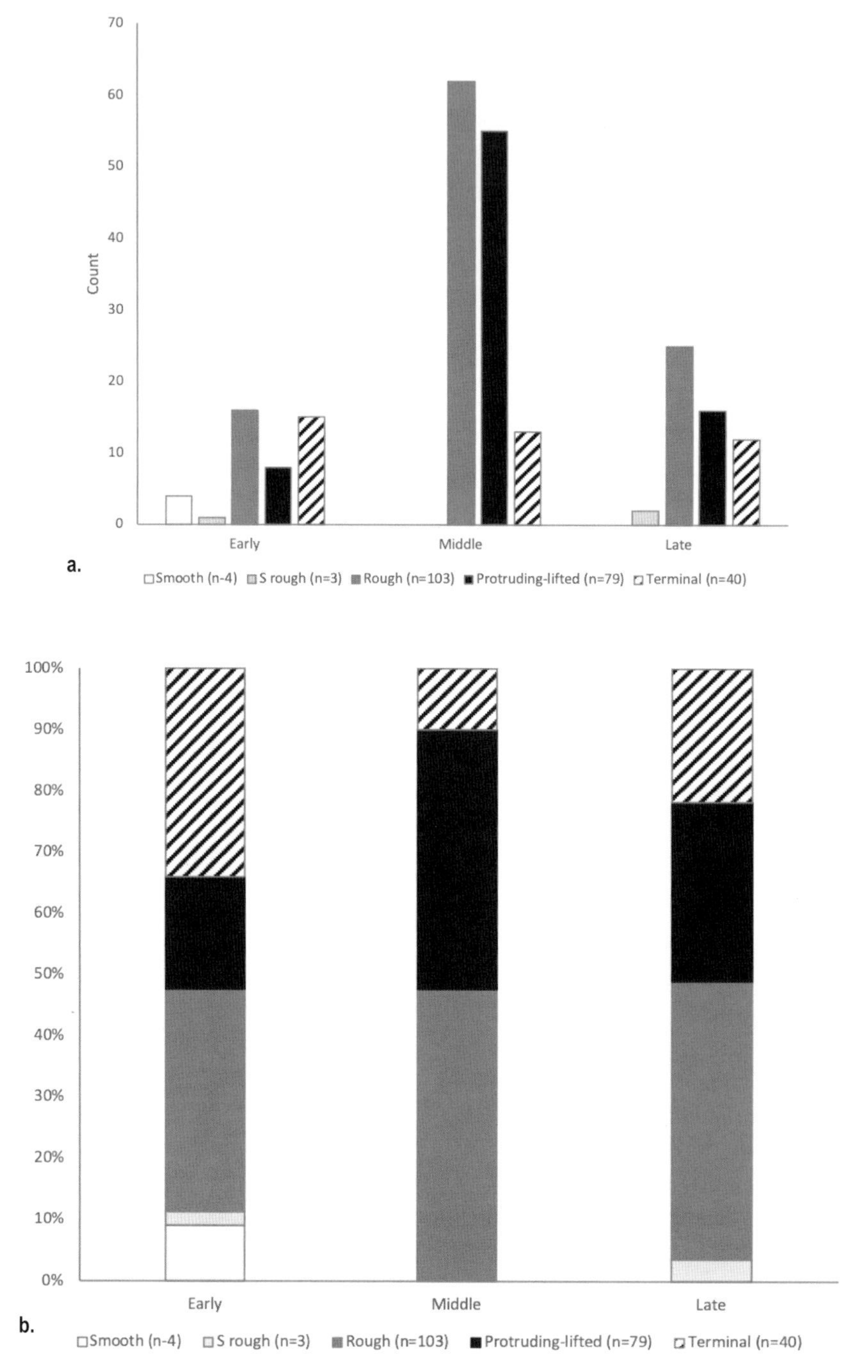

Figure 6.6. Summary of scar types in Çatalhöyük emmer spikelet forks: (a) counts of scar types by site period; (b) percentages of scar types by site period.

through time. In other words, the domestication process for the NGW was still ongoing at this time. This morphological trend is accompanied by an increase in NGW chaff through time and its dominance over emmer especially through the Late Neolithic Levels (Bogaard et al. 2017; this volume, Chapter 5). Thus, emphasis on the NGW as a favoured cereal crop over emmer was associated with fixation of its domesticated morphology.

	Smooth	*Slightly rough*	*Rough*	*Protruding*	*Terminal*	*Torn*	*Total*
Late	43	124	94	105	83	166	615
Middle	47	63	49	27	16	1	203
Early	8	6	4	5	2	3	28
							846

Table 6.3. Summary of scar types in Çatalhöyük NGW spikelet forks.

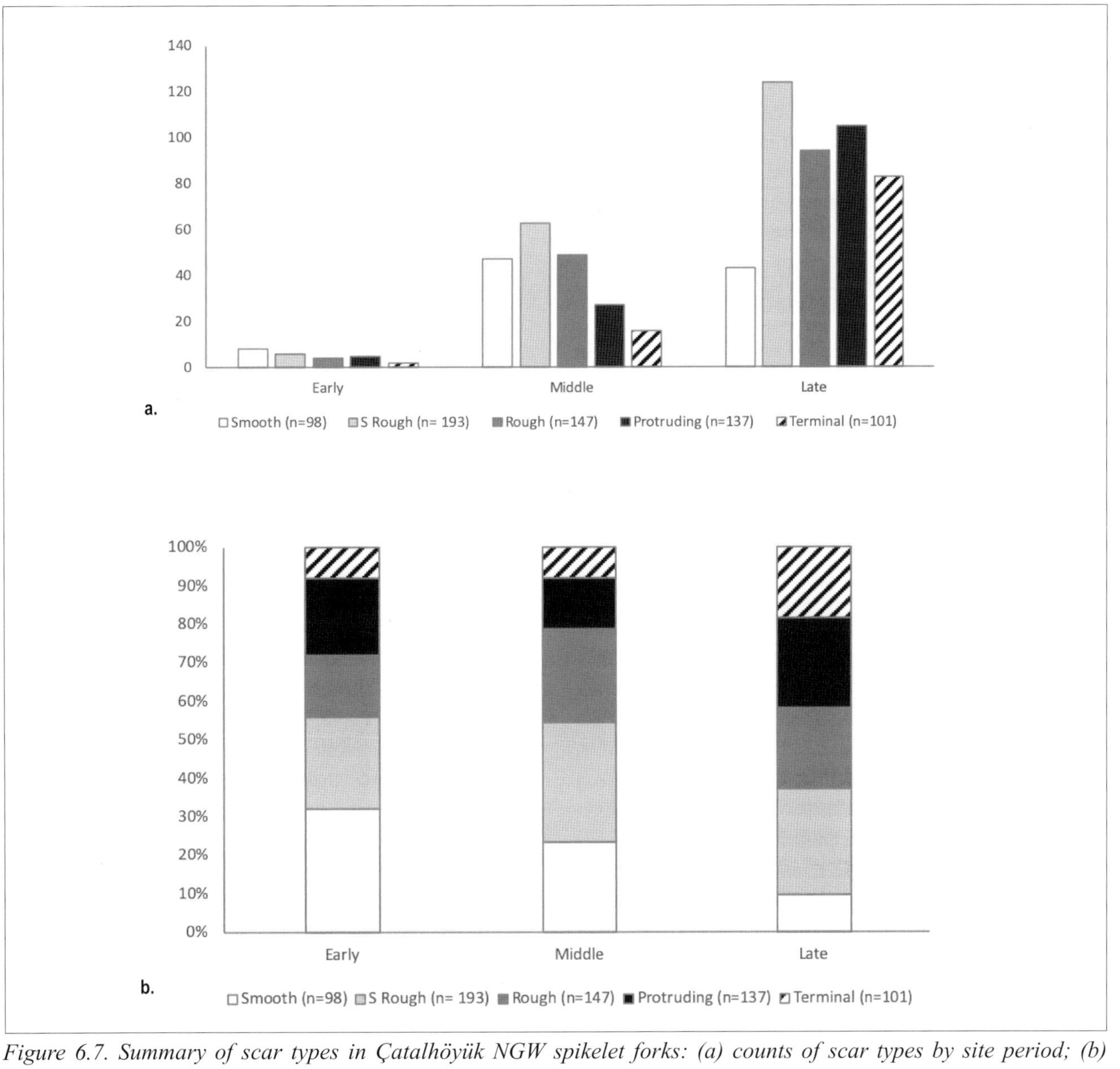

Figure 6.7. Summary of scar types in Çatalhöyük NGW spikelet forks: (a) counts of scar types by site period; (b) percentages of scar types by site period.

Grain and pulse seed size through time

Figure 6.8 summarises the average size of major cereals and pulse crops from the Early to Middle and Late periods at Çatalhöyük. It is notable that most taxa increase in size from Early to Middle period levels. This tendency includes crops that increased in relative frequency through time (NGW, pea, naked barley, free-threshing wheat) as well as one that decreased (bitter vetch) (Chapter 5). Other crops (lentil, emmer) remained the same or decreased in size from the Early to Middle periods. The implication is that size increases may bear some relationship to economic importance. Rainfall increases from the Early to Middle periods of the Neolithic sequence in the wider region (see Chapter 2) may also play a role in size increases in prevalent crops.

The shifts in size from the Middle to Late levels are more variable. It is clear that barley increases in size throughout the sequence, though it should be noted that a shift from six- to two-row naked barley (that is, from a mixture of central/lateral grains to only central grains) likely contributed to this trend, at least from Early to Middle Neolithic (Bogaard et al. 2017; this volume, Chapter 5). In other cases, increases from the Early to Middle periods were not sustained in the Late period, perhaps in part due to decreasing regional precipitation towards the 8.2K event (Chapter 2).

Comparison of the Çatalhöyük assemblage with those documented by Fuller et al. (2014a) from across Western Asia show a close fit in grain thickness and pulse seed length (fig. 6.9). Barley is the exception to this pattern, possibly due to the Çatalhöyük barley being naked and the shift from six- to two-row barley through time at the site. Nevertheless, it is notable that Early period barley of the early seventh millennium BC is distinctly smaller than contemporary barley in the Levant. Eventually this observation may be understandable in terms of the origins of Central Anatolian naked barley. The steady increase in barley size through time at Çatalhöyük is also notable in that it parallels the shape of earlier size increase(s) in hulled barley. Otherwise the variation in Çatalhöyük grain/seed sizes through time more or less fits within ranges among contemporary assemblages.

Discussion and conclusions

If the co-evolutionary, niche construction perspective on domestication (Rindos 1984; Zeder 2015; Fuller, Stevens 2017) is to be taken seriously, it means that domestication was a process that continued through and indeed beyond the Neolithic. In this sense it is no surprise that even unquestionably domesticated crops at Çatalhöyük, like naked barley and free-threshing wheat, showed some morphological change through time, including the steady grain size increase observed in

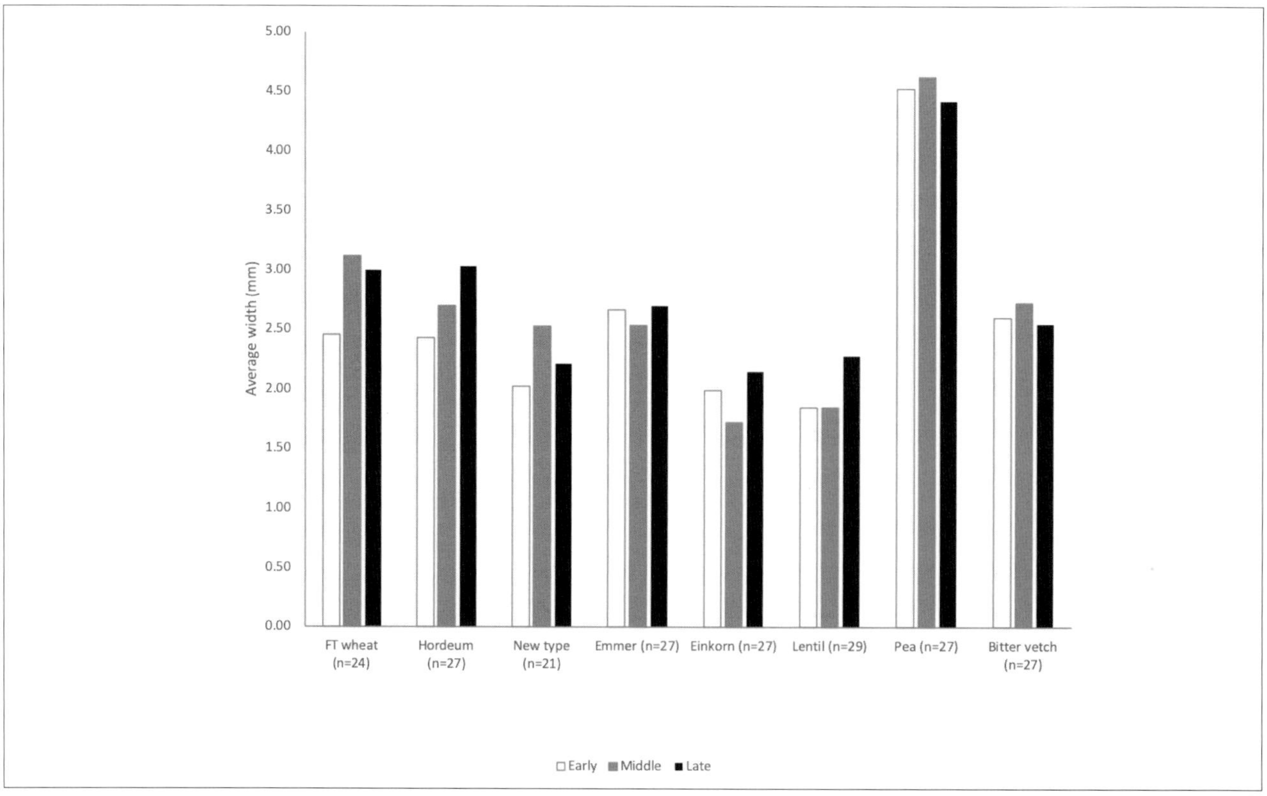

Figure 6.8. Summary of average grain/seed thickness (mm) per site period in cereal and pulse crops at Çatalhöyük.

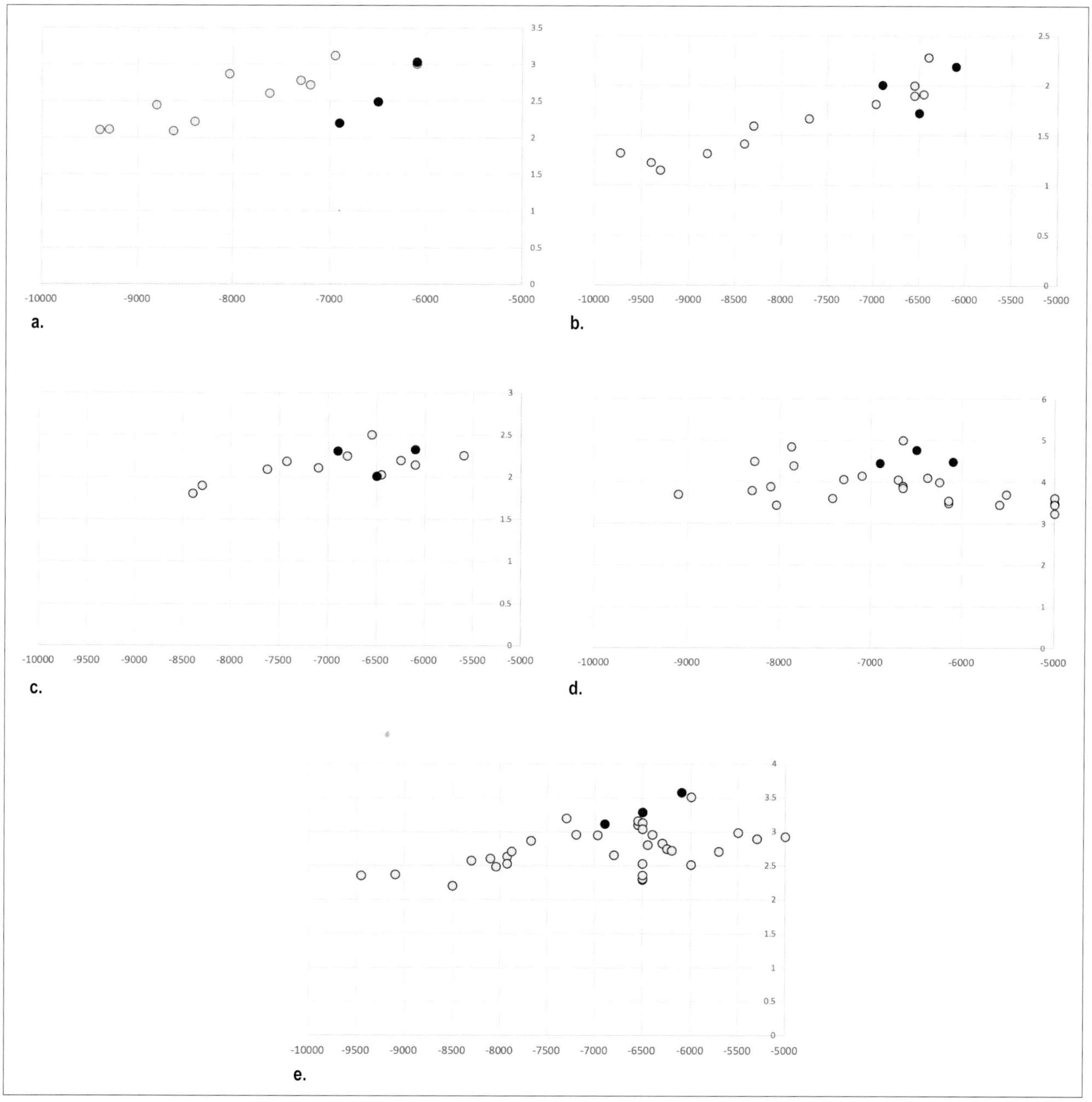

Figure 6.9. Comparison of average grain/seed thickness of Çatalhöyük crops (filled circles) per site period (Early, Middle and Late Neolithic) with metrical data summarised by Fuller et al. (2014a) from across Western Asia (open circles). (a) barley, (b) einkorn, (c) emmer, (d) pea, (e) lentil.

barley. This chapter presents only a first pilot study in domestic crop traits at Çatalhöyük, but an emerging challenge is to assess how grain/seed size change relates to climate change, quantitative shifts in preferred crops and agroecology through time (Chapter 2 and Chapter 5). The high-resolution archaeobotanical documentation and thousand-year sequence at Çatalhöyük East make it an ideal case study for assessing how domestication relationships articulated with these other factors within a single local sequence.

This study has also raised new questions about an increasingly domesticated/non-shattering status of the NGW glume wheat through the occupation sequence. There are grounds for thinking that the NGW was a distinctively Anatolian Neolithic crop. It is not currently reported from other regions of Western Asia, though this may in part be due to its recent definition as a morphological type (Jones et al. 2000). It appears likely that it occurred at Early-Middle PPNB Cafer Höyük (reported as 'machoid emmer') (De Moulins

1997; cf. Jones et al. 2000). NGW is well attested at Aşıklı Höyük, though emmer was the dominant glume wheat (Ergun et al. 2018). This dominance of emmer continued at Early period Çatalhöyük, with a distinctive shift towards NGW first in the North Area of the settlement in the Middle period, and by the Late period in the South Area (Bogaard et al. 2017; this volume, Chapter 5).

The preliminary results reported here suggest that the increasing preference for NGW over emmer was associated with a detectable shift from a semi-shattering to predominantly non-shattering form. The immediate cause of such a shift was presumably a sickle-harvesting routine that favoured the collection of non-shattering ears. Sickle harvesting was well established throughout the Levant before this time (Maeda et al. 2016), and use-wear traces of cereal harvesting on sickle blades are well attested at Çatalhöyük (see Volume 14, Chapters 14 and 15). By the Late period <10% of NGW chaff was of the shattering/smooth scar type, bringing it into line with the status attained by the earlier domesticated Western Asian cereals (barley, einkorn, emmer) prior to the onset of occupation at Çatalhöyük. Though the results reported here need to be explored further through a larger scale study, it appears that Çatalhöyük should no longer be regarded as simply an 'end point' after the Western Asian plant domestication episodes, but was still engaged in the domestication process of a new Anatolian wheat. The beginning and end of the domestication episode of the NGW both appear to be later than the cereals that were brought into cultivation in the Levant. These results also bolster the case for regarding Central Anatolia as a distinctive region within the broader multi-regional process of domestication (cf. Fuller et al. 2011, 2018; Stiner et al. 2014; Bogaard et al. 2017).

7. Following the plant pathways: a synthesis of phytoliths and starch grains analyses at Çatalhöyük

Carlos G. Santiago-Marrero, Javier Lara-Recuero, Carla Lancelotti and Marco Madella

Introduction

Phytoliths are microscopic structures made of opal silica that accumulates within the cells and tissues during a plant's lifespan (Piperno 2006). These bodies are particularly abundant in grasses and are incorporated into the soil following plants' decay, surviving in sediments for thousands of years. Their morphology, being a cast of the plant cells in which they formed, gives information about plant taxonomy and ecology. Another valuable quality of phytoliths is the possibility they provide to distinguish between leaves, culms, fruits and inflorescence, helping identify the specific use of plant parts. Due to these characteristics, phytoliths have been widely integrated into archaeological and environmental sciences (Piperno 2006; Ball et al. 2015). On the other hand, starches are microscopic grains made of amylose and amylopectin, two types of glucose produced by plants. These grains are present in different plant tissues for storing energy and are particularly abundant in seeds and storage organs such as roots, tubers, rhizomes and bulbs. Despite being biodegradable, starches can survive for thousands of years incrusted within stone tools' crevices and pores or trapped in dental calculus and sediments (Henry et al. 2011; Henry 2012; Pagán-Jiménez et al. 2016). The study of starch grains has been a valuable tool for identifying plant foods that do not produce phytoliths and are not represented in the macrobotanical record. In addition, observation of starch degradation by anthropic activity such as grinding, cooking and fermentation processes has evidenced patterns that allow us to infer such activities in the past by comparing archaeological starches with modern ethnobotanical and experimental reference collections (Henry et al. 2009).

The value of microbotanical studies has not been overlooked at Çatalhöyük, as evidenced by the integration of phytoliths analyses in 1998 (Rosen 2005). Since then, different researchers have carried out phytolith and starch analysis as part of an integrated and multivocal research strategy (Hodder 2000b; Matthews et al. 2000; Hardy et al. 2013; Shillito 2017a). Following these initial implementations, microbotanical studies have been used to interpret the inhabitants' relation with their landscape, food processing, artisanship, the use of space and different aspects of Neolithic daily life (Rosen 2005; Rosen, Roberts 2005; Ryan 2011; 2013; Shillito et al. 2011a; Ryan, Rosen 2016). These microfossils have been the footprint of plant processing activities and the varied networks of actions that plant materials undergo to become part of the material culture repertoire. However, the integration of micro-botanical studies at Çatalhöyük has been subject to multiple challenges. In this chapter we address those challenges, present a synthesis of the data and illustrate what future developments we foresee for the microbotanical studies at the site.

A brief history and key moments in the development of phytoliths and starch analyses at Çatalhöyük

Microbotanical studies started at Çatalhöyük in an effort to provide a holistic view of the human-plant interactions at the settlement by integrating a group of researchers from different fields within the archaeobotanical sciences (Hastorf, Near 1998). This group was put together to combine different data sets while working on common goals from their respective specialisations (Hodder 2000b; Shillito 2017a). Phytolith analysis was tested on the recurrent impressions of baskets, mats and construction materials (Rosen 1998). Those first examinations evidenced the abundance of silica skeletons from cereal husks, reeds and sedges, highlighting the ability of microbotanical remains to reveal unknown aspects of plant uses and taxa. Starch granules analysis was pioneered by Hardy et al. (2013) with a sampling programme aiming to investigate the broad distribution of starch across selected areas of the site.

Despite their potential, phytolith studies were implemented as a complementary approach stemming from other research aims or as a supporting analysis. One reason for this was probably related to the fact that phytolith studies are time-consuming and require specialised equipment and facilities. This meant that microbotanical studies were not always able to provide, as other areas of archaeobotany were, a stream of on-site data. Nevertheless, this downside did not stop phytolith specialists from developing their own methodological and theoretical approach, which was later evidenced by the multiple implementations of phytoliths studies (Shillito 2011; Shillito et al. 2011a; Wendrich, Ryan

2012). Microbotanical remains proved to be an effective tool for inferring plant use at the site and at the same time, this tool was flexible enough to be easily integrated into the diverse research themes that characterised the excavations of Çatalhöyük. These methodological and theoretical developments were the basis for the subsequent smoother integration of starch grain studies at the site almost a decade later, which amplified the information on plant use and the identification of otherwise invisible plant foods.

In sum, the microbotanical studies approach and focus at Çatalhöyük changed over time and saw the implementation of distinctive sampling strategies, providing the basis for a better integration of phytoliths and starches into the site's research programme for the years to come. Here we present key moments and strategies that exemplify the development and integration of microbotanical studies and how these became part of the discussions of the site's complexities.

Target sampling. The sampling strategies here presented are those that, in our opinion, give phytoliths and starch a more relevant position within the site's research framework. In target sampling, samples are chosen because of the visual identification of silicified remains and imprints during field excavation and for those contexts identified as *priority units* (Rosen 1998; Ryan 2011). These samples often originate from primary contexts, such as storage bins, fire installations, basketry, burials, ovens, and construction materials (Ryan 2013, 2011). This same strategy was implemented for starch grain analysis. These first tests focused on understanding, and they finally proved, starch survival in different contexts throughout the site (Hardy 2007), providing the basis for the development of this line of investigation.

Spatial sampling. The peculiar life cycle of houses at Çatalhöyük, characterised by a cleanliness during their use and often a complete rebuilding after their abandonment (Russell et al. 2014), makes microremains more suited for understanding the use of space than macroremains, due to their higher preservation and more difficult displacement once they are embedded in the floors. In addition, domestic spaces at Çatalhöyük seem to be well defined, based on architectural installations such as rooms for storage and cooking areas with ovens and hearths. This situation creates a favourable scenario for the long-term accumulation of plant microremains produced by repetitive activities performed in such areas. In this sense microbotanical remains became part of the floors and surfaces through the lifespan of the house, making them better suited for understanding the spatiality of plant-related activities.

Spatial sampling was applied from the very beginning of microremains investigations (Rosen 1998; Rosen 2005), first through point samples aimed at collecting sediments representative of a specific floor or surface (Ryan 2009) and then as grid sampling fully or partially covering the floor of a house. This strategy attempted not only to understand plant distribution, but also to unify and bring together target contexts from across the houses.

The grid sampling for phytoliths included a variety of surfaces, such as platforms and areas next to dumping spots, hearths, ovens, bins and baskets within the same building (Rosen 2005; Ryan 2013). Similar sampling was carried out in parallel for the starch grains (Locht, Hardy 2009) and was aimed at creating a starch grains distribution pattern to bring to light the use of starchy plant/organs at Çatalhöyük (Locht, Hardy 2009). The results of these first spatial analyses highlighted links between the distribution of starch and some of the various plant-related activities in buildings, in some cases where macrobotanical remains were absent (Hardy et al. 2013).

Multiproxy sampling. Multiproxy sampling is the third strategy that allowed phytoliths and starches to have a stronger voice within the site's multivocality. Here we define multiproxy samples as those taken with the aim of being approached through a multiple set of analyses and by multiple teams. This strategy allows having comparative data from the same contexts and increases the interpretative possibilities. The first application of this strategy was related to the sub-sampling of sediments or other materials previously analysed by other specialists. Here we present some cases that exemplify the development of multiproxy approaches at Çatalhöyük, using the study of middens as an example.

Middens can be a mix of primary, secondary or tertiary deposits with input of plant-based materials from very different origins. Middens in Çatalhöyük are more than areas of disposal: they are active spaces and important activity areas throughout the history of the site (Bogaard et al. 2014b). These deposits contain rich amounts of phytoliths and other microremains associated with multiple moments of deposition, representing daily or seasonal activities (Matthews 2005a; Shillito et al. 2011a; 2013a; Hardy et al. 2013). Due to these characteristics, middens are often sampled with a multi-sampling approach aimed at geochemical, micromorphological and archaeobotanical analyses (Shillito 2011; Shillito et al. 2009; 2011a). This approach allows production of directly comparable data about plant use, depositional dynamics and taphonomic processes with (often) a short temporal perspective (Rosen 2005; Rosen, Roberts 2005; Roberts, Rosen 2009; Shillito 2011a; Ryan 2013; Shillito et al. 2013a).

The last five years of plant microremains work at Çatalhöyük

The 2014 season marked the start of the new microremains team and a pilot study seeking to integrate the research of phytoliths and starches (and geochemical signatures) into the communities of practice approach (Madella, García-Granero 2014). The first step was to construct a theoretical framework defined by a set of research objectives that aimed at understanding foodways and pathways of plant material use by subsistence farmers (Madella, García-Granero 2014; Lancelotti et al. 2016). Learning from previous research strategies, such as the ones described above, the team designed a working method that allowed a full integration into Çatalhöyük's multivocality, enhancing the dialogue with other archaeobotanical specialists on site in order to produce comparable data from the same contexts. This approach allows for amplifying and cross-validating both macro- and microbotanical data. The first strategy put in place was to carry out a sub-sampling for each sediment sample brought to the flotation station, to create an exhaustive body of directly correlated samples (Madella, García-Granero 2014). The second strategy was the implementation of a spatial grid sampling from living surfaces and from the material culture lying on or related to such surfaces (Madella, García-Granero 2014). Grid sampling was carried out in Buildings 52, 80, 89, 131, 132 and 160 and from open spaces and middens (Madella et al. 2015; Lancelotti et al. 2016). Finally, a target sampling strategy was carried out on sediments from plant-related contexts and macroscopically visible silica deposits such as basketry and woven artefacts (Madella, García-Granero 2014; Madella et al. 2015; Lancelotti et al. 2016).

Evidently, the way in which researchers have implemented microbotanical analyses at Çatalhöyük is more complex than the strategies and examples provided. This can be attested by the multiple times that microbotanical evidence has been used to address various aspects of plant and human relationships. However, the sampling strategies here presented are those that, in our opinion, give microbotanical specialists a more relevant position within the site's research framework, allowing the development of transdisciplinary approaches to the use of plants at the site for the years to come.

On-site contributions

Our presence at the site focused only on three seasons. Nevertheless, and beyond the sampling exported for proper analysis (a total of 2,331 sediment samples were finally exported in 2017 to the Laboratory for Environmental Archaeology at the Universitat Pompeu Fabra), various contributions were made to direct the excavation and to understand the excavated contexts (mostly priority units and other plant-related contexts) for a total of 238 samples. Phytoliths need to be separated from other mineral and organic components, concentrated and then mounted on microscopy slides before a full analysis can be undertaken. However, samples were quickly prepared at the field laboratory by mounting on a microscopy slide whole sediments with ordinary bleach (commercial sodium hypochlorite) to discolour them. Once discolouration occurred, the slides were quickly examined at the microscope to allow a first assessment of the phytolith composition. These analyses allowed the excavators to have a better understanding of the context in question and to make informed decisions on how to proceed and what samples to collect. The quick assessments also contributed to a first understanding of the plant evidence, especially for the grave contexts where different plant materials were identified even when not directly visible.

With you until death: further insights on burial basketry

Previous and current studies on Çatalhöyük basketry include the analysis of the manufacturing technique and possible use (Wendrich 2005; and see Volume 14, Chapter 16) and of the raw materials employed for their production (Rosen 2005; Ryan 2011). Furthermore, Wendrich and Ryan (2012) carried out an integrated approach that 'enhances the possibility to construct the life history of objects and object groups'. The microscopic observation of phytoliths and especially silica skeletons from basketry *ghosts* (*sensu* Hastorf 2001), allowed for the concurrent identification of materials and techniques, including the reconstruction of the *chaîne opératoire*, life histories (Wendrich, Ryan 2012) and biographies of objects (Tringham 2015) in relation to basketry (see also Ryan 2011). Phytoliths studies have shown that the plants used to produce basketry in Çatalhöyük are local and wild, collected both in the marshes surrounding the site and along the primary channel and the distributaries of the Çarşamba River (Rosen 2005; Ryan 2011; Charles et al. 2014).

Our team's recent work on basketry aimed at expanding the information on the raw materials used for the diverse plant-made objects recovered from burials, but also allowed for confirming or extending the knowledge of the manufacturing technique (for in-depth discussion on this, see Volume 14, Chapter 16). Two mats, four coiled baskets, one plaited element (described as a bundle during the excavation) and a possible mat wrapping a skeleton are part of this example of phytolith analysis. The artefacts were recovered from eight burials from Buildings 5, 77, 132 and 161 excavated in 2015 and 2016.

The mat found under the skeleton from B.5, unit (22194) consists of common reed culms (*Phragmites* sp.), as does the plaited element recovered from under an obsidian blade. Other samples from the same burial (F.3808), excavated in 2016 and described at that time as possible linen fibres (see Lancelotti et al. 2016), seem also to pertain to *Phragmites* culm. What these preliminary results highlight is the widespread use of a species (common reed) and a specific part of the plant (culms) for the manufacture of diverse products. Ryan (2011) and Wendrich and Ryan (2012) already described the use of common reed culms, especially for mats. The woven tabby-plaited mat from B.161, unit (32608), on the other hand, is composed of unknown plant fibres that do not appear to belong to reeds, sedges or cereals and could be a soft fibre of dicotyledonous origin.

The coiled baskets examined were produced using two different species of plants: one for the bundles and another for the wrapping strand – to stitch the bundles (Wendrich 1999). The basket from B.77, unit (21682) shows an abundance of silica skeletons that Rosen (2005) and Ryan (2011) identified as a wild panicoid grass (fig. 7.1). However, there are also silica skeletons that can be attributed to rush culm (*Juncus* sp.) (Tsartsidou et al. 2012). From the low proportions of these elements, we suggest that they represent the more flexible wrapping strand(s), while the bundle was made of the wild panicoid grass. The basket from B.132, unit (21688) also has a double composition but with *Phragmites* as fibres (more abundant) and *Juncus* (less abundant) silica skeletons. On the other hand, the baskets from B.5, (22196) and B.132, (21634) are composed of only one type of silica skeleton, *Phragmites* culm that underwent some treatment to release the fibres. These observations highlight that coiled baskets in Çatalhöyük were sometimes constructed using two different plants for the bundle and the wrapping strand. Of interest is the utilisation of *Phragmites* for the bundles, as this plant has been traditionally used mostly to produce mats and roof covers (Kiviat, Hamilton 2001; Köbbing et al. 2013). It is the first time that common reed culms are identified as raw material for coiled basketry, a practice that, nowadays, does not seem to be particularly widespread (Cirujano, Morales 1997; Kiviat, Hamilton 2001; Ertug 2006; Gerritsen et al. 2009; Köbbing et al. 2013; Ehinmore 2014). To produce the fibre-rich and flexible material needed for the coiled basketry bundles *Phragmites* culms need to be soaked until they are softened and subsequently smashed to eliminate the rest of the soft tissues and separate the fibres (Shaikh et al. 2016).

An interesting example of how phytoliths can help understanding of burial practices is the case of burial F.3808 from B.5, units (22194) and (22196). The burial contains two superimposed secondary depositions that correspond to two consecutive moments. First, an adult cranium and mandible (22196) was deposited in the grave within a basket made of fibres from *Phragmites* culms (fig. 7.2). The cranium and mandible seem to have been plastered and covered with red pigment, probably cinnabar, also seen on parts of the basket. A possible textile/matting, so interpreted during excavation, which covers the cranium, is also part of the coiled basket, as shown by the manufacturing technique and the same raw material. A plaited item was recovered from under an obsidian blade deposited with the cranium and mandible. This element was elaborated from discrete parts of

Figure 7.1. (a) Coiled basket from (22196) Building 5; (b) silica skeleton from Phragmites *culms associated with fibres. These elements can be identified by their fusiform morphology (scale=200µm).*

Phragmites culms. The second deposition was a partially disarticulated infracranial skeleton (22195) within fill (22194). The absence of a cranium and mandible for this individual, as well as the lack of clear anthropological differences between cranium and mandible (22196) from same feature, suggests that (22196) and (22195) may represent the same individual even if they were deposited at different times. The skeleton was deposited on and wrapped in a plaited mat of *Phragmites* culm. Burial F.3808 with two deposition events contained different plant-made objects related to the burial practices, the plant raw material was the same (*Phragmites*) but underwent different treatments (fibre extraction versus simple use of the culm) and different manufacturing techniques (coiling versus matting). Furthermore, on-site analysis documented the presence of phytoliths from *Stipa* sp. (esparto grass) and chaff from cereals (wheat/barley). The cereals seem to suggest some kind of grain deposition associated with the burial (Lancelotti et al. 2016: 132).

Scratching the surface: insights from the ground stones

Numerous studies have demonstrated the survival of plant micro-remains on ground stones and their value to inferring plant processing activities in the past (Piperno et al. 2004; Aranguren et al. 2007; Nadel et al. 2012). However, previous studies on ground stones at Çatalhöyük have raised numerous questions related to the functions of such tools that can be answered with the implementation of a multi-proxy approach as suggested by Hardy et al. (2013). The aim of exploring the multiple functionalities was developed further by the current microremains team while sampling several tools in an integrated effort with the ground stone team members (Madella, García-Granero 2014). This approach consisted of recovering the sediment trapped and accumulated along the several surfaces of use of the artefact, within the stones' porous surfaces, cracks and crevices. This process was carried out in a cleaned and isolated subset of the archaeobotany laboratory, to avoid any possible contamination from modern starch. The extraction process consisted of: 1) identification of the grinding stone's used surfaces, 2) dry brushing of the chosen area, followed by 3) wet brushing to recover all sediments, from which the starch and phytolith extraction was carried out following the methods described by García-Granero et al. (2015). A total of 18 stone tools were sampled from Buildings 80, 52 and 44, producing 70 samples.

Five of these grindstones (hand stones and querns from Buildings 80 and 52) have already been analysed, while the others are currently under study as part of a doctoral project. These ground stones come from the building with the most thorough sampling, which means that the data produced by the stones will be contrasted with results from spatial analyses on floors and surfaces from where such tools were recovered and other contexts within the same building to clarify whether there is a

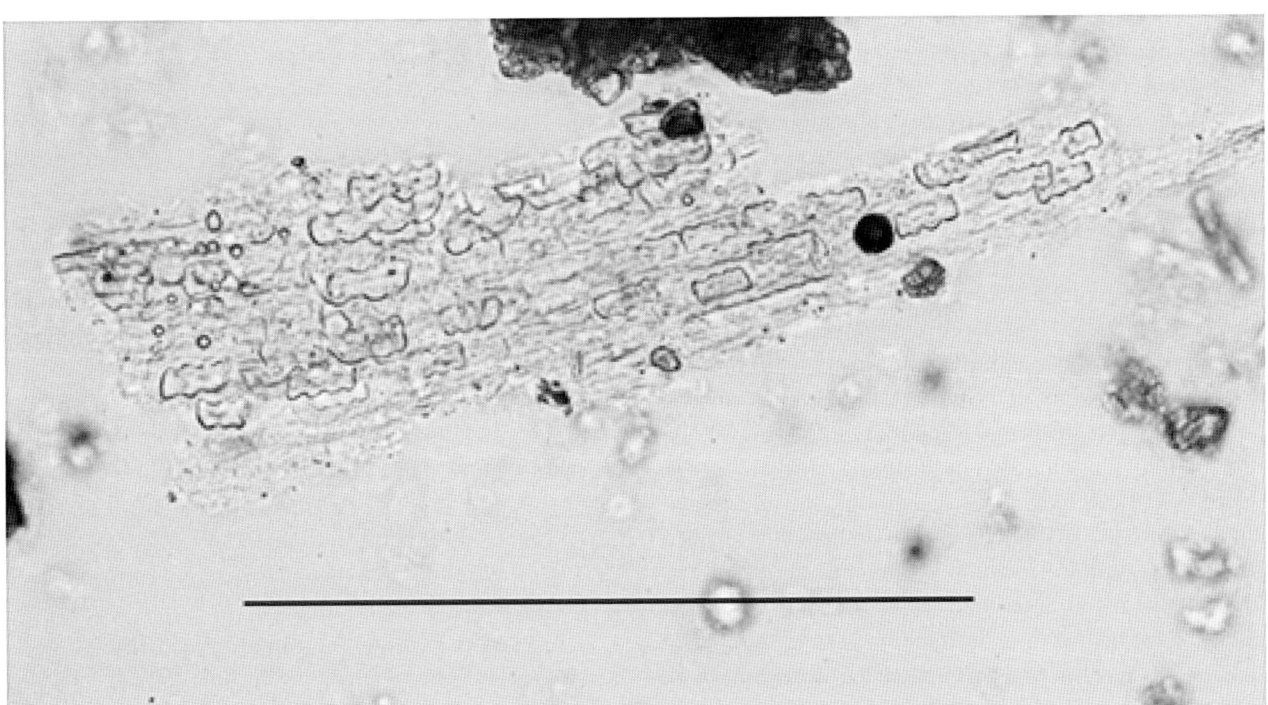

Figure 7.2. Silica skeletons with bilobates and polilobates from a Panicoid found in (21682) Building 77 (scale=200µm).

correlation between the material culture and the continuous use of a specific space for a particular plant-related activity. Preliminary results have highlighted the abundance of archaeological starches from the stone artefacts as well as the use of alternative starchy resources such as tubers/rhizomes (Santiago-Marrero et al. in preparation) (fig. 7.3). In addition, the analyses found a positive correlation between phytoliths and starch ratios that may imply an association between tools and an inferred function.

However, there are still some unresolved issues before the contribution of plant resources at Çatalhöyük can be properly understood, such as the creation of a starch reference collection and ethnographic work on food processing. The need for a reliable reference collection for starches has been a challenge since starch analyses were introduced at the site (Hardy et al. 2013). In fact, reference collections are a constant problem for starch analyses everywhere, partly because comparative starch research is time-consuming (Pagán-Jiménez 2015; Mercader et al. 2018). This problem is even greater for Anatolia and the Middle East where, contrary to other areas of the world (Lentfer 2009; Messner 2011; Pagán-Jiménez 2015; Mercader et al. 2018), there are no published or open access reference collections. We have currently created a list of economically important plants based on the abundant ethnographic data on wild edible and useful plants in the region (Dogan et al. 2004; Kreziou et al. 2016). The criteria used to create the list considered the plants' current range of distribution, ecological niche and traditional use for food or beverage preparation. Furthermore, past environmental settings were considered in order to add economically important plants that are not found nowadays but might have been present in Neolithic Çatalhöyük (plants from marshes). Finally, wild relatives of modern domesticates were also included. The list contains more than 300 species and the collection is currently under development.

These observations have also been made by Hardy et al. (2013). In fact, those previous works served as a precedent to corroborate some of the starch morphotypes that have been found in the current research and to have an idea about the nature of starch presence, survival and distribution. Current and forthcoming works will go deeper on these matters by integrating phytoliths and starches into the spatial analyses to add another level of resolution to the interpretation of microbotanical distribution patterns on household levels. In sum, these first studies have helped to understand the starch presence at the site, to have a first insight on the different starch morphologies and to highlight the benefits of integrating both analyses.

Learning from the past: reflections for an integrated micro (and macro) botanical approach

The plant microremains work at Çatalhöyük is an opportunity to reflect on and rethink the theoretical and methodological approach for properly understanding plant exploitation, plant-related activities and how plant materials arrive at and move within the site. The first assumption we make is that the majority of botanical microremains found at different contexts are equivalent to plant-processing leftovers (Lancelotti, Madella 2012). Phytoliths have been successfully used to investigate agricultural and processing models in the past, while identifying the different plant components at each stage (Harvey, Fuller 2005). These models are based on the notion of *chaîne opératoire*, which is sometimes conceptualised as a linear development (Harvey, Fuller 2005; Lancelotti, Madella 2012; Hastorf 2016). This means that such models are usually employed to synthesise and explain the steps plant material goes through to achieve a specific outcome (for example, crop processing and by-product obtention/uses). However, operational systems and materials processing are not straightforward and plants can often undergo treatments such as part separation even before arriving at the site (Fuller et al. 2014b; Hastorf 2016), with separated parts heading to different *chaînes opératoire*. This dynamic view of plant processing has been implemented in foodways studies to address the complex processes, entanglements and decisions in food production (Hastorf 2016; Peres 2017). We suggest that *chaînes opératoire* (and the models based on them) can be seen as entangled components of more complex models. To archaeologically address these past dynamics is without doubt one of the biggest challenges in the study of plant-processing activities in the past. In such an approach, we can give relevance to the network of activities and decision-making, as well as conceptualising plant by-products not as leftover (or discharged) elements but as main components of other *chaînes opératoire*, diverging onto different paths as part of the material culture as a whole in the form of food or crafts (fig. 7.4). Here we refer to this model as plantways, a term that simultaneously englobes all the aspects related to plants, including food and craft productions. This at the same time exemplifies the kind of subjects that our team intends to explore under the current research programme, following the different plant pathways through the buildings (what plants arrive at the houses and what happens afterwards).

This proposal brings multiple challenges that need to be tackled by a wider, more articulated use of plant microremains, as well as a shared interpretation for plant macro- and microremains. Previous studies

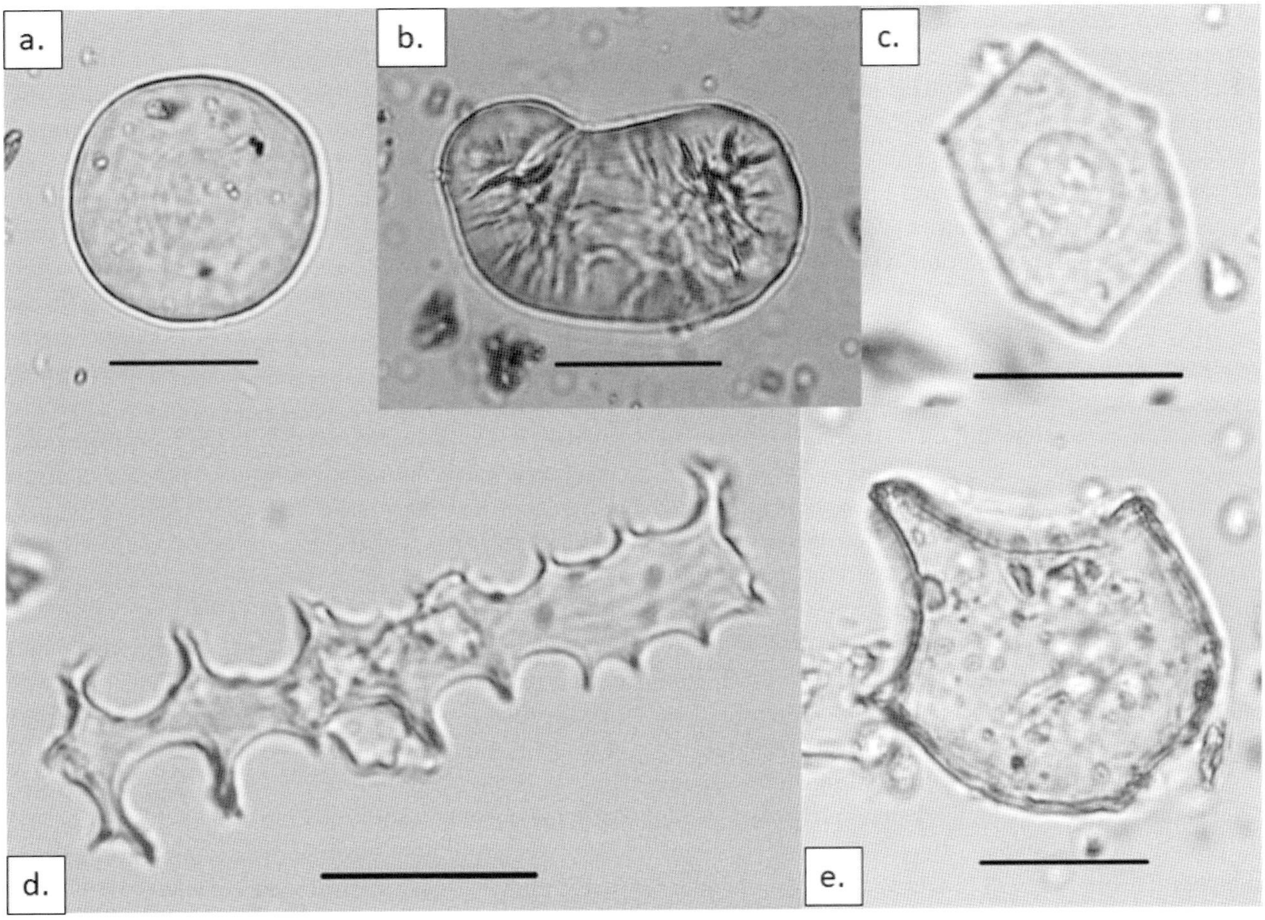

Figure 7.3. Starches and phytoliths recovered from the ground stones: (a) round-shape starch with visible lamellae, this type is commonly found on cereals; (b) 'kidney'-shape starch with visible lamellae, this shape is commonly found on legumes; (c) polygonal cone phytolith from Cyperaceae; (d) dendritic long cell; (e) bulliforms (scale=20µm).

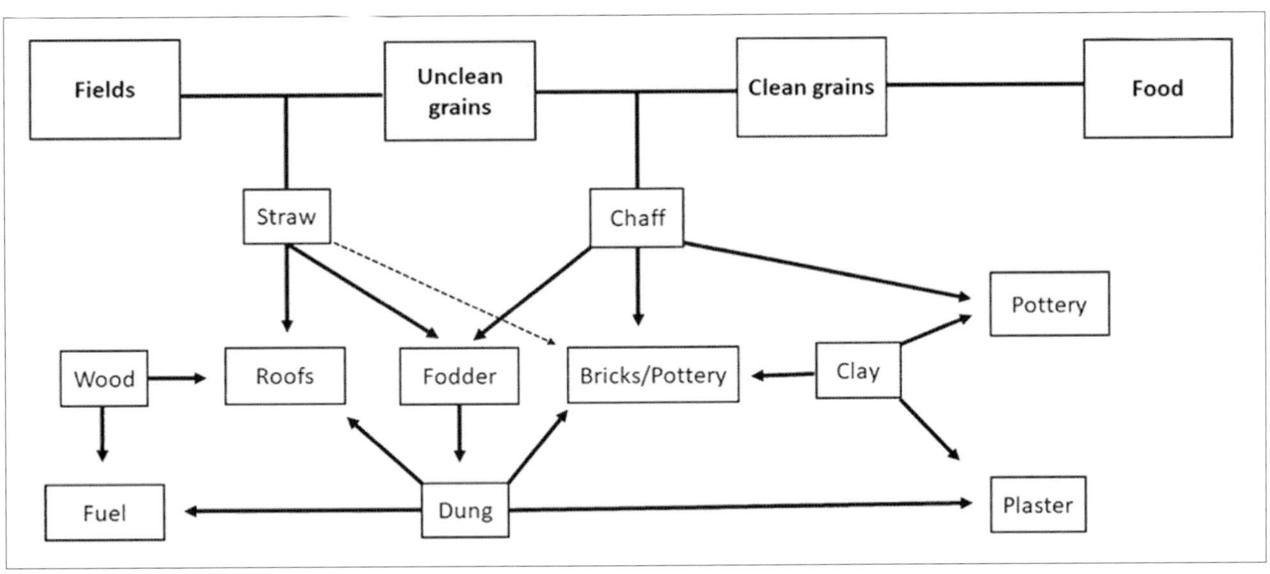

Figure 7.4. Example of a simplified chaîne opératoire *on crops processing and the entanglements emerging from the multiple plants' by-products. Each part of the plant is utilised for various means, becoming new plant-based products. The paths that these plants follow and the final outcome is not always recognisable in the macrobotanical record, since sometimes the plant itself is not visible anymore.*

emphasised the importance and benefits of using microbotanical remains to recognise crop by-products such as straws and husks at different processing stages (Harvey, Fuller 2005; Dal Corso et al. 2017; Out, Madella 2017). Understanding how these systems work at Çatalhöyük is crucial to refining methodological approaches to reveal some of the entanglements between plants and the multiple aspects of the inhabitants' daily life.

Remarks: reflections for the years to come

We would like to emphasise that this chapter should be taken as a reflection on how the microbotanical remains at Çatalhöyük can contribute to revealing the relationships between the inhabitants, their landscape and the plants that moved alongside them. For this reason, we have presented some key strategies implemented by previous research that opened the path for the integration of microbotanical analyses into the wider research programme at the site. These first implementations have also shown how microbotany can make valuable contributions to other specialists as well. Following this, one of the goals for our current team is to contribute to the work previously done while unifying phytoliths and starch into the current analyses. This in turn will produce more accurate interpretations of the use of plants at the site, adding another level of dynamism to our understanding of the past inhabitants.

Certainly, Çatalhöyük has become an exemplary site for archaeologists, not only for its remarkable preservation and richness, but for the influential figures that have emerged from it as well. In this sense, this chapter falls short of addressing the various methodological and theoretical contributions made by the multiple specialists through the 25 years of the project. Here we have highlighted some of them and how the valuable lessons left by previous specialists can become a platform for future researchers. This has been evidenced in the preliminary results here presented on past and ongoing work. Indeed, we believe that Çatalhöyük can serve as a model for the contributions that microbotanical studies can make to enhance our knowledge of past use of plants within archaeobotany and other fields within the archaeological sciences as well.

Acknowledgements

We would like to thank Ian Hodder for his continuous support and all the Çatalhöyük field team for helping with the many needs of the micro-remains sampling. Especially, we are immensely grateful to Bilge Küçükdoğan for her help with the management and above all with the samples export. Also, we want to thank Juan José García-Granero, Abel Ruiz-Giralt and Judit Barastegui for overseeing the fieldwork tasks and collecting the samples currently being analysed. Finally, Javier Lara-Recuero would like to thank A. Delgado and S. Biagetti for their support during his Masters course and MA thesis defence.

8. Macromammals of Çatalhöyük: new practices and durable traditions

Katheryn C. Twiss, Jesse Wolfhagen, G. Arzu Demirergi and Jacqui A. Mulville

Introduction

This chapter reports on 83,424 macromammalian specimens recorded during the Çatalhöyük Research Project (ÇHRP)'s final tranche of excavation and analysis in 2009–2017. These specimens, which all come from the North and South Areas of the settlement, span a significant chronological range. A minority can be associated with specific houses; many derive from middens and fills that were probably produced mostly by people living nearby, allowing us to examine animal use across neighbourhoods as well as through time.

The data presented in this chapter expand our understanding of animal production, management and use at Çatalhöyük, building upon the research presented in Russell and Martin (2005), Russell (2012) and Russell et al. (2013) (as of this writing, faunal analyses of the West Mound and TP/TPC projects have yet to be published). Of particular importance are this publication's contributions with respect to (a) exploitation of animals other than sheep and goats and (b) the economy of the site during the Middle period occupation.

The efforts made in this tranche of research to record specimens other than caprines (so well documented in previous publications) are outlined in the Methods section of this chapter. As for our chronological contributions, figures 8.1 and 8.2 (and online table S8.1) summarise the data that underlie each of the three major faunal chapters from the ÇHRP. Please see online appendix S8.2 for raw measurement data and online appendix S8.1 for a wide variety of additional information about the 2009–2017 macromammalian assemblage. All of the data reported on in this publication will also be made available in OpenContext, joining those covered in Russell et al. (2013) and Russell and Martin (2005).

The ÇHRP introduced Hodder Levels in 2008 in order to improve analyses of the village's chronological development (Farid 2008; 2013b). Russell and Martin (2005) thus present their data in terms of Mellaart levels; Russell et al. (2013) and this publication both use Hodder levels. The two sets of levels cannot be directly correlated. However, both sets can be grouped into broad chronological phases. Figure 8.1 illustrates the extent to which each of the three major Çatalhöyük faunal publications speaks primarily to different eras of the site's occupation. Russell and Martin (2005) provides extensive information about the Early period occupation, incorporating far more data from the mound's lower levels than either of the subsequent two publications. Russell et al. (2013), in contrast, is rich in information about Çatalhöyük's later developments. Heading into this research tranche, therefore, the faunal team knew considerable amounts about the earlier and later occupations of Çatalhöyük, but data from the centuries in between were sparser. This chapter fills in that gap, relying on a preponderance of specimens dating to the Middle period.

Changing methodologies

Faunal recording strategies at Çatalhöyük changed in 2012 and again in 2014. Curtailing some elements of our analyses and adding new ones enabled us (a) to analyse a higher proportion of the site's fauna; (b) to quantify assemblages in new ways; and (c) to provide basic quantitative characterisations of units that would not be fully studied. Data recorded using our new methodologies are comparable to data gathered prior to 2012 in most but not all ways. Key differences between the faunal methodologies used to produce Russell and Martin (2005) and Russell et al. (2013) and those used to produce this chapter are summarised in online appendix S8.1. Of particular relevance to this chapter are a new method of recording portions of elements present ('Cardiff zones') and a system aimed at maximising data on taxa other than caprines ('2014 Recording,' in which only rare taxa and unusually informative cattle/caprine specimens are individually recorded). We also developed Tier 1 (T1) recording, which can, unlike 2014 Recording, be used to analyse taxonomic proportions at Çatalhöyük. Tier 1 recording resembles the Long (L) form recording used by Russell and Martin (2005) and Russell et al. (2013) but includes Cardiff Zones and excludes individual specimens' weights.

We stress that *analysts must compare and contrast taxonomic proportions from comparably analysed units only*. Future research using the Çatalhöyük Project faunal database *must* begin with a check of the recording status of each unit as noted in its Faunal Unit Description entry.

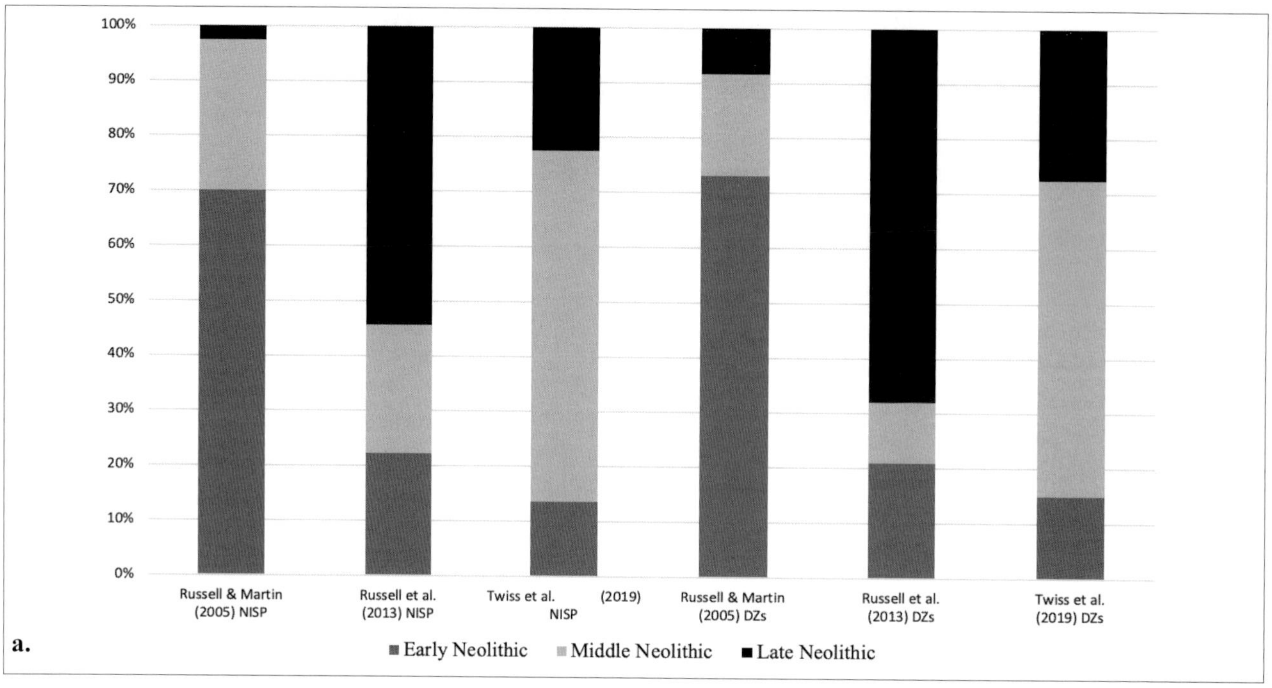

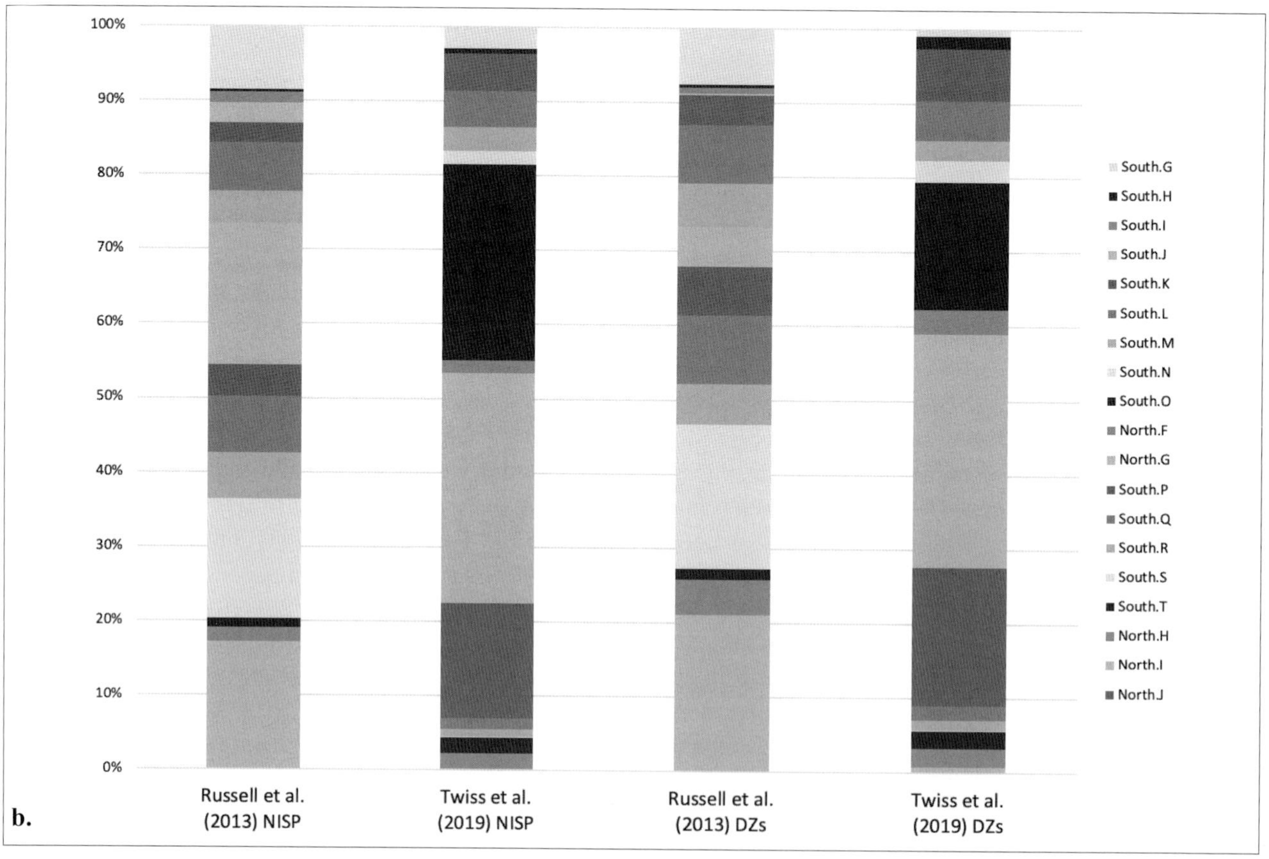

Figure 8.1. Chronological distribution of specimens published in previous ÇHRP macrofaunal reports: (a) Summary of data described in Russell, Martin (2005), Russell et al. (2013) and this paper, all three publications by occupation period; (b) specimens published in this chapter and in Russell et al. (2013), by Hodder Level. The figure includes only specimens assigned to (i) family, genus and/or species, and (ii) a specific level. Here and throughout this chapter, we include specimens cautiously attributed to specific levels (that is, deriving from spaces with no direct stratigraphic link to the site chronological 'backbone') with the securely attributed specimens. Previous publications did likewise.

Taxonomic proportions

Table 8.1 and figure 8.2 summarise taxonomic proportions at Neolithic Çatalhöyük by level. Table 8.2 reports all of the specimens recorded during this tranche of research. It cannot be used to assess taxonomic proportions at the site, as it includes selectively recorded units.

We turn now to discussing the individual taxa represented in Çatalhöyük's macromammalian assemblage. For each taxon we summarise salient information about skeletal element representation, mortality patterns (that is, age and sex-specific culling) and inferred role in Çatalhöyük society. Please see online General and Measurements Appendices for considerable additional information. Our goal is to characterise how the inhabitants of Çatalhöyük produced and consumed different kinds of animals.

Cattle (*Bos* sp. cf. *primigenius*)

We recorded 6,119 Bos fragments (Diagnostic Zones [DZ]=961.5) during this tranche of research; 18,518 indeterminate fragments were assigned to cattle-sized animals (table 8.3 shows NISP and DZ totals for fully-recorded contexts).

Understanding the domestication status of cattle and the potential timing of the introduction of domesticated cattle to Çatalhöyük East has been one of the research foci for the faunal team throughout the recent project (Russell, Martin 2005; Russell et al. 2005; Twiss, Russell 2010; Russell et al. 2013). Russell and Martin (2005) demonstrated that the *Bos* in the early occupation of the site were morphologically wild; Russell, et al. (2013) determined that morphologically domestic cattle were likely present in the later occupation levels of the South Area (South P-T) and in the North Area. We revisit this question with the full set of Çatalhöyük East *Bos* remains from the North and South Areas: please see Chapter 9 for expanded discussion of the biometric analyses.

Biometry

We follow previous Çatalhöyük publications in using the log size index (LSI) technique to interpret *Bos* measurements from across the body, but we use the natural log ($\log_e$) rather than $\log_{10}$ to rescale our measurements because differences in $\log_e$ are more directly interpretable (that is, they more closely match percentage differences in the original scale) and transforming measurements with $\log_e$ transforms the data from log-normally distributed to normally distributed (*contra* Meadow 1999), which aids modelling. Wolfhagen et al. (this volume, Chapter 9) detail the modelling of LSI values using a Bayesian multilevel mixture model of 477 postcranial specimens.

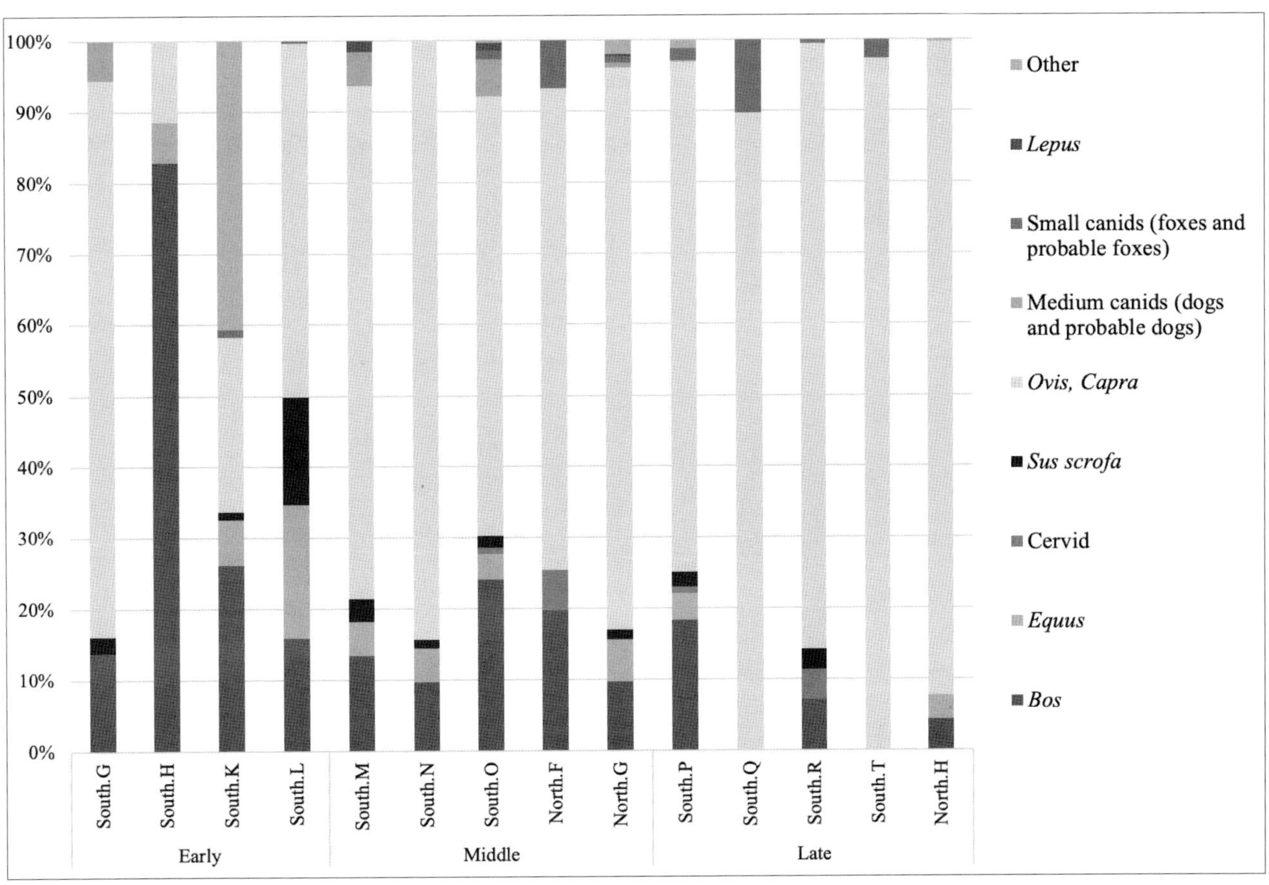

Figure 8.2. Faunal taxonomic proportions, major taxa only, Çatalhöyük 2009–2017. By Diagnostic Zones (DZs); no size-classed specimens. L and T1 unit data only.

		North.F	North.G	North.H	South.G	South.H	South.K	South.L	South.M	South.N	South.O	South.P	South.Q	South.R	South.T	Total
Size classes	Aurochs-sized															
	Cattle-sized		1.5							2	1	0.5				5
	Boar-dog-sized?											1				1
	Boar-sized										1					1
	Sheep-sized		1	2						4	0	2		0		9
	Hare-sized															0
	Bos primigenius		5						1.5	0	7.5					14
	Bos sp.	3.5	29.5	2.5	3	14.5	24.5	11	7	4	77	61.5		2.5		240.5
	Bos taurus											1.5				1.5
	E. ferus									1						1
	E. hemionus/ hydruntinus															
	Equus sp.		21	2		1		12			4	11				51
	Large equid						4			1	4					9
	Small equid						2	1	3		5	2				13
	Cervus elaphus	1	0.5								1	0.5				3
	Large cervid										2	0.5				2.5
	Dama dama											2				2
	Small cervid													1.5		1.5
	Sus scrofa		5		0.5		1	10.5	2	0.5	5.7	7		1		33.2
	Capra		17		1		1.5	3	2		7.5	2		1	4.5	39.5
	Ovis	3.5	104.5	10	3.5	1	11	15	14.5	9.5	65.5	75.5		17	8	338.5
	Ovis/Capra	8.5	161.5	43.5	12.5	1	10.5	16.5	29	25.5	144	169	3.5	12	24	661
	Medium artiodactyl															0
	Small artiodactyl										0					0
	Ursus											2.2				2.2
	Canis familiaris								1		4	1				6
	Canis sp.		2.2		1.2						13.4	0				16.8
	Medium canid		0.4						2		1.2					3.6
	Vulpes	1.2	3.2				1				4.4	5.6	2	0.2	1	18.6
	Small canid							0.2			1.2	0				1.4
	Medium felid		1.2				35.8									37
	Medium carnivore															0
	Meles meles		3.2				1				1	2				7.2
	Small mustelid		3	0.2												3.2
	Talpa europaea															
	Small carnivore						1.2				0.2	0.2				1.6
	Lepus		1						1		4					6
	Erinaceus europaeus										1					1
	Homo										1					1
	Total	17.7	360.7	60.2	21.7	17.5	92.5	69.2	63	47.5	357.6	348	4.5	35.2	37.5	1,532.8

Table 8.1. Taxonomic representation by Level, 2009–2017. DZs, T1 and L form data only. Specimens originally assigned to Ovis/Capra/Capreolus *(n=5) are here included in the* Ovis/Capra *totals.*

Early								*Middle*									*Middle/Late*	*Late*							*Unphased*		*Unknown*							**Taxon**
South.G	*South.H*	*South.H, South.I*	*South.I*	*South.I, South.J*	*South.K*	*South.K, South.L*	*South.L*	*North.F*	*North.F, North.G*	*North.G*	*South.M*	*South.M, South.N*	*South.M,South.N*	*South.N*	*South.N,South.O*	*South.O*	*South.O,South.P*	*North.H*	*North.I*	*South.P*	*South.Q*	*South.R*	*South.S*	*South.T*	*South.L, South.M*	*North.G, North.H*	*South. Unknown*	*North. Unknown*	*Total NISP*	*Early total*	*Middle total*	*Middle/ Late total*	*Late total*	
										1	1					3				3							0		**8**		*5*		*3*	**Microfauna (smaller than rabbit)**
10		14					4	5		25	1	1		5		11		1		54		1	1				2		**135**	*28*	*48*		*57*	**Hare-size (rabbit to medium dog)**
5602	53	2065			762	5	172	593	157	14555	470	132	71	660	96	6337	465	1877	3	9127	11	780		1002	1		1783	934	**49253**	*13511*	*23073*	*1580*	*12800*	**Sheep-size (medium dog to medium sheep)**
																				1							0		**1**				*1*	**Medium dog to wild boar**
	1	11			4		115	28	5	108	4			8		13				7	1	1					1	5	**312**	*142*	*166*	*2*	*9*	**Pig-size**
										1				1		58						1					1		**62**		*60*		*1*	**Medium sheep to medium cattle**
25	327	740		1	390		344	141	39	2338	594	29	2	85	6	4012	500	227	1	2939	5	58		91		1	150	419	**13464**	*2696*	*7248*	*691*	*3321*	**Cow-size (cattle/red deer/horse)**
							1																				0		**1**	*1*				**Aurochs/ bison-size (larger than medium cattle)**
																1											0		**1**		*1*			**Small artiodactyl**
																1				1							0		**2**		*1*		*1*	**Medium artiodactyl**
							2			3						1								1			1		**8**	*3*	*4*		*1*	**Large bovid**
1							3			1		1				5			1	12							0		**24**	*4*	*7*		*13*	***Ovis/Capra/Capreolus***
168	14	149		1	112	115	165	64	21	1967	218	57	10	110	8	1332	250	222	13	930	67	86	4	207	1		298	81	**6670**	*970*	*3787*	*322*	*1529*	***Ovis/Capra***
14	5	17		1	22	1	42	26	2	375	24	9	3	40	6	168	9	20	2	138	15	28		41	1		70	17	**1096**	*119*	*653*	*26*	*244*	***Ovis***
2	2	2			4		30	2	2	56	4			1		52	1			9		1		8			9	1	**186**	*41*	*117*	*3*	*18*	***Capra***

Table 8.2. Taxonomic representation by Level, 2012–2017. By NISP, all data.

Early								Middle									Middle/Late	Late							Unphased		Unknown		Total NISP	Early total	Middle total	Middle/ Late total	Late total	Taxon
South.G	South.H	South.H, South.I	South.I	South.I, South.J	South.K	South.K, South.L	South.L	North.F	North.F, North.G	North.G	South.M	South.M, South.N	South.M,South.N	South.N	South.N,South.O	South.O	South.O,South.P	North.H	North.I	South.P	South.Q	South.R	South.S	South.T	South.L, South.M	North.G, North.H	South. Unknown	North. Unknown						
133	47	35	1	24	196		132	53	10	1167	133	4		35	4	164	100	18		684	40	17	7	13	2		39	29	**4527**	*655*	*3011*	*113*	*779*	**Bos sp.**
					1		1	1		9	2	1				2	1			6	2			1			1		**28**	*3*	*15*	*1*	*9*	***Bos taurus***
							1	1		13	4			9		55	1			3						2	2		**91**	*1*	*82*	*1*	*3*	***Bos primigenius***
					6			8		13						2								1			1		**31**	*6*	*23*		*1*	**Small cervid**
	6	1			11		6	6		131	6					57	2	3		21	3						0		**253**	*26*	*200*	*2*	*27*	**Large cervid**
																						2					2		**4**				*2*	***Capreolus capreolus***
2	2				3		3	3	4	43	1			3	2	23				5	3						0		**97**	*12*	*79*		*8*	***Cervus elaphus***
								1												2							0		**3**	*1*	*1*		*2*	***Dama dama***
32	1	9	1		20		125	13		132	15	1		7		51	5	2	2	65	1	1	2	2			5	7	**509**	*195*	*219*	*7*	*85*	***Sus scrofa***
8					13		47	13		44	21	3		12	1	31	5	1	12	39	14			1			1	1	**267**	*68*	*125*	*6*	*67*	**Small-medium equid**
					16		7			2	1			6		8	2			10	2			1			1	1	**57**	*24*	*17*	*5*	*13*	**Large equid**
2							1									2	1										0		**6**	*3*	*2*	*1*		***Equus hemionus/ hydruntinus***
							1	3						2		3		5		8							0		**22**	*1*	*8*		*13*	***Equus ferus***
21	10	27		1	2		66	10	2	188	1			11	2	30	17	6		49	3		2	1	2	1	6	7	**487**	*166*	*244*	*19*	*61*	***Equus sp.***
	1	2			4			1		17	1					5				4	1						0		**36**	*10*	*24*		*5*	**Small carnivore**
		1			2			2		8						2								1			1		**17**	*4*	*12*		*1*	**Medium carnivore**
					247			1		14																	0		**262**	*247*	*15*			**Medium felid**

Table 8.2 (continued). Taxonomic representation by Level, 2012–2017. By NISP, all data.

Early								Middle									Middle/Late	Late							Unphased		Unknown		Total NISP	Early total	Middle total	Middle/ Late total	Late total	Taxon
South.G	South.H	South.H, South.I	South.I	South.I, South.J	South.K	South.K, South.L	South.L	North.F	North.F, North.G	North.G	South.M	South.M, South.N	South.M,South.N	South.N	South.N,South.O	South.O	South.O,South.P	North.H	North.I	South.P	South.Q	South.R	South.S	South.T	South.L, South.M	North.G, North.H	South. Unknown	North. Unknown						
		1								4								1									0		**6**	*2*	*4*		*1*	**Small mustelid**
										1																	0		**1**	*1*	*1*			**Large mustelid**
							1														1						0		**2**	*1*			*1*	***Martes***
										1											1						0		**2**		*1*		*1*	***Mustela***
					3		1			11				4	1	2				3			1				1	1	**28**	*4*	*18*		*4*	***Meles meles***
																						1					1		**2**				*1*	***Martes martes***
										1																	0		**1**		*1*			***Martes foina***
		1					1	11			1				1	11				12	3						0		**41**	*3*	*24*	*4*	*15*	**Small canid**
					1			1		9	8				2	11		3		3		1		1			2		**43**	*2*	*31*		*8*	**Medium canid**
11					10		3	1		29	1			5		84		1		8	1		4				4		**162**	*25*	*120*		*14*	***Canis* sp.**
							24			6	3					18				64	2			4			4		**125**	*25*	*27*		*70*	***Canis familiaris***
					1					8				2	2	3		2		17	1			3			3		**42**	*1*	*15*		*23*	***Vulpes* sp.**
6		8			8		6	2		15	2			1	1	20	6		1	29	17	3	3				6	3	**137**	*34*	*41*	*6*	*53*	***Vulpes vulpes***
										1						1				4							0		**6**		*2*	*1*	*4*	***Ursus***
								1			1					2				1							0		**5**	**2**	*4*		*1*	***Erinaceus europaeus***
																1											0		**1**		*1*			***Talpa europaea***
										3	2					2				1							0		**8**		*7*		*1*	**Rodent**
										1										2							0		**3**		*1*		*2*	**Lagomorpha**
1					1		2	1		2	1	4			2	9			1	7	2		1	1			2	1	**38**	*5*	*19*		*12*	***Lepus***
	1							1		11					29	42	18		1	8	3	1					1		**116**	*2*	*83*	*18*	*13*	***Homo***

Table 8.2 (continued). Taxonomic representation by Level, 2012–2017. By NISP, all data.

	Early				*Middle*				*Late*			
	Bos		*Cattle-sized*		Bos		*Cattle-sized*		Bos		*Cattle-sized*	
Element	*NISP*	*DZ*	*NISP*	*DZ*	*NISP*	*DZ*	*NISP*	*DZ*	*NISP*	*DZ*	*NISP*	*DZ*
Cranial	9	0	1	0	152	0	672	1	91	0	641	0
Maxilla	2	0	0	0	7	1	3	0	19	3	55	0
Mandible	15	1	0	0	51	14	17	0	87	11	60	0
Horn Core	4	0	38	0	2,380	0	116	0	252	0	37	0
Atlas	9	9	0	0	14	11	0	0	9	7	1	0
Axis	3	2	0	0	10	3	30	0	2	1	1	0
Other Vertebrae	3	0	83	0	45	0	440	0	78	0	255	0
Sacrum	1	1	0	0	7	3	4	1	4	2	1	0
Ribs	0	0	128	0	7	0	431	0	18	0	870	0
Scapula	16	7	7	0	74	31	331	1	44	20	299	2
Humerus	17	8	0	0	30	10	5	0	41	12	6	0
Radius	13	4	0	0	38	12	2	1	32	5	3	0
Ulna	16	8	2	1	32	13	2	1	39	13	2	0
Carpals	22	21	0	0	28	19	0	0	32	25	2	0
Metacarpal	15	10	1	0	15	7	0	0	15	8	0	0
Pelvis	7	2	3	0	46	16	42	0	59	14	147	2
Femur	11	4	3	0	34	10	1	0	31	4	2	1
Tibia	14	8	0	0	21	7	2	0	22	8	7	1
Os malleolare	3	3	0	0	4	4	2	0	3	2	0	0
Astragalus	2	2	0	0	13	13	0	0	7	7	0	0
Calcaneus	7	5	0	0	15	7	2	0	6	8	0	0
Other Tarsals	7	4	0	0	6	0	0	0	14	2	0	0
Metatarsal	13	7	2	0	9	7	0	0	49	17	0	0
Phalanx I	17	1.5	0	0	47	15	3	0.5	49	12	6	1
Phalanx II	16	4	0	0	48	19	1	0	50	14.5	0	0
Phalanx III	4	2	0	0	20	12	0	0	26	10	1	0
Total	246	113.5	268	1.0	3,153	234.0	2,104	5.5	1,040	195.5	2,396	7.0

Table 8.3. Bos *and cattle-sized skeletal elements distribution 2009–2017. Includes only T1 and L-form data.*

Modelling of eight other Anatolian assemblages inferred to contain either aurochs or domesticated cattle remains informed probability calculations of Çatalhöyük subassemblages' fits with cattle-herding assemblages.

Bos measurements from Early period Çatalhöyük are consistent with Central Anatolian aurochs (fig. 8.3; Wolfhagen et al. this volume, Chapter 9). While the average size of cattle decreases throughout the site's occupation, all Middle and most Late period North and South Area *Bos* assemblages are likewise consistent with aurochs rather than domesticated cattle. Modelling suggests that one or two of the Late period subassemblages probably include morphologically domestic *Bos*, but they are far from consistent presences in Late Çatalhöyük levels: most subassemblages are consistent with wild aurochsen. Final period and West Mound deposits conform to morphologically domestic cattle (Russell et al. 2014a; Pawłowska in press); we infer that cattle herding did not become commonplace until these periods.

Body part representation

We recorded 543 new *Bos* Diagnostic Zones (table 8.3) from T1 and L-form units. Figure 8.4 reveals relatively even body part representation throughout the occupation, though scapulae are over-represented (as also indicated by Cardiff Zones data and in previous analyses; Russell et al. 2013). Low-food-value elements such as cranial remains and lower limb bones are generally more common than meat-rich upper limb elements (attrition can influence this).

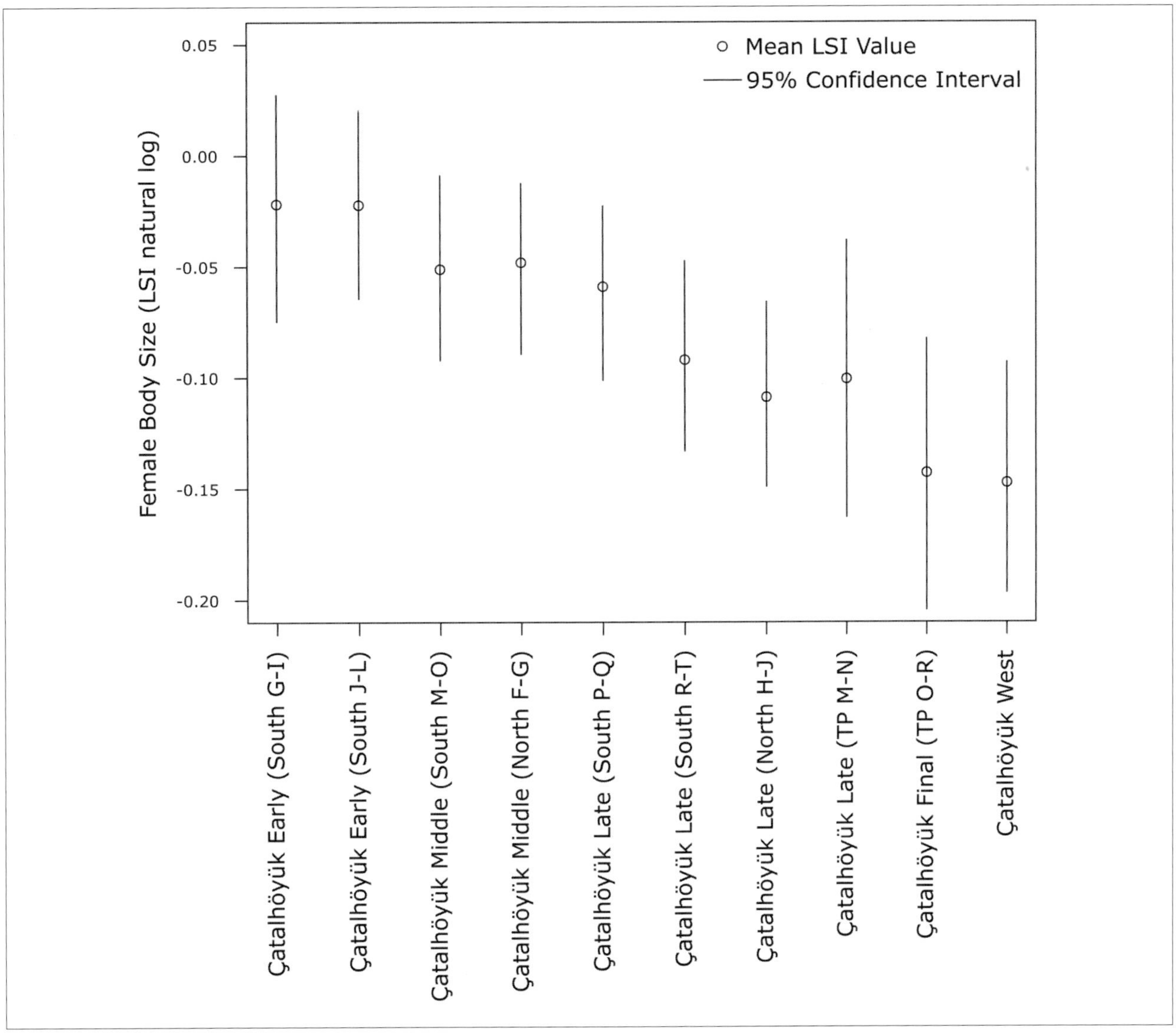

Figure 8.3. Estimated size of female Bos *at Çatalhöyük through time. Data from this project, Marciniak (2013) and Orton et al. (2013).*

Middle and Late period %DZ values are relatively even except for scapulae and astragali, which may have been curated for non-dietary use (Russell et al. 2013; 2014b).

Culling patterns

We recorded 20 new sexable *Bos* pelves, bringing the total of sexable pelves from Early–Late period Çatalhöyük to 39. Late period *Bos* pelves are significantly female-dominated (table 8.4); small sample sizes preclude discerning sex ratios in the earlier periods.

We recorded 20 new *Bos* mandibles with cheek tooth eruption and wear data; we now have 57 total for the North and South Area Neolithic occupations (online table S8.5). Of these, 22 mandibles could be assigned to a single Halstead (1985) stage; the remaining mandibles were probabilistically assigned to stages based on the estimated age ranges for these stages (after Jones, Sadler 2012: table 7) (fig. 8.5). The single-stage mandibles provide largely similar age trends to the multiple-stage mandibles: adult and elderly cattle (Stages G–I) predominate, though neonatal and infantile animals (Stages A–B) are present. We infer that animals were taken at a wide range of ages; we see no specifically focused hunting or management strategy.

We recorded 493 *Bos* specimens with epiphyseal fusion data. Adding these to previously reported specimens (total N=1,261) enables diachronic and spatial comparison of *Bos* fusion stages (fig. 8.6). North and South Area culling strategies do not appear to differ. Subadult and adult fusion rates are lower in later Çatalhöyük levels than in earlier ones, suggesting a decline in the proportion of animals that lived into adulthood.

Pathologies. We recorded two noteworthy pathological *Bos* specimens in this tranche of research (online table S8.19). Multiple pieces of obsidian embedded in a healed scapular fragment (17383.F143: fig. 8.7) testify to an aurochs's escape from hunters – at least its first escape, as the animal's remains ended up inside the settlement, in a Level South O fill deposit. A second scapula from the North Area (20988.x11) bears healed glenoid damage

Area	*Phase*	*2009–2017*		*Whole Project*			*Probabilities for θ*		
		Female	*Total*	*Female*	*Total*	*Posterior Mean θ*	*θ < 67%*	*33% < θ < 67%*	*θ < 33%*
South	Early	1	3	3	8	42% (14–72%)	6%	63%	31%
	Middle	1	3	1	5	31% (2–63%)	3%	39%	58%
	Late	6	8	9	12	72% (50–93%)	72%	28%	0%
North	Middle	3	4	7	10	68% (41–91%)	56%	44%	1%
	Late	2	2	4	4	85% (58–100%)	89%	10%	0%

Table 8.4. Proportions of male and female Bos *based on pelvic morphology. Bars represent 95% highest posterior density intervals. Proportions and intervals derived as per Wolfhagen et al. (this volume, Chapter 9).*

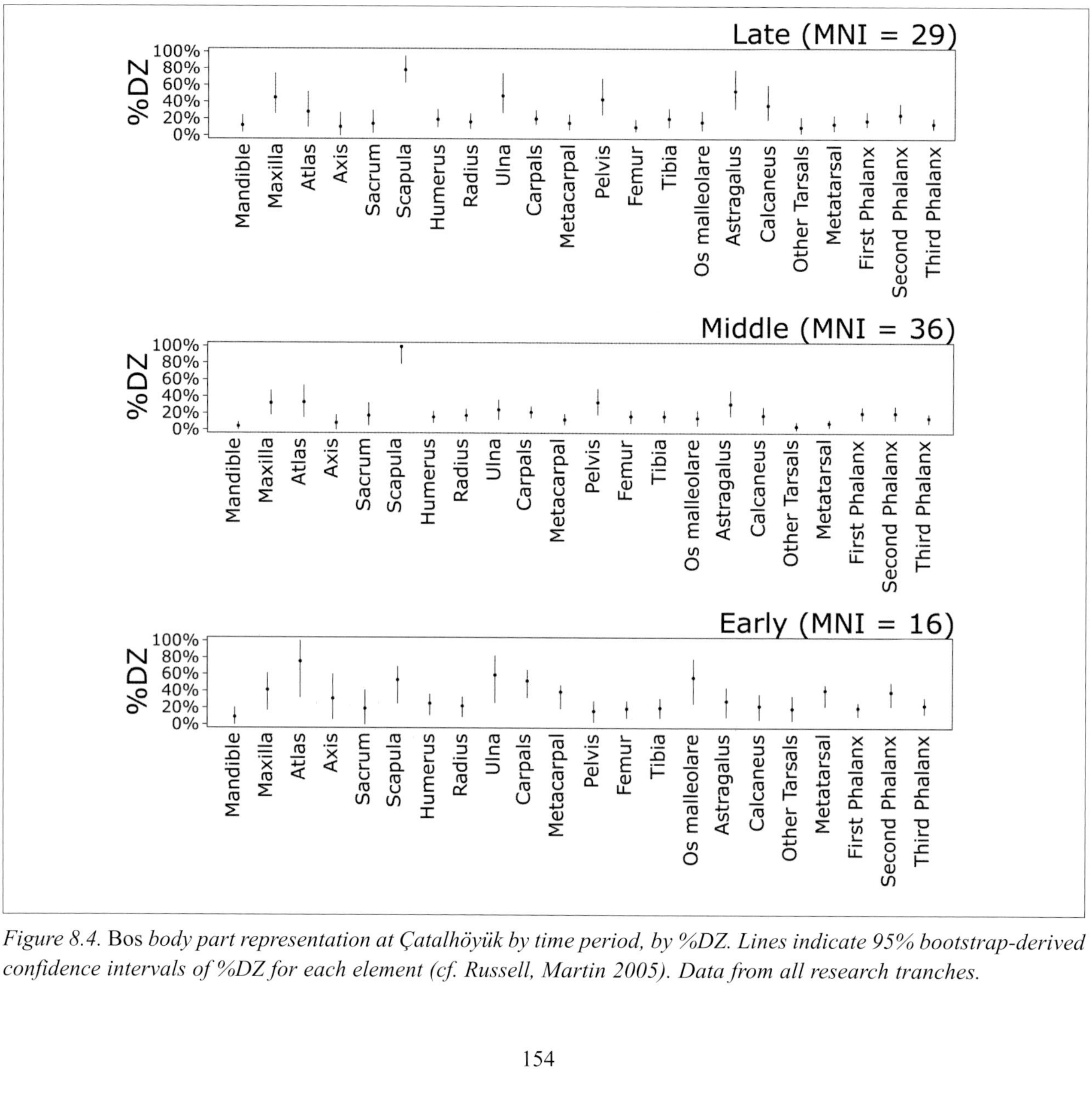

Figure 8.4. Bos *body part representation at Çatalhöyük by time period, by %DZ. Lines indicate 95% bootstrap-derived confidence intervals of %DZ for each element (cf. Russell, Martin 2005). Data from all research tranches.*

(resembling osteochondrosis dissecans). In later periods such damage is commonly interpreted as traction-induced, but in the absence of other evidence suggesting that Çatalhöyük's villagers were using cattle for labour, we do not assume a similar origin here. The bone is likely an adult male aurochs based on its size. Most of the other bovine pathologies recorded in this tranche of research are related to age (for example, osteoarthritis) or minor trauma.

Discussion

Gradual reduction of *Bos* body size throughout the Middle and Late periods at Çatalhöyük is inconsistent with a relatively sudden introduction of morphologically domesticated cattle into villagers' economic repertoire (cf. Arbuckle, Makarewicz 2009; Arbuckle 2013a). Rather, gradual size depression might reflect overhunting (Marom, Bar-Oz 2009) and/or increasing human interference in the lives of local aurochs populations (Hecker 1982; Zeder 2012). We deem the latter possibility particularly plausible, as relatively even body part representation and non-selective culling profiles suggest that local *Bos* populations were not being extirpated to the point that hunters needed to travel further from the site to find aurochsen.

We conclude, therefore, that the combination of long-term size depression and apparent local abundance reflect increasing entanglement of human activities with local aurochs populations, wherein interventions used to maintain local aurochs populations created feedback loops that required greater interventions in later generations (this volume, Chapter 9; Fuller et al. 2016). These interventions may have affected which animals were encouraged to live near the community and were thus accessible for slaughter.

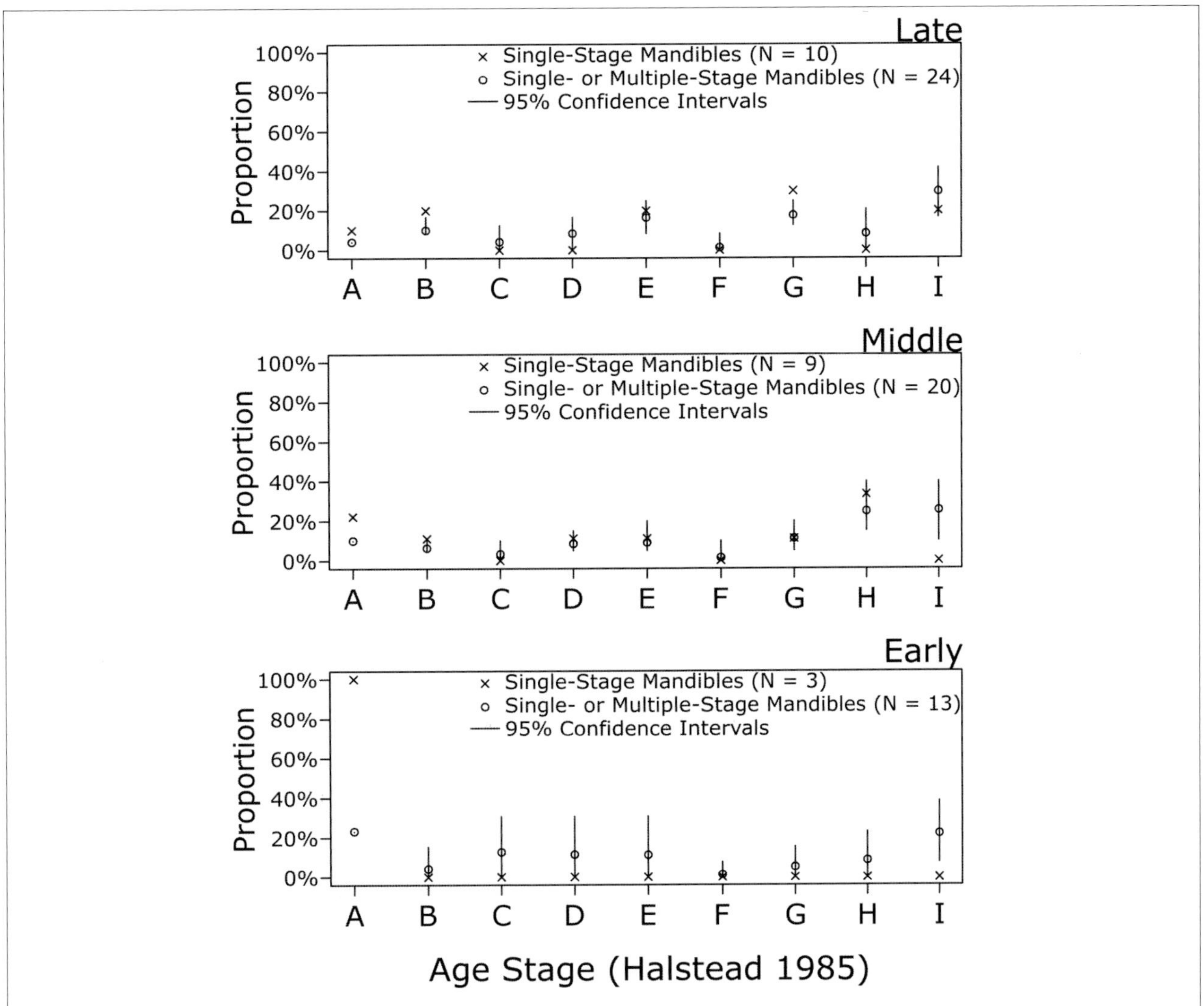

Figure 8.5. Proportions of Bos *Age Stages. After Halstead (1985). 95% confidence intervals defined by simulating mandibles into specific stages based on the probabilistic assignments.*

Evidence for potential *Bos* penning deposits in the Late period (Shillito, Ryan 2013; this volume, Chapter 14), along with evidence for increased reliance on young animals, dovetails with biometric evidence suggesting that some Late period deposits probably contain morphologically domestic *Bos* but that herding of biologically domesticated cattle did not take hold at the site until its Final period and West Mound occupations.

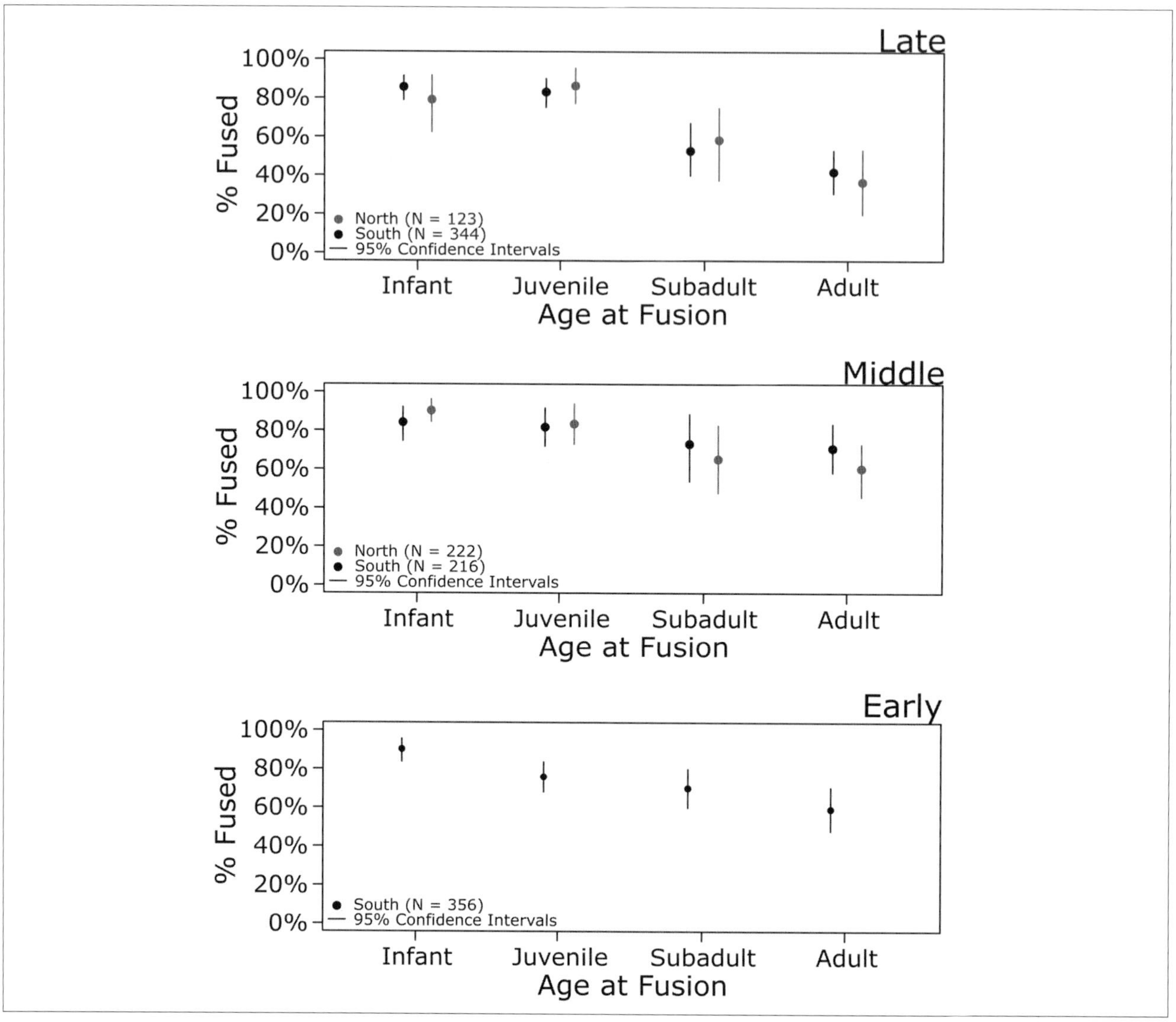

Figure 8.6. Proportions of fused elements with bootstrap-derived 95% confidence intervals in each of the age stages for different occupation phases.

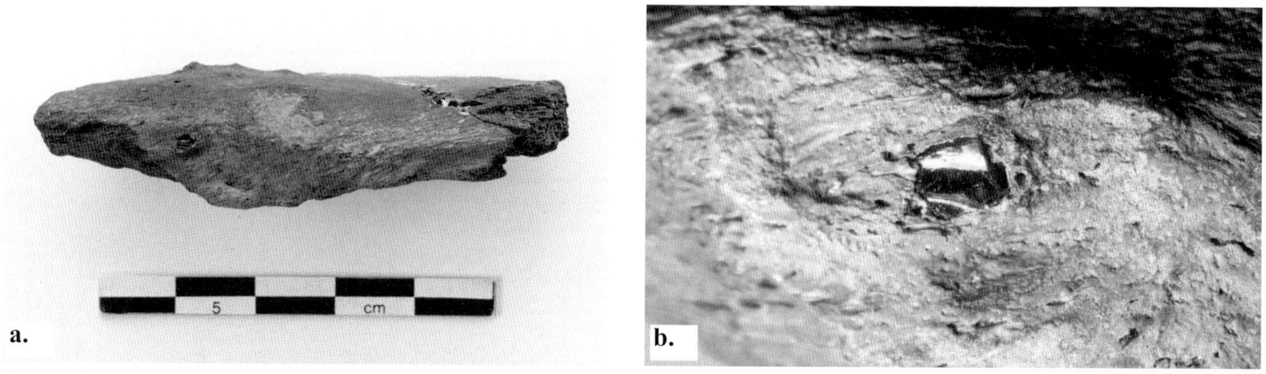

Figure 8.7. (a) Bos *scapular fragment with embedded obsidian (17383.F143); (b) close-up of obsidian fragment.*

Equids (*E. ferus*, *E. hemionus/hydruntinus*)

We recorded 816 new wild equid specimens (271 DZs) in this tranche of research. Of these, we originally assigned 11% (Number of Identified Specimens [NISP]=86; 33 DZs) to large equids (*Equus ferus*), 35% (283 NISP; 126.5 DZs) to small-medium equids (*E. hemionus/hydruntinus*) and the remaining 55% to Equus sp. We assign all loose teeth to *Equus* sp. (see Twiss et al. 2017) (table 8.5).

It remains clear that the inhabitants of Çatalhöyük hunted primarily small-bodied equids, surely the onagers/wild asses (*E. hemionus/hydruntinus*) that have been identified at numerous sites across Neolithic Southwest Asia. Their bones and teeth constitute 34% of our equid specimens, while only 9% of specimens derive from wild horses (*E. ferus*/large equids). When feasible, we discuss these two populations of equids separately, as their different body sizes and ethologies raise the possibility that they were hunted, butchered, transported and consumed in distinct ways.

As noted previously by Russell et al. (2013), equid proportions peak mid-sequence and then drop precipitously in the Late period. In our data the Middle period peak is driven by 21 equid DZs from assorted Level North G units plus 12 from a single midden (2739) in Level South L. No other Middle period levels have high equid proportions in fully recorded units. No equid DZs postdate South P. (If we use NISP, the mid-sequence peak is attributable to sharp elevations in equid representation in South L (equids=2.7% of speciated NISP) plus South M (equids=2.4% of speciated NISP, against a sitewide equid baseline of 0.6% speciated NISP). Since NISP is subject to fragmentation issues, we rely on DZs). We have only one equid specimen from Level South Q and none at all from Levels South R-T (total NISP in these three levels=2,224).

Body part representation

All body parts are present for both large and small equids (table 8.6; given the limited size of the large equid assemblage, we group all levels of the site for the MNEs based on DZs). We see neither selective transport nor significant differences in processing between larger and smaller animals. If we group all of the 2009–2017 equid remains together and look at element representation by occupation phase (online fig. S8.20), we can see that throughout the site sequence hunters regularly hauled both low- as well as high-value body portions (for example, complete cut-up carcasses) home with them.

However, high proportions of scapulae and pelves and low proportions of phalanges in the Middle period deposits suggest that hunters of that era often brought home only the valuable portions of their equid prey. While this inference is based on a relatively small sample, it conforms with Russell et al.'s (2013: 224) analyses of equid remains from largely Middle period samples in the erstwhile 4040 (now North) Area. As virtually all of our Middle period scapula and pelvis specimens come from middens and fills, we do not think that the rise in high-value elements is merely a function of deposit type: it is not driven by an overrepresentation of special deposits in the Middle period faunal assemblage. All Early period and 85% of Late period equid remains derive from the South Area. Only 34% of Middle period equid remains come from the South Area. Middle period North Area deposits are strikingly rich in equid scapulae and pelves: in these deposits, shoulder blade and pelvic remains total 19.3% (31 out of 161) of all equid specimens. As we describe below, and as Russell et al. (2013: 217) also mention, equid scapulae are frequent presences in feasting and abandonment deposits.

Cull patterns

A small sample of equid tooth rows testify to equids being taken at ages ranging from very young to elderly (online table S8.6). These data suggest that Çatalhöyük's hunters were far from selective when it came to the ages of the equids they would take. In contrast, age profiles based on epiphyseal fusion (fig. 8.8) suggest that a preponderance of equids were taken as adults. We grouped elements that fuse at roughly similar times into sequential 'age stages' (not directly equated to actual ages at death); the percentage of fused specimens in

Taxon	*NISP*	*% equid NISP*	*DZs*	*MNI (zones-based)*
Small-medium equid	278	*34.1*	88.5	4
E. ferus	13	*1.6*	4	2
Large equid	64	*7.8*	28	2
Equus sp.	461	*56.5*	150.5	
Total	816	100	271	

Table 8.5. Equid taxa, 2009–2017. All loose teeth assigned to Equus sp., *even if originally assigned to a species based on occlusal morphology. All recording types.*

Element	*Early*	*Middle*	*Late*
Atlas	2	2	1
Axis	2	3	1
Scapula	2.5	8.5	1
Humerus	1.5	2.5	2
Radius	3.5	2.5	5
Ulna	1.5	4.5	1.5
Intermediate carpal (semilunar)		0.5	
Metacarpal	3	4	1.5
Pelvis	1.5	7	3
Sacrum		1	1
Femur	2	2	2
Tibia	1.5	1.5	1.5
Astragalus	2	2	2
Calcaneus	2	2.5	1
Central tarsal (scaphoid/navicular)		0.5	0
Tarsal III (external cuneiform)	0.5		
Metatarsal	1.5	3	
1st phalanx	3		1.25
2nd phalanx	3	1.5	
3rd phalanx	0.75	0.75	0.75
Total	33.75	49.25	25.5

Table 8.6. Equid anatomical relative abundances (MNE/# of times in body) for each phase. Tallied using diagnostic zones (DZs, after Martin, Russell 2005; Russell et al. 2013). Data from 2009–2017 excavations.

each stage estimates the proportion of animals that survived into and perhaps past the relevant age range. In all but the latest-fusing elements large majorities of specimens are fully fused. We think that what appears to be a drop-off in survivorship during young adulthood (that is, age stage FG) is the product of sampling bias, as only three specimens are appropriate for assessing the latest stages of epiphyseal fusion. Limited sample sizes also make it difficult to tell whether Çatalhöyük's early inhabitants took more young animals than did later villagers, as figure 8.8 suggests. Clearly, however, our data counterweigh Russell et al.'s (2013: 217) noting of more younger animals *later* in the sequence. Russell et al. suggested that rising proportions of young equids during Levels South P-T might reflect the effects of hunting pressure on the regional equid population: data recorded during this research tranche do not support such an inference.

Discussion

We continue to see the Çatalhöyük equids as game, pursued (unlike bovids) with a preference for prime-aged adults. Our data contain no evidence for humans having placed significant hunting pressure on the accessible equid population. Hints of selective transport of high-value elements (that is, scapulae and pelves) exist only in the Middle period.

Deer (*Cervus elaphus*, *Dama dama*, *Capreolus capreolus*)

While fallow and perhaps roe deer are present in small numbers, red deer dominate Çatalhöyük's cervid remains (table 8.7). Among the remains recorded in this research tranche, *Dama dama* is represented by four chronologically and spatially scattered bones: tibial and scapular fragments found in a Level South P pit, a calcaneus from South H-K and another tibial fragment from North F. Only two bones, a second phalanx and a metacarpal from a single South R unit, were recorded as perhaps deriving from *Capreolus capreolus*. Previous publications have reported very small numbers of *C. capreolus* remains at Çatalhöyük, but their rarity, combined with the tentative nature of our species assignment, means that we prefer to categorise our two as small cervid rather than roe deer. Thirty-one other specimens were also assigned to small cervid (that is, roe or fallow deer): 28 pieces of antler, two carpals and another second phalanx. Çatalhöyük's residents rarely exploited roe or fallow deer.

Red deer appeared more frequently on the menu, although venison was far from common at the site. Ninety-nine red deer specimens (DZ=34) and 257 large cervid specimens (DZ=4.5) come from across the site and Early through Late periods: Levels South K? through South Q and North F through North H?. Red deer specimens derive from all parts of the animals' bodies, although lower limbs and hooves – which probably arrived on site attached to hides – far outnumber upper limbs, axial elements and cranial remains other than antler (table 8.8). Similar proportions of red deer body parts characterise East Mound assemblages from earlier research tranches and contrast with those found in the offsite KOPAL Area, where meaty upper limb bones were well represented and antler and hoof bones scarce (Russell, Martin 2005; Russell et al. 2013).

We echo Russell and Martin (2005) and Russell et al. (2013) in concluding that Çatalhöyük's primary interest in cervids appears to have been their antler and their hides. Seventy-three per cent of the 393 cervid specimens identified in 2011–2017 are red deer and large cervid antler fragments and small cervid antler is another seven per cent. As antler has a tendency to crumble, the number of specimens is not an accurate measure of

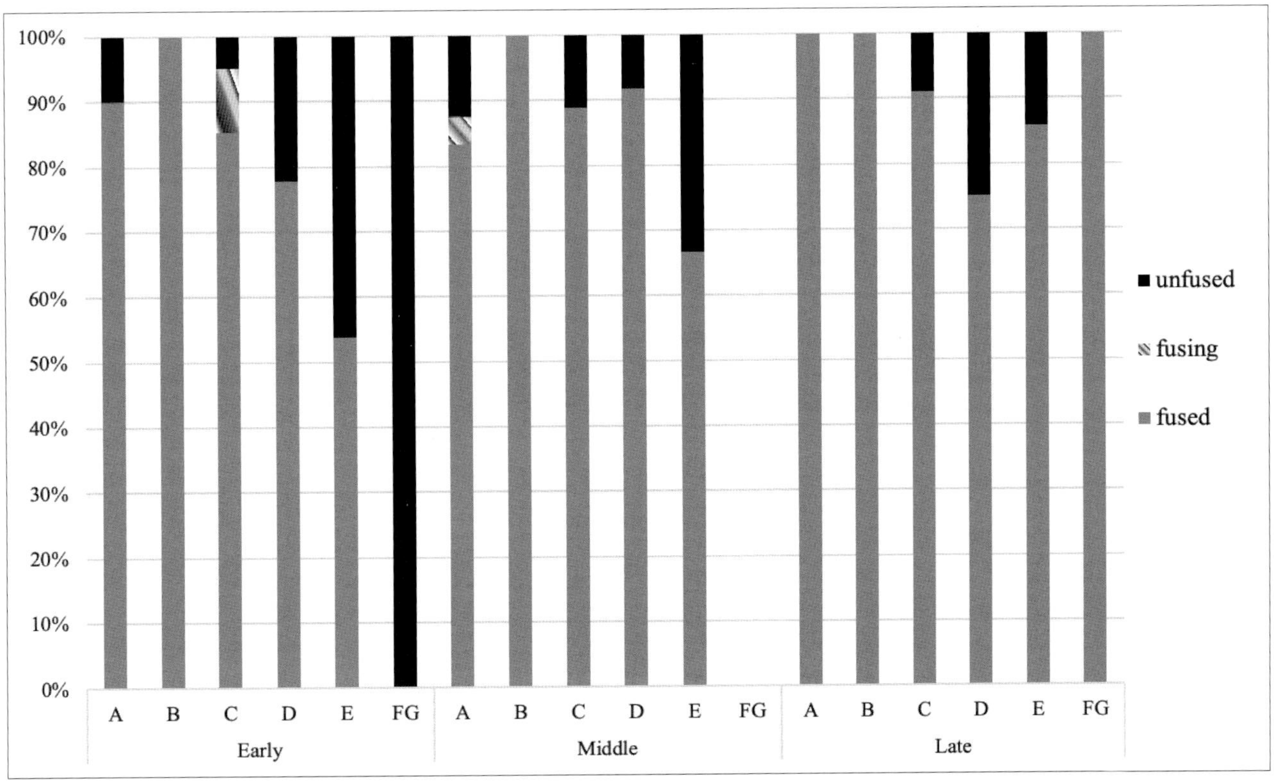

Figure 8.8. Equid epiphyseal fusion. Each stage represents a group of epiphyses that fuse at a similar age, based on Silver (1969). The percentages of fused specimens can thus be seen as estimates of the proportion of animals that survived at least into the age group represented by each stage (stage labels following Zeder (2006) for sheep/goat postcranial fusion). Fusion groups are: (A) pelvis, proximal radius; (B) scapula, distal humerus; (C) phalanges; (D) distal tibia, distal metapodia; (E) distal radius, proximal ulna, proximal and distal femur, proximal tibia, calcaneus; (FG) proximal humerus.

Taxon	*NISP*	*DZs*
Cervus elaphus	99	3
Dama dama	4	2
Large cervid	257	2.5
Small cervid	33	1.5
Total	393	9

Table 8.7. Cervid specimens at Çatalhöyük 2009–2017. Two specimens were identified as possible Capreolus capreolus. *As the species assignment is uncertain we include them among the 33 'small cervid' specimens.*

	C. elaphus		*Large cervid*	
	NISP	*DZs*	*NISP*	*DZs*
Antler	46	0	240	0
Cranial	7	1	1	0
Axial skeleton	2	2	2	1
Upper & mid-limbs	4	1		
Lower limbs	11	8	4	1
Hooves	23	19	5	1.5

Table 8.8. Red deer / large cervid body part distribution. Cranial excludes antler. Axial skeleton=vertebrae, pelvis; upper and mid-limbs=scapula, humerus, radius, ulna, femur, patella, tibia, malleolus; Lower limbs=carpals, tarsals, metapodia; hooves=phalanges.

villagers' appreciation for the material. Its presence throughout the site and particularly in special deposits (Russell et al. 2009; 2013), however, makes its value to Çatalhöyük's residents clear. The extent to which foot bones outnumber other postcranial remains testifies to repeated imports of hides with the feet attached. The paucity of red deer remains on site suggests that such imports were infrequent, but we have no way of knowing if footless hides were more regularly used.

Looking at cervid exploitation patterns diachronically, Russell and Martin (2005) proposed that after Çatalhöyük's earliest occupation, villagers brought hides and antler to the site almost exclusively. Russell et al. (2013) modified that statement to note that body part distributions evened out again to some extent in Late period Çatalhöyük levels. We have too little Late period data to revisit the latter finding in a meaningful way, but of our 29 *C. elaphus* and large cervid specimens from Levels South P and South Q, all but one scapular fragment are antler (n=23) and phalanges (n=5) (the sample is again far too small to be reliable, but the five Late period fallow deer and small cervid specimen include only one antler fragment and one phalanx, plus scapula, tibia and metacarpal pieces).

In sum, our data suggest limited exploitation of red deer hides and antlers and only rare consumption of venison. Fallow and roe deer were uncommon at Çatalhöyük.

Boar (*Sus scrofa*)

Boar remains are scarce in the Çatalhöyük assemblage (online table S8.8). This tranche of research provided no resolution to the ongoing question of why the villagers of Çatalhöyük ate so little pork; the animals were locally available, but at no time during the site's occupation did they provide more than a tiny fraction of its meat (see also Russell, Martin 2005; Russell et al. 2013; contrast with Baird et al.'s (2012) reporting on nearby Boncuklu). We found again no demonstrable change in boar metrics over time (see online appendix S8.2); Çatalhöyük suids' measurements (excluding extreme juveniles) average around 10% larger than those of the female boar described in Hongo and Meadow (2000).

Body parts

All suid body parts are present on site (fig. 8.9, online table S8.9). Although sample sizes are small, figure 8.9 suggests that relative proportions of different body segments might change through time, with a particular rise in meaty limb bones relative to far-less-appetising feet.

Cull patterns

Neonatal remains appear frequently in the suid assemblage; presumably hunters were taking litters when they could (online table S8.10). Although samples are small, epiphyseal fusion data suggest that hunters took few adult boars.

Sixteen specimens could be sexed on the basis of morphology. All but one are teeth and mandibles. Males slightly outnumber females, with eight males and one questionable male to six females and one questionable female. If we exclude the worked teeth as probably subject to different curation than other elements (as per Russell et al. 2013), the sex ratio evens, as three of the four excluded artefacts are male lower canines. No chronological patterning is apparent in either case (table 8.9). Mixture model analysis of suid postcranial measurements from all research tranches (fig. 8.10) also suggests a relatively even sex ratio.

Discussion

Boar hunts were infrequent events at Çatalhöyük and collecting litters of piglets seems to have been as common as dramatic pursuits of dangerous adult animals. Some people wore beautifully carved boar tusk collars, some wore beads carved from tusks and some resided in homes with boar remains installed in the walls (Russell, Meece 2006; Twiss 2006; Russell, Griffitts 2013), but no one ate pork very often and some may never have eaten it at all.

Caprines (*Ovis, Capra*)

The 2014 decision to focus recording on taxa other than the already well-documented sheep naturally caused the proportion of caprines in the analysed assemblage to decline in comparison with previous research tranches

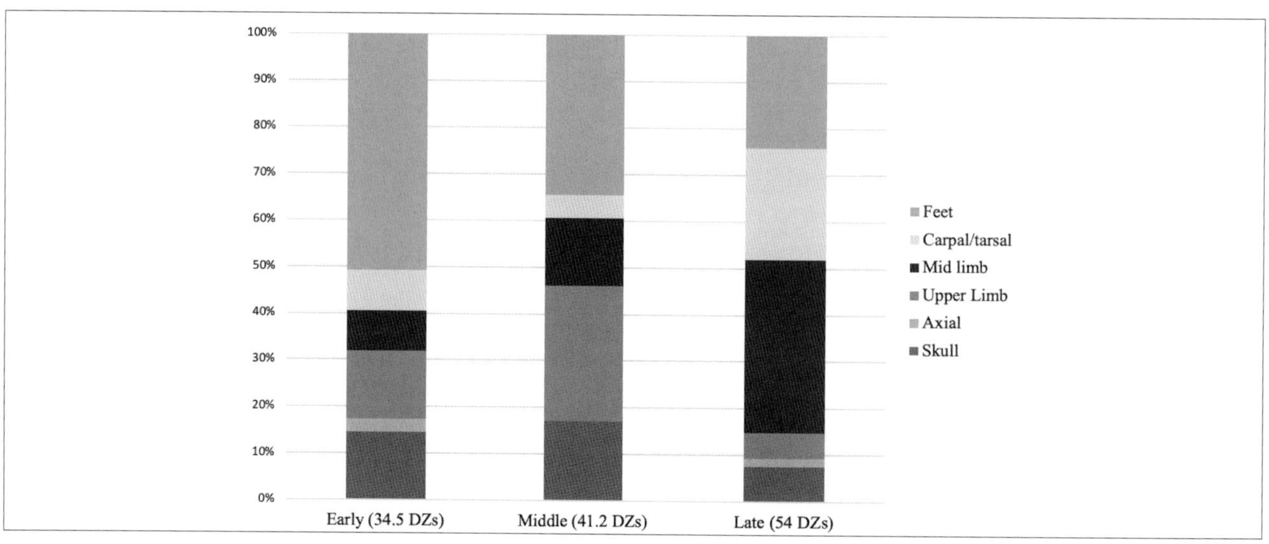

Figure 8.9. Suid anatomical proportions through time.

Occupation Phase	Including worked teeth					Excluding worked teeth				
	M	*?M*	*F*	*?F*	*% (M and ?M)*	*M*	*?M*	*F*	*?F*	*% (M and ?M)*
Early			1		0			1		0
Middle	5	1	2	1	67	2	1	2	1	50
Late	2		2		50	2		2		50
Middle/Late			1		0			1		0
Unknown	1				100					n/a

Table 8.9. Morphologically sexed suid specimens. ?M=possible males, ?F=possible females.

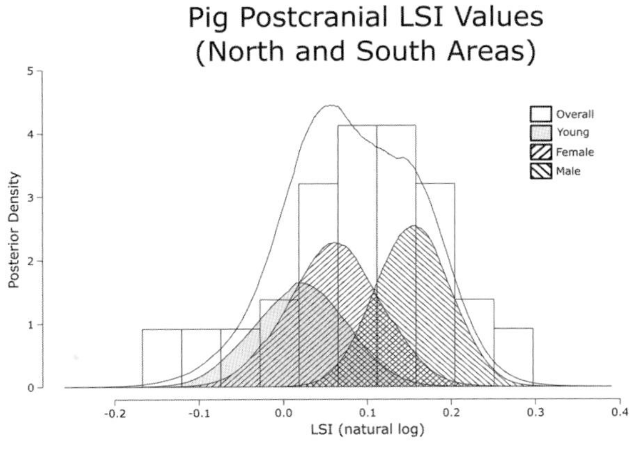

Figure 8.10. Estimated posterior mixture model analysis of North and South Area suid postcranial measurements (all research tranches).

(fig. 8.3, table 8.10). We recorded 8,184 caprine specimens (1,940.5 DZs), of which 15.9% were assignable to either sheep or goat. The ratio of sheep to goats is roughly 6:1 using NISP or 11:1 using DZs, broadly comparable to the proportions published by Russell and Martin (2005) and Russell et al. (2013).

Body part representation

We examine the relative abundance of body parts for *Ovis* and *Capra* as one group (fig. 8.11). In Early period Çatalhöyük levels, all *Ovis/Capra* body parts are present but meaty upper limb bones predominate, suggesting a focus on the best meat-bearing elements. We infer on-site/nearby slaughter and preparation, with entire animals consumed and disposed of at the site. There may have been more selectivity about body parts brought on site during the Middle period, as high-value elements such as meaty upper limb bones and metapodia (known to be valued in tool manufacture: Russell 2005) predominate and there are relatively few horn cores or lower-meat-value lower limb and toe elements. During the Late Çatalhöyük occupation, meat-bearing upper limb bones continue to predominate and mandibles and metapodia remain common.

Cull patterns

Tooth eruption and wear. Fifty-six mandibles could be assigned Payne (1973) mandibular wear stages (table 8.11). Nothing suggests differential mortality between sheep and goats, although *Capra* sample sizes are very small. Peak mortality occurs during wear stages C-F; these wear stages commonly occur during animals' later juvenility and early adulthoods (ca ages 1–4 years), although wear rates vary across populations. Russell et al. (2013) suggested that caprine cull patterns shifted during the later occupation of the site, with fewer animals surviving into adulthood. We have too few mandibles (particularly from the early phases of occupation) to determine whether mortality changed through time, but the limited data we have are consistent with more, rather than fewer, animals surviving infancy during the Late Çatalhöyük occupation. To explore caprine culling at Çatalhöyük further, we turn to epiphyseal fusion data.

Epiphyseal fusion. Figure 8.12 shows the proportions of fused, unfused and fusing sheep, goat and *Ovis/Capra* specimens from different phases of the site's occupation. Based on our admittedly small sample, it appears that virtually all goats survived into physical maturity during Çatalhöyük's Early period. Beginning in the Middle period villagers culled a minority of juveniles during age stage C (roughly 12–18 months) and significant numbers during age stages D and then E (which ends around 48 months of age). The Late period Çatalhöyük goat sample is too small for meaningful inferences.

	NISP	*DZs*	*%NISP*	*%DZs*
Ovis	1,109	636	13.6	32.8
Capra	189	71	2.3	3.7
Ovis/Capra	6,862	1,227.5	83.8	63.3
Ovis/Capra/Capreolus	24	6	0.3	0.3
Total	8,184	1,940.5	100.0	100.0

Table 8.10. Caprine specimens reported on in this chapter.

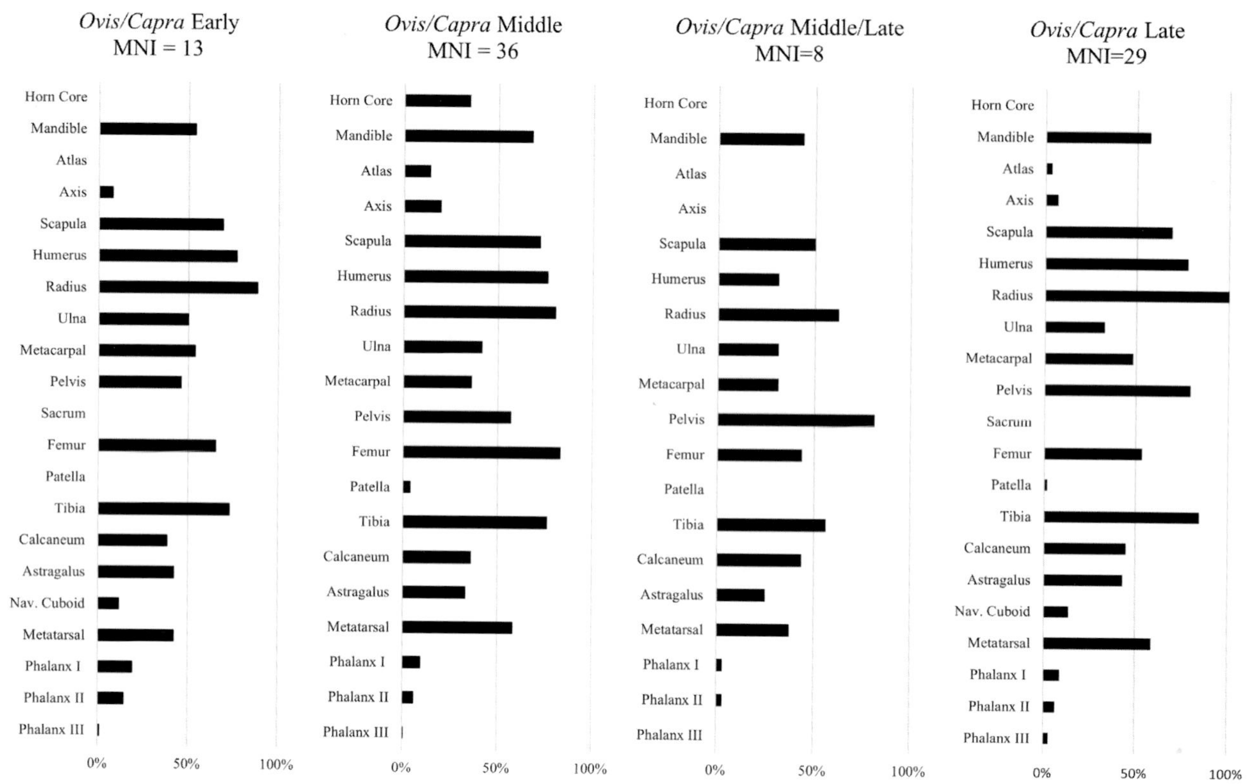

Figure 8.11. Relative abundance of body parts, Ovis *and* Capra *(grouped).*

	By taxon				*By occupation phase*				
Payne Wear Stage	Ovis/Capra	Ovis	Capra	*Total*	*Early*	*Middle*	*Mid/Late*	*Late*	*Total*
A	1			1		1			1
B		2		2		2			2
(B?)	3	2		5	1	2		1	3
C	1	7	1	9	3	4		2	6
(C?)		1		1				1	1
D	5	5	1	11		4		7	11
E	1			1				1	1
F	4	5	2	11		6	1	4	11
G		6	1	7	2	2		3	5
(G?)		1		1	1			1	1
(G or H)	1	1		2				1	1
H		3		3		2		1	3
I	1	1		2			1	1	2

Table 8.11. Caprine mandibular wear stages. After Payne (1973). NISP. All levels.

Sheep faced similar culling strategies. During the Early and Middle Çatalhöyük periods, more than 80% of lambs survived their first year and a half of life; survivorship dropped precipitously during age stage E (ca 30–48 months). It's possible that the three-year cull claimed fewer Late period sheep than had previously been the case. We assume that a large majority of '*Ovis/Capra*' specimens were sheep, however, and the overall caprine sample suggests heavier culling throughout the animals' lives than is apparent in either the diagnostic goat or sheep samples. The difference is presumably due to the fact that unfused specimens are harder to assign to species, as most caprine long bone diagnostic features a) are found at the epiphyses and b) cannot be evaluated in very young lambs or kids.

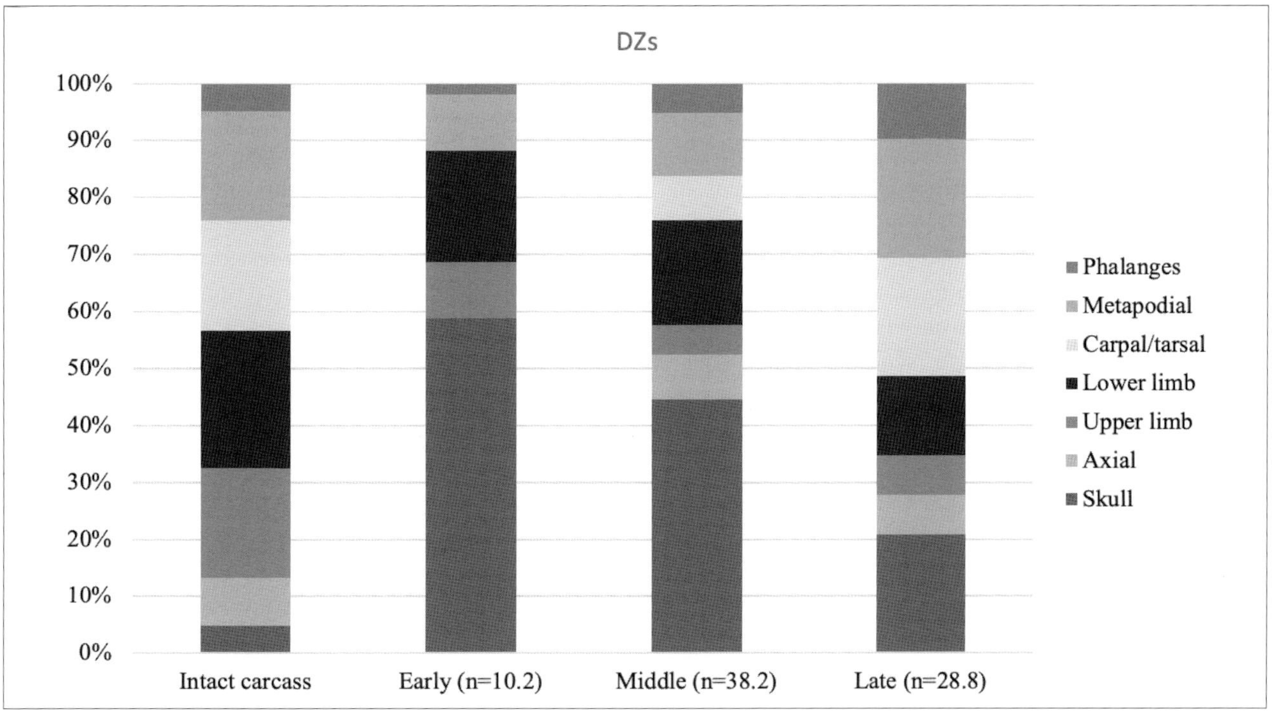

Figure 8.13. Dog/medium canid body part representation by occupation phase.

Metrical analysis. We, like previous researchers (Russell et al. 2013: 238–39), find dog sizes to be more variable in Late Çatalhöyük deposits than in earlier ones (online appendix S8.2). Late Çatalhöyük dogs are also somewhat smaller than those dating to earlier periods. These findings are consistent with Russell and colleagues' idea that in the Late/Final period Çatalhöyük dogs started to be used to fulfil a variety of roles.

Mortality and pathology. In contrast to previous publications, we recorded mostly subadult and adult dog remains. Aside from bones from the young animal (19814) found in Level South O inter-wall fill, we only recorded one unfused dog or medium canid bone, a South O scapular fragment (online table S8.12). Of 20 dog, medium canid and *Canis* sp. mandibles with teeth, only (19814)'s contained deciduous dentition and only two others contained less-than-fully-erupted permanent teeth: both P4s, which come into wear relatively late. These findings are in sharp contrast with the preponderance of puppies reported by Russell and Martin (2005) and Russell et al. (2013). We do not know why the pattern appears different in this tranche of research, nor can we explain why there would be more adult animals than young ones, given that dogs have litters and we would expect relatively high infant mortality in a semi-urban canid population receiving limited human care.

Most of the dog pathologies are mouth anomalies (online table S8.22). We recorded a resorbed canine tooth root on a mandible recovered from a deposit of mixed debris in Level South O (18545.F2) and a South T? mandible with no M3 alveolus (perhaps the dog was congenitally missing its third molar). The dog skeleton in South P (18625) lacked its left upper first molar; two of its premolars broke during the animal's life. A cusp broke off the lower left P4, and the upper right P4's break exposed its pulp cavity but did not create an abscess. (Elements from this skeleton are shaded grey in online table S8.22.) The only postcranial pathology is a variably fused and swollen metacarpal from the juvenile dog in between the walls of B.76 and B.80 in South O (19814).

Discussion. Our data are consistent with previous findings that dogs were regular and largely unremarkable presences in and around Çatalhöyük. Scavengers and cleaners, they roamed open-air middens and spaces throughout the settlement. Their numbers were probably never high, as not only do we have limited numbers of dog bones in the Çatalhöyük assemblage, but we also see limited amounts of carnivore gnawing and digestion in most midden assemblages (Russell, Twiss 2017). Past finds have suggested that at least a few dogs were eaten and that dog heads occasionally featured in symbolic activity, but we have no evidence of either among the bones recorded since 2009.

Foxes (*Vulpes* sp.)

Çatalhöyük's hunters took foxes regularly but not frequently (table 8.2 and online table S8.13). They tended to bring home whole carcasses, almost all of adult animals. Villagers used the animals for meat and for skins; there are no indications that foxes held any particular symbolic significance at Çatalhöyük as they did elsewhere in Neolithic Southwest Asia.

Previous publications on the Çatalhöyük East fauna reported that head elements were overrepresented among fox remains, perhaps because Neolithic villagers were sometimes bringing whole carcasses to site and sometimes bringing home only skins. Russell et al. (2013) note that the proportion of cranial remains drops during the later occupation of the site and speculate that either foxes were being killed relatively close to the village or fox skins lost value. We perceive no such patterning in the fauna recovered in this tranche of research. Rather, looking at the %NISP of fox elements in Early, Middle and Late period deposits, we see a rough consistency through time (online table S8.14).

There are no infants and no elderly animals in the fox assemblage (online table S8.15). Perhaps hunters were targeting subadults and adults, but it is at least as possible that fox hunting was largely opportunistic, young kits usually remained in dens, and few foxes survived to be old enough that their bones are recognisably geriatric. This seems particularly likely because a) there are five infantile bones among the 43 specimens identifiable only as small canid and b) infantile and elderly foxes have been reported at Çatalhöyük during previous tranches of research (Russell, Martin 2005). Even if villagers preferred prime-age meat and fur, their hunting methods yielded, and they were willing to consume, younger and older animals as well. Online table S8.23 summarises fox and small canid pathologies.

Badgers (*Meles meles*) and other mustelids

The mustelid assemblage from this tranche of research is quite small. Three quarters of our 36 mustelid specimens are badger remains. The remaining nine represent other small mustelids. One specimen is probably marbled polecat (*Vormela peregusna*), one was identified as a weasel (*Mustela nivalis*) and the rest may be martens (*Martes* cf. *foina*), polecats or weasels. We identified only one pathological specimen, a badger mandible with a possibly resorbed M2 found in a Level North G midden (32128).

A majority of the specimens derive from Middle period Çatalhöyük deposits, particularly Level North G (online table S8.16). The North G bones were found in several units and few articulate: the level's mustelid remains derive from multiple animals, not just one or two relatively complete skeletons. This, combined with the fact that their surface conditions accord with those of other bones in their units, indicates that the North Area badgers are not intrusive but generally reflect Neolithic villagers' use of the animals. We do not know what proportion of the small mustelid remains reflect human activity. Small mustelids surely came on the site in pursuit of the rodents that infested households' grain stores. As mustelids are nocturnal hunters, we think that they were living near the settlement, entering at night to hunt, and staying away during the day: this would explain the scarcity of small mustelid remains in the village in combination with the presence of small carnivore gnawing and scat (Jenkins, Yeomans 2013; this volume, Chapter 10).

Badger and small mustelid bones come from all parts of the animals' bodies (online table S8.17). Russell and Martin (2005: 83) infer that an articulated badger paw found in Building 17 floor sweepings was trimmed off a hide or from an animal being butchered; we assume that the three complete and articulating badger metacarpals found in a Level North G midden (19564) come from a similar amputation. All of our mustelid remains come from middens, fills or construction deposits and none shows traces of burning, gnawing or modification. We see no evidence of special treatment of mustelids.

Cats (*Felis silvestris*)

Prior to this tranche of research, felid remains at Çatalhöyük were rare and since virtually all derived from heads or lower limbs and feet, we thought that wild cats entered the site almost exclusively as skins. Only one upper limb bone fragment had been recorded, from a concentration of animal bone on the final floor of B.64 that might represent either an abandonment deposit or a pocket of midden material (Russell, Martin 2005; Russell et al. 2013).

It is with excitement, therefore, that we report a nearly complete kitten skeleton found in (32496), a foundation deposit (F.7847) near the northern wall of B.160 in the South Area. We believe the cat to have been *F. silvestris* rather than *F. chaus*, based on its mandibular carnassial morphology. The skeleton, which was missing the anterior portion of its skull, was found in partial articulation (possibly due to deposit slumping). Epiphyseal fusion data (variable fusion of proximal phalanges, unfused lateral condyle of the distal humerus) indicate that the animal was around 3–4 months old at the time of its death (following Smith 1969). Although the kitten would not have been fully grown, its postcranial measurements are well within the range of European *Felis silvestris* specimens (Morales, Giannini 2013). Indeed, the right os coxa LA (length of acetabulum) measurement is 14.67mm: this is large for modern European *F. silvestris*, so the animal might have been larger than modern wildcats if it had reached adult size. We identified no skinning or consumption marks on the animal.

The kitten's placement in the foundation deposit appears intentional. It lay on its right side with its back towards the wall. Other material in the foundation deposit (clay balls, ground stones) lay arranged in clusters; the cat skeleton was positioned to rest on its own.

Lepus & *lagomorph specimens*			*'Hare-sized' specimens*	
Body part	*NISP*	*DZs*	*Body part*	*NISP*
Axial	3	2	Axial	54
Feet	15	6.2	Feet	4
Lower limb	3	3	Indet. limb	6
Upper limb	11	6	Upper limb	1

Table 8.14. Lagomorph, Lepus *and hare-sized body part representation. 'Hare-sized' bones surely derive from foxes as well as hares.*

Hares (*Lepus* sp.)

Çatalhöyük's hunters caught hares throughout the Neolithic sequence, but in very low numbers (see also Russell et al. 2013). We assigned 40 specimens to either *Lepus* sp. or a broader lagomorph category. Every level had at least one such specimen, but a peak is apparent mid-sequence in the South Area. Level South O yielded nine specimens, South P seven and South M-N four; every other level yielded either one or two.

Specimens identified as either lagomorph or *Lepus* are heavily weighted towards foot elements (foot NISP=15; limb NISP=14; axial NISP=3). We have too few DZs for a secure evaluation of Russell et al.'s (2013: 84) suggested bias towards meaty upper limb bones, although we do have modestly more upper than lower limb specimens (table 8.14). We identified few *Lepus* axial specimens, but there are high proportions of axial elements among specimens that could be identified as hare-sized but no further. Many of these are surely fox bones, however, so we cannot assess whether whole hares were coming on site. Arguably the dearth of lagomorph cranial elements indicates that the animals were decapitated prior to being brought on site. We do not identify either meat or fur as hare hunters' priority.

Hedgehogs (*Erinaceus concolor*)

Once again the faunal team found few hedgehog remains (tranche NISP=8, MNI=4; online table S8.18) in the masses of bone analysed; all were mandibular or maxillary. Russell and Martin (2005) reported finding one humeral fragment, but apart from that lone specimen, every single fragment of hedgehog found at Çatalhöyük in a quarter-century of research has been from an animal's head and specifically from its upper or lower jaw. The finds come from both North and South Areas and from Early, Middle and Late period deposits, Levels South G to South P. All were found in mixed deposits such as middens and fills. Only one bears traces of possible carnivore gnawing or digestion. It is highly unlikely that such a biased set of remains derive from natural deaths onsite. Rather, people occasionally brought hedgehog mandibles or maxillae into the village. The scarcity of specimens and the aggregated nature of their contexts makes it impossible to narrow down potential motives for their import: we may imagine them as children's toys, as valued talismans, as art equipment, or as all of the above during different phases of their use-lives.

Macromammals in buildings

Since the last round of publications, the ÇHRP has completed excavations of burnt Building 77 and investigated Buildings 131 and 132 in the North Area and Buildings 80, 89, 97 and 160 in the South Area. Leaving aside B.77's previously reported above-floor remains (Russell et al. 2012), only burnt B.80 yielded large quantities of remains associable with house founding, use or abandonment (table 8.15; see also the online appendix S8.1 for expanded discussions of building contents).

We compared the proportions of different kinds of animals' remains in different spaces (table 8.16). Figure 8.14a demonstrates that the relative proportions of different size classes of fauna vary between buildings; figure 8.14b shows that relative proportions are usually reasonably consistent within each building (for example, across spaces/rooms). This is not true of B.131, where Sp.500 is a clear outlier, but that is because the space contains such a high proportion of unidentifiable remains. If we exclude unidentifiable remains as well as B.77's exclusively sub-floor deposits from consideration, B.131 as a whole stands out far less; its main room (Sp.500) continues to look atypical but that may simply be because sample sizes are small (fig. 8.14c and d).

Returning to figure 8.14a and b, we can see that the North Area buildings generally contained more unidentifiable fragments than did their South Area peers. Building 77, where we examined only the sub-floor deposits (notable burial fills) is easily explained, but what might be happening in B.131? We suspect that it contains so many small and unidentifiable fragments for two reasons. First, the building burned: high-temperature burning distorts, fragments and can destroy bones.

Area	*Building*	*Hodder phase*	*Wholly excavated?*	*Spaces*	*Occupation deposits**	*Associated special deposits*
North	77	North.G	Y	336 (main) 337 (side)	All fauna from above floor level (that is, from the building's burning phases) are reported in Russell et al. (2013). It is unclear what of these fauna reflect normal occupation vs ceremonial activity. The only primary or secondary faunal remains recovered from B.77's pre-burning phases are bone personal adornments in burials.	
	131	North.G		500 (main room) 504 (eastern side room) 556 (northern side room)	Burned rodent bones inside a basin. Pendant fragment on Space 504's floor possibly related to the room's earliest use.	Two left cattle scapula blades embedded into opposing sides of niche (F.7986).
	132	North.F		511 (side room) 531 (main room) 633 (earliest occ phase in main room)	Possibly one bone: a cattle scapula fragment found at floor level.	None
South	80	South.O		135 (main room) 373 (southern room)	*All finds are in Space 135.* Stone and bone cluster (18955) atop the oven included a fragment of (gnawed, not fresh) unworked red deer antler – perhaps raw material for toolmaking. Cluster of worked and possibly worked burnt bones found inside a post retrieval pit: not normal storage, but perhaps specimens were household tools.	Faunal installations in Space 135: 1. Unburned horned bucranium set into bench 2. Horned goat head probably on western wall of the building (A dismantled cattle skull installation found in a dump deposit cannot be linked to Building 80's occupation. Ditto an unburned cattle skull fragment and atlas in a wall collapse unit. Several other fill and collapse units also contained horn cores and cranial fragments.) Fills and dumps are rich in scapulae (often equid): all that could be sided are lefts. Notably, 5 left scapulae (2 cattle, 3 equid, all but one unburned) = deliberate placements in room fill. This fill related to one in neighbouring Building 79 (Space 134). Building 79's fill also held a left equid scapula. Taphonomically similar human cranial fragments found in both fills. Collection of heavily burned bones beneath the house entrance: perhaps household waste and/or perhaps symbolic deposit. Location echoes that of Building 77 fish bone/pea cluster in Building 77 (House 2013: 496).

*Table 8.15. Faunal remains plausibly linked to the occupations and closures of Buildings 77, 131, 132, 80, 89, 97 and 160. Excludes personal adornments from sub-floor burials. * This category includes only remains believed to represent activities in the house. Floor deposits and construction materials almost always contain bits of redeposited bone; we omit such accidental inclusions from consideration. ** An additional 123 specimens were assigned to either Sp. 365 or Sp. 469.*

Area	*Building*	*Hodder phase*	*Wholly excavated?*	*Spaces*	*Occupation deposits**	*Associated special deposits*
South	89	South.N, South.?N	N	379 (main)	One worked antler base deposited on dirty floor, possibly soon after plastering (potentially linked to occupation).	Plastered bucranium in Sp. 379 (damaged and covered by multiple floor layers).
				565 (annex)		
	97	South.O, South.?M	N	365 (main room; truncated by Mellaart excavations)	One worn cattle scapula buried in the side of a bin; link to B. 97's occupation is tenuous.	Ritualised activity in/near abandoned B.97: feast remains in the fill above the floors of Sp. 365.
				469 (side room)		Closure/abandonment, probably.
				503 (early phase of B.97)		
	160	South.K	Y	551 (main)	None recorded.	Abandonment deposits: articulating aurochs bones placed in three pits in Sp. 551. MNI = probably 2–3 aurochsen.
				552 (side)		Foundation deposit: wild cat skeleton and antler placed under north wall of Sp. 552.

*Table 8.15 (continued). Faunal remains plausibly linked to the occupations and closures of Buildings 77, 131, 132, 80, 89, 97 and 160. Excludes personal adornments from sub-floor burials. * This category includes only remains believed to represent activities in the house. Floor deposits and construction materials almost always contain bits of redeposited bone; we omit such accidental inclusions from consideration. ** An additional 123 specimens were assigned to either Sp. 365 or Sp. 469.*

Second and more importantly, the deposits we studied from Sp.500 were (as in B.77) mostly burial fills and construction material, both of which commonly contain only small bone fragments that were inadvertently caught up in intentional soil deposits. All of the other buildings represented in figure 8.14 had a broader variety of context types analysed, notably including middens and fills. Buildings 97 and 160 are perceptibly richer in cattle remains than any of the other buildings. Let us now discuss the contents of each of these buildings in order to explain why this might be.

Aside from B.80, the buildings analysed in this tranche of research contained largely fills and middens, with relatively few remains testifying to life inside a Neolithic

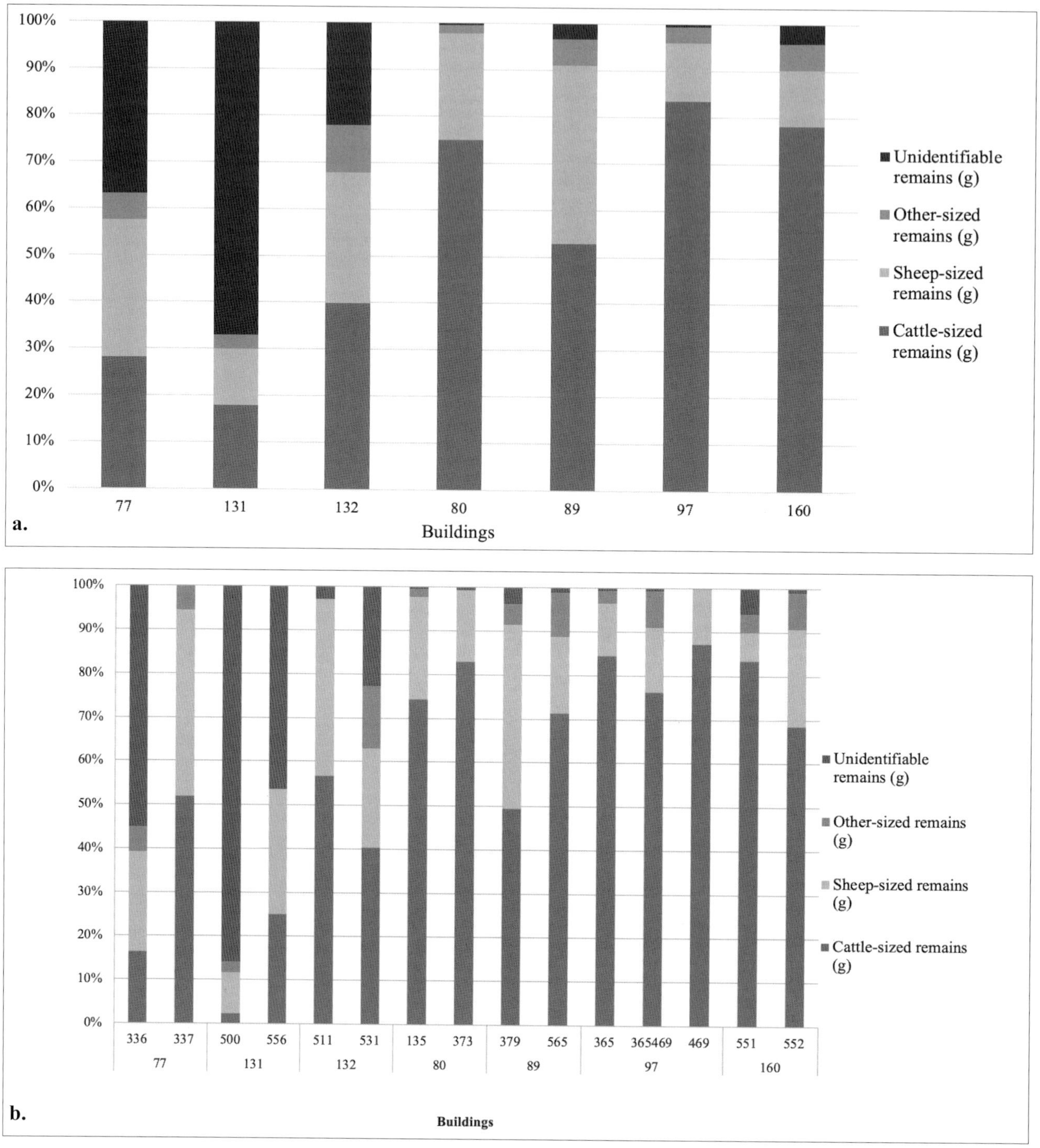

Figure 8.14. Proportions of different size classes of animals in different buildings, by weight (g). Weights by (a) buildings; (b) spaces within buildings; (c) buildings (identifiable specimens only, B.77 sub-floor deposits excluded); (d) spaces (identifiable specimens only, B.77 sub-floor deposits excluded). Building 77 data in charts (a) and (b) include only the sub-floor deposits recorded during this tranche of research.

house. Exceptions include two left cattle scapula blades (23074.x1 and x2) embedded into the sides of a niche in B.131's main room (Sp.500). One of the scapulae comes from an aurochs; the other is smaller and might come from either a female aurochs or a domesticated animal. Building 89's main room contained a large worked antler base (21952.x1) – potentially abandoned near a working surface and quern stone cluster and plausibly used as a pounder or a soft hammer – and a damaged plastered bucranium (21968.x1; a female aurochs). The latter lay in floor deposits and it is unclear whether it had been dismantled or remained installed. A worn *Bos* scapula lay buried in the edge of a B.97 bin (19665.x1); we do not know when or why it was deposited in the bin's siding.

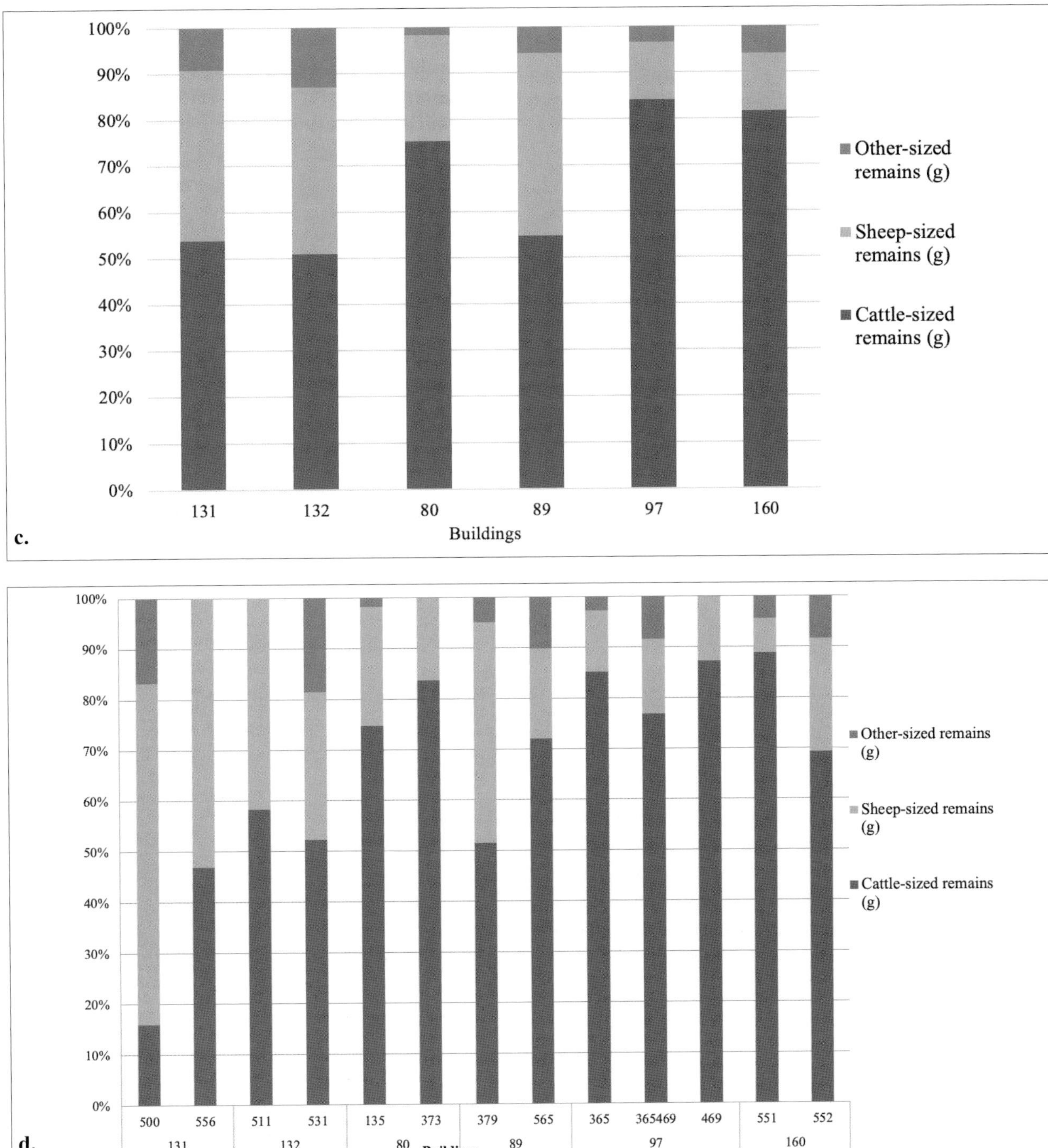

Figure 8.14 (continued). Proportions of different size classes of animals in different buildings, by weight (g). Weights by (a) buildings; (b) spaces within buildings; (c) buildings (identifiable specimens only, B.77 sub-floor deposits excluded); (d) spaces (identifiable specimens only, B.77 sub-floor deposits excluded). Building 77 data in charts (a) and (b) include only the sub-floor deposits recorded during this tranche of research.

Building	Space	Cattle-sized remains (g)	Sheep-sized remains (g)	Other-sized remains (g)	Unidentifiable remains (g)
77		371	388	73	487
	336	144	202	49	487
	337	227	186	24	
131		252	172	43	950
	500	22	92	23	847
	500, 556	176	19	20	3
	556	54	61	0	100
132		1,382	975	352	763
	489, 511	327	331	14	143
	511	160	114	0	8
	531	895	498	318	503
	531, 633	0	32	20	109
80		19,091	5,823	431	123
	135	17,757	5,559	430	113
	135, 373		3		
	373	1,334	261	1	10
89		3,041	2,199	325	192
	379	2,424	2,048	237	183
	565	617	151	88	9
97		18,951	2,825	790	136
	365	15,186	2,160	505	116
	365, 469	2,568	492	283	20
	469	1,197	173	2	
160		13,453	2,049	981	705
	551	9,298	704	474	650
	552	4,155	1,345	507	55

Table 8.16. Weight of bone from various animal size classes in Buildings 77 (sub-floor deposits only), 131, 132, 80, 89, 97 and 160.

New evidence for house founding and abandonment activities is as follows. A foundation deposit lay under the north wall of B.160 (Sp.552): a young cat (see *Felis silvestris* above), a large piece of antler, clay balls, a wood plank and limestone pebbles. After its last residents left B.160, two or perhaps three aurochsen's remains were deposited in three pits (F.7831, F.7832 and F.7833) in the main room of the house. Each pit contained multiple, deliberately placed aurochs bones, with articulations linking all three fills. It is possible that other aurochs bones (the pairs 32453.x3 and 32453.F1, 32467.F1 and 32467.F2) also derive from the same event, as there are other articulating sets of aurochs remains in contexts that have been interpreted as abandonment deposits. The butchered aurochsen probably included at least one male and at least one female based on the sizes and fusion states of the specimens, and the bones found in B.160's pits are meaty ones: vertebrae, scapulae and limb bones, the latter broken up to retrieve the fatty, delectable marrow inside. Aurochs bones aside, the fill of the three pits resembles the rest of the building's infills.

Excavators found a collection of bones (18647) associated with the floors of B.97's main room (Sp.365). Here, several large pieces of cattle bone – coming from more than one animal – lay together with equid, sheep, fox and dog remains. Not all of the bones would have been meaty ones, but those that were would have provided enough meat for at least ten people to have eaten a pound of meat each. We infer that a feast took place in or near the house, close enough to the time of its abandonment that the leftover bones lay directly atop its floor, to burn when the structure did (also burning inside Sp.365, but not in direct contact with fire, were a cattle horn core and sacrum. These bones were recovered together with a cluster of seeds (19238) and might either have been stored with them or lain near them on the floor of the house. It is also entirely possible, however, that one or both derive from the surrounding fill).

We turn now to burnt B.80, whose main room (Sp.135) contained multiple faunal installations as well as a cache of osteological material that may be related to the house's occupation.

A handful of grave goods and caches testify to practices predating the burning and abandonment of B.80. A stone and bone cluster (18955) found atop the building's oven included a fragment of burnt red deer antler. Unworked, it may have been collected as raw material for toolmaking (that it was not a fresh specimen is indicated by the presence of gnawing marks). A cluster of worked and possibly worked burnt bones (19194) lay inside a post retrieval pit: their location is inconsistent with regular storage across the lifespan of the house, but their presence at its termination suggests that they were household tools in some fashion or another. One specimen was a needle and the others – rib fragments, possibly unfinished or rejected equipment – may have been needles, burnishers or pottery polishers (after Russell, Griffitts 2013: 279–80).

Building 80 burned with a collection of bones (18964) lying beneath its entrance ladder. The calcined bones' fine surface conditions and lack of rodent or carnivore gnaw suggest that they were relatively fresh when they burned. Sheep-size specimens represent meaty body parts such as the torso and haunches (vertebrae and ribs, upper limb bones), whereas larger animals are represented by less-meaty distal limb specimens (two cattle metacarpal fragments, an equid distal radius with its carpals). Perhaps these latter bones were intended for boiling? The excavator of this cluster suggested that the bones might have been waste, piled near the entrance for future disposal; it seems equally likely that some or all of the deposit was symbolically intended. Its location echoes that of the fish bone/pea cluster found beneath the entry to B.77.

Building 80 also contained at least two and perhaps several more installations at the time of its destruction. Set into a bench sat a horned bucranium (22430.x1), unburnt apart from its left horn core and somewhat crushed (fig. 8.15). A horned goat head probably protruded from the western wall of the building; excavators recovered a burnt *Capra* frontlet (anterior cranium, 18576.x4), set in clay, from a collapse deposit. It retained two fragmentary horn cores (we believe the sheaths to have been present in the original installation), and two pole-shaped impressions at its base suggest how the installation was supported. We think that a horned cattle skull found upside down in a dump deposit (18543.x1) was a dismantled installation, but we cannot link it to B.80's occupation. Nor do we know the relationship between B.80's occupation and an unburnt aurochs anterior skull fragment and atlas (plausibly once articulating) found in a wall collapse unit (18531). Several other fill and collapse units in the building also contained horn cores, often with cranial fragments; their presence in such numbers is reminiscent of similar high proportions of cattle skull remains in burnt Buildings 52 and 77 in the North Area.

Building 80's fills and dumped deposit units in this building also contained numerous scapulae – and all of those for which symmetry could be determined came from the left sides of animals' bodies. This attention to symmetry is reminiscent of fauna from burnt B.52 (Twiss et al. 2008). In contrast to Buildings 77 and 52, however, many of B.80's scapulae came from equids rather than cattle.

Room fill (17342) contained five left-hand-side scapulae: only two came from cattle. Two others came from large equids (*E. ferus*) and the last came from a small/medium equid. Only one cattle scapula was burnt. We believe these scapulae to have been deliberate placements in the fill. The use of bones from five different animals echoes practices in Buildings 3, 131 and 65, and the choice of scapulae in particular accords with the selection of these elements for a variety of symbolic deposits across the site (Russell et al. 2013: 214). This fill unit probably lay above and related to a fill (18508) in neighbouring B.79 (Sp.134). Building 79's fill also held a complete, left-hand-side equid (*E. ferus*) scapula, and taphonomically similar human cranial fragments (plausibly derived from one person) were found in both fills.

Finally, one B.80 dump deposit held a cattle or large equid scapula showing signs of use wear along its cranial edge (18947.F3), and another held one bovine and one large equine scapula (18543.x2 and x4) whose symmetry

Figure 8.15. Damaged bucranium 22430.x1, set into a bench in Building 80 (photo: Mateusz Dembowiak).

could not be determined. This unit also contained a partial dog skeleton (1854). The dog bones (vertebrae, pelvis and femur) were burned a uniform chocolate colour on all surfaces, suggesting that they were defleshed at the time they burned. We infer that this was a secondary, or redeposited, dog burial.

Building 80's fills were unusual in ways that extend beyond their richness in scapulae and equid remains. Unlike most fills and wall/roof collapse deposits at Neolithic Çatalhöyük, those in B.80 were often rich cattle-size specimens (for example, 18519, 18531, 18544). Some units contain bones consistent with feasting, perhaps in situ. The collapse deposit that contained the *Capra* frontlet (18576) also held a variety of cattle and caprine remains, roughly 90% of which were burnt. A series of cattle vertebrae may derive from a single animal; we think it plausible that they and the rest of the unit derive from a single event (alternatively, perhaps they were stored goods).

Butchery

This section provides a brief summary of North and South Area cut mark data recorded from the beginning of the Çatalhöyük Research Project through 2017. A fuller review of the data is available in online appendix S8.1 under the heading 'Butchery marks of the North and South Areas'. The total number of Neolithic faunal specimens with butchery marks is 1,953; only 0.2% of Neolithic specimens deriving from the two excavation areas show evidence of butchery.

The frequency of cut-marked bones in various types of deposits accords with the frequency of all bones in such deposits: bones with visible butchery marks were not deposited differently from unmarked bones. This is not surprising, as we assume that most remains on site derive from butchered animals and that the scarcity of marks on the bones reflects a combination of ancient skill and reliance on very sharp cutting tools (Russell, Martin 2005: 85; Dewbury, Russell 2007). These two circumstances might also contribute to the fact that the North and the South Areas have broadly similar proportions of cut-marked bones within each animal size category. In both areas small (hare-sized) remains are more frequently marked than large (cow-sized) ones, and they most frequently bear evidence of filleting, while large animals' bones bear traces of dismemberment and consumption. Perhaps skilled butchers found it easier to fillet beef than hare without nicking underlying bones? Or perhaps hares were caught, butchered and eaten ad hoc – by relatively unskilled or careless butchers – whereas cattle, symbolically as well as socially and economically important, were filleted with care? In any case, dismembering something as large as an aurochs was hard enough work that sharp obsidian cutting edges could have dulled or bigger chert tools (Carter, Milić 2014: 447) been deployed to wrest apart immense joints.

In general, the inhabitants of the North and the South Areas probably used similar stone tool types and butchery methods and the overall intensity of their meat processing remained fairly consistent through time (the incidence of cut marks declines slightly through time in the South Area but holds relatively steady in the North Area). The two areas had similar proportions of various cut mark types, with dismemberment, consumption and filleting cuts dominant. We infer that meat-drying, boneless meat stewing and bone-in meat baking were all practised with some regularity at Neolithic Çatalhöyük.

Relative proportions of these kinds of cooking may have changed differently through time in the different Areas. A Middle-to-Late period increase in dismemberment and filleting marks is apparent in both the North and the South; in the North Area consumption cuts also rise, but in the South Area they decrease in frequency. Perhaps the Late period Çatalhöyük denizens of the North Area preferred to keep baking their meat as their great-grandparents had done, while their neighbours in the South Area liked stews more?

Meal sharing

Four fire spot deposits in North Area Spaces 631, 610 and 85 (online table S8.3; Sp.631 and Sp.610 are Level North G, Sp.85 is Level North H?) probably represent coherent individual events. We use the weight of bone in these units to estimate that between 250g and 3,965g of meat was served at each event (online table S8.4, calculations following Demirergi 2015: 156–57). If each participant received 250–500g of meat, between two and 16 people – the inhabitants of one to three houses? – shared each meal (see also online appendix S8.1, 'Meal sharing').

An abandonment deposit (32418) recovered from pit F.7832 in B.160 (see 'Macromammals in buildings') testifies to a far larger event. Excavators recovered 3,372g of cattle and cattle-sized remains from this pit, representing in the order of 17kg of edible meat. This would have fed between 33 and 67 people (four to 17 houses' inhabitants?), at 250–500g of meat each. Two other pits in this building (F.7833 and F.7831) yielded remains testifying to approximately one third to one half as much meat: still plenty to feed the residents of multiple houses. It is also possible that the three pits derive from a single event, especially as we found multiple plausible (although not certain) articulations involving cattle bones from different pits. A single feast large enough to produce all three pit fills could easily have involved a hundred people or more.

Meat serving sizes surely varied, and our estimates of diner numbers may be high, low or both in alternation. However, we clearly see meal sharing at a variety of scales. If our estimates of participant numbers are reasonably accurate, the fire spot remains plausibly testify to meals shared among people affiliated with one or two houses (perhaps B.114 and/or B.131, which abut the relevant spaces). In contrast, an entire neighbourhood or kin group might have contributed to the closure of B.160. Demirergi (2015) suggests that most Çatalhöyük feasts probably seated between seven and 34 people (perhaps one to seven houses). We infer events both smaller and larger than this average. We see these diverse meals as reflecting what Hodder (2014e: 153) characterises as a 'dispersed overlapping mosaic of relationships': as evidence of people's participation in household, neighbourhood and myriad other social spheres.

More and more sheep

We checked to see if the inferred rise in sheep herding during the Late period (Russell et al. 2013) was borne out by new data. Using our newly recorded estimates of %NISP (see 'Changing Methodologies' above), we found a dramatic rise in sheep proportions during the site's later occupation (online fig. S8.19; see also fig. 8.2 above). This rise is not due to a change in the size of units recorded; that is, we are not seeing it because we are looking at a higher proportion of small units such as burial fills, which tend to contain numerous relatively small pieces of bone.

Changing relative proportions of different animal size classes do not, of course, reveal changes in the total amounts of mutton or beef that people were eating: to assess those we need to look at bone densities per litre of soil.[1] Figure 8.16a indicates a sharp rise in the Late period after a drop in the Middle period across all kinds of meat. Interestingly, the soaring ratios of the Late period derive entirely from the South Area (fig. 8.16b). Meat quantities actually decrease slightly from the Middle to the Late period in the North Area (a caveat: our Late period Çatalhöyük North Area derive from only 13 units). The Late period rise is not attributable to a particular type of deposit (fig. 8.16c), as we see increased bone densities in middens, fills and activity areas (floor use deposits' faunal contents are almost always small, crumbly and unintentionally present – they are what people did not toss out or sweep up and we do not consider their faunal proportions reflective of general meat consumption).

Conclusion

What do the findings of this third tranche of faunal research at Neolithic East Çatalhöyük add to our overall understanding of human-animal relations at the site? Let us review some key expansions of information.

First, we have deepened and expanded our diachronic data set, adding in particular information about the animal economy of Middle period Çatalhöyük (fig. 8.1). As this was the era during which the inhabitants of Çatalhöyük modified not only their economy but also several other aspects of their centuries-old ways of life (see Hodder 2014a), expanding the Middle period faunal data set enhances our understanding of how economic practices did and did not change as site populations, technologies and built environments shifted.

We have significantly improved our grasp of when and how the people of Çatalhöyük integrated cattle management into their economy (for further discussion, see this volume, Chapter 9). We have developed new evidence for human entanglement with aurochsen over the course of centuries: biometric data reveal people interfering with local aurochs populations at least during Levels South M-O (that is, Middle period) and this interference continued/intensified throughout the rest of the North and South Area occupations. Cattle in these levels were *not* definitively domesticated, but their decreasing size suggests that their lifeways were changing, likely due to human interference through hunting or (more likely) attempts to manage the population (Marom, Bar-Oz 2009; Zeder 2015). Geoarchaeological evidence for cattle stalling in Levels South R and S (Shillito, Ryan 2013) suggests that human interference, in the later stages of this process, may have been similar to attempts to maintain populations of minimally managed sheep at Aşıklı Höyük (Stiner et al. 2014).

Recent models of domestication posit long-term interactions between humans and animals that eventually result in morphological change. These models emphasise that the nature and intensity of such interactions and the pacing and duration of the resultant domesticatory processes vary across taxa and through time (e.g., Peters et al. 2015; Allaby et al. 2017; Zeder 2016; 2018). Our inferences are entirely consistent with such views of domestication – or adoption of domesticates – as an extended and complex process. However, they also complicate current models of how and when domesticated cattle spread across Neolithic Anatolia and of what happened after they arrived.

---

1 This follows Russell et al.'s (2013: figs. 71–74) methodology, but please note that they examined bone densities by taxon rather than size class. We included all cattle- and sheep-sized specimens as we wished to a) base our conclusions on as large a data set as possible, specimens and b) include specimens that cannot be assigned to a particular species on the basis of their morphology (for example, ribs, vertebrae). Large majorities of the cattle-sized specimens surely derive from *Bos*, and sheep-sized specimens from *Ovis*, supplemented by *Capra*.

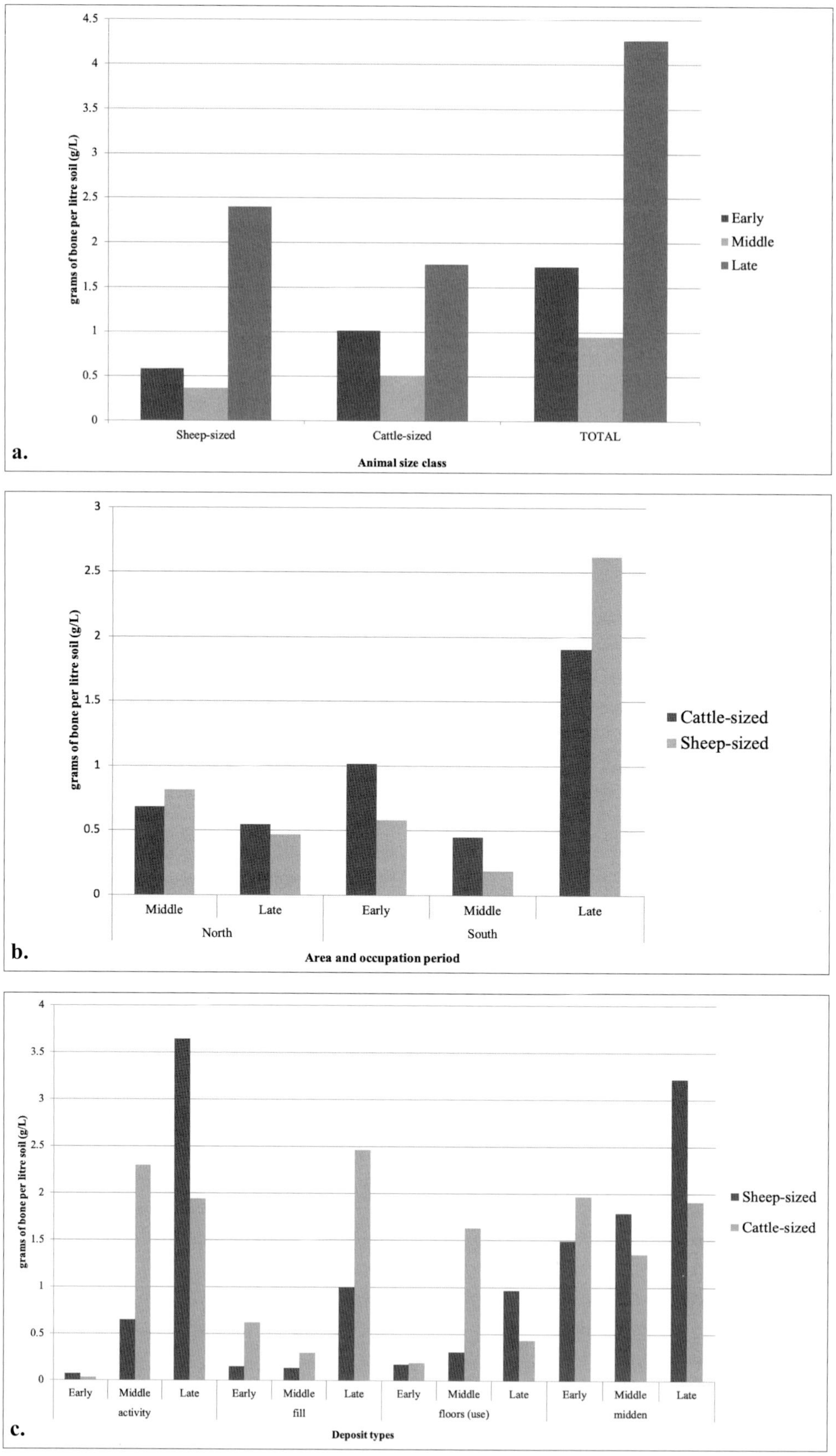

Figure 8.16. Faunal bone densities through time: grams of faunal bone per litre soil (g/L), by time period and animal size class. (a) all areas; (b) North vs. South Areas; (c) by deposit type. Based on Faunal Unit Description weight totals recorded for T1, L and 2014 units with soil densities greater than zero. 757 units total: 145 Early, 514 Middle, 98 Late periods (see online fig. S8.19 for changes in remains assigned specifically to caprines).

Modern analyses of the Çatalhöyük macrofauna have universally concluded that cattle were morphologically domesticated elsewhere and then incorporated into the site socioeconomy (e.g., cf. Perkins 1969; Russell et al. 2005; Arbuckle, Makarewicz 2009; Arbuckle et al. 2014). Some publications have further proposed that Çatalhöyük's people resisted adopting morphological domesticates, which were exploited elsewhere in Central Anatolia for centuries prior to their inferred appearance in the Çatalhöyük assemblages (Arbuckle, Makarewicz 2009). These papers raise the possibility that the site's later levels include the remains of large animals as well as smaller ones because – for ideological reasons – Çatalhöyük's herders preferred to mate their cows with wild bulls, producing managed herds that remained phenotypically wild for centuries (Arbuckle, Makarewicz 2009; Peters et al. 2015).

People were probably managing cattle during the Late period at Çatalhöyük, but we are not convinced that they were more reliant on herded animals than on wild ones (for further discussion, see this volume, Chapter 9). We can see neither resistance to domesticates nor an eventual decision to accept the smaller animals being raised elsewhere. It is true that Late period slaughter patterns are female-dominated while earlier culls were relatively balanced between male and female animals (a pattern produced by some management strategies), but *Bos* stable isotopic signatures do not become more constrained (as we might expect if people were habitually controlling grazing of domestic livestock) (this volume, Chapter 11; cf. Peters et al. 2015). A modest increase in infant and juvenile culling does not indicate a dramatic change in culling strategies (cf. Russell et al. 2013). Organic residues testify to dairy consumption at Neolithic Çatalhöyük, but milk proteins dating to the later Çatalhöyük West Mound occupation – when there is clear evidence for domesticated cattle – derive primarily from sheep and goats (Hendy et al. 2018). We therefore believe caution is warranted when discussing Late period cattle herding at Çatalhöyük and we encourage people to avoid assuming that most of the site's beef came from domesticated cows or calves.

On another front, we have increased the number of houses for which we have in-depth faunal data. Of particular importance is burnt B.80, which joins burnt Buildings 52 and 77 as the symbolically richest and arguably most socially informative structures excavated by the ÇHRP.

We have provided additional support for some previous inferences about animals at Neolithic Çatalhöyük. Unsurprisingly, our data on caprine husbandry accord with the inferences (based on immense amounts of data) put forth in previous publications. Culling strategies suggest eight centuries of unspecialised herd management, although more animals survived into their fourth years during the Late period. However, densities of faunal bone indicate that later-period residents of the South Area consumed distinctly more meat, particularly sheep meat, than did either their neighbours in the North Area or earlier villagers. Pathologies on three Late period sheep horn cores are consistent with malnutrition: perhaps herd conditions declined as sheep populations increased in and around Çatalhöyük (see also Henton 2012; Pearson et al. 2015).

We have also provided information inconsistent with other previously suggested inferences. Whereas previous publications reported a preponderance of puppies, we recorded mostly subadult and adult dogs. We identified high proportions of fully fused bones in equid remains from all phases of occupation, *contra* Russell et al.'s (2013: 217) theory that increasing consumption of young equids during Levels South P-T was due to the effects of hunting pressure on the regional equid population. That said, equid numbers do decline in the Late period. Consumption of these game animals waned; we do not know whether they became less available, or less desirable.

With these data added to what was previously known, we arrive at an ever-richer understanding of animal exploitation across eight centuries (ca 7100–6300 BC) of life at Neolithic Çatalhöyük. We see a pastoral economy centred on the herding of domesticated sheep, with some goats scattered throughout the flocks. These animals were reared using a generalised strategy that balanced villagers' desires for herd security, meat and fat production, some dairying (it is unclear when this started) and the myriad other products derivable from caprine carcasses such as bones, hides and tendons. Dogs roamed the outdoor areas of the settlement, scavenging what they could to eat and dropping gnawed bones and faeces in middens and yards (Russell, Twiss 2017). After centuries in which brave hunters regularly hauled home the immense carcasses of wild aurochsen, midway through the site's occupation people began managing (morphologically wild) cattle as well. These managed animals did not fully replace the aurochsen in either symbolic activity (which the aurochsen had long dominated) or economic utility, but they changed people's labour, foodways and perhaps landscape exploitation (Wolfhagen 2019).

As for wild animals other than aurochsen, antlers were regular presences on site and deer hides useful if probably less-than-common resources. Deer canines became personal adornments and sometimes people shaped other osteological materials to resemble them. Venison was consumed very rarely. Pork was likewise scarce, coming mostly from piglets that people presumably took as litters. Bear remains arrived at Çatalhöyük only occasionally, in the form of pelts. Smaller wild carnivores (foxes and badgers) and hares provided meat as well as fur. Some small carnivores (weasels, for example) may have lived in the village, hunting the rodents that liked to infest household grain supplies. Perhaps small populations of wild cats were among these hunters, but most of Çatalhöyük's cats probably arrived on site dead, as pelts. Hedgehog heads came on site, but not the rest of the animals.

We now know much about the ways in which Neolithic villagers raised, hunted, butchered, cooked, ate and displayed animals in and around Çatalhöyük. We see a fundamentally stable but not static animal economy and a richly complex set of traditions involving food sharing, symbolic deployment of animal remains and ritualised house establishment and closure. Alive, animals were constant presences not just around the settlement but inside it; dead, they filled bellies, adorned fingers and necks and protruded from walls and benches. At Çatalhöyük, 25 years of faunal research have revealed centuries of human-animal entanglements.

Acknowledgements

We thank David Orton, Julie Daujat, Moussab Albesso, Kamilla Pawłowska, Hayley Foster, Richard Madgwick, Julia Best, Jennifer Jones and Adrienne Powell for their analytic and conversational contributions: this report is much richer for their input. We have benefited immensely from the hard work of previous faunal team members, including Nerissa Russell, Louise Martin, Levent Atıcı, Denise Carruthers, Sheelagh Frame, Robert Symmons and Lisa Yeomans. Non-zooarchaeologist members of the ÇHRP have sharpened our thinking and enhanced our understanding of human-animal interactions at Neolithic Çatalhöyük. Finally, we thank Ian Hodder, Bilge Küçükdoğan and the staff of the ÇHRP dig house: without your help none of our research would have been feasible. Our analyses have been supported by Cardiff University, the Wenner-Gren Foundation (Grant No. 9192) and the National Science Foundation (Grant No. 0647131).

Supplementary material

For supplementary material related to this chapter, please visit https://doi.org/10.18866/BIAA/e-13. It comprises appendix S8.1 (general appendix, with 20 figures: S8.1–S8.20 and 23 tables: S8.1–S8.23) and appendix S8.2 (measurements appendix).

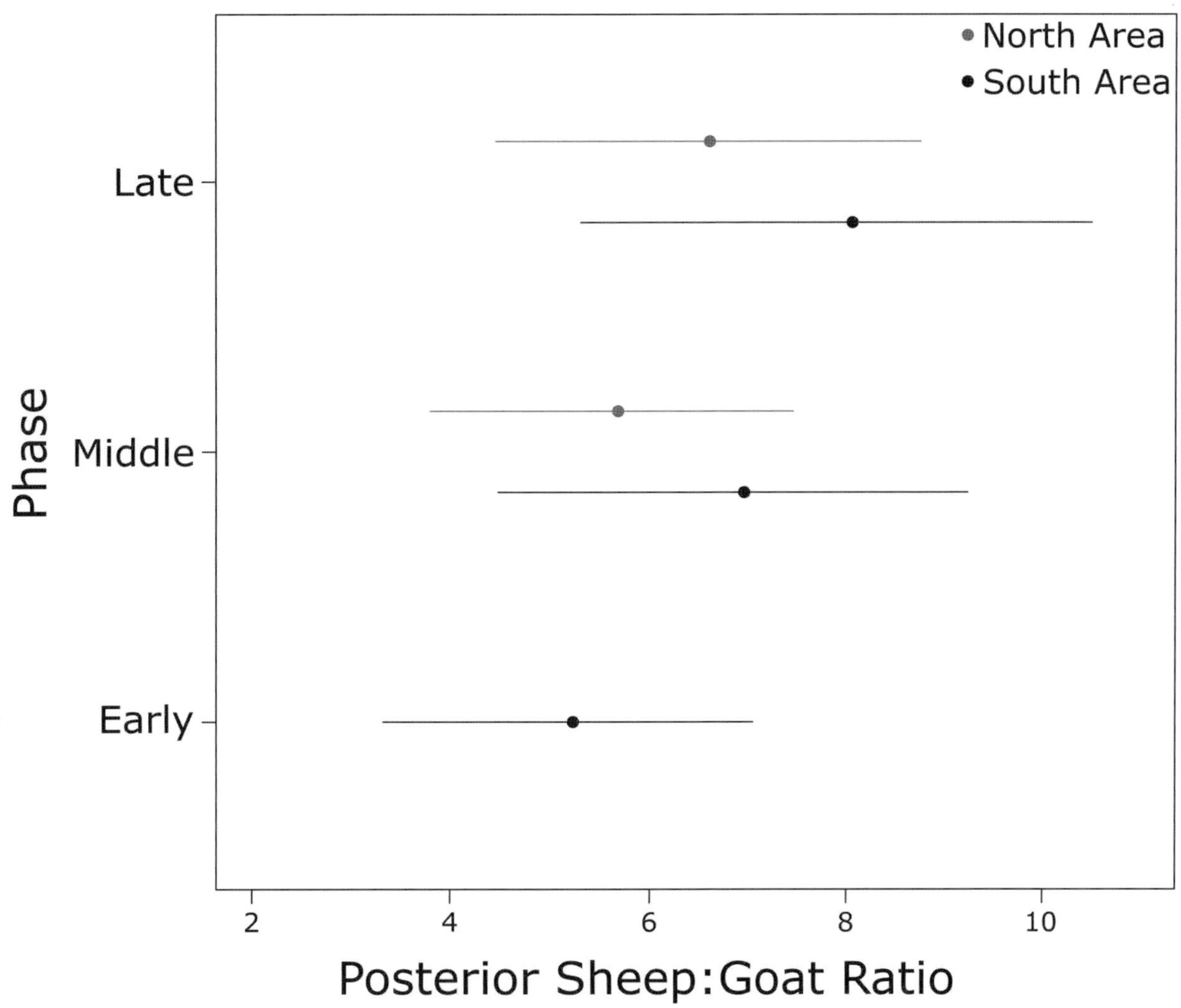

Figure 9.2. Sheep:goat ratios in the North and South Areas through time. Periods are Early, Middle and Late. Ratios are defined as the number of sheep per one goat. Points are posterior mean values, while lines denote 95% HPDIs (McElreath 2016b). North Area assemblages are shown above contemporary South Area assemblages.

Early–Late Çatalhöyük periods, though some fluctuations suggest potential differences in herding intensity (Manning et al. 2015). The proportion of young sheep is low (<20%) but relatively stable in Early–Middle Çatalhöyük, but in Late Çatalhöyük (Levels South R–T, North H–J) drops below 10% on average. This drop in the proportion of young animals is not reflected by any change in the adult sex ratio, which remains relatively female-dominated throughout the Early–Late Çatalhöyük periods. On average, females constitute 67% of the adult populations (95% HPDI: 56–77%).

We applied the same modelling approach (with the same sub-assemblage definitions) to remains identified specifically as *Capra* in order to estimate goat management at Çatalhöyük East. Table 9.3 tallies goat measurements from the sub-assemblages, while table 9.4 shows the measured estimates of the demographic parameters (π_{young} and θ_{female}).

While the *Capra* model results are based on a dataset that is only roughly an eighth the size of the *Ovis* dataset, we still gain insight into Çatalhöyük goat herding practices. The biometric model traits suggest that there were no major changes to goat management from the Early–Late Çatalhöyük periods (fig. 9.4). Goats may have trended towards larger body sizes throughout the occupation, which could indicate that as time passed fewer goats were giving birth at early ages (Özgül et al. 2009; Manning et al. 2015). In contrast to sheep, we see

Measurement	*South G-I*	*South J-L*	*South M-O*	*North F-G*	*South P-Q*	*South R-T*	*North H-J*	*Standard value*
Scapula GLP	3	3	10	21	10	17	35	33.0
Scapula LG	3	4	10	22	11	18	38	26.0
Humerus Bd	6	22	26	60	36	60	58	33.0
Humerus BT	6	24	27	61	38	63	59	29.5
Radius Bp	2	8	9	19	39	38	28	33.5
Radius BFp	2	8	10	17	39	37	26	30.5
Radius Bd	1	3	4	12	15	32	28	31.0
Metacarpal Bp	0	0	11	26	6	3	9	25.0
Metacarpal Bd	4	9	11	38	14	47	28	26.5
Femur DC	1	4	8	6	8	8	5	21.1
Femur Bd	0	0	0	3	0	1	0	39.5
Tibia Bp	0	0	1	4	0	2	1	42.5
Tibia Bd	8	12	14	51	46	60	84	26.5
Astragalus Bd	17	14	17	47	41	74	183	19.6
Metatarsal Bp	2	4	6	14	4	5	4	22.5
Metatarsal Bd	5	9	5	27	19	20	30	26.0
Individual specimens	49	91	114	307	237	363	493	Total specimens: 1,654

Table 9.1. Ovis *measurement counts for Çatalhöyük periods. Measurements defined by von den Driesch (1976). Standard values are the measurements of an* Ovis orientalis *female (FMC 57951) (Uerpmann, Uerpmann 1994: table 12). 'Astragalus Bd' measurement originally listed as 'Astragalus BC'. Individual Specimens excludes multiple measurements per specimen.*

Sub-assemblage	π_{young} *Measurement*	θ_{female} *Measurement*
Early Çatalhöyük (South G–I)	0 / 15	5 / 7
Early Çatalhöyük (South J-L)	5 / 21	5 / 11
Middle Çatalhöyük (South M-O)	5 / 50	16 / 21
Middle Çatalhöyük (North F-G)	23 / 126	25 / 31
Late Çatalhöyük (South P-Q)	16 / 77	12 / 29
Late Çatalhöyük (South R-T)	5 / 170	18 / 35
Late Çatalhöyük (North H-J)	10 / 226	32 / 43

Table 9.2. Demographic data to estimate Ovis π_{young} *and* θ_{female} *for Çatalhöyük sub-assemblages. All data come from fully recorded (T1 or L-form) contexts from the Hodder levels designated in sub-assemblage. For* π_{young}, *the numerator is the count of unfused first and second phalanges and the denominator is phalanges with any fusion data. For* θ_{female}, *the numerator is the number of morphologically female pelves and the denominator is the number of pelves assigned to a sex (either 'Female', 'Probably Female', 'Male' or 'Probably Male').*

an even sex ratio among adult goats, which suggests that both male and female animals were living into adulthood. This accords with the low proportion of young goats (mean: 6%) throughout the Early–Late Çatalhöyük periods. Allowing animals to survive into adulthood and keeping relatively similar proportions of male and female adults are consistent with Redding's (1981) model of herd security, though these data are also consistent with a variety of herding motivations (Marom, Bar-Oz 2009).

Because the sheep and goat mixture models rely on specimens assigned to either *Ovis* or *Capra*, they likely underestimate the proportion of young animals. As noted above, young specimens are unlikely to preserve the relevant morphological traits or those traits may be less well-defined in young animals (Zeder, Lapham 2010). Could the Late Çatalhöyük decline in the proportion of young sheep be an artefact of recording unfused specimens regularly as Sheep/Goat rather than trying to

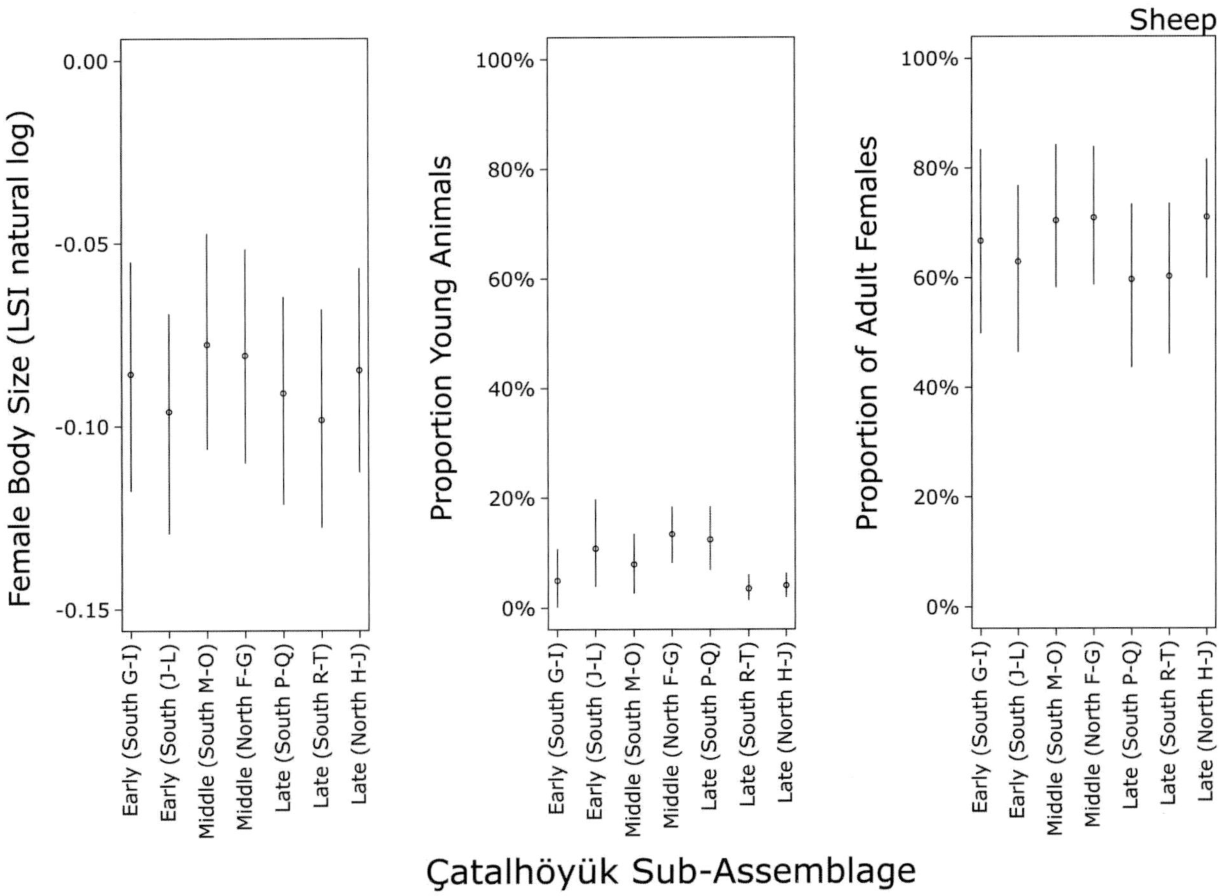

Figure 9.3. Ovis *modelled demography through time. Posterior estimates of female body size (μ_{female}), proportion of young animals (π_{young}) and adult sex ratio (θ_{female}) across different Çatalhöyük sub-assemblages. Points are posterior mean values, while lines denote 95% HPDIs (McElreath 2016b). Sub-assemblages are ordered in roughly chronological order, with South Area sub-assemblages occurring before contemporary North Area sub-assemblages. Female body size is in LSI units ($\log_e$ scale).*

speciate them? As most remains assigned only to Sheep/Goat are likely sheep (see fig. 9.1), we undertook a new mixture model using both *Ovis* and Sheep/Goat specimens (relevant measurements and demographic estimates in tables 9.5 and 9.6).

Adding almost 1,000 specimens to the *Ovis* dataset does not materially alter our sub-assemblage estimates of μ_{female} and θ_{female} (although these estimates do become more precise). We take this as confirming that most Sheep/Goat remains derive from sheep (fig. 9.5). Including Sheep/Goat material does, however, dramatically increase our estimated proportions of young animals in all sub-assemblages: the average proportion of young sheep rises to 26%, compared to an average of 7% when only using *Ovis* remains. We also see the proportion of young sheep declining gradually over the course of the Early–Late Çatalhöyük periods. Young sheep make up ~35% of Early Çatalhöyük Ovis and Sheep/Goat remains, on average; this drops to 27% in Middle Çatalhöyük levels and to 19% in Late Çatalhöyük levels. Levels South R–T still have a significantly lower proportion of young sheep (mean: 12%) than do earlier levels of the site.

Discussion

Caprine management at Çatalhöyük was focused on sheep; sheep and goats were probably herded collectively, and the small number of goats may have been included to improve the efficiency of sheep herding and long-term survival of herds (Dahl, Hjort 1976). Sheep stable oxygen isotopic data and dental microwear suggest that sheep herding at Çatalhöyük required neither regular long-distance movement outside of the alluvial plain nor supplemental foddering (Henton 2013). On-site evidence for sheep penning, based on micromorphological evidence of compacted dung lenses and calcitic spherulites (for

Measurement	*South G-I*	*South J-L*	*South M-O*	*North F-G*	*South P-Q*	*South R-T*	*North H-J*	*Standard value*
Scapula GLP	1	1	0	4	3	6	7	35.7
Scapula LG	1	1	1	6	3	5	7	28.5
Humerus Bd	1	2	0	7	4	3	7	35.0
Humerus BT	1	3	0	7	4	3	7	34.3
Radius Bp	1	2	0	3	4	7	7	35.5
Radius BFp	1	1	0	3	4	7	6	34.4
Radius Bd	1	1	0	3	3	5	3	33.3
Metacarpal Bp	0	0	0	2	0	1	3	27.5
Metacarpal Bd	1	0	0	0	2	3	2	30.5
Femur DC	0	0	0	2	0	1	0	23.0
Femur Bd	1	0	0	0	0	0	0	42.5
Tibia Bp	0	0	0	0	0	0	0	45.8
Tibia Bd	1	1	0	9	7	11	24	28.0
Astragalus Bd	1	3	1	6	8	7	20	20.3
Metatarsal Bp	0	0	0	1	1	0	3	22.5
Metatarsal Bd	0	0	0	0	2	6	6	28.5
Individual specimens	8	11	2	39	34	47	80	Total specimens: 221

Table 9.3. Capra *measurement counts for Çatalhöyük periods. Measurements defined by von den Driesch (1976). Standard values are the average measurements of a male and female* Capra aegagrus *(BMNH 651M, BMNH 651L2) (Uerpmann, Uerpmann 1994: table 14). 'Astragalus Bd' measurement originally listed as 'Astragalus BC'. Individual Specimens excludes multiple measurements per specimen. 'Tibia Bp' listed for completeness but not included in the model (no archaeological specimens).*

Sub-assemblage	π_{young} *Measurement*	θ_{female} *Measurement*
Early Çatalhöyük (South G–I)	0 / 2	0 / 1
Early Çatalhöyük (South J-L)	1 / 7	0 / 0
Middle Çatalhöyük (South M-O)	1 / 5	0 / 4
Middle Çatalhöyük (North F-G)	1 / 9	2 / 2
Late Çatalhöyük (South P-Q)	1 / 7	1 / 5
Late Çatalhöyük (South R-T)	2 / 18	2 / 5
Late Çatalhöyük (North H-J)	2 / 19	1 / 1

Table 9.4. Demographic data to estimate Capra π_{young} *and* θ_{female} *for Çatalhöyük sub-assemblages. All data come from fully recorded (T1 or L-form) contexts from the Hodder levels designated in sub-assemblage. For* π_{young}, *the numerator is the count of unfused first and second phalanges and the denominator is phalanges with any fusion data. For* θ_{female}, *the numerator is the number of morphologically female pelves and the denominator is the number of pelves assigned to a sex (either 'Female', 'Probably Female', 'Male' or 'Probably Male').*

example, Sp.620 in Level South G), proves that at least some sheep lived within the community and that dealing with sheep herds would have been a regular part of life for portions of the Çatalhöyük community (Chapter 14).

Biometric modelling suggests that goat management at Çatalhöyük was oriented towards resiliency and long-term survival. We have no evidence for management aimed at specific products. Demographic parameters (π_{young} and θ_{female}) are consistent with limited culling of young animals; so are goat epiphyseal fusion patterns, though both fusion and biometric data likely underestimate the number of young animals killed, due to the lack of Sheep/Goat specimens (Twiss et al. this volume, Chapter 8: fig. 8.12.ii). The even adult sex ratio does not appear to be an artefact of wild goat hunting, as our estimated male and female body sizes are both significantly smaller than

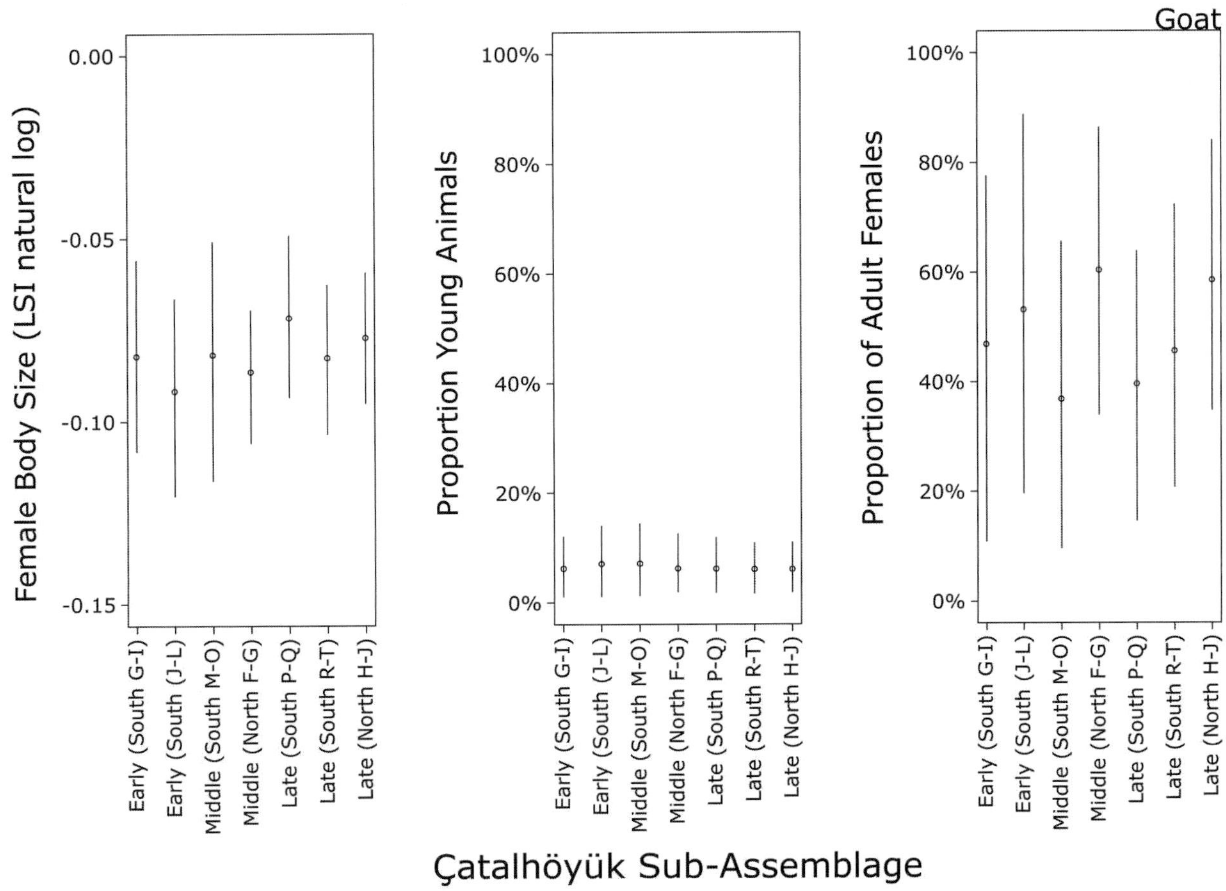

Figure 9.4. Capra *modelled demography through time. Posterior estimates of female body size (μ_{female}), proportion of young animals (π_{young}) and adult sex ratio (θ_{female}) across different Çatalhöyük sub-assemblages. Points are posterior mean values, while lines denote 95% HPDIs (McElreath 2016b). Sub-assemblages are ordered in roughly chronological order, with South Area sub-assemblages occurring before contemporary North Area sub-assemblages. Female body size is in LSI units (log$_e$ scale).*

the standard animal values based on modern wild goats. We see lower proportions of goats among postcranial remains than cranial remains. It appears that wild goats did not contribute much at all to the meat diet at Çatalhöyük through the Early–Late periods; perhaps hunters valued wild goats more for their horns than for their meat.

The paleo-demographic structures of sheep sub-assemblages are distinct from those of contemporary goat assemblages at Çatalhöyük, suggesting that the motivations and goals of sheep herding were distinct from those of goat herding. Adult sex ratios were skewed towards females, indicating that more males were killed at a young age – a strategy that could point towards a general interest in increasing herd sizes (Zeder, Hesse 2000; Arbuckle, Atıcı 2013). While the high proportions of young sheep could be due to a strategy of young male culling, it is noteworthy that the proportions decline from the Early through Late Çatalhöyük periods. Almost all faunal material from Çatalhöyük South G–I (Early Çatalhöyük) comes from contexts associated with penning activity (Spaces 181, 198, 199, 620 and 628), where we might expect relatively higher proportions of immature animals and neonates. It is also possible that the high proportions of young sheep in Early Çatalhöyük indicate higher neonatal/infant mortality among sheep populations. If so, this may reflect husbandry practices that were not conducive to infant survival: cramped conditions in on-site stalls, for example (e.g., Gillis et al. 2014).

Late period Çatalhöyük sheep were herded across a wider variety of environments than earlier flocks, according to stable isotopic data (Pearson et al. 2015; Pearson this volume, Chapter 11). This may mean that herds were spending less time on or near the site. The lower proportions of young animals in the Late period – coupled with the skewed adult sex ratio – might also indicate that not all animals were living near the site year-round (e.g., Bréhard et al. 2010). Comparisons with the rock shelter Pınarbaşı B, occupied at roughly the same time as Çatal-

Measurement	*South G-I*	*South J-L*	*South M-O*	*North F-G*	*South P-Q*	*South R-T*	*North H-J*	*Standard value*
Scapula GLP	3	7	14	30	17	26	62	33.0
Scapula LG	3	11	16	32	19	28	78	26.0
Humerus Bd	6	24	31	64	49	83	72	33.0
Humerus BT	6	26	32	68	57	89	78	29.5
Radius Bp	3	11	14	20	46	42	31	33.5
Radius BFp	3	9	13	18	42	40	29	30.5
Radius Bd	2	5	6	12	19	37	31	31.0
Metacarpal Bp	1	1	15	43	34	45	40	25.0
Metacarpal Bd	4	9	13	43	21	50	34	26.5
Femur DC	1	9	14	35	26	35	42	21.1
Femur Bd	0	2	1	4	2	10	8	39.5
Tibia Bp	0	0	3	10	5	8	10	42.5
Tibia Bd	17	16	39	74	92	89	115	26.5
Astragalus Bd	21	24	25	54	54	89	195	19.6
Metatarsal Bp	4	5	9	32	25	36	27	22.5
Metatarsal Bd	5	12	6	31	22	20	33	26.0
Individual specimens	67	131	186	436	417	571	716	Total specimens: 2,524

Table 9.5. Ovis *and sheep/goat measurement counts for Çatalhöyük sub-assemblages. Measurements defined by von den Driesch (1976). Standard values are the measurements of an* Ovis orientalis *female (FMC 57951) (Uerpmann, Uerpmann, 1994: table 12). 'Astragalus Bd' measurement originally listed as 'Astragalus BC'. Individual Specimens excludes multiple measurements per specimen.*

Sub-assemblage	π_{young} *Measurement*	θ_{female} *Measurement*
Early Çatalhöyük (South G–I)	56 / 150	15 / 23
Early Çatalhöyük (South J-L)	48 / 120	11 / 27
Middle Çatalhöyük (South M-O)	48 / 155	34 / 57
Middle Çatalhöyük (North F-G)	94 / 362	38 / 66
Late Çatalhöyük (South P-Q)	70 / 255	40 / 77
Late Çatalhöyük (South R-T)	78 / 375	59 / 120
Late Çatalhöyük (North H-J)	63 / 526	107 / 160

Table 9.6. Demographic data to estimate sheep and sheep/goat π_{young} *and* θ_{female} *for Çatalhöyük sub-assemblages. All data come from fully recorded (T1 or L-form) contexts from the Hodder levels designated in sub-assemblage. For* π_{young}*, the numerator is the count of unfused first and second phalanges and the denominator is phalanges with any fusion data. For* θ_{female}*, the numerator is the number of morphologically female pelves and the denominator is the number of pelves assigned to a sex (either 'Female', 'Probably Female', 'Male' or 'Probably Male').*

höyük's Middle and Late periods and with cultural ties to Çatalhöyük (Baird et al. 2011), indicate a general change in sheep herding strategies on the Konya Plain. Pınarbaşı B has a very high proportion of young sheep. This, along with increased variability in sheep stable carbon and nitrogen isotopic values, suggests that sheep herding in Late Çatalhöyük may have been more extensive than in earlier periods, with fewer young sheep living in or near the Çatalhöyük settlement itself (Carruthers 2006; Pearson 2013).

Cattle (*Bos primigenius* and *Bos taurus*)

Domestic cattle are sources of long-term security and wealth in modern agropastoral communities thanks to their greater resilience and lifespan compared to smaller stock, factors that would have also been important in the past (Dahl, Hjort 1976; Russell 1998). As a reliable source of manure, milk and potentially labour, cattle drastically impact not only animal economies but the agricultural opportunities for communities who have

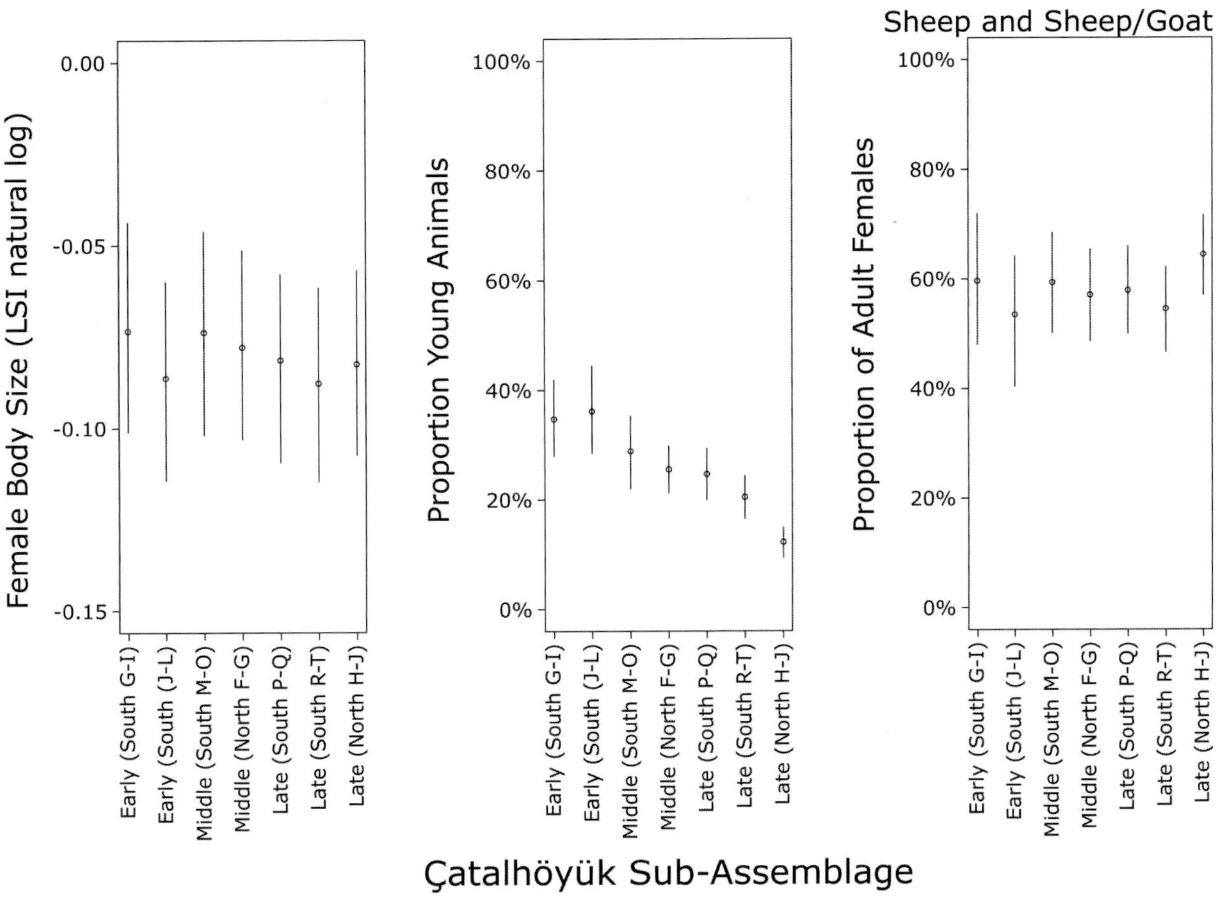

Figure 9.5. Ovis *and sheep/goat modelled demography through time. Posterior estimates of female body size (μ_{female}), proportion of young animals (π_{young}) and adult sex ratio (θ_{female}) across different Çatalhöyük sub-assemblages. Points are posterior mean values, while lines denote 95% HPDIs (McElreath 2016b). Sub-assemblages are ordered in roughly chronological order, with South Area sub-assemblages occurring before contemporary North Area sub-assemblages. Female body size is in LSI units ($\log_e$ scale).*

access to them (Borgerhoff Mulder et al. 2010). The origins and spread of domestic cattle in southwest Asia have been studied extensively, generally in site-specific studies that explore the impacts of cattle keeping on local communities (cf. Grigson 1989; Helmer et al. 2005; Arbuckle et al. 2014, 2016).

We investigate the appearance of domestic cattle at Çatalhöyük using a multilevel mixture model of animal bone measurements to estimate cattle biometry through time. This allows us to evaluate the likelihood of different scenarios of cattle domestication and/or adoption at the site. We compare cattle body sizes during different occupation periods at Çatalhöyük with the sizes of ancient Anatolian cattle confidently inferred to have been either wild or domestic. Because comparative data indicate that domesticated cattle are significantly smaller than wild cattle, we model these body size estimates with a separate mixture model to define the probability that assemblages are consistent with domestic cattle (Monchot, Léchelle 2002).

Modelling cattle exploitation in Neolithic central and southwest Anatolia

We model cattle body sizes and demographic variables for Çatalhöyük and eight other assemblages from central and southwest Anatolia (table 9.7) using our multilevel mixture model. The model uses the same finer chronological and spatial sub-assemblages as the caprine datasets but supplements them with TP Area (divided into Late Çatalhöyük and Final Çatalhöyük periods) and Çatalhöyük West data, as these assemblages are probably dominated by domestic cattle (Russell et al. 2013b; Pawłowska in prep.) (table 9.8). The other assemblages, spanning roughly 3,000 years of cattle exploitation, contextualise the scale of biometric changes during the Çatalhöyük occupation. The central Anatolian assemblages encompass sites where cattle were definitively wild (Aşıklı Höyük and Musular: Russell et al. 2005) and two sites where the status of the cattle was less clear but may have included domestic animal herding (Erbaba

Höyük and Kosk Höyük II-III: Arbuckle 2013a). The southwest Anatolian sites were included because they are broadly contemporaneous with Çatalhöyük's occupation and have clear evidence for domestic cattle herding throughout the occupation sequence (Arbuckle et al. 2014). Metric and demographic data for most of the sites come from Open Context, except for data for Höyücek, Bademağaci Höyük, Aşıklı Höyük and Musular (Duru, De Cupere 2003; Russell et al. 2005; De Cupere et al. 2008; Arbuckle 2013b; 2013c; Çakırlar 2013b; Galik 2013; Orton et al. 2013; Marciniak 2013).

The Bayesian multilevel mixture model estimated the biometry and demographic proportions of the assemblages. Figure 9.6 shows the posterior estimates for these demographic parameters for the Çatalhöyük sub-assemblages. Unlike previous analyses of cattle body sizes at Çatalhöyük (e.g., Arbuckle, Makarewicz 2009), our model suggests that cattle became gradually smaller throughout the site's occupation. Body sizes are generally stable within occupation periods, though cattle from Levels South P–Q are more similar to cattle from Middle Çatalhöyük levels than to those from Late Çatalhöyük levels. While the proportion of young animals stays relatively stable in the face of decreasing body size, θ_{female} increases gradually. Table 9.9 shows the average adult sex ratio (females per male) across different periods; Early and Middle Çatalhöyük cattle sex ratios were relatively even, in contrast to Late and Final/West sex ratios, which were definitively female-dominated.

In isolation, this pattern appears to fit an 'autochthonous' domestication process, paralleling patterns at Çayönü Tepesi or Mehrgarh (Meadow 1984; Hongo et al. 2009). However, when we consider the Çatalhöyük sequence in the context of broadly contemporary central Anatolian and southwest Anatolian data (fig. 9.7, table 9.7), this image becomes more complicated. Figure 9.7 shows that Early and Middle Çatalhöyük

Site	*Date estimate (cal BCE)*	*Specimens*	*Measure-ments*	π_{young} *Measurement*	θ_{female} *Measurement*	*Reference*
Aşıklı Höyük*	8400–7400	67	85	75 / 283	10 / 13	Russell et al. 2005
Musular*	7600–7200	236	268	2 / 58	13 / 16	Russell et al. 2005
Early Çatalhöyük (South G–I)*	7100–6700	26	31	9 / 36	2 / 4	OSF**
Early Çatalhöyük (South J-L)*	7100–6700	97	112	23 / 83	1 / 2	OSF
Middle Çatalhöyük (South M-O)*	6700–6500	70	81	15 / 78	1 / 5	OSF
Middle Çatalhöyük (North F-G)*	6700–6500	89	129	7 / 47	7 / 10	OSF
Late Çatalhöyük (South P-Q)	6500–6300	73	93	9 / 50	4 / 5	OSF
Late Çatalhöyük (South R-T)	6500–6300	32	38	7 / 46	5 / 7	OSF
Late Çatalhöyük (North H-J)	6500–6300	51	62	8 / 47	4 / 4	OSF
Late Çatalhöyük (TP M-N)	6500–6300	11	14	2 / 15	0 / 1	Marciniak 2013
Final Çatalhöyük (TP O-R)	6300–5950	11	13	5 / 16	0 / 1	Marciniak 2013
Çatalhöyük West	6000–5500	17	19	2 / 12	0 / 0	Orton et al. 2013; Orton et al. 2018
Erbaba Höyük	6500–6000	84	99	9 / 67	1 / 1	Arbuckle 2013a
Köşk Höyük	6200–5500	74	78	14 / 113	0 / 0	Arbuckle 2013b
Höyücek†	6400–6200	30	35	2 / 16	0 / 0	Duru, De Cupere 2003
Bademağacı Höyük†	6500–6000	40	46	0 / 0	0 / 0	De Cupere et al. 2008
Çukuriçi Höyük†	6200–6000	12	14	5 / 17	2 / 2	Galik 2013
Ulucak Höyük†	6500–5660	25	28	7 / 50	2 / 2	Çakırlar 2013

Table 9.7. Counts of unfused Bos *phalanges and female* Bos *pelves for Çatalhöyük sub-assemblages, Central Anatolian Sites and southwest Anatolian Sites. Specimens – the number of unique measured specimens. Measurements – the total number of included measurements. For π_{young}, the numerator is the count of unfused first and second phalanges and the denominator is phalanges with any fusion data. For θ_{female}, the numerator is the number of morphologically female pelves and the denominator is the number of pelves assigned to a sex (either 'Female', 'Probably Female', 'Male' or 'Probably Male'). Reference is source of measurement, sex, age and dating data; date estimates for Çatalhöyük East sub-assemblages from online database.*

** Site used to define the 'hunting' group; † Site used to define the 'herding' group*

*** Associated Open Science Framework (OSF) page: https://osf.io/mz97t/*

Measurement	*South G-I*	*South J-L*	*South M-O*	*North F-G*	*South P-Q*	*South R-T*	*North F-G*	*Standard value*
Scapula GLP	2	5	7	29	16	2	7	89.0
Scapula LG	4	7	9	31	14	2	9	70.0
Humerus Bd	3	3	4	7	3	3	2	97.0
Humerus BT	3	7	5	12	7	2	2	89.0
Radius Bp	0	10	1	7	5	3	2	100.0
Radius BFp	0	9	1	8	5	2	3	90.0
Radius Bd	0	4	5	5	2	2	2	92.0
Metacarpal Bp	2	10	4	4	3	1	2	74.0
Metacarpal Bd	4	13	10	4	8	5	4	73.0
Femur DC	1	10	9	3	3	2	5	57.0
Femur Bd	0	1	1	0	0	1	0	116.0
Tibia Bp	0	3	0	2	0	0	0	126.0
Tibia Bd	1	11	6	7	3	7	6	78.0
Astragalus Bd	5	4	11	7	19	3	8	52.0
Metatarsal Bp	3	4	5	2	4	3	2	62.0
Metatarsal Bd	3	11	3	1	1	0	8	68.0
Individual specimens	26	97	70	89	73	32	51	Total specimens: 438

Table 9.8. Bos *measurement counts for Çatalhöyük North and South Area sub-assemblages. Measurements defined by von den Driesch (1976). Standard values are measurements of a female* Bos primigenius *(Ullerslev) (Degerbøl 1970). 'Humerus Bd' measurement updated following Grigson (1989: table 1). Individual Specimens excludes multiple measurements per specimen.*

cattle are broadly consistent with wild aurochsen from Aşıklı Höyük and Musular – and Çatalhöyük cattle from Late Çatalhöyük through Çatalhöyük West are similar in size to those from Erbaba Höyük and Köşk Höyük (Arbuckle, Makarewicz 2009; Arbuckle 2013a). We are uncertain whether these latter cattle are morphologically domestic: they are larger than contemporary southwest Anatolian herded cattle (e.g., De Cupere et al. 2008; Çakırlar 2012).

What might be driving the larger body sizes in central Anatolian cattle? The size difference between central Anatolian and southwest Anatolian cattle could be due to environmental differences between the two regions (Zeder 2001; Wright, Viner-Daniels 2015). Environmental differences could also explain why Çatalhöyük sheep are larger than sheep from southwest Anatolian sites like Ulucak Höyük (Çakırlar 2012, 2013b). It is also possible that the larger body size could be due to the continued presence of some wild cattle in these assemblages (Orton 2010; Russell et al. 2013b). If aurochs remains were present in an assemblage, the estimates of both μ_{male} and μ_{female} would increase because the model would likely consider aurochs as males, thus some male domesticates would be interpreted as females.

Evaluating scenarios of when domestic cattle appeared at Çatalhöyük

Mixture modelling easily separates central Anatolian wild aurochsen from southwest Anatolian domestic cattle in female body size (μ_{female}) values (fig. 9.7). Early and Middle Çatalhöyük cattle are entirely consistent with aurochsen from earlier central Anatolian sites Aşıklı Höyük and Musular (Russell et al. 2005). Cattle μ_{female} values from Late–Final Çatalhöyük and Çatalhöyük West do not fit neatly into either domestic cattle herds from southwest Anatolia or aurochs populations from earlier central Anatolian assemblages. Do these assemblages represent small aurochsen or large domestic cattle? We can answer this question probabilistically by mixture modelling cattle μ_{female} values from the 'known' sites (Monchot, Léchelle 2002).

Mixture models use categories populated by 'known' data in order to estimate membership probabilities (π_{site}) for unknown data. Here π_{site} means the probability that an assemblage accords with southwest Anatolian herding assemblages. To evaluate π_{site} for the Late, Final and West Çatalhöyük μ_{female} values, we defined two 'known' categories: Aurochsen Hunting (μ_{female} from Aşıklı Höyük, Musular and Early–Middle Çatalhöyük) and

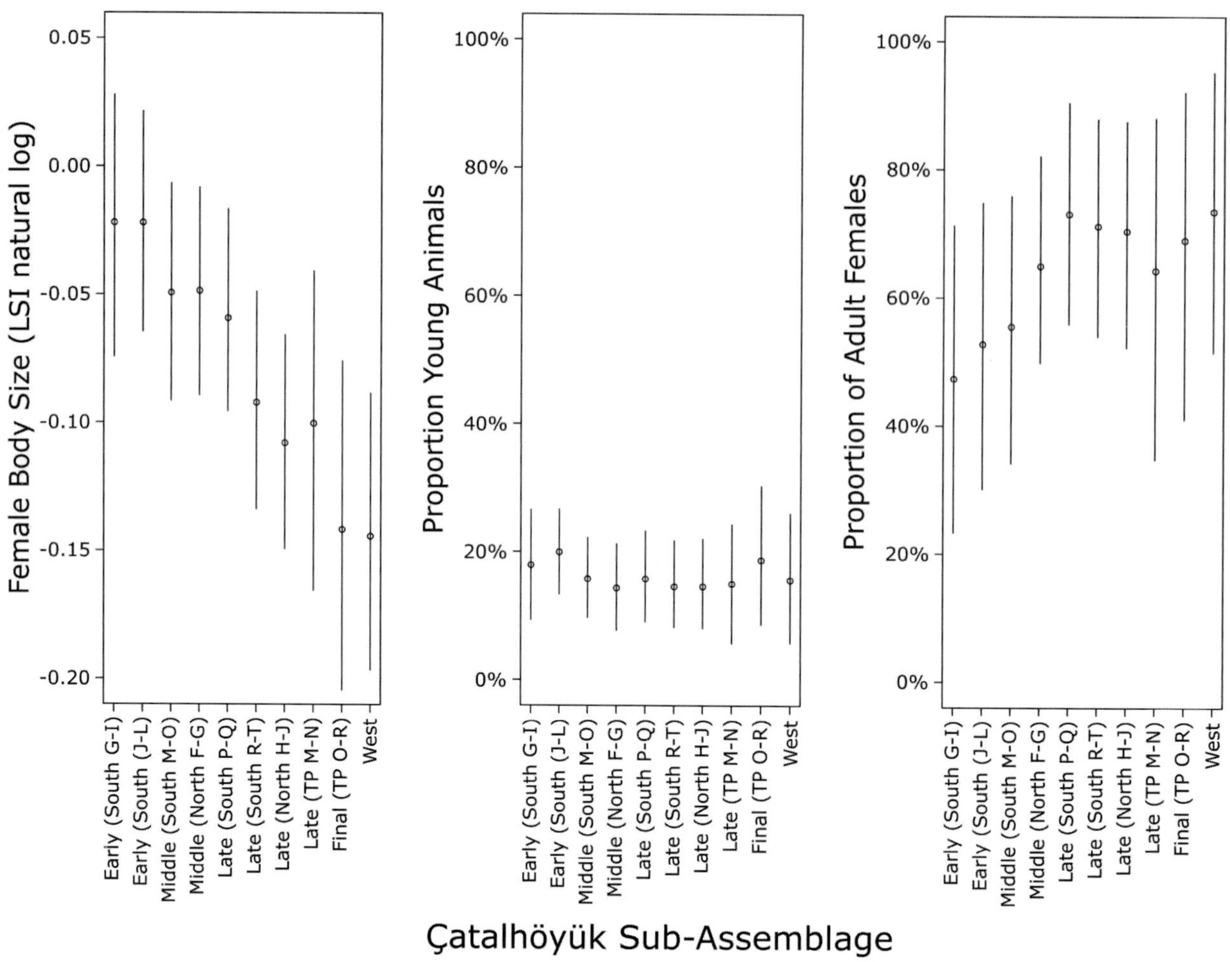

Figure 9.6. Bos *modelled demography through time. Posterior estimates of female body size (μ_{female}), proportion of young animals (π_{young}) and adult sex ratio (θ_{female}) across different Çatalhöyük sub-assemblages. Points are posterior mean values, while lines denote 95% HPDIs (McElreath 2016b). Sub-assemblages are ordered in roughly chronological order, with South Area sub-assemblages occurring before contemporary North Area sub-assemblages. Female body size is in LSI units (log_e scale).*

Cattle Herding (μ_{female} from Höyücek, Bademağacı, Çukuriçi Höyük and Ulucak Höyük). We modelled each group with normal distributions; these distributions define the likelihood that an assemblage with a specified μ_{female} came from that assemblage (fig. 9.8). The following formula defines an unknown site's probability of being a herded site (π_{site}), where x_i is the μ_{female} value for that site:

$$\pi_{site}[i]=\pi * Normal(x_i \mid \mu_{herding},\sigma_{herding}) + (1 - \pi) * Normal(x_i \mid \mu_{hunting},\sigma_{hunting})$$

We defined the overall mixture probability π as 0.4 because the sample sizes of the two known distributions are uneven (4 herding sites / 10 'known' sites).

Occupation phase	*Mean θ_{female}*	*Mean sex ratio (females per male)*
Çatalhöyük Early	50%	1:1
Çatalhöyük Middle	60%	1.5:1
Çatalhöyük Late	70%	2.33:1
Çatalhöyük Final	69%	2.23:1
Çatalhöyük West	74%	2.85:1

Table 9.9. Estimated adult Bos *sex ratios in different Çatalhöyük sub-assemblages. Mean θ_{female} values calculated by averaging across relevant sub-assemblages. θ_{female} – the female proportion of adult cattle.*

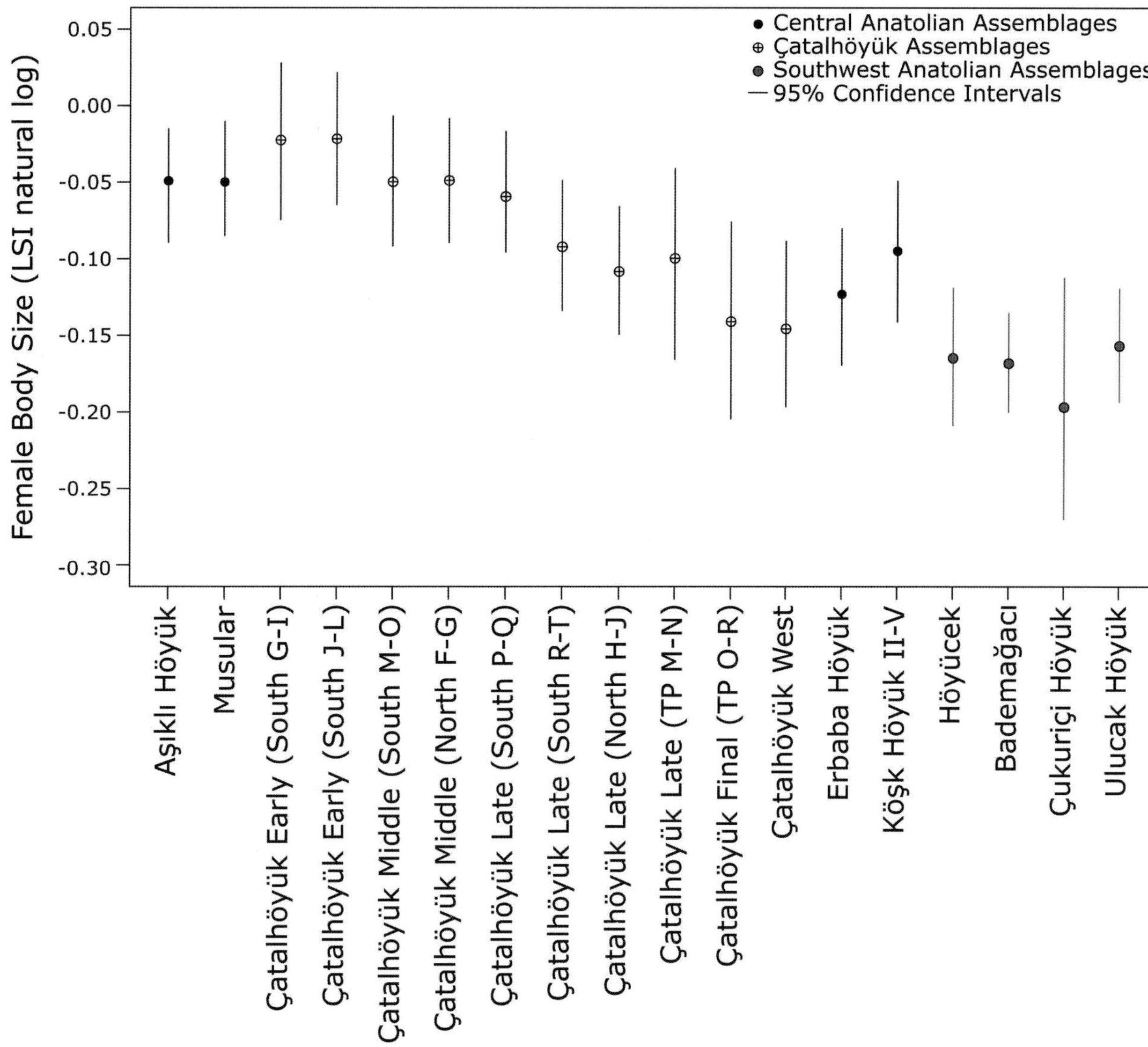

Figure 9.7. Comparison of posterior Bos *female body size for Çatalhöyük sub-assemblages, Central Anatolian sites and southwest Anatolian sites. Points are posterior mean values, while lines denote 95% HPDIs (McElreath 2016b). Assemblages are ordered in roughly chronological order, with South Area sub-assemblages occurring before contemporary North Area sub-assemblages. Female body size is in LSI units (log$_e$ scale).*

Table 9.10 shows the mean and median π_{site} values for the Late, Final and West Çatalhöyük assemblages. These π_{site} values are independent of one another but we can multiply them together to calculate the probability of different cattle domestication/adoption scenarios (table 9.11). We assume that once domesticated cattle were introduced to Çatalhöyük they remained available, so we exclude scenarios wherein domestic cattle phase in and out of use across occupation periods. We do include the scenario wherein Late, Final and West Çatalhöyük villagers kept no domesticated cattle. We calculate that it is probable that domestic cattle were herded in at least one of the Late Çatalhöyük sub-assemblages (~75% probability); at the same time, we calculate that it was highly unlikely that domestic cattle were herded in all four of the Late Çatalhöyük sub-assemblages (~1% probability). Levels South P–Q have a low probability of fitting with the herding distribution; South R–T also has a moderately low probability. We infer that villagers began keeping domesticated cattle during the Late Çatalhöyük period, but that cattle herding was not necessarily well established as a practice until the Final and West Çatalhöyük occupations.

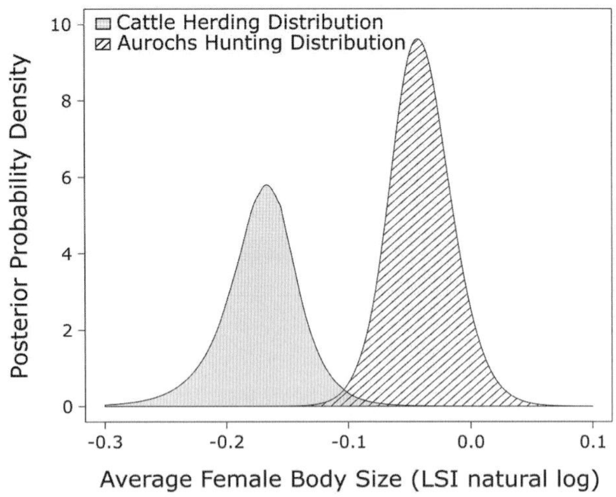

Figure 9.8. Distributions of cattle herding and aurochs hunting female body sizes. Distributions used for the mixture model to calculate π_{site} *values. 'Aurochs Hunting' distribution is based on the average* μ_{female} *values from Aşıklı Höyük, Musular and Early–Middle Çatalhöyük. 'Cattle Herding' distribution is based on the average* μ_{female} *values from Höyücek, Bademağacı, Çukuriçi Höyük and Ulucak Höyük. Height of distributions shows the relative likelihood that an assemblage with the corresponding* μ_{female} *value is consistent with that exploitation group.*

Discussion

Our results complicate and enrich earlier interpretations of how and when cattle herding began to be practised at Çatalhöyük. First, we find that cattle body sizes and adult sex ratios changed gradually throughout the site's occupation: there was no dramatic shift with the introduction of a new population (cf. Arbuckle, Makarewicz 2009). Local aurochs populations changed over the course of centuries, presumably in response to sustained hunting pressure and landscape modifications; hunting techniques may have been modified in efforts to maintain the viability of local herds (Hecker 1982; Marom, Bar-Oz 2013). Plausibly Çatalhöyük's inhabitants were engaged in cattle 'proto-domestication' (*sensu* Ducos 1988), investing more and more effort over time to maintain and attract local aurochs herds. Analyses on the ritual uses of cattle remains – including artistic representations of cattle in figurines and wall art – find evidence for changes in the nature human-cattle interactions over the course of the site's occupation (e.g., Twiss, Russell 2009; Martin, Meskell 2012; Johnsen 2019). We can also see, in our mixture modelling results, evidence for Çatalhöyük's inhabitants' deepening entanglement with local aurochs populations (Fuller et al. 2016). The continued importance of these human-aurochs interactions may partly underlie the continued presence of aurochs hunting in the later central Anatolian assemblages (Arbuckle 2013a).

Second, while our probabilistic estimates of when cattle herding was established at Çatalhöyük are broadly concordant with those of Russell et al. (2013a), our results provide increased chronological and spatial resolution that highlights the complexity of the process. Domesticated cattle were probably first herded in the Late Çatalhöyük period – but this introduction was variable across different occupation areas. The low probability that Levels South P–Q deposits contain domesticated cattle suggests that cattle herding is introduced to the community *after* the Middle–Late Çatalhöyük transition and associated changes to social organisation and geography (cf. Hodder, Doherty 2014). We have no clear evidence for domesticated cattle in periods with inferred or potential penning areas (South Area Sp.339: Shillito, Ryan 2013; North Area Sp.636: Chapter 14): perhaps these areas were not used for cattle keeping or perhaps they represent early attempts to manage morphologically wild cattle. That domesticated cattle were probably present in Levels North H–J but not in South P–Q highlights the possibility that the North and South Areas employed different economic and/or land-use strategies. Such differentiation is consistent with agricultural and cooking data as well (Bogaard et al. 2017; González Carretero et al. 2017; Volume 15, Chapter 14), exciting thoughts of distinct agropastoral norms in different – potentially contemporary – portions of the settlement.

Conclusions

Our biometric models of the major animal groups exploited at Çatalhöyük – caprines and cattle – produce new insights into how people interacted with these animals through the Early–Late Çatalhöyük periods. Modelling sheep:goat ratios across different body elements underscores the importance of sheep herding at the site while also highlighting a potential focus on the heads and horns of goats. Mixture modelling *Bos* data clarifies the long-term trajectory of human-cattle relations at Çatalhöyük. The higher chronological and spatial resolution provided by the mixture model also allows us to probabilistically explore different scenarios of when and how domesticated cattle were introduced to the community.

Biometric modelling of caprine remains shows that management strategies for sheep and goats may have differed drastically. Mixture modelling suggests that goat postcranial remains were predominantly domestic: the morphologically wild goat horn cores found on site may be the result of people hunting wild goats but not returning their whole bodies. While goat management at Çatalhöyük does not appear to have been focused on producing any specific products, sheep herding may have been focused on producing meat. Female-dominated adult populations and high proportions of young animals

Sub-assemblage	*Mean π_{site}*	*Median π_{site}*
Çatalhöyük Late (South P–Q)	2%	0%
Çatalhöyük Late (South R–T)	32%	5%
Çatalhöyük Late (North H–J)	59%	82%
Çatalhöyük Late (TP M–N)	46%	25%
Çatalhöyük Final (TP O–R)	87%	100%
Çatalhöyük West	93%	100%

Table 9.10. Mean and median π_{site} values for Late, Final and West Mound assemblages. π_{site} represents the probability that a sub-assemblage is consistent with domestic cattle herding rather than aurochs hunting. Distributions of π_{site} values calculated for each sub-assemblage for each posterior sample. π_{site} calculated using the mixture model equation based on the sub-assemblage's μ_{female} value.

Scenario of domestic cattle's introduction	*Average weighted probability*
Present in one Late Çatalhöyük sub-assemblages	28%
Present in two Late Çatalhöyük sub-assemblages	28%
Present only in Final and West Çatalhöyük	17%
Present in three Late Çatalhöyük sub-assemblages	16%
Present only in West Çatalhöyük	7%
No domestic cattle in Neolithic Çatalhöyük (through West)	3%
Present in all four Late Çatalhöyük sub-assemblages	1%

Table 9.11. Probabilities of different scenarios. Scenarios listed in order of weighted probability, from most likely to least likely. Probability of each scenario calculated by taking the product of independent π_{site} values for each sub-assemblage. In scenarios wherein domestic cattle were introduced in the Late period, domestic cattle were also considered present in Final and West Mound periods.

suggest that Çatalhöyük's herders practised young male culling (Arbuckle, Atıcı 2013). Proportions of young sheep decline in the Late Çatalhöyük period based on the mixture modelling, though caprine dental eruption/wear data show a parallel trend: only 5% (9/169) of Late Çatalhöyük caprine mandibles are Payne Stages A–D (under ca 2 years old: Payne 1973) compared to 11% (63/587) of Early and Middle Çatalhöyük caprine mandibles (see also Twiss et al. table 8.12). The decline in young sheep proportions may reflect a reorganisation of sheep herding wherein sheep travelled more extensively across the Konya Plain and fewer sheep were born near Çatalhöyük (Pearson 2013); such reorganisation might correspond to the use of nearby herding stations like Pınarbaşı B (Baird et al. 2011).

Over the course of Çatalhöyük's occupation, aurochs hunting was not always the same activity. Local aurochs populations adapted to changing local landscapes and hunting pressures, while hunters refined their hunting strategies to avoid extirpating local herds. This long-term trend can be seen as 'cultural manipulation' (*sensu* Hecker 1982: table 1a). When accounting for how Çatalhöyük's residents manipulated local habitats and environments more broadly (Asouti, Kabukcu 2014; Kabukcu 2017), we can see human-aurochs interactions at Çatalhöyük as cultural niche construction (O'Brien, Laland 2012; Smith 2007). In terms of the lived experience of human-cattle interactions, Çatalhöyük's inhabitants may have been entangled in local cattle lifeways throughout the site's occupation. However, cattle remains are not consistent with cattle herding assemblage until sometime in the Late Çatalhöyük period. Earlier cattle penning areas in the North and South Areas may testify to these entanglements, as people invested more and more effort into maintaining access to local animals. Adoption or integration of morphologically domestic cattle from elsewhere does not appear to have materially changed practices established in earlier periods, as aurochs remains appear in *Bos* assemblages well after the adoption of cattle herding (Arbuckle 2013a).

The models we present in this chapter represent baselines in human-animal interactions to further explore connections with other materials and aspects of animal lives (Sykes 2014). The integration of sheep and goat herding likely allowed herders to experiment with new strategies of sheep herding with goats as a resilient fallback (Mace 1993); this resilient management strategy allowed herders to innovate and new strategies to spread when advantageous (Bogaard et al. 2017). Mixture modelling cattle remains suggests that cattle management should not be seen as a particular on/off invention, but rather as part of a long-term process of adaptation to local aurochs populations. The results of this long-term process can be seen in other arenas of social and economic life like landscape use and task scheduling. Together, these biometric models show that human-animal interactions at Çatalhöyük were always under renegotiation.

Acknowledgements

We thank David Orton, Hayley Foster, Richard Madgwick, Julia Best, Jennifer Jones, Adrienne Powell, Julie Daujat, Moussab Albesso, Virginia Garcia-Díaz and Kamilla Pawłowska for their contributions as both analysts and colleagues. This research would not have been possible without the foundations set by previous faunal team members, such as Nerissa Russell, Louise Martin, Levent Atıcı, Denise Carruthers, Sheelagh Frame, Robert Symmons and Lisa Yeomans. Finally, we thank Ian Hodder, Bilge Küçükdoğan and the staff of the ÇRP dig house for making all our research possible. Heather Lynch and Jeroen Smaers provided crucial statistical advice and support. The development of these analyses was supported by the Wenner-Gren Foundation (Grant No. 9192).

10. The Çatalhöyük Microfauna

Michelle C. Feider and Emma Jenkins

Introduction

Microfauna are often viewed as non-cultural material and therefore used for palaeoenvironmental reconstruction only (Edwards, Martin 2007). However, as well as being useful for reconstructing palaeoenvironments, microfauna can be used to explore how archaeological sites were inhabited (O'Connor 1988). A shift in changes from anthrophobic to anthropophilic species can help identify increasing levels of urbanisation and the reverse, eventual abandonment of sites, as species represented can reflect recolonisation due to human habitation, or lack thereof. As such, microfauna can be a useful tool in understanding more about archaeological sites, rather than simply in interpreting past environments. They can be used to explore their potential use as part of a broad-spectrum economy, in line with current ethnographic examples today (Peršič 1998; Crittenden, Zes 2015) and may have played a role in ritual activities on sites (Jenkins 2012a). Microfauna can also be used to understand the transition from mobile or semi-sedentarising groups to those who occupy permanent settlements in the landscape (Hesse 1979; Tchernov 1991; Weissbrod 2017).

Microfauna can enter sites through different means, as live animals or brought in by an avian or mammalian carnivore such as an owl, a fox or a cat. Predation can potentially be seen in the microfaunal assemblage as digestion evident on the bones and teeth of the prey, or gnaw marks on the bones, as well as skeletal element representation and fragmentation (Andrews 1990) due to hunting strategy. Much work has been done on recognising the taphonomic signatures of predation on small mammals (Mellet 1974; Mayhew 1977; Andrews 1990; Fernandez-Jalvo, Andrews 1992; 2003; Terry 2007); however, other microvertebrate species, for example, herpetofauna, have not seen the same amount of study (Denys et al. 2017).

At Çatalhöyük, microfauna have been studied since the initial phases of excavation by James Mellaart in the 1960s and were reported by Brothwell (1981). Since excavation work at the site was renewed in 1995, the previous microfaunal assemblages were analysed by Jenkins (2005; 2009) for the 1995–1999 seasons, Jenkins and Yeomans (2013) for the 2000–2008 seasons, and Jenkins (2012b) for the BACH Area.

Materials and method

Recovery

To date, a total of 7,412 specimens have been analysed and recorded from 19 of the available 214 priority units (table 10.1) that were recovered during this final excavation period.

Microfauna at Çatalhöyük are recovered from the heavy residue as part of the flotation process and sieved through 4mm, 2mm and 1mm stacking sieves. Percentages of the heavy residue samples are then sorted by local women who remove 'shaped' and potentially diagnostic bones. No record was kept of unidentifiable specimens, as those were also mixed with small fragments of macrofaunal remains. Due to the varying size of samples taken at Çatalhöyük, sub-sampling during the heavy residue process can vary from 6.25% of a large sample to 100% of smaller samples. This decision is at the discretion of the person responsible for the heavy residue processing (see Volume 14, Chapter 2). Material examined for this analysis largely came from the 2mm and 1mm flotation samples, and the percentages to which they were sorted can be found in table 10.2. Identifiable elements and taphonomically interesting specimens were then sorted by Jenkins prior to export to the UK for analysis. This was necessary because export of any material from Çatalhöyük is problematic and the number of specimens selected for export has to be kept to a minimum.

The assemblage was sorted and analysed using a Brunel Stereomicroscope and a comparative collection of Turkish microfauna, held at Bournemouth University and curated by Jenkins. Scanning Electron Microscopy (SEM) was undertaken using a JEOL JSM-6010LV SEM at Bournemouth University. This is an Environmental SEM, which means that specimens do not need to be coated, and hence destroyed, for analysis.

Taxonomic and taphonomic analysis

Micromammal remains were identified to species or genus level where possible, using the cranial elements, while herpetofauna were identified taxonomically using diagnostic elements, such as the ilia, scapula and humeri, with the assistance of Dr Chris Gleed-Owen, a herpetofauna specialist. Sub-species of the genus *Mus* have previously been identified from Çatalhöyük using both

Unit	*Category*	*Interpretive category*	*Building*	*Space*	*Area*	*Mellaart level*	*Hodder level*
21367	Layer	Infill	119	512, 513	North	Unassigned	North F
21573	Layer	Oven superstructure	119	513	North	Unassigned	North F
23215	Layer	Plaster floor	131	500	North	Unassigned	North G
30543	Layer	Dirty floor		17	North	Unassigned	North ?G
30554	Layer	Dirty floor		17	North	Unassigned	North ?G
30591	Layer	Dirty floor patches, occupational surface	132	511	North	Unassigned	North F
32717	Layer	Burial infill	132	531	North	Unassigned	North F
32782	Layer	Dirty floor	132	531, 633	North	Unassigned	North F
32793	Layer	Occupational sediment	132	633	North	Unassigned	North F
18523	Layer	Burnt collapse	79	134	South	Unassigned	South O
18578	Layer	Dump room infill	80	135	South	Unassigned	South O
19802		Fill between walls	76, 80	135, 137	South	Unassigned	South O
21814	Layer	Burial fill	17	170	South		South K
32403	Layer	Burial infill	160	551	South	IX	South K
32611	Layer	Arbitrary infill of burial	161	605, 606	South	IX	South J
32616	Layer	Roof collapse, primary room infill	161	605, 606	South	IX	South J
32632	Layer	Oven superstructure	161	605, 606	South	IX	South J
30217	Layer	Infill layer	110	486	TPC	Unassigned	TPC N
30269	Layer	Infill	110	486	TPC	Unassigned	TPC N

Table 10.1. Details of units analysed for microfauna.

skeletal identification following Harrison and Bates (1991) and geometric morphometrics (Cucchi, pers. comm. December 2018). Analysis has shown that *Mus musculus domesticus* is present and, as it outcompetes *M. macedonicus* in a commensal niche, an assumption has been made that all *Mus* spp. therefore belong to *Mus musculus domesticus*. Other species were identified to genera only, as due to the limited quantity recorded, identification to species level would not have altered the results of the analyses. Species-level identification will be confirmed at a later date. In some cases, where diagnostic portions of elements were present, it was possible to distinguish between rodent and insectivore post-crania and these were recorded as such. Where no diagnostic features were present, post-cranial remains were then recorded simply as micromammal. If it was unclear as to whether the specimen was mammalian or herpetofauna, then it was recorded as microfauna; however, this category was rarely used. Insectivore is used here to mean a small mammal that feeds primarily on insects, worms and other invertebrates formerly in the obsolete order *Insectivora* (now largely replaced with *Eulipotyphla*).

Taphonomic processes affecting the specimens were also recorded, which included fragmentation and breakage, digestion, gnawing, weathering and burning. The methodology for taphonomy followed Andrews (1990), Fernandez-Jalvo and Andrews (1992), Pinto Llona and Andrews (1999), Jenkins (2009) and Fernandez et al. (2017) so as to be compatible with previously recorded microfaunal assemblages.

Incisor digestion followed the methodology of Fernandez-Jalvo and Andrews (1992) and murid molar digestion using diagrams from Fernandez et al. (2017: 143, fig. 3), with post-cranial digestion following Andrews (1990) and Jenkins (2009).

Digestion has been analysed on a unit-by-unit basis for those with a Number of Identified Specimens (NISP) greater than 100. Generally, units should not be combined when looking at digestion, as units may have had different taphonomic pathways. As well as levels of digestion, the presence of gnawing was noted and levels of fragmentation explored in order to understand the potential predator, or predators, responsible for accumulation. SEM was used to examine taphonomy and measure puncture marks.

Unit	Flot no.	NISP	% sampled (HR)	Corrected NISP for 100%	Volume of sample (L)	NISP per litre
18523	8973	4	100	4	24	0.17
18578	9122	3	100	3	24	0.13
19802	9825	228	100	228	28	8.14
21367		9	Hand collected?			
21573	11610	78	100	78	24	3.25
21814	13007	47	25	188	473	0.4
23215	13256	1	100	1	2.5	0.4
30217	10356	1	50	2	37.5	0.05
30269	10528	1	100	1	27	0.04
30543	10624	2	100	2	4	0.5
30554	10629	10	100	10	27	0.37
30591	10777	2	100	2	35	0.05
32403	12823	24	100	24	263	0.09
32611	13008	836	75	1115	105	13.5
32611	13008	301	100	301	105	N/A
32616	12940	1265	100	1265	40	31.6
32632	13138	1779	87.5	2033	1	4547
32632	13138	2514	100	2514	1	N/A
32632	13038	239	100	239	24	10
32717	13163	6	100	6	28	0.21
32782	13376	8	25	32	75	0.53
32782	13376	4	50	8	75	N/A
32782	13378	19	100	19	49	0.4
32782	13386	11	100	11	10	1.1
32782	13387	10	100	10	25	0.4
32782	13379	1	100	1	60	0.02
32793	13392	3	100	3	28	0.11
32793	13394	4	100	4	35	0.11
32793	13396	2	100	2	25	0.08

Table 10.2. Number of Identified Specimens (NISP) by unit and corrected NISPs per litre for comparison.

Quantification

NISPs were calculated for each unit and the term 'specimen' included both complete elements and identifiable fragments of elements (O'Connor 2000). The NISPs were then adjusted by multiplying the numbers up to represent the number expected if 100% of the sample had been available, and a NISP per litre was calculated so units could be compared.

Minimum Numbers of Elements (MNEs) were calculated by taking into account the portion of the specimen recorded, for example, proximal versus distal, as well as siding of the element (O'Connor 2000). These numbers were used to both calculate Minimum Numbers of Individuals (MNIs) and examine body part representation in all units that had a NISP greater than 100.

Results

NISP and species composition

Whole assemblage. NISPs for each unit, corrected for sampling strategy, and NISP per litre can be found in table 10.2. Only four units had a NISP greater than 100, and this included combined NISPs for all units within a space or building. They were (19802), fill between walls of Buildings 76 and 80; (36211), arbitrary infill of a burial in B.161; (32616), primary room infill of B.161 and; (32632), oven superstructure in B.161.

Species composition for the whole assemblage (fig. 10.1) shows that rodents dominate the assemblage, with the next largest category being micromammals. Herptofauna and insectivores together make up only 1% of the assemblage, where rodents account for at least 64%. When the larger groupings are removed, the assemblage is still dominated by rodents, with *M. m. domesticus* accounting for 96% of species within the assemblage (fig. 10.2). Anurans and insectivores make up 1.4% and 2.4% respectively. *Meriones* spp. and *Microtus* spp. were represented by loose molars only, and *Natrix* spp. only by vertebrae, and each forms a very small part of the overall assemblage.

Building 161. Three of the four units that have a NISP greater than 100 were excavated from B.161 in the South Area. Unit (32632) comes from phase 1.3 of the building, whereas (32616) and (32611) come from Phase 2, meaning they are potentially not contemporary accumulations.

Unit (32632). This unit was interpreted as oven superstructure and contained the highest NISP per litre of the assemblage at 4,547 specimens per litre.

Of the identifiable species in the assemblage, *M. m. domesticus* once again dominates, with 97.7%, followed by *Crocidura* spp. at 2.2% and *Microtus* spp. at 0.1%.

Units (32616) & (32611). Units (32611) and (32616) are associated with the primary room infill of B.161 and had a NISP per litre of 13.5 and 31.6 respectively. Again, these assemblages were dominated by rodents, with unit (32611) being comprised of 64% rodent and unit (32616), 75% rodent; both units had anurans and insectivores present in low numbers. Unit (32611) also contained a single vertebra of *Natrix* spp. and unit (32616), a single *Meriones* spp. molar.

Microfauna by unit category

Fill. In total, ten fills were analysed, including (19802), (32611) and (32616), which were amongst the largest units recorded. Unit (21367) was composed of testudines carapace fragments, which had been hand collected during excavation rather than recovered through flotation. Because of this, there is no NISP per litre information for this unit.

The average adjusted NISP per litre for this interpretive category is 6.02; however, if the three larger units are removed the NISP per litre becomes 0.15, which appears to better reflect the other units in this category.

This category contained a broad range of microfauna, including *M. m. domesticus*, *Meriones* spp., *Crocidura* spp., *Natrix* spp., *Pelophylax* spp., *Pelobates* spp. and *testudinae* spp. The majority of the rodent species were from the larger fills, (19802), (32611) and (32616), with the others having a higher number of anura, even though MNIs remained low (table 10.3).

Construction/make-up/packing. Three units were analysed from this category and they included (32632), which was the largest unit recorded here. The adjusted NISP per litre was calculated as 1,140, however, this was inflated by the inclusion of (32632), which was a dense cluster of microfauna recovered from the packing around an oven and interpreted by the excavators as 'foundational deposits'. When (32632) is excluded from the calculation, the adjusted NISP per litre is 1.7, which again better reflects the other units in the category.

This category contains rodents, anura and insectivores, with the only incidence of *Microtus* spp. being recorded from (32632) and consisting of a single molar.

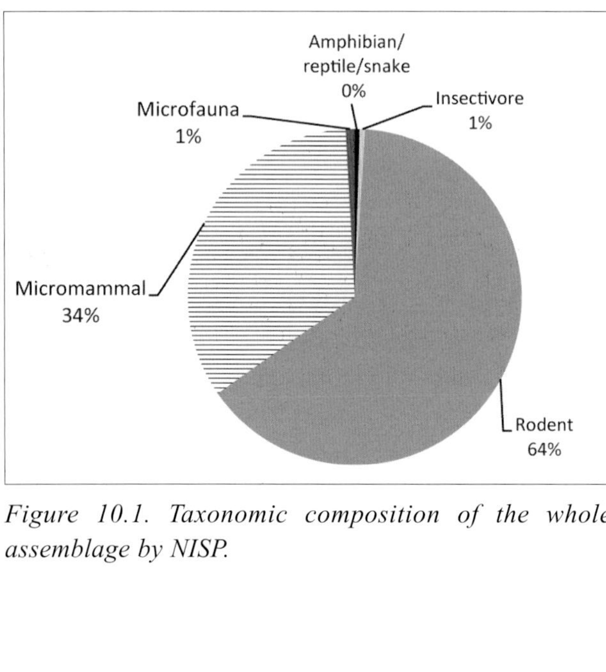

Figure 10.1. Taxonomic composition of the whole assemblage by NISP.

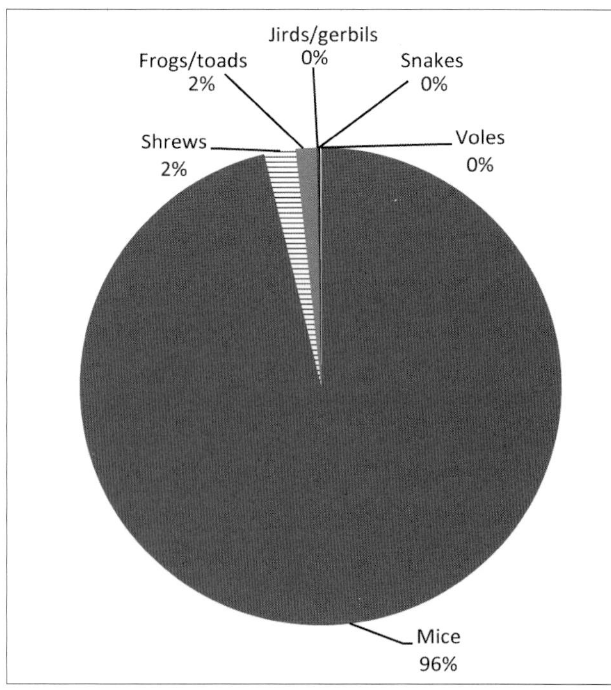

Figure 10.2. Taxonomic composition of the whole assemblage by NISP with rodent, micromammal and microfauna excluded.

	19802	*21573*	*21814*	*23215*	*32403*	*32611*	*32616*	*32632*	*32717*	*32782*	*32793*	*21367*	*18578*	*18523*	*30554*	*30269*	*30543*	*30217*	*30591*
Microfauna				1						1	1								
Micromammal	3	3			1	14	20	85		1	1								
Rodent	17	3	5		2	61	101	239	1	3	1				1				1
Insectivore						1	1	1											
Mus sp.	17	2	2		2	42	70	177	1	3									
Crocidura sp.	1		1		1			6											
Anuren						2	1	1				1		1	1		1	1	
Meriones	1						1												
Microtus								1											
Testudine												1							
Natrix						1							1						
Pelophylax ridibundus													2						
Ranidae														1					
Bufo viridis															1				
Pelobates sp.																1			

Table 10.3. Minimum Numbers of Individuals (MNI) per unit, based on actual NISP counts for MNE, and not corrected for sample volume.

Floors. This is the only unit category that did not contain a microfaunal cluster and therefore gave an adjusted NISP per litre of 0.3 for the six units analysed. It contained rodents and anura, but no other genus was recorded.

Body part representation

Body part representation was calculated for all units with a NISP over 100 (fig. 10.3), with both loose and in situ molars and incisors included in the calculations. This was to determine if there was any underrepresentation of certain body parts that may have suggested selective hunting strategies by predators or processing by humans.

Generally, elements are represented to varying degrees, however, teeth seem to represent the highest fractions, except for (32611), where vertebrae are the most common element. Despite the fact that there are high numbers of tibiae, femora are less abundant than humeri suggesting that there is no bias in fore or hind limb representation (fig. 10.3).

The vertebra were mostly caudal, rather than from the entire length of the vertebral column. This is in accord with the results of Brothwell (1981), who focused on a microfaunal assemblage recovered from a human burial. His conclusion was that due to the lack of non-caudal vertebrae, scapula and ribs, the assemblage was produced from skinning practices (Brothwell 1981). No evidence of cut marks on the bones, however slight, has been recorded by us, and in many instances we see evidence of digestion, as well as gnaw marks from a small mammalian carnivore, suggesting that the accumulation was most likely due to predators hunting on site.

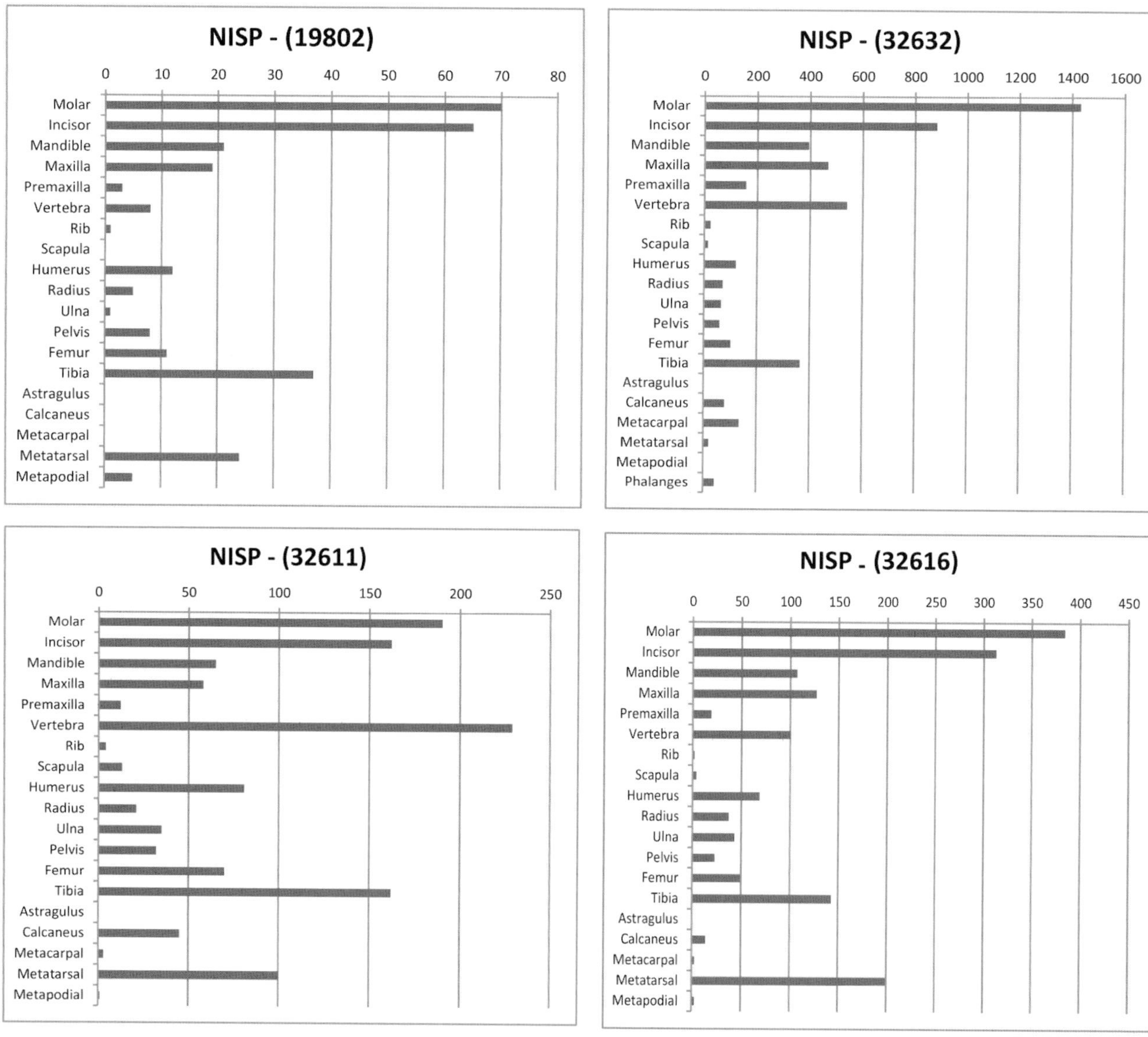

Figure 10.3. Relative proportions of elements for each unit with a NISP greater than 100.

Fragmentation

Fragmentation patterns for the units with a NISP of over 100 were analysed in order to understand the potential predator that may have been involved in assemblage formation (table 10.4). Very few complete elements were recovered, with the exception of the ulna from (32632). Shafts were also low in number, but this could be due to a lack of diagnostic features leading to an under-identification, rather than lack of presence. In the case of proximal and distal ends, the ones most represented are those that fuse early, for example the distal versus the proximal humerus. The distal end of the humerus fuses before the proximal end, and as many of the proximal humeri were recorded as being unfused, the difference in representation could be due to bone survivability, with unfused ends potentially being more susceptible to either breakage or digestion. The same pattern is seen in the femur, with the proximal end being represented in greater numbers because it fuses before the distal end.

Greater numbers of shafts of the tibia were recorded; however, this could be due to the tibia having a greater degree of diagnostic features all the way along the bone, rather than mainly at the articular ends as is the case for many of the other long bones.

	(19802)		*(32611)*		*(32616)*		*(32632)*	
	No.	*%*	*No.*	*%*	*No.*	*%*	*No.*	*%*
Humerus								
Complete			1	1.25	3	4.3	9	9
Proximal			19	23.75	1	1.4	19	19
Shaft	2	16.7	6	7.5	13	18.8	8	8
Distal	10	83.3	54	67.5	52	75.4	64	64
Ulna								
Complete			1	2.9			11	33.3
Proximal	1	100	4	11.4	26	70.3	21	63.6
Shaft			30	85.7	1	2.7		
Distal			0		10	27	1	3
Femur								
Complete	1	10	1	1.8	5	10.4	6	9.2
Proximal	8	80	33	58.9	26	54.2	35	53.8
Shaft			11	19.6	11	22.9	6	9.2
Distal	1	10	11	19.6	6	12.5	18	27.7
Tibia								
Complete	1	2.9			3	2.1	5	1.4
Proximal	2	5.7	16	9.9	5	3.5	13	3.7
Shaft	3	8.6	35	21.7	30	21.3	54	15.3
Distal	29	82.9	110	68.3	103	73	280	79.5

Table 10.4. Post-cranial breakage for units with a NISP greater than 100. Epiphyses not counted (micromammals only, including those identified to Genus).

As well as analysing which parts of the long bones were best represented, the percentage remaining was also recorded (table 10.5). In nearly all instances the fraction of bones that fell in the ¼ to ½ category was the most represented. This shows that the levels of breakage at the site were high, with few complete bones recorded. The numbers in the 'less than ¼' category are lower than expected, given the fragmentary nature of the assemblage, however this could be due to identification issues rather than an absence, potentially meaning that the assemblage is in fact more fragmentary than it appears in the analysis.

Maxillary (table 10.6) and mandibular breakage (table 10.7) was high, with 100% of maxilla and 72.5% of mandibles being broken to the greatest extent. Mandibles are more robust than maxilla and skull fragments, which may account for the slight differences in levels of breakage; however, it is interesting to note than none were recorded as complete.

Predator-induced modifications

Gnawing. Table 10.8 shows the percentage of gnawing on each element for each of the four units with a NISP greater than 100.

Unit (32632) has the highest percentage of gnawing, with 6.5% of the specimens displaying evidence of isolated or multiple puncture marks. This unit had the broadest range of elements being gnawed, with 14 different elements affected, including one basisphenoid skull fragment not included in table 10.8, compared to nine elements from (32611), eight elements from (32616) and six elements from (19802). Of all the gnawed elements in (32632), the tibiae were the most affected with 30% of all gnawed bone being tibiae and 24.4% of all tibiae being gnawed. The next most affected element was the mandible, 10.5% of all gnawed elements for this unit, which showed a consistent pattern of puncture marks on the lateral surface below the first and second molars (fig. 10.4). The femora were also highly affected, comprising 8.1% of all gnawed elements, and with 37.3% of all recorded femora being gnawed.

Unit (32616) had the next highest number of elements gnawed and showed a similar pattern to (32632), with the tibiae and mandibles most affected (tables 10.8 and 10.9). 6.2% of the unit had either isolated or multiple puncture marks and the patterning of gnaw marks on the mandible was similar to those in (32632).

Of the specimens in (32611), 2.7% were gnawed; however, the femora and the tibiae were the most affected (table 10.9). In addition, 15.4% of all mandibles, and 14.3% of all femora were gnawed (table 10.8).

		(19802)		(32611)		(32616)		(32632)	
Element	*Category*	*N*	*%*	*N*	*%*	*N*	*%*	*N*	*%*
Humerus	Less than 1/4			26	32.5	3	4.3	27	23.7
	1/4 – 1/2	7	58.3	34	42.5	39	56.5	38	33.3
	1/2 – 3/4	5	41.7	10	12.5	21	30.4	13	11.4
	More than 3/4			3	3.75	3	4.3	13	11.4
	Complete			1	1.25	3	4.3	9	7.9
	Epiphysis			6	7.5			14	12.3
Ulna	Less than 1/4			7	20	6	16.2	2	6.1
	1/4 – 1/2			14	40	15	40.5	10	30.3
	1/2 – 3/4	1	100	11	31.4	13	35.1	5	15.2
	More than 3/4			2	5.7	3	8.1	5	15.2
	Complete			1	2.9			11	33.3
	Epiphysis								
Femur	Less than 1/4			30	44.1	11	22.4	19	20.2
	1/4 – 1/2	6	54.5	18	26.5	22	44.9	26	27.7
	1/2 – 3/4	3	27.3	7	10.3	9	18.4	9	9.6
	More than 3/4					1	2	5	5.3
	Complete	1	9.1	1	1.5	5	10.2	6	6.4
	Epiphysis	1	9.1	12	17.6	1	2	29	30.9
Tibia	Less than 1/4	2	5.7	26	16.1	17	12.1	39	10.9
	1/4 – 1/2	24	68.6	92	57.1	72	51.1	227	63.2
	1/2 – 3/4	7	20	38	23.6	34	24.1	66	18.4
	More than 3/4	1	2.9	4	2.5	15	10.6	15	4.2
	Complete	1	2.9			3	2.1	5	1.4
	Epiphysis			1	0.6			7	1.9

Table 10.5. Post-cranial breakage categories for units with a NISP greater than 100 (micromammals only, including those identified to Genus).

In Unit (19802), 5.2% of elements were gnawed; however, as the NISPs were quite low, the percentages of all gnawed elements were broadly similar. The humeri and femora, however, did appear to be more affected, with 18.1% of all femora and 16.7% of all humeri showing evidence of isolated or multiple puncture marks.

A small sample of the gnawed mandibles from (32632) was examined using an SEM and measurements were taken of the puncture marks. The preliminary results for this line of analysis demonstrate that the average puncture size is approximately 0.69mm x 0.49mm; however, further work needs to be done to ensure these measurements are statistically sound. These puncture marks are larger than the ones found in previous analysis, if only marginally. No measurements were taken regarding the distance between multiple puncture marks on any single element.

Digestion. Table 10.10 shows digestion categories for incisors, both mandibular and maxillary, murid molars, distal humeri and proximal femora. None of the non-murid molars exhibited signs of digestion; however, this could be due to small sample size.

Outside of the four units with a NISP greater than 100, only two other units showed any evidence of digestion. Due to the small size of these units, the percentages of digestion are inflated and so do not represent the typical patterns seen in the larger units. In total, for the four units, 3.5% of incisors and 1.9% of molars were digested. The majority of the specimens were only lightly digested; there was some moderate and heavy digestion, primarily on the molars, but this was at very low levels. In contrast, the percentages of post-cranial digestion are higher, with 16.7% of humeri and 3.4% of femora being digested.

	A Complete, with zygomatic region		*B Complete with zygomatic but break to base of skull*		*C Broken with zygomatic intact*		*D Maxilla fragment lacking the zygomatic process*	
Unit	*N*	*%*	*N*	*%*	*N*	*%*	*N*	*%*
19802							19	100
32611							58	100
32616							118	100
32632							467	100

*Table 10.6. Maxillary breakage for all units with a NISP greater than 100 (*M. m. domesticus *and rodents only).*

	A Complete		*B Broken ascending ramus*		*C Ascending ramus missing*		*D Ascending ramus missing and inferior border broken*	
Unit	*N*	*%*	*N*	*%*	*N*	*%*	*N*	*%*
19802					4	20	16	80
32611					13	20.3	51	79.7
32616			7	6.5	26	24.3	74	69.2
32632			16	4.2	92	24	275	71.8

*Table 10.7. Mandibular breakage for all units that have a NISP greater than 100 (*M. m. domesticus *and rodents only).*

	(19802)			*(32611)*			*(32616)*			*(32632)*		
Element	*NISP*	*No. gnawed*	*% gnawing*	*NISP*	*No. gnawed*	*% gnawing*	*NISP*	*No. gnawed*	*% gnawing*	*NISP*	*No. gnawed*	*% gnawing*
Molar	70			190			384			1431		
Incisor	65	1	1.5	162	1	0.6	313	3	1	883	36	4.1
Mandible	21	2	9.5	65	10	15.4	107	28	26.2	394	31	7.9
Maxilla	19			58			127			467	1	0.2
Premaxilla	3			12			19			158	5	3.2
Vertebra	8			229	6	2.6	100	6	6	541	33	6.1
Rib	1			4			2			23		
Scapula				13			4			14	5	35.7
Humerus	12	2	16.7	81	1	1.2	69	11	15.9	120	22	18.3
Radius	5			21			37			71	8	11.3
Ulna	1			35			43	5	11.6	65	14	21.5
Pelvis	8	2	25	32	1	3.1	23	1	4.3	59	22	37.3
Femur	11	2	18.1	70	10	14.3	49	9	18.4	102	24	23.5
Tibia	37	3	8.1	162	3	1.9	143	15	10.5	365	89	24.4
Astragulus												
Calcaneus				45	3	6.7	14			79	4	5.1
Metacarpal				3			3			135		
Metatarsal	24			100			199			21		
Metapodial	5			1			3			1		
Phalanges										42		

Table 10.8. Percentage of gnawed elements from all units with a NISP greater than 100 (all microfauna included).

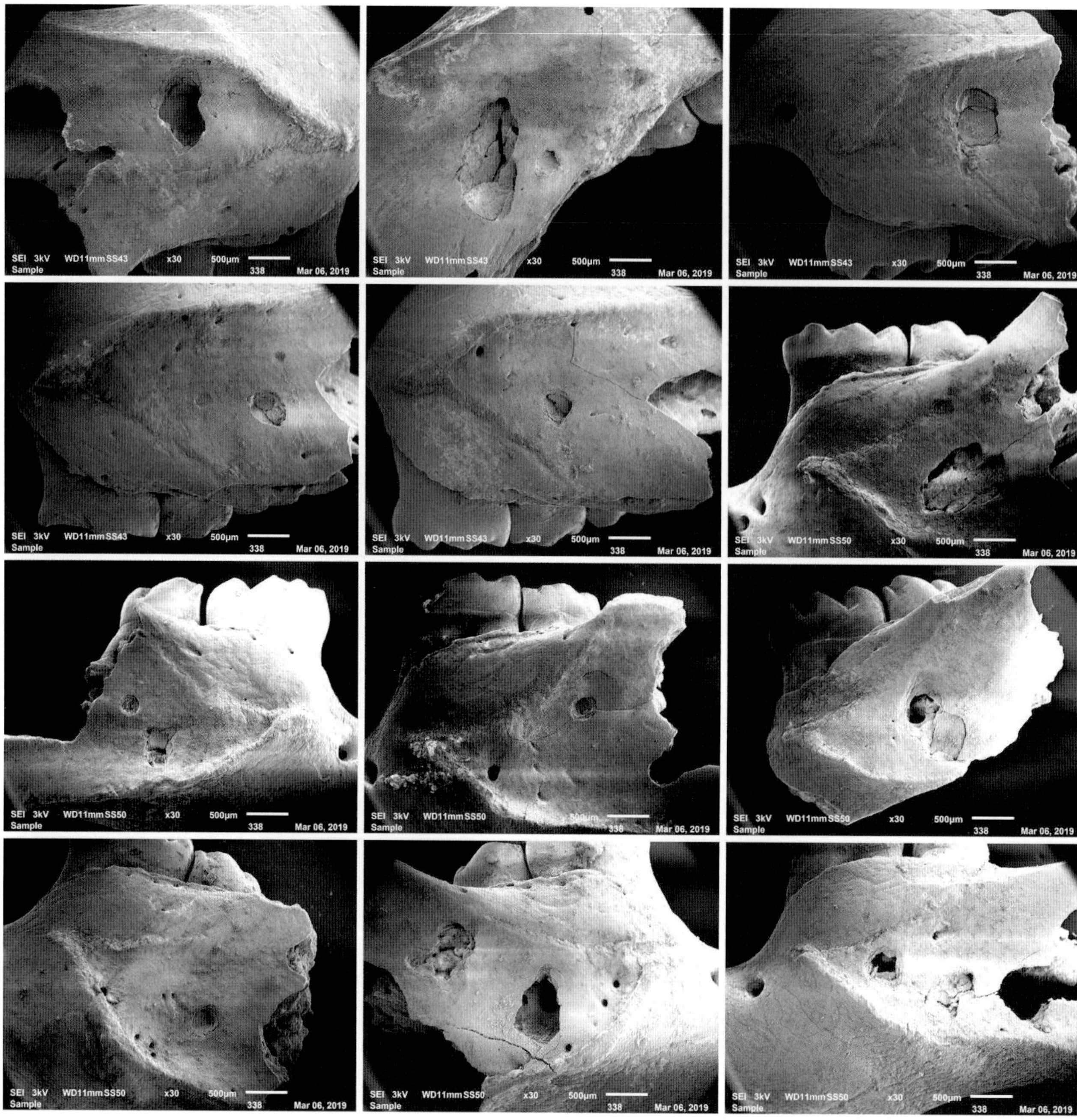

Figure 10.4. SEM micrographs displaying mandibles from (32632) with isolated and multiple puncture marks.

Discussion

Palaeoenvironmental reconstruction using the microfauna from Çatalhöyük shows that the immediate environment was very much an urban, or proto-urban, one, with settlement dense enough to encourage commensal species to the site and support them in reasonable numbers. Habitat preference for all species recovered on site can be found in table 10.11; however, due to the limited number of these species recovered on site, they are unlikely to be representative of the local environment. The domination of the assemblage by *M. m. domesticus*, including in the early levels, reflects how quickly commensal mice adapted to human-altered environments and excluded most other species. Other early sites, such as Tepe Ganj Dareh in western Iran (Hesse 1979), and cave sites such as Iraq el-Dubb in Highland Jordan (Edwards, Martin 2007), show early commensal species, however, they are also accompanied by a wide variety of other microfauna (Tchernov 1991).

Predators at Çatalhöyük

The results for the analysis of the 2009–2017 seasons show a species composition that appears to have been accumulated by a predator, rather than one in which

	(19802)		*(32611)*		*(32616)*		*(32632)*	
Element	*No. gnawed*	*% gnawing*	*No. gnawed*	*% gnawing*	*No. gnawed*	*% gnawing*	*No. gnawed*	*% gnawing*
Lower incisor	1	8.3	1	3.2	3	3.8	23	7.8
Upper incisor							13	4.4
Mandible	2	16.7	10	32.3	28	35.9	31	10.5
Maxilla							1	0.3
Premaxilla							5	1.7
Caudal vertebra			5	16.1	6	7.7	21	7.1
Vertebra			1	3.2			12	4.1
Scapula							5	1.7
Humerus	2	16.7	1	3.2	11	14.1	22	7.5
Radius							8	2.7
Ulna					5	6.4	14	4.7
Pelvis	2	16.7	1	3.2	1	1.3	22	7.5
Femur	2	16.7	10	32.3	9	11.5	24	8.1
Tibia	3	25	1	3.2	15	19.2	89	30
Calcaneus			1	3.2			4	1.4
Total	12	100	31	100	78	100	294	100

Table 10.9. Showing percentages of the gnawed assemblage for all units with a NISP greater than 100.

multiple taphonomic pathways were in effect. This is borne out by the high predator selection bias in the assemblage, with 96% *M. m. domesticus* being present with few anurans or insectivores. The 1995–1999 assemblage, with 96.2% *M. m. domesticus* (Jenkins 2005; 2009) and the microfauna assessed from the human burial excavated by Mellaart (Brothwell 1981), were derived from carnivore scats and therefore represented the diet of the predator, rather than the generic microfauna of the area. However, the assemblage recorded in the second phase of excavation, 2000–2008 (Jenkins, Yeomans 2013), had a much higher percentage of amphibians and snakes, with 72% *M. m. domesticus* and 19% amphibian species, and was similar in composition to the microfauna recorded from the BACH excavation, with 78% *M. m. domesticus* and 16% amphibian species (Jenkins 2012b), suggesting a mixed taphonomic pathway for assemblage accumulation.

The species of predator, whilst still unknown, appears to be consistent throughout all phases of excavation so far, as evidenced by the patterning of puncture marks, particularly on the lateral surfaces of the mandibles.

Levels of breakage for this assemblage were compared against those caused by potential predators (Andrews 1990). The most likely suggestion would be a mammalian carnivore, for example, pine marten or fox, and although this is borne out by the presence of puncture marks on the bones, the fragmentation pattern does not conform. For example, in the breakage of major skeletal elements in Andrews (1990: 51, table 3.3) the distal tibia was not recovered in examples of scats from any of the mammalian carnivores. Coyote, Red Fox and Pine Marten produced no complete elements, 67–90% of proximal tibiae, and between 10 and 33% of shaft fragments, with no distal ends recorded. In contrast, the tibiae recorded from the Çatalhöyük assemblages show few complete elements, only 1.4–5%, with 3.5–9.9% being comprised of proximal tibiae, in stark contrast to the mammalian carnivores above. 8.6–21.7% of shaft fragments fit more closely with what would be expected. However, distal tibiae in the Çatalhöyük assemblage equate to 68.3 to 82.9% of the fragments, compared to none recorded for the mammalian carnivores (Andrews 1990).

Evidence of gnawing was not restricted to any elements in particular; however, some exhibited more consistent gnawing than others. For example, in all four of the larger units studied, the mandibles, humeri, femora and tibiae had high percentages of gnawing. The mandibles also showed a consistent pattern of puncture marks on the lateral surface, below the first and second molar, which could be indicative of a predator hunting strategy; for example, killing prey by biting off the head. Fifteen individual gnaw marks from 12 mandibles from (32632) were examined under a SEM and the puncture marks had a mean puncture size of 0.69mm x 0.49mm. Previous studies on the canine teeth from weasels, which are the smallest predators found at Çatalhöyük, gave an

		Incisor		*Molar*		*Humerus*		*Femur*	
Unit	*Category*	*N*	*%*	*N*	*%*	*N*	*%*	*N*	*%*
21814	Light			2	22.2			1	20
	Moderate								
	Heavy								
	Extreme								
32782	Light							1	100
	Moderate								
	Heavy								
	Extreme								
19802	Light	2	3.1			1	8.3	1	9.1
	Moderate								
	Heavy								
	Extreme								
32611	Light	20	12.3	16	8.4	14	17.3	1	1.4
	Moderate			1	0.5	2	2.5		
	Heavy								
	Extreme								
32616	Light	26	8.3	10	2.6	5	7.2		
	Moderate			3	0.8	7	10.1		
	Heavy			2	0.5				
	Extreme								
32632	Light	1	0.1	7	0.4	14	11.7	5	4.9
	Moderate	1	0.1			3	2.5	1	
	Heavy					1	0.8		
	Extreme								

Table 10.10. Digestion categories for each of the units with a NISP greater than 100.

Species name	*Common name*	*Diet*	*Habitat prefences*
Mus musculus domesticus	House mouse	Omnivorous, preference for cereals	Commensal, human habitation
Crocidura	White-toothed shrew	Insectivorous	Pastureland, cultivated fields
Meriones	Jird	Omnivorous	Arid regions including steppe
Microtus	Vole	Herbivore	All but dense forest
Testudine	Tortoise/turtle	Omnivorous	Variable depending on species
Natrix	Grass snake	Carnivorous	Open woodland, woodland borders, field margins. Good swimmers so may be found in association with bodies of fresh water
Pelophylax ridibundus	Marsh frog	Insectivorous	Water dweller
Bufo viridis	Green toad	Insectivorous	Steppe, semi-desert, urban areas
Pelobates sp.	Spadefoot toad	Insectivorous	Sandy soils, burrower. Lays eggs in temporary pools

Table 10.11. Showing the species recorded, their diet and habitat preferences.

average width of 1.85mm for the upper canines and 2mm for the lower (Jenkins 2012b). As the puncture marks could have been made by only the tip of the tooth, the size ranges for the teeth of the predator and the puncture marks seen on the assemblage fall within the same range. It is therefore likely that small mustelids could be the potential predator responsible for the creation of the scat assemblages.

However, despite the gnaw marks being potentially indicative of small mustelids, levels of digestion on the bones are low, which is more indicative of an avian predator, for example, a Barn owl, rather than a mammalian carnivore. A feeding experiment of a captive population of small mustelids, amongst other potential predators, would prove useful, especially as captive animals are more likely to be fed murids as opposed to arvicolids, and therefore would be of much greater value for the interpretation of predator selection and effects of predation on the assemblage at Çatalhöyük. For example, are the low levels of digestion that we see at Çatalhöyük a reflection of the availability of food for the predator? If food is abundant, then the prey may not spend as much time traversing the digestive tract, which could lessen the effect digestive acids have on bones and teeth (Denys et al. 2017). Similarly, fragmentation is not as high if the prey items are smaller than those reported in Andrews' (1990) study, mice being smaller than voles, which would require less mastication. Murid molars are also more resistant to digestion than microtine molars (Fernandez et al. 2017), lacking the salient edges for digestive attack (Williams 2005); however, incisors of both genera should be affected similarly, due to the similar structure of those teeth.

The percentages of *M. m. domesticus* and species composition in the predatory deposits show that most, if not all, hunting, was being done on the site, demonstrating that the levels of this commensal species were high enough to sustain predators without the need for off-site hunting to supplement the diet. Indeed, commensal weasels in Egypt have been reported to catch and kill mice within houses, as well as being seen during daylight hours despite being predominantly nocturnal (Osborn, Helmy 1980: 109).

Building 161

The vast majority of microfauna analysed for this study came from units excavated from B.161. This building is located in the South Area, in one of the early levels – South J. There were several phases of construction/occupation/abandonment of the building, and the largest concentration of microfauna, (32632), comes from Phase 1.3, one of the final occupation phases of the structure. Units (32611) and (32616) come from Phase 2 of the building, the abandonment phase.

Unit (32632). The microfauna were originally interpreted as being commemoration offerings, due to it being a discreet, high-density cluster, or a foundational deposit for the oven structure (fig. 10.5), as it was found in association with macrofaunal remains, including a large scapula, and a deposit of shells, as noted by the excavators:

> During the removing process of that superstructure we found what we think that there were (sic) foundation deposits. We detected three different deposits, one in the north part of the superstructure that was a line of micro fauna (S3)...
>
> After [we] remove[d] the remains of the superstructure that was abutting the building walls... We also were able to see that between the superstructure and the walls in the E side of the structure there were more remains of micro fauna, which probably were related with all this commemoration offerings (sic)...
>
> The oven was composed by a light orange dome around 0.15cm thick which was preserved at least 0.45cm. In a second phase, probably related with a reparation (sic), the space between the dome and the walls corner was filled with mud bricks and clay to give more consistency to the structure. Relates (sic) with this process, as a commemorative offers they placed micro fauna, bones, shell...(x1 to x5 finds) between the dome and the walls (CB 2017).

The taxonomic and taphonomic analysis of the unit, however, found evidence for both digestion and small mammalian carnivore gnawing on the bones, leading to the conclusion that the microfauna were introduced into the unit as scats, rather than fully fleshed animals. What remains unclear is why scats were incorporated into the fabric of a utilitarian structure. Other remains, including macrofauna and a quantity of shells were also excavated in close proximity, leading to the conclusion that the incorporation was intentional. But what meaning could be given to carnivore scats? Were the scats used so that the high density of microfaunal remains could be used to temper the clay and add strength to the structure? If so, this appears to be a very rare practice, as the construction/make up/packing category does not usually contain high-density clusters. Bone has been used as a temper in pottery in Neolithic Europe (Stilborg 2001); however, the bone tends to be finely ground rather than incorporated as actual elements and botanical tempers are more frequently used, such as barley straw, which is soft and makes good temper (Jenkins et al. 2017). As many structures are found intact prior to excavation, and if this inclusion was a routine practice, we would expect the unit category to contain a much higher percentage of microfauna than has been found to date.

Figure 10.6 illustrates mice running around a house and climbing on an oven structure similar to that found in B.161; however, we have to question whether this would have been a common sight at Çatalhöyük, or whether mice would have been more restricted in their movements and limited in the space they occupied when alive. It is highly unlikely, given the taphonomic signatures on the specimens, that mice in B.161 fell behind the

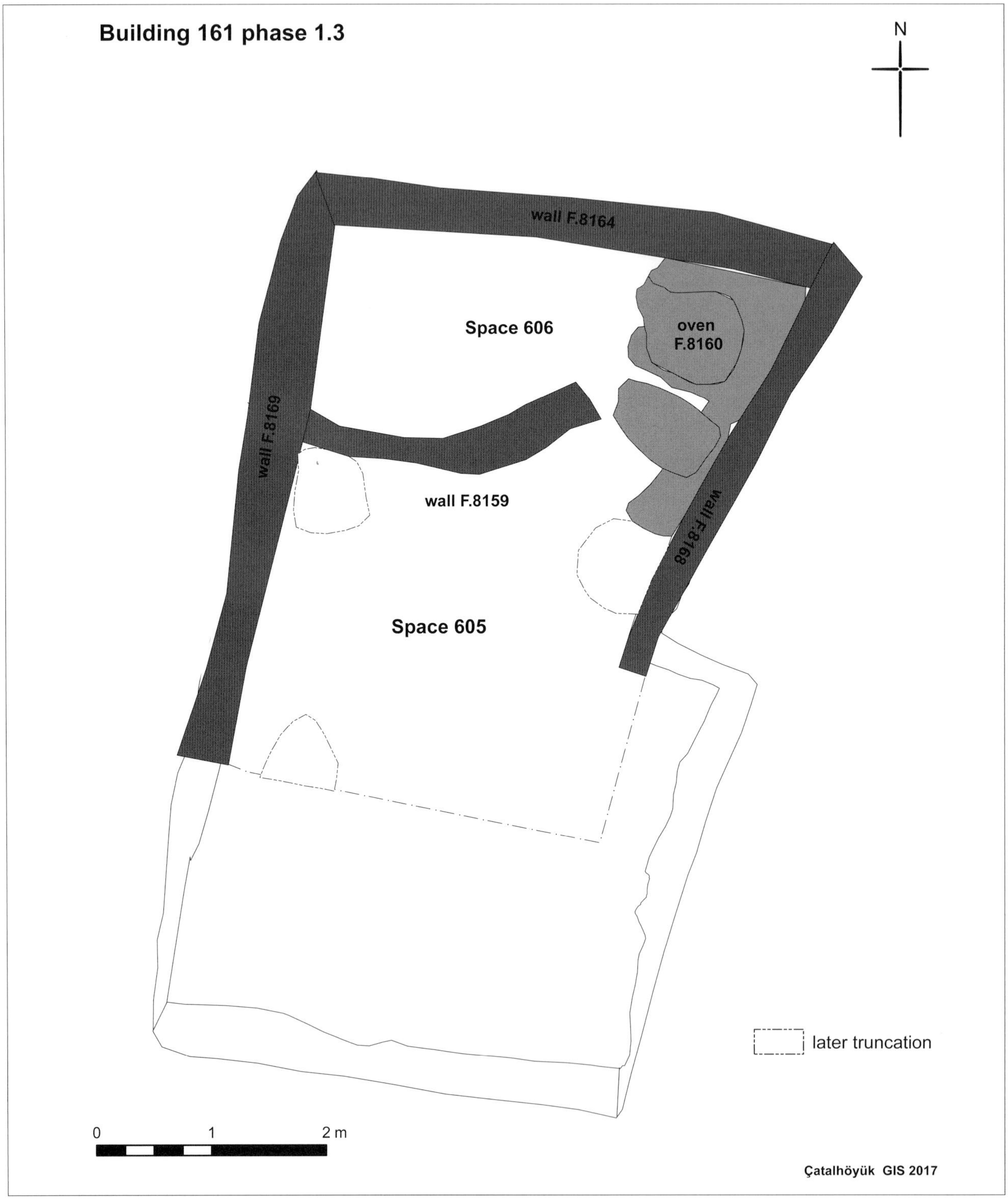

Figure 10.5. GIS building plan of Phase 1.3 of B.161, showing location of oven F.8160 in the northeast corner which contained unit (32632) (plan: Camilla Mazzucato).

oven and died in high enough numbers to provide an MNI of over 200. Not only are mice excellent climbers, but it also would have been obvious to the occupiers of the dwelling and in all likelihood the small carcasses would have been removed, or potentially burned. As such, it would seem most likely that scats were, for some unknown reason, deliberately incorporated into the superstructure of the oven.

Unit (32616). This unit was described as primary room infill and was noted by the excavators as including a large quantity of collapsed material, including roof material, bricks, mortar and oven superstructure.

It is interesting to note that this fill contained a high proportion of microfauna, with a NISP per litre of 31.6, and a MNI of over 100. The abandoned space could have been used as a latrinal area by a small carnivore, or alternatively, if the incorporation of microfauna as foundation deposits or the use of scats as a packing material for oven superstructure is indicative of this building, the inclusion of oven superstructure in the fill could be the source of the microfauna cluster. The southernmost part of the building was truncated by a later dwelling, so there may have been additional ovens in that part of the building that were destroyed during the abandonment phases and included in the primary room fill.

It is also worth noting that the percentage of gnawing in this unit, which is stratigraphically below that which includes the human remains from this phase of the building's lifespan, was greater than the percentage of gnawing from the unit recovered around the human remains (32611), showing scats present earlier in the building's depositional history. Despite the assumption that the scats were included with the body intentionally, as appears to have been the case in previous burials at Çatalhöyük (Jenkins 2005; 2012a), in this instance, they may have just been incorporated with the general fill rather than as part of an intentional depositional event associated with the human remains.

Unit (32611). This unit is comprised of the fill excavated around human remains (32608) that did not appear to have their own grave cut (fig 10.7). As such, an arbitrary cut was assigned by the excavator so that the finds directly around the body could be distinguished from those in the wider fill. With regards to the microfauna the excavators noted:

Figure 10.6. Illustration showing mice around the oven in B.161 (illustration: Kathryn Killackey).

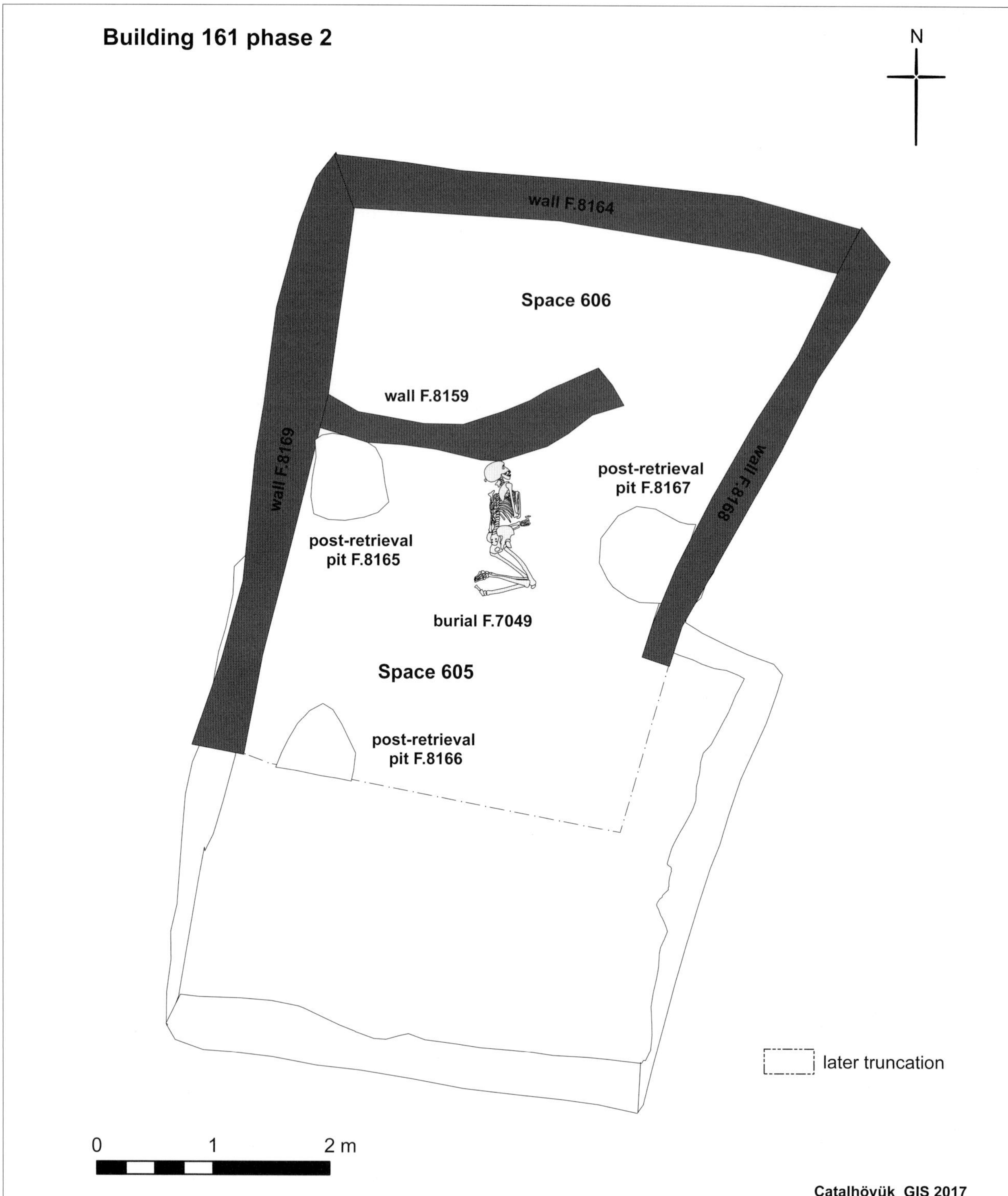

Figure 10.7. GIS building plan of phase 2 of B.161, showing location of human remains associated with units (32611) and (32616) (plan: Camilla Mazzucato).

One of the things that we detected is [that] around the body there was quite a lot of spots of microfauna, but we found also this kind of spots outside the line that we define[d] as a burial infill. I[t']s true that the concentration of [the] microfauna remains was mostly in the central part of the building, where the skeleton was found. Have to be define (sic) if this microfauna was deposited on [purpose] and is related to the body, or was just [a] natural process (CB 2016).

Unfortunately (32606), the unit stratigraphically above the skeleton and thought to be the same as (32611) by the excavators, as noted on the unit sheet, was not available for export and therefore it is difficult to confirm whether the microfauna were limited to the human remains or were just part of the general fill, albeit in patches.

The fact that unit (32616) is directly below the human remains and has a higher degree of microfauna, with a NISP per litre of 31.6 to 13.5 for (32611), indicates that the microfauna could be part of the general fill rather than something intentionally deposited on top of and around the human remains.

The human remains themselves (fig. 10.8) do not conform to previous burials in which large concentrations of microfauna were found (Jenkins 2005). The others were more typical of the standard burial practices on site. Instead of being in a discrete grave cut, skeleton (32608) appears to have been thrown onto the primary room infill during the abandonment and filling of the building, with the body following the south to north inclination of the fill below and the head abutting the partition wall. The haphazard nature of the burial is not typical of those seen at Çatalhöyük and in addition, these are also the only human remains recorded with perimortem injury, which involved a broken mandible (Volume 15, Chapter 15). Phytolith evidence suggested that the body was lying on or possibly wrapped in matting, and there was evidence of delayed burial, with some elements being found in disarticulation, including a tooth among the ribs and the left scaphoid next to the left acromion process of the scapula.

Figure 10.8. Human skeletal remains (32608) thought to be associated with microfauna.

The discrete nature of the microfaunal accumulations as mentioned by the excavators could indicate that the abandoned building was being used as a latrinal area for small mammalian carnivores. There were no indications of gnawing on the human skeletal remains, which may have been expected if a small carnivore and scavenger were using the space regularly and came upon a body, especially as it would take time to build the scat accumulation to the level where we would see a MNI of over 60 for this unit; however, the matting around the body may have been enough to protect it.

As the human remains excavated in B.161 are so unlike others found in association with microfauna, it is unlikely that the concentrations were intentionally placed with the remains. The body was left in an abandoned building in a haphazard fashion, unlike other burials at the site. High levels of microfauna are also recorded in the fill below, and it seems that the abandoned space may have been used by a small mammalian carnivore as a latrinal area. It is incongruous, but not impossible, that remains would be treated so haphazardly and yet show evidence of mortuary practices in the form of incorporation of scats. It is more likely that the inclusion is accidental and the microfauna are simply part of the fill.

Conclusion

Species composition of the assemblage from the 2009–2017 seasons shows that *M. m. domesticus* dominate, with only a fraction of the assemblage given over to other species, most notably *Crocidura* spp. and herpetofauna. Whilst insectivores and herpetofauna at other sites have been classed as potentially commensal, due to the attraction of increased insect activity around waste accumulation or middens (O'Connor 2013), these taxa are both found in such low numbers at Çatalhöyük that this is unlikely to be the case here. The low numbers could be the result of spatial patterning, with these species being limited to the periphery of the site, which we cannot ascertain with only a small portion of the site having been excavated, or there could be a natural dearth, with better habitats being available around the site rather than in it.

Palaeoenvironmental reconstruction using the microfauna shows Çatalhöyük to be an early urban environment that was quickly exploited by the commensal house mouse. Other early sites that contained high levels of *Mus musculus domesticus*, such as the cave site of Iraq el-Dubb, showed *M. m. domesticus* in association with an array of other microfauna, which is not mirrored here. As such, it is clear that *M. m. domesticus* exploited the commensal niche to such a degree that other species were outcompeted.

Digestion and gnawing in the assemblage for this phase of excavation indicate accumulation of microfauna by a predator, and this is supported by previous analyses. The 2000–2008 phase of analysis (Jenkins, Yeomans 2013) and the BACH phase (Jenkins 2012b) found a greater degree of variation in species composition, including higher numbers of amphibians, representing a mixture of predator-accumulated and natural death assemblages, whilst the 1995–1999 phase (Jenkins 2005, 2009) is dominated by *M. m. domesticus*, indicating a high predator selection bias.

As in previous studies (Jenkins 2005; 2009; 2012b; Jenkins, Yeomans 2013), there appears to be a low level of background microfauna across all levels and feature types; however, levels of *M. m. domesticus* were high enough to encourage small predators onto the site. The domination of the assemblage by a single species would indicate that most, if not all, hunting was being done on site, with little contamination of 'external' microfauna, suggesting the populations of *M. m. domesticus* at Çatalhöyük were enough to sustain the predators without the need to supplement the diet elsewhere.

There is little evidence, however, to suggest the predators were living on site, as small mustelid, felid and canid remains have been recovered in very low numbers and the felid remains are thought to be associated with skins rather than live animals (Russell, Martin 2005).

Despite the microfaunal assemblage being accumulated by predators, the high-density clusters appear in specific anthropogenic contexts; for example, being incorporated into construction material or into burials. It is clear that the people at Çatalhöyük must have known they were sharing space with these commensal rodents, however, the commensal relationship balance between rodents and humans was not positive/neutral. Mice would have been eating and fouling stored food, as evidenced by mouse remains in bins within burned houses (Jenkins, Yeomans 2013) and the discovery of charred mouse pellets in bins containing food (Twiss et al. 2009). As the recorded numbers of small carnivores at Çatalhöyük are low, we cannot conclusively determine if they were being encouraged onto the site in order to control mouse populations. However, the incorporation of their scats into human environments, for example, construction materials and burials, could show some deferment to the animals. Are small carnivores being associated with the control of pest species, or could the meaning be more ephemeral, like an association with abandonment and closing of buildings, as they potentially used these empty spaces as latrinal areas?

That the house mouse was so ubiquitous even in the early levels at Çatalhöyük shows that this species took advantage of the commensal niche rapidly and could quickly have become an issue for the inhabitants of the site. Overall, the Çatalhöyük microfaunal assemblage represents one of the largest and earliest 'urban' or 'proto-urban' microfaunal assemblages found to date. The relationship of the people of Çatalhöyük with these animals is unclear. Evidently, the people of Çatalhöyük attributed some symbolic or practical significance to the small carnivore scats which incorporated the microfauna. Whether the significance is with the scats, the small carnivores that produced them, or the microfauna themselves remains unknown.

11. Multi-isotope evidence of diet (carbon and nitrogen) and mobility (strontium) at Neolithic Çatalhöyük

Jessica Pearson, Angela Lamb and Jane Evans
(with contributions by Claudia Engel, Nerissa Russell, Mehmet Sömel, Wim Van Neer and Wim Wouters)

Introduction

The past 20 or so years of stable isotope analysis at Neolithic Çatalhöyük have focused on reconstructing human diet through carbon and nitrogen isotope analysis of bone collagen almost exclusively (see also Styring et al. 2016). Here, this chapter provides the final carbon and nitrogen data for the remaining houses at the site not yet measured and supplements these (and previous data) with a large-scale study of strontium isotope analysis throughout the duration of the site's occupation. Despite previous extensive analyses resulting in the sampling of almost 1,300 bones and results of over 900 measured individuals (see also Pearson 2013), several questions related to diet remain, chiefly: how important were particular domestic and wild resources, how do houses compare through the site's occupation and can differences be identified between houses with different features and elaboration? For the strontium isotope analysis there are two new considerations: did the population of Çatalhöyük originate from the local area, and if not, can we find evidence for patrilocal or matrilocal kinship practices?

Previous work

Earlier work (Pearson et al. 2007; Pearson 2013) has shown that animal carbon and nitrogen isotope values are surprisingly diverse, reflecting natural and anthropogenic feeding regimes that result in large ranges (up to 9.5‰ variation) in isotope values even amongst animals with narrow, herbivorous diets. A summary of the faunal data from Pearson (2013) can be found in table 11.1.

All species show small to moderate consumption of C_4 plants in some animals, but overall the mean carbon isotope values for fauna suggest the impact of these plants was relatively small, with many more animals observing a C_3 plant-dominated/exclusive diet. These interpretations are based on C_3 plant carbon isotope measurements averaging around −23‰ for wild and domestic plants (Pearson 2013); and slightly higher (ca −22.5‰) for crops alone (Pearson 2004; Wallace et al. 2015). Therefore an exclusive C_3 plant-consuming animal that eats a mixture of wild plants from grazing and some fodder from crop waste will measure around −17.8‰. Individual animals with more C_4 plants in their diets tended to be species of sheep, goats and the later specimens of *Bos* spp. suggesting that the role of these plants in animals' diets was a feature of their management by humans and the mosaic landscape visited by them, together with changes in these habitats over time, with the drier episodes offering optimal growth to C_4 plants. That some equids had a C_4 component to their diet suggests these plants (likely grasses) were available in hunting and perhaps also in

Genus	*N*	*δ¹³C‰ (VPDB)*					*δ¹⁵N‰ (AIR)*				
		Min	*Max*	*Range*	*Mean*	*s.d.*	*Min*	*Max*	*Range*	*Mean*	*s.d.*
Ovis sp.	154	−20.2	−13.6	6.6	−17.5	1.2	4.9	14.4	9.5	10.1	1.6
Capra sp.	22	−19.8	−12.5	7.3	−17.7	1.6	5.2	12.2	7.1	9.0	2.0
Bos spp.	79	−19.8	−14.6	5.1	−18.1	1.1	6.0	13.8	7.7	9.8	1.7
Equus spp.	45	−19.9	−14.8	5.1	−19.0	1.0	5.1	11.3	6.2	8.3	1.4
Sus sp.	28	−21.3	−15.3	6.0	−19.5	1.2	5.9	11.9	6.4	8.0	1.6
Vulpes sp.	8	−19.0	−16.9	2.2	−18.2	0.7	8.2	15.0	6.8	10.8	2.7
Canis spp.	19	−21.3	−16.3	5.1	−18.5	1.6	7.7	13.2	5.5	11.2	1.5
Canis lupus	1	-	-	-	−14.3	-	-	-	-	16.8	-
Cervus sp.	8	−20.1	−14.1	6.0	−18.7	2.2	6.7	13.8	7.2	8.7	2.3
Lepus sp.	3	−20.4	−18.7	1.7	−19.3	1.0	3.5	6.6	3.1	4.9	1.6

Table 11.1. Summary carbon and nitrogen isotope data from fauna bone collagen (data taken from Pearson 2013).

grazing zones. Equids have a relatively smaller range in values, which as wild species (*Equus ferus* and *Equus hemionus*) may indicate the average isotope values of grasses in the local area that were available to free-to-roam wild animals. Data from suids alongside human isotope values have been used elsewhere to determine successfully whether these animals were consuming the human meal leftovers and thus indicating proxy evidence of early domestication (for a review see Rowley-Conwy et al. 2012). Table 11.1 shows mean values for boar at Çatalhöyük were $\delta^{13}C$ −19.5‰ and $\delta^{15}N$ 8.0‰, which considered alongside the mean values for the inhabitants of Çatalhöyük ($\delta^{13}C$ −18.7‰ and $\delta^{15}N$ 12.7‰) strongly suggests that the majority of animals did not eat discarded food waste and instead points to these animals being wild boar living in their natural habitat. In contrast, the canids ($\delta^{13}C$ −18.5‰ and $\delta^{15}N$ 11.2‰) are more similar to humans, suggesting these animals had access to leftovers. Only a single wolf at the site has been measured successfully, which provided a nitrogen isotope value of 16.8‰ and provides one example of what might be expected in consumers with meat-dominated diets.

Previous work on the human remains from the site has shown while C_4 plants play an important role as food for some animals, there is no evidence for the passing on or direct consumption of these resources in humans ($\delta^{13}C$ −18.7‰). In terms of the importance of houses as locations for food-sharing, Pearson et al. (2015a) show how the isotope data generated from the middens of specific houses do not seem to be part of the same foodweb as the adults buried beneath the house floors. The reasons for this could be because people spent some time away from the site, foods were not strictly shared, food was consumed in different locations or houses, people came from outside of the area originally, or people were not permitted to consume some of the foods available. It was also possible that whole food groups, which had not been measured, might account for some of the discrepancy. In particular, birds/eggs, reptiles and fish are found at the site, but their role in human diet so far is unknown isotopically. Whatever the reason for this variation, it suggests that food production, consumption and disposal were the result of carefully structured agentive processes. Comparing values across the inhabitants revealed some surprising data: male and female diet was probably the same, as indicated by virtually identical isotope values (female: $\delta^{13}C$ −18.8‰, $\delta^{15}N$ 12.6‰; males: $\delta^{13}C$ −18.6‰, $\delta^{15}N$ 12.7‰). Instead the population differentiated between social groups through age (Pearson, Meskell 2013; Pearson et al. 2015b), a finding echoed in the figurine corpus and grave goods (Nakamura, Meskell 2009; 2013a; 2013b).

The principles of stable isotope analysis

Carbon isotope analysis and dietary reconstruction

Carbon in animal tissues originates from atmospheric carbon dioxide where it is fixed by plants during photosynthesis. There are three methods of photosynthesis used by terrestrial plants: C_3, C_4 and crassulacean acid metabolism (CAM). Each pathway discriminates against the heavier carbon isotope ^{13}C in different ways resulting in characteristic values for each form of photosynthesis (See O'Leary 1981): C_3 plant $\delta^{13}C$ values range between −34‰ and −22‰; C_4 plants from −20‰ to −7‰ and CAM plants from −23‰ to −11‰. C_3 plants account for most plants globally, including cereals such as wheat and barley, most trees, fruit, nuts, pulses, all leguminous plants and tubers. Typical C_4 plants include maize, millet and sugarcane and arid-adapted grasses and chenopods but only account for ~8,000 (<5%) of terrestrial plants. CAM plants (such as cacti) form a smaller percentage but have little nutritional importance in Southwest Asia during the Neolithic, in contrast to C_4 plants, which have been identified among charred plant taxa recovered from the site (Fairbairn et al. 2005a; Bogaard et al. 2013), including in sheep dung (Bogaard et al. 2013). Isotope values in plants can also be affected by global climatic, micro climatic and environmental conditions and anthropogenic activities (e.g., van der Merwe, Medina 1989; Tieszen 1991; Glagoleva, Chulanovskaya 1992; Tieszen, Fagre 1993; Heaton 1999). Plant water-use efficiency also influences plant carbon isotope ratios, causing lower carbon isotope values in well-watered plants. Nutrient impoverishment (lack of nitrogen or phosphorus) and cooler temperatures also lower carbon isotope values in C_3 plants (Tieszen 1991). Phenomena that increase carbon isotope values include aridity, salinity and high altitudes (Tieszen 1991; Stewart 1995; Araus et al. 1997; Long 1999; 2001).

C_4 plant consumption by animals raises an interesting question, because the photosynthetic structures of C_4 plants, comprised of more silica and fibre than their C_3 counterparts, makes their palatability to animals inferior (Heckathorn et al. 1999). Thus, the relative consumption of C_3 and C_4 plants in animal tissues is unlikely to be representative of the standing plant biomass available to free-to-roam animals because, given a choice, C_4 plants will be avoided unless they are the only available food seasonally (that is, in late summer) or because they are used as fodder. Ruminant animals have a greater tolerance for such plants on account of their digestive physiology, so normally C_4 plants would be avoided by non-ruminants such as equids and suids, especially if those species are considered to be wild. Previous work by Pearson et al. (2007) has demonstrated the consumption of both types of plants at Çatalhöyük and elsewhere, and how this may be indicative of herding practices.

Carbon is passed from plants through the food chain at two trophic levels: a ~5‰ shift between plants and herbivores, and a ~1‰ shift between herbivore and carnivore (DeNiro, Epstein 1978) along with any environmental or physiological influences. Other resources that have an impact on reconstructing human diets in particular include the consumption of aquatic foods such as fish, aquatic mammals and molluscs. These resources tend to cause much higher carbon isotope values for marine foods and slightly lower values for freshwater resources in consumers. While marine resources are unlikely to have made a major contribution to human diet at Çatalhöyük, even if some inhabitants travelled to the Mediterranean coast (ca 200km away), freshwater resources are a distinct possibility. Fish bones have been found on site originating from small cyprinids and loaches (under 15cm in length) (Van Neer et al. 2013; Hamilton-Dyer this volume, Chapter 12) and the identified bird bones reflect a range of local aquatic species, some of which are shown to have been eaten (Russell, McGowan 2005; Russell 2019a; Best et al. this volume, Chapter 13).

Nitrogen isotope analysis and dietary reconstruction

Nitrogen also enters the food chain, either as adsorbed nitrogenous compounds in soil or, in the case of leguminous plants, having been fixed directly from the atmosphere. The enrichment factor between diet and a body (the so-called trophic level effect) is ~4‰ (DeNiro, Epstein 1981; Ambrose, Norr 1993; Hedges, Reynard 2007). Nitrogen isotope values are also influenced by a large number of specific environmental factors, most of which cause values to rise. In plants these include aridity (e.g., Heaton 1987; Ambrose 1991; Cormie, Schwarcz 1994; Gröcke et al. 1997; Schwarcz et al. 1999), salinity (e.g., Handley, Scrimgeour 1997; Robinson et al. 2000; van Groenigen, van Kessel 2002), high temperatures (Ambrose 2000), ruminating (Macko et al. 1982; Sealy et al. 1987; Wattiaux, Reed 1995; Cormie, Schwarcz 1996) and starvation (Barboza et al. 1997; cf. Ambrose 2000; Faber 2003; Fuller et al. 2004; Stewart, Smith 2005). More recently, Bogaard et al. (2007) and Fraser et al. (2011) have shown that anthropogenic inputs from crop manuring have the same effect and may influence how accurately the trophic level of an individual can be determined.

Strontium isotope analysis and mobility

Strontium (Sr) is a naturally occurring trace element present in geological substrates, organic organisms and water sources. It is present as four stable isotopes: ^{84}Sr, ^{86}Sr, ^{87}Sr and ^{88}Sr. The isotope ^{87}Sr is formed by the radioactive decay of rubidium (Rb), specifically ^{87}Rb. This isotope is also present in geological substrates, and hence substrates of different ages will have different Rb/Sr ratios and $^{87}Sr/^{86}Sr$ profiles (Faure, Mensing 2005; Bentley 2006). Strontium in the local water, plants and fauna is deposited into the skeleton and tooth enamel by substituting for calcium in apatite crystals formed in consumer tissues. Adult tooth enamel in humans forms during childhood and does not remodel, thus the food and water ingested at the time of tooth enamel mineralisation reflects the $^{87}Sr/^{86}Sr$ profile of where people were living at the time (e.g., Bentley et al. 2004). In order to identify an individual as local or non-local, a baseline for the local geological substrates must be produced. This can be generated using the bone apatite values of the same individual (which equilibrate with local soil during burial) or local plants and soils from the site where the individual was found. If the $^{87}Sr/^{86}Sr$ values of an individual's tooth enamel do not match values of local plants or the bone apatite of the individual themselves, then they are likely to have spent their childhood (the age range depends on the tooth analysed) in a different location.

Materials and methods

The following section details the methodology by which samples were identified to species, chosen for sampling, prepared and assessed for sample integrity, homogeneity and analytical error. All samples discussed here were taken from seasons 1996–2017.

Carbon and nitrogen stable isotope analysis

As discussed elsewhere (Mellaart 1967; Andrews et al. 2005; Molleson et al. 2005; Boz and Hager 2013; Haddow et al. this volume, Chapter 15), burial practices at Çatalhöyük included single inhumations as well as multiple and secondary burials, many of which were frequently disturbed. The choice of individuals sampled follows Pearson (2013), but briefly: all individuals were sampled except those that were burnt or represented by only a few elements or fragments; secondary burials, specifically crania, were sampled from the least diagnostic fragments that were clearly associated; pathological conditions were avoided; for multiple burials, care was taken only to sample from elements clearly associated to a particular individual. Age and sex estimation for all human remains follows previous work (Andrews et al. 2005; Molleson et al. 2005; Hillson et al. 2013; Larsen et al. 2013). In contrast to the human remains, animals were sampled at a maximum of c. ten samples per taxa per point of interest (that is, midden or level or building) following Pearson and Grove (2013). Details regarding the species recovery and identification can be found in Russell and Martin (2005), Russell et al. (2013) and most recently, this volume, Chapter 8. All reagents used were Analytical (AR/Analar) Grade or above and all water used was of ultrapure (18.2 MΩ) quality. All bone samples were taken as ~500mg chunks using a diamond powder-coated cutting wheel from the broken diaphysis, avoiding glue, id number, pathological conditions, articulations and cut marks. To prevent lipid contamination, fish samples were subjected to a chemical liquid lipid extraction using 2:1 chloroform:methanol, following the method described in Guiry et al. (2016), prior to collagen extraction. Collagen extraction for all samples follows the modified Longin (1971) method (Brown et al. 1988): samples were immersed whole in a test tube of 0.5M $HCl_{(aq)}$ and refrigerated at 4°C. Once the bone was fully demineralised (2–10 days), it was rinsed with ultrapure H_2O to neutral pH (three rinses). Samples were gelatinised by adding pH3 H_2O (HCl) and heated at ~70°C for 48 hours in a heater block. The gelatinous solution was then filtered through an 8μm EZee filter, transferred to a plastic test tube and sealed with Parafilm, and frozen before freeze-drying until dry (~48 hours). Once dried the new samples reported here were loaded into tin capsules and combusted online to CO_2 and N_2 in an elemental analyser (Flash/EA) coupled to a ThermoFinnigan Delta Plus XL isotope ratio mass spectrometer via a ConFlo III interface (running in continuous flow mode) which measures the $\delta^{13}C$ ratio of CO_2 and the $\delta^{15}N$ of N_2. M1360p (powdered gelatine, British Drug Houses) with expected $\delta^{13}C$ values of −20.32‰ (calibrated against CH7, IAEA) and $\delta^{15}N$ values of +8.12‰ (calibrated against N-1 and N-2, IAEA) was used as the internal reference standard material. The long-term 1σ reproducibility for internal reference standards is normally better than ±0.1‰ for $\delta^{15}N$ and $\delta^{13}C$, which represents the measurement precision for this study. We used the following measures to indicate collagen integrity: Atomic C:N ratios between 2.9 and 3.6 (DeNiro 1985; Ambrose 1993); at least 1% collagen yield and elemental yields of 15.3%–47% for carbon and 5.5%–17.3% for nitrogen (Ambrose 1990). Particularly poor preservation was noted amongst the secondary burials, amongst crania especially, and hints at different post-mortem treatment and taphonomic processes, as seen previously (Pearson 2013).

Strontium stable isotope analysis

All humans were potentially available for sampling; however, there were several caveats that reduced the sample size: samples were to be taken from individuals with a fully mineralised, erupted and little-worn second (M2) or third molar (M3) tooth where the antimere was also present. These criteria confined the sampling to adolescent and adult individuals. Teeth with pathological conditions or developmental disruptions (caries, obvious signs of linear enamel hyperplasia) were also avoided. This approach allowed for a total of 80 individuals to be selected for strontium isotope measurement. The majority of teeth sampled were the second molar (n=78) and rarely a third molar (n=2). This sample set represents one of the largest studies currently undertaken of adult mobility from a single site and is the largest from Southwest Asia in any time period. Teeth were not exported whole, instead, using a diamond-coated cutting wheel, the roots were completely removed and the crown cut in half. Each tooth was photographed on the occlusal and buccal/lingual surface before sampling. Samples exported, prepared and measured therefore comprised a half crown of an M2 or M3 tooth. Samples were restricted to the Neolithic occupation of the site and were sampled in both the North Area (n=38) and South Area (n=41); a single sample came from the KOPAL environmental trenches excavated between 1996 and 2001 (Boyer et al. 2006). Within this sample, 23 were males/possible males, 33 were females/possible females and 24 have no recorded sex estimation, either because they are too young, or because they were too poorly preserved to enable an accurate assessment.

The enamel deposited on permanent adult human M2 and M3 teeth covers a period of 2.5–8.5 years for M2 and 8.5–14.5 years for M3 (AlQahtani et al. 2010). The first molar (M1) was not selected because of the overlap with breastfeeding, which for the purposes of strontium isotope analysis changes the source water consumed by the individual being analysed, being instead formed from the water ingested by the individual providing breast milk to the nursing infant, which may not be the biological mother. Since breastmilk is formed over a relatively short period of time, breastfeeding might also pass on evidence of group logistical mobility as the mother engages with community activities (see Humphrey 2014 for a review).

Sample preparation follows the method described in Evans et al. (2006). Briefly, tooth enamel was cleaned using high-purity water in an ultrasonic bath and rinsed twice to remove loosely adhered material. The enamel surface of the tooth crown half was abraded to a depth of >100μm and all dentine and cementum removed using a tungsten carbide dental burr. Thin enamel slices were then cut from the tooth using a flexible diamond-edged rotary dental saw. These samples were then transferred to a clean (class 100, laminar flow) working area and cleaned ultrasonically in high-purity water to remove dust, rinsed twice before being dried down in high-purity acetone and weighed into pre-cleaned Teflon beakers. A known amount of ^{84}Sr tracer solution was added to each sample and dissolved in Teflon-distilled 8M HNO_3. Samples were then converted to chloride using quartz

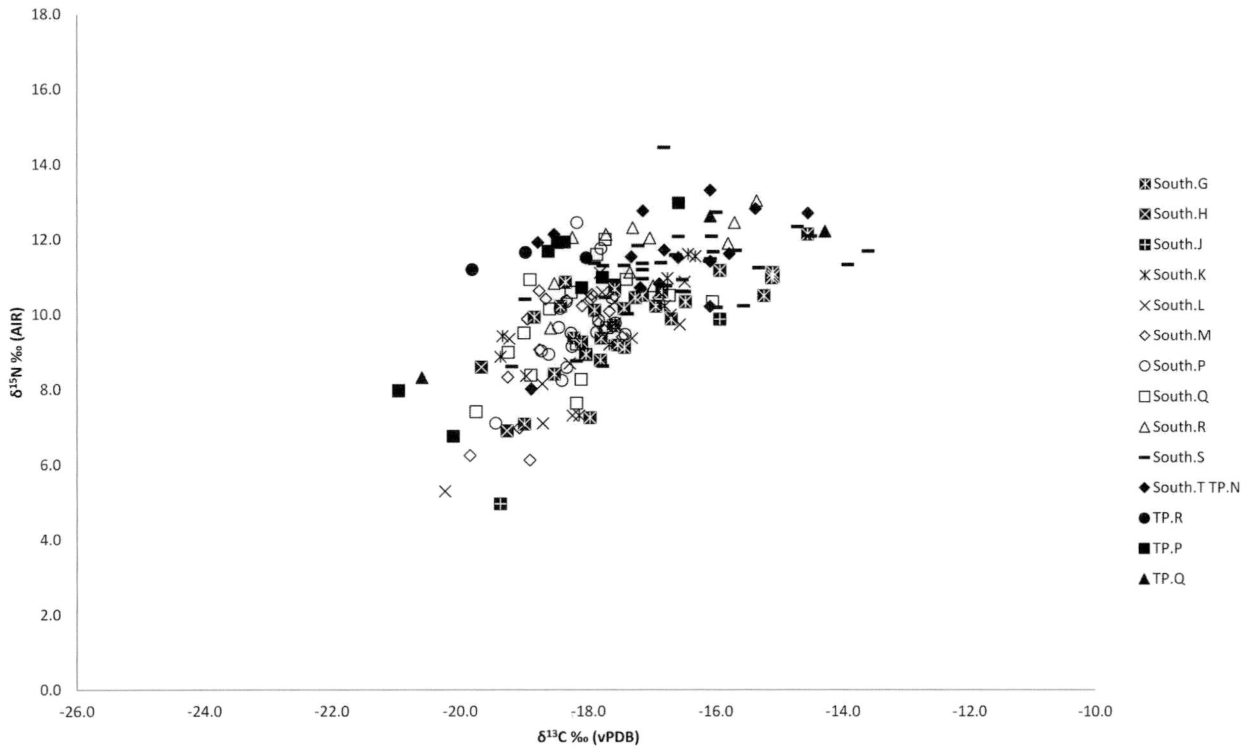

Figure 11.5. Ovis *spp. plotted according to level in the South and TP Areas (n=184).*

Capra *spp*. In addition to the 22 individuals previously measured, a further 15 specimens of goat have been added (fig. 11.6). Goats remain a rarely exploited species at the site, being vastly outnumbered by sheep in terms of identified specimens, and cattle/aurochs too when meat volume is considered (Twiss et al. this volume, Chapter 8). These additional samples shift the mean, maximum, range and standard deviation values by just 0.1‰, although the range is no longer the largest at the site (table 11.4). The number of C_3/C_4 plant consumers has increased, with a large quantity of C_4 plants identified in a small number of additional animals. The nitrogen isotope values have been affected by the additional samples, with the range now covering between 5.2‰ and 15.2‰, a range of 10‰. These data also increase the mean to 9.5‰, which brings this species closer to *Bos* and *Ovis* species averages and subsequently has not alleviated the difficulty in establishing the relative importance of all three species to human diet. The temporal variation is based on only a small number of samples in the Middle period, but again these data seem to show lower values for both isotopes, with none of the individuals from this period seeming to consume C_4 plants (fig. 11.6).

This suggests a similar scenario to the sheep, depending upon whether individual animals are domestic or wild and whether the Middle period represents responses to climatic, environmental or socio-political change, altering how herders managed and/or hunted the animals they consumed. The individuals from the Middle period that are affected by a transition in behaviours are one individual from Level South M and three from Level North G.

Equids. A further 11 equids can be added to the original 45 specimens (fig. 11.7). It should be noted that this group is represented by two species (*Equus ferus* and *Equus heminonus*), which cannot be distinguished between based on postcranial bone sampled here. Again the extra samples make only a small difference to the previously reported values (Pearson 2013). The $\delta^{13}C$ values are from −19.9‰ to −14.5‰ (5.4‰ range), and in $\delta^{15}N$ from 4.8‰ to 11.3‰ (6.5‰ range), with mean values of −19.0‰ and 8.4‰ in carbon and nitrogen, respectively. The two taxa comprised of 56 specimens still have a smaller range in isotope values compared to key ruminants. The diets of these animals remain dominated by C_3 grasses, in stark contrast to the sheep, goats and cattle. This might give a clue as to the 'natural' grazing available for animals free to roam and point to the small but consistent role of C_4 plants in animal management and herding.

The three animals with a small C_4 plant contribution and one other with a larger portion in their diet were reported previously (Pearson 2013). In terms of considering the management of the other species, there is no shift in diet in the Middle period for equids, suggesting that the shifts seen in the sheep, goats and cattle may point to any change in this period of occupation being the result of anthropogenic factors. In contrast, the equids show a different dietary pattern in the Late period of occupation, which may reflect a shift to different equid hunting grounds, where a greater quantity of C_4 grasses were consumed by animals living there. The C_4 consumers appear in Levels South S and R, with no C_4 plants in the diet in the early levels (unlike the sheep).

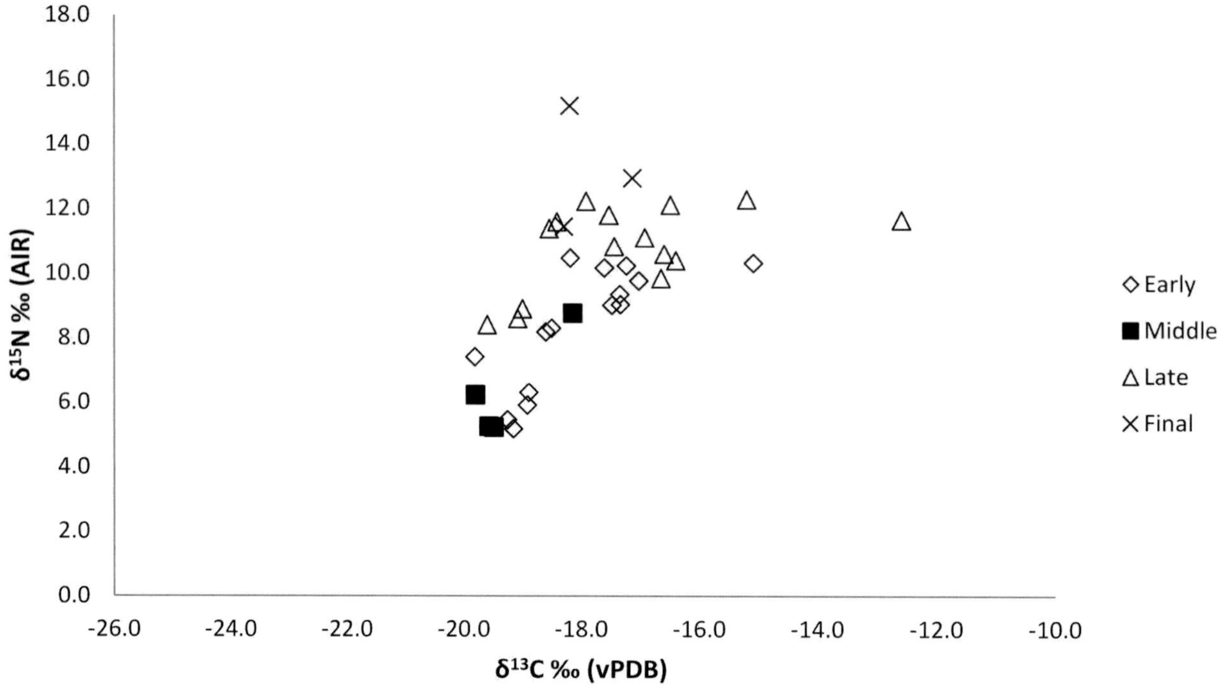

Figure 11.6. Capra *sp. plotted according to broad occupation period in all areas (n=37).*

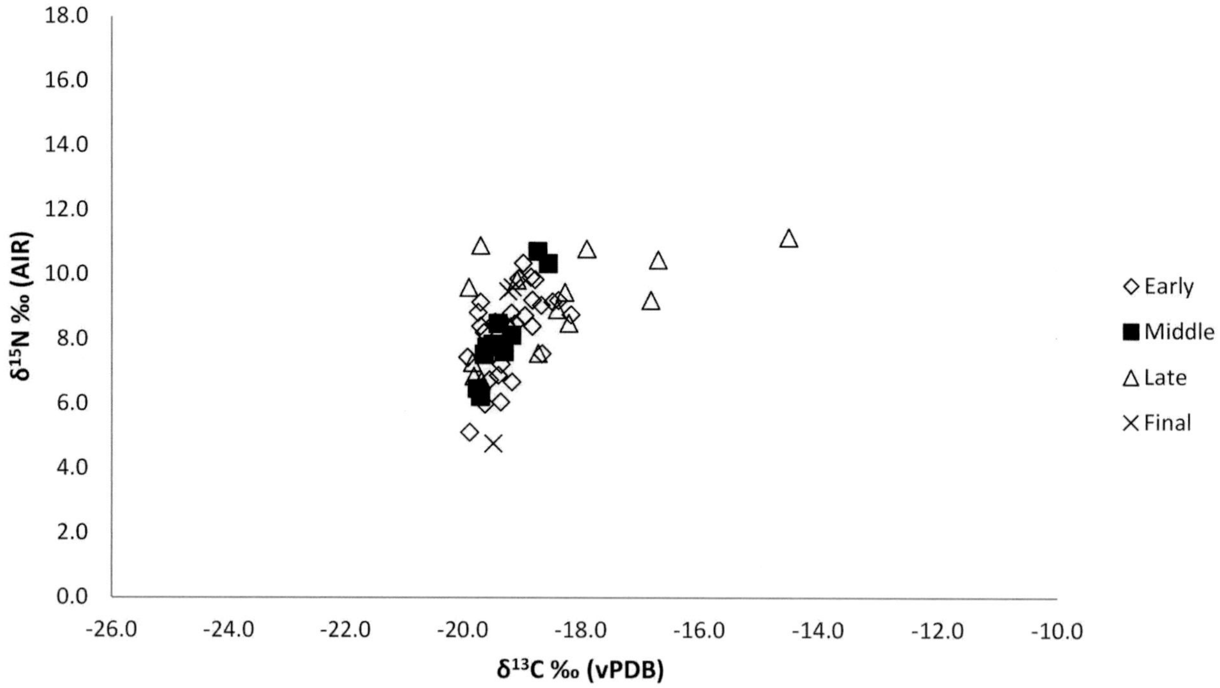

Figure 11.7. Equids plotted according to broad occupation period in all areas (n=56).

Boar. A further four samples can be added to the existing dataset, leaving an overall total of 32 measured specimens (fig. 11.8). Most of the samples came from the South Area. The carbon isotope values for the taxa remain unchanged: −21.3‰ to −15.3‰ (6.0‰ range), but the nitrogen isotope values have a new lower end of 5.1‰, leaving an overall range of 5.1 to 11.9‰ (6.8‰ range), with no effect on the mean values, which remain −19.5‰ and 8.0‰ in carbon and nitrogen respectively.

The presence of C_4 plants in wild boar diet remains an unexpected finding. C_4 plant taxa are almost exclusively restricted to the grass and sedge families and have a relatively higher silica and fibre content compared to C_3 plants, as described earlier. The preferred habitat of wild boar in wetter and shaded areas of the landscape would seem to exclude zones where C_4 plants might flourish – more open, drier areas – from being areas that wild boar might visit for food, and yet some of the specimens have a significant component of this resource in their diet. This apparent signature may have been acquired through the consumption of a lower/adjacent-trophic level C_4 plant consumer, which may have included leftovers from human foods or an as yet unmeasured resource. The appearance of C_4 plants only occurs during the Late period (Level South S), similar to equids described above. However, since there are no C_4 plants eaten by humans present at the site, the hypothesis that the C_4 input in boar was through leftover human foods seems unlikely. Similarly, the mean $\delta^{15}N$ value of 8.0‰ is 4+‰ lower than humans and thus instead seems to point to a food not consumed at the site.

Carnivores. The results of 46 samples including 18 new samples from the Carnivora family (*Vulpes vulpes* n=21, *Canis* sp. n=23, and *Canis lupus* n=1, *Ursus* sp. n=1) are given in figure 11.9.

These species show a broad range of both carbon and nitrogen isotope values, suggesting that diets varied between the different species. Overall the mean values for this order are −18.1‰ and 11.4‰ with a range of −21.3 to −14.3‰ (7‰ range) in carbon and 4.6 to 16.5‰ (11.9‰ range) in nitrogen. These ranges serve as an illustration for the potential range of different meat-eating taxa and help to bookend the values that might be expected in humans. The mean nitrogen isotope value for the order overall is surprisingly low given the combined mean for sheep, goats and cattle (9.9‰) and adult humans (12.3‰). This suggests a range of trophic level foods in the diet of the site's carnivores, and while the lower nitrogen isotope value relative to humans might suggest *more* plants in their diet, the very high nitrogen isotope values in the site's cereals (Bogaard et al. this volume, Chapter 5) may be the cause of high nitrogen

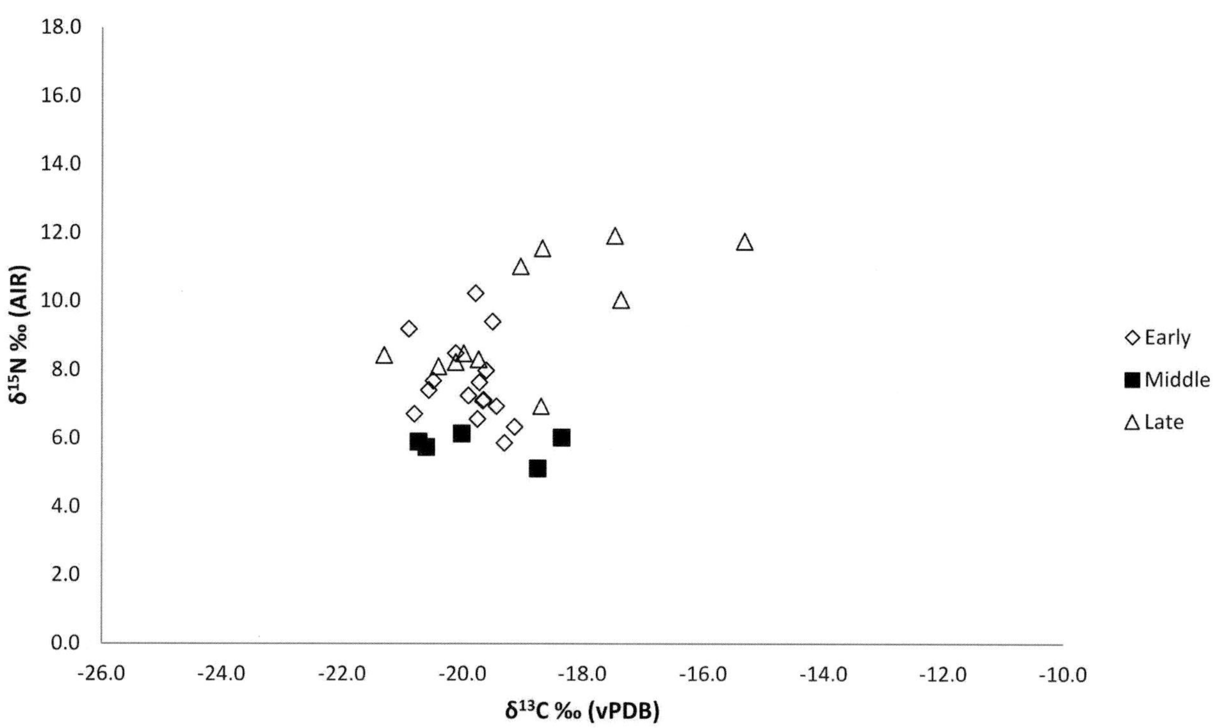

Figure 11.8. Boar plotted according to broad occupation period in all areas (n=32).

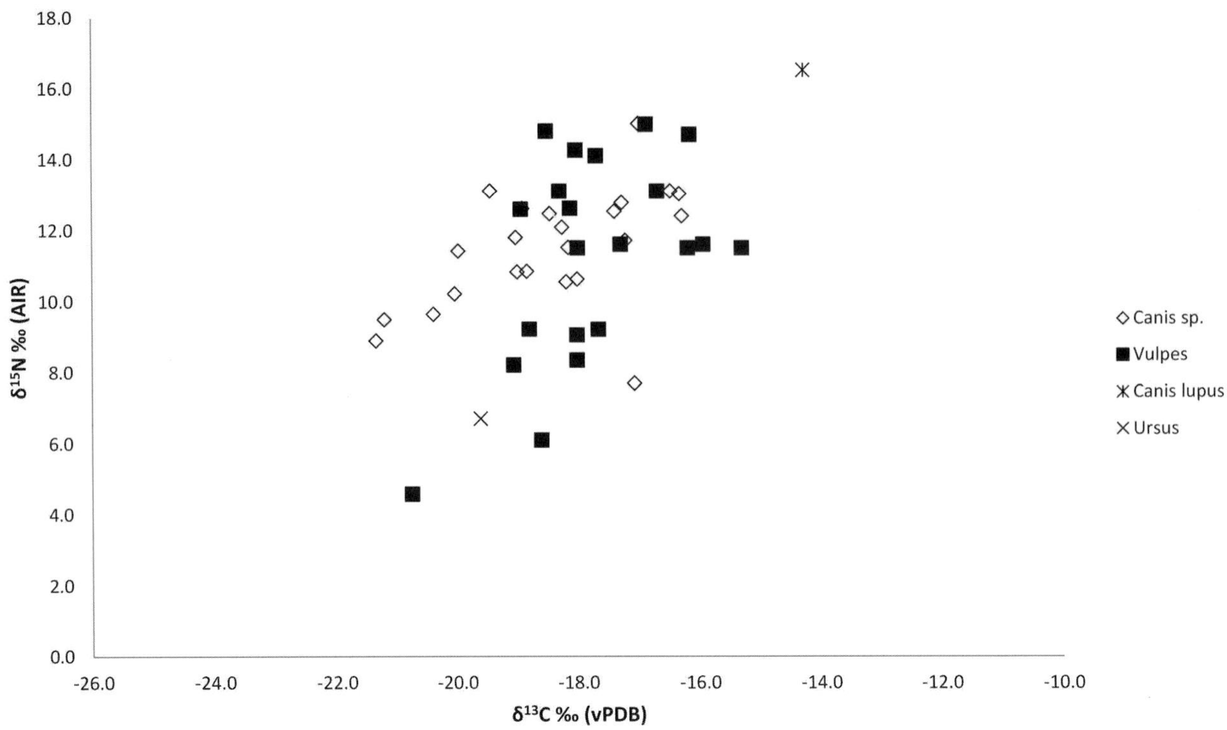

Figure 11.9. Carnivora plotted according to species in all areas (n=47).

isotope values in humans. The carnivore carbon isotope values are higher than for humans by 0.7‰, almost one trophic level, which suggests a different source of low trophic-level foods in their diet. This perhaps indicates consumption of the meat-based human foods, perhaps parts of animal carcasses considered less desirable, but not human plant-based foods such as breads or gruels. As discussed elsewhere, dogs, wolves, foxes and bears will consume smaller animals such as insects, smaller mammals, birds and eggs, which may also explain the difference between these animals and humans, despite their omnivorous physiology. For individual genera and species, the additional five *Canis* sp. specimens result in no changes to the $\delta^{13}C$ value, which remains −21.3‰ to −16.3‰ (5‰ range) and mean of −18.4‰.

The $\delta^{15}N$ values now range from 7.7‰ to 15.0‰, a new range of 7.3‰ and a new mean of 11.5‰. For *Vulpes vulpes* with 13 new values, the range has now increased in $\delta^{13}C$ value from −20.7 to −15.3‰ with a new mean of −17.8‰, and in nitrogen the new lower end of the range is 4.6‰ with the same upper end 15.0‰, resulting in a slightly different new overall mean of 11.3‰. In terms of temporal change, figure 11.10 illustrates these taxa according to the broad occupational periods. Again, as illustrated with other fauna, the consumption of foods with the highest carbon and nitrogen values occurs in the later periods.

Other mammals: Cervus *and* Lepus. Five additional deer (four *Cervus* sp. and one *Dama* sp.) can now be added to the 11 samples previously discussed (Pearson 2013). Although based on a small number of individuals, these taxa offer an insight into the diets of true wild herbivores. There are now 12 *Cervus* sp. samples (fig. 11.11), which have a slightly larger range than the previous work (Pearson 2013).

The carbon isotope values now range from −21.2‰ to −14.1‰ (7.1‰ range) with a new mean of −19.0‰, and in nitrogen isotope values from 4.0‰ to 13.8‰ (9.8‰ range) with a new mean of 7.9‰. The *Dama* sp. specimen ($\delta^{13}C$ −18.9‰; $\delta^{15}N$ 6.2‰) sits towards the lower end of the nitrogen isotope range for *Cervus* sp. The *Lepus* sp. samples reported previously (Pearson 2013) remain the species with lowest $\delta^{15}N$ values.

All animals except two of the *Cervus* specimens have C_3-plant dominated diets. The presence of C_4 plants is surprising given their ecological preference for a wooded environment, although perhaps this illustrates the presence of C_4 plants in mixed zones of woodland and grassland vegetation, or relatively close to woodland. Figure 11.12 shows these taxa according to the broad chronological periods at the site. Again, animals with a significant C_4 component to their diet came from the later periods of occupation at the site (Levels South Q and South S).

For the final phase of analysis, a further 251 human individuals were sampled, of which 96 failed to yield well-preserved collagen (38% failure rate). The failures were the greatest in the North Area of the site, with 69 of the 96 failed samples originating from this area. Across the site, the mean adult human carbon and nitrogen values are −18.8‰ and 12.3‰ respectively. For adult males and females, the means remain strikingly similar, with females measuring: δ^{13}C −18.8‰; δ^{15}N 12.3‰; n=57, and males measuring δ^{13}C −18.9‰; δ^{15}N 12.3‰; n=45. This suggests the diets were virtually the same across the sexes, although these new data have reduced the mean nitrogen isotope values for both sexes by 0.4‰ relative to Pearson (2013). The adult mean isotope values of δ^{13}C −18.8‰ and δ^{15}N 12.3‰ indicate that the average value for food consumed (plants and animals) will be approximately 8.3‰, with an average carbon isotope value of −19.8‰ (using $\Delta^{15}N_{consumer\text{-}diet}$ value of 4‰ and $\Delta^{13}C_{consumer\text{-}diet}$ value of 1‰ trophic level shift discussed above), which would have been generated from a combination of plants and animals. The most common animal protein element of the diet would have been sheep (δ^{13}C −17.6‰; δ^{15}N 10.2‰) and cattle (δ^{13}C −18.2‰; δ^{15}N 10.1‰) (fig. 11.16).

However, since there is still an offset between the values of animals and the expected human diet, this suggests that other resources with lower carbon and nitrogen isotope values likely made up a significant proportion of the diet. As a farming community, cereals and legumes seem the most likely sources of additional food and would have made a major contribution to the diet. Fish can be ruled out as having been a major resource, since the carbon isotope values for that resource are 3‰ lower than the value expected for the whole diet. However, it should be noted this interpretation is based on measurements from only two samples (although each sample comprised tens of specimens). Aquatic and terrestrial birds cannot be ruled out as dietary contributors. The mean values for these two groups correspond well to the carbon isotope values, in terms of their likely contribution to the human diet, but the relatively higher nitrogen isotope values in the terrestrial birds suggests that these were not a dominant component of the whole diet. The extensive range in nitrogen isotope values amongst the humans indicates that some of the measured fauna would have been more important to some individuals than others. These data suggest the possibility, for instance, that some groups organised either as households or other social groupings consumed significantly more specific resources such as birds, cattle, and the rarer species such as equid, suid and deer. Potential groups or possible environmental effects and their relationship to diet are discussed in more detail below.

Figure 11.17 illustrates adult human diet by age group and sex. Comparing diet in individuals of different age as well as different sex reveals some differentiation in their diets.

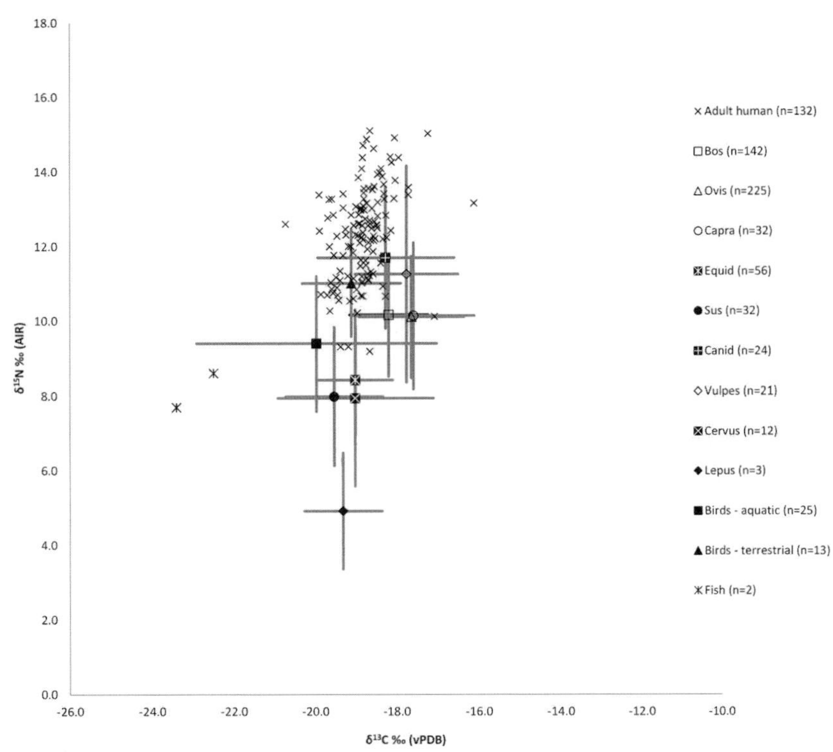

Figure 11.16. All Neolithic faunal species plotted as mean and 1SD error bars against all Neolithic adult humans.

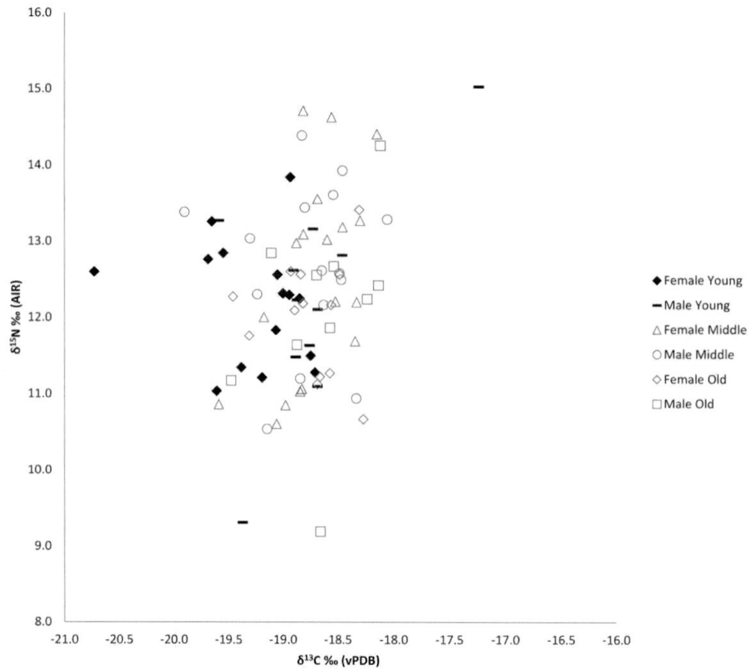

Figure 11.17. Adult humans plotted by age and sex.

These data show a significant shift in carbon isotope values between young and old females (p=0.005) and to a lesser (but still significant) extent comparing young females with young males (p=0.02), suggesting that age-based diet variations have a greater impact on the isotope evidence of diet than do sex-based variation. However, comparing young females against old males gives the greatest significance (p=0.004). This suggests that the main driver in dietary differentiation is age-related, but that a smaller but sustained impact can be attributed to sex and likely was inextricably linked to it. Therefore, the age shift in diet seems to have played out differently according perhaps to gendered foodways. Both groups of younger women and older men had similar mean nitrogen isotope values (12.3‰ versus 12.1‰ respectively), which broadly suggests similar trophic level foods, but the differences in carbon isotope mean values (−19.3‰ versus −18.6‰ respectively) suggests that younger women would have been more likely to have consumed foods with lower isotope values. Such resources would have been wild animal species including equids, suids and birds, plus plant foods with relatively higher nitrogen isotope values, whereas older men are more likely to have eaten similar foods as the rest of the occupants at the site, since their mean values are close to the site means. A possible biological factor is the difference in isotope values in the tissues of fertile women. However, modern studies of pregnant women show small shifts (up to 1‰ lower) are frequently observed in nitrogen rather than carbon (Fuller et al. 2004; D'Ortenzio et al. 2015), allowing this to be ruled out here.

Area-specific diets. A comparison of the groups of individuals living in different parts of the mound, chiefly the North (fig. 11.19) and South (fig. 11.18) Areas, is shown below. The poor preservation in the North Area has somewhat hampered a robust comparison in terms of sample size, but some preliminary data for the North Area suggests large variation in the diets of animals, although the large standard deviation values could be due to small sample sizes in some fauna.

The South and North Area sheep, which are the most statistically robust sample populations, suggest that there is a difference in terms of the diets of the animals from these two areas. However, the North Area animals only span the Early and Middle periods of occupation, whereas sheep for the South Area span Early, Middle, Late and Final periods. Thus, it seems likely that the difference between these two areas is a function of time depth sampling, with the North Area not being represented by the Late and Final period fauna when the C_4 plant 'reflux' described above in a number of fauna occurs. Other differences between the two areas appear in the birds; the grey heron mean carbon isotope value for the South Area ($\delta^{13}C$: −20.0‰) suggests a C_3 diet foundation, in contrast to the North Area ($\delta^{13}C$: −16.5‰), which suggests some C_4 input in the local food chain (since this species is largely residential). A closer inspection of these data indicates that the South Area animals date to the Early period of the occupation, whereas the North Area animals date to the Late period, suggesting a possible chronological effect. Many of the fauna with the highest nitrogen isotope values seem to occur in the South Area, but again this seems to be

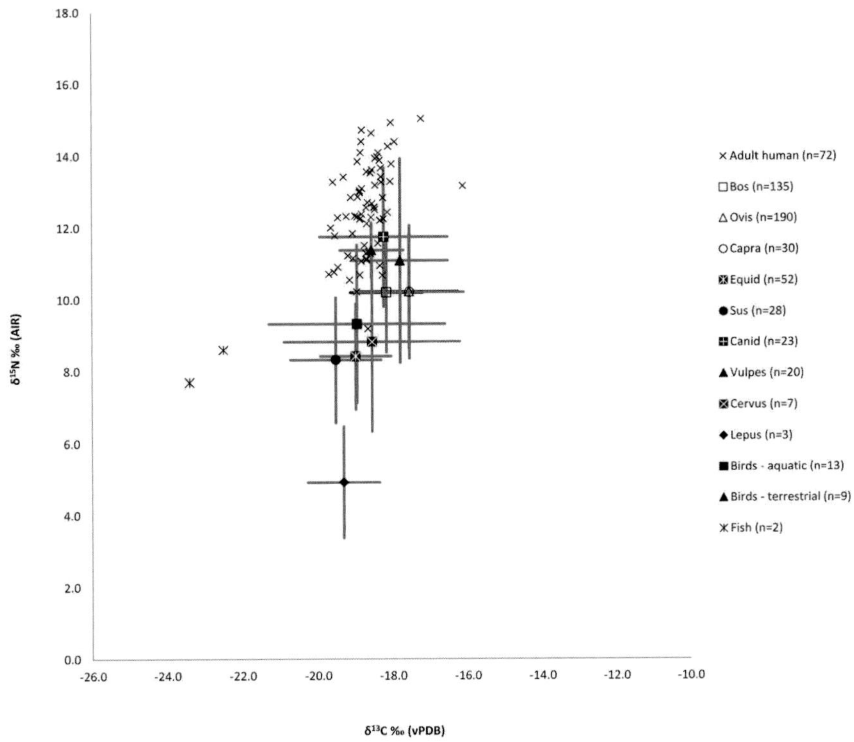

Figure 11.18. South Area faunal species plotted as mean and 1SD error bars against South Area Neolithic adult humans (n=584).

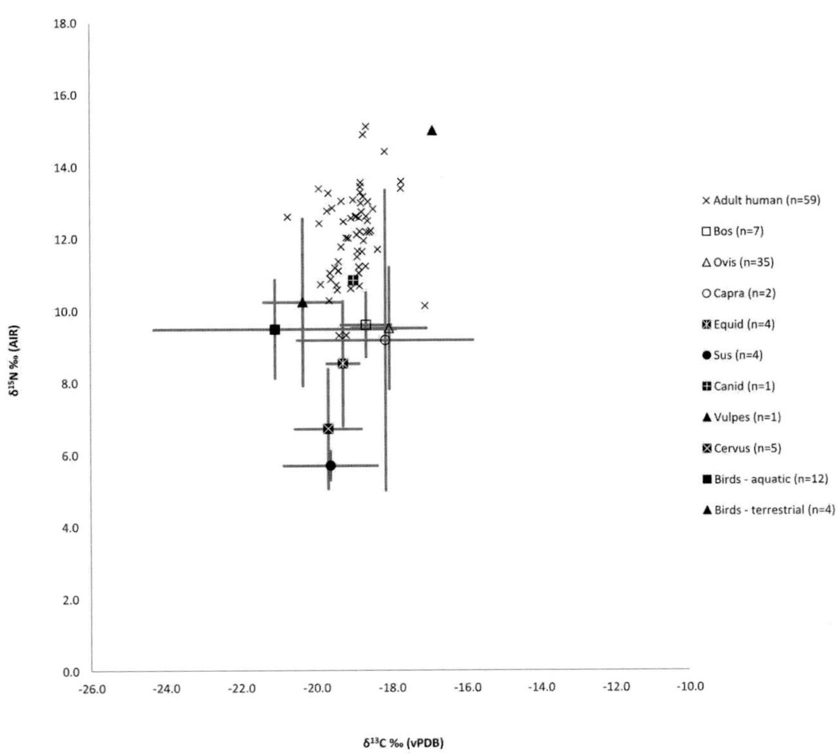

Figure 11.19. North Area faunal species plotted as mean and 1SD error bars against North Area Neolithic adult humans (n=134).

a feature of sample size and sampling across the different levels of the site, since the adult humans are broadly similar. Statistical analysis using an independent samples t-test of the Middle period of occupation, where the sample population for humans is largest, confirms no significant difference between the diets of humans from these areas ($p=0.944$). The human carbon and nitrogen isotope values are discussed below building by building.

Period-specific diets. Figure 11.20 shows the nitrogen element of the 'reflux' first identified in the sheep in Pearson (2013) and confirmed above for sheep (fig. 11.4), goat (fig. 11.6), and boar (fig. 11.8), is also present in the human diet, suggesting that this shift likely represents either an environmental impact, or a shift in how the landscape was used for hunting, gathering and farming.

The Middle period witnesses a sizeable shift downwards in nitrogen isotope values, which may indicate a greater focus on plant resources, although what might cause this (that is, population pressure, environmental change) is unclear. However, a stress response concerning availability of resources would normally result in elevated nitrogen isotope values. These shifts in values, though, are identified in both domestic and wild resources, suggesting a more fundamental shift at the environmental level. There continues to be little impact of C_4 plant consumption in animals on human carbon isotope values. Although many of the elevated nitrogen isotope values in the animals discussed above appear to correlate with increased C_4 plant consumption, that effect does not appear to feature here. One reason for this may be the C_4 plant increase was itself a response to local environment changes. This suggests C_4-eating animals generally made very little contribution to the lifetime diet of adult humans. This apparent shift in values over time across a number of species also complicates the identification and comparison of building-specific diets and consideration of human diet and its relationship to building attributes discussed below, which can only be compared within occupation periods: Early, Middle, Late and Final.

Building-specific diets (with contributions by Claudia Engel). The adult human data in figure 11.21 are plotted according to the building underneath which they were found. It should be noted that these numbers of

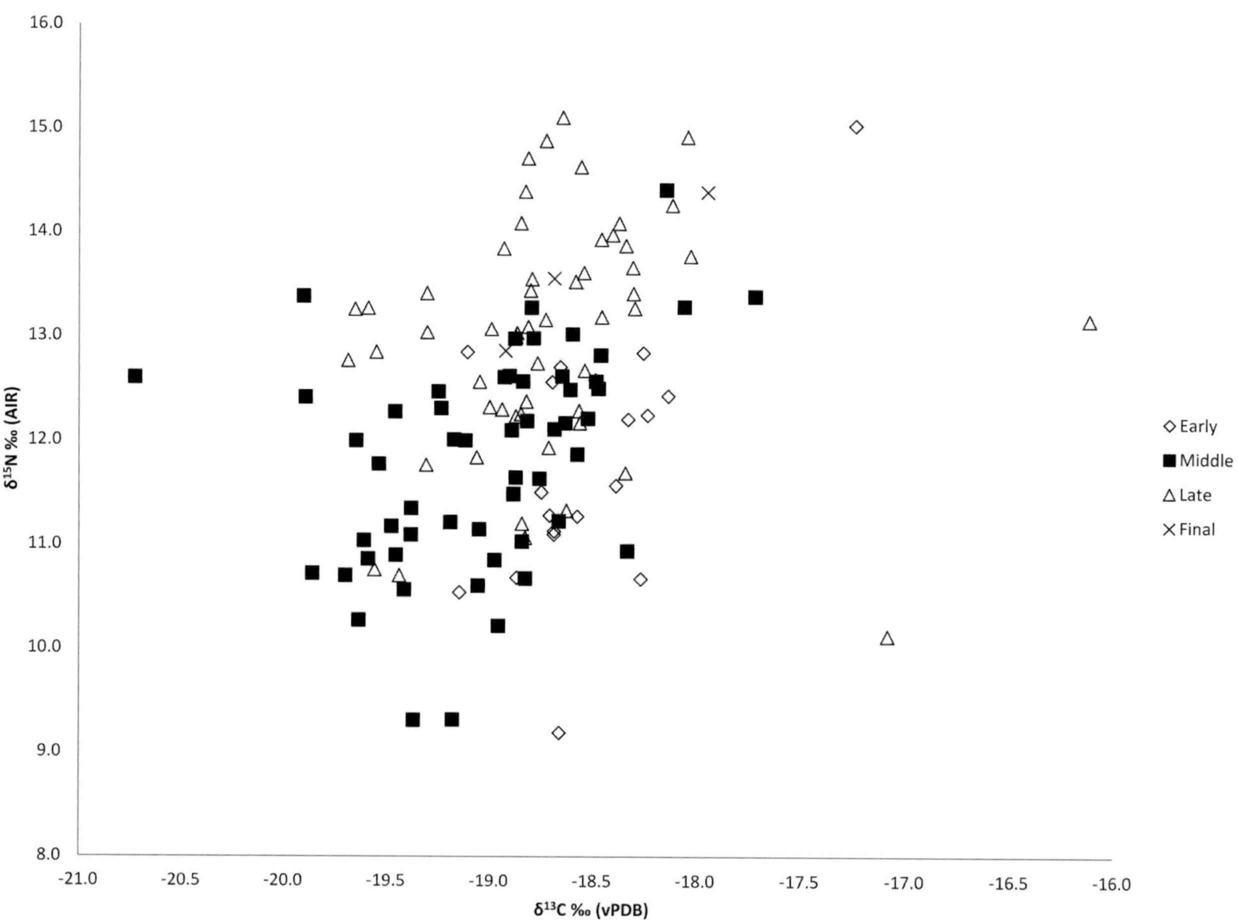

Figure 11.20. Adult humans plotted according to Early, Middle, Late and Final occupation periods.

individuals only represent those that had good collagen preservation, and in most cases there were other adults also buried in the building; therefore, these data do not give a complete picture of the diet of a given building. Only buildings with at least two individuals with well-preserved collagen are included in order to facilitate comparison.

These data show variation in diet within buildings, which is to be expected, as individuals undertook different roles and performed their required tasks as part of the community. Such tasks would have necessitated that individuals spent differing amounts of time away from the site in a range of environments, which would lead to the consumption of different foods where these activities occurred on a regular or even seasonal basis. However, a central question is whether diets within houses are still more similar than across houses in spite of these activities. In other words, are the diets within a house more similar than might be expected if diets (and thus the burying of individuals) were randomly distributed across houses. To test this robustly, considering the small sample size for each house, and given both the likelihood these data are not normally distributed and the need to compare as many as possible, permutational multivariate analysis of variance (PERMANOVA) and permutational analysis of multivariate dispersions (PERMDISP) were used to examine the centroids and dispersal. The centroids are of clusters of values relating to a location, in this case a house, and dispersal indicates whether the individuals from one house had widely spread diets in comparison to other houses. Comparing houses with at least four individuals, PERMANOVA was significant at 0.001 across the occupation but PERMDISP was not significant. These tests demonstrate that there is a location (house) effect, which confirms diets differed between houses with no statistical difference in the degree of dispersion within houses (fig. 11.22).

When chronological effects of the major occupation periods are taken into consideration, dietary patterning by house does not reach significance in the Early period ($p=0.549$) but does in the Middle ($p=0.033$) and Late ($p=0.013$) periods, perhaps suggesting increasing clustering over time. Since houses are also part of neighbourhoods, groups of houses were also compared to examine whether there were distinct neighbourhood diets. When six house clusters were compared overall, regardless of period, the groups are significantly different ($p=0.001$). However, when these are further separated into Middle and Late periods, they do not show significant differences within a particular period (fig. 11.23).

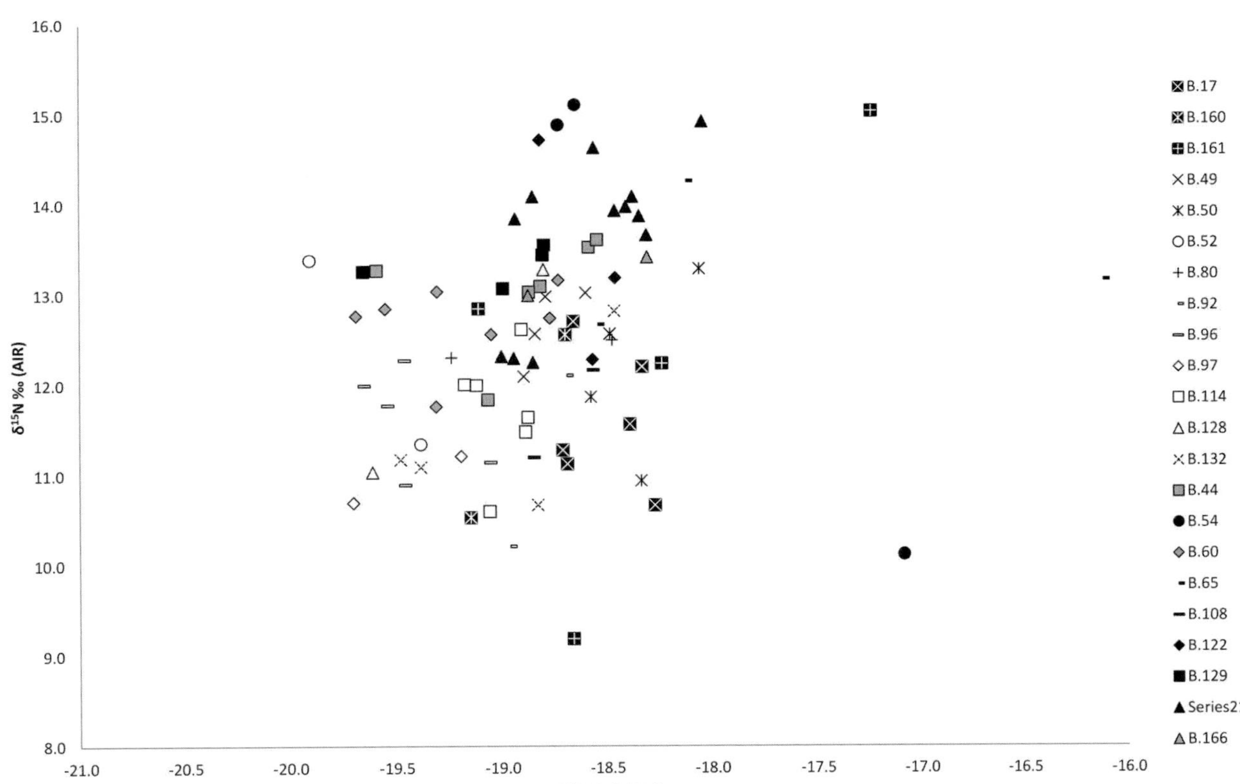

Figure 11.21. Adult humans plotted according to building and occupation period (Early: white on black symbols; Middle: black line/unfilled symbols; Late: filled black and grey symbols).

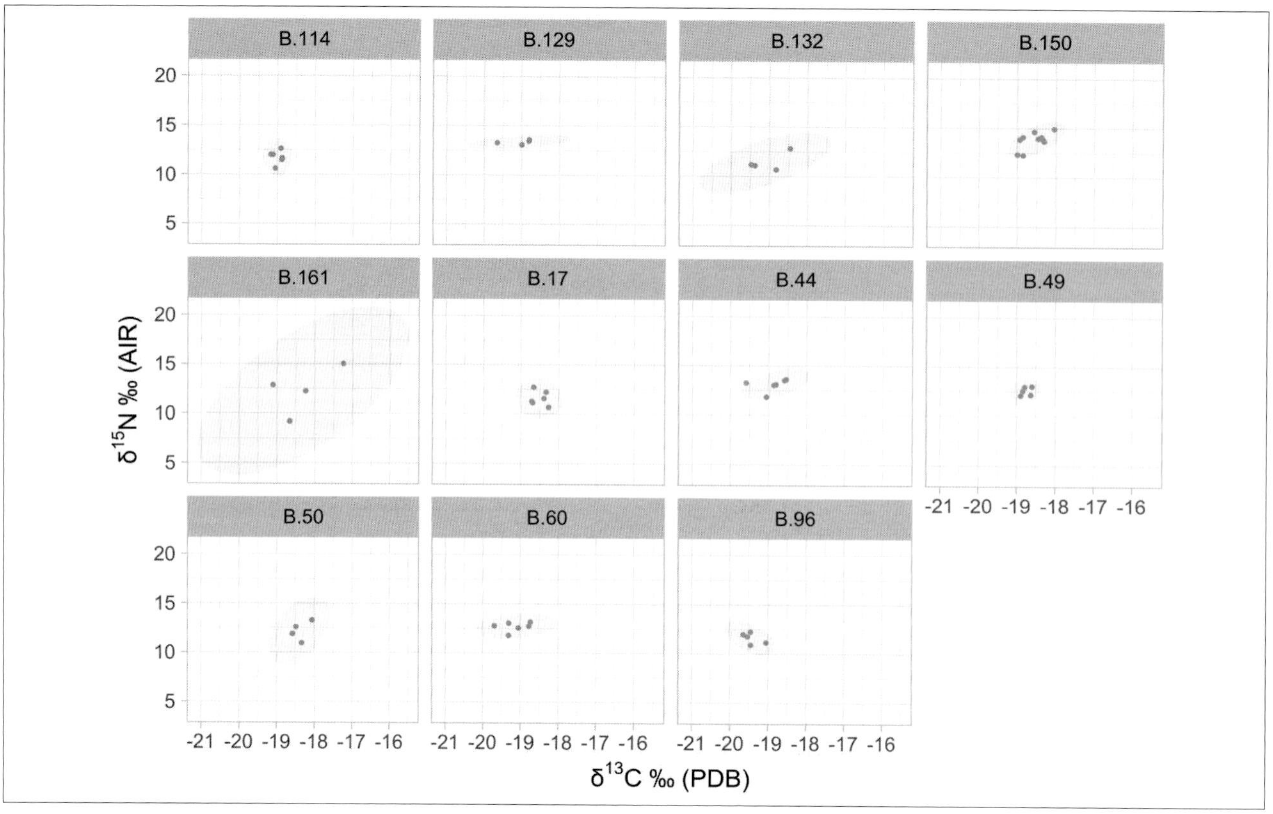

Figure 11.22. Buildings and spaces with >3 individuals with 95% confidence ellipses.

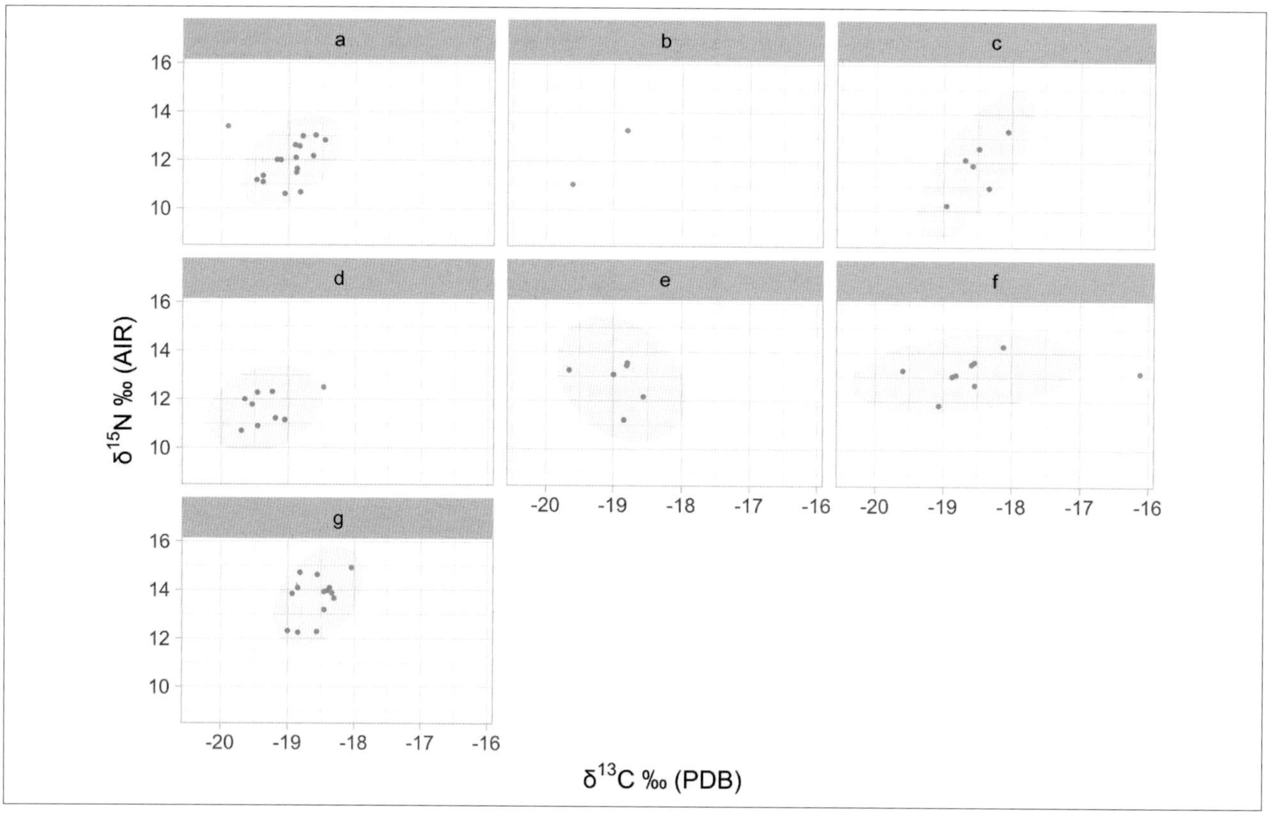

Figure 11.23. Neighbourhood spatial clusters with 95% confidence ellipses: (a) Middle period North (B.132, B.52, B.114, B.49); (b) Middle period North B.128, Sp.17; (c) Middle period South B.92, B.50; (d) Middle period South B.80, B.96, B.97; (e) Late period North B.129, B.108; (f) Late period South B.44, B.65; g) Late period South B.122, B.150, B.165.

Finally, neonates were compared across houses in an attempt to identify clearer evidence of household diets by hypothesising that neonatal bone provides a short-term summary of female diets as they enter the last trimester of pregnancy. Three buildings (fig. 11.24) had at least four neonatal individuals with well-preserved collagen. These data show significantly different values between the buildings (p=0.001) suggesting they may indeed represent evidence of the house-based diets for females. However, there are several caveats that should also be borne in mind: 1) the error margin of neonates is two months in utero to five months post-partum, and older individuals may have had some breastmilk represented in their bone; 2) because pregnancy causes metabolic stress, the isotope values may not reflect true dietary values; and 3) the isotope values of these neonates might also reflect the impact of the pathological condition which they suffered prior to death.

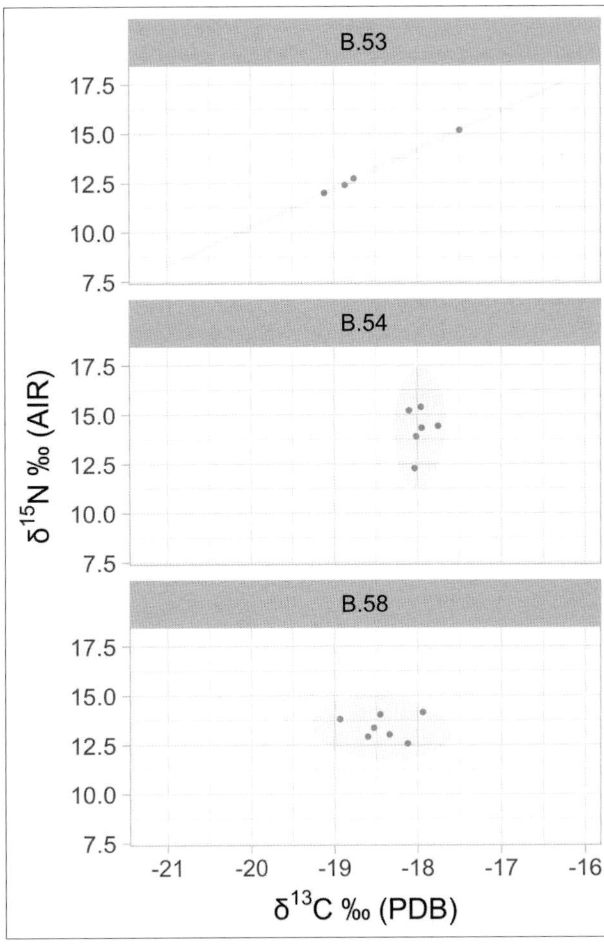

Figure 11.24. Buildings and spaces with 3+ neonatal individuals with 95% confidence ellipses.

Human diet and building attributes. A final source of variation in diet concerns whether specific buildings or areas of buildings were used for particular individuals: specifically, whether the diet of the people of Çatalhöyük correlated according to platforms under which they were buried, or the numbers of burials/elaboration of houses. Platforms in the Early, Middle and Late periods were compared where ages were mixed so as to avoid age-based differences already identified, and no patterning was found. This suggests that individuals under specific platforms were not afforded similar diets. Comparisons were also made using the Elaboration Index (Ei) and Minimum Number of Individuals (MNI) burial data calculated for the site (see Chapter 1). Only buildings where at least 75% of the structure had been excavated were considered. The following linear correlations were explored in order to highlight potential relationships, but since most demonstrated no correlation between any of the building attributes, further statistical analyses were not undertaken (table 11.5).

A total of 29 isolated skulls/crania were also identified for sampling in order to test for a potential difference in diet according to primary and secondary burial. The mean values differ from the overall population by 0.1‰ for carbon and 0.4‰ for nitrogen, suggesting that there was no major difference. Secondary burial treatments of crania and mandibles also feature most prominently in the Late period of occupation, where relatively higher isotope values are seen, suggesting that this difference is not unexpected. Comparisons between the skulls/crania also illustrated there was no dietary difference between the sexes (0.2‰ and 0.1‰ differences were observed in carbon and nitrogen, respectively).

Phase	*Attribute - Isotope*	*R2*
Early	Ei-Carbon	0.4630
Early	MNI-Carbon	0.4630
Early	Ei-Nitrogen	0.0060
Early	MNI-Nitrogen	0.0060
Middle	Ei-Carbon	0.0035
Middle	MNI-Carbon	0.0691
Middle	Ei-Nitrogen	0.0030
Middle	MNI-Nitrogen	0.0526
Late	Ei-Carbon	0.1372
Late	MNI-Carbon	0.0354
Late	Ei-Nitrogen	0.0332
Late	MNI-Nitrogen	<0.0001
ALL	Ei-Carbon	0.0437
ALL	MNI-Carbon	0.0050
ALL	Ei-Nitrogen	0.0139
ALL	MNI-Nitrogen	0.0140

Table 11.5. R^2 data for comparisons of carbon and nitrogen isotope data and building attributes Elaboration Index (Ei) and Minimum Number of Individuals (MNI) burial data.

Weaning patterns. In Pearson (2013), using a dataset of 38 subadults aged between birth and age stage five years, the nitrogen isotope data suggested that the first onset of weaning began around 18 months of age, and was complete by three years, giving an overall weaning duration of 18 months. The carbon data were inconclusive.

Adding a further 75 subadults in the interval between birth and five years of age, the nitrogen isotope data now suggest that the start of weaning began before the first year of age (fig. 11.25), much earlier than previously suggested, but that breastfeeding had ceased by three years of age. This shift in the interpretation of these data may have been the result of small sample sizes in the individual age groups. The carbon isotope data also suggest the first introduction of complementary foods began at some point before the first year of age and the wholesale shift to this food was completed by three years of age (fig. 11.26). Some caution should still be exercised, as these new interpretations are based on a small number of individuals with relatively higher values, which may also be a sign of stress in these particular infants.

Strontium isotope evidence of residential mobility

In order to identify individuals that did not spend their childhood in the local area, we used a $^{87}Sr/^{86}Sr$ measurement baseline map previously published by Bogaard et al. (2014a) that was constructed using modern plants sampled from the centre of the Konya Plain alluvial basin across to the nearby limestone terraces. These data showed only a small overlap between the two geological substrates, with the terrace $^{87}Sr/^{86}Sr$ ratios ranging from 0.7069 to 0.7076 and those in the alluvial plain from 0.7076 to 0.7081. Therefore, these data indicate $^{87}Sr/^{86}Sr$ measurement constraints of 0.7076 to 0.7081 as evidence for a childhood spent at Çatalhöyük. The $^{87}Sr/^{86}Sr$ measurements show that the majority (n=72) of the individuals sampled spent their childhood (for most individuals between 2.5 and 8.5 years of age, and for two, 8.5–14.5 years) on the Konya Plain, suggesting they are local individuals that grew up at Çatalhöyük. There are eight individuals that fall outside the range for a childhood spent on the alluvial plain. One individual (14753) has $^{87}Sr/^{86}Sr$ ratios indicative of a childhood spent on the limestone terraces, and seven individuals have $^{87}Sr/^{86}Sr$ ratios indicative of a childhood (or part of a childhood) spent in a more geologically radiogenic zone, likely towards the Taurus Mountains. It is possible that the individuals closest to the higher ratio cut-off for the plain likely spent part of their childhood, perhaps on a seasonal basis, away from the site. Since the Konya Plain alluvium is comprised of soil generated from the Taurus Mountain region (Bogaard et al. 2014a), it is possible that the higher strontium isotope ratios show short-term travel between

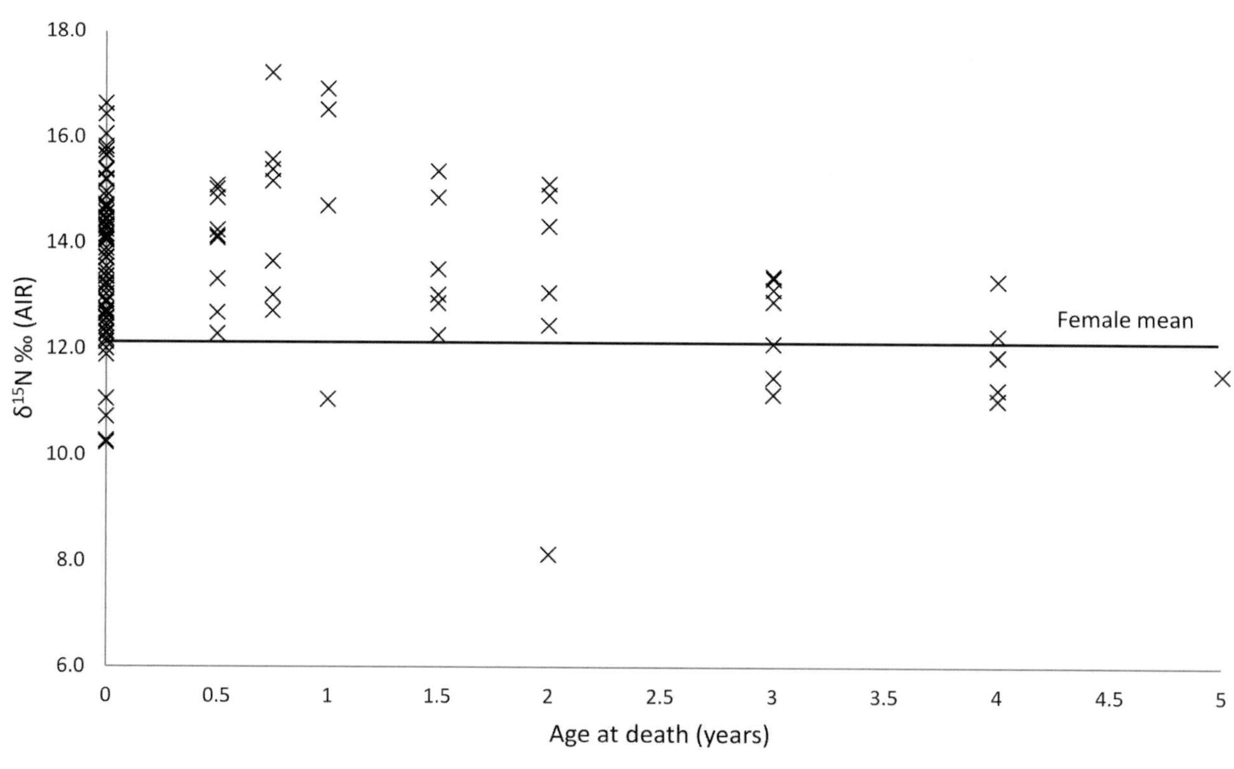

Figure 11.25. Nitrogen isotope values for subadults assigned to age stages birth–5 years age stage. The solid black line indicates the female mean.

12. The fish remains

Sheila Hamilton-Dyer

Introduction/previous work

Until 2008, fish remains found at Çatalhöyük were not analysed in detail; bones of small fish had been noted but thought largely incidental, arriving in the mud used for building. Some artefacts of fish-hook shape had been found, but they are too large for catching fish of this small size and the remains of large fish seemed very few. In the 2008 excavation season a study was undertaken to assess the occurrence of fish remains at the site and elucidate whether some could be the remains of food and of deliberate fishing (Van Neer et al. 2013). For that analysis a selection of 120 units was made out of the 1,287 excavated units that contained high concentrations of faunal remains and other finds. In addition, a study was made of all the samples from B.77, a burnt house in the North Area. The finds from this building included an unusual concentration of burnt material that included fish bones and peas.

An examination of the density of the fish finds found few bones in mortar but quite a number in the soft clay floors. The highest concentrations occurred in the same midden and fill deposits as the other faunal remains and were thought highly likely to be food remains.

As a result, many thousands of bones were examined (18,693 specimens, Van Neer et al. 2013: table 15.1), of which 11,955 were identified at least to superfamily. The identified remains are mainly of Cyprinidae (carps, minnows) with a small percentage of Cobitoidea (loaches). The cyprinids that could be identified to genus are mostly of *Pseudophoxinus* sp. and *Capoeta* sp. Although the exact species and their distribution is imprecisely known, in general the ecology of the species within the genus is likely similar. It was proposed that the fish assemblage represents exploitation of small fish from shallow, marshy waters. These include the adults and young of small taxa such as the loaches and *Pseudophoxinus*, and the juveniles of larger fish such as the *Capoeta*. The remains of young fish suggest that the fishing season lasted for a few months over late spring to early summer. Capture from the shallow waters could have been by the use of scoop baskets in small pools, static baskets and traps, or fine meshed nets. It is also possible that ichthyotoxic plants could have been used to stupefy the fish for easy gathering. The fish could have been used immediately or sun-dried on matting on the roofs, for example.

The relative quantity of fish in the midden deposits was found to decline over time; perhaps through hydrological changes or a changing focus on food procurement, for example, intensification of sheep herding.

Current study

For the present study, samples were examined from 150 units. These were a random selection from across the site and had been collected from 1996 to the final excavation season; half of the units were excavated between 2009 and 2017. Eleven units are from the Early occupation period in the South Area; the other units from this area are split roughly equally between the Middle and Late periods. There is a similar number of samples from the Middle period in the North Area and a smaller number from the Late period.

Sampling and method

The material was generated from the standard protocol; multi-component soil samples were taken onsite, generally for water flotation. At least 10l of soil was floated from each unit and the macrofaunal material was recovered over a >4mm mesh. Any fish material observed during sorting and examination of the >4mm residue fraction was separated out for analysis. A further small number of fish remains were collected from dry sieving of the remainder of soil from the unit over a 4mm screen.

Identification and recording

Most of the recognisable bone fragments are of the pharyngeal plates, basioccipitals, or are vertebrae; there are very few other diagnostic elements present. Indeterminate rays, spines and fragments of cranial bones are relatively few. All elements that could be identified are exclusively of Cyprinidae. This is a very large family containing many morphologically similar species, and most bones are difficult to discriminate; vertebrae, for example, other than the first and second vertebrae, are not at present able to be discriminated and were recorded only as cyprinid. An overall total of 567 specimens was recorded, of which 279 were identified to sub-family or genus, 219 only as Cyprinidae and 69 as indeterminate fish.

Given the difficulties of identification of this group, recording concentrated on the more diagnostic pharyngeal plates and basioccipitals – together with vertebrae, these in any case accounting for the vast majority of the present material.

As discussed in Van Neer et al. (2013), the freshwater fauna of Anatolia is not fully known, with descriptions of new species and systematic changes frequent. The present distribution of species is also likely to be different to that of the past, as the hydrology has been much altered. It is therefore difficult to identify the precise species represented from archaeological bones and unsafe to make identifications beyond genus level in most cases. For some elements and broken bones only the family or subfamily level is appropriate.

The majority of the remains identified beyond the level of family Cyprinidae are of the Leuciscine cyprinid *Pseudophoxinus* (91) or, in the case of incomplete bones, are very closely comparable with this genus (52) and included as this in tables. The bones attributed to this genus are all pharyngeals and basioccipitals. They appear to represent up to three morphotypes, but no attempt was made to separate them; Van Neer et al. (2013) also observed at least three types but did not attempt discrimination.

A few bones are of a different genus within the same Leuciscinae subfamily; these match *Leuciscus/Squalius* cf. *cephalus*.

No other species were identified; the group of cyprinid bones classed as Leuciscine because they are definitely not of *Capoeta* or of the Cobitoidea (loach superfamily) are assumed to be all of *Pseudophoxinus*, as in the previous analysis where other Leuciscine species were identified only very rarely. Online table S12.1 lists the taxon counts for each sample/unit in phase order; table 12.1 gives a summary total by area and phase.

Distribution

As expected from the results of the previous study, most of the fish bones were recovered from midden deposits and fills. Lesser amounts were recovered from floors and only a few remains from other contexts (table 12.2).

In particular, the majority of the pharyngeals and basioccipitals of *Pseudophoxinus* were recovered from midden fills. The present material does not contain any special concentrations such as that reported from B.77, which contained pieces of burnt material with peas and fish (Van Neer et al. 2013: 320).

Area/Period	Pseudophoxinus	Leuciscus/squalius cephalus	*Leuciscinae*	*Cyprinidae, indet.*	*Fish, indet.*	*Total NSP*
North						
Middle	78	1	51	71	28	229
%	34.1	0.4	22.3	31.0	12.2	
late	3		10	17	6	36
%	8.3	0	27.8	47.2	16.7	
Unphased			2			2
Total	**81**	**1**	**63**	**88**	**34**	**267**
%	30.3	0.4	23.6	33.0	12.7	
South						
Early	8	1	10	23	2	44
%	18.2	2.3	22.7	52.3	4.5	
Middle	16	6	33	60	18	133
%	12.0	4.5	24.8	45.1	13.5	
Late	39	1	20	45	15	120
%	32.5	0.8	16.7	37.5	12.5	
Total	**63**	**8**	**63**	**128**	**35**	**297**
%	21.2	2.7	21.2	43.1	11.8	
GDN Late TP				2		2
IST Unphased				1		1
Total Taxon	**144**	**9**	**126**	**219**	**69**	**567**
%	25.4	1.6	22.2	38.6	12.2	

Table 12.1. Fish taxon by excavation area and major occupation period.

Taxon/Context type	*Unclassed*	*Activity*	*Cluster*	*Construction/ Makeup/Packing*	*Cut*	*Fill*	*Floors (use)*	*Midden*	*Total*
Pseudophoxinus	0	1	2	8	0	23	16	94	144
%	0	0.7	1.4	5.6	0	16.0	11.1	65.3	
Leuciscus/Squalius cephalus	0	0	1	0	0	5	0	3	9
%	0	0	11.1	0	0	55.6	0	33.3	
Leuciscine	1	6	1	6	2	36	36	38	126
%	0.8	4.8	0.8	4.8	1.6	28.6	28.6	30.2	
Cyprinidae	3	2	4	29	0	64	42	75	219
%	1.4	0.9	1.8	13.2	0	29.2	19.2	34.2	
Fish, indeterminate		8	5	1	0	19	21	15	69
%	0	11.6	7.2	1.4	0	27.5	30.4	21.7	
Total	**4**	**17**	**13**	**44**	**2**	**147**	**115**	**225**	**567**
%	0.7	3.0	2.3	7.8	0.4	25.9	20.3	39.7	

Table 12.2. Fish species by context type.

The vast majority of individual units contained only one or two fish bones, with just 13 units contributing over ten specimens (see online table S12.1).

The largest amounts from combined units from buildings are 40 and 54 bones from Buildings 49 and 132 (North Area) and 38, 27 and 26 bones from Buildings 89, 96 and 97 (South Area) respectively, all from the Middle period. Online table S12.2 gives a summary of the species and elements by area, phase and building.

The highest frequency from an individual unit came from midden deposit 19131 (South Area), which contributed 42 bones, 18 of which could be identified as *Pseudophoxinus*, a similar midden deposit (21660) (North Area) had 29 bones, including 27 of *Pseudophoxinus*, and of the 38 total from B.89 (South Area), 29 bones came from fill unit (18780). No *Pseudophoxinus* were identified in this particular deposit (18780), but it does contain at least four bones of significance. The remains are fragments of a pair of pharyngeal plates from a cyprinid matching *Leuciscus/Squalius cephalus* (chub) of about 45cm SL (standard length). Also in this context are a Leuciscine quadrate bone and the first vertebra of a similar sized fish – these elements are less diagnostic than pharyngeals but are a very good match for chub, confirmed with advice from Wim Wouters (Royal Belgian Institute of Natural Sciences (RBINS)). Although not diagnostic to species, there are also three cyprinid precaudal vertebrae and three caudal vertebrae that could easily be from the same individual fish. Other remains in the context were more typical of the rest of the assemblage.

Unit (18648) in B.97 (South Area) contributed 11 caudal vertebrae of similar size and type; even though not identifiable to species, they do represent a similarly large fish.

These finds are important as the majority of the other fish remains and those in the previous material were from small and very small fish, with individuals of 10cm SL or less. In the previous material, only one chub bone was identified and this was from a fish of less than 5cm length.

Size of *Pseudophoxinus*

A limited suite of measurements was taken on the bones including the inferior pharyngeal plates of the *Pseudophoxinus*. The chord length measurements of complete pharyngeals have been compared with measurements of *Pseudophoxinus handlirschi* and *Phoxinellus anatolicus* of known length in the comparative skeletal collections at the RBINS. The chord length range of the archaeological bones is 6.4mm to 13mm with a mean of 10.6mm. Assuming that the modern species have a similar bone to fish length relationship, these measurements indicate that the smallest fish represented would have been about 9cm SL and that the closest represented by the mean value would have been of about 13cm SL. The RBINS does not contain fish comparable with the largest bones but these are likely to have been fish of around 16cm SL (Wim Wouters pers. comm.) (online table S12.3). In comparison with the previous, very large dataset, the fish mainly represent the two largest size classes of 10–15cm SL and 15–20cm SL. There are few of the 5–10cm size class represented and none of the fish of less than 5cm SL. These two smaller size classes accounted for the vast majority of the remains in the previous study. This difference is almost certainly due to that dataset incorporating material from smaller sieve sizes. The limited data on *Pseudophoxinus* species in Anatolia indicates that some species may attain lengths up to 22cm TL (total length) but most are smaller (Bogutskaya 1992). The remains in this present material therefore probably represent adult fish.

Discussion

This small assemblage contains bones exclusively of Cyprinidae, with diagnostic elements mainly of the swamp minnow genus *Pseudophoxinus*. A few bones could be identified as comparable with *Leuciscus/Squalius cephalus*. Most of the less diagnostic or broken elements are comparable with the Leuciscinae cyprinid sub-family and are likely to be mainly of *Pseudophoxinus*. No other species were identified.

The current assemblage totals less than 3% of that analysed by Van Neer et al. (2013) at 567 specimens versus 18,693. The absence of the minor taxa, for example, *Gobio* and *Chondrostoma*, identified in small numbers in that collection, is therefore to be expected.

The current assemblage contained no bones of the cyprinid genus *Capoeta*. This taxon was common in the previous material, being slightly more frequent than finds of *Pseudophoxinus*, and its absence in the current assemblage needs explanation. This genus is from a different non-Leuciscine group of cyprinids and has very distinctive pharyngeal plates and basioccipitals. While it is accepted that the previous work was carried out to a higher level of distinction and a few elements may have been missed, this does not explain the lack of pharyngeals and basioccipitals, elements that dominate the present material. In the previous assemblages it was calculated that that 89.9% of the *Capoeta* bones were from fish that would have been less than 5cm SL (standard length) when alive. The bones from fish of this size will be extremely small and the most likely explanation for their absence is that they were not present in the <4mm sieve residues, whereas the previous work also included material recovered from the smaller, 2mm and 1mm mesh, sieve sizes. Indeed, unless the density of remains is much lower in the present selection of samples, which seems unlikely, the majority of the fish may have been of the smaller sizes considering that the number of selected units is similar.

No Cobitoidea (loaches) were identified either; in the previous material 9% of the identified remains were of this superfamily, therefore only a few would be expected from a much smaller collection and, again, as these are very small fish, with over half of the previous finds being bones from fish under 5cm SL, this negative evidence is not likely to be significant, as they are also unlikely to have been recovered from the <4mm residues.

Of significance are the bones identified as chub *Leuciscus/Squalius cephalus*. The only bone recovered from the previous material was that of a very small juvenile fish of less than 5cm SL. The bones from the current samples represent much larger fish, probably of adult breeding size. Hooks that have been found at the site are clearly too large to be used to catch the tiny *Pseudophoxinus* and *Capoeta*. If these artefacts were indeed used in fishing rather than being decorative items, the chub could represent a possible target. According to anglers, adult chub can be caught in streams as shallow as 50cm, but they are generally found in deeper running waters. Sometimes they may enter flooded areas out of river channels and it is possible the few larger fish represented may have been trapped as the water receded.

Temporal changes, such as the decrease in numbers of fish remains seen in the later phase (Van Neer et al. 2013), are not clearly visible in this small assemblage. The numbers of *Pseudophoxinus* are quite high in the Late phase of both the North and the South Areas, but the units from these phase groups include most of the midden deposits where these were frequent. The few bones of larger fish appear to be more evenly distributed.

Conclusion

All of the identifiable fish remains are of Cyprinidae, mainly the swamp minnow *Pseudophoxinus*, which was also frequent in the previously studied material. As this assemblage is from a similar number of excavation units and includes material from areas where much of the previous material was recovered, the lack of *Capoeta*, the other common genus in the previous study, and of the extremely small specimens in general, from the units in the present study indicates that the current assemblage is subject to collection or processing bias. Apart from the lack of the smallest remains from the lower sieve sizes, the results are broadly similar to those from the much larger study – there are very few remains of fish over 20cm SL despite the fact that the absent *Capoeta* can reach at least 40cm and at that size the bones would be readily recovered from the larger mesh sizes and even by hand. The bones matching chub *Leuciscus/Squalius cephalus* of 45cm SL are therefore rather unusual finds at Çatalhöyük. It is likely that most of the fishing activity was concentrated on catching small spawning species and juvenile fish in shallow water on the floodplain in late spring, perhaps by wading with scoop baskets.

Acknowledgements

With thanks to Wim Wouters for measurements of specimens in the RBINS and assistance with identifications.

Supplementary material

For supplementary material related to this chapter, please visit https://doi.org/10.18866/BIAA/e-13. It comprises appendix S12.1 (data) and three tables (S12.1–S12.3).

13. The avian remains

Julia Best, Beatrice Demarchi and Samantha Presslee

Introduction

This chapter reports on the avian bone remains from the Çatalhöyük Research Project's final series of excavation and analysis in 2009–2017. These specimens, which all come from the North and South Areas of the settlement, span a significant chronological range, and there are limited numbers represented in each period and subdivision. Therefore, due to the relatively small assemblage size, the bird bones are not considered in such spatial detail as the mammalian remains, but where possible their temporal and locational distribution is discussed.

The chapter also briefly reports on the preliminary analysis of the eggshell recovered from this phase of excavation and also includes a number of pre-2009 samples. At present a small sample of eggshell has been analysed using an integrated approach that combines high-powered microscopy with mass spectrometry. The pilot research has proved successful, but challenging. Additional analysis is planned to produce a larger and more representative dataset.

Bird bone

Assemblage summary

The bird bone assemblage is small compared to the mammalian one. Only 925 fragments have been recovered from the 2009–2017 tranche of excavation and recording in the North and South Areas. The recovery of small bird bone fragments suggests that this is not entirely a recovery bias and instead indicates that wild bird capture was not conducted very regularly. Although small, the avian assemblage still provides useful information on resource use, seasonality, landscape use, faunal diversity and ritual activities.

Avian bone identification often needs a larger reference collection than mammalian zooarchaeology. Therefore, over half of the 925 bird bones reported on here were exported to Cardiff University for further analysis. As a result, not all of the assemblage has been analysed to the same degree, with the material that was not available for export in some cases being recorded in less detail. As with the mammalian bone assemblage, recording practices changed in 2012. Therefore, some care should be taken in directly comparing different units. However, in large the information for birds is comparable and mainly focused on species and element abundance.

The exported birds were recorded using Cohen and Serjeantson's (1996) zoning and metric systems and identified using multiple physical reference collections. To ensure comparability with earlier analyses, the morphological and metric criteria guides used by Russell (as outlined in Russell 2019b: 6395) were also employed to aid identification. Habitat and seasonality data are based on Kirwan et al. (2008) supplemented by Snow and Perrins (1998), unless otherwise stated. Latin species names are given in table 13.2 for key species, or at first reference beyond this.

More avian material was recovered from the North Area (57% of the bird bone) than the South Area (43% of the bird bone). Middle Neolithic deposits produced the largest avian assemblage in this tranche of research, totalling 607 fragments (66%), which mainly derived from the North Area (table 13.1). Similar to the mammalian bone assemblage, previous work has provided detailed information about the Early period of the site and to some degree the later one. The material considered in this report helps to shed some light on the important Middle period. The majority of the material derived from middens and fills, with a smaller amount being associated with specific houses and floors, or with burial fills.

Taxa

Of the 925 recovered fragments in the 2009–2017 assemblage just 17% (154 fragments) were identifiable to species or a useful taxonomic group (table 13.2). A further 366 were assigned to an informative grouping or size category (such as very large bird, or cf. rallidae). The remaining 405 fragments, totalling 44% of the assemblage, were unidentifiable. However, this identification rate is lower than it should be as a result of the bones that were not exported to the UK only being identified to size category in the field. Around 30% of the exported bones were identified to species or close taxonomic grouping.

	Early	*Middle*	*Late*	*Indeterminate*	*Total*
North	0	502	23	1	526
South	40	105	237	17	399
Total	40	607	260	18	925

Table 13.1. Avian assemblage breakdown: number of fragments shown by period and area.

This is still lower than has been seen in other studies of the bird bone (Russell, McGowan 2005; Russell 2019a), and may be partially taphonomic. Many of the bird bones considered in this tranche were highly fragmentary and exhibited a mixture of modern and ancient breakages.

The assemblage is relatively diverse, with at least 24 species represented (table 13.2). This pales in significance compared to the 79 species identified in previous analyses, although admittedly that diversity comes from a larger assemblage where a far greater proportion of the remains were able to be exported for analysis (Russell 2019a). The number of species in this assemblage is in reality likely to be far higher, since a large proportion of the bone was fragmented and missing the diagnostic zones and measurement points that would have allowed more detailed identification. Recent work into duck and geese measurements has shown that metrical work is key to separating many species (e.g., Poland 2018) and as such the Anseriformes in this assemblage have been treated with caution where lacking clear morphological or metrically distinguishing features, and often assigned to broad identification categories.

The exploitation of birds is very heavily focused on waterfowl (fig. 13.1), with smaller contributions from wading shorebirds, raptors and passerines. The NISP is dominated by geese, ducks and herons, and their relative proportions are very similar to the pattern identified in other avian material from the site (Russell, McGowan 2005; Russell 2019a). The range of species identified is again similar to the material already published, but with some small variations.

Ducks, geese and swans (Anatidae). Geese are the most well-represented birds, accounting for just over 33% of the NISP (table 13.2, fig. 13.1). Greylag geese are present, as are other *Anser* species, most of which align best with greater white-fronted geese. A small number of lesser white-fronted geese may also be present. Two bones match best with a *Branta* species – likely the red-breasted goose. Greylag geese would have bred in the area, and still do so today, but it is probable that many of the geese were taken in the winter, when migratory birds from more northern areas increased the numbers present in the Konya Plain. The thickness of fragments in the eggshell assemblage potentially indicates that a greater number of large geese (such as the greylag) were breeding in the area than is represented in the bone remains, which, when all analyses are considered, are more focused on the white-fronted geese who can be found wintering in Anatolia in significant numbers (see eggshell section below and Russell 2019a). Subadult geese were recorded, and around 50 further bone fragments were identified as juvenile large birds that bore most similarity to geese and herons.

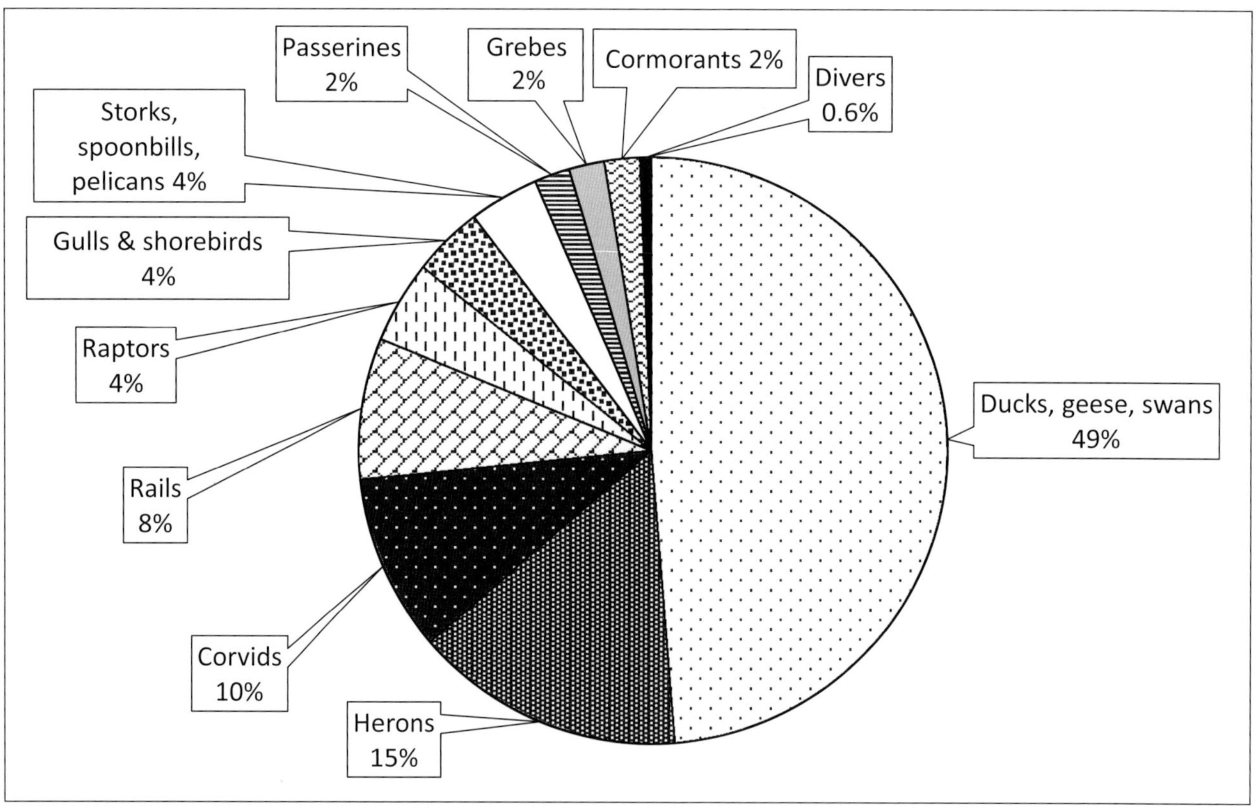

Figure 13.1. NISP by taxonomic groupings.

Order	*Family*	*Taxon*	*Common name*	*NISP*	*% NISP*
Anseriformes	Anatidae		Goose	5	3.2
Anseriformes	Anatidae	*Anser anser*	Goose cf. greylag	18	11.7
Anseriformes	Anatidae	*Anser albifrons*	Goose cf. greater white-fronted	21	13.6
Anseriformes	Anatidae	*Anser erythropus*	Medium goose cf. lesser white-fronted	4	2.6
Anseriformes	Anatidae	*Branta ruficollis*	Small goose cf. red-breasted	2	1.3
Anseriformes	Anatidae		Goose/duck	1	0.6
Anseriformes	Anatidae		Goose/swan	1	0.6
Anseriformes	Anatidae		Large duck	4	2.6
Anseriformes	Anatidae		Medium duck	7	4.5
Anseriformes	Anatidae		Small duck	1	0.6
Anseriformes	Anatidae	*Aythya sp.*	Diving duck	4	2.6
Anseriformes	Anatidae	*Anas platyrhynchos*	Mallard	6	3.9
Anseriformes	Anatidae	*Anas crecca/querquedula*	Teal/garganey	1	0.6
Podicipediformes	Podicipedidae		Grebe	1	0.6
Podicipediformes	Podicipedidae	*Tachybaptus ruficollis*	Little grebe	2	1.3
Ciconiiformes	Ciconiidae	*Ciconia cf. ciconia*	Stork cf. white stork	3	1.9
Pelecaniformes	Threskiornithidae	*Platalea leucorodia*	Eurasian spoonbill	2	1.3
Pelecaniformes	Ardeidae		Heron	1	0.6
Pelecaniformes	Ardeidae		Large heron	5	3.2
Pelecaniformes	Ardeidae		Small heron	2	1.3
Pelecaniformes	Ardeidae	*Ardea cinerea*	Grey heron	12	7.8
Pelecaniformes	Ardeidae	*Egretta garzetta*	Little egret	3	1.9
Pelecaniformes	Pelecanidae	*Pelecanus onocrotalus*	Great white pelican	1	0.6
Suliformes	Phalacrocoracidae	*Phalacrocorax carbo*	Great cormorant	3	1.9
Accipitriformes	Accipitridae		Accipitridae cf. hawk	1	0.6
Accipitriformes	Accipitridae	*Gyps fulvus*	Vulture cf. griffon	1	0.6
Accipitriformes	Accipitridae	*Circus sp.*	Harrier	1	0.6
cf. Accipitriformes	cf. Accipitridae		Raptor cf. harrier	1	0.6
Gruiformes	Rallidae		Rail	2	1.3
Gruiformes	Rallidae	*Fulica atra*	Coot	10	6.5
Charadriiformes	cf. Scolopacidae		Medium wader	2	1.3
Charadriiformes	cf. Scolopacidae		Small wader	3	1.9
Charadriiformes	Laridae	*Hydrocoloeus minutus*	Little gull	1	0.6
Strigiformes	Strigidae	*Asio flammeus*	Short-eared owl	3	1.9
Passeriformes	Corvidae	*Corvus cornix*	Hooded crow	8	5.2
Passeriformes	Corvidae	*Corvus cornix/frugilegus*	Crow/rook	5	3.2
Passeriformes	Corvidae		Large passerine cf. corvid	2	1.3
Passeriformes			Small passerine	3	1.9
Gaviiformes	Gaviidae		Gavia Sp.	1	0.6
			Very large bird	70	
			Large bird	84	
			Large + bird	63	
			Medium bird	121	
			Small bird	28	
			Bird	405	
			Total	925	

Table 13.2. Bird bone assemblage by NISP (number of identified specimens), %NISP and counts of unidentified material.

Twelve of these fragments are likely to represent a large *Anser* goose – supporting the interpretation that the greylag geese were breeding in the area. Greylags, their young and their eggs may thus have been exploited in the summer, whilst winter fowling included a greater focus on the arriving white-fronted geese. The greylags of course could have been taken year-round, and we cannot assume that Çatalhöyük's fowlers differentiated between what we identify as separate species. Instead they may have categorised birds by their habitats, seasonality, morphology and behaviour. These birds, and many other species, could have been taken by net, noose, snare or projectile.

Ducks form 15% of the NISP and are also represented by multiple species including a range of dabbling and diving ducks. Mallard is most commonly identified. Teal or garganey is represented by a single bone. Morphology and the limited suite of metrics indicate that both tufted ducks (*Aythya fuligula*) and common pochards (*Aythya farina*) are present in the diving duck category. Most of these birds would have been present in some form all year round and could have been taken at any point, however, some such as the common pochards may have been more common in the area during winter, when flocking and migrating. Whilst many of the ducks were not currently identifiable to species, the large common shelduck (*Tadorna tadorna*) was not present, despite being identified in small numbers in previous assemblages (Russell 2019a). Although ducks may have been more numerous in the winter months, the eggshell data (see below; Sidell, Scudder 2005; Best et al. 2015) indicates that some of these birds were also being targeted in the summer breeding season, with likely candidates including the mallard.

Herons (Ardeidae). Herons are the next most common avian grouping, accounting for 15% of the NISP. Grey heron, a resident species, is most numerous, but the purple heron (*Ardea purpurea*) also appears to be present in the 'large heron' material (although it could not be identified with confidence). Little egret was represented by only three bones and likely was present year-round and may have bred in the area. Immature and subadult large heron bones were present. An immature large heron coracoid and a probable articulating scapula were recovered from unit 11376 (a Late period midden layer above a room fill). This unit also contained part of an articulated adult heron (see below). Russell (2019a) also recorded a number of young herons. Herons and other waterfowl may nest together, and as such it is probable that many of the young waterfowl birds could have been taken from the same areas. The herons mainly come from midden deposits, but two are from floor-use surfaces. Herons appear to be one of the species used for food but also for feathers. Interestingly, grey herons are today one of the earliest species to lay eggs in Turkey, sometime commencing their first lay as early as February (Kirwan et al. 2008).

Pelicans (Pelicanidae). A single bone from a great white pelican was identified – a proximal ulna fragment. The rarity of these birds in the assemblages indicates that even though they breed in the area they were not a regularly targeted species and that they may not have been present in huge numbers (Russell 2019a).

Spoonbills (Threskiornithidae). Eurasian spoonbill was represented by two coracoids and possibly by a small fragment of the beak. They may have been captured when breeding or whilst migrating through Central Anatolia, and the remains from previous analyses have indicated that they were used for food but also held wider significance (Russell 2019a). The unusually shaped beak may have been used as an item – either practical or ritual.

Storks (Ciconiidae). Three stork bones (cf. white stork) were identified: two tarsometarsus fragments (not from the same bone) and a carpometacarpus. These bones all come from one unit (19342), a dump associated with oven material and bone discard, and may belong to the same individual. It is possible that these birds were eaten and that these limb extremities were discarded (there will be no meat on these elements); however, it is also possible that the carpometacarpus is discard from feathery portions brought to site. One of the fragments was sub-adult, indicating that, like today, species such as the white stork were also breeding in the region. Considering the number of waterfowl exploited overall, and that today storks migrate through the area in large numbers, it is surprising that they are not better represented; Russell (2019a) has suggested that there may be taboos against their capture.

Cormorants (Phalacrocoracidae). Cormorant is represented by just three bones from one individual (a radius, ulna and carpometacarpus) that were recovered from a midden layer above a room fill. Feathers may have been the primary reason for capturing these birds on occasion. Cormorants, although present year-round, were probably taken whilst breeding in summer, when they occur in larger numbers.

Rails (Rallidae). Rails are common in the assemblage, accounting for 8% of the NISP. Of these most are coot, but two bones come from a smaller species, potentially the corn crake (*Crex crex*). Five of the coot bones were recovered from one unit (20472) and very probably represent one individual. Another unit (11376) produced remains from at least two coots. The coot is resident to the area and may have been captured year-round alongside other waterfowl, which may explain its frequency. An additional nine very fragmented specimens were recorded as cf. rail, again showing that waterbirds were an important part of the avian assemblage.

Cranes (Gruidae). Cranes are conspicuous by their rarity, or possibly absence, in this assemblage. Two probable crane specimens have been recognised but were unable to be confirmed (see special finds below). This is unusual, as they formed around 3% of the NISP in the material analysed by Russell (2019a). Their rarity here may indicate that their usage was focused in specific areas of the site and again suggests that although some were eaten, their capture was potentially more symbolic and had special significance.

Bustards (Otididae). One coracoid fragment assigned only to a size category was possibly a small fragment from a great bustard. However, beyond this one possible specimen none were identified. The absence of this species is one of the only large variations from the previously analysed material. It may be that these birds were not present in large numbers and only occasionally hunted.

Raptors (including Accipitridae and Strigidae). Raptorial birds account for 4% of the assemblage. A diverse array of species appears to be present for the number of bones recorded, including a harrier, a vulture, a possible hawk and owls (table 13.2). One of the harrier bones fits well with the large marsh harrier (*Circus aeruginosus*), but the other, an isolated talon, was not identified further. A single vulture bone was identified and probably represents a griffon vulture, although the fragment was too small to assign with confidence. This is unusual in comparison to previous analyses, where vultures formed over 2% of the NISP (Russell 2019a). Russell has suggested that the use of vultures at the site was highly ritualised and had many symbolic associations (Russell 2019b). It is therefore possible that their absence in this assemblage, and the rarity of some other species such crane, may result from fewer 'special deposits' being analysed in this tranche. No eagles were identified in this assemblage, although they were present in small numbers in the other material (Russell 2019a).

The three owl bones have been identified as short-eared owl based on metrics (Emslie 1982: 157). These birds may have been found around the tell site or areas of marshland nearby.

These raptorial birds are unlikely to have been a food source and may have been killed for their feathers, to protect livestock, or for other special uses (Lockie, Stephen 1959: 43–50).

Galliformes. No galliformes species, such as partridge or quail, were identified in this assemblage. These remains are usually distinctive even when fragmented and so their absence is unlikely to result from taphonomic/identification biases. Previous analyses (Russell 2019b) have also noted how rare these birds are at Çatalhöyük. Since galliformes should have been relatively common in the steppe areas of the landscape, it is probable that they were, for some reason, actively avoided. The relative rarity of steppe species in general suggests that birds in this environmental zone were not favoured to the extent of waterbirds (fig. 13.2).

Grebes (Podicipedidae). Grebes are not as well represented as in the previous analyses, only forming 2% of the NISP. The little grebe would have been taken during the summer, but other species would have been available in winter.

Passerines (Passeriformes including Corvidae). Small passerines are only represented by three bones, but corvids make up over 9% of the NISP. Hooded crows and rooks are the most common passerines recorded, eight bones of which come from an articulated crow and five of which were not identifiable to exact species due to fragmentation (Tomek, Bocheński 2000). Both of these species are present in the area year-round. These birds are sociable scavengers who may have come to the site of their own volition and been taken locally.

Gulls and waders (Charadriiformes including Laridae and Scolopacidae). One little gull was identified, and like today, they were probably rare visitors overwintering in the vicinity of the site. At least two wader species (also known as shorebirds) are present, indicating that some shorebirds were being taken from water's edges, presumably as food.

Habitats

Nearly all of the birds could have been sourced from wetland environments (fig. 13.2). These may have included a range of habitats such as: open water, the shoreline, marshland and wet grassland. Some birds such as the harrier may have been taken from steppe areas, and others such as the probable griffon vulture could have been found in more mountainous areas or as a vagrant in others. Many other raptor species could have inhabited varied habitats around the site. It is likely that the passerines, including the corvids, were found in and around the settlement, or in ecotonal environments. Scavenging birds, such as the corvids, may have been captured whilst feeding on midden deposits. The small passerine assemblage probably includes some birds who died of natural causes and were incorporated non-anthropogenically into the assemblage.

This profile suggests that the site inhabitants were familiar with a range of habitats and chose to focus largely on waterbirds in wetland environments that could have been found near to the site. The avian assemblage analyses in this tranche show no distinct seasonality, with evidence for the exploitation of resident species, winter visitors, and birds of the summer breeding season. When the presence of juvenile birds (such as herons and geese) is also combined with the sizable eggshell assemblage (below), it indicates that the site inhabitants were using birds year-round in both planned and opportunistic fowling activities. Birds would have moved around and through this landscape at different points of the year. For example, summer-breeding greylag geese would have been joined by larger numbers of wintering greylags moving onto the Konya Plain, whilst a range of other birds (including raptor, shorebird and passerine species) may only have been seen whilst passing through the area on migration (Kirwan et al. 2008; Russell 2019a). The movement of birds around the landscape at different points of the day and in different seasons may have held symbolic significance in marking time and space, or even transcending boundaries in worldviews, such as in the movement between life and death, or between human and animal.

Body part representation

The birds at Çatalhöyük are mainly represented by wing elements and bones from the shoulder girdle (table 13.3 and fig. 13.3). When all birds are considered together, the most commonly occurring individual element is the corocoid, closely followed by

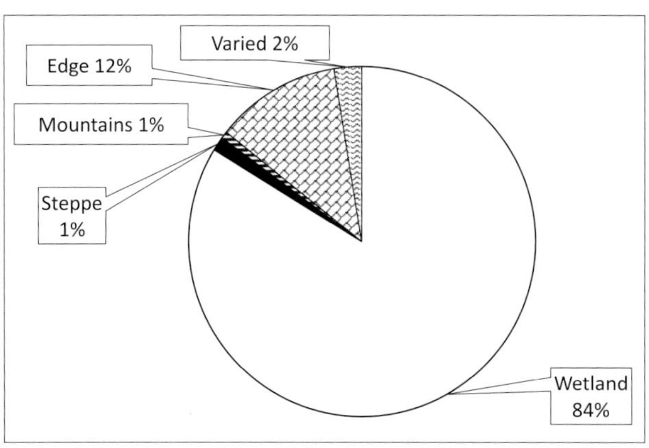

Figure 13.2. NISP by broad habitat area.

the humerus, ulna and tibiotarsus. However, the majority of the wing elements come from geese and herons, which both have a large number of wing bones and body elements (particularly those of the shoulder), with an underrepresentation of legs. It is likely that this pattern reflects a dual usage, with these birds being both used as food and processed for feathers. Almost all of the head and foot elements in this tranche were only assigned to size category, due to fragile elements such as the skull being highly fragmented and bones such as foot phalanges being tricky to assign to exact species for many birds.

Coots and other rails are largely represented by elements of the body, with a fairly balanced ratio of wing and leg bones suggesting that they were brought to site whole and then consumed. Extremities of the wing are represented in the birds assigned to size categories such as medium bird cf. rallidae, supporting this interpretation.

Ducks also seem, from their elemental distribution, to have been consumed, with leg elements better represented than for other Anseriformes such as geese. Wing elements are underrepresented in comparison to both geese and herons (fig. 13.3). There is a lack of limb extremities for ducks, with no tarsometatarsii, only two carpometacarpii and no wing digits. This suggests that the meaty elements of the bird are represented here and that waste elements may have been disposed of elsewhere pre-cooking and consumption. These birds may still have been used for feathers, but it does not appear to be their primary purpose in this tranche. Interestingly, in the material analysed by Russell (2019a) the meaty elements are underrepresented. It is possible that these birds were being used differently through time, or that the feathery parts received differential processing and were underrepresented in the material considered here.

Raptors are well represented by leg elements and in particular the tibiotarsus, although the assemblage size is small. The pattern may result from specific elements entering the site for bone working or ornamentation. Alternatively, feathery elements such as the ulna may have been preferentially removed and thus be missing from this assemblage. It is plausible that other elements from the skeletons of the individual birds recorded here may have been distributed across the site in special deposits already recorded and reported on elsewhere (e.g., Russell, McGowan 2005; Russell 2019a; 2019b).

Corvids have a very equal representation of body, leg and wing elements, indicating that they were whole on site, and may have been caught very locally and minimally processed. The absence of ulnae may result from these elements being removed for feather processing.

Element	All birds	Geese	Ducks	Herons	Rails	Corvids	Raptors
Head/neck	14	0	0	0	0	0	0
Wing	138	26	5	10	2	5	0
Body	118	21	10	10	8	5	2
Leg	81	3	8	3	2	5	5
Foot	41	0	0	0	0	0	1

Table 13.3. Element frequency (by NISP) for all birds and by main taxonomic groupings.

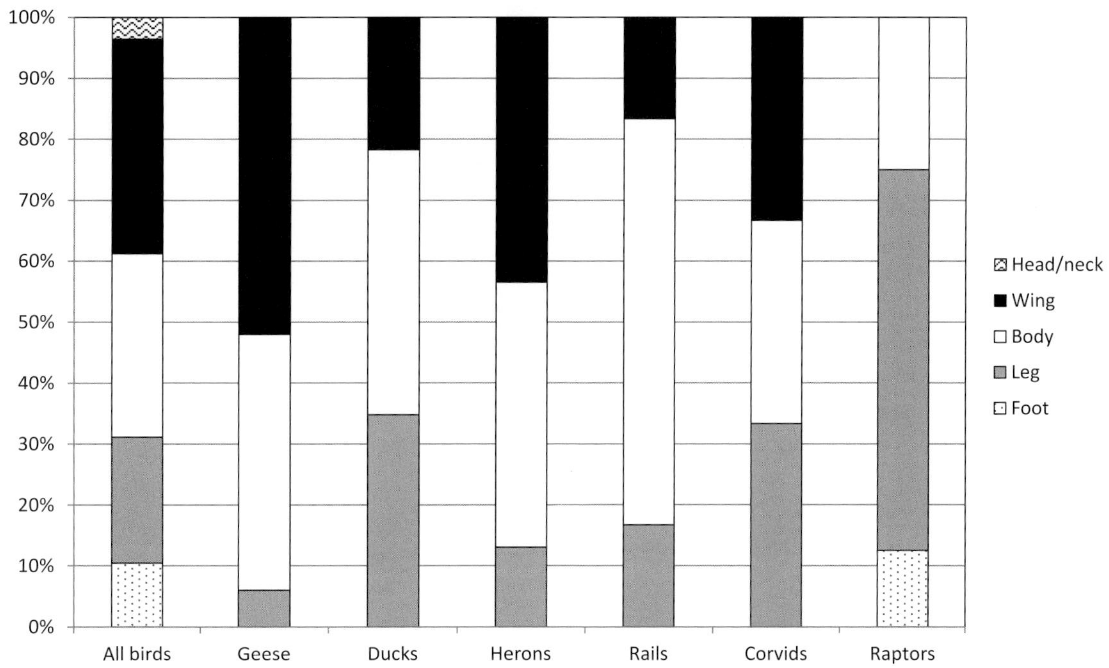

Figure 13.3. Element frequency for all birds, and by main taxonomic groupings, as % NISP.

Taphonomy

Butchery marks are very scarce, a pattern which was also seen in the mammalian data. Only two knife cuts were recorded, and one possible cut mark was noted. A large duck scapula had a small scooping scrape characteristic of light contact during flesh removal. A large goose furcula exhibited two fine cut marks indicative of disarticulation – this specimen articulates with a coracoid and may represent food waste. Both examples are consistent with very sharp flint or obsidian blades. Another large goose carpometacarpus had a possible knife cut and belonged to part of an articulated lower wing which may have been systematically removed. It is likely that this low frequency of butchery marks is partially caused by skilled butchers processing animals without leaving traces, but it also may indicate that birds were not processed heavily for cooking. Birds can, however, be butchered without knives and one large goose coracoid exhibited peeling marks associated with overextension of the articulation or twisting.

Carnivore gnawing occurs on only five bones; one duck, three geese and one unidentified large bird. Two other goose bones exhibited possible gnawing marks, including one coracoid that may have a human tooth mark. A final goose specimen had been digested. There were no instances of rodent gnawing. The results indicate that some bird remains were accessible in the middens to scavengers, primarily dogs, but also some smaller carnivores.

Burning occurs on 32 bones. The majority of these were charred or heat stained, but one was calcined. Many of the burnt bones are unidentified beyond size categories, but a range of birds are represented, including ducks, geese, small waders, rails and a crow/rook. This burning seems to represent a combination of cooking and post-depositional burning. One unidentified bird bone was recovered from the highly burnt burial fill of unit (21113) and may have been an intentional inclusion.

A further 12 bones exhibited staining that may be a result of indirect heat modification from the burial environment. Only one bone exhibited erosion and there were no weathered specimens, indicating that in general the bone was not exposed for long. Two specimens had trampling striations.

Special finds and associated bone groups

Previous work on the bird bones from Çatalhöyük has identified that whilst avian meat resources were consumed, bird use often goes beyond subsistence and includes a comparatively large number of special or structured deposits that show birds also played a ritual or symbolic role.

A large number of associated bone groups (ABGs) and possible articulations were recorded in this tranche – at least 11. In some instances, clusters of these ABGs are represented in the same unit. Unit (20472) contained what appears to be an articulated hooded crow, but only the right side of the body. One bone, the tarsometatarsus, was charred and slightly calcined, indicating that this was not a natural mortality and that this bird had been intentionally processed. The same unit also produced a rallidae ABG, namely a coot. This bird had bones from both sides of the body present and seems to be another partially articulated skeletal deposition, potentially from food waste, as again one bone is charred – in this case the sternum. Part of an associated goose shoulder girdle was also retrieved from the same unit. This large goose did not exhibit burning or butchery but had been gnawed by a carnivore.

Unit (11376), a Late period midden layer above a room fill, produced what appears to be a partially articulated heron ABG. The key elements consist of two humerii (a left and a right) broken in antiquity in the same place on the shaft (fig. 13.4). Two heron coracoids (again a left and a right, which metrics show are probably the same individual) and a scapula were also present. The symmetrical breakage could suggest that the lower wings of this bird were intentionally removed; potentially as waste elements, but more likely for their feathers or use as an item. Russell (2019a; 2019b) has noted the importance of feathers at Çatalhöyük. It is possible that the left and right bone groups come from different individuals, in which case paired articulations may have been specifically chosen for deposition. Probably associated with this grouping are eight vertebrae and a synsacrum.

Other unusual groupings have been noted, such as the seven juvenile tibiotarsii from unit (21660). These were not identifiable to species due to their porosity, but the proportions match large waterfowl such as geese. It is likely that these birds were taken from the nest in a targeted trip that selected multiple individuals. The lack of other skeletal elements in this unit may suggest these leg elements were consumed together.

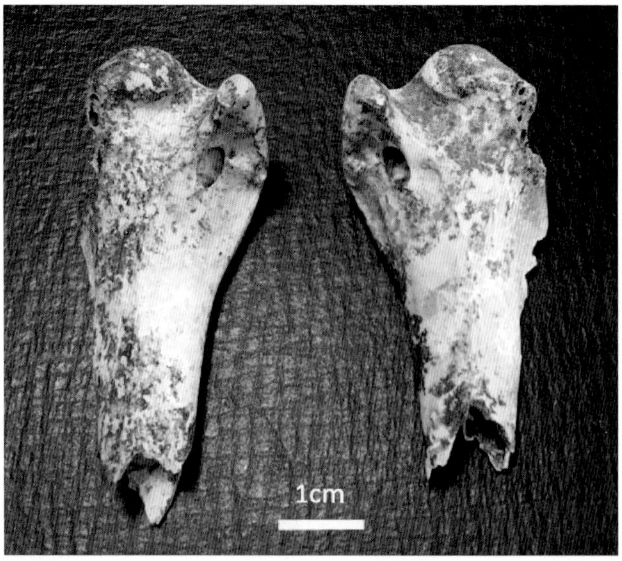

Figure 13.4. Pair of grey heron humerii with matching breaks.

Unfortunately, one specimen of potential ritual significance from an abandonment deposit in Middle period unit (20863) was not permitted for export and as such no further analysis or species confirmation has taken place. An ulna of a very large bird, thought to be a crane or a stork, was recovered from this unit alongside other animal remains. Considering the proposed importance of such species at Çatalhöyük (e.g., Russell, McGowan 2002), the selection of this item for deposition may have had cultural and ritual significance. It is also worth noting that the well-known crane wing ABG was associated with a construction deposit of B.1 (e.g., Russell, McGowan 2002; Russell 2019a). It may be that certain birds played a significant role in starting and ending events.

A small ulna fragment from a very large bird, probably crane, has been identified as a bead-making remnant due to polish on the shaft – unit (19383), Level South M, a mixture of dump and midden deposit. A very large bird long-bone fragment with polish on the shaft was also recovered from unit (13183) and is probably a remnant of bead or point making. Another medium bird long bone (unit (15909), floor surface, Level North G) was identified as a tubular bead using a cut-and-break technique (this specimen was not exported and has not been identified further).

Eggshell

Assemblage summary

The Çatalhöyük eggshell assemblage is unparalleled in size and preservation, in both its geographic and its temporal situation, but also represents the largest known eggshell assemblage currently recovered archaeologically (Sidell, Scudder 2005). The assemblage has the capacity to inform on human-avian interactions in prehistory, and it is a valuable dataset to be examined alongside the bone record.

Over 940 units have produced eggshell, providing thousands of fragments (totalling at least 5 kilograms) and resulting in a substantial assemblage for analysis (Mulville et al. 2014). Eggshell has been recovered mainly through flotation, although there are a number of groups of hand-collected material. There are some large collections of eggshell, with several individual contexts revealing numerous fragments, including some of substantial size. Preservation is generally very good. Prior to this study only a sample of material from a range of feature types and areas relating to two occupation levels from the 1995–1999 excavation seasons had been analysed using SEM (Scanning Electron Microscopy). Sidell and Scudder's (2005) results revealed a predominance of duck and goose eggs (Anseriformes), a small number of which appeared to have hatched. Other material remained unidentifiable but was thought to be seabird (possibly gulls, spoonbills or storks), although the remains of these birds are rare on the site.

Eggshell analysis has until recently been limited to microscopic SEM analysis, which although providing valuable information, when used alone is restricted by the subjective nature of identification, limited by any damage on the eggshell, used to require expensive, destructive coating of the samples, and is time consuming. Now microscopy (SEM and high-powered digital microscopy) can be used in conjunction with the analysis of eggshell proteins by mass spectrometry (MALDI-TOF-MS and LC-MS/MS), which allows identification of family or taxon-specific protein markers for avian species, based on comparison with known reference datasets (Presslee et al. 2017; Demarchi et al. 2020). This palaeoproteomic approach is beneficial since it increases the number of samples that can be analysed and identified even where morphologically diagnostic features of the eggshell have been damaged by preservation or by hatching.

As such, 90 new samples have been analysed using palaeoproteomics and microscopy as a pilot study to test the preservation of the proteins and to identify deposits that would benefit from further analysis (building on Best et al. 2015. For palaeoproteomic method details see Demarchi et al. 2020). The analysis involved identifying the fragments taxonomically by both methods, and recording changes to the internal surface of the shell (on the mammillae). These changes can be caused either by chick development (resorption caused by the chick mobilising calcium from the eggshell), or by taphonomic processes (such as acidic corrosion). These markers can be very difficult to separate. However, where there is no evidence for damage it indicates that the egg was freshly laid or infertile. Meanwhile, changes in the mammillae surface may reflect the stage of chick development/hatching (Beacham, Durand 2007) or inform on taphonomic processes in the burial environment (Sichert et al. 2019). This analysis (online table S13.1) uses 'no resorption' to indicate a lack of developmental activity, and 'corrosion' where developmental and taphonomic signals cannot be conclusively separated. Corrosion is described by its regularity/irregularity and intensity of mammillae destruction (minimal, moderate, high, very high). Where corrosion is very regular, it is more likely to result from chick development, whereas patchy, irregular corrosion is often taphonomic in nature (Morel 1990; Sichert et al. 2019). Eggs were selected from midden, burial and house deposits from the North and South Areas. Where possible Early, Middle and Late period samples were chosen, but the small sample size restricted more detailed distributional analysis.

Species

The combined mass spectrometry and microscopy results reveal that Anseriformes or very probable Anseriformes dominate the avian eggshell assemblage (online table S13.1; fig. 13.5), accounting for over 68% of the samples.

Mass spectrometry-based identification of Anseriformes at a lower taxonomic level is hampered by the fact that the main goose eggshell protein (ansocalcin) is also present in duck eggshell. Unfortunately, the exact sequence of ansocalcin is currently unknown and therefore at present we cannot use potential amino acid mutations to distinguish between duck and goose. The eggshell thickness, however, indicates that geese are well represented, and microscopic features of the eggshell suggest that multiple species of both ducks and geese are likely to be present. It is possible that swan may also be represented (online table S13.1). This combined approach has enabled identification of more fragments than either technique could do in isolation. Mass spectrometry allowed fragments with damaged mammillae (one of the key diagnostic features in microscopy) to be identified as Anseriformes; these could then be studied further via other characteristics such as pore count and thickness to reveal diversity within the Anseriformes assemblage (fig. 13.6).

There are a range of potential geese species that could be represented. The bird bones analysed in this report and previous analyses demonstrate that greylag geese are present in the assemblage in regular numbers.

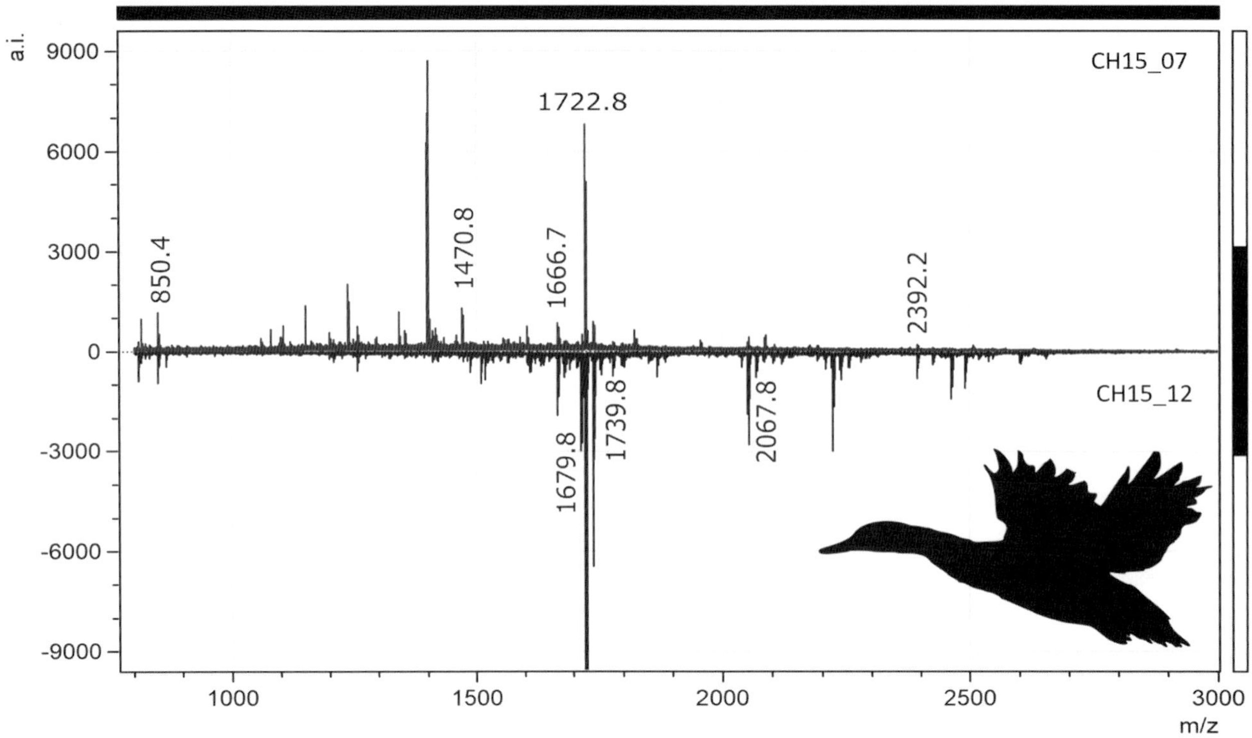

Figure 13.5. MALDI-TOF-MS spectra for Anseriformes samples CH15_7 and CH15_12.

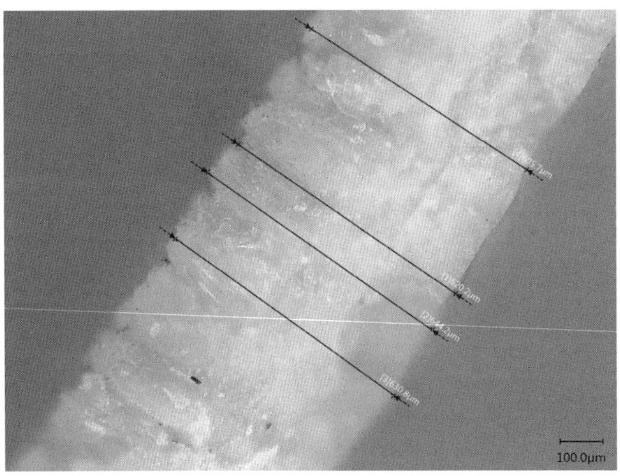

Figure 13.6. Large Anseriformes eggshell fragment, likely Anser anser *(CH19_11).*

Whilst the avian bone assemblages, and in particular those analysed by Russell (2019a), indicate that greylag are not the most common goose species by bones, it is important to note that they are the only species of the geese identified that is likely to have bred locally in the past and which still breeds in the area today (Snow, Perrins 1998: 175–81). The thickness of the Anseriformes eggshell analysed is at the top end of the ranges associated with *Anser*, and as such greylag geese are a likely candidate for much of the eggshell material. If egg taking was a key part of avian exploitation, it stands to reason that the adult birds would not have been killed in large numbers, as this would then deprive the fowlers of the valuable egg resources. However, the range of eggshell thicknesses and variation in the mammillae suggest that multiple goose species are represented. Potentially this may include other species commonly identified in this and previous bone assemblages (Russell 2019b; Russell, McGowan 2005) such as the greater white-fronted goose and the lesser white-fronted goose. Interestingly, although neither of these breed in the area today, the lesser white-fronted goose does breed in Eastern Turkey, Syria and Greece, which may suggest that breeding distributions extended to the Konya Plain in the past. At present other related birds such as black geese (*Branta*) cannot be eliminated using mass spectrometry or microscopy.

Other species have been tentatively identified (online table S13.1) including several large Gruiformes (cf. crane) fragments from units including an Early period midden layer (30625), as well as potentially other birds, such as herons and bustards (Demarchi et al. 2020). Crane, heron and bustard bones have all been identified in the Early, Middle and Late period bone assemblages (e.g., Russell, McGowan 2005; Mulville et al. 2014; Russell 2019a). The eggshell analyses may suggest that these birds were nesting in the vicinity. The presence of young geese and herons as represented in the bone data (see above) also supports this interpretation. The potential presence of only small quantities of heron eggshell in this assemblage indicates that eggs from these birds were not as frequently exploited as those of Anseriformes. However, several juvenile herons are represented in the bone assemblage. This suggests that the inhabitants of Çatalhöyük may have avoided taking large numbers of heron eggs if they instead wished to take young birds.

Developmental stage

Analysis of internal structures of the eggshell revealed that the majority of the samples analysed were derived from unhatched eggs (online table S13.1). Just over half of the Anseriformes and probable Anseriformes (31 of 61 samples) exhibited no resorption, indicating that many of the eggs were taken when recently laid. A further six specimens had very patchy corrosion which is likely to be taphonomic damage. The remaining 24 fragments had mostly uniform corrosion, which could represent either chick development, taphonomic damage, or a combination of the two (fig. 13.7). The regularity of the corrosion suggests it is caused by the growth of a chick within the egg, demonstrating that these eggs were fertilised and developing; however, taphonomic processes can also sometimes cause similar profiles (Sichert et al. 2019). Of the corroded fragments, 11 had high levels of corrosion (including duck samples CH19_15 and CH19_25, and a goose CH19_17). These probably come from eggs that are very late in the developmental sequence, indicating that a small number of live birds may have been hatching in the vicinity of the site. This is consistent with the presence of marshes near the site. The non-Anseriformes samples exhibited a similar pattern, with most exhibiting no resorption, but some fragments revealing corrosion (online table S13.1).

Eggshell discussion

This study suggests that egg exploitation largely focused on Anseriformes, with a range of duck and geese species present. This correlates well with the bone data and suggests that eggs of these waterfowl were the key avian resources gathered. Considering the relatively small size of the avian bone assemblage it is possible that the (presumably wild) Anseriformes were being systematically exploited to maximise egg production, rather than large numbers of birds being killed for meat. Whilst the wetland environments around the site would have provided suitable nesting environments for waterfowl including Anseriformes, Gruiformes and some Pelecaniformes in summer, it is

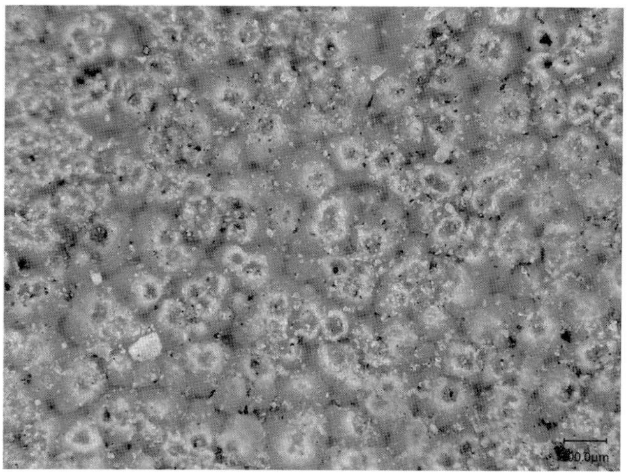

Figure 13.7. Anseriformes eggshell with some uniform corrosion indicating that this egg may have been mid development.

possible that some egg resources may have been brought from further afield or traded, either as food or special items, given the excellent portability of eggs and their durability as food staples.

Smaller but significant egg contributions were provided by other bird taxa, including Gruiformes, potential Pelecaniformes and some taxa which at present cannot be identified. Although additional work, better genomic references and a larger number of samples are needed to refine these identifications, their presence is not surprising considering their representation in the overall bone assemblage. The potentially special and ritual role held by crane at Çatalhöyük should also be recognised. It is possible that like the birds themselves, their eggs may have also been used in both mundane and special ways, as two eggshell fragments of a large Gruiformes cf. crane were recovered from burial fills. Symbolic roles may have also existed for other more common taxa. Whilst geese and duck eggshell fragments were most numerous in midden deposits, they were also represented in burial fills, although unintentional incorporation cannot be excluded for these or other birds (online table S13.1).

Developmental evidence suggests that the eggs, being largely unhatched, would have provided a valuable food resource, but may also have been used as a material in pigment binding, and in some instances have held ritual significance.

The quantity of avian eggshell present at Çatalhöyük could potentially indicate early management of waterfowl in the surrounding landscape where egg collecting was prioritised over meat exploitation; this would be of international significance. This work has demonstrated identifiable proteins are well preserved in the Çatalhöyük eggshell remains, and with ongoing developments in the genomic and proteomic reference databases, more detailed information may be attained in future. Due to the size of the assemblage a larger body of material should be analysed in order to obtain a representative and accurate picture of the taxonomic makeup and to fully explore the roles that different birds may have played at the site.

Summary of the avian bone and eggshell

It is clear that bird meat did not form a large proportion of the food consumed at Çatalhöyük, but that these animals were used year-round in small numbers. Some species such as ducks and coots were taken primarily for food, but other birds, particularly geese and herons, appear to have been targeted for their feathers and for their meat. It is probable that a number of the birds exploited held symbolic roles and would have been part and parcel of inhabiting the site, both as a living element of the landscape (such as on migration) and as resources to be used.

Although bird meat would not have been a regular dietary addition, eggs may have been used on a larger scale. A similar range of species were exploited, and Anseriformes such as geese and ducks in particular provided egg resources that may have been used for food, as ritual items and in other forms such as pigment stabilisers.

When combined with previous analyses, very similar patterns can be seen, which enable this avian assemblage to be considered as a whole, and demonstrate that whilst not numerous by bone count, birds were integral to life and death in Çatalhöyük.

Supplementary material

For supplementary material related to this chapter, please visit https://doi.org/10.18866/BIAA/e-13. It comprises one table (S13.1).

14. Micromorphology: exploring micro-contextual traces of settled life at Çatalhöyük

Aroa García-Suárez, Wendy Matthews and Marta Portillo

Introduction

This chapter reports the results of the micro-contextual investigation of archaeological deposits in buildings and open areas excavated between 2009 and 2017 at Çatalhöyük. The main goal of this research is the contextual analysis of forensic-scale traces of occupation. This approach brings us closer to the micro-historical scales at which consciousness of place and social relations are shaped in the everyday, in specific events and in life cycles (Matthews 2005b). To examine these traces and timescales, we have conducted high-resolution analyses of intact microstratigraphic sequences of floors and accumulated residues in large resin-impregnated thin-sections and spot samples by applying integrated micromorphological, opal phytolith and dung spherulite analyses. Through this methodology we investigate the origin and use of interior and exterior areas and human-plant-animal interactions to study key issues in the creation of settled life and communities. In addition, we aim to explore ways in which the case study of Çatalhöyük can provide new deep-time perspectives on current global challenges (Isendahl, Stump 2019), including sustainable energy and construction, communities, health and the built environment, and the nature of inequalities and socio-economic independence.

Firstly, this investigation aims to assess the functional independence of houses through the micro-contextual examination of floors and surfaces in and around buildings. Plaster floors and accumulated occupation residues are viewed here not only as records of architectural choices, activities and uses of space, but also as an important category of material culture that embodies social practices, technological knowledge and ecological relations (Boivin 2000; Karkanas, Efstratiou 2009; Anderson et al. 2014; Matthews 2018). Particular attention has been paid to small built environments, those smaller than ca 10m^2, usually interpreted as annexes to larger buildings nearby and considered too small to have functioned as autonomous units, whose microstratigraphic investigation has been widely overlooked in previous studies (Düring 2001; García-Suárez 2017; 2018).

Secondly, this study considers the issues of health and the built environment through the high-resolution analysis of occupation surfaces and accumulated residues in buildings and open areas within the settlement in order to explore concepts of space, discard practices, sanitation and health and the built environment (López 2012; Larsen et al. 2019). The composition, depositional pathways and possible recurrent uses of dung and coprolites, specifically, are deemed here to be of key significance for the investigation of potential disease reservoirs and vectors and how the design, construction and maintenance of the built environment might have influenced these.

Thirdly, human-animal interactions are explored through the microscopic identification of dung matter, particularly in penning areas located within the settlement and indicative of close human-animal relations representative of, or leading to, domestication and herding practices.

Fourthly, this research seeks to review community sustainability as a critical concept to the comprehensive interpretation of ecological strategies, uses of space and social dynamics within the large settlement at Çatalhöyük. To this end, this research concentrates on examining accumulated occupation residues and the material remains of houses, middens, and suspected remains from roofs/upper stories, placing the emphasis on their composition and multiple temporalities to throw light on settlement dynamics involving economic practices, household organisation and units of consumption.

Integrated micro-contextual analyses from building and open area stratigraphies in the North and South Areas of Çatalhöyük are included in this chapter. Micromorphological results on occupation sequences excavated in the TPC and GDN Areas and corresponding to the Final period are reported in García-Suárez (Marciniak (ed.) forthcoming) and in Barański et al. (Volume 15, Chapter 5) respectively.

Materials and methods

This study integrates high-resolution micro-contextual approaches, including micromorphology and microfossil plant and dung analyses, in order to identify the nature of archaeological deposits and materials, their precise depositional histories and contextual associations, and their ecological and sociocultural significance more widely (Weiner 2010; Anderson et al. 2014; Matthews et al. 2014). The sampling strategy of this study has been critically developed attending to the specific research questions addressed, the composition of the depositional layers and the type of analysis. Attention has been paid to variations in the occurrence and preservation status of specific archaeological materials in different contexts and areas, which has significantly increased the integration of results from the different analyses.

Sampled locations

Integrated micro-contextual sampling during this final stage of the Çatalhöyük Research Project was carried out during the 2013, 2014, 2015 and 2017 excavation seasons. Sediment samples (micromorphological blocks and spot samples for microfossil and chemical analyses) were collected from a number of built environments and open areas spanning the Early, Middle and Late periods of occupation in the North and South Areas of Çatalhöyük. The samples collected are listed in table 14.1.

Sample	*Area*	*Blg.*	*Sp.*	*Mellaart level*	*Hodder level*	*Context*
20627-4	North	114	87	Unassigned	North.G	Building fill and clean plaster floors
20627-5	North	114	87	Unassigned	North.G	Building fill and hearth rake-outs/occupation residues
20628-3	North	114	87	Unassigned	North.G	Clean plastered platform and floors
21175-3	North	114	87	Unassigned	North.G	Floors and fills of fire installation feature 7345
21364-3	North	114	87	Unassigned	North.G	Lining of fire installation feature 7345 and dirty floors
21535-3	North	114	87	Unassigned	North.G	Floors and fills of fire installation feature 7607
30596-3	North	114	87	Unassigned	North.G	Clean plastered platform and floors
21572-1	North	114	87	Unassigned	North.G	Floors of unexcavated early fire installation
21572-2	North	114	87	Unassigned	North.G	Clean floors to the east of the space
30022-3	North	114	87	Unassigned	North.G	Core of burial platform feature 638 and floors below
30032-2	North	114	87	Unassigned	North.G	Plasters on the south wall
20473-4	North	77	336	Unassigned	North.G	Oven floors and fills of feature 7108
22087-1	North	77	336	Unassigned	North.G	Clean floors of platform feature 6062
22087-2	North	77	336	Unassigned	North.G	Clean floors of platform feature 6062
30136-1	North	77	336	Unassigned	North.G	Clean floors of platform feature 6051
30161-2	North	102	18	Unassigned	North.G	Sequence of floors and accumulated deposits
21562-4	North	119	513	Unassigned	North.F	Oven floors of feature 7322
20988-3	North	132	511	Unassigned	Post-North.F	Collapsed construction materials
20988-4	North	132	511	Unassigned	Post-North.F	Collapsed construction materials
32149-5	North	-	636	Unassigned	North.F	Open area/midden deposits
32150-5	North	-	636	Unassigned	North.F	Open area/midden deposits
23616-1	North	-	630	Unassigned	North.F	Open area/midden deposits, penning and in situ burning
23616-2	North	-	630	Unassigned	North.F	Open area/midden deposits
19835-3	South	89	379	Unassigned	South.N	Clean floors
19839-1	South	89	379	Unassigned	South.N	Dirty floors
19839-2	South	89	379	Unassigned	South.N	Dirty floors
19393-4	South	-	470	Unassigned	South.M	Occupation floors and activity residues
23249-3	South	-	628	XI, XII	Unassigned	Open area/midden deposits
23214-10	South	-	620	XI, XII	Unassigned	Open area and penning deposits
23214-11	South	-	620	XI, XII	Unassigned	Open area and penning deposits
32689-3	South	162	621	XI, XII	Unassigned	Building floors
32688-3	South	162	621	XI, XII	Unassigned	Building floors
32653-31	South	162	621	XI, XII	Unassigned	Floors

Table 14.1. Micromorphological samples collected and analysed for this study.

Thin-section micromorphology

In the field, stratigraphic sequences were cleaned with artist's palette knives and air-blowers, and subsequently mapped and photographed at high-resolution. Block samples measuring ca 15×7×7cm, were then cut out from the field sections for micromorphological examination in large resin-impregnated thin-sections.

After collection, the sediment blocks were carefully packed and transported in rigid boxes to avoid any physical disturbances that could alter the morphology of the samples. Once in the laboratory, the blocks were oven dried at 40°C, impregnated with epoxy resin under vacuum and hardened at 70°C for 18h. After cooling, a slice of approximately 1cm thickness was cut from the face of interest of each of the hardened block samples using a diamond saw, and subsequently trimmed. Each cut slice was then impregnated on a hot plate at 40°C with a mixture of Araldite epoxy resin, temporarily mounted on glass media, and placed in a Brot grinder-polisher to obtain perfectly planed sample surfaces before permanent mounting to a glass slide. Samples were then ground to a thickness of 30μm and cover-slipped.

Micromorphological slides were analysed using a Leica DMLP microscope at magnifications from ×25 to ×400 under plane-polarised (PPL), cross-polarised (XPL) and fluorescent light. Photomicrographs were captured using a DFC420 camera and enhanced through the LeicaV2.3 image analysis software. The samples were described using the standard terminology developed by Stoops (2003).

Phytolith and dung spherulite analyses

In the field laboratory, research included microscopic scanning of plant-related contexts and concentrations of plant and faecal microfossils including phytoliths, wood ash pseudomorphs and calcitic dung spherulites from suspected faecal deposits. Spot samples were also collected above and/or below layers associated with these dung contexts, which served for comparison as controls. A total of 18 non-extracted spot samples were mounted in non-permanent slides using clove oil. Slides were examined at 400× magnification with an optical microscope under cross-polarised light (XPL) for spherulite analysis, whereas phytoliths and ash pseudomorphs were examined in plane-polarised light (PPL) using a Nikon SMZ645 instrument. These pilot microfossil results refer to 'non-processed' sediments that did not undergo chemical extraction, and all observations reported in the 2017 Archive Report are to be taken with caution (Matthews, Portillo 2017). These results are re-evaluated here based on the subsequent laboratory analysis of 19 extracted spot samples and 19 additional sub-samples collected from the micromorphological blocks prior to thin-section manufacture (coded as ss) at the University of Reading, following the methods of Katz et al. (2010). A minimum of 200 phytoliths with diagnostic morphologies were counted at 200× and 400× magnifications. Morphological identification was based on modern plant reference collections and standard literature (Twiss et al. 1969; Brown 1984; Mulholland, Rapp 1992; Rosen 1992; Twiss 1992; Albert, Weiner 2001; Piperno 2006; Tsartsidou et al. 2007; Albert et al. 2008; 2016; Portillo et al. 2014; Neumann et al. 2019). Dung spherulite analysis followed the methods of Canti (1999). Spherulites were counted in random fields at 400× magnification. Both microfossil assemblages were compared to modern reference ethnoarchaeological and ethnobotanical datasets of modern dung remains and dung-products from the Central Anatolian Plateau (Anderson, Ertug-Yaras 1998; Portillo et al. 2019; Portillo, Matthews 2020) as well as geo-ethnoarchaeological assemblages from the Mediterranean region (Tsartsidou et al. 2008; Portillo et al. 2012; 2014; 2017).

Results and interpretation

Micromorphological analyses allowed investigation not only of the mineral and organic components in the deposits as indicators of access to resources but also, as importantly, the fabric of these inclusions, both with regard to themselves and to other components, as well as the microstructural characteristics of the sediments. The study of layer composition, texture and fabric at the microscale was undertaken in order to develop more robust interpretations on the origin of the materials, their depositional pathways and any modifications they underwent post-depositionally. In the following sections, the results emerging from the micro-contextual observations of sampled sequences at Çatalhöyük are reported in three sections. The first part considers the arrangements of components in building contexts as represented by distinct and recurrent groups of similar composition and organisation, designated as deposit types. The second part examines the contextual variation and the socio-ecological implications of open area deposits and sequences. Finally, the last section evaluates the type and impact of post-depositional alterations on occupation deposits at Çatalhöyük.

Variation and classification of deposit types in Middle and Late period buildings

Deposits from building sequences from Middle and Late occupation levels have been classified according to diagnostic microscopic attributes, mainly microstructure, particle size, sorting, c/f ratio, c/f related distribution, and colour. The classification of accumulated residues has been made according to the abundance of specific inclusions and aggregates in the deposits examined in this study, such as charred plants and faecal aggregates. This strategy has allowed the identification of similar deposits occurring in different buildings across the site.

Occupation floors and surfaces. A range of samples of floor plasters from Buildings 77, 89, 102, 114 and Space 470, in addition to collapsed roof/upper storey sequences found in Sp.511, have been analysed in order to study variations in the selection of floor materials at Çatalhöyük, their use in specific contexts, and periodicity. Both 'dirty' areas surrounding fire installations, and 'clean' contexts from the central and northern areas of buildings, including raised platforms used for sitting and sleeping, have been selected to delineate variations in spatial boundaries and activity areas.

The first deposit type identified, 1A, corresponding to pale grey plasters, is characterised by an extremely fine particle size (clay grade), and the low occurrence of inclusions and sediment aggregates. Heavily tempered with plant stabilisers (ca 10–15% abundance), only two occurrences of these plasters, ranging in thickness from 0.1 to 2cm, have been found, specifically in the clean areas of Buildings 89 and 114.

Deposit Type	*Description*	*Interpretation*	*Blg./Sp.*
1A	Calcareous plasters (pale grey)	Extremely thin and fine-grained layers of calcareous sediments tempered with ca 10–15% chaff and almost completely devoid of finds.	B.114/Sp.87 B.89/Sp.379
1B	Marl plasters (grey)	Poorly sorted silty clay floors of slightly oxidised marl sediment displaying ca 5–15% plant stabilisers.	B.114/Sp.87 B.77/Sp.336 B.132/Sp.511 B.89/Sp.379
1C	Silty clay plasters (orangish brown)	Moderately sorted deposits of lightly to strongly oxidised sediment showing abundant calcareous and non-calcareous minerogenic aggregates (ca 15–30%).	B.114/Sp.87 B.77/Sp.336 B.102/Sp.18 B.132/Sp.511 B.89/Sp.379
1D	Silty clay loam plasters (light brown)	Moderately sorted plasters often displaying a strong parallel orientation and a highly variable temper content (2–20%).	B.114/Sp.87 B.102/Sp.18 B.132/Sp.511 B.89/Sp.379
1E	Clay loam plasters (dark brown)	Poorly sorted coarse plasters, usually lightly tempered (ca 5% plant pseudomorphs) and containing abundant charred flecks (ca 10–15%).	B.114/Sp.87 B.77/Sp.336 B.102/Sp.18 B.132/Sp.511
1F	Loamy sand (brownish grey)	Moderately sorted sandy deposits, mainly formed by quartz minerals and displaying an intergrain microaggregate structure. Likely used as levelling units.	B.114/Sp.87
1G	Sandy loam (light greyish brown)	Moderately sorted and unoriented coarse plasters containing abundant sand-sized particles.	B.114/Sp.87
1H	Mixed clay loam plasters (dark greyish brown)	Poorly sorted coarse deposits, frequently over 1 cm in thickness and apparenly used as floor base/packing material. High occurrence of sediment aggregates and re-utilised architectural materials.	B.114/Sp.87 B.77/Sp.336 B.102/Sp.18 B.132/Sp.511 B.89/Sp.379 Sp.470

Table 14.2. Floor plaster deposit types in buildings from Middle and Late occupation levels at Çatalhöyük.

Deposit type 1B, formed by slightly oxidised grey marl sediments constituted by fine silty clay and ca 10% of quartz, amphibole, plagioclase and chert mineral grains, has been detected in multiple floor sequences across Buildings 77, 89 and 114, in addition to suspected roof/upper storey plasters found in secondary deposition within Sp.511. These layers, which frequently contain embedded calcareous and clayish sediment aggregates (ca 5% abundance), in addition to charred plant flecks and mollusc shells (ca 2–5%), display a variable content of plant temper, as inferred from the occurrence of plant pseudomorphic voids, which ranges between 5 and 15%.

Orangish-brown silty clay plasters have been classified under deposit type 1C. These apparently oxidised deposits vary in colour from reddish brown to light orangish-brown and are frequently identified as thick (>1cm) layers within clean floor sequences from the central and eastern areas of houses. The plasters classified as 1C usually contain abundant calcareous and clayish sediment aggregates (ca 15–30%), and occasional fragments of recycled wall plasters (ca 2–5%). These deposits appear to have been lightly tempered, with plant pseudomorphic voids displaying a low abundance (ca 5%) and charred plant flecks representing 5% of these units.

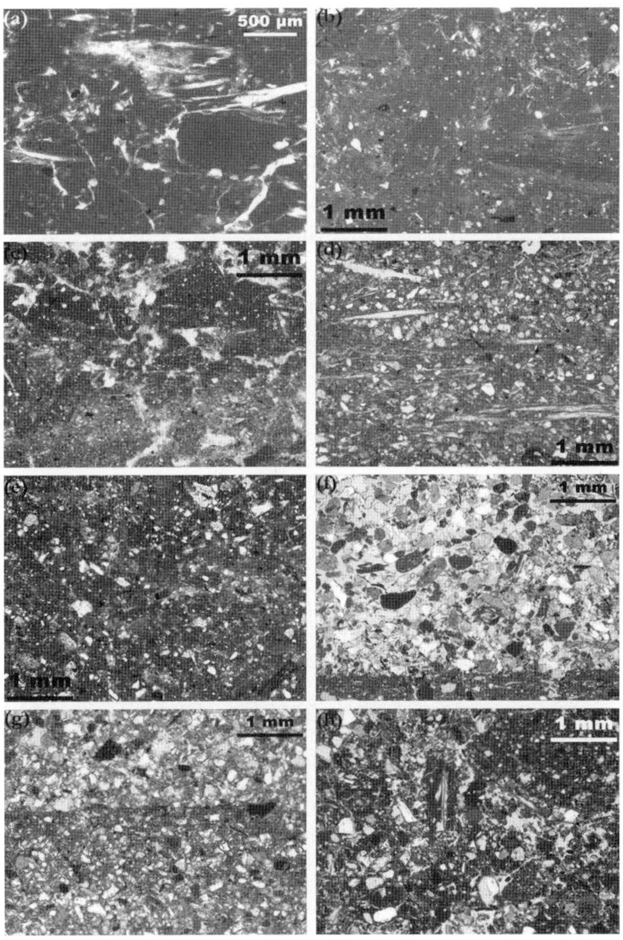

Figure 14.1. Photomicrographs of plaster floor types: (a) 1A, pale grey, calcareous plaster in B.89, PPL; (b) 1B, marl-based plaster in B.89, PPL; (c) 1C, orangish-brown silty clay deposit in B.114, PPL; (d) 1D, silty clay loam floor in B.89, PPL; (e) 1E, clay loam plaster in B.114, PPL; (f) 1F, loamy sand layer part of central platform sequence (F.7114) in B.114, likely used as a levelling unit, PPL;(g) 1G, sandy loam floor in B.114, PPL; (h) 1H, mixed plaster in B.77, PPL.

The light brown floor plasters classified as deposit type 1D are the most common in the Çatalhöyük sequences analysed for this study. These units have been identified in Buildings 114, 102 and 89, and the suspected roof/upper storey plaster sequence found in secondary deposition in Sp.511. These layers, consisting of moderately sorted silty clay loam sediments frequently displaying a strong parallel orientation and a highly variable temper content (2–20% of plant chaff), include a number of oxidised clayish sediment nodules (ca 5–15%), some of which display parallel graded bedding, suggesting that these aggregates originated from an alluvial source.

Deposit type 1E, consisting of dark brown clay loam plasters identified in the floor sequences of Buildings 114, 77 and 102, and the collapsed remains found in Sp.511, is characterised by the presence of poorly sorted sub-rounded minerals, including quartz, amphibole, plagioclase, chert, muscovite and calcite. The incidence of plant-pseudomorphic voids derived from added stabilisers is generally very low in these deposits (<5%), although fragments of charred plants are more abundant than in the other plaster types, ranging between 10 and 15%. The occurrence of sediment aggregates is also lower than in other types of mud plasters, forming 5–10% of these layers.

Loamy sand deposits, classified as type 1F, have only been documented in B.114, as part of the sequence forming the central platform F.7114. These unusual plasters are constituted by moderately sorted rounded and sub-rounded mineral grains of quartz, but also chert, amphibole, calcite and plagioclase (ca 80% abundance), and a small number of calcareous and silty clay aggregates (ca 5%). Due to their high content of sands, it is unlikely that these units were used as occupation surfaces. Instead, the deposit detected in B.114 could represent a base or levelling unit for the platform.

Deposit type 1G, referring to greyish brown sandy loam plasters, has also been uniquely identified in B.114. This unit, although constituted by a high proportion of sand-sized particles (>50%), displays a more compacted microfabric than the layers classified as 1F due to the higher clay content. Anthropogenic inclusions are extremely rare, with the exception of fragments of wall plasters (2% abundance). Other components include natural aggregates of calcareous and silty clay sediments (ca 5% abundance).

The last plaster category identified, deposit type 1H, consists of coarse mixed units of poorly sorted clay loam sediment, usually between 1 and 3cm in thickness, frequently found towards the base of the stratigraphic sequences studied. These layers have been detected in Buildings 114, 77, 102, 89, Space 470 and the collapsed roof/upper storey sequence unearthed in Sp.511. Interestingly, although these units usually display continuous accumulations of occupation residues along their upper boundaries and were therefore used as living surfaces, they are frequently classified as packing/make-ups in the field due to their thickness and overall heterogeneity. In fact, 1H plasters are constituted by a high proportion of various types of calcareous and clayish sediment aggregates (ca 30–40% abundance), and re-utilised construction materials, predominantly wall plasters and fragments of baked oven floors (ca 5–10%). The occurrence of charred plant flecks, ca 10–15% abundance, is significant when compared to the percentage of anthropogenic inclusions of organic origin found in other plaster types, but while their presence in these deposits was possibly intentional, to act as stabilisers, it was more probably caused accidentally at the place of manufacture, likely a midden/open space.

Fire installation floors. The fire installations considered in this study correspond to the two hearths of B.114 (Features 7607 and 7645), the unexcavated baked floors by the south wall of B.114, suspected to correspond to an early oven feature that was dismantled, the oven of B.77 (F.7108) and the oven of B.119 (F.7322), all located in the North Area of the site. Five types of plaster have been identified based on key micromorphological attributes, including particle size, microstructure and sorting (table 14.3). Colour has not been taken into account in this categorisation as it has been found to be too dependent on post-depositional alterations related to firing temperatures. These were considerably higher in the ovens, with plasters that were completely baked and hardened, than in the hearths, which display substantial rubefaction and scorching of their upper surfaces but were not hardened by heat (fig. 14.2).

The first deposit type identified, 2A, corresponding to poorly sorted silty clay loam plasters, often displays a well-developed crumby microstructure towards its upper boundary, likely an effect of high burning temperatures and erosion caused by maintenance practices, such as the regular sweeping of these surfaces. This type of plaster, tempered with ca 5% plant chaff, is the most common oven/hearth floor in the samples studied. The deposits classified as 2A contain between 40 and 60% calcareous and clayish sediment aggregates.

Deposit Type	*Description*	*Interpretation*	*Blg./Sp.*
2A	Silty clay loam plasters	Poorly sorted deposits, often displaying well-developed crumby microstructures towards the upper boundary, and containing abundant calcareous and clayish sediment aggregates (ca 40–60% abundance).	B.114/Sp.87 B.77/Sp.336 B.119/Sp.513
2B	Sandy clay loam plasters	Moderately sorted coarse plasters formed by >40% sand-sized quartz minerals.	B.114/Sp.87 B.77/Sp.336
2C	Silty clay plasters	Well-sorted units displaying ca 5–15% plant pseudo-morphic voids and few (ca 10–15%) non-calcareous sediment aggregates.	B.114/Sp.87
2D	Sandy clay plasters	Poorly sorted coarse plasters frequently displaying a crumby microstructure caused by heat impact and sweeping practices. High occurrence of re-used architectural materials such as mudbrick (ca 20–30% abundance) and wall plasters (1%).	B.114/Sp.87
2E	Hearth lip (sandy clay loam)	Completely unsorted and unoriented deposit rich in clayish sediment aggregates (ca 50–60%).	B.114/Sp.87

Table 14.3. Fire installation plaster deposit types in buildings from Middle and Late periods at Çatalhöyük.

Deposit type 2B, corresponding to sandy clay loam plasters formed by over 40% quartz grains, has only been identified in the oven sequence of B.77 and the unexcavated oven of B.114. Sediment aggregates (ca 5–10% abundance) and anthropogenic inclusions (2–5%) are rare in these deposits, which appear to have been substantially tempered with plant materials (ca 15–20% plant pseudomorphic voids). These layers are frequently black to very dark brown in colour due to the intensity of firing temperatures and the repeated use of the oven. In spite of their high porosity (approximately 25–30% voids), these units were found to be extremely hardened in the field and were described as 'baked oven floors' during the excavation. Occasionally, vesicle voids have been detected in these plasters, likely produced by the effect of high temperatures on the plant materials used as stabilisers, which appears to have caused them to burst.

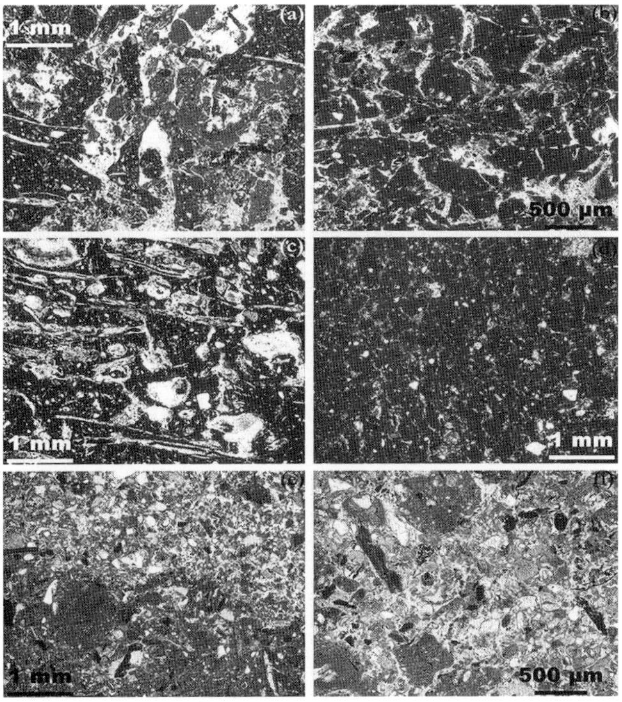

Figure 14.2. Photomicrographs of plastered surfaces found within fire installations in buildings: (a) 2A, heterogeneous layer formed by different sediments in oven F.7108, B.77, possibly acting as a levelling unit, PPL; (b) 2A, silty clay loam plaster within hearth F.7607, B.114, PPL; (c) 2B, baked oven floor displaying abundant plant pseudomorphic voids and vesicles, the latter caused by high firing temperatures, PPL; (d) 2C, silty clay deposit part of hearth sequence F.7345, PPL; (e) 2D, sandy clay plaster within hearth F.7607, B.114, PPL; (f) 2E, fabric of hearth lip deposit surrounding fire installation F.7345 in B.114, PPL.

Orangish-brown silty plasters have been classified as deposit type 2C. These fine-grained floors, which display a marked rubefaction gradient with scorched upper boundaries, have been found as part of the hearth sequence F.7345 in B.114. Well sorted and moderately tempered with approximately 5–15% plant stabilisers, these units often contain embedded alluvial aggregates of oxidised silty clay and silty clay loam (ca 10–15% abundance).

Coarse, sandy clay hearth plasters have been identified within the sequence of fire installation F.7607, in B.114. These deposits, classified as type 2D, were poorly tempered (ca 5% plant stabilisers), which appears to have been the cause for the crumbliness and fragmentation observed in their upper boundaries, damaged by heat impact and sweeping practices. In addition, the top of these plasters contains re-precipitated ashes and abundant complex packing voids along their upper boundary. Deposit type 2D displays a high degree of compositional heterogeneity overall, although it includes recycled architectural materials such as mudbricks (ca 20–30% abundance), baked floors (ca 5%) and wall plasters (ca 1%).

The sandy clay loam lining of hearth F.7345 in B.114, classified as deposit type 2E, is constituted by a range of unsorted and unoriented components, which include coarse silty clay loam and silty clay sediment aggregates (approximately 50–60% abundance) and re-cycled fragments of mudbrick. Although no plant pseudomorphic voids were identified, the small amount of fragmented charred plant remains found embedded in this unit could potentially have acted as stabilisers.

Accumulated residues. Units of accumulated occupation residues in buildings have been encountered in floor areas around fire installations as re-deposited rake-outs, and as thin 'dust' layers of trampled materials on the surfaces of fire installations and building floors (table 14.4). These accumulations are often rich in charred residues of plant origin and ashes, and they can vary in thickness from 200µm to several millimetres. In the case of rake-outs, these units are usually formed by superimposed microlaminations of charred materials, each of which seems to constitute a single depositional event of oven/hearth discards (fig. 14.3). In the clean areas of buildings at Çatalhöyük, mainly used for sitting and sleeping (Matthews 2005a), accumulated deposits are often thin and sparse, comprising microlenses of silt-sized organic and inorganic materials that were gradually deposited under mats and soft furnishings.

The first deposit type in this category, 3A, consists of highly organic accumulations rich in silicified plant materials, which represent up to 70% of these units, and other inclusions such as charred plant fragments

Deposit Type	Description	Interpretation	Blg./Sp.
3A	Accumulations rich in silicified plant materials	Accumulated residues on occupation floors and fire installations consisting of >50% phytoliths, ca 5–15% charred plants and ca 15–25% calcitic ashes. These layers frequently display a parallel orientation relative to the base of the unit, and appear to represent in situ decay of plant remains, possibly related to fuel, craft, or cooking activities.	B.114/Sp.87 B.119/Sp.513
3B	Accumulations rich in charred plant materials	Highly organic deposits formed by charred plant materials (>50% abundance) mixed with dispersed dung inclusions (ca 5–10%), and sediment aggregates (ca 10–20%), likely representing mixed plant/dung fuels.	B.114/Sp.87 B.89/Sp.379
3C	Accumulations rich in faecal matter	Millimetric layers containing abundant inclusions of reworked herbivore dung (ca 20–30%) found accumulated on building floors and fire installations.	B.114/Sp.87 B.102/Sp.18 Sp.470
3D	Mixed accumulations (no dominant inclusions)	Heterogeneous deposits containing mainly unoriented minerogenic aggregates (ca 10% abundance), calcitic ashes (ca 5%), dung inclusions (ca 5%), and plant materials (ca 20–30%).	B.114/Sp.87 B.132/Sp.511 B.89/Sp.379

Table 14.4. Deposit types of accumulated living residues in buildings from Middle and Late occupation levels at Çatalhöyük.

(ca 5–15% abundance), calcitic ashes (ca 15–25%) and, less frequently, phosphatic aggregates of suspected faecal origin (ca 1%). Small particles of inorganic materials in the form of rounded aggregates of clayish sediment (ca 5–15%) have also been found in these layers. Deposits classified as 3A have only been detected in hearth sequences, displaying strong parallel orientations that are indicative of in situ deposition. Thus, these were probably fuel remains of herbaceous/mixed origin.

Deposit type 3B comprises layers rich in charred plants (>50% abundance), often poorly sorted and highly fragmented, and other components of both organic and inorganic origin that are present in lower quantities, such as phytoliths (ca 5%), herbivore dung (ca 5–10%), bone fragments (ca 2–5%) and calcitic ashes (ca 5%). Burnt sediment aggregates are especially abundant in these units, including swept floor particles and rounded inclusions of calcareous and clayish materials, some of which appear to be fragments of oven plaster. This type of deposit, often consisting of multiple microlayers that display a moderate parallel orientation, are especially frequent in the dirty area of buildings, adjacent to fire installations and locations for food preparation.

Deposit type 3C is distinguished by the relatively high proportion (ca 20–30%) of faecal lenses of suspected herbivore origin occurring in these layers. These units are only a few millimetres thick and appear substantially reworked by trampling. Interestingly, deposit type 3C has been identified in two very different contexts: as re-deposited rake-outs from the fire installations in B.114, and as gradually accumulated and severely trampled layers in Sp.470 (Barański et al. 2015). In the former case, 3C deposits are constituted by burnt herbivore dung and other charred components, including rounded sediment aggregates (ca 5% abundance) and anthropogenic inclusions such as charred plants (ca 5% abundance), and dispersed fragments of bone, shell, and obsidian flakes. Here, the abundance of faecal aggregates and their contextual associations suggest the use of dung as fuel, although given the scarcity of these layers, this was probably an unusual practice in buildings. By contrast, in Sp.470, in situ herbivore dung lenses and fine sediment particles constitute deposit type 3C, which was considerably altered by trampling, a depositional pattern reminiscent of penning deposits identified elsewhere on site through thin-section micromorphology (Matthews et al. 1996).

The last deposit type, 3D, has been described as mixed accumulations, due to the absence of clearly dominant inclusions. These heterogeneous units, the most common in buildings at Çatalhöyük, contain a wide range of unoriented and randomly distributed components, including burnt plant materials (ca 20–30%

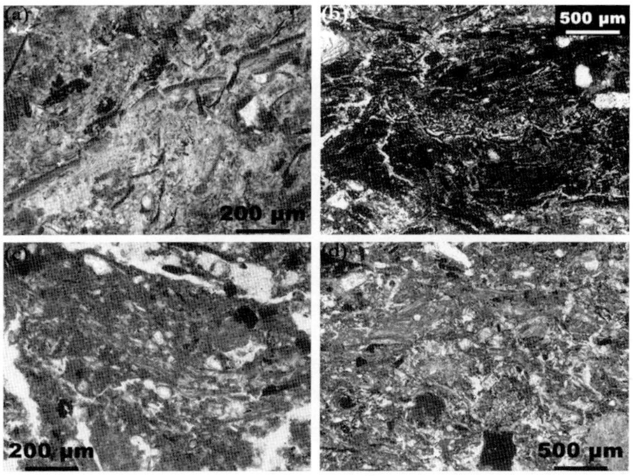

Figure 14.3. Photomicrographs of deposit types of accumulated materials found on building surfaces: (a) 3A, hearth fill dominated by silicified plant materials in B.114, PPL; (b) 3B, rake-out layer rich in charred plant materials identified within the northwestern quadrant of B.114, PPL; (c) 3C, discontinuous accumulations of herbivore dung on the floors of B.114, PPL; (d) 3D, mixed accumulations of charred and silicified plants, calcitic ashes, dung and sediment aggregates in B.89, PPL.

abundance), heated bone and shell fragments displaying varying degrees of charring (ca 5%), burnt faecal matter (ca 5%), calcitic ashes (ca 5%), and approximately 10% rounded and subrounded minerogenic aggregates. These re-deposited units seem to represent mixed hearth/floor sweepings that were gradually accumulated in the dirty area of buildings through daily maintenance practices.

Variation of deposits in Early period buildings

The focus of this analysis was on B.162, Sp.621, to compare the two core contrasting interior activity areas separated by a curb (F.8186): a well-plastered area in the south, and an area with an oven and ashy rake-outs in the north.

In order to examine the micro-history of the white plastered floors in the south of B.162, a micromorphology block was collected from the section face. Under the microscope, it became patent that these floors were laid on a thick layer of greyish-brown silty clay packing that had impressions from plant temper that would have reduced shrinkage on drying of these smectite-rich clays (Roberts et al. 1999). While the first plaster was manufactured from loamy sand sediments, the overlying floors were comparatively thick and made of whitish silty clay. Some of the thin accumulated lenses on these surfaces include organic aggregates, probably omnivore coprolites, while others had traces of dung ash and charred flecks, possibly derived from the oven in the north of this room (F.8183). The floor surface in this area was raised a second time by the deposition of a thick, more heterogeneous packing. The later floors were made of much thinner whitewash/silty clay plaster.

Of particular note is the lens of ashy occupation deposits with red pigment scattered throughout it, which includes charred flecks and bone fragments visible both in the field and under oblique incident light. This lens accumulated at, or close to, the closure of the building, and it suggests increased ritual activity at a key transitional period in the life of the individuals associated with it, towards the end of this house.

To study an oven area and adjacent floors and occupation deposits from the north of B.162, a spot sample of in situ fuel was collected from oven F.8183, and both block and spot samples were collected from adjacent floors. In thin-section, the floors in front of the large oven were covered in successive accumulations of oven rake-out residues rich in charred plant remains (including some wood fragments), phytoliths, dung ash, and rounded floor sweepings integrated by plaster, charred bone fragments and omnivore coprolites, some of which had been burnt perhaps as a sanitation measure. The latest levels were particularly rich in omnivore coprolites and may represent a latrine or an area in which latrine waste was discarded.

The collapsed wall plaster in this area has traces of soot on some surfaces and is less frequently and less elaborately plastered than later levels at the site, suggesting a fairly short-lived building occupancy.

Five additional spot samples were collected from Sp.620, from oven contexts (23212), (23215) and (23216). These samples were also examined under the microscope in the field laboratory (Matthews, Portillo 2017). Calcitic microfossil associations, including dung spherulites, ash pseudomorphs resulting from the burning of wood to at least 450°C and phytoliths, likely relate to the use of mixed fuel remains (fig. 14.7a). Phytoliths showing evidence of partial melting, which results in deformations due to high temperatures, have been found to correspond to reeds, sedges and Pooid grasses, as the morphology of original cells was preserved and identifiable in most cases (fig. 14.7b-c). These pilot results have been confirmed by further laboratory analyses (Portillo et al. 2019).

Variation of deposits in open areas

The sections below report the integrated micromorphological and microfossil results for open spaces in the North (spaces 630 and 636) and South (spaces 620 and 628) Areas of Çatalhöyük, corresponding to midden layers, penning deposits and activity areas.

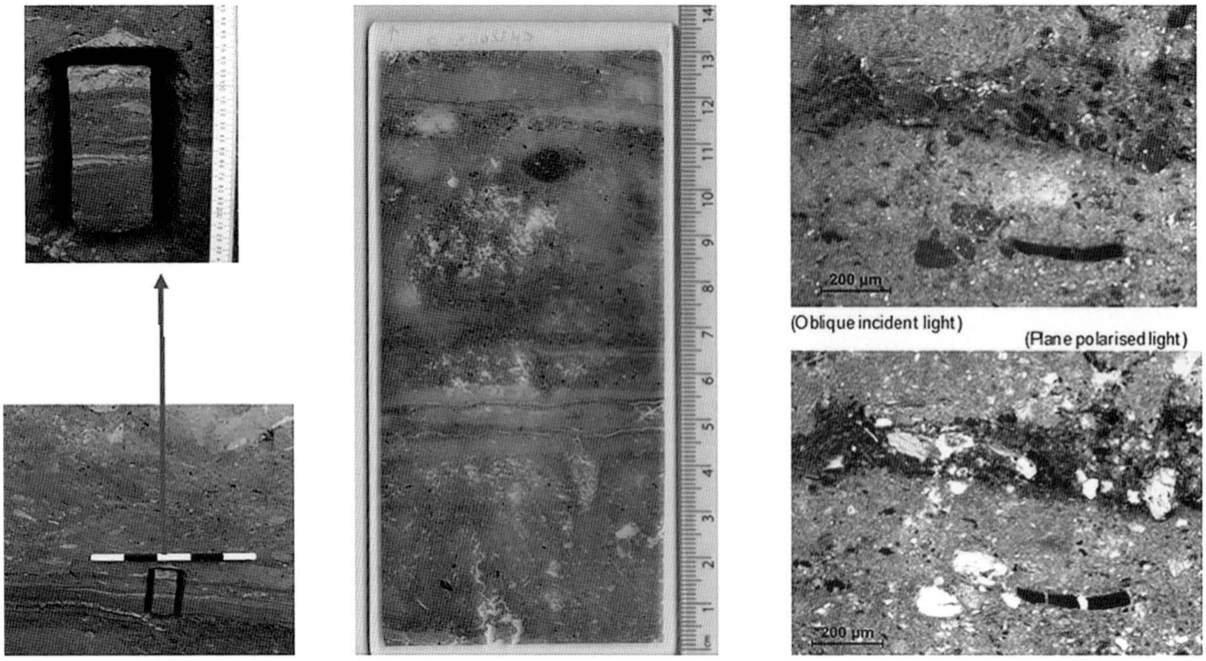

Figure 14.4. Micromorphology block sample and thin-section corresponding to early plastered floors in the south of B.162 (left and centre), and photomicrographs of red pigment (right).

Open areas and animal pens in the North Area.

Open area/animal pen: Space 630. As outlined in the 2017 archive report by Klimowicz (2017: 20), Sp.630 was an 'open, unroofed and widely accessible space of a unique character' that differed significantly from other open areas constituted by midden deposits. Little evidence was recovered in the field indicative of its purpose, apart from possible post placements for a light roof, low and poorly constructed partitions to the north and south of the space, several fire-spots across the area, and finely laminated deposits.

In thin-section, this sequence comprises heterogeneous midden deposits overlain by >2cm of compacted in situ banded herbivore dung with calcitic spherulites from a period of animal penning. The overlying in situ burnt surface comprises traces of dung mixed with diverse residues, including burnt bone and charred plant remains. Mixed layers of dung and sediments form the uppermost deposits in this space, together with traces of a later episode of in situ burning that includes dung, which had similar blackened and rubefied underlying layers (fig. 14.5).

This evidence suggests that this area was used as a pen for ruminants, at least periodically and for sustained periods, as evident in the banded sequence. The results from integrated phytolith and calcitic microfossil analyses from extracted block sub-samples indicate variable concentrations of spherulites in association with the leaves and culms of Pooideae grasses and reeds, suggesting that these plants may have been used as fodder and, therefore, also as bedding (fig. 14.7f). This plant composition contrasts with that of the animal pen from Sp.620 in the South Area, representative of a dicotyledonous-based diet (Portillo et al. 2019).

The excellent preservation of the finely banded deposits suggests that this area may have been at least periodically roofed. The fire-spots would have helped to heat this area, pointing to its use as shelter in colder months, perhaps for birthing/young animals (Anderson et al. 2017).

Open area/midden: Space 636. To contribute to micro-excavation and research on exterior spaces by Issavi (2017), a large section through a deep sequence of open area/midden deposits in Sp.636 was cleaned, recorded and sampled. In thin-section, this sequence comprises, from the basal levels upwards: intact omnivore coprolites; lenses with abundant omnivore faecal materials; bands of deposits rich in ashes, charred seeds, charcoal fragments, dung aggregates, burnt bone and sub-angular fragments of plasters; and layers with well-preserved large fragments of charred dicotyledonous wood, burnt aggregates and remains of construction materials with traces of omnivore coprolites and burnt bone.

The upper part of this sequence is characterised by alternating layers of aggregates of re-deposited construction materials and lenses of ashy rake-outs with well-preserved charred dicotyledonous wood fragments. Overlying these deposits is a layer rich in omnivore

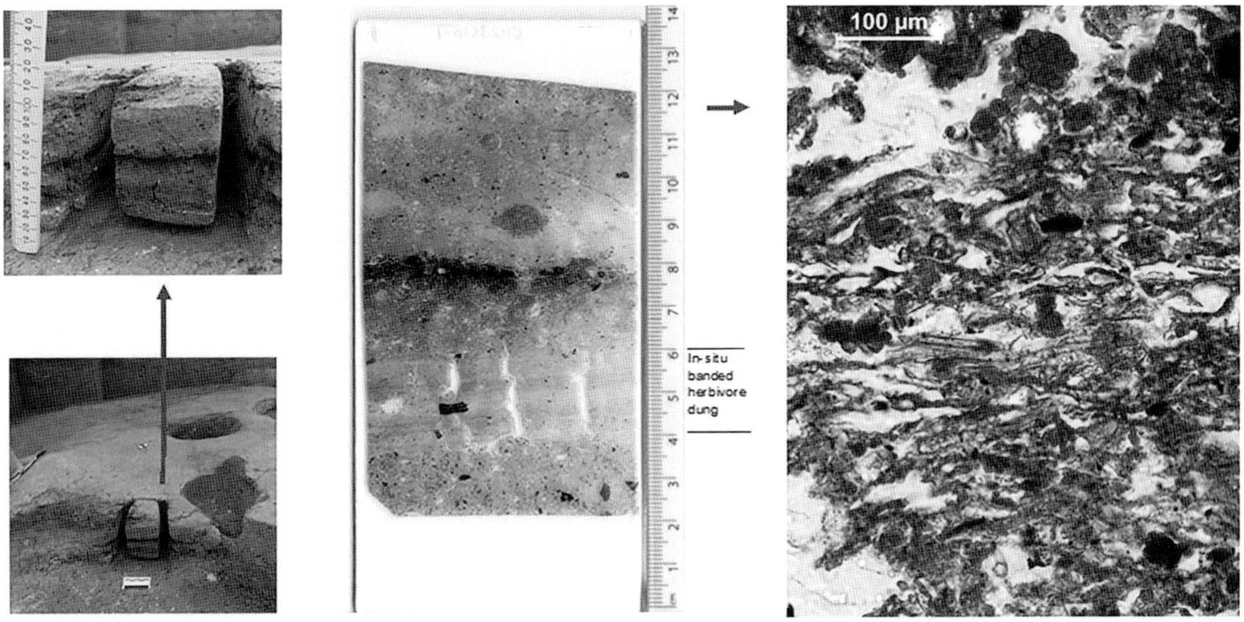

Figure 14.5. Surfaces, fire-spots and penning deposits in Sp.630, North Area.

coprolites, some of which display digested bone inclusions, and a thick lens of intact omnivore faecal matter with partially burnt hackberry endocarps preserved. An undisturbed layer of burnt plant remains with articulated phytoliths and charred cereal grains was identified towards the top of this sequence, overlaid by layers formed by discarded construction materials, burnt aggregates, charred wood, and lenses of omnivore coprolite fragments mixed with ashy rake-outs.

Open areas and animal pens in the South Area. In the South Area, the early history of the settlement is characterised by large open areas where midden deposits accumulated. These extend spatially across more than 60m from Sp.181, excavated in the Deep Sounding in 1999 (Matthews 2005a), to Spaces 620 and 628, recorded in 2017 (Taylor 2017, Archive Report). In Sp.181 these deposits accumulated to a current depth of ca 4m. In both Spaces 181 and 620, microstratigraphic evidence points to their use for high temperature burning, arguably for the production of fired-lime that was used to plaster interior surfaces of buildings from the earliest levels of site occupation until ca Level XII (Matthews 2005a; Matthews et al. 2013; Cessford 2007a).

Midden-like deposits and high-temperature burning: Space 628. In thin-section, Sp.628 comprises a sequence of heterogeneous accumulated deposits and surfaces, many less than 5mm thick. Of particular interest are lenses with well-preserved omnivore coprolites, compacted herbivore dung with calcitic spherulites probably derived from transitory penning practices, abundant rock fragments likely from craft/grinding activities, and rounded floor plaster fragments from sweeping in a built environment. The lenses of dung precede and arguably presage the thicker sequences of compacted dung from the penning areas identified across this sector of the South Area discussed above, indicating increased proximity to animals within the site at this time. The good preservation of the omnivore coprolites and articulated phytolith lenses, in particular, suggests that whilst this area was the locus of residues from a diverse range of activities, many of these were conducted elsewhere, and no construction surfaces were laid to facilitate the performance of activities in this location.

Early animal pen and animal management: Space 620. Space 620 was used to pen animals in a yet undefined area with ca 5cm of compacted laminated dung deposits sometime prior to the construction of B.17, as attested by a 3–5cm deep sequence of compacted animal dung in the south-west corner of this space (fig. 14.6).

Preliminary microfossil observations in the field laboratory allowed the identification of abundant calcitic spherulites in these deposits, indicative of herbivore dung accumulations (Matthews, Portillo 2017). Interestingly, well-preserved articulated multi-celled phytoliths derived from dicotyledonous leaves were dominant in these samples, as confirmed by further laboratory analyses (fig. 14.7d-e). This is especially noteworthy given that dicotyledonous plants are minor phytolith producers. Their presence in dung may indicate a diet that is either based on, or includes, an important component of dicotyledonous matter. A diet enriched in dicotyledonous leaves

Figure 14.6. Micromorphological sample of livestock penning in the South Area (Sp.620).

may suggest human manipulation of livestock fodder (Rasmussen 1993; Macphail et al. 1997; Halstead, Tierney 1998), or it may also be the result of a selection by obligate browsers, such as caprines (Tsartsidou et al. 2008) and, therefore, it can as well reflect some degree of seasonality (Portillo et al. 2019).

Open area feasting deposit: Space 629. An exceptional phase of intervening activity in an open area possibly associated with feasting was examined in Sp.629, comprising the transition between B.162 and the later structure constructed on this plot, B.161. A spot sample (32675.s3) was collected from the in situ fuel remains found in the fire-pit/fire installation F.8177, and the sequence in the adjacent floors was sampled for micromorphology. Phytolith results from the fuel ashes point to a mixed component of grasses and woody materials, including dicotyledonous leaves. In contrast with many firing deposits, reeds and sedges were absent, whereas large concentrations of multicellular phytoliths from the floral parts of grasses derived mainly from the husks of wheat were identified. This occurrence may be due to the presence of agricultural by-products such as those derived from cereal processing, and it could potentially constitute a seasonality indicator, although plant material collected in multiple seasons was stored at the site (Anderson, Ertug-Yaras 1998).

In thin-section, many of these layers are heterogeneous in composition, displaying burnt and unburnt bone fragments, phytoliths, charred plants and calcitic ashes. The deposits associated with the 'feasting deposit' include an unusually thick brown silt loam plaster with charred plant remains inclusions that was laid to provide a surface to surround the sunken fire installation. The top boundary of this deposit is irregular and appears to have been trampled. Some of the bone fragments in these layers, like most from the early occupation levels at Çatalhöyük, show a dark coating of an undetermined material, the significance of which needs to be investigated with analytical chemistry, as it might be indicative of the presence of burnt meat on the bone when buried, similar to the fat-derived char materials formed from the burning of flesh or animal fat documented in Palaeolithic sites and experimental studies (Goldberg et al. 2009; Miller et al. 2010).

Post-depositional alterations

The most common post-depositional features of natural origin observed in thin-section include bioturbation, recrystallised salts and organic staining. Post-depositional processes of anthropogenic origin comprise trampling, abrasion, sweeping, burning, compaction below mats, and truncation of surfaces and deposits.

Post-depositional alterations are particularly marked in open sequences. The most conspicuous of these processes is bioturbation caused by insect and small mammal burrowing, also comprising the deposition of excremental pellets and root action. The latter is distinguished by traces of modern roots, often displaying ferruginous coatings, partially degraded cellulose tissue and associated calcitic hypocoatings.

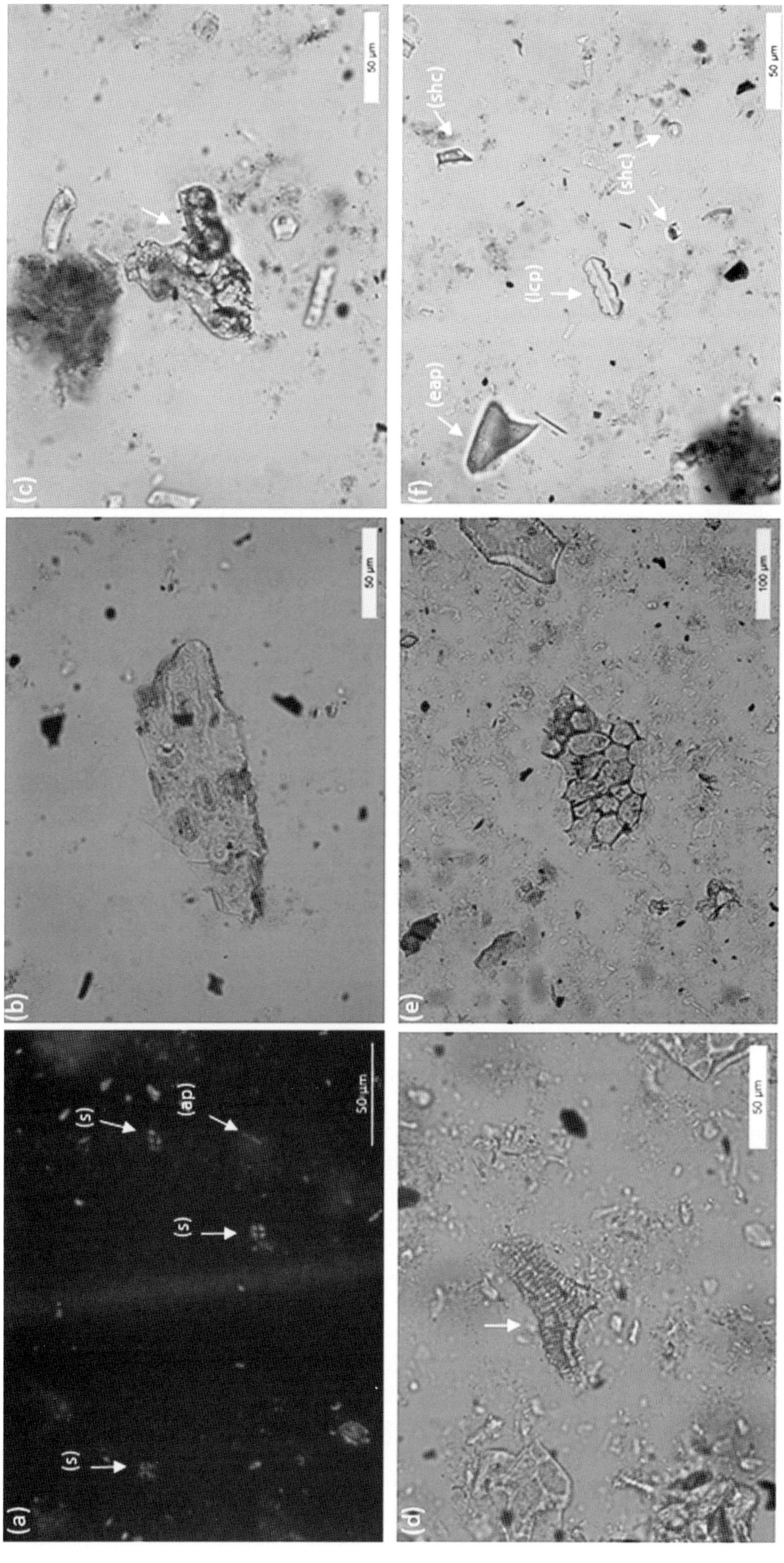

Figure 14.7. Photomicrographs of faecal and plant microfossils identified in the studied samples (at 400x or 200x): (a) dung spherulites [s] and ash pseudomorphs [ap] from Sp.630 (23616.s1.ss5) (XPL); (b) multi-celled phytoliths with stomata cells from reed leaves/stems from Sp.630 (23616.s1.ss3); (c) melted phytoliths from Sp.630 (23616.s1.ss5); (d) tracheary from dicotyledonous leaves from Sp.620 (23214.s12); (e) dicotyledonous epidermal leaf tissues from Sp.620 (23214.s12); (f) grass phytoliths: [eap] epidermal appendage acute bulbosus (prickle); [lcp] elongate (long cell) polylobate; [shc] short cell rondel from Sp.630 (23616.s1.ss5).

Precipitation of gypsum salts is common in building contexts, occurring as poorly developed lenticular, xenotopic or, more rarely, needle crystalline formations infilling channels, sometimes visible macroscopically in the field as white nodules up to several millimetres in size. The formation of these crystals often causes the surrounding groundmass to become disaggregated. In open areas, gypsum has been observed to occur as well-developed lenticular and hypidiotopic pedofeatures in voids produced by biological agents, and also as dispersed crystals within the deposits. Gypsum precipitation occurs when the concentrations of calcium and sulphate ions in the soil are high. These ions can derive from the dissolution of calcium sulphates from rocks, or from the alteration of limestone and sulphate-bearing formations (Poch et al. 2010). Gypsum formations are more abundant in arid and semi-arid regions, where soil moisture evaporates easily and salts from solution are deposited (Herrero, Porta 2000). This phenomenon occurs both within the groundmass and in voids created by plant roots and faunal activity, which provide a pathway for the flow of gypsum-saturated solutions, ultimately leading to the precipitation of these salts in voids.

Ferruginous features occur mainly in the form of typic and dendritic nodules within the matrix of building plasters, likely representing inherited features that formed in the natural sediments from which the architectural materials were manufactured. Localised iron staining of bone, plant and faecal inclusions has been observed in some deposits, especially in middens, an alteration associated with microbial action and the breakdown of organic matter.

Human and animal trampling is a conspicuous anthropogenic phenomenon in floor and accumulated deposits at Çatalhöyük, which results in the compaction of the groundmass close to the upper boundary of these layers and the in situ fragmentation of inclusions, in particular of fragile plant remains (Matthews et al. 2010).

Discussion

The final phase of micro-contextual research has proved particularly important in addressing the key hypotheses highlighted at the beginning of this chapter, as well as new research on global challenges. This study has built on the micromorphological records and reference collections that span the entire duration of the 1993–2017 excavation seasons, and on the wider interdisciplinary research that is the hallmark of the Çatalhöyük Research Project. These concluding sections examine in turn the significance of integrated micromorphological and microfossil approaches and the contributions that they have made to understanding this settlement at high-resolution timescales: house independence, health and the built environment, as well as human/animal interactions represented by the rarely emplaced animal pens. The final section discusses the contribution to sustainability.

House independence

The microstratigraphic evidence from building sequences examined in this study demonstrates the existence of more similarities than differences between houses in terms of the selection of architectural sources and concepts of space. The main rooms of buildings at Çatalhöyük display a marked division of space into clean areas (usually towards the north) and dirty areas (southern part of the house where fire installations are commonly located), both of which consist of a series of floors and raised platforms of variable heights coated with multiple layers of plaster. As the micromorphological categorisation of deposits has demonstrated, the origin of these floor plasters, dependent on the selected natural sediments used for their manufacture, was variable. In general, all plasters were prepared by mixing alluvial or lacustrine sediments with water and stabilisers, usually plant chaff (by-products of crop processing such as cereal husks but also reeds and grasses from the surrounding marshes) and occasionally also charred flecks and re-utilised fragments of architectural materials such as wall, floor, and oven plasters. Surprisingly, herbivore dung does not appear to have been considered an appropriate temper for floor construction materials by the residents of this settlement, as the scarcity of this material in the sequences analysed appears to suggest. The prepared mixture was then applied to the floor area, carefully compacted to produce a smooth and even surface, and left to dry (Houben, Guillaud 1994; Matthews et al. 1996; Matthews 2005a).

It has been observed, based on the assessment of plaster sequences from the main rooms of Buildings 114, 77, 102 and 89, that mud-based plasters are generally more frequently selected than calcareous, marl- or soft-lime-based plasters. Building 89, in particular, shows a very consistent choice of architectural sources, especially in its clean area. Here, cyclical patterns of floor construction involving the exact same sediment materials have been identified. Overall, only six calcareous floors out of a total of 26 occupation surfaces have been identified, again demonstrating a preference towards mud plasters.

It is here acknowledged, however, that this pattern can be due to the perceived properties and symbolic status of specific materials, possibly indicating that the use of calcitic sediments as construction material was linked to marking important events during the life-history of the building (Boivin 2000). In this regard, the analysis of sequences in the earliest building excavated by the Hodder team, B.162, has shed new light on how consciousness of place and the functional independence of household groups evolved at Çatalhöyük, and whether these early buildings represented a blueprint that continued to be emulated. Although the location of the two core areas in this early building was diametrically opposite to that of many later buildings, the concepts of interior places with cleaner areas defined by whitish plasters and separated by raised areas and/or curbs from cooking areas was clearly established at this early date. In addition, red ochre was used to mark particularly significant events in arguably ritual practices, such as the end of occupancy. Interestingly, there was also a clear cycle in the life-history of this building and a devolution in its use from habitation to a probable latrine before infilling, as observed in later buildings that were not immediately rebuilt (Matthews 2005a).

Many of the floor and hearth/oven plasters observed throughout the Çatalhöyük sequence were manufactured from a variable mixture of sediments, including marl, lower alluvium and backswamp clays, resulting in a wide diversity of materials that is reflective of the diversity of the local environment (Ayala et al. 2017; this volume, Chapter 2). Further, anthropogenic refuse from middens and hearth rake-outs, in addition to recycled materials such as fragments of wall plasters and baked floors, were often added to coarse clay loam plasters during the manufacture process, as attested by micromorphological observations, which greatly complicates the sourcing of these architectural materials.

This overall patterning suggests that similar practices were taking place across different buildings and it shows that, although production was done at a house-based level and likely in different temporalities, the knowledge of this production was probably shared across households. Similarly, plaster sources, while numerous even within a single house, show considerable continuity through the different buildings studied here. The production of plasters, then, although carried out at a household level, may well have been implemented through shared networks and collaboration.

Health and the built environment

Although the transition to sedentary agriculture 10–12,000 years ago, on which modern life is based, potentially increased food security and human fertility, it also increased exposure to major diseases from domesticated animals and commensals, and has been argued to represent a major epidemiological transition (Harper, Armelagos 2013; Larsen et al. 2019). In particular, penning inside the settlement during the Early period (Portillo et al. 2019; 2020; and this chapter), as well as milk consumption, documented from 6000 BCE in the West Mound (Hendy et al. 2018), would have been major risk factors for brucellosis and tuberculosis. In responding to these challenges, the creation and use of the built environment at Çatalhöyük offered its occupants ways for controlling disease reservoirs, vectors and pathways, whether consciously or not, as disease is not only biologically but also socially, economically and culturally constructed (López 2012).

The identification of animal pens within the settlement in its early history and the widespread use of dung fuel both within buildings and in open areas (Matthews 2016; Bogaard et al. this volume, Chapter 5), provide important evidence of the proximity of humans and animals and the potential spread of zoonoses. Much of the settlement during the Middle and Late periods, however, does not appear to have been used for penning animals, particularly ruminants, suggesting that for a variety of reasons – ecological, social and cultural, as well as perhaps sanitary – animals were penned away from the community (Henton 2012; 2013; Wolfhagen et al. this volume, Chapter 9). Further evidence suggests that this activity took place at increasing distances later in the occupation of the settlement (Pearson et al. 2007; 2015; this volume, Chapter 11; Roberts, Rosen 2009; García-Suárez et al. 2020).

The extensive long-lived open areas with midden-like deposits would have also created potential disease reservoirs, due to the accumulation of living refuse such as discarded food remains and, particularly, faecal matter, throughout the history of the settlement. Significantly, a wide range of deposits within these open spaces were burnt, including coprolitic materials such as those identified in Sp.636, suggesting that fire may have been used as one way of reducing the disease risks from these sources, whether consciously or not. Microbial traces such as eggs of faecal parasites and fungal spores, however, are present in these and other deposits across the site in thin-section and are currently being investigated in more detail to evaluate health implications.

The interiors of buildings were generally kept scrupulously clean and re-plastered frequently with highly calcitic wall and floor plasters that would have provided some anti-microbial as well as insect-repellent properties (Panagiotakopulu et al. 1995). The surfaces on raised platform areas for sitting/sleeping, as well as burial, were plastered exceptionally frequently with calcitic plasters and whitewashes. These were the cleanest areas within buildings, often covered in soft furnishings with few traces of dust on their surfaces (Matthews et al. 1996, 2013). The hygro-thermal and health properties of surfaces and earthen materials are currently being investigated in a collaboration between the School of Archaeology, Geography and Environmental Science and the School of the Built Environment of the University of Reading with Dr Zhiwen Luo, and in discussion with Prof Peter Walker and Dr Dan Maskell, BRE Centre for Innovation in Construction Materials, University of Bath.

Although the interiors of buildings at Çatalhöyük were generally very clean, traces of amorphous organic material, closely resembling omnivore coprolites and human coprolites in particular, have been identified within some of the early buildings at the site, including Sp.621 in B.162. Human faecal matter has been identified in a wide range of contexts across the settlement (Bull et al. 2005; Shillito et al. 2011b; 2013a; 2013b). Other rare traces of omnivore coprolites within buildings include a possible walled latrine area in the northeast corner of B.1, Sp.111 in the North Area (Matthews 2005a), and lenses on surfaces in B.2, South Area and B.114, North Area (Farid 2007b; Matthews 2005a; García-Suárez 2017).

It has long been recognised that oven areas were the dirtiest areas within building interiors at Çatalhöyük, often placed in the south. This was probably for functional airflow, as well as for cosmological reasons, as part of a habitus (Bourdieu 1977), and due to continuity in social roles and relations (Jenkins 2004; Bloch 2010). Accumulations of micro-refuse in these areas, however, were still often only millimetric in scale after each rake-out and sweeping, in contrast to the thick accumulating deposits in open areas/middens. Ovens and hearths within interior areas would have adversely affected indoor air quality and impacted health by introducing smoke, particulates and pollutants, as attested on successive interior wall plasters, particularly during the winter months when most cooking was conducted within buildings (Matthews 2005a; 2005b; 2012; 2016). The use of biofuels and health at Çatalhöyük is currently being investigated in more detail (Shillito 2017b; Shillito et al., thematic volume, forthcoming).

Human/animal interactions

The 2017 excavation season resulted in the identification of two out of four animal penning areas within the site as determined through thin-section micromorphology (Matthews, Portillo 2017; Portillo et al. 2019; 2020). The two recently discovered pens are Sp.620 in the South Area and Sp.630 in the North Area, both comparatively early in the history of the settlement. The two other areas of penning identified in thin-section are in the South Area. One of these consists of successive walled pens in Spaces 199 and 198, Level XII-XI, less than seven metres from, and probably contemporary with, those in Sp.620 (Cessford 2007a; Matthews 2005a). The other sequence is formed by deep, compacted penning deposits in Level VIII/X.25 (Matthews et al. 1996). The coprolite-rich deposits in the fill of B.2, South Area, have been occasionally depicted as representing an animal pen, but these have been identified by biomolecular analyses as human in origin (Shillito et al. 2011b). This evidence supported initial micromorphological interpretations of these faecal remains as indicative of human latrine deposits mixed with large lime-rich aggregates at the end-life of the building (Matthews 2005b).

The creation of two possibly contemporary areas of penning at the end of a long history of open area/midden use in the South Area, as observed in Sp.620 and Spaces 199-198, suggests that greater proximity to, and management of, animals in pens within the boundaries of the site shortly preceded settlement expansion, perhaps representing increased population influx and/or need to protect or be close to livestock and their products (Matthews 2018).

Community sustainability

One of the most frequent characterisations of Çatalhöyük is that of a 'mega-site', commonly using the words 'earliest city'. The site is certainly large, at more than half a kilometre in length, covering an area of 37ha. One of its most striking characteristics is its longevity, spanning more than 1,000 years from ca 7100 to 5950 cal BC, creating a mound over 20m high. The surveys and research on adjacent sites and past environment by a wide range of researchers, notably Ayala and colleagues (2017; this volume, Chapter 2), Baird (2012) and Roberts et al. (1999), provide an important local, regional and chronological context for examining sustainability and resilience at Çatalhöyük.

A wide range of fuels was used at Çatalhöyük throughout the history of the settlement, including animal dung burnt as fuel within buildings, as observed in B.162 through micromorphology, and not just in exterior areas, *contra* Bogaard et al. (2014b) (Matthews 2016; Portillo et al. 2020). Animal dung provided an important and sustainable second generation fuel source, as the bulk of the combustible materials in dung have already supported a biomass – animals are thus vital to sustained settled life, as argued by Hesse (1984). In addition, the by-products of fuel rake-outs in midden materials were periodically excavated and used as stabilisers in mortars (Matthews et al. 1996) as well as fertilisers to enrich soils for crops together with manure (Bogaard et al. 2017), providing important re-cycling opportunities and further enhancing their sustainable applications.

Finally, the buildings at Çatalhöyük were constructed from locally available earthen materials and could stand for over ca 70 years. They were well maintained and frequently plastered and modified. Particularly where abutting, buildings are likely to have provided well-insulated interior spaces, reducing energy costs. The regularity with which these interior places were laid out and their adaptability to the life cycle of their occupants are likely to have contributed to the sustainability of the community at Çatalhöyük, providing both predictability and representations of continuity in social roles and relations (Jenkins 2004; Bloch 2010), in addition to enhanced health.

Conclusions

The results of this research, focused on the integrated analysis of microstratigraphic sequences and small-sized material residues of human activities at the microscopic level, and therefore closer to the short timescales that created the data, have made important contributions to our understanding of household and community practices. The value of this evidence, however, transcends its ability to inform us on the precise loci and nature of domestic practices across time and space, to provide key insights into behaviour and relationships between social actions and the material, thus contributing to our understanding of how the community at Çatalhöyük was articulated.

Firstly, a rich body of data on building floors and the diverse range of activities on and around them has been documented. The results show evidence of associations between larger and smaller buildings, manifested in the range and repeated occurrences of source materials used for plaster manufacture across different households, as well as the functional independence of small built environments (Sp.87, B.114) and the differences in the longevity of individual houses.

Secondly, the micro-contextual examination of the earliest building excavated to date at the site, B.162, has provided evidence of the performance of religious and ritual activities early in the occupation of the settlement. The spread of red pigment on the floors of this house at its closure appears to mark the transition from one building to another and the social transformations associated with this event.

Finally, this study has provided clear evidence for the widespread penning of animals inside the settlement in the South Area in the early phases of occupation and approximately 250 years later (A. Bayliss, pers. comm.) in the North Area in Sp.630 (Portillo et al. 2019; 2020). Although the high incidence of calcitic dung spherulites in these deposits points to the ruminant origin of these faecal remains (Canti 1999; Shahack-Gross 2011), further species determination is not possible with the methods used in this research.

Acknowledgements

We are very grateful to all the Çatalhöyük team members and to the Project Director Ian Hodder for their kind collaborations. We would also like to thank the Turkish Ministry of Culture and Tourism and its representatives for permission to conduct this research at Çatalhöyük. We are indebted to Burcu Tung for her continuous guidance and feedback throughout the several excavation seasons at the site. We also wish to thank Marek Barański and Arek Klimowicz for their help and insight in correlating our microstratigraphic samples with field data and further collaborative work; Jason Quinlan for his detailed photographs of the microstratigraphy and Yıldız Dirmit and Bilge Küçük-doğan for their support. Special thanks are due to Matthew Britten for his hard work in the field. Scott Haddow, Marta Perlińska, Camilla Mazzucato, Ceren Kabukcu, Lara González Carretero, Patrycja Filipowicz, Jędrzej Hordecki, Maciej Chyleński, Chris Knüsel, Arek Marciniak, Maurizio Forte, Alex Bayliss, Dorian Q. Fuller, Justine Issavi, Cristina Belmonte, David Orton, Dominik Lukas, Zhiwen Luo and many other team members are thanked for their helpful discussions and support. The micromorphological thin-sections were produced at the Micromorphology Preparation Unit of the University of Reading with the invaluable help of John Jack. Thanks are also extended to Ann-Marie Catten and Kate Dudgeon, who helped in the phytolith laboratory of the University of Reading.

This research has been funded by the Arts and Humanities Research Council (Project No. 1227582), the University of Reading, the British Institute of Archaeology at Ankara, the Research Center for Anatolian Civilizations (ANAMED) of Koç University, the European Union's Horizon 2020 research and innovation programme under the Marie Skłodowska-Curie grant agreement No. H2020-MSCA-IF-2015-702529 (Project MICROARCHEODUNG), and the Templeton Foundation (Award No. 52003 to I. Hodder).

Supplementary material

For supplementary material related to this chapter, please visit https://doi.org/10.18866/BIAA/e-13. It comprises colour versions of figures 14.1–14.7.

15. Funerary practices I: body treatment and deposition

Scott D. Haddow, Eline M.J. Schotsmans, Marco Milella, Marin A. Pilloud, Belinda Tibbetts, Barbara Betz and Christopher J. Knüsel

Introduction

Neolithic Çatalhöyük is striking for its seeming cultural stability over an occupation that lasted 1,100 years (Bayliss et al. 2015). Seen on a macroscale, a considerable preoccupation with conformity, continuity and repetition is apparent in many aspects of life – and death (Hodder, Cessford 2004; Hodder 2006c). This propensity for continuity can be seen, for example, in the buildings at Çatalhöyük, which tend to conform to a similar layout consisting of a large central room with smaller rooms typically located to the west or north of the central room, a pattern sustained through most of the life of the settlement. Except for houses from the earliest levels, ovens and hearths are typically located along the southern wall, with the ladder and roof entrance in the southeast corner. The central rooms are divided between 'clean' platform areas in the north and 'dirty' floors in the southern half of the building near the ovens and hearths. Wall paintings tend to occur in the same locations, usually along the north and east walls (Czeszewska 2014), while platforms, animal installations and walls are regularly re-plastered and painted during the lifecycle of a building. After the abandonment and demolition of buildings, new structures are often rebuilt on the same footprint as the previous ones, sometimes over several construction sequences (Matthews 2005b; Farid 2014; Russell et al. 2014). A similar preoccupation with continuity and repetition is also evident in the funerary practices observed on site.

During much of the occupation at Çatalhöyük, the overwhelming majority of burials occurred within houses, typically beneath the northern and eastern platforms of the central room, although perinate, neonate and infant burials are found in more variable locations within the house (Andrews et al. 2005; Boz, Hager 2013; Tibbetts 2017). Most intramural burials took place during the occupation phase of the house, although a smaller number of burials have been found that correspond with construction and abandonment phases. Perinates, neonates and infants also account for a large proportion of burials that occurred during the construction phase of houses, but are rarely found in abandonment contexts (Boz, Hager 2013).

Previous studies of funerary practices at Çatalhöyük

Given the long-running history of the Çatalhöyük Research Project, the leadership, personnel structure and research directions of the human remains team (as with other specialist labs at Çatalhöyük) have changed considerably over time. From 1995–2003, Theya Molleson and Peter Andrews were the joint heads of the Human Remains Laboratory. In 2004, Clark Spencer Larsen and Simon Hillson followed. In 2011 there was another major changeover with Simon Hillson and other long-term team members departing. In 2012, Christopher Knüsel joined Clark Spencer Larsen as co-director of the lab, along with a number of additional researchers, some of whom are co-authors of this chapter.

Over the course of 25 years, the focus of the human remains team has been directed primarily towards fundamental bio-cultural questions of health, diet, physical activity and biological kinship – these research avenues produced a plethora of data that contributed to understanding of early settled life in the Near East. Until quite recently, however, a systematic approach to the study of funerary practices at Çatalhöyük has been over-shadowed, as the focus tended to prioritise the biological aspects of the human skeleton over more site-specific questions of socio-cultural behaviour. Previous discussions of funerary practices at Çatalhöyük were founded on patterns observed early on in the Hodder excavations, at a time when the number of burials was relatively small and diachronically restricted (Hamilton 1998a; 2005; Andrews et al. 2005; Moses 2012). Much of this early work focused on re-evaluating Mellaart's claims regarding aspects of funerary practice at Çatalhöyük; for example, vulture excarnation, secondary burial and age/sex segregation of burials within houses (Hamilton 2005: 301). Subsequent work, based on larger sample sizes, remained largely descriptive in nature, while continuing to employ a similar interpretative framework (Boz, Hager 2013; 2014). Other recent studies have focused on specific examples of funerary treatment (e.g., Patton, Hager 2014; Haddow et al. 2016a; 2019; Pilloud et al. 2016a; Haddow, Knüsel 2017), or have approached the subject from a particular philosophical viewpoint (Palaver 2019).

In 2013, with the development of a specifically funerary archaeological component of the laboratory, attention shifted to emphasise archaeothanatological (that is, field-based) observations of the disposition of skeletal remains and archaeological scientific analyses of the grave context to better understand funerary taphonomic aspects of the treatment and deposition of the once fleshed body (see Knüsel, Robb 2016) in what has recently been termed the '*chaîne opératoire funéraire*' (Valentin et al. 2014; 2016; Sellier 2016). This shift in focus occasioned the introduction of a revised field recording system for human remains based on adherence to standard anatomical terminology (Knüsel 2014a) and recording of the state of skeletal articulations and positioning of skeletal remains, articulated skeletons and partially articulated as well as disarticulated skeletal elements, to better understand the state of the corpse upon deposition. Increased communication and collaboration with the other specialist labs aided the development of fuller integration with the overall archaeological context of the site to better address project research questions, as well as refinement of the specific links between osteological indicators such as age, sex, pathological conditions (see Chapter 16), and grave inclusions (see Chapter 17), including floral and molluscan accompaniments (see Volume 14, Chapter 14), and items of personal adornment, including colourants and residues (see Volume 14, Chapter 16 and Volume 15, Chapter 16). This type of collaborative work underlines an ambitious undertaking for a site of this size with excavation extending over 25 years. This has sometimes resulted in unevenness in the reporting of results, but provides a more detailed understanding of '*gestes funéraires*', funerary gestures/actions as part of multi-stage and variable-length funerary ritual sequences.

Demographic considerations. While adolescents (aged 12–20 years) are technically subadults from the perspective of skeletal growth and development, for questions of cultural practices such as funerary treatment the remains of these individuals are included with those of adults in this chapter, as they were likely to have been considered full members of the community, having achieved sexual maturity and attendant adult social and labour responsibilities (Crawford 1991; 1993; Gowland 2006; Sofaer 2011; Fibiger 2013). The complete demographic profile for the Çatalhöyük primary skeletal assemblage is presented in Chapter 16, along with a discussion of the methods employed to assess age and sex.

Burial deposition categories

In order to assist in the interpretation of the funerary practices at Çatalhöyük, each burial excavated by the Hodder team has been assigned to one of six deposition categories that reflect the variable contexts from which skeletal remains have been recovered (table 15.1). **Primary** burials, defined as the inhumation of an articulated corpse, are split between three depositional subcategories: (1) *primary undisturbed*: articulated skeletons found in their original interment location that have not been disturbed by subsequent Neolithic activity; (2) *primary disturbed*: the in situ remains of articulated skeletons found in their original location that have been disturbed by subsequent Neolithic activity, and (3) *primary disturbed loose*: disarticulated or partially articulated skeletal elements found in grave fills that derive from disturbed primary burials. **Secondary** burials are defined as a partial or complete skeleton, typically in a disarticulated or partially disarticulated state, which has been intentionally moved by Neolithic people from a previous location to a subsequent interment location. The **tertiary** deposition category is reserved for isolated, disarticulated, or partially articulated skeletal elements found outside of typical burial contexts (that is, underneath house floors and platforms). Such non-interment contexts can include middens and other external spaces, as well as house construction and abandonment layers. Lastly, human remains in the **unknown** deposition category are typically recovered from heavily eroded surface layers or have been displaced by animal burrowing, such that their original deposition cannot be determined with certainty; in addition, highly commingled remains whose deposition cannot be clearly distinguished (for example, secondary versus primary disturbed loose) are also assigned to the unknown deposition category.

The stratified skeletal assemblage

The Hodder-era excavations at Çatalhöyük between 1995 and 2017 recovered at least 816 individuals from stratified Neolithic contexts (while the number of individuals from primary contexts represents the total number of individuals, the number for the secondary and tertiary contexts represents a minimum number of individuals (MNI)). Individuals from the primary disturbed loose category are excluded from the total in order to avoid the potential for counting individuals twice. Figure 15.1 presents the number of stratified individuals by deposition category. In this chart, individuals from unknown deposition contexts (n=75) are excluded. As such, the total number of individuals studied in this chapter is reduced from 816 to 741.

Fully articulated skeletons, along with disarticulated and partially articulated skeletal elements, were commonly observed under house platforms and floors during both the Mellaart and Hodder excavation projects (Haddow et al. 2016a). However, the interpretation of this commingling has evolved since Mellaart's day.

1. Primary undisturbed	An undisturbed, articulated skeleton, complete or nearly complete, found in its original place of interment
2. Primary disturbed	The in situ skeletal remains of a primary articulated skeleton found in its original burial location but subsequently disturbed/truncated by Neolithic people
3. Primary disturbed loose	Loose, scattered, disarticulated skeletal elements found in grave fills and deriving from a disturbed primary burial
4. Secondary	Partial or complete skeleton moved by Neolithic people from a previous location and deposited in a final interment context
5. Tertiary	Isolated, disarticulated or partially articulated skeletal elements found in non-interment contexts (for example, room fill, midden)
6. Unknown	Unable to determine original deposition of bones

Table 15.1. Description of burial deposition categories employed by the Çatalhöyük Research Project.

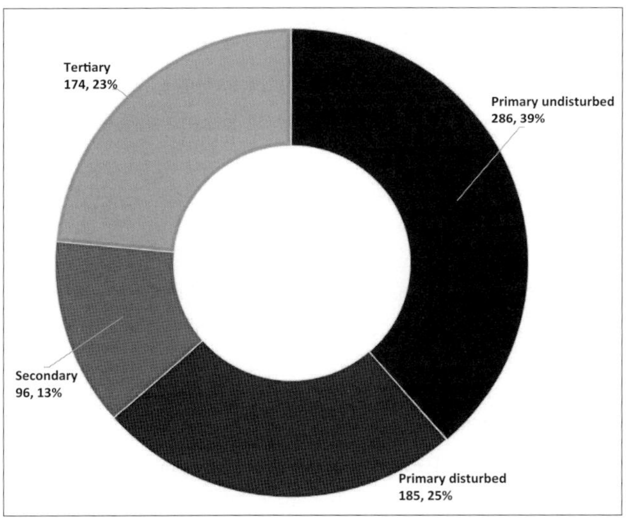

Figure 15.1. Distribution of individuals by burial context at Neolithic Çatalhöyük. Secondary and tertiary context individuals are represented by MNI (individuals from primary disturbed loose and unknown depositions are excluded).

Mellaart interpreted the presence of disarticulated and commingled remains as evidence for secondary burial practices involving the exposure and excarnation of bodies prior to intramural burial: 'These burial habits … strongly suggest that secondary interment was practised involving the reburial of parts of the skeletons after complete or partial decomposition of the flesh' (Mellaart 1962: 51–52). Given the extreme flexion of some of the articulated skeletons, Mellaart also believed that these, too, represented secondary burials: 'Evidently care was taken to preserve the skeleton intact in anatomical position...' (Mellaart 1967: 204).

In contrast to Mellaart's view, the predominant narrative of the Hodder excavations, until quite recently, has been that the majority of burials are primary depositions, rather than secondary in nature (Andrews et al. 2005; Andrews, Bello 2006). In this reading, the observed commingling of skeletal remains is interpreted as mainly the result of sequential disturbances of earlier primary interments by later ones within house platforms (Boz, Hager 2014).

> [T]he majority of depositions are primary single interments. Burials containing multiple individuals interred in a single event rarely occur. Multiple event interments where primary burials are subsequently disturbed by new interments are common and result in the many primary disturbed skeletons at Çatalhöyük (Boz, Hager 2013: 415).

Recent archaeological observations, in combination with newly excavated material from the earliest and latest occupation levels, have prompted a re-evaluation of the funerary data at Çatalhöyük. Based on these analyses it now appears that funerary practices at the site were far more complex and variable than previously considered, likely involving extended, multi-stage funerary treatments for at least some members of the Neolithic population. Such findings are especially relevant to understanding the ways in which Neolithic societies were structured and changed through time. The evidence and interpretation of such diverse funerary practices at Çatalhöyük are presented in detail below via a series of case studies, followed by a discussion of how these practices varied both spatially and temporally.

Bodies of evidence: recent insights and new interpretations

The starting point for a re-evaluation of funerary practices at Çatalhöyük can be traced to a number of recent archaeological findings, including the excavation of a series of burials uncovered from buildings and spaces in the North Area of the site. In light of these findings, a number of patterns have emerged that provide fresh insights into many earlier observations that did not fit previous interpretative models. Furthermore, the addition of burial data from the earliest and latest levels of the site reveals patterns of funerary behaviour that were much less consistent throughout the occupation of the site than previously thought.

Case study 1: Building 129

Between 2012 and 2015, several primary and primary disturbed sub-floor burials in Building 129 (Level North H) were found in association with loose crania and large amounts of disarticulated and partially articulated infracranial remains (Knüsel et al. 2012; Haddow, Knüsel 2017). The skeletal remains, representing a minimum of 20 individuals from seven distinct burial events, were located in Space 77 (fig. 15.2), the eastern platform area of the central room (Tung 2014). The varied dispositions of the human remains recovered here offer a panoramic view of the range of variation in funerary practices encountered elsewhere at Çatalhöyük.

Southernmost burial sequence.

Feature 3643. Feature 3643 represents the primary disturbed burial of a child (19451) aged 8 years (+/−2 years) at death (fig. 15.3). This burial is the earliest in the southern sequence of interments. Only the cranium and articulated mandible and the cervical and superior-most thoracic vertebrae and ribs were in articulation; other skeletal elements were found scattered throughout the fills of subsequent burials F.3645 and F.3686. Given the labile nature of the cervical vertebrae found in situ (Duday et al. 1991; Duday 2009), this individual was probably disturbed early in the decomposition process and its remains dislodged and scattered by subsequent burials in the same location. Feature 3643 was partially truncated at its western end by the later grave cut for Feature 3645.

The incomplete and disarticulated remains of a neonate (19578) were found in the grave fill of F.3643. The bones of this individual may derive from an earlier burial located in this area and subsequently disturbed by later interments. However, because this individual is represented solely by cranial elements, as well as the right scapula and humerus (no other neonatal skeletal elements were recovered from any of the subsequent grave fills in this sequence), it may represent a secondary deposit of skeletal remains made at the time skeleton (19451) was interred. No artefacts were found in association with this burial.

Feature 3645. Feature 3645 contained the primary disturbed burial of a middle adult male (20457). Only the feet, *ossa coxae*, forearms and hands were found articulated and in situ (fig. 15.4). The rest of the skeleton appears to have been disinterred when the grave cut for the subsequent interment of skeleton (20430) (F.3686; see below) was dug. Isolated skeletal elements found in the fill of F.3686 and also the fill of the later cut, F.3639, likely belong to skeleton (20457). The skeleton appears to have been placed in a flexed position on its left side with the head oriented to the west. No artefacts were found in association with this burial.

Feature 3686. Directly above F.3645 lay F.3686, which contained the primary disturbed burial of a middle adult male (20430) (fig. 15.5). The body was placed on its back (supinely but leaning slightly toward its left side) in a tightly flexed position with the head oriented to the west. No artefacts were found in association with this burial. The cranium, mandible, atlas and axis were not recovered, but the presence of the hyoid bone and ossified thyroid cartilage in anatomical position indicates that the body must still have been intact at the time the missing bones were removed. Furthermore, a lack of cut marks on the remaining cervical vertebrae and no apparent displacement of the infracranial skeleton suggest that the ligaments and other soft tissue structures holding these elements in place were decayed sufficiently to allow the removal of the skull (cranium and mandible) and first two cervical vertebrae with little effort. Based on this evidence, the missing skeletal elements of skeleton (20430) appear to have been intentionally removed at some point after the body was initially interred, likely via a retrieval pit (see below).

Feature 3639: skull retrieval pit. Feature 3639 (fig. 15.6) consists of a circular pit lined with the disarticulated and poorly preserved skeletal remains of at least two individuals represented by long bones, additional infracranial elements, and two crania, (19450) and (19493). Skeleton (19450) is a poorly preserved adult cranium (possibly male) found on the northwest edge of the cut. The first cervical (atlas) vertebra was found under the cranial base, but there were no cut marks present, which would seem to indicate that these elements were moved after initial decomposition of the articulation between the atlas and axis. Skeleton (19493) is represented by a poorly preserved cranium of an adult of indeterminate sex. It was found isolated on the southern edge of the cut. No artefacts were recovered from this feature.

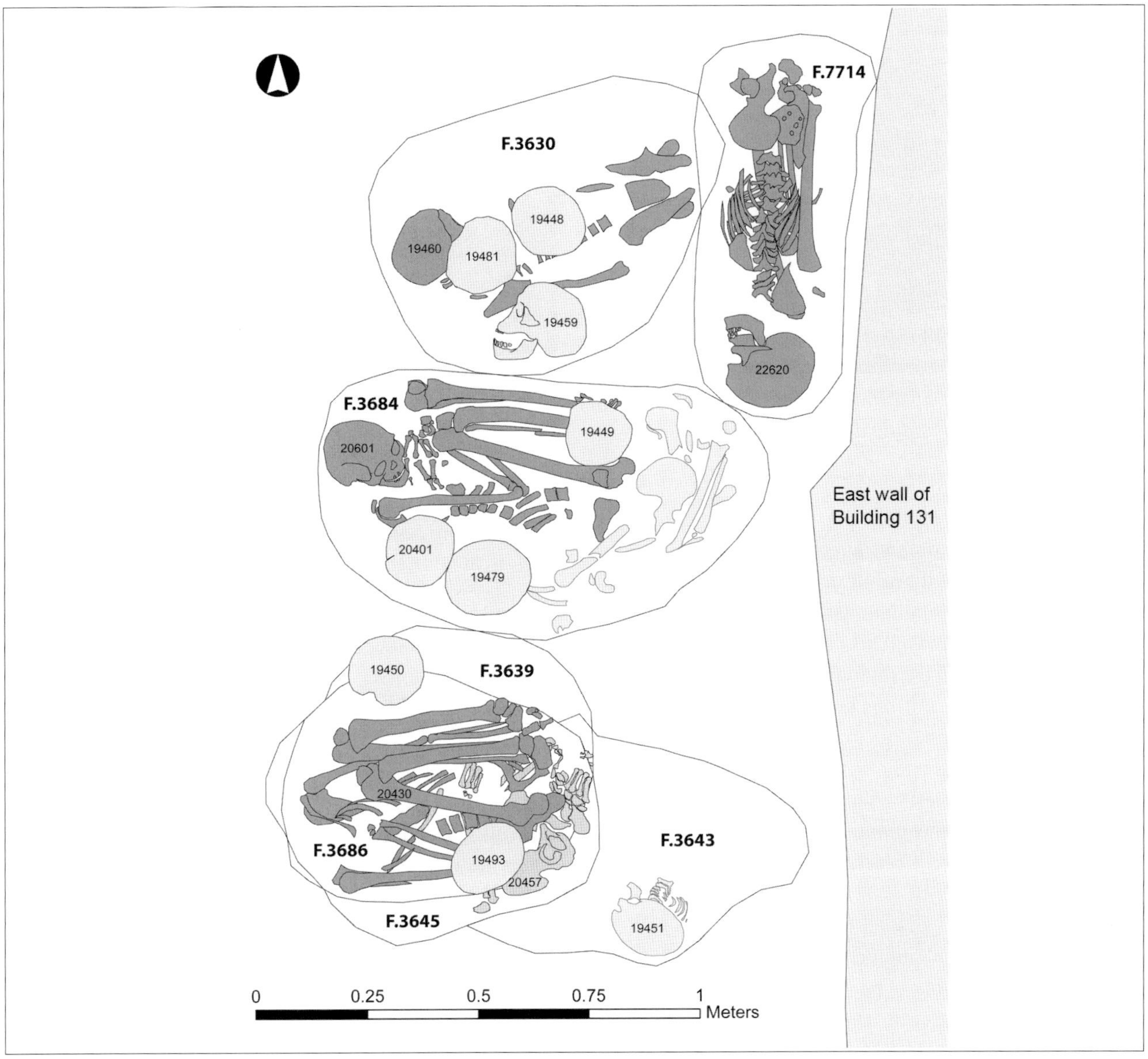

Figure 15.2. Plan of Space 77, Building 129, showing locations of individual burial features.

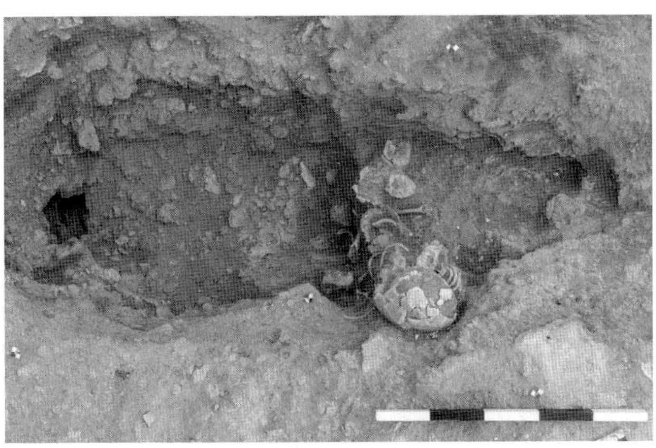

Figure 15.3. Feature 3643: primary disturbed burial of a child (8 years +/−2 years of age at death) (19451).

Figure 15.4. Feature 3645: primary disturbed burial of a middle adult male (20457).

Figure 15.5. Feature 3686, which contained the primary disturbed burial of a middle adult male (20430).

Figure 15.6. Feature 3639: circular pit lined with the disarticulated and poorly preserved skeletal remains of at least two individuals represented by two crania, (19450) and (19493), long bones and additional infracranial elements.

Structure from Motion 3D modelling techniques enabled the identification of a retrieval pit located above the primary burial of the adult male, (20430), that facilitated removal of the cranium and mandible (Berggren et al. 2015; Haddow, Knüsel 2017). The cut for this pit can be seen clearly above the region of the head of the underlying primary interment of skeleton (20430) in F.3686 and appears to represent a 'skull retrieval pit' that was dug in order to access and remove the skull of this earlier burial (fig. 15.7, left). The disarticulated skeletal elements found in the grave fill of F.3639 (fig. 15.7, middle) were carefully placed within the cut after the retrieval of the skull of (20430) (fig. 15.7, right). This type of retrieval pit has long been hypothesised to have been the means of obtaining crania at many Near Eastern Neolithic sites, but this is one of the few times, and the first at Çatalhöyük, that it has been confirmed archaeologically (see Goring-Morris, Horwitz 2007 for another case), due to difficulties in identifying smaller secondary pits stratigraphically.

Central burial sequence.

Feature 3684. Burial F.3684 was found immediately to the south of F.3630 and contained the primary interment of a robust, middle adult male (20601), along with the disarticulated remains of at least two additional individuals (fig. 15.8). Skeleton (20601) was tightly flexed on its left side and oriented east to west, with the head to the west. The head was flexed on the upper torso facing east. Both lower limbs were tightly flexed against the torso, with the knees close to the chin. The right upper limb lay on top of the thorax and was flexed at the elbow at roughly 45°. The right wrist was palmar flexed under the chin. The left forearm of this individual was found between the right and left lower limbs, an observation consistent with extreme flexion and close-packing of the body within the grave cut. Several stone beads (20492.x1-x9, and x11-15), as well as a single copper bead (20492.x10), were recovered from the grave fill.

The incomplete, disarticulated remains of at least two additional individuals were found in the grave fill of burial F.3684, placed above the primary skeleton. The first individual, (19479), is represented by a cranium belonging to an older adult of indeterminate sex. It was found on the southeast edge of the grave cut, above the right shoulder of skeleton (20601), facing south and oriented north to south, with its vertex uppermost. This cranium lay next to another isolated cranium (20401) located immediately to the east and belonging to a young adult female. It was found in the southwest part of the grave cut to the west of and facing (19479). This cranium was oriented east to west resting on its cranial base. A large amount of disarticulated adult infracranial remains from at least two individuals were also recovered from the grave fill of burial F.3684, the majority of which were recovered from the eastern end of the grave cut. Due to poor preservation, it is unclear if the two loose crania were associated with the disarticulated infracranial remains. As with burial F.3630, the grave cut for skeleton (20601) does not appear to have disturbed any earlier burials in this location. Thus, the disarticulated bones found in the grave fill are more likely to represent a secondary deposition of skeletal remains brought from elsewhere to accompany the primary interment.

Northernmost burial sequence.

Feature 3630. Feature 3630 consists of the primary interment of a child (19460) aged 11 years (+/−2.5 years) at death (fig. 15.9). The tightly flexed body of this individual was found prone and oriented east to west,

Figure 15.7. False colour 3D models showing stages of the excavation of F.339: (left) upper burial layer showing circular fill of skull retrieval pit (dotted circle) above a primary flexed adult burial; (middle) middle burial layer after removal of soil from skull retrieval pit, revealing disarticulated skeletal elements placed in cut for skull retrieval pit; parts of the underlying primary flexed adult skeleton are now visible; (right) lower burial level after full exposure of the primary flexed adult 'headless' skeleton.

Figure 15.8. Feature 3684: primary interment of a middle adult male (20601), along with the disarticulated remains of at least two additional individuals.

Figure 15.9. Feature 3630: primary interment of a child (19460) aged 11 years (+/−2.5 years) at death. Two obsidian 'mirrors' (19447.x3, 19447.x4) were also recovered from the fill units of this burial.

with the head towards the west and facing north. The lower limbs were strongly flexed at the hip and knee, so strongly that the tibiae and fibulae were found beneath the femora bilaterally. The right upper limb ran parallel to the torso and flexed at the elbow such that the distal end of the forearm and hand were beneath the torso. The left upper limb was in a similar position on the left side. Numerous stone (19447.x1, x7, x8, 20450.x1-5) and shell (19447.x2, x6) beads were found in the grave fill and around the ankles (19447.x2), chest (19447.x1), neck and head region (19447.x6) of this individual. Bone beads (19460.F1-3, 20450.F1-13) were also recovered from the fill (see Chapter 17), as well as an obsidian blade (20450.x9). Two obsidian 'mirrors' (19447.x3, 19447.x4), one placed near the head of (19460), were also recovered from the fill units of this burial, although it is unclear whether they are directly associated with the primary burial.

The incomplete remains of three additional individuals were found in the upper grave fill of burial F.3630. The first individual consists of a poorly preserved cranium belonging to an adult of 20+ years of age, (19448), found directly above the thoracic region of skeleton (19460). Sex could not be determined due to the poor state of preservation. The second individual consists of an articulated adult cranium and mandible belonging to a possible male (19459). It was found lying on its right side facing west, and oriented north to south, vertex to the north. It was found in the upper grave fill above the left shoulder of skeleton (19460). The third individual, (19481) (not shown on plan), represents an agglomeration of disarticulated subadult skeletal remains found in the upper grave fill above the neck region of skeleton (19460). The incomplete nature of the remains precludes a more precise age estimate. Additional bones recovered from the fill include an adult articulated left forearm and hand, as well as disarticulated vertebrae, ribs and other loose skeletal elements.

Because the grave cut for skeleton (19460) does not appear to have disturbed any earlier burials in this location, the remains of the three additional individuals found in the fill of F.3630 may represent a secondary deposition of skeletal remains brought from another location to be deposited above skeleton (19460). Given the variable bony constituents and states of articulation of these secondary remains, it would seem that they were in different stages of decomposition at the time of their re-deposition.

Feature 7714. Excavated in 2015, F.7714 contained the poorly preserved primary disturbed remains of an adult (22620) along with disturbed disarticulated bones from at least one additional adult (22655) recovered from the grave fill (fig. 15.10). This feature was located immediately to the east of F.3630, excavated in 2012. It appears that the two individuals ((22620) and (22655)) were placed in the grave cut (22626) at the same time, as some elements from skeleton (22655) rest directly on skeleton (22620). Due to the extremely friable and poorly preserved state of the bones, a more precise age estimate could not be obtained for either individual. Although the *ossa coxae* of skeleton (22620) were poorly preserved, preliminary examination of the sciatic notch in the field indicated that the individual may possibly have been female.

Skeleton (22620) was lying in a prone, tightly flexed position with the abdomen resting directly over its tightly flexed lower limbs. Nearly all of the skeletal elements of skeleton (22620) were present, except for some elements of the right upper limb and the left forearm, which may have been disturbed by F.3630. The cervical vertebrae were completely absent, suggesting that the skull had been intentionally placed in a position approximating anatomical articulation. Overall, the preservation and the extremely hyperflexed and partially articulated positioning of the remains give the impression of a tightly bundled package of bones. A corroded piece of beaten native copper (22623.x4) (perhaps a ring) was found under the right fingers. Rows of carnelian beads were found around both wrists (22623.x1, x2) and stone and shell beads were found around the ankles (22623.x3, x5). Over 200 stone, shell and clay beads were also recovered from the grave fill, some in the neck region of skeleton (22620).

Figure 15.10. Feature 7714: poorly preserved primary disturbed remains of an adult (22620), along with disturbed disarticulated bones from at least one additional adult (22655) recovered from the grave fill.

Discussion. The final stage in a series of funerary acts carried out in the southernmost area of Sp.77 entailed the deliberate targeting of the skull of the primary interment in F.3686. Unlike some cases of missing crania in primary burials observed at Çatalhöyük, the disturbance to F.3686 did not result from the cutting of a grave for a subsequent burial, but rather from an excavated pit expressly targeting the region of the head of skeleton (20430). After the cranium and mandible and superiormost cervical vertebrae had been removed, partially and completely articulated remains of at least three individuals (including cranial and infracranial remains) were placed in the pit before it was filled in.

It has been postulated previously at Çatalhöyük that skeletal elements removed from a burial context are often redeposited in the same location at a later date, often at the end of a house occupation or platform use-life (Haddow et al. 2016a). It is possible that one of the crania and some of the infracranial remains originate from the first burial in this sequence, that of skeleton (20457) in F.3645; much of the skeleton of this individual was removed during the interment of skeleton (20430). Due to the poor preservation of the skeletal elements lining the retrieval pit, it is impossible to confirm this scenario. However, even if two of the seven isolated crania/skulls found in B.129 derive from the primary disturbed skeletons ((20457) and (20430)) found there, five additional crania/skulls remain for which infracranial skeletons have not been accounted.

Case study 2: Building 52

Between 2013 and 2014, two distinctive skeletal assemblages were uncovered within adjacent platforms in B.52 (Level North G) (Knüsel et al. 2013; Haddow et al. 2014; 2016a; 2017) (see Volume 12, Chapter 21 for additional detail). The two assemblages exhibit various degrees of commingling and represent the outcome of divergent funerary practices: one characterised by long-term, successive inhumations and the other by a single interment episode consisting of multiple individuals – a rare occurrence at the site. Each of these skeletal assemblages can be linked to the occupational history – or life course – of B.52, which culminated in a dramatic closure event characterised by an intentional fire that burned most intensely in the central area (Sp.94) of the building (Twiss et al. 2008; Harrison et al. 2013; Tringham 2013). By analysing the manner in which the two sets of commingled remains were deposited, it is clear that each assemblage reflects distinctive funerary behaviours that correspond to particular ways of entangling individuals (or parts of individuals) within the structural modifications of B.52 at different occupation stages.

The northeast platform. The skeletal remains of at least five individuals (three adults and two subadults), representing at least three separate burial events, were recovered from the northeast platform (Haddow et al. 2016a; Knüsel et al. 2013). These three interments occur at distinct phases in the use-life of the northeast platform, during which the height of the platform was successively raised. The uppermost (last) burial in the sequence (F.7112) is that of a middle adult female (20655) placed in a tightly flexed supine position and leaning slightly to her left with the head to the west and feet to the east (fig. 15.11). Phytolith bands running across the ankles, right proximal femur, and left ilium suggest that the body was tightly bound with reed cordage when it was placed in the grave. Despite being the final and most shallow interment within this platform, this individual appears to have been unaffected by the fire that consumed B.52. No heat-related colour changes were apparent on the bones, and no traces of carbonised soft tissue were found within the endocranium or anywhere else on the skeleton (as has been found in other sub-floor burials within burnt houses at Çatalhöyük), suggesting that this individual was buried long before the fire took place and with sufficient time for full decomposition of the soft tissues of the body. A disarticulated cranium and mandible (20661) belonging to a possible young adult male were also found within the grave fill of this burial.

Located immediately below burial F.7112 was the primary inhumation (F.7120) of a middle adult male (30522) and an infant (30523) (fig. 15.12). This burial was cut into an earlier surface of the platform. Both of these individuals appear to have been buried in a single event. The adult was placed on his back in a tightly flexed position and oriented with the head to the west and the feet to the east. The infant, tightly flexed on its right side, was placed directly above the torso of the adult. While the remains of the lower-lying adult male were undisturbed, the remains of the infant were

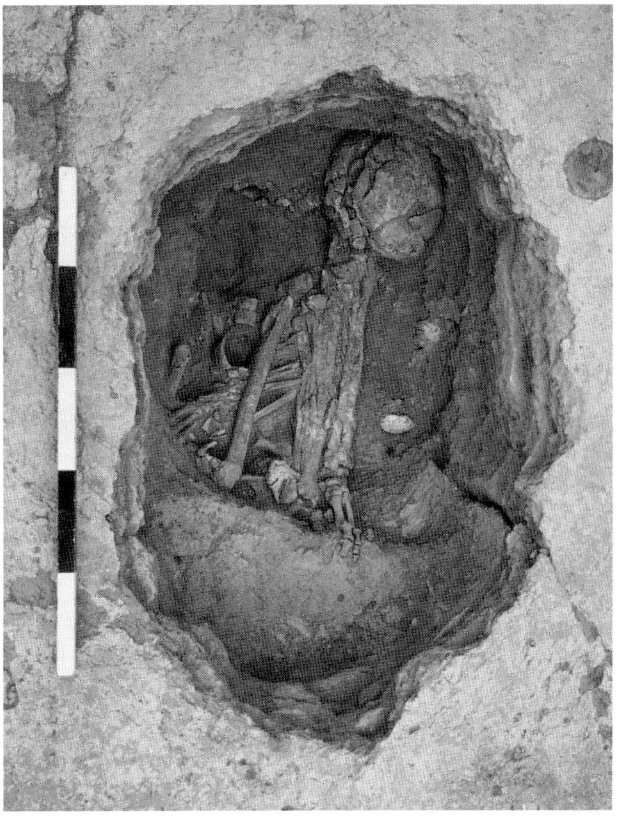

Figure 15.11. Building 52, F.7112: latest burial in the northeast platform, contained a middle adult female (20655) placed in a tightly flexed supine position.

disturbed and partially disarticulated by the grave cut for burial F.7112 described above. Three rows of small stone disc beads in various colours were found on the abdomen of the infant, along with an additional string of similar beads around the left ankle. In addition, traces of red pigment were observed on the frontal bone and two green stone beads (possibly serpentinite) were also found, one on either side of the temporal area of the cranium. The disarticulated remains of a fifth individual were found within the lower grave fill of the grave cut. This individual is represented by the cranium and mandible of a child (30521). A set of disarticulated long bones, including left and right tibiae and femora, as well as a right humerus and fibula were also recovered from the fill. Based on their size, these likely belonged to the same individual skeleton (30521).

Burial F.7606 (fig. 15.13), located immediately below burial F.7120, represents the earliest interment in the northeast platform burial sequence (Haddow et al. 2014; 2016a; Tung 2014); in fact, it may predate the construction of this platform, for which it may have been a foundation deposit, and was cut into an older platform belonging to an earlier phase of B.52. The grave cut

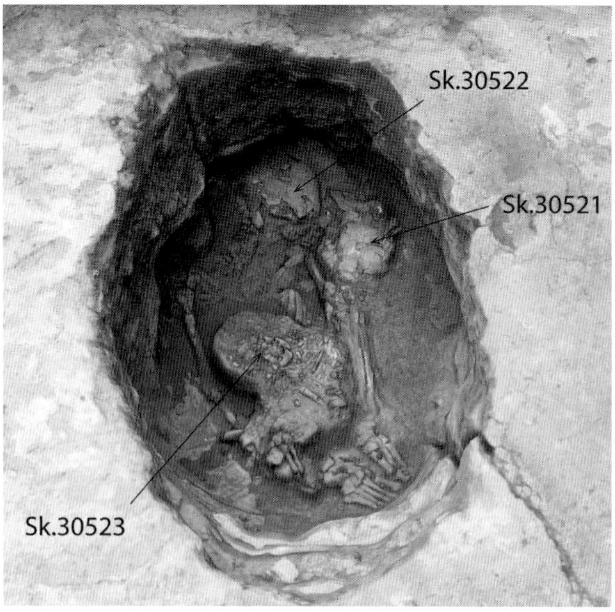

Figure 15.12. Building 52, F.7120: primary inhumation of a middle adult male (30522) and an infant (30523) in northeast platform. Disarticulated remains of a child found within the grave fill were also recovered and matched with remains from earlier burial F.7606.

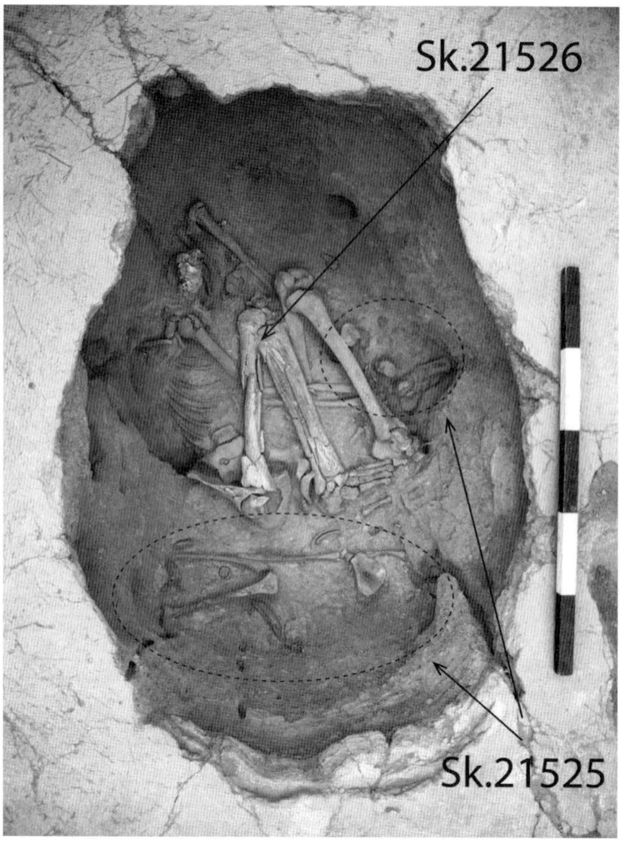

Figure 15.13. Building 52, F.7606: primary disturbed skeleton of a young adult possible female (21526) and the disarticulated infracranial remains of a child (21525) found scattered throughout the grave fill. These remains were matched with disarticulated remains recovered from the grave fill of F.7120. This burial represents the earliest interment in the northeast platform burial sequence.

contained the primary disturbed skeleton of a young adult possible female (21526) and the disarticulated infracranial remains of a child (21525) found scattered throughout the grave fill. The young adult (21526) was placed on its left side in a flexed position, with the head oriented to the west and the feet to the east. The cranium and mandible of this individual were missing, although all of the cervical vertebrae were recovered. It appears that the cranium and mandible were removed during the subsequent interment of the middle adult male (30522) and infant (30523) in F.7120.

Based on developmental similarities (that is, dental development and diaphyseal lengths), the disarticulated cranium and mandible of the child (30521) and disarticulated infracranial remains found in the grave fill of F.7120 were matched in the lab with the subadult remains found loose in the grave fill of earlier burial F.7606. It is also likely that the loose adult cranium and mandible (20661) found in the grave fill of the latest burial F.7112 may belong to skeleton (21526) from F.7606, as several loose teeth found in the grave fill of F.7606 fit within the maxillary tooth sockets of the cranium (20661). The occurrence of loose subadult remains from a single individual within two separate burial features has two potential explanations: (1) the skeletal remains of the child ((30521)=(21525)) represent an earlier primary burial predating this platform and belonging to the preceding phase, which was then completely disturbed by the subsequent burial of the young adult possible female (21526). In this scenario, the original grave cut was completely obliterated by the grave cut for skeleton (21526) and the bones of the child were re-deposited with the grave fill of the young adult male; or (2) the disarticulated subadult remains represent a secondary deposit placed in the grave F.7606 at the same time as the primary burial of the young adult possible female (21526). The latter scenario appears more likely, however, as there is no trace of an earlier grave cut for the child, nor were any of its bones found in a primary in situ position, as is the case with the many disturbed burials at Çatalhöyük (Boz, Hager 2013; 2014). In both scenarios it appears that the grave cut for the subsequent interment event, F.7120, dislodged the cranium and mandible of the primary young adult possible female (21526), along with a large amount of the disarticulated child skeleton, including the cranium and mandible. The bones of the child were then re-deposited in the grave fill of F.7120 – with special care taken to place the cranium and mandible alongside that of the middle adult male (30522). Meanwhile, the young adult possible female cranium and mandible (20661) likely belonging to skeleton (21526) appear to have been retained above ground before being reburied with the final interment (F.7112) in the platform. Intriguingly, this individual has a healed blunt force injury on the occipital region of the cranial vault. It seems this cranium and mandible once belonged to an individual associated with violence (Volume 15, Chapter 15); could this have been a reason for its removal and retention?

The northwest platform. The northwest platform of B.52 contained three separate burial features (F.8508, F.8510 and F.7127). A fourth burial (F.8516) recovered at a much deeper level than the other inhumations may predate construction of the platform. Feature 8508 contained the undisturbed primary burial of an approximately 34-week (in utero) prenate (23469), while F.8510 contained the primary disturbed remains of an infant (23477). Feature 8510 was truncated by the last burial in the northwest platform, F.7127.

Feature 7127 contained the primary undisturbed remains of a middle adult male (30514) and at least six subadults ((30510), (30511), (30512), (30513), (30514) and (30524)) in various states of articulation. Disarticulated subadult bones were also found scattered in the uppermost levels of the grave fill. With the exception of a few elements likely deriving from earlier subadult burial F.8510, it appears that all of these individuals were interred in a single burial event. The adult male was the first individual to be placed in the grave cut (fig. 15.14), in a supine and flexed position with the head to the west and feet to the east, and the subadults were placed on top of his body. Carbonised brain tissue was recovered from within the cranial vault of the adult male, and the bones ranged in colour from yellowish-orange to greyish-brown as a result of heat transference through the platform during the fire that consumed B.52. Two large red-painted mollusc shells (30503.x6) were placed beside the right knee of the adult; the internal shell surfaces contained a brownish organic material. A small, shiny, flat piece of metallic mineral (30503.k1) (likely hematite), roughly 20mm by 15mm and perforated at one end, was recovered from the grave fill just below the cranium of skeleton (30514) and may have been worn as a pendant.

The first subadult to be placed in the grave cut with the adult male was a child (30524) aged 4 years (+/−1 year) at death (see fig. 15.14). The body, placed on the left humerus of the adult male, appears to have been at least partially decomposed at the time of burial, as the skeleton was not completely articulated and certain elements, such as those of the right upper limb and the left *os coxae*, were missing. The next subadult to be interred was a child (30513) aged

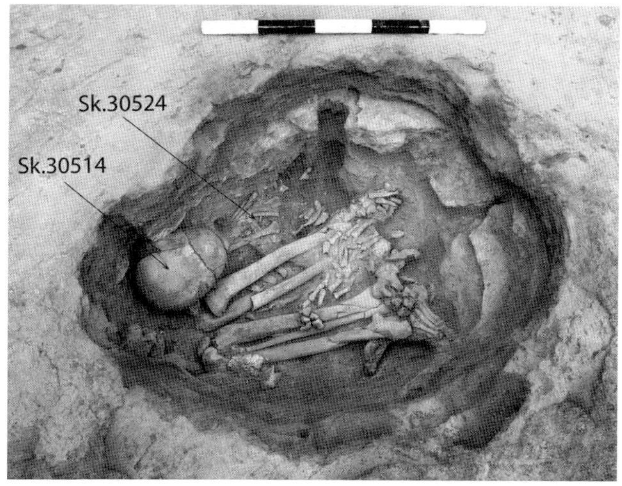

Figure 15.14. Building 52, F.7127: adult male (30514) and partial remains of child skeleton (30524) above the left humerus of the adult.

3 years (+/−1 year) at death (fig. 15.15). The occipital of this child was in direct contact with the frontal bone of the adult male, and a relatively well-preserved circular wooden object (30513.x1), possibly a bowl, was placed on top of the cranium. This individual was missing the left upper limb, as well as the left tibia and fibula. As with the adult male, carbonised brain tissue was recovered from the endocranium of this child, and the bones exhibited heat-related colour changes. An infant (30511) aged 6 months (+/−3 months) at death (fig. 15.16) was placed on top of this child along the northern wall of the grave cut, with a large amount of well-preserved linen textile separating the two bodies (see textile chapter (Volume 14, Chapter 11) for details). Carbonised brain tissue was recovered from the endocranium, and additional carbonised soft tissue was found in the abdominal region. The bones from the upper body of this infant were orange-brown in colour, while the bones of the lower body were blackened and partially calcined, likely due to the proximity of these remains to the platform surface. Unlike the other subadults, this individual was fully articulated and largely complete.

The partially complete remains of a child (30510) (see fig. 15.16) were recovered just to the south of the infant and child previously described. Only the axial skeleton of this individual was in articulation, and the bones of the upper and lower limbs were dispersed in the upper grave fill. Given this partially articulated state, it would appear that this child was in an advanced stage of decomposition at the time of burial. Traces of textile cordage were found running from the top of the cranium to the underside of the mandible, possibly as a means of binding these elements together when they were in a fleshed state. Only a small amount of carbonised brain tissue was recovered from the endocranium.

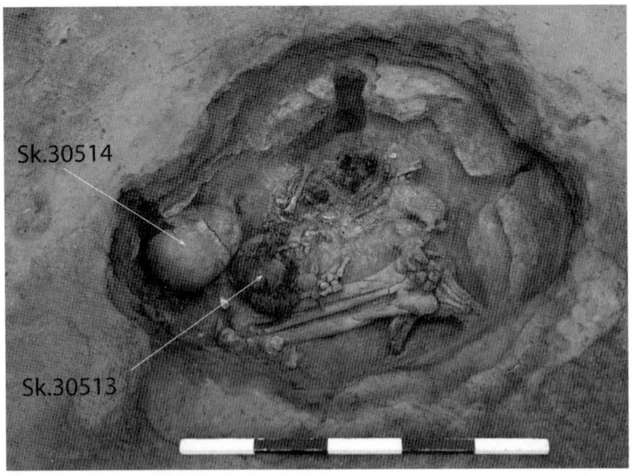

Figure 15.15. The next subadult to be interred within F.7127 was a child (30513) aged 3 years (+/–1 year) at death placed above the torso of the adult skeleton (30514). A well-preserved circular wooden object (30513.x1) was placed on top of the child's cranium.

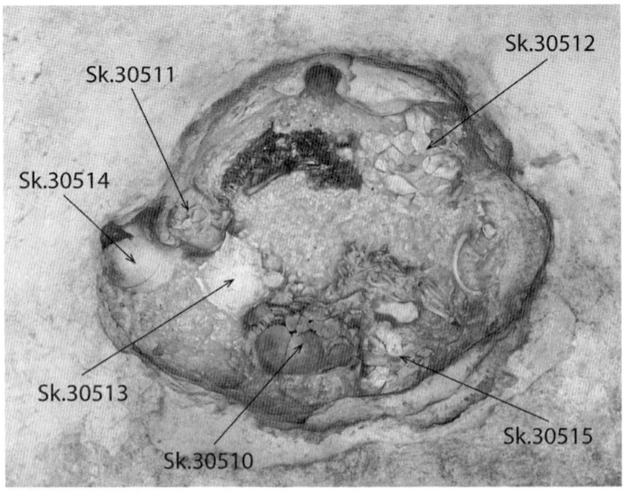

Figure 15.16. Building 52, F.7127: location of skeletons (30510), (30511), (30512) and (30515) in upper grave fill.

An isolated cranium of a child (30512) (see fig. 15.16) was recovered in the northeast corner of the grave cut. Unlike the others in this burial, the cranium of this child was crushed flat and shattered, with all single-rooted teeth missing post-mortem. Without soft tissue to support it, the cranial vault collapsed under the weight of the overlying grave fill, and only a very small amount of carbonised brain tissue was recovered. This suggests that the cranium was completely skeletonised when it was deposited and may have been retrieved from a primary interment in another location before being re-deposited here. A partial cranium and mandible of a 5-year-old (+/–16 months) child (30515) (see fig. 15.16) were also recovered from the fill against the south wall of the grave cut. No brain tissue was recovered from this individual. Once the adult male and partially articulated subadults were placed in the grave, a layer of loose subadult bones was deposited above them before the grave cut was sealed.

While defining the bottom of the grave cut after the human remains had been removed, a complete marble 'bracelet' (30503.x8) (see this volume, Chapter 17 and Volume 14, Chapter 10) was found close to a rodent burrow along the north wall of the grave cut. It was not directly associated with any particular skeleton, but may have been moved from its original location by rodent activity. However, while bioturbation of skeletal elements due to rodent activity often occurs at Çatalhöyük, there is no indication that burrowing is responsible for the disarticulation and commingling of the subadult remains in burial F.7127.

In attempting to associate the loose, disarticulated bones recovered from the grave fill with the partially articulated subadults described previously, it became apparent that there may be at least two additional subadults interred within this burial feature; these are represented by two sets of right and left femora and several bones of the upper limb whose shaft length measurements provide age estimates that are at least two years younger than any of the partially articulated skeletons. If these elements represent two additional individuals, this would bring the total of subadults interred with the adult male to eight, with only one being fully articulated (and thus having only recently died) when they were buried in the platform. The others appear to have died much earlier and partially decomposed elsewhere, having either been kept aside or retrieved from another burial (or burials) before being interred with the adult male. The incomplete skeletal representation of many of the subadults and the recovery of little or no carbonised brain tissue from the crania of these partially disarticulated individuals, in comparison to the adult male and articulated subadults, strengthens the interpretation that these individuals represent secondary depositions.

Case study 3: External Space 602

During the 2015 excavation season a series of intriguing inhumations was uncovered within the upper infill layers of abandoned B.132 (external Sp.602) (Haddow et al. 2015). At the time these burials took place, only the outer walls of B.132 remained and its successor, B.77, had not yet been built. Two adult females (F.7863, skeleton (21685); F.7634, skeleton (21672)), one adult male (F.7632, skeleton (21636)), an adolescent (F.7864, skeleton (21691)) and an infant (F.7633, skeleton (21676)) were recovered. In contrast to the majority of grave cuts at Çatalhöyük (typically ovoid and just large enough to accommodate the body), the cuts for the Sp.602 burials were very large and circular in shape.

Except for F.7634, located centrally within the space, their placement within the infill of B.132 anticipated the layout of the northern and eastern platforms of subsequent B.77, which was constructed within the footprint of the earlier B.132.

While no artefacts were found in association with the adolescent and infant, the adults were buried with an unusually large and diverse number of grave goods/inclusions, including – variously – beads, obsidian and chipped stone blades, projectiles and scrapers, as well as a number of worked animal bone tools and pigment, some of which appears to have been applied directly to the bone (see this volume, Chapter 17).

Although these burials maintained a semblance of anatomical articulation, the bones themselves were highly fragmented and evinced multiple dry fractures. Additionally, many skeletal elements were out of place or missing entirely. In F.7632 for example, the left forearm of the young adult male (21636) was detached at the elbow and found tucked under the right humerus (fig. 15.17); this must have occurred before the body was placed in the grave. In addition, the cranium was crushed flat, the base of the sacrum was rotated inferiorly and many of the thoracic vertebrae were no longer in articulation. The skeleton was very tightly flexed and traces of phytoliths on the bones indicate that the body had been tightly bound (Haddow et al. 2015).

The Sp.602 burials are among the very few primary inhumations found at Çatalhöyük that occurred in what would have been, at the time, an external space. The majority of burials at Çatalhöyük are typically associated with specific houses. These intramural burials occur most often during the occupation phase, but also in smaller proportions during the construction and abandonment phases (see discussion of spatiotemporal variation below for further detail). Other examples of burials in external spaces at Çatalhöyük include a series of neonate and infant burials in Sp.144, an external area that preceded the B.65-B.56-B.44 house sequence (Regan 2014: 143) in the South Area, and the burial of an adult male in a midden (Sp.115) (Farid 2007c; Milella et al. 2016).

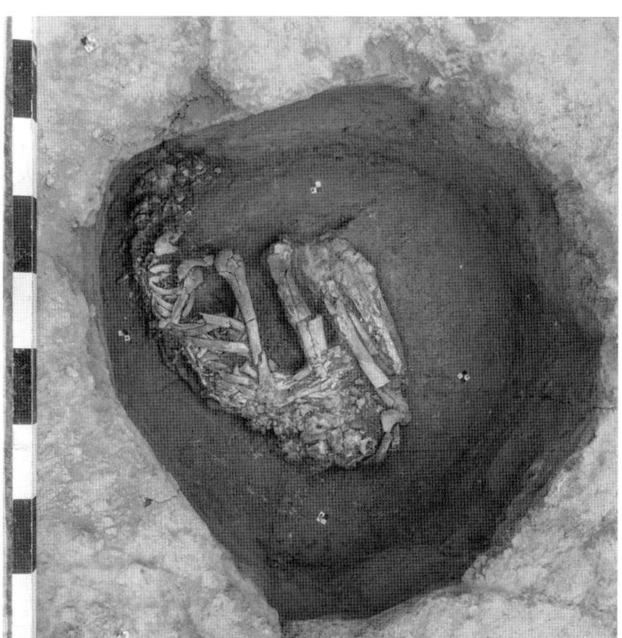

Figure 15.17. Burial F.7632, skeleton (21636) in Space 602.

Preservation of organic materials

While James Mellaart uncovered a great deal of organic material, including textiles, cord and wood from burnt houses at Çatalhöyük (Mellaart 1967), it was not until 2009 that the Hodder-led team fully excavated such buildings. Since then, five completely burnt houses have been excavated in their entirety, and each of these contained sub-platform burials with exceptional preservation of organic material, including wooden bowls, animal hide, cord, textiles, reed matting and basketry and, in some cases, carbonised human soft tissue (see Volume 14, Chapter 16). In unburnt burial contexts, matting (in the case of adults) and lidded baskets (in the case of perinates, neonates and young infants) are attested via the presence of phytoliths (Rosen 2005; Boz, Hager 2013). Brown residues observed in some burials may suggest the presence of degraded animal hide, perhaps as clothing items or used to wrap the body. In other cases, smaller amounts may indicate their use as pouches (Boz, Hager 2013: 423). The occurrence of these 'packaging' materials in burials are discussed in more detail in Chapter 17, while textiles and animal hides are specifically analysed in Volume 14, Chapter 16.

With the exception of phytolith impressions and brown residues, such findings are rare in unburnt contexts and furnish a glimpse of what were likely customary burial accompaniments at Çatalhöyük. The use of such materials suggests that great care was taken to tightly wrap and package the corpse, and this may have been done in order to retain the anatomical integrity of the body for a period of time before burial.

Putting the pieces together: delayed burial?

These recent observations have prompted a re-examination of previously excavated primary burials for similar anomalies. On closer inspection many primary skeletons appear to show signs of an extended interval between death and final interment. Some skeletons are very tightly flexed; such positions would likely be difficult to achieve if the body were fully fleshed at the time of interment. This suggests the removal or reduction of fleshy mass prior to burial. In addition, persistent joints such as the lumbar vertebrae, the knee, the ankle and the sacro-iliac joint, as well as the humero-ulnar joint (as described in burial F.7632 above), are often disarticulated. Some bones are often out of place or missing entirely (for example, the basicranium, small bones of hands and feet). There are also differences in preservation between burials located in the same building and, at times, within the same platform. Lastly, the presence of packing and wrapping materials is associated with many of these tightly flexed inhumations.

Such observations suggest a number of potential behaviours, in particular the practice of delayed burial, in which the bodies of at least some members of the Neolithic population were retained as part of a multi-stage funerary ritual that culminated in an intramural interment only after a variously extended interval above ground, possibly seasonal in nature but perhaps much longer. During this interval, soft tissue mass appears to have been reduced or removed without substantially compromising the anatomical integrity of the skeleton. If delayed burial had been practised at Neolithic Çatalhöyük, a number of questions arise: firstly, with regard to the manner in which corpses had been processed, and secondly, if there were spatio-temporal patterns underlying these and other secondary funerary practices. A number of potential means of removing or reducing soft tissue mass were investigated, including the removal of flesh using tools or via exposure of the corpse to raptor scavenging, as well as rudimentary forms of natural mummification. The answers to these questions, especially regarding temporal changes, have important implications for understanding changes in social structure and organisation through time and the emergence of inequality in the Neolithic and subsequent periods.

Body treatment

Exposure and excarnation (anthropogenic versus animal). The removal of flesh from a corpse as part of a delayed burial regime is a well-documented phenomenon across various geographic regions and time periods (e.g., Robb et al. 2015). The manner in which this can be achieved, however, is variable.

Moore and Molleson (2000) proposed that secondary burial involving exposure and excarnation was the dominant funerary treatment at early Pre-Pottery Neolithic (early ninth to the late eighth millennium B.P.) Abu Hureyra in the Middle Euphrates Valley in Syria. This interpretation is based on several archaeological observations: (1) the occurrence of single or collective burials with incomplete skeletons (Ibid., 278); (2) elements placed tightly together as if secured in a bag or by matting (Ibid., 284); (3) the presence of 'headless' skeletons (Ibid., 279), and (4) disarticulated crania found singly or in groups (Ibid., 280). Evidence for the manual removal of flesh at Abu Hureyra is minimal and derives from a single humeral diaphysis fragment bearing multiple cutmarks (Ibid., 287). In another building at the same site, human remains were placed on the floor prior to the eventual interment of some of them beneath the floors of structures.

Anthropogenic excarnation. Manual excarnation, or the removal of flesh from a corpse using tools of some kind has been observed in Neolithic contexts (see, for example, Erdal 2015; Robb et al. 2015) and may have been used at Çatalhöyük as a first stage in preparing a corpse for an extended period above ground. Verification of manual excarnation is typically made via the identification of linear markings on the diaphyseal surface of bones, either with the naked eye or microscopically (Fernández-Jalvo, Andrews 2016).

In Anatolia, clear cutmarks associated with flesh removal have been observed on a small percentage of human bones at PPNA Körtik Tepe (3.6% of 281 individuals) (Erdal 2015: 8); if such practices occurred at Çatalhöyük, they should also be observable. However, while cutmarks associated with bodily disarticulation (typically found in the vicinity of articular or peri-articular bone surfaces of joints) have been detected in a small number of cases (four individuals (2%) from primary disturbed contexts, one individual (0.6%) from a tertiary context), particularly pertaining to the separation of the head from the neck, there is an apparent lack of cutmarks associated with flesh removal. The overall frequency of cutmarks in the Çatalhöyük faunal assemblage is also low, however, and has been attributed to the use of extremely sharp obsidian tools, which lessens the need for deep cuts that would potentially leave marks on bone (Russell et al. 2013: 241–42).

Animal excarnation. One of the more intriguing finds of Mellaart's excavations was a series of wall paintings depicting vultures attacking what look like headless bodies (Mellaart 1967: 166). Mellaart interpreted these scenes as depicting the process of excarnation prior to burial; as such, they were a major influence on his reconstruction of the funerary practices at Çatalhöyük. During the current excavations, however, the idea of vulture excarnation playing a role in the funerary practices at Çatalhöyük was largely abandoned (Andrews et al. 2005: 265). While the occurrence of secondary burials continued to be recognised at Çatalhöyük, these were seen as rare in comparison to primary burials. As such, the narrative surrounding funerary practices moved towards a more simplified model of death followed by immediate burial of a fully fleshed corpse (Boz, Hager 2013). However, based on recent forensic taphonomic studies of raptor scavenging (Domínguez-Solera, Domínguez-Rodrigo 2011; Spradley et al. 2012), Pilloud et al. (2016) revisited the 'vulture hypothesis' at Çatalhöyük. Raptorial birds are capable of stripping flesh from a corpse while leaving the body fully articulated, and they may do so while leaving very few, if any, marks on bones. They may also remove and scatter articulated bones. While vulture excarnation at Çatalhöyük cannot be established definitively in the absence of observable taphonomic evidence, the pattern of skeletal flexion and part representation observed among certain burials is not inconsistent with more recent observations of raptor scavenging.

Mummification/desiccation. Another means of preparing and preserving a corpse for an extended period above ground is by mummification or desiccation. Such techniques need not be as sophisticated as those practised by the Ancient Egyptians. At the Bronze Age site at Cladh Hallan in the Western Isles of Scotland, for example, radiocarbon dating of the skeletal remains of two individuals revealed their remains to be several hundred years older than the building in which they were interred (Parker Pearson et al. 2005). One skeleton was revealed to be a composite of three different individuals, and histological analyses of the bones showed that the decomposition process had been arrested soon after death via some form of mummification (Parker Pearson et al. 2005). A subsequent study has shown that rudimentary mummification practices were likely more widespread in the British Bronze Age (Booth et al. 2015) than previously considered. The Cladh Hallan 'mummies', apparently kept in circulation for well over a hundred years, represent an extreme case of delayed burial, but the results of such a process could explain the condition and appearance of some Çatalhöyük skeletons that show signs of an interval between death and burial; for example, the burials from Sp.602 described above.

Possible methods for desiccating a corpse at Çatalhöyük may have included air-drying, smoking and other desiccating surface treatments such as the application of salts or ochre to the skin, or more advanced techniques such as the removal of internal organs to help reduce putrefaction. Actualistic studies using fresh corpses are currently being conducted to better understand these treatments (Schotsmans et al. 2017; see also Mickleburgh, Wescott 2018). Combinations of these postmortem treatments are cross-cultural in nature and are documented *inter alia* for ancient Egypt (Aufderheide 2003; Papageorgopoulou et al. 2015), modern-day Papua New Guinea (Beckett et al. 2011; Beckett, Nelson 2015) and medieval Europe (e.g., Weiss-Krejci 2005; Knüsel et al. 2010). Other methods for preserving bodies may be indicated by a plaster burial found in B.131, which is similar to those from PPNA Körtik Tepe (Erdal 2015), and the burial of a juvenile from Abu Hureyra – especially regarding the presence of matting/textile in both cases (Moore, Molleson 2000: 284).

Use of colourants. The remains of some individuals at Çatalhöyük were treated with pigments. The most common pigment found in burials is red ochre (haematite – Fe_2O_3), although the manner in which it was applied is not always clear. Sometimes ochre was found associated with reed matting, in other cases the ochre appears to have been sprinkled over the body, or applied directly to the body prior to interment. The occurrence of pigments in burials is presented in further detail in Chapter 17 and detailed in Schotsmans et al. Volume 15, Chapter 16.

Another red colourant, cinnabar (vermilion – HgS), was uniquely found on crania of a small number of individuals, as well as in primary, primary disturbed and secondary burial depositions and often in co-occurrence with other grave goods/inclusions. Red ochre and cinnabar were also present in shells found in burials (and other contexts, see Volume 14, Chapter 14 and Schotsmans et al. forthcoming). A very small number of blue and green colourants in the form of lumps or loose powders, such as azurite ($Cu_3(CO_3)_2(OH)_2$ and malachite ($Cu_2CO_3(OH)_2$) were also recovered in burials (See this volume, chapter 17 and Schotsmans et al. Volume 15, Chapter 16).

Analytical approaches for the study of delayed burial and initial results

Histotaphonomy. In the absence of preserved soft tissue, cutmarks or other surface modifications of bone, microscopic analyses of bone can be used to characterise taphonomic changes associated with the burial of fleshed and unfleshed bodies (Bell 2012). Wedl tunnelling, for example, is a form of microbial bone alteration typically associated with exogenous fungal attack (Machiafava et al. 1974; Hackett 1981). Furthermore, some researchers believe that endogenous gut bacteria released during the early stages of decomposition are responsible for a particular pattern of bone microstructure alteration (Jans et al. 2004; White, Booth 2014; Booth 2016; although see for example Turner-Walker 2019 for counter-arguments). As such, reduced levels of bacterial degradation of bone may indicate an intentional intervention by Neolithic people to reduce or halt the decomposition process, either through the removal (that is, excarnation) or reduction (that is, desiccation) of soft tissue body mass. Other histotaphonomic indicators that have the potential to identify particular types of body treatment, as well as burial environments, include collagen preservation (birefringence), microfissures, staining and the inclusion of foreign materials in bone pores (Hollund et al. 2012; Brönnimann et al. 2018).

During the 2016 and 2017 excavation seasons, sections of femoral cortical bone from 52 Neolithic individuals were obtained for analysis at Koç University in Istanbul and Middle Eastern Technical University (METU) in Ankara. Five additional individuals from post-Chalcolithic contexts at Çatalhöyük were also sampled and analysed as control specimens. Thin sections ranging between 40 and 125 micrometres in thickness were produced for each sample and analysed using standard light microscopy and backscatter scanning electron microscopic (BSEM) imaging techniques. Each thin section was scored using the Oxford Histological Index (OHI), which grades the preservation of bone microstructure on an ordinal scale from 0 (microstructure completely obliterated) to 5 (microstructure fully intact) (Hedges et al. 1995; Millard 2001). Levels of bacterial alteration, Wedl tunnelling, collagen birefringence, microfissures and thermal alteration were also recorded, using the indices compiled by Brönnimann and co-workers (2018: 48, table 2). Staining of bone tissue and foreign inclusions were also recorded.

While the cortical bone thin sections varied considerably in terms of Wedl Type 2 tunnelling, collagen birefringence, microfissures and other histotaphonomic indicators, only one Neolithic sample from Çatalhöyük appeared to show evidence of bacterial bioerosion under light microscopy: a middle adult female (11659) from B.44 in the South Area whose cranium and mandible appears to have been intentionally removed. However, subsequent BSEM analysis revealed that the observed bioerosive features lack the characteristic redeposition of hyper-mineralised bone typically associated with bacterial tunnelling. Instead, these features may represent Wedl Type I tunnelling and may imply that the body was exposed (or re-exposed) for a period above ground. While Wedl Type I tunnels are present in other samples from Çatalhöyük, this is the only individual in which the microstructural alterations affect more than 50% of the bone area. Based on the apparent absence of bacteria-linked bioerosion, it appears that the natural process of decomposition was arrested in all of the study samples; the explanation for why this occurred, however, is not entirely clear. Further work is required in order to differentiate potential natural phenomena from intentional human interventions. This histotaphonomic study has now been expanded to include other Neolithic sites in Anatolia, including Boncuklu Höyük, Barcın Höyük and Aşıklı Höyük. Although this comparative research remains ongoing, preliminary analyses of the Boncuklu and Barcın material indicate much greater variability in terms of microbial alteration levels than observed in the Çatalhöyük material. In addition, young adults of both sexes appear to have the lowest rates of bacterial alteration at Boncuklu and Barcın.

Spatio-temporal trends

Burial deposition

There are substantial differences in the proportion of burial deposition categories between the three main excavation areas on the Neolithic East Mound at Çatalhöyük: North, South and TP/TPC/GDN. While the proportion of individuals from primary burials is nearly identical between the North and South Areas (68% and 66%, respectively), primary burials in the TP/TPC/GDN Areas make up only 37% of the total number. Individuals from secondary contexts account for 13% of the proportion of burials in the North Area, and 3% of the proportion in the South Area. In the TP/TPC/GDN Areas, however, secondary burials account for 43% of the total number of individuals. Much – but not all – of the variation between excavation areas is due primarily to the occupation periods under excavation in each of these areas. For example, the South Area has the deepest stratigraphy, with material from the Early, Middle and Late occupation periods; the North Area contains only material from the Middle and Late periods, while the TP/TPC/GDN Areas have focused entirely on the Late and Final periods. The spatial locations of skeletal remains in secondary contexts vary substantially across the site. In fact, two North Area buildings, B.131 and B.129, account for 40% of the total number of individuals found in secondary contexts. Building 131 contained the remains of 24 individuals (including 17 crania) in association with three separate primary burials, while Building 129 contained the remains of 13 individuals (including nine crania) in association with three separate primary burials. Building 129 was built on the footprint of the earlier Building 131 after it was abandoned and intentionally burnt. These two buildings are linked in other ways, including the very rare occurrence of obsidian mirrors in both buildings in association with human remains.

Plotting changes in the proportion of burial deposition categories over time at Çatalhöyük (fig. 15.18) reveals that primary burials dominate the first three occupation periods at the site. Within the primary burial category, however, the proportion of undisturbed primary burials decreases substantially through time as disturbances to primary burials increase. These disturbances are usually the result of earlier burials being truncated by subsequent burials in the same house platform. As the proportion of primary burials decreases over time, the proportion of secondary deposits of human remains begins to rise. From the Early occupation period, where no secondary burials have been found, the proportion increases steadily over time, eventually overtaking primary burials as the most common type of burial context in the Final occupation period. In this period, sub-floor burials during the occupation of houses no longer took place. Instead, it appears that rooms or spaces were created within abandoned buildings for the accumulation of loose and partially articulated skeletal remains (see, for example, discussion of Spaces 248 and 327 in Marciniak, Czerniak 2007; Marciniak et al. 2015). Lastly, the proportion of human remains found in tertiary contexts (that is, middens, construction/abandonment infills) shows a slight increase over time, although it remains relatively stable in comparison with the primary and secondary deposition categories.

The primary burial assemblage. Individuals recovered from primary burial contexts, both disturbed and undisturbed (n=471, 64%), account for the majority of the Neolithic skeletal assemblage at Çatalhöyük. Multiple individuals are commonly recovered from beneath house platforms and floors; in the past, such assemblages have been misleadingly referred to as 'multiple burials', but they are more accurately defined as collective burials, as defined by Duday (2008) and applied by Crevecoeur et al. (2015), for example. Knüsel and Robb (2016: 657) define collective depositions as 'human remains deposited successively over time rather than in a single episode, often inferred from variations in completeness and articulation among the remains.' Of 378 primary burial features excavated, only 34 (9%) contained more than one primary individual (this count does not include primary burials that contain additional human remains in secondary depositions (see below)). In terms of demography, these multiple primary interments include four burials containing two adults each, two burials containing three adults and one burial containing two adults and an adolescent; there are 13 burials containing an adult or adolescent with one or more subadults and 14 burials containing only subadult groupings. Of the adults and adolescents interred with subadults, there are seven

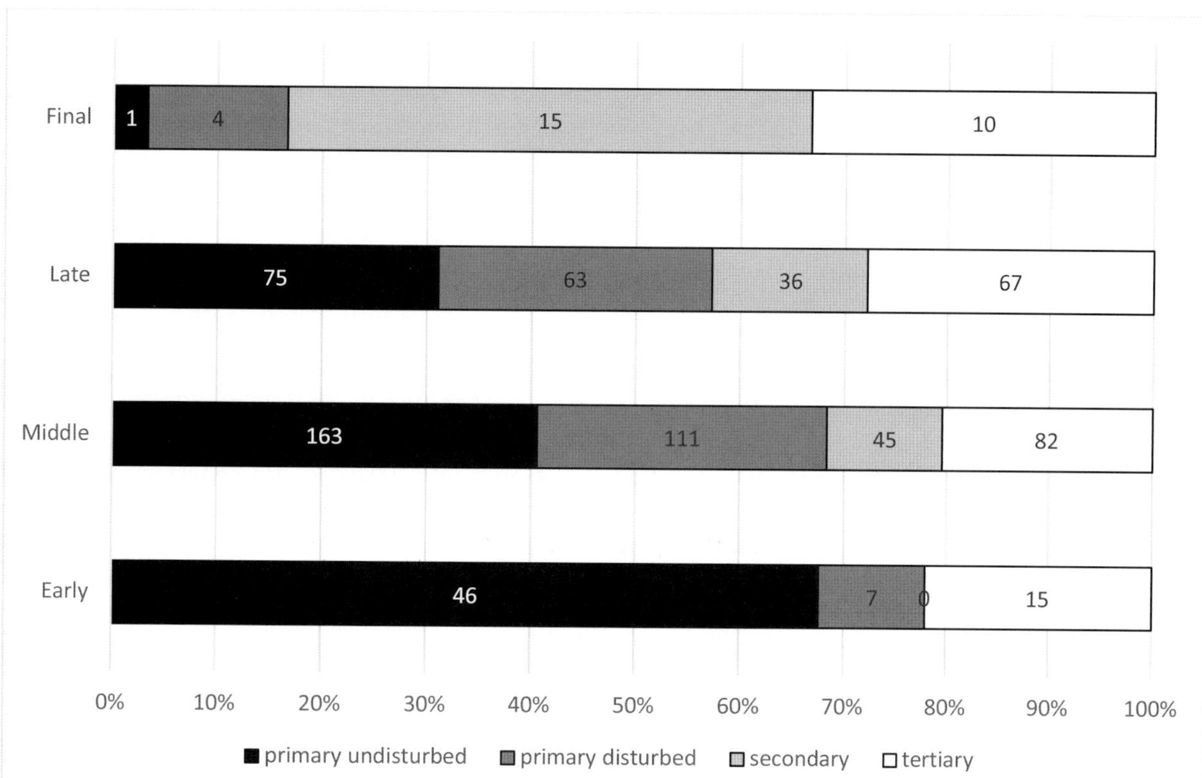

Figure 15.18. Temporal distribution of burial deposition categories through major occupation periods at Çatalhöyük.

females, four males and two of indeterminate sex. Two of the females appear to have died in childbirth. See Chapter 16 for a full discussion of the primary burial demographic profile.

In terms of spatiotemporal variation, burials containing multiple primary individuals occur throughout the occupation of the site and in all excavation areas. However, a group of Late period buildings and external spaces in the North Area (Buildings 54, 57, 58, 88 and Space 279) account for a substantial proportion of the multiple subadult primary individuals recovered. In addition to three single primary inhumations and a triple adult burial, B.54 contained one burial containing four neonates (F.2037) and another containing one neonate and one infant (F.2159); B.57 contained a burial containing three primary neonates (F.1585); in addition to three single primary inhumations, B.58 contained a triple neonate burial (F.2020) and a double neonate burial (F.2028); B.88 contained a double neonate burial (F.2559) in addition to four single primary inhumations; Sp.279 contained a double neonate burial (F.2031) (along with an additional single primary burial of a neonate).

The secondary deposition skeletal assemblage. While primary burials dominate the Çatalhöyük skeletal assemblage, it is the rarer secondary burials (n=96, 13%) that are often the most intriguing. One of the most obvious examples of secondary burial treatments at Çatalhöyük can be seen in the retention and redeposition of elements of the skull (that is, cranium and mandible) (see below for further detail). However, secondary burial practices also involve the infracranial skeleton.

Secondary deposits of human remains at Çatalhöyük typically occur in association with primary inhumations. Out of 26 recorded burial features containing secondary skeletal remains, only eight were recorded in isolation (that is, unassociated with a primary burial). Some of these burial features contained a single primary skeleton associated with the secondary remains of multiple individuals; for example, three separate burial features in B.131 (F.7961, F.7962 and F.7963) contained a total of three primary skeletons and the remains of at least 25 individuals in secondary contexts (Haddow et al. 2015, 2016b; 2017). The nature of the relationship(s) between these primary individuals and their accompanying secondary remains is currently unclear, although it is hoped that the ancient DNA analyses currently being conducted (this volume, Chapter 18) will help shed light on this question. In some ways, the secondary remains associated with primary burials at Çatalhöyük could be interpreted as another category of grave inclusion, albeit one with important implications for the social organisation of the site.

In terms of age structure, the secondary burial assemblage does not follow the same pattern as the primary assemblage. While the primary burial assemblage has a higher subadult (prenates to children) to adolescent/adult ratio (264: 207), a pattern that is not inconsistent with pre-industrial mortality profiles (Hillson et al. 2013: 358; Knüsel et al. this volume, Chapter 16), subadults occur much less frequently (36:60) in the secondary burial assemblage; this pattern suggests an age-based selection bias. There are no secondary burials in the Early period, but during the Middle period the subadult to adolescent/adult ratio for the secondary assemblage is nearest to being equal (20:25), while the Late and Final periods skew higher towards adolescents/adults (11:25 and 5:10, respectively) – the equivalent ratios for the primary assemblage are: Early (31:22), Middle (147:127), Late (83:56) and Final (3:2). As with the secondary burial assemblage, the Middle period primary assemblage has the most balanced subadult to adolescent/adult ratio, while the Late and Final periods skew higher towards subadults. The sex distribution of the secondary assemblage is roughly evenly split between males and females. There are nine males vs seven females in the Middle period, four males vs six females in the Late period and one male vs two females in the Final period. Thus it would appear that sex did not play a role in selection for secondary burial treatment.

Intriguingly, the increase in the proportion of individuals found in secondary contexts mirrors a trend observed across multiple data sets that shows an increase in the mobility of the Çatalhöyük population over time as part of an ever-widening use of the landscape and its resources (e.g., Pearson et al. 2007; Charles et al. 2014; Larsen et al. 2015). Interpreted from this perspective, these secondary remains may represent individuals who died at some distance from the site and were not immediately returned for burial. As such, this temporal trend may be better characterised as a practical rather than cosmological shift in behaviour, although both may be implicated. As resource-acquisition regimes, availability and access tend to fluctuate seasonally, the trend is also more consistent with the practice of pre-scheduled burial rites rather than a standardised liminal period. However, increased mobility does not sufficiently explain the rapid decrease in the proportion of primary burials between the Late and Final occupation periods. It is likely that there are multiple behaviours and intentions in operation here.

The tertiary deposition skeletal assemblage. As described earlier in this chapter, the tertiary deposition category comprises isolated, disarticulated or partially articulated skeletal remains found outside of typical burial locations such as beneath building platforms and floors. These contexts include middens and other external spaces, as

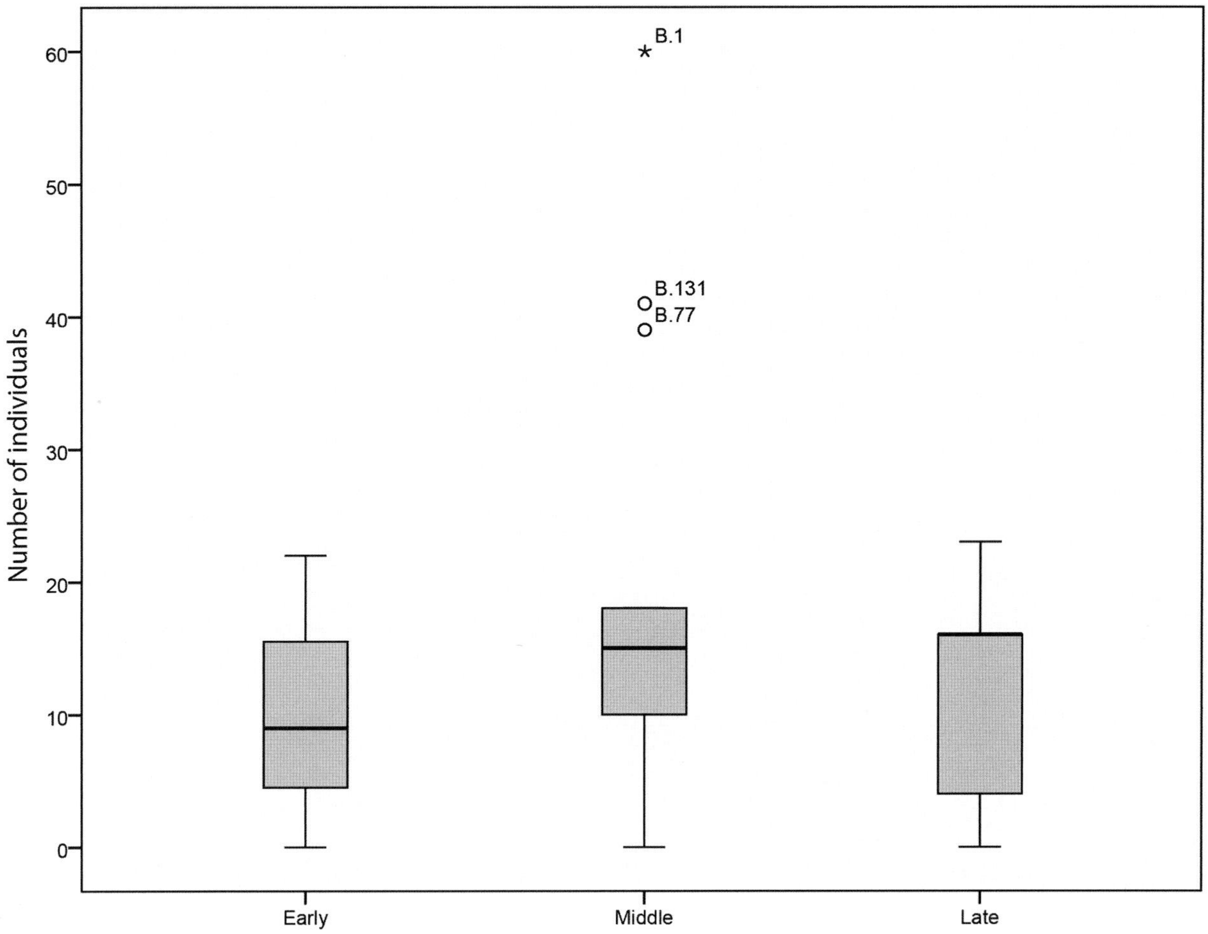

Figure 15.22. Box plots showing variation in number of individuals per building in the Early, Middle and Late periods.

floors in the north and clean floors in the south. This architectural inversion appears to have affected the placement of burials as well, with burials along the western and southern walls occurring in higher frequencies than in later periods. In addition, burials in side rooms are less frequent than in later periods.

During the Middle period, burials in the northern and eastern areas of the main room became increasingly common (fig. 15.25). Burials under floors became much less frequent than in the Early period, while burials in the southern area of the main room remained roughly the same. Lastly, side room burials increased slightly during this period.

In Late period houses, side room burials occurred in substantially higher proportions than in previous periods, while burials in the southern area of the main room and under main room floors occurred much less often (fig. 15.26). Over the course of the occupation at Çatalhöyük the east platform gradually became the preferred location for burials; in the Late period it received nearly half of all primary and secondary inhumations, while other locations within houses were used less frequently.

Age and sex breakdown by location. Figures 15.27 and 15.28 plot the variation in location by age categories for primary and secondary burials, respectively. While there are no areas of houses that are used exclusively for one age group as opposed to another, considerable spatial variation does occur: adult (including adolescents) primary burials occur more frequently than subadult primary burials in the east, northeast and west platforms, as well as under central floors, whereas subadults predominate within side rooms, north and northwest platforms, or the south area of the central room (fig. 15.27). Secondary deposits of adult individuals occur more often than those of subadults in north and east platforms, and exclusively in the south and west areas of the central room. Secondary deposits of subadult individuals occur most often in side rooms and northwest platforms. Secondary deposits of human skeletal remains are never found in the southern area of the central room or under central room floors (fig. 15.28). Previous studies have clearly demonstrated that the location of burials within houses was not sex-segregated (Hamilton 2005; Boz, Hager 2013).

Area	*Building*	*Level*	*Period*	*MNI*	*% excavated*	*MNI per period*	*Total houses per period*	*Total houses with full burial set*	*Avg MNI per house*
South	118	South H	Early	0	25	**56**	**13**	**3**	**10.3**
South	18	South J	Early	1	25				
South	162	South J	Early	0	50				
South	23	South J	Early	2	50				
South	16	South K	Early	0	25				
South	160	South K	Early	4	50				
South	*17*	*South K*	*Early*	*22*	*75*				
South	*2*	*South K*	*Early*	*0*	*75*				
South	161	South K?	Early	8	50				
South	*43*	*South L*	*Early*	*9*	*75*				
South	130	South L	Early	0	25				
		South L	Early	0	50				
		South L	Early	10	25				
North	5	North F	Middle	5	50	**350**	**33**	**14**	**19.0**
North	139	North F	Middle	0	25				
North	119	North F	Middle	0	75[1]				
North	*132*	*North F*	*Middle*	*10*	*75*				
North	48	North G	Middle	1	25				
North	*59*	*North G*	*Middle*	*1*	*100*				
North	66	North G	Middle	0	25				
North	167	North G	Middle	4	<25				
North	*3*	*North G*	*Middle*	*8*	*100*				
North	*49*	*North G*	*Middle*	*15*	*100*				
North	*114*	*North G*	*Middle*	*15*	*75*				
North	*51*	*North G*	*Middle*	*0*	*100*				
North	*52*	*North G*	*Middle*	*18*	*85*				
North	*77*	*North G*	*Middle*	*39*	*100*				
North	*131*	*North G*	*Middle*	*41*	*100*				
North	*1*	*North G*	*Middle*	*60*	*100*				
North	82	North G	Middle	0	50				
North	112	North G?	Middle	0	50				
North	101	North G?	Middle	2	<25				
North	128	North G?	Middle	12	50				
North	102	North G?	Middle	36	50				
South	24	South M	Middle	0	25				
South	40	South M	Middle	0	<25				
South	100	South M?	Middle	1	25				
South	91	South M?	Middle	4	25				
South	92	South M?	Middle	5	25				
South	*50*	*South M?*	*Middle*	*15*	*75*				

*Table 15.4. Minimum number of individuals (primary and secondary depositions) per building, with percentage of building excavated. Buildings that have been excavated over 75%, or where we are confident that all burials were recovered, are italicised. * Not excavated below floors.*

Area	*Building*	*Level*	*Period*	*MNI*	*% excavated*	*MNI per period*	*Total houses per period*	*Total houses with full burial set*	*Avg MNI per house*
South	*89*	*South N?*	*Middle*	*14*	*75*				
South	76	South O	Middle	5	50				
South	79	South O	Middle	0	25				
South	96	South O	Middle	9	25				
South	*97*	*South O*	*Middle*	*14*	*100*				
South	*80*	*South O*	*Middle*	*16*	*75*				
North	12	North H	Late	1	15	**179**	**27**	**4**	**14.8**
North	45	North H	Late	3	25				
North	57	North H	Late	3	25				
North	88	North H	Late	7	25				
North	13	North H	Late	0	15				
North	55	North H	Late	0	25				
North	58	North H	Late	9	25				
North	108	North H	Late	9	25				
North	64	North H	Late	0	25				
North	67	North H	Late	0	25				
North	60	North H	Late	10	25				
North	*129*	*North H*	*Late*	*23*	*25***				
North	54	North I	Late	12	25				
North	71	North I	Late	0	25				
North	47	North J	Late	0	25				
South	75	South P	Late	3	25				
South	53	South Q	Late	8	50				
South	*65*	*South Q*	*Late*	*16*	*100*				
South	*56*	*South R*	*Late*	*4*	*100*				
South	42	South R	Late	8	50				
South	*44*	*South S*	*Late*	*16*	*100*				
South	*10*	*South T*	*Late*	*0*	*75*				
TPC	*122*	*TPC M*	*Late*	*12*	*50?*				
GDN	140		Late	1	50?				
GDN	142		Late	2	50?				
TPC	150		Late	26	50?				
TPC	166		Late	6	50?				
TP	33	TP R	Final	1	?	**1**	**6**		
TP	34	TP R	Final	0	?				
TP	61	TP R	Final	0	?				
TP	73	TP P	Final	0	?				
TP	62	TP P/TP Q	Final	0	?				
TP	72	TP O	Final	0	?				

*Table 15.4 (continued). Minimum number of individuals (primary and secondary depositions) per building, with percentage of building excavated. Buildings that have been excavated over 75%, or where we are confident that all burials were recovered, are italicised. ** All subfloor burials recovered.*

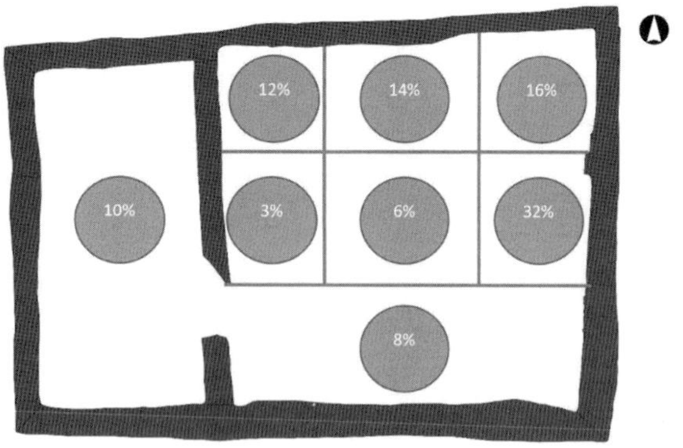

Figure 15.23. Spatial variation of burials within buildings (all periods combined).

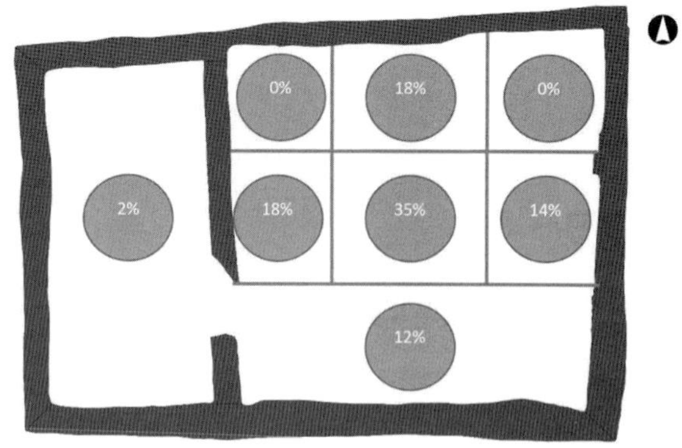

Figure 15.24. Spatial variation of burials within Early period buildings.

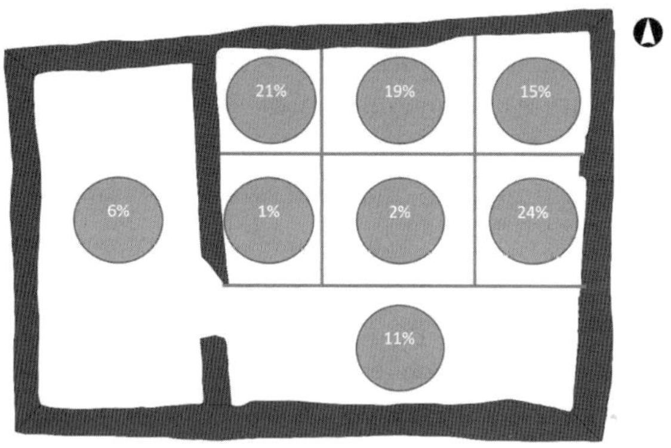

Figure 15.25. Spatial variation of burials within Middle period buildings.

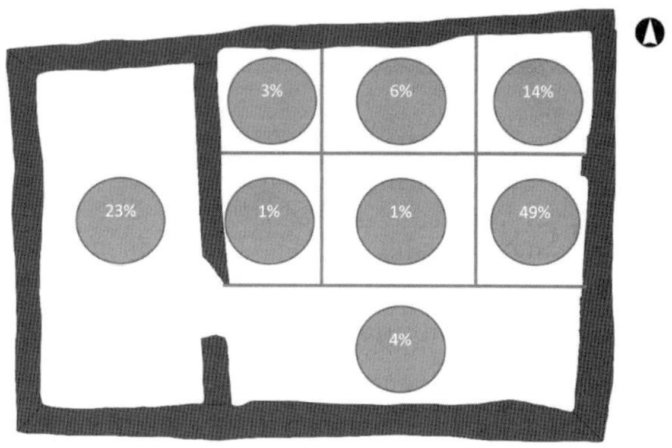

Figure 15.26. Spatial variation of burials within Late period buildings.

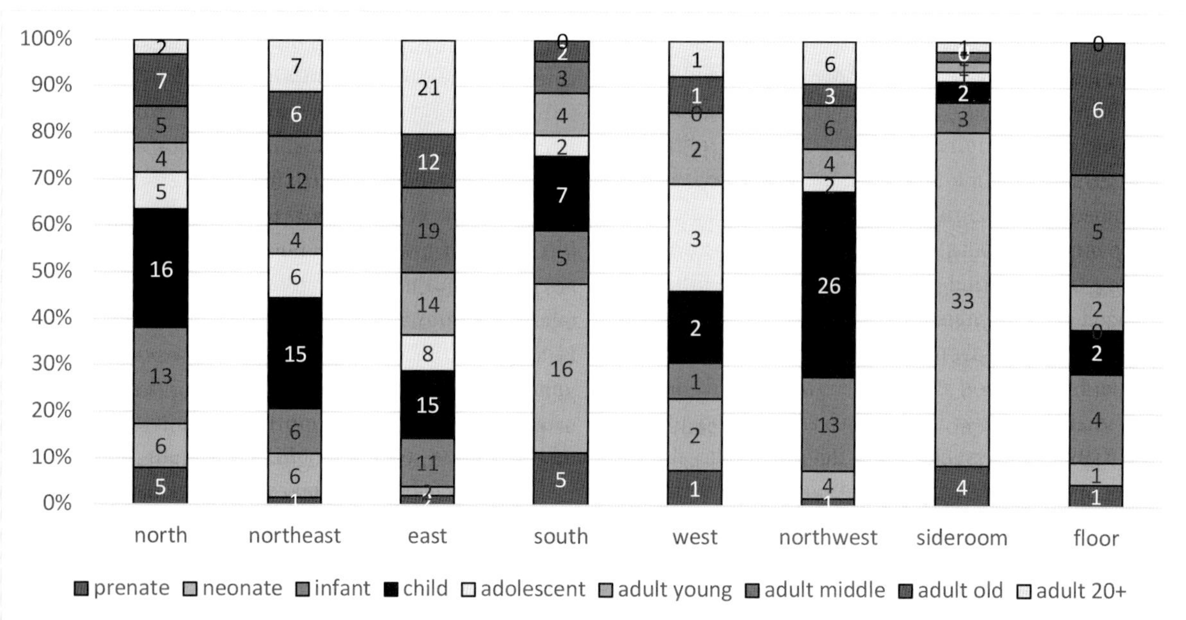

Figure 15.27. Spatial location of primary burials within houses by age category. Data presented for all occupation phases, including construction and abandonment.

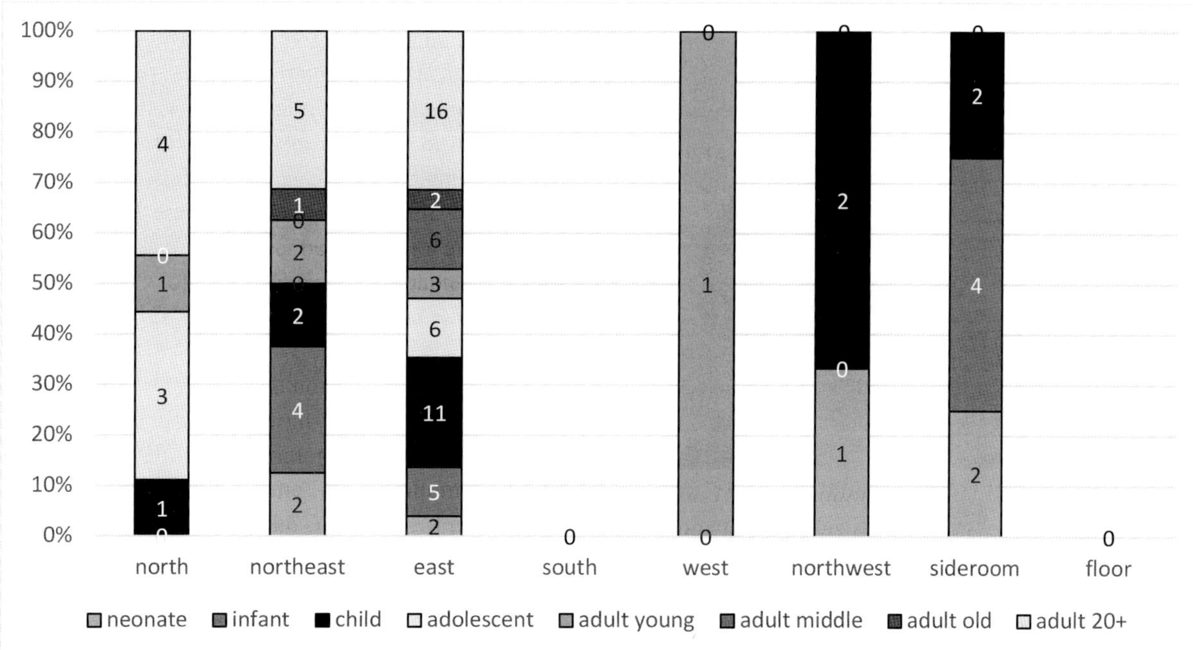

Figure 15.28. Spatial location of secondary burials within houses by age category. Data presented for all occupation phases, including construction and abandonment.

Deposition type and age breakdown by Building occupation phase. Figure 15.29 shows the proportion of primary and secondary burials across occupation phases of houses at Çatalhöyük. Here, a clear trend is visible, with the proportion of secondary burials increasing over time throughout the life of the house.

The majority of intramural burials at Çatalhöyük, both primary and secondary, occur during the period in which the house is occupied. A smaller number of inhumations occur during construction and abandonment phases. Figures 15.30 and 15.31 plot the age distribution of primary and secondary burials, respectively, by house occupation phase. Here, we see that primary subadult burials occur more often during the construction and early occupation phases of houses, while the reverse is true for secondary subadult burials; here, they occur more often during late occupation and abandonment phases.

Foundation/construction burials. Previous studies at Çatalhöyük noted an increase in the number of individuals recovered from foundation/construction contexts over time (Boz, Hager 2013; Nakamura, Meskell 2013a). With the addition of burial data from the 2009–2017 excavations, however, it now appears that the reverse is true. As a proportion of the total number of individuals recovered from primary and secondary contexts, the Early period has a 19% frequency (10/53) of foundation/construction burials. The Middle period has a 7% frequency (22/319) and the Late period a 6% frequency (11/174). There are no foundation/construction burials in the Final period. Figure 15.32 plots the proportion of foundation/construction burials by age category for the Early, Middle, Late and Final periods.

In terms of age distribution, subadults (prenates, neonates, infants and children) represent 67% of the foundation/construction burial assemblage and this is roughly consistent with the proportion calculated in previous studies (Boz, Hager 2013). When the assemblage is examined temporally the proportion of subadult to adult burials fluctuates over time, but subadults consistently predominate (fig. 15.33). In the Early period, subadults represent 80% of foundation/construction burials, in the Middle period, 58% and in the Late period, 73%.

A previous study of the sex distribution of adult foundation/construction burials noted the complete absence of males (Boz, Hager 2013: 420), but in the final excavation season in 2017 one male and one possible male were recovered in the construction layers of Early period Building 17 and two possible males from the construction layers of Middle period Building 132 (see excavation volume). In comparison, four females and four possible females have been recovered from foundation/construction contexts since the Hodder project began.

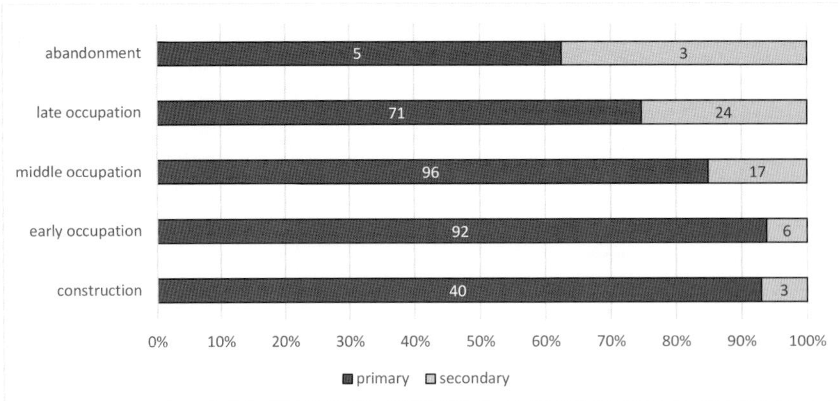

Figure 15.29. Proportion of primary and secondary burials during house occupation phases.

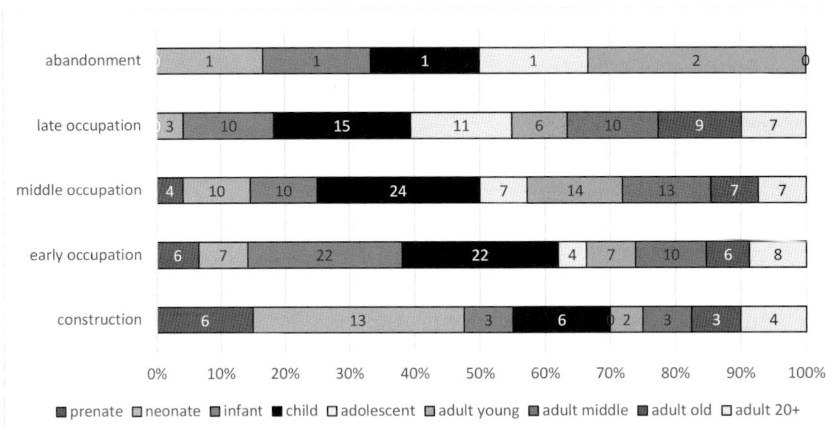

Figure 15.30. Age breakdown of primary burials within houses by occupation phase.

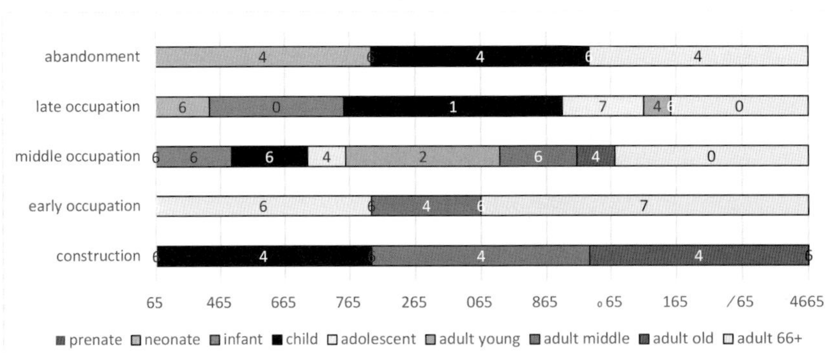

Figure 15.31. Age breakdown of secondary burials within houses by occupation phase.

Orientation of the body

Previous studies at Çatalhöyük demonstrated that the majority of primary burials were oriented with their heads towards the west (36%), followed by southern (20%) and eastern (17%) orientations (Boz, Hager 2013: 420–21). When Early, Middle, Late and Final period data are combined for the current sample the previously observed trends hold, although with a stronger emphasis for heads oriented to the west (45%), followed again by southern (16%) and eastern (12%) orientations (fig. 15.34). When analysed diachronically, however, it becomes clear that skeletal orientation did not remain static throughout the

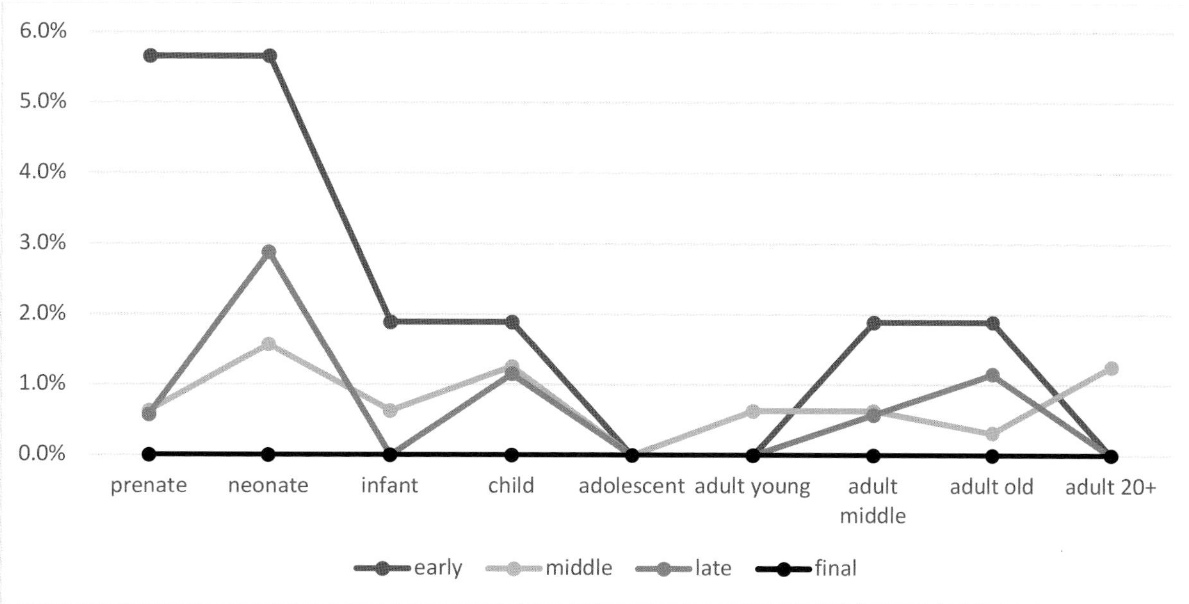

Figure 15.32. Proportion of foundation/construction burials per total number of individuals by age category for the Early, Middle, Late and Final period primary and secondary burials.

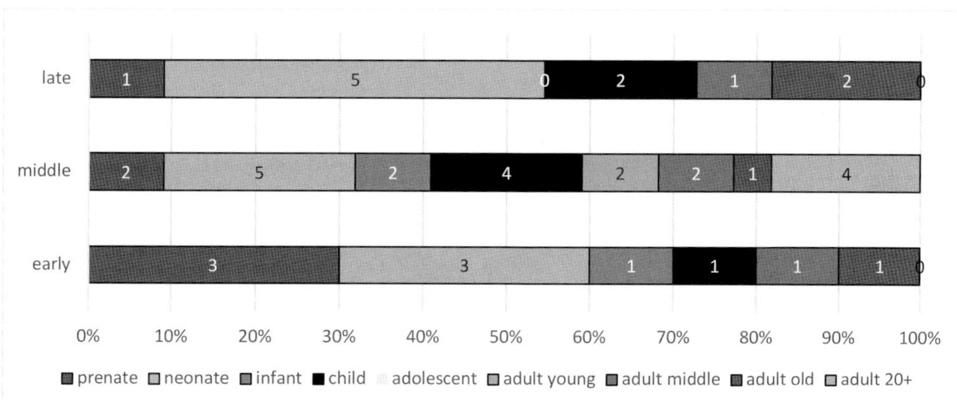

Figure 15.33. Age distribution of foundation/construction burials for Early, Middle and Late period primary and secondary burials.

occupation of Çatalhöyük (fig. 15.35). During the Early period heads oriented to the west represent 38% of the total number of primary burials, followed by an equal number of northern and eastern orientations (16%). During the Middle period heads oriented to the west represent 41% of the total number of primary burials, followed by southern (20%) and eastern orientations (11%). In the Late period heads oriented to the west make up 57%, followed by southern (13%) and eastern (12%) orientations. As seen in figure 15.34, while a majority of individuals were oriented with their heads toward the west throughout the occupation sequence at Çatalhöyük, their predominance increased over time, with variations in skeletal orientation becoming less common. Furthermore, in the Early period heads oriented to the north were also common, but they became increasingly rare over time. Perhaps the switch from northerly to southerly orientations is linked to changes in the orientation of houses during the Early period, when ovens and dirty floors were commonly located in the north part of the house and platforms in the south. This arrangement was inverted in subsequent periods.

Skeletal flexion and orientation

Another notable temporal trend is the increase in the degree of skeletal flexion over time (Haddow et al. in press). Many skeletons from the earliest levels (for example, South Area buildings B.17, B.160, B.161) appeared to have been placed in a more loosely flexed, less compact position than skeletons in subsequent levels. In order to quantify and test this general observation, a 'skeletal flexion index' was created for each primary skeleton by recording the angle of the knee, hip and elbow joints and taking the average (primary disturbed skeletons were also included if the angles of the knee, hip and elbow joints were still observable). Prenatal and neonatal individuals, as well as infants, were excluded from this analysis because they are often placed in non-standard positions/orientations due to their small body size and reduced limb proportions. The coordinate geometry (COGO) tool in ArcGIS 10.4 was used to record joint angles from the digitised skeleton plans within the Çatalhöyük Research Project GIS. These measurements permit the definition of operational terms to characterise the level of skeletal flexion observed on site:

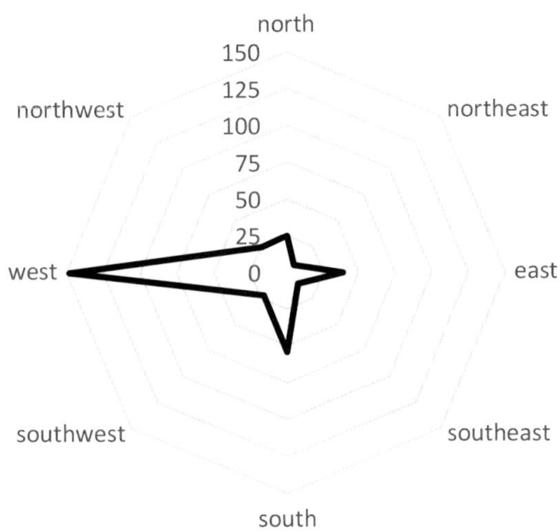

Figure 15.34. Skeletal orientation (head placement) of Çatalhöyük primary burials (N=330), all occupation periods combined.

- Hyperflexion: average knee/hip/elbow angle less than 15 degrees
- Tight flexion: average knee/hip/elbow angle between 15 and 30 degrees
- Medium flexion: average knee/hip/elbow angle between 30 and 45 degrees
- Loose flexion: average knee/hip/elbow angle greater than 45 degrees

Because skeletons excavated before 2000 are not yet part of the GIS, these individuals could not be measured. However, there are a sufficient number of digitised skeletons recorded between 2000 and 2017 to sufficiently represent each temporal period except for the Final, in which only a few primary inhumations occurred; only one adult had joint angles that could be recorded.

The skeletal flexion index was calculated in two ways: (1) using all three joint angles (knee, hip and elbow) and employing an average of all three, and (2) using only an average of the knee and hip angles. This was done because it was observed that while the lower limbs of some skeletons were in tightly flexed positions, the upper limbs were sometimes extended alongside the trunk of the body or loosely flexed across the chest. As the compactness of the skeleton is not typically affected by whether or not the upper limbs are extended along the body or flexed against the thorax, measuring the flexion of the knee and hip joints alone may provide a better representation of overall skeletal flexion/compaction. Table 15.5 presents the number of flexion types by occupation period using all three joints and using only the knee and hip joints.

Table 15.6 presents the average skeletal flexion index for each occupation period using both indices (all three joints versus only the knee and hip joints). Because only one individual in the Final occupation levels could be measured, average skeletal flexion indices are not available for this period, and it is excluded from subse-

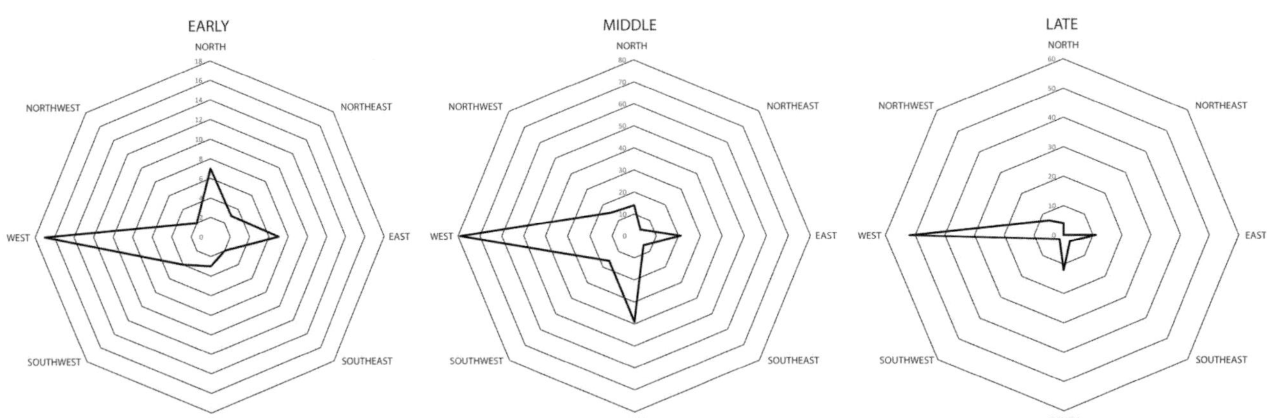

Figure 15.35. Skeletal orientation (head placement) of Çatalhöyük primary burials across temporal periods.

	Knee/hip/elbow joints				*Knee/hip joints*			
	Early	*Middle*	*Late*	*Final*	*Early*	*Middle*	*Late*	*Final*
Hyperflexion	0	10	4	0	1	38	11	1
Tight flexion	2	28	14	0	3	36	19	0
Medium flexion	7	30	7	1	7	7	2	0
Loose flexion	11	18	13	0	9	5	6	0

Table 15.5. Count of skeletal flexion categories by occupation period using three joints (knee, hip and elbow) and two joints (knee and hip).

	Knee, hip, elbow	*Knee, hip*
Early	56°	44°
Middle	35°	20°
Late	37°	25°

Table 15.6. Temporal variation in average skeletal flexion values using three joints (knee, hip and elbow) and two joints (knee and hip).

quent statistical analyses. It is clear from this table that, on average, primary burials in the Early period are more loosely flexed than primary burials in the Middle and Late periods. An independent samples Student's t-test using all three joints shows a significant difference ($\alpha = .05$) between Early period flexion averages and both Middle ($p = .001$) and Late ($p = .003$) period averages. There is no significant difference between Middle and Late period flexion averages ($p = .645$). There is still a significant difference between Early flexion averages and Middle ($p < .001$) and Late ($p = .002$) averages when analysing only the knee and hip joints. There is also no significant difference between Middle and Late flexion averages ($p = .127$). These results confirm the original, anecdotal observations made in the field and may indicate that less emphasis was placed on tightly wrapping the body during the early occupation of Çatalhöyük, perhaps because interments occurred shortly after death while the body was still fully fleshed. By this criterion, burials in the Middle and Late periods appear to have been prepared differently, perhaps because the removal/reduction of soft tissue body mass facilitated the creation of a more tightly bundled corpse, or perhaps for other reasons, such as the necessity to conserve space in order to place a greater number of individuals under house platforms.

Discussion

Given the excess of loose crania, mandibles and other disarticulated skeletal elements found in association with primary burial deposits and in other contexts, it seems likely that some segments of the population may never have been accorded a primary burial. Rather, their bodies, or in rare cases parts of their bodies (primarily crania and sometimes mandibles) were retained for a time before eventually being deposited in a variety of contexts. Whether this practice occurred as part of a sharing/redistribution scheme for the purposes of social cohesion or stress alleviation, as Kuijt (1996, 2000a) has argued, is debatable. In fact, the presence of partial skeletons or disarticulated bones in secondary contexts at Çatalhöyük does not necessarily imply that bodies were 'partible' or that body parts were re-distributed in the Neolithic, as has been previously postulated, to indicate that a different understanding of the individual existed at this time (Fowler 2004). The ratio of detached crania to 'headless' skeletons, for example, suggests that the retrieval and circulation of crania was a relatively rare occurrence. Thus, it may be time to de-emphasise the 'skull cult' at Çatalhöyük and replace it with an interpretative model in which bodies themselves appear to have circulated. In some cases, the loose or partially articulated skeletal elements found accompanying primary burials may actually represent what was left of an articulated corpse (originally held together by textiles, hides, matting and cord), as well as dried connective tissues, such as tendons and ligaments, that was kept above ground for some time before gradually fragmenting and eventually being buried. In Medieval European contexts the cranium (and sometimes long bones) often served as a funerary synecdoche: the part that represents the whole, the head defining the place of burial (Binski 1996: 55). Thus, many of the disarticulated bones found in secondary and perhaps tertiary contexts may represent the final deposition of a fragmented and skeletally incomplete individual, not the redistributed parts of a 'dividual'.

The important point here – and noted by other researchers (e.g., Croucher 2010; Goring-Morris, Belfer-Cohen 2014; Bocquentin et al. 2016) – is that a one-size-fits-all model likely oversimplifies a situation in which multiple actions and intentions were at play, and that there was not a standard funerary treatment applied to all (cf. Knüsel, Robb 2016).

Pre-scheduled/seasonal rites versus prescribed liminal period?

One interpretation, offered originally by Mellaart (1964: 92–94), is that intramural burials took place only at certain times of the year, when disruptions caused by opening holes in the platforms did not completely disrupt daily activities, perhaps during the spring or summer months. Other researchers on the current project (e.g., Fairbairn et al. 2005b; Matthews 2005b) have raised this possibility, but it has never been systematically investigated, until recently, from the perspective of the funerary treatment of human remains. If inhumations were scheduled on a seasonal basis, the variations in the degree of flexion and skeletal articulation observed among burials would reflect the amount of time between the death of an individual and the deposition of the corpse itself.

Alternatively, there may have been a prescribed liminal period between death and interment. Many social groups employ a very short liminal period, for example, in Jewish and Islamic funeral traditions, while other groups exercise a much longer interval, even lasting decades in some instances (Metcalf 1981: 576). If a prescribed liminal period was employed at Çatalhöyük, a standard level of flexion/articulation across all burials would be expected, but the observed variation argues against this interpretation. It could be, however, that the liminal period was longer for some individuals than it was for others. The challenge is then to decode the selection criteria employed by the Neolithic inhabitants of Çatalhöyük. Obvious possibilities include age and sex groupings, membership in sodalities or in other social groups, or the social circumstances in which death occurred.

A delay in burial for some individuals might also reflect more practical considerations. Some inhabitants of Çatalhöyük likely spent a portion of the year away from the site as part of a seasonal cycle of resource management/acquisition; for example, hunting, harvesting timber, sheep/goat pastoralism, and obsidian and ground stone procurement (Fairbairn et al. 2005b; Asouti 2013; Charles et al. 2014). Thus, some secondary burials may represent individuals who died away from the community and were eventually returned for burial after the body had partially/fully decomposed or became desiccated naturally. Variation in the degree of flexion and skeletal articulation might then reflect the amount of time elapsed before a body could be returned to Çatalhöyük for burial. Perhaps communal burial rites were scheduled at times during the year when the bulk of the population was most likely to be on site (cf. Matthews 2005b).

Other considerations include pragmatic concerns for maintenance of hygiene associated with living in proximity to decomposing corpses, such as escaping gases and insect infestation. The burial of a fully fleshed corpse under the floor of a house would also create structural instabilities caused by soil subsidence as soft tissue body mass was lost during the decomposition process.

Regional comparisons

It is during the Final period on the East Mound at Çatalhöyük that the burial practices begin to resemble those encountered in certain Late PPNB and Early Pottery Neolithic sites in the Levant and Mesopotamia, in which there appear to be very few burials within settlements (e.g., Banning 1998; Rollefson 2001; Bienert et al. 2004; Goring-Morris 2005). Some researchers have interpreted this apparent selection for particular individuals as representing a form of early social differentiation (Rollefson 2002). Furthermore, Rollefson (Ibid.) observes a similar increase in the proportion of secondary burials during the PPNC levels at 'Ain Ghazal (Jordan). There are also similarities between Final period Çatalhöyük and Late Neolithic Tell Sabi Abyad in Syria, where burials were found in abandoned rather than occupied houses (Akkermans 2008). Lastly, the burnt structure at LPPNB Bouqras in northern Iraq has been interpreted as a charnel structure for the storage and processing of bodies before final interment (Merrett, Meiklejohn 2007), as has a similar structure at Abu Hureyra (Moore, Molleson 2000). These structures, as well as the 'collective burials' found at Late PPNB Ba'ja in Jordan (Gebel et al. 2006), bear a resemblance to the 'burial chambers' (Spaces 248 and 327) found in abandoned houses during the Final period at Çatalhöyük (Marciniak, Czerniak 2007; Marciniak et al. 2015).

Conclusions

In terms of temporal change, the biggest shift in funerary practices at Çatalhöyük is observed during the Late and Final periods. During this time individuals in sub-floor primary burial contexts decline in proportion to the number of individuals found in secondary contexts. In the very latest occupation levels of the East Mound it seems that sub-floor primary inhumations ceased entirely. Instead, modified spaces within abandoned buildings were used to contain the accumulated remains of secondary and disturbed primary skeletal material (Marciniak, Czerniak 2007; Marciniak et al. 2015).

While an earlier round of inquiries relating to consciousness in the Neolithic theorised that an emerging sense of bounded identity and personal autonomy should be linked with a corresponding decrease in secondary burial associated with the fading of the so-called 'dividual' (Haddow et al. 2020), the Çatalhöyük funerary data do not appear to support this argument. In fact, the increase in the proportion of human remains found in secondary and tertiary contexts observed at Çatalhöyük is exactly the opposite of what one would expect, as bodies actually appear to become more fragmented over time, until 'whole bodies' (for example, undisturbed primary burials) nearly disappear from the site. This trend continues in the early Chalcolithic settlement established on the West Mound at Çatalhöyük, where human remains in any context within the settlement are almost completely absent (Biehl 2012; Anvari et al. 2017; Orton et al. 2018).

If the original argument is turned on its head, however, the increase in human remains found in secondary/tertiary contexts (and the corresponding reduction in primary burials) during the Late and Final occupation periods at Çatalhöyük could be interpreted, instead, as representing a shift from an earlier egalitarian/collective mentality towards one in which certain people were singled out for secondary/delayed treatments as part of an emerging focus on particularly significant individuals or groups – perhaps as part of a trend towards increased social differentiation. Perhaps, then, primary burials at Çatalhöyük East represent a standard/collective mode of burial, while the secondary remains and delayed primary burials represent treatments reserved for 'special individuals' – or 'special groups'. Indeed, while the primary burial assemblage, with its excess of subadult individuals, is thought to reflect a normal mortality profile (Hillson et al. 2013; see also this volume, Chapter 16), the secondary and tertiary burial assemblages are skewed towards adolescent and adult individuals, a pattern that suggests an age-based selection bias. As secondary treatments increase through time, this could be interpreted as an increased focus on particular individuals/groups, although increased mobility and off-site deaths cannot be excluded here.

Furthermore, the reduced number of primary burials over time may represent a reduced emphasis on collectivism and social memory tied to houses, which is demonstrated in other Çatalhöyük data sets: the reduction in large-scale feasting activities, houses becoming more autonomous both spatially and in terms of production, and symbolic systems previously linked with the house, such as wall paintings and animal installations, starting to become portable, with the emergence of stamp seals and eventually painted pottery on the Chalcolithic West Mound (Marciniak, Czerniak 2007; Hodder 2014d; Hodder, Doherty 2014; Marciniak et al. 2015a). Perhaps bodies (and increasingly only particular bodies or parts of bodies) become more portable as well, and less tied to specific houses.

In shifting from a macroscale view to a more fine-grained analysis of the funerary practices at Çatalhöyük, subtle patterns likely reflecting a much wider range of intentional behaviours than previously conceived begin to emerge. Furthermore, the addition of data from the earliest and latest levels of the site reveals unequivocal evidence of temporal changes in burial practices. It is clear that the increasing occurrence of disturbed, secondary and commingled remains at the site masks a variety of pathways that bodies, whole or fragmented, seem to have taken before their final deposition under house floors, external spaces or other contexts. For some individuals, however, the journey appears to have lasted longer than for others. Decoding the motivations behind these practices and distinguishing between the equifinal behaviours that culminate in the archaeological record requires careful attention to the stratigraphic links between burial sequences and grave fills, as well as meticulous osteological and taphonomic analyses. In order to adequately address higher-level questions of socio-cultural behaviour and cognitive change, it is crucial that these fundamental archaeological issues are sufficiently resolved beforehand.

Acknowledgements

This research received financial support from French State via the 'Investments for the Future' framework programme, Initiative d'Excellence de l'Université de Bordeaux (IdEx), (Award No. ANR-10-IDEX-03-02 to CJK), the Research Center for Anatolian Civilizations (ANAMED) of Koç University (to SDH), the British Academy Newton Advanced Fellowship (Award No. NAFR1180202 to SDH), France-Stanford Collaborative Projects for Interdisciplinary Studies (to CJK and Ian Hodder), the Templeton Foundation (Award No. 52003 to I. Hodder), and a National Geographic Society Project Grant (to CS Larsen and CJK). Scott Haddow is grateful to Dilşad Sila Votruba (Archaeology and History of Art Department, Koç University) for training and assistance with the preparation of bone thin sections; Rana Özbal (Archaeology and History of Art Department, Koç University) and Fokke Gerritsen (Netherlands Institute In Turkey) for providing comparative bone samples from Barcın Höyük; Jessica Pearson and Douglas Baird (Liverpool University) for providing comparative bone samples from Boncuklu Höyük; Barış Yağcı (KUYTAM, Koç University) for providing the SEM analyses of the bone thin sections; Thomas Booth

(Natural History Museum, London) for assistance with interpreting the bone thin sections and Peter Gethin (Liverpool University) for SEM training. A number of other Project members deserve mention for their time and inspirational investment that helped to realise this chapter, including Numan Arslan, Åsa Berggren, Bonnie Glencross, Lisa Guerre, Rémi Hadad, Katarzyna Harabasz, Kristina Jonsson, Clark Spencer Larsen, Ashley Lingle, Selin Nugent, Jason Quinlan, Joshua Sadvari and Milena Vasić.

Supplementary material

For supplementary material related to this chapter, please visit https://doi.org/10.18866/BIAA/e-13. It comprises colour versions of figures 15.18, 15.19, 15.20, 15.21, 15.27, 15.28, 15.29, 15.30, 15.31, 15.32, 15.33, 15.34 and 15.35.

16. Bioarchaeology at Neolithic Çatalhöyük: indicators of health, well-being and lifeway in their social context

Christopher J. Knüsel, Marco Milella, Barbara Betz, Irene Dori, Evan Garofalo, Bonnie Glencross, Scott D. Haddow, Marissa Ledger, Evilena Anastasiou, Piers Mitchell, Jessica Pearson, Marin A. Pilloud, Christopher B. Ruff, Joshua W. Sadvari, Belinda Tibbetts and Clark S. Larsen

Introduction: a great transition and its aftermath

The transition to farming and its immediate aftermath is frequently associated with a decrease in health, a result of the greater proximity of people to animals, proximity of living quarters to one another and to waste disposal areas, and higher settlement population densities. All these factors, together with a narrower dietary niche, are today interpreted as responsible for the sudden spread of infectious and parasitic diseases among early Holocene agricultural communities (Cohen, Armelagos 1984; Larsen 1995; Armelagos et al. 2005). Bioarchaeology has significantly contributed to understanding this key moment in the past, a process often referred to as 'Neolithisation', which commenced in the preceding Epi-Palaeolithic period, dating to roughly between 20,000 and 10,000 BC in Anatolia (Düring 2011a: 31–32). This process reached its apogee with the emergence of a number of large settlements, termed 'megasites' (Bogaard, Isaakidou 2010; Der, Issavi 2017), throughout the Near East towards the end of the Pre-Pottery Neolithic period (that is, Late PPNB – 8500–6500 cal B.C.) (Kuijt 2000b). Features of the Late PPNB include the full domestication of plants and animals and the advent of large multi-room structures. As one of only two megasites in Anatolia (the other being Aşıklı Höyük), Çatalhöyük provides the opportunity to examine the health and well-being of a very large and well-established sedentary Late Neolithic farming community. Because of its long occupation sequence from 7100 to 5950 BC (Bayliss et al. 2005) and size (13ha, equivalent to roughly 30 football fields), Neolithic Çatalhöyük furnishes an ideal candidate for exploring the effects of diachronic changes in population size and their effects on the health, well-being and way of life of the sexes and for age and social groups through time. Results of previous analyses suggested a diachronic change in infectious disease prevalence, a likely result of increased population density and a related exposure to pathogens (Larsen et al. 2015). Conversely, estimates of growth trajectories for stature and body mass did not seem to deviate from those expected for 'healthy' populations (but see below). These contrasting results suggest the need for further analyses of the socio-cultural environmental challenges present at the site through time and their biocultural effects on the inhabitants.

This chapter addresses measures of health (palaeopathology), well-being and lifestyle (way of life) that provide insights into the social context of the Neolithic population sample excavated at Çatalhöyük. The term 'well-being' is defined here as the ability to attain the full genetic potential for growth and development in the face of stress, the non-specific response of the body to physiological demands placed on it (cf. Selye 1936), a measure of individual and population adaptation to social, economic, political and psychological circumstances. Goodman et al. (1988) defined a group of osteological indicators of stress or 'stressors', 'that which produces stress' (Selye 1936: 51, 78–79), for the field of biological anthropology, which has been expanded since but without exhausting the list of indicators or the application to population samples in subsequent research to date. Although not all of these indicators of stress could be included in this treatment, the studies described below include those based on changes in the palaeodemographic composition of the population sample, especially of the very young. Constituents and quality of the diet and lifestyle are assessed through studies of dental enamel hypoplasia, fluctuating asymmetry and disease; growth (stature and body mass); violence-related trauma; long bone cross-sectional geometric properties; and dietary isotopic ratios. Cross-sectional geometric properties are also employed in conjunction with strontium isotopic ratios and patterns of dental non-metric traits to address residential mobility and migration. In addition, parasitic infections are considered from individual grave contexts and human faecal remains. The results of this multi-faceted research programme both confirm and expand previously published insights into the health and way of life of the Neolithic population of Çatalhöyük (see Larsen et al. 2015; Hillson et al. 2013; Larsen et al. 2013).

Palaeodemography of Çatalhöyük: diachronic changes in age-at-death, sex composition and fertility

Recovery of substantial numbers of human remains during the 2009–2017 field seasons provides the opportunity to refine and update previous palaeodemographic estimates for the East Mound at Çatalhöyük. In particular, the incorporation of new data from the lower stratigraphic levels provides a more informed discussion of the demographic composition of the Çatalhöyük community during the earliest occupation of the site. This, together with a more fine-grained estimation of subadult age-at-death, facilitates a more informed contextualisation of the biological and social patterns observed in this Neolithic setting. Due to the potential for double counting of secondary/tertiary burials, the present discussion considers only primary inhumations. The analysis of demographic patterns associated with various funerary treatments is provided in the funerary archaeology chapter (Chapter 15).

Sex determination was based on the scoring of cranial, mandibular and pelvic sexually dimorphic features following standard protocols (Buikstra, Ubelaker 1994). Cranial features included the size of the mastoid process and the robusticity and rugosity of the supramastoid crest, temporal line, glabella, supraorbital ridge, superior nuchal line and external occipital protuberance. Mandibular features were also scored, including the shape of the coronoid process, ascending ramus, gonial angle and mental eminence, as were features of the *ossa coxae*, including the shapes of greater sciatic notch and ventral arc, sub-pubic concavity and inferior ischio-pubic ramus ridge of the pubis. Apart from the last three (recorded as present or absent), all dimorphic features were scored on a 5-point ordinal scale (from distinctly female to distinctly male, with the third score corresponding to indeterminate sex). Subadult age-at-death was estimated on the basis of skeletal and dental developmental and eruption milestones (Schour, Massler 1941) and for the youngest individuals, neonates and young infants, the dental development and eruption sequence proposed by AlQahtani et al. (2010). Multiple methods were applied to estimate age-at-death from subadult skeletal remains, including the presence and morphology of ossification centres, element and epiphyseal fusion, dimensions of specific skeletal elements, and dental development. Subadult age estimates were divided into the following categories: pre-term (<38 weeks gestation); neonate (0–2 months); infant (2 months–3 years); child (3–12 years); and adolescent (12–20 years). Adult age-at-death was estimated on the basis of the morphological changes of the pubic symphysis (Brooks, Suchey 1990) and auricular surface of the ilium (Lovejoy et al. 1985). The resulting age estimates were used to divide the sample into three adult age categories: young adults (20–34 years old), middle adults (35–49 years old), and old adults (≥50 years old). Adults for whom skeletal preservation did not allow inclusion in a specific age category were recorded as 'adults' (≥20 years old).

Chronologically, the site is divided into stratigraphic levels corresponding to the main excavation areas on the East Mound: North, South, TP and TPC. Because some stratigraphic levels contain very few individuals, groups of levels have been combined into four major occupation periods in order to increase sample sizes (see Chapter 1 for details of site chronology): Early (7100–6700 BC), Middle (6700–6500 BC), Late (6500–6300 BC) and Final (6300–5950 BC). Furthermore, for this chapter, the Late and Final periods were grouped together, owing to the extremely small number of primary burials retrieved from the Final period of occupation. The final total sample of 471 individuals was used to address three main research questions:

1. What is the age composition of each period of the settlement, and how does it vary through time?
2. What is the proportion of males versus females in the population of Çatalhöyük, and was this ratio stable or prone to fluctuations through time?
3. What inferences can be made from the above patterns regarding the interaction between the population of the site and their biological and social environment?

Age

Figure 16.1 and table 16.1 show the relative frequency of each age category by occupation period. From this, it appears that the frequency most prone to vary through time is that of pre-term individuals (a dramatic decrease after the Early occupation period), and children (whose frequency reaches a peak during the Middle occupation period). It is worth noting that, even if the frequency of the three adult age categories appears to be relatively constant, especially for the Middle and Late/Final periods, the relatively large number of individuals not assignable to a specific adult age category is likely to bias the overall picture. Figures 16.2 and 16.3 provide a more detailed picture of the relative proportion of each age category, respectively, for adults and subadults. Table 16.2 illustrates the age distribution of the subadult sample. For adults – even though biased by 'uncategorised' adults – the clearest trends are expressed by the middle and old adult categories. Middle adults show a marked increase in frequency from the Early to the Middle period, with no changes after the latter. Conversely, the relative frequency of old adults decreases from the Early to the Middle period, with no changes during the Late/Final periods. Results for subadults are more variable than those for adults. Apart from the adolescents, which tend to present a relatively homogenous frequency through time, the other age categories show marked diachronic variability. This can be summarised by a decrease in pre-term individuals after the Early period, and a peak in the relative frequency of children during the Middle period and of neonates during the Late/Final period. Running an ordered logistic model with age category as an independent variable and occupation period, settlement area, and their interaction as predictors (when testing settlement area only Middle and Late/Final periods were considered) demonstrates that the above changes are statistically independent from the area of the settlement sampled (table 16.3).

Pre-term mortality exhibits a marked decrease between the Early and Middle occupation periods despite evidence for increased population density (fig. 16.3). This decrease suggests that the risk of pre-term spontaneous delivery was reduced through an improvement in maternal physiological condition and/or a diminished impact of extrinsic factors affecting foetal viability. The increase in pre-term mortality during the Late/Final period may reflect delayed, trans-generational, epigenetic insults (see Gowland 2015) that affected maternal condition during the Middle period, particularly the subsequent generational effects of subadult female health during development (Roseboom et al. 2001; Barker 2004). This is also the case for neonatal mortality during the Late/Final occupation period, but the increase is considerably more marked, rising from 15% to 56% of the total subadult sample. As with pre-term mortality, neonatal mortality decreases between the Early and Middle occupation periods, again supporting an interpretation of a general improvement in maternal condition and foetal viability into the Middle occupation period. This trend appears to have been reversed in the Late/Final period, suggesting diminishing beneficial conditions within the population and, in particular, for reproductive females.

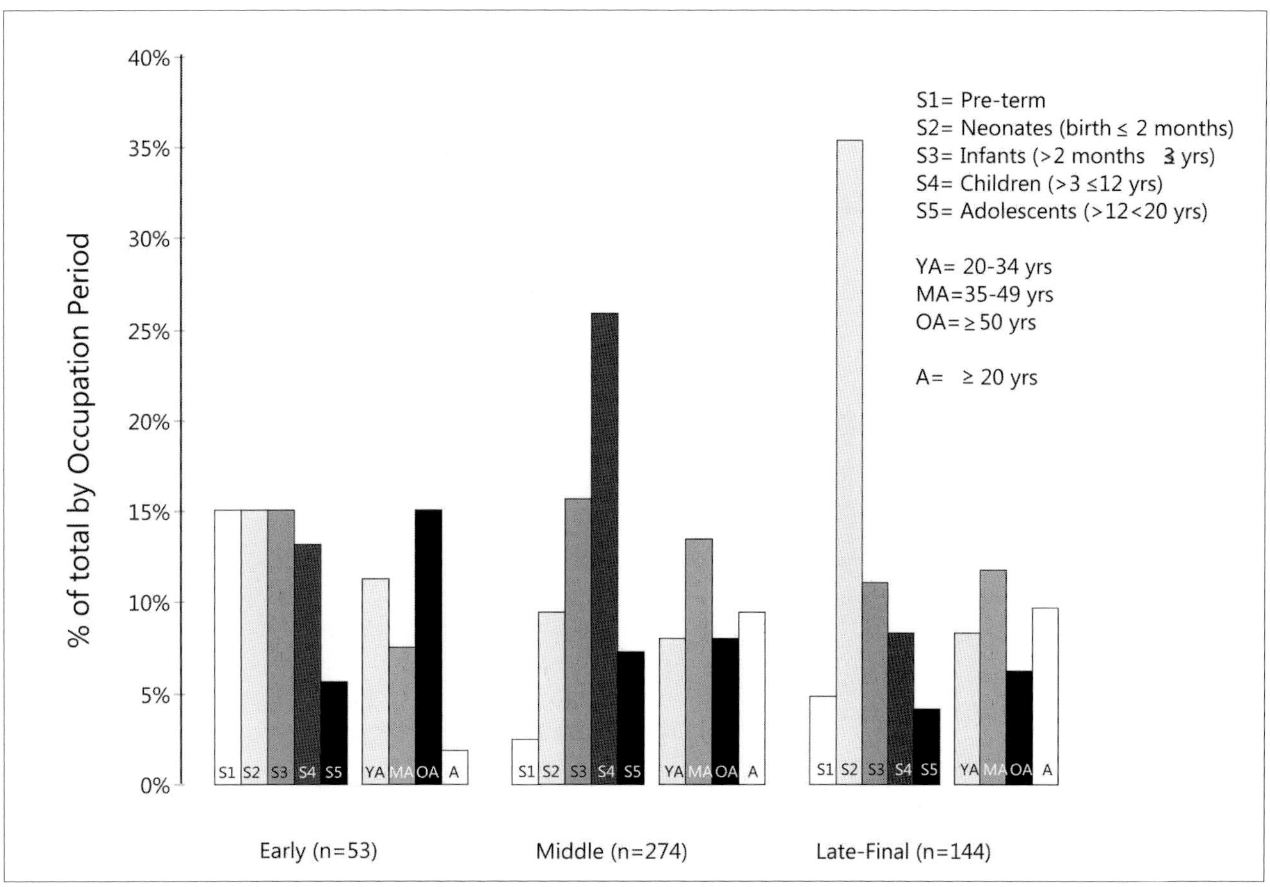

Figure 16.1. Age categories of the Çatalhöyük human skeletal sample (only primary and primary disturbed inhumations included). Frequencies refer to the number of individuals in each age category divided by the total sample size for each period of occupation, expressed in percentages. YA= Young adult, MA= Middle adult, OA= Old adult, A= Adult.

(a)

	Early (n=53)		Middle (n=274)		Late-Final (n=144)	
Age category	*n*	%	*n*	%	*n*	%
Pre-term	8	15.1	7	2.6	7	4.9
Neonates (birth≤ 2months)	8	15.1	26	9.5	51	35.4
Infants (>2 months ≤ 3 yrs)	8	15.1	43	15.7	16	11.1
Children (>3≤ 12 yrs)	7	13.2	71	25.9	12	8.3
Adolescents (>12<20 yrs)	3	5.7	20	7.3	6	4.2
YA (20–34 yrs)	8	15.1	22	8.0	9	6.3
MA (35–49 yrs)	4	7.5	37	13.5	17	11.8
OA (≥ 50 yrs)	6	11.3	22	8.0	12	8.3
≥ 20 yrs	1	1.9	26	9.5	14	9.7

(b)

	Early				Middle				Late-Final			
	F (n=8)		M (n=11)		F (n=57)		M (n=35)		F (n=24)		M (n=20)	
Age category	*n*	%	*n*	%	*n*	%	*n*	%	*n*	%	*n*	%
Adolescents (>12<20 yrs)	0	0.0	1	9.1	2	3.5	1	2.9	1	4.2	0	0.0
YA (20–34 yrs)	2	25.0	3	27.3	13	22.8	8	22.9	7	29.2	4	20.0
MA (35–49 yrs)	2	25.0	2	18.2	20	35.1	15	42.9	7	29.2	10	50.0
OA (≥ 50 yrs)	3	37.5	5	45.5	15	26.3	5	14.3	4	16.7	4	20.0
≥ 20 yrs	1	12.5	0	0.0	7	12.3	6	17.1	5	20.8	2	10.0

Table 16.1. The demographic sample divided by age category (a) and sex (b). Only stratified primary and primary disturbed individuals included.

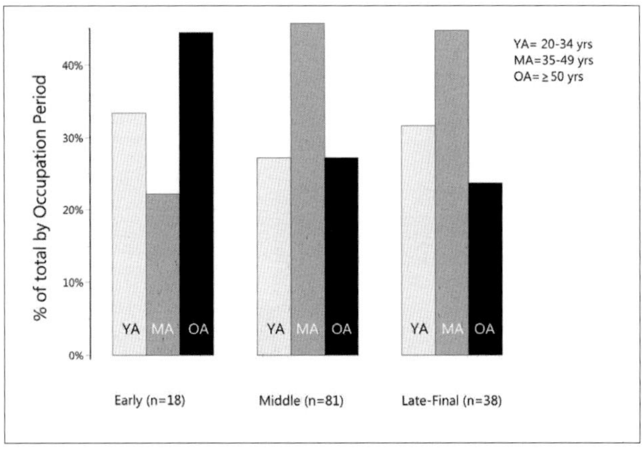

Figure 16.2. Age divisions of adults. Frequencies refer to the number of individuals for each age category divided by the total sample size for each occupation period, expressed in percentages. YA=Young adult, MA=Middle adult, OA=Old adult, A=Adult.

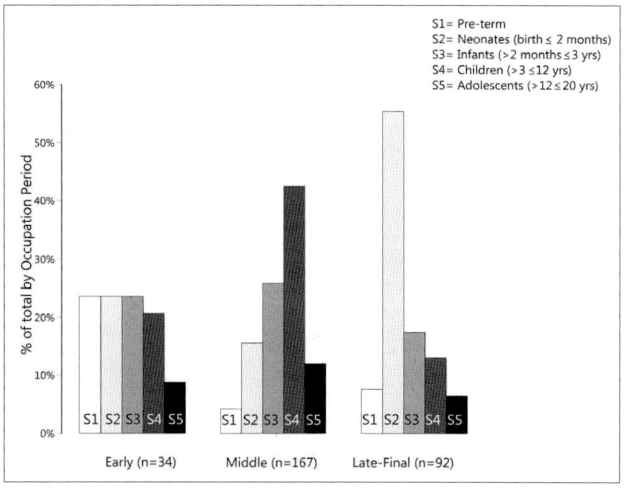

Figure 16.3. Age divisions of subadults. Frequencies refer to the number of individuals for each age category divided by the total sample size for each occupation period, expressed in percentages.

Age	*Early*	*Middle*	*Late-Final*
Pre-term	23.5 (8)	4.2 (7)	7.6 (7)
Neonates	23.5 (8)	15.6 (26)	55.4 (51)
Infants	23.5 (8)	25.7 (43)	17.4 (16)
Children	20.6 (7)	42.5 (71)	13 (12)
Adolescents	8.8 (3)	12 (20)	6.5 (6)

Table 16.2. Absolute (in brackets) and relative frequencies of subadult age categories. Relative frequencies are calculated based only on the subadult sample.

(a)	*N params*	*DF*	*ChiSq*	*p (Bonferroni-corrected)*
Occupation Period	2	2	17.93	<.0001
(b)				
Occupation Period	1	1	19.24	<.0001
Settlement Area	1	1	0.00	0.9967
Occupation Period x Settlement Area	1	1	4.90	0.0538

Table 16.3. Ordinal logistic models with age categories as an independent variable, and (a) period of occupation, (b) occupation period, settlement area, and their interaction as predictors (but considering only Middle and Late-Final periods).

Age (months)	*Early*	*Middle*	*Late-Final*
2–6	0	14.6 (6)	13.3 (2)
6–12	37.5 (3)	48.8 (20)	60 (9)
12–18	25 (2)	17.1 (7)	13.3 (2)
18–24	37.5 (3)	7.3 (3)	0
24–36	0	12.2 (5)	13.3 (2)

Table 16.4. Absolute (in brackets) and relative frequencies of individuals in the infant age category by month-specific ranges.

Infant mortality appears relatively constant from the Early into the Middle occupation period but displays a substantial decrease in the Late/Final period. This age category spans a considerable range (two months to three years). Further refinement of the ages at death within this category can be found in table 16.4. The small number of infant deaths during the Early period limits the interpretation of this age category. Higher mortality within the 6–12 month age range could be associated with the introduction of foods at the commencement of weaning, which isotopic analyses suggest started prior to 12 months of age (see 'Weaning', below), as well as greater infant mobility and interaction with the environment and exposure to pathogens. This age range is associated with increased mortality risk for those individuals that had survived the first six months of life. A reduction in the number of infant deaths in the Late/Final period, despite a consistent ratio compared with the Middle period, may suggest a cultural response to mitigating infant mortality, but a much larger sample would be required to test this hypothesis.

Child mortality effectively doubled from the Early to Middle periods and then decreased by three-quarters in the Late/Final period. The peak in child mortality during the Middle period could indicate a negative shift in adult health and well-being towards the Late/Final occupation period. It can reasonably be argued that the health and development of surviving children may also have been compromised during the Middle period. As such, evident peaks in older subadult (child and adolescent) mortality could prove to be useful indicators for general population health and well-being in subsequent periods.

Sex

Figure 16.4 presents the relative frequency of each sex for each occupation period, as well as the diachronic trajectory of the ratio between males and females expressed by the difference between females and males divided by their sum multiplied by 100. These analyses consider all the individuals for whom it was possible to determine sex. Accordingly, the sample includes, in addition to a majority of adults, some adolescents. The most prominent pattern is represented by an inversion of the ratio of females to males from the Early period, with relatively more males, to the Middle period and a marked majority of females. During the Late/Final period, the frequencies of the sexes are more similar, even if females are relatively more numerous. However, when testing them with a logistic model, neither occupation period, nor area of the settlement (North Area versus South Area) demonstrates a significant majority of males or females (table 16.5).

Age and sex

Testing the effect of sex and occupation period on age category by means of an ordinal logistic model indicates a lack of sex-specific age trends through time (table 16.6). This is further illustrated in figure 16.5, which presents the relative frequencies of each age category by sex through time. Note that the few (n=5) adolescents and all generic 'adults' (n=21) are not considered, due to, respectively, their small number and the lack of a precise age category.

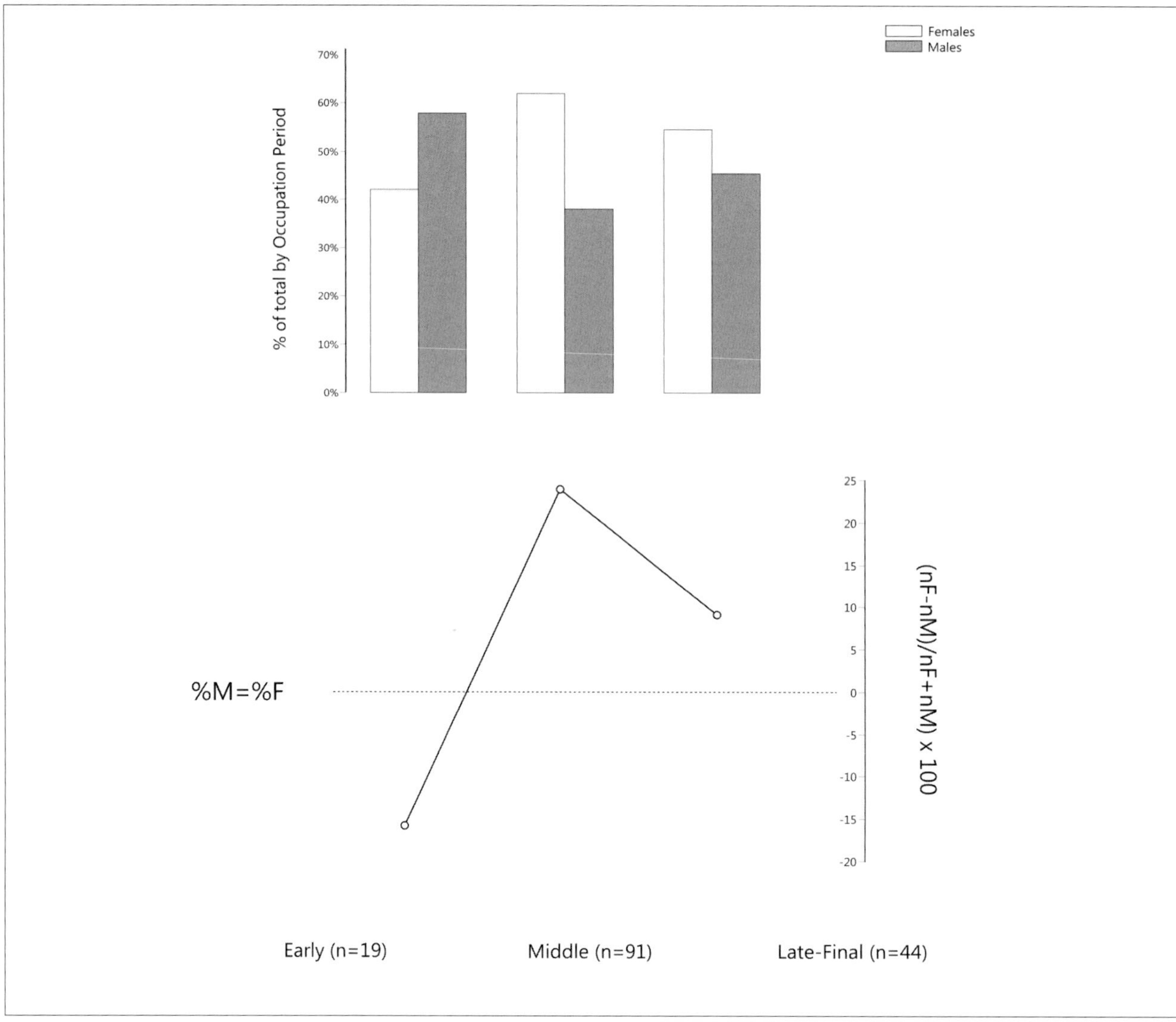

Figure 16.4. (a) Relative frequency of each sex by period of occupation; (b) changes in the sex ratio through time. Positive values indicate more females, negative ones more males.

	N params	*DF*	*ChiSq*	*p (Bonferroni-corrected)*
(a)				
Occupation Period	2	2	2.73	0.2558
(b)				
Occupation Period	1	1	0.19	1
Settlement Area	1	1	0.51	1
Occupation Period x Settlement Area	1	1	0.02	1

Table 16.5. Logistic models with sex as an independent variable, and (a) occupation period, (b) period of occupation, settlement area, and their interaction as predictors (but considering only Middle and Late-Final periods).

The overlap of age trajectories between males and females is exemplified by the middle adult (MA) and old adult (OA) age categories. The frequency of the MA tends to increase for both males and females from the Early to the Middle period, with no marked change during the Late/Final period. There is also a clear decrease in frequency of OA for both sexes from the Early to the Middle occupation period.

Juvenility index: population change in temporal perspective

Previous estimates of the ratio between juveniles and adults at Çatalhöyük have been used as a proxy for fertility rates in this population and found to reflect a diachronic change consistent with population growth driven by an increase in fertility from the Early to the Middle occupation periods, followed by a decrease (Hillson et al. 2013; Larsen et al. 2015). Re-calculation of the juvenility index – the ratio between individuals aged between 3 and 19 years and the total of individuals above 3 years of age – produces results that are similar to those of Larsen and colleagues (2015) for the total sample. The index follows a precise diachronic trend (Early=9/29 [.31], Middle=96/218 [.44], Late/Final=20/77 [.26]). The changes represent a growth in population in the Middle period relative to the Early period – growth driven by an increase in fertility. The reduction in the ratio in the Late/Final period is consistent with a decline in population represented by other variables, including but not limited to marked reduction in the number of houses and house density in the Late/Final period (Hodder 2013c).

Wrapping it together: age, sex, time and demographic changes

As already mentioned for Çatalhöyük, in particular, and the Neolithic, in general, the transition from foraging to farming with the adoption of agriculture had several interwoven effects on the biology of past human populations (Larsen 1995; 2006; Bocquet-Appel 2008; Larsen et al. 2015). The increasing availability of carbohydrates fostered a dramatic positive shift in fertility and birth rates, leading to large demographic changes (Bocquet-Appel 2002; 2008; 2011). These changes were coupled with an increase in infant mortality, one of the negative effects of a crowded environment in close proximity to animals and waste disposal areas (Bocquet-Appel 2008). The data from Çatalhöyük expand the discussion about the biological correlates of the Neolithic way of life by providing a glimpse into the demographic changes through time characterising a specific Late Neolithic population.

An important caveat in the interpretation of the demographic patterns at Çatalhöyük (especially of subadults) is that only a relatively small proportion of the East Mound at the site has been excavated. The unexcavated areas could well contain specific burial zones with different treatments of the burial population compared to those that have already been analysed. In general, these findings indicate that the population of Çatalhöyük underwent profound demographic change throughout more than a millennium of settlement occupation. In particular, the age composition of the population tended to vary through time, with the more marked changes during the Middle period representing

	N params	*DF*	*ChiSq*	*p (Bonferroni-corrected)*
Occupation Period	2	2	1.04	1
Sex	1	1	0.04	1
Sex x Occupation Period	2	2	1.40	1

Table 16.6. Ordered logistic model with age category as the independent variable, and occupation period, sex, and their interaction as predictors.

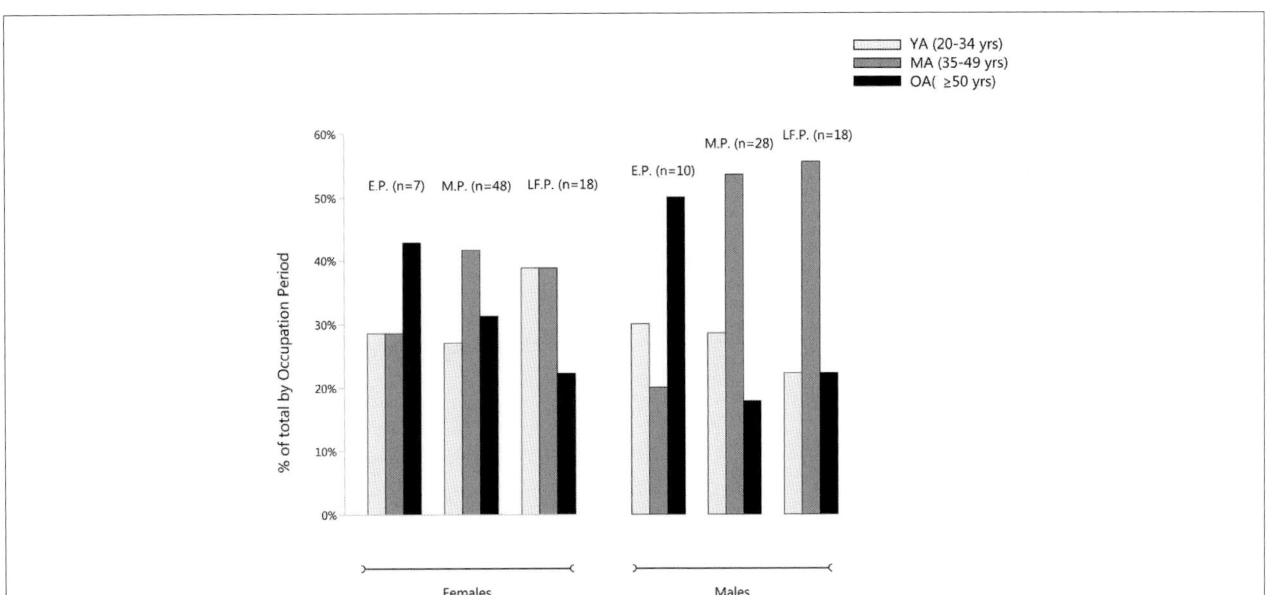

Figure 16.5. Age categories of the Çatalhöyük human skeletal sample partitioned by sex. Frequencies refer to the number of individuals for each age category divided by the sample size of each sex for a specific occupation period, expressed in percentages.

increased birth rates and fertility (fig. 16.6). During this time, the frequency of pre-term individuals drops suddenly and frequencies of infants and fertility reach a peak, whereas old adults are less frequent. All these trends depict a scenario consistent with previous reconstructions envisioning fertility as the driving force underlying an increase of population size at the site (Larsen et al. 2015; Milella et al. 2018). Similarly, even if differences in the relative proportion of sexes through time fail to reach statistical significance, the inversion of the ratio between the frequencies of males and females during the Middle period, when females become (and remain) the more numerous sex, may suggest a concomitant role of immigration of women (that is, in the context of patrilocal social systems) in the fertility estimates at the site (cf. Knipper et al. 2017). The lack of significance of these changes may suggest that, even if this hypothesis is correct, female immigration was probably not so pronounced, relatively homogenous through time, and, possibly, counterbalanced by outward mobility of part of the female population of Çatalhöyük.

The specific demographic profile characterising the Middle period may also be indicative of the type of social and environmental challenges facing the population under study. In particular, the peak in neonates in the Late/Final period points to relatively high mortality before two months of age. Considering the stability of fertility estimates across occupation periods, this result suggests that living conditions during the late occupation of the settlement were probably not markedly improved, even in the context of a decrease in population size and density.

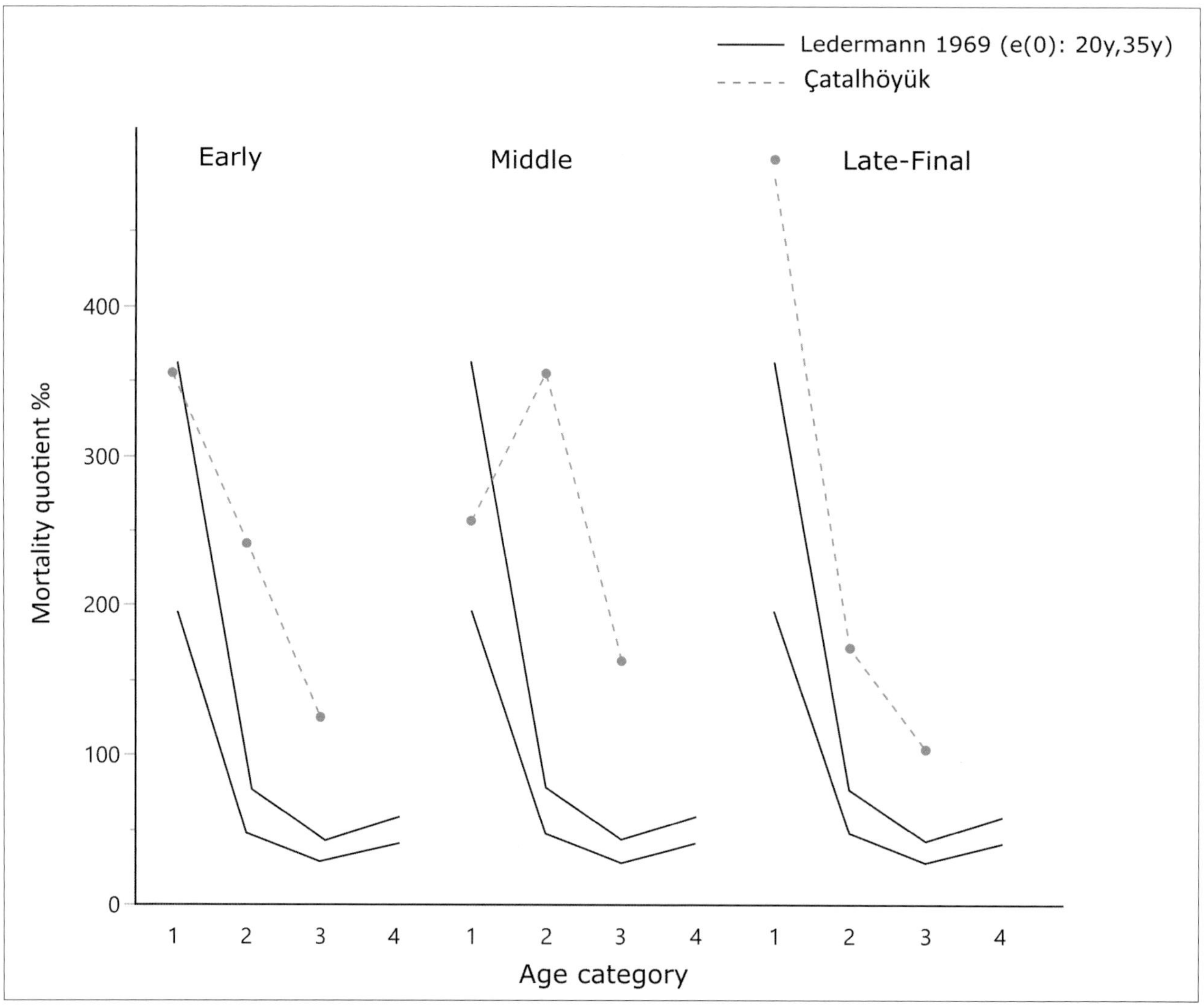

Figure 16.6. Mortality quotients by age category for the youngest members of the Çatalhöyük community (grey dotted line) and expected values for non-epidemic populations (black continuous line, from Ledermann 1969) with life expectancies at birth of 20 and 35 years. Note that age categories do not overlap completely: 1: 1–4 years old (Ledermann), 1–3 years old (Çatalhöyük); 2: 5–9 (Ledermann), >3–12 (Çatalhöyük); 3: 10–14 (Ledermann), >12–19 (Çatalhöyük); 4: 15–19 (Ledermann).

Subadults. Subadult mortality at Çatalhöyük appears to support an interpretation of growing stability in resource availability among the members of the living population and the relatively successful management of population health from the Early period into the Middle period of occupation, although the rise in childhood mortality suggests the presence of independent variables specific to this age category (fig. 16.6). There is evidence for the general improvement of maternal health and well-being into the Middle period, with a strong decline in the Late/Final period reflected in the increase of pre-term and neonatal mortality coupled with declining fertility. The potential for socio-cultural buffering against extrinsic factors affecting foetal viability and reducing pre-term and neonatal mortality is also supported by the mortality profile of the Middle period. There is evidence across the occupation periods for a socio-cultural transition around the age of six months that increased the mortality risk of individuals who had survived birth and early infancy (cf. Chávez et al. 2000; Christian et al. 2015).

Shifts in health and well-being of the adult population may also have influenced later subadult mortality profiles. However, it is almost certainly the case that the peak in child mortality during the Middle period is a precursor to a negative shift in adult health and well-being towards the Late/Final occupation period. The mortality profiles presented here provide evidence for the generational effects of subadult health and well-being on subsequent generations, as well as the impacts of health and environmental stress on subadults, including factors leading to either early death or survival to adulthood.

Finally, the relative proportions of age categories provide the opportunity to test for the presence of epidemic episodes during the Late/Final period at the East Mound, possibly responsible for the abandonment of that part of the site. Comparing the age composition of skeletal samples from plague burial sites with those expected from non-epidemic contexts, Castex and Kacki (2016) highlight in the former a marked under-representation of very young infants (up to 4 years old) and a high number of individuals aged between 5 and 14 years old. Mortality quotients at Çatalhöyük do not show this type of shift, overall fitting in the range expected for non-epidemic contexts (fig. 16.6). This result militates against the presence of major epidemic outbreaks during the life of the settlement, and excludes a sudden, anomalous mortality shift at the end of the occupation of the East Mound.

Growth: stature and body mass

Environment and health conditions greatly influence the process of growth that results in the attainment of stature and body mass (Bogin 1999). Given socio-economic changes occurring during the occupation of Çatalhöyük, this section explores the temporal changes in body size of infants and children recovered from the site. Stature and body mass were estimated for a sample excavated since 1993 associated with dental remains that also met infra-cranial preservation criteria required for cross-sectional geometric analyses (table 16.7) (see below). To provide a context for these results, Çatalhöyük growth profiles are compared to growth profiles from two relatively skeletally healthy (that is, lacking a high prevalence of bone lesions) populations: the Denver Growth Study, a longitudinal mid-20th century US middle-class sample (McCammon 1970) and a sample of Arikara from the Sully sites, a proto-historic indigenous North American cemetery group from South Dakota (Jantz, Owsley 1984). Table 16.8 provides descriptive statistics for these samples. Dental age was assigned to Çatalhöyük and Arikara individuals based on tooth length regression equations and dental development (Smith 1991; Liversidge 1994; 2008; Liversidge, Molleson 2004; AlQahtani et al. 2010). Stature estimations for ages at death $\geq$0.5 years were calculated from femoral length and body mass (ages at death $\geq$1 year) from distal femoral metaphyseal breadth or femoral head supero-inferior breadth using age-specific equations (Ruff 2007). Mean adult values for stature (Ruff et al. 2012) and body mass (Ruff et al. 2018) indicate an 'end point' for growth. Residuals were calculated from polynomial regressions fit through the pooled Çatalhöyük sample to compare growth among occupation periods. Body size residuals for each Neolithic occupation period: Early, Middle and Late (where the Final period is collapsed into Late) were compared using Kruskal-Wallis ANOVA tests with Bonferroni adjusted post-hoc multiple comparison test of mean ranks between groups.

	Pooled Çatalhöyük	
	Femur	*Humerus*
Age range (years)	*N*	*N*
Perinate to 0.9	16	15
1 to 2.9	9	8
3 to 4.9	12	8
5 to 8.9	10	6
9 to 12.9	2	1
13 to 17.9	6	4
18 to 21	8	5
Subadult total	63	47
Adult total	71	48

Table 16.7. Adult and subadult pooled cross-section sample by element.

	Çatalhöyük means				*Arikara means*				*Denver means**			
	Stature (cm)		*Body mass (kg)*		*Stature (cm)*		*Body mass (kg)*		*Stature (cm)*		*Body mass (kg)*	
Age range (years)	n		n		n		n		n		n	
0.5 to 1	7	65.6			2	64.3						
1 to 2.9	12	77.4	12	9.5	4	69.6	4	8.5	20	78.7	20	10.3
3 to 4.9	14	93.0	14	13.4	6	89.7	6	12.9	20	97.7	20	14.6
5 to 8.9	11	114.6	11	21.5	6	103.7	6	16.8	20	117.7	20	21.4
9 to 12.9	2	135.9	2	28.4	6	135.8	6	35.9	20	140.8	20	34.0
13 to 17.9	6	140.0	6	39.4	3	158.9	3	53.6	20	165.8	20	55.5
18 to 21	11	153.6	12	53.9	1	166.9	1	72.0				
Subadult total	63		57		28		26		20		20	
Adult total	122	158.2	123	58.7	9	159.6	9	62.4	20	173.4	20	64.5

*Table 16.8. Growth profile for Çatalhöyük by dental age, categorised by temporal period. Yearly means from the Denver Growth Study (McCammon 1970; means derived from Ruff, unpublished data). Mean adult male and female values are plotted at 23 years of age (a) for stature (Çatalhöyük stature (cm)= 61.4925+8.8439*x-0.2063*x^2; Arikara stature (cm)= 54.044+9.7884*x-0.2197*x^2; Denver stature (cm)= 63.9238+9.8858*x-0.2192*x^2); (b) for (log) body mass (kg) (Çatalhöyük (log) body mass= 1.8917+0.2038*x-0.0063*x^2+6.905E-5*x^3; Arikara (log) body mass= 1.8334+0.1825*x-0.002*x^2-6.9381E-5*x^3; Denver (log) body mass= 2.1337+0.1442*x+0.0004*x^2-0.0001*x^3).*

Figure 16.7a presents stature growth profiles from the Çatalhöyük, Arikara and Denver samples, each fit with sample-specific second order polynomial regressions. While the Denver growth profile demonstrates a greater mean stature throughout the growth process, all profiles overlap considerably prior to 6 years of age. Arikara and Çatalhöyük stature profiles are similar into adulthood and individuals from these groups are shorter than those of the Denver sample. Body size differences that appear during later childhood and adolescence may imply emerging genetic body size differences rather than health disturbances (Johnston et al. 1976; Frisancho et al. 1980; Bogin 1999; WHO 1999), but could also indicate that the later stages of pubertal maturation were delayed, perhaps due to environmental stress (Shapland, Lewis 2013). Temporal comparisons within the Çatalhöyük sample indicate there is no significant difference in stature residuals.

Growth in body mass is shown in fig. 16.7b with third order polynomial regressions fit to each sample. As also found with stature, body mass overlaps extensively in earlier childhood in the three samples, but the growth curves diverge in adolescence. It is notable that the majority of the Çatalhöyük individuals aged 13 to 21 years of age were assigned female sex (F: n=10; M: n=3; U: n=5), which is driving the divergence in mean body size estimates for that age range. While adolescent Çatalhöyük body mass estimates are lower than those of the Arikara and Denver samples, comparisons between Arikara and Çatalhöyük adult body mass estimates do not differ statistically. Temporal differences in body mass residuals are not significantly different between periods.

It is thought that growth during infancy and early childhood is more sensitive to environmental and nutritional disruption than growth occurring in adolescence, when genetic effects are increasingly influential (Habicht et al. 1974; Johnston et al. 1976; Frisancho et al. 1980; Bogin 1999; WHO 1999). Taken together, while body size differences are present among adolescents and adults, the growth profiles for all three samples are quite similar during early growth years, suggesting a pre- and post-natal environment where conditions (nutritional, metabolic, infectious, etc.) do not appear to have impaired the growth process. Similarly, the ontogenetic series complements adult body size analyses that do not identify secular trends at any time in the occupation of the settlement.

Dental bioarchaeology

The study of teeth is an important component of bioarchaeological research due to the amount of information they can provide, and because they are the skeletal elements that preserve best. Contrary to the rest of the skeleton, teeth do not undergo remodelling throughout life; their characteristics are set during development. Alterations in their structures are thus permanent and provide insight into disturbances occurring during growth, insults that reflect on the well-being of infants and children (cf. El-Najjar et al. 1978; Goodman, Rose 1990; Skinner, Goodman 1992; Hillson 2014: 41), as well as health status of the oral cavity, the deciduous and permanent dentitions,

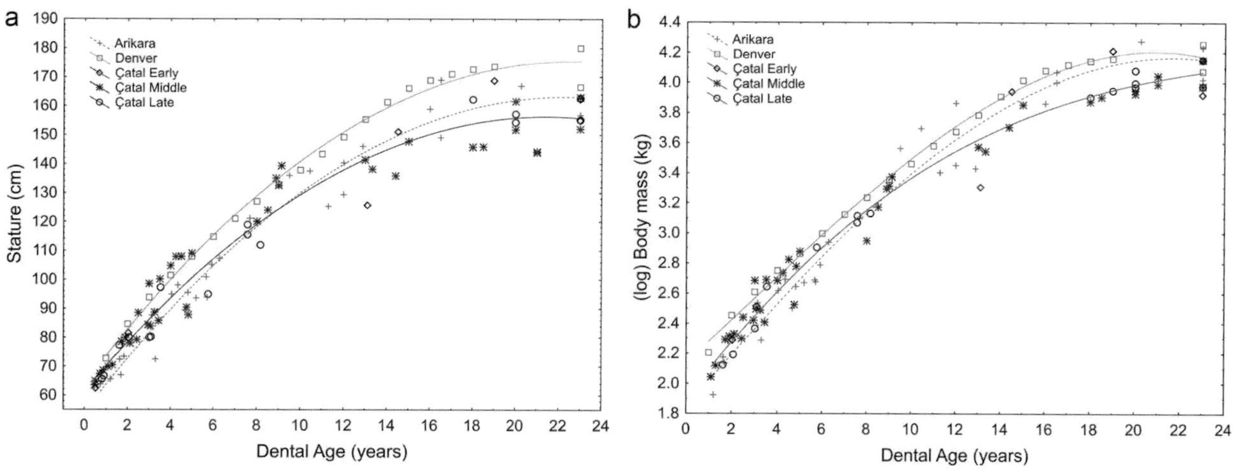

Figure 16.7. Growth profiles by (a) stature and (b) body mass for Çatalhöyük by dental age, categorised by temporal period and compared with the Denver Growth Study and Arikara samples. Yearly means from the Denver Growth Study (McCammon 1970; means derived from Ruff, unpublished data). Mean adult male and female values are plotted at 23 years of age (a) for stature (Çatalhöyük stature (cm)= $61.4925+8.8439*x-0.2063*X^2$; Arikara stature (cm)= $54.044+9.7884*x-0.2197*X^2$; Denver stature (cm)= $63.9238+9.8858*x-0.2192*X^2$) and (b) for (log) body mass (kg) (Çatalhöyük (log) body mass= $1.8917+0.2038*x-0.0063*X^2+6.905E-5*X^3$; Arikara (log) body mass= $1.8334+0.1825*x-0.002*X^2-6.9381E-5*X^3$; Denver (log) body mass= $2.1337+0.1442*x+0.0004*x^2-0.0001*X^3$).

diet and culturally mediated dental modifications of past populations (cf. Molnar 1971; 1972; Smith 1984). The sections below consider insights into population health and well-being of the Neolithic inhabitants of Çatalhöyük through considerations of dental fluctuating asymmetry, linear enamel hypoplasia (LEH), and diet from the perspective of dental disease (caries). Further, social structure and mobility are explored through an analysis of dental metric and morphological variation.

Dental fluctuating asymmetry

One means by which to test the effect of stress on the human body and its structures is through random deviations from symmetry in bilateral traits, a phenomenon referred to as 'fluctuating asymmetry' (Van Valen 1962; Klingenberg 2003). Several studies have adopted fluctuating asymmetry of dental, craniofacial and skeletal structures as a tool for exploring patterns of developmental stress in living and archaeological populations (Perzigian 1977; Doyle, Johnston 1977; Costa 1986; Kieser et al. 1986; Albert, Greene 1999; Hoover et al. 2005; DeLeon 2007; Hoover, Matsumura 2008; Kujanova et al. 2008; Barrett et al. 2012; Gawlikowska-Srokaet al. 2013; 2017). Here, the results of the analysis of dental fluctuating asymmetry (henceforth DFA) at Neolithic Çatalhöyük (Milella et al. 2018) are presented and discussed.

Materials and methods. Analyses of DFA were performed on a sample of 259 individuals representing all occupation periods at Çatalhöyük, including both sexes and all age categories (table 16.9). Bucco-lingual (BL) and mesio-distal (MD) diameters of the permanent canines (C), first molars (M1) and second molars (M2) of both maxillae and mandibulae were measured twice. The presence of significant DFA was tested for each measurement by means of a two-way mixed-model analysis of variance (ANOVA) with individuals as the random factor and sides as fixed to assess the relative amount of variance due to DFA versus measurement error (ME). DFA for each trait was quantified by means of the index $|(R-L)|$ (Palmer, Strobeck 2003) for each individual. Differences in DFA between periods, age groups and sexes were then tested by means of Kruskall-Wallis and Kendall Tau correlation tests. The Late and Final occupation periods were combined in a single chronological period due to the extremely small sample size of the Final period group.

Results. Five of the measurements, the mesio-distal and bucco-lingual diameters of the upper second molar (UM2MD and UM2BL, respectively), the mesio-distal diameter of the lower first molar (LM1MD) and lower canine (LCMD), and the bucco-lingual diameter of the lower second molar (LM2BL), showed significant fluctuations in asymmetry and were thus targeted for analysis. The comparison between occupation periods indicates a general decrease in asymmetry with time (fig. 16.8a), although this result is significant only for females for LM2BL (fig. 16.8b). When comparing asymmetry between subadults (<

	Early			*Middle*			*Late/Final*			*Unstratified Neolithic*			*Total*
	F	M	I	F	M	I	F	M	I	F	M	I	
Infant (2 months–3 y)	0	0	0	0	0	3	0	0	1	0	0	0	4
Child (3–12 y)	0	0	5	0	0	58	0	0	13	0	0	5	81
Adolescent (12–19 y)	0	1	1	3	2	17	2	1	5	0	1	3	36
Young adult (20–34 y)	2	2	1	13	12	1	5	4	0	2	0	2	44
Mature adult (35–50 y)	2	2	0	13	10	0	7	7	1	2	1	1	46
Older adult (> 50 y)	2	2	0	7	2	1	0	3	0	0	0	0	17
Adult (> 20 y)	1	0	0	6	6	2	3	6	3	1	1	2	31
Total	7	7	7	42	32	82	17	21	23	5	3	13	259

Table 16.9. Sample size for fluctuating asymmetry by sex, age and period of occupation. y=years; M=males; F=females; I=indeterminate sex.

20 years old) and adults (≥ 20 years old) subadults show greater asymmetry in three out of five measurements (LM2BL, LM1MD, and LCMD). Conversely, no significant difference was found between sexes.

Discussion. The results of these analyses indicate a subtle decrease in asymmetry with time, a trend consistent with the expected decrease in population density from the Middle period resulting in lower environmental stress. Note, however, that in only one case is this pattern significant, in females for LM2BL. Three main factors may be responsible for the observed patterns:

1. Contrasting developmental stability between different tooth classes (Bailit et al. 1970; Garn et al. 1966; 1967; Sofaer et al. 1971; Garn, Bailey 1977; Perzigian 1977; Harris, Nweeia 1980; Hershkovitz et al. 1993; Guatelli-Steinberg et al. 2006).
2. The presence of a moderate amount of environmental stress throughout the entire occupation of the site, *even* in a situation of decreased population density, though mitigated perhaps by a shared diet between the sexes. This would explain the lack of differences between sexes. It is indeed possible that the exposure to relatively high developmental stress would have obscured sex differences in immuno-competence and fragility.
3. A non-linear relationship between population size and population density. The extent of occupation of Çatalhöyük probably varied during the 1,000 years plus of occupation (cf. Cessford 2005). Occupied areas of larger and smaller size would have led to population density variation, given a specific population size.

The difference in asymmetry between adults and subadults suggests an inverse correlation between the amount of developmental stress and age at death. In other words, greater asymmetry is significantly associated with premature mortality. This pattern not only supports an interpretation of dental FA as a marker of developmental stress but is also in agreement with the age distribution and relative prevalence of periostosis at Çatalhöyük, and is in line with the postulated weakened immuno-competence of juveniles in this population (Larsen et al. 2015).

Linear enamel hypoplastic (LEH) defects

Another means of assessing well-being from the dentition comes in the form of linear enamel hypoplasia (LEH). These are developmental defects visible on tooth crown surfaces as lines of decreased enamel thickness caused by a disruption of enamel secretion during crown formation (Goodman et al. 1984). Because enamel does not remodel after crown formation, these defects provide an enduring record of physiological stress experienced by an individual during tooth crown development, which spans infancy and childhood (cf. Klaus, Tam 2009). These defects result from systemic metabolic insults, such as stress episodes caused by high fevers, infectious agents, malnutrition, weaning and hormonal changes (Goodman et al. 1980; Goodman, Armelagos 1985a; 1985b; Dobney, Goodman 1991; Guatelli-Steinberg 2001; Goodman, Martin 2002).

LEH defects were present on permanent mandibular canine teeth of 122 of 123 (99%) individuals studied from Neolithic Çatalhöyük. These data were collected from fully developed canine tooth crowns with Smith (1984) dental wear scores of five or lower using a 10x hand lens. Defects were only counted if they could be matched on at least two canines per individual (Smith 1984; Goodman, Rose 1990; Buikstra, Ubelaker 1994; Steckel et al. 2006).

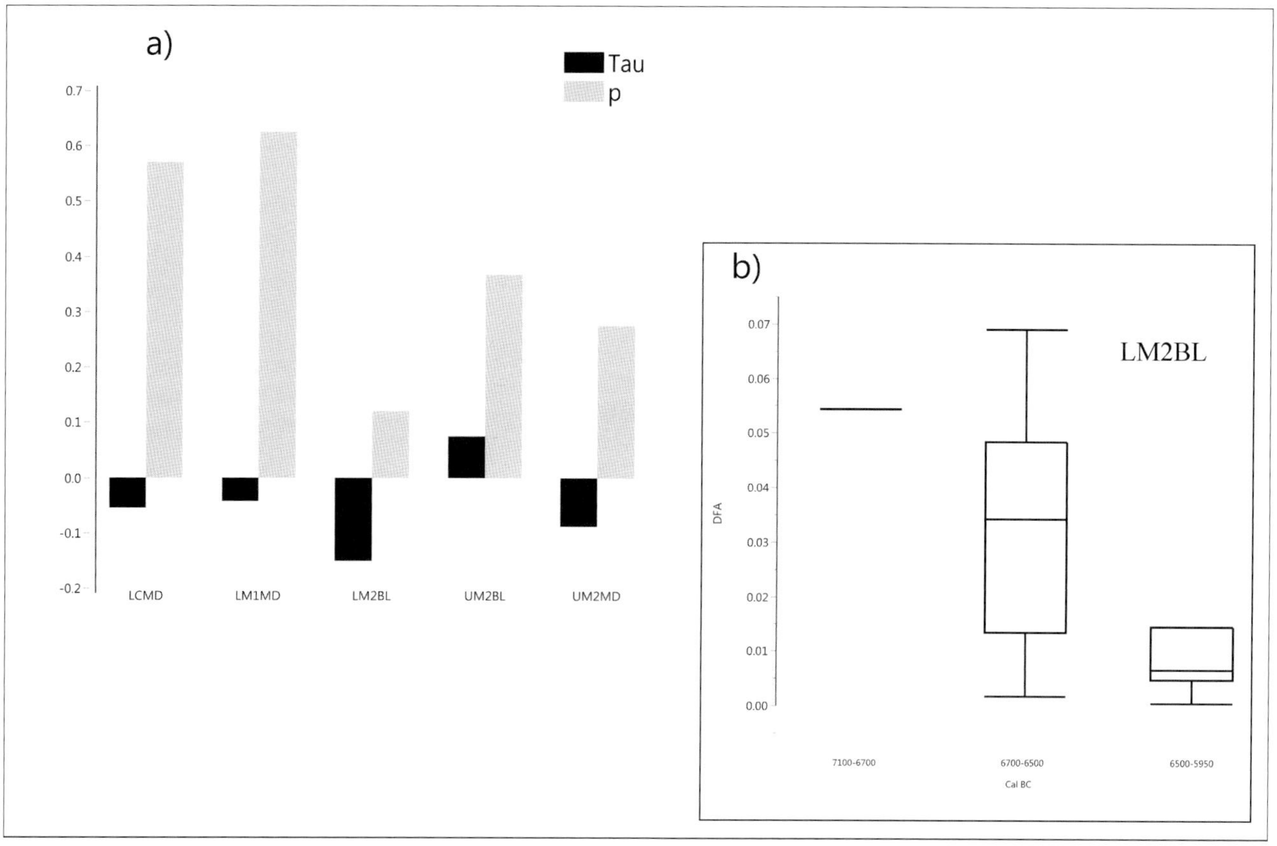

Figure 16.8. Patterns of dental fluctuating asymmetry through the occupation periods at Çatalhöyük: (a) correlation between DFA and occupation periods, combined sexes. Note that four out of the five measurements show a negative (but not significant) correlation, indicating decreasing asymmetry through time; (b) a box-plot of DFA of bucco-lingual diameter of the lower second molar in females through time. In this case, the decrease in asymmetry is significant.

These results are consistent with those of a previous study that found 100% LEH prevalence in a sample of maxillary and mandibular central and lateral incisors from 54 individuals from Neolithic Çatalhöyük (Bocaege 2015).

Comparisons between occupation periods illustrate that LEH defect prevalence was high throughout the Neolithic occupation of the site (fig. 16.9). In the Early period (n=11), 100% of individuals had at least one enamel defect on their mandibular canines, and ten individuals (90.9%) had two or more defects. In the Middle period (n=77), 98.7% of individuals had at least one enamel defect on their mandibular canines, and 72 individuals (93.5%) had two or more defects. In the Late and Final periods (n=22), 100% of individuals had two or more defects on their mandibular canines. While there are no statistically significant differences between LEH prevalence during different occupation periods, it is interesting to note that during the Early and Middle periods there are at least a few individuals who experienced only one stress episode leading to the formation of an enamel defect, whereas in the Late period all individuals experienced multiple defect-causing stress episodes.

This record of physiological stress is consistent with expectations for a community living a relatively sedentary way of life in a setting characterised by high population size and density (Cohen, Armelagos 1984; Goodman et al. 1988; Armelagos, Goodman 1991; Larsen 2015). Such living conditions are often associated with dependence on domesticated plants and animals and tend to expose community members to insults to well-being relating to diet, sanitation, pathogen exposure, and other factors that promote stress in general (Larsen 1995; Armelagos et al. 2005; Bocquet-Appel 2008). Research employing more fine-grained enamel microstructure analyses of LEH defect timing and duration is ongoing and should lead to additional insights into the distribution of risk of experiencing physiological stress episodes during childhood within the Neolithic population at Çatalhöyük. While these developmental assessments of well-being from disrupted growth provide insight into the life experiences in the very youngest members of the Çatalhöyük community, palaeopathological analyses of oral health from the study of adult dentition lend insights into the exposure to pathogens in later life based on dietary constituents.

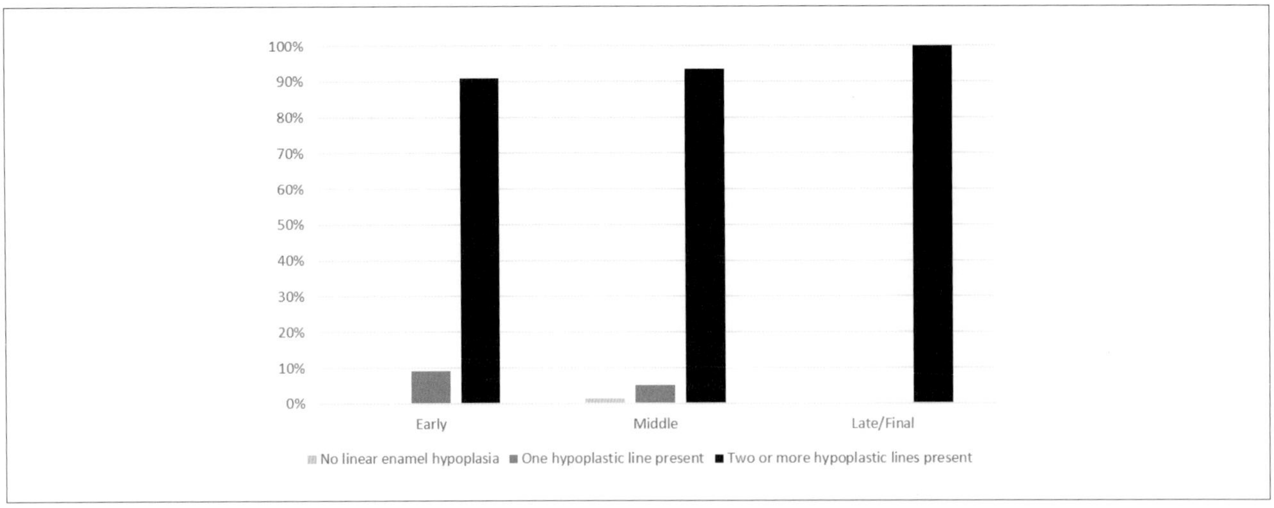

Figure 16.9. Prevalence of linear enamel hypoplasia (LEH) defects per individual on permanent mandibular canine teeth within the Early, Middle, and Late occupation periods at Neolithic Çatalhöyük.

Oral health: dental caries at Neolithic Çatalhöyük

Tooth decay or caries, an oral infectious disease, involves focal demineralisation of the dental hard tissues – enamel, dentine and cementum – by acids produced by bacterial fermentation of dietary carbohydrates, sugars and starches (Aas et al. 2008; Larsen 2018). The demineralisation of dental tissues ranges from small opacities in the enamel to cavities that expose the underlying dentine and pulp that may eventually destroy the crown and the root of the tooth. This infectious disease is commonly associated with low pH (acidic) saliva, poor oral hygiene, genetic and hormonal factors, crown morphology, extent of wear, and age-related increase in prevalence (cf. Rowe 1982; Powell, 1985; Hara, Zero 2010). There is a close relationship between the prevalence of carious lesions and the types of food consumed, their carbohydrate content and food-preparation techniques. Analyses of caries can therefore provide insights into environmental stressors and the diet of past populations that shed light on diachronic, sex and age-related patterns in lifestyle.

This study analysed prevalence of carious lesions in the permanent dentition of a large sample of individuals (n=137), taking into account sex, age-at-death and occupation period (tables 16.10 and 16.11). Given that demographic pressure/crowded conditions are considered to correlate with decreased health and well-being, an increase in dento-alveolar pathologies and sex-related differences during the period of greatest site complexity and population density is expected during the Middle occupation period (6610–6250 cal BC) (cf. Cessford 2005) (Larsen et al. 2015; Milella et al. 2018).

Crown and root carious lesions were assessed macroscopically following Hillson (2001). For the purposes of this study, five scores were used to describe the nature of the lesions: 1) lesions penetrating only the enamel of the crown; 2) lesions penetrating the enamel and the dentine; 3) lesions reaching the pulp chamber; 4) lesions on the root surface; and 5) gross caries comprised of large cavities, in which it is impossible to determine the site of origin, and multiple lesions on the same tooth. Observations were scored both by individual and by affected tooth. Scores for upper and lower teeth and left and right sides were combined.

Table 16.11 compares the presence/absence frequencies of caries in individuals and by tooth in males, females and those of indeterminate sex by age group and period. Overall, some 50.0% of individuals have at least one tooth affected by caries, for a total of almost 8.0% of teeth with lesions. It is apparent that the disease is strongly age-related, with a significant increase in the older age category (p-value: <0.001). When considering subsamples by sex and pooling age categories, there is no statistically significant difference between males and females in the individual prevalence of caries (F=65.5%, M=75.0%; p-value: 0.20) or by tooth (F=11.8%, M=10.2%; p-value: 0.20). Conversely, within some age groups, a difference between sexes can be observed. In particular, mature adult males have a slightly higher frequency of caries per individual when compared to females in the same age category (F=74.2%, M=93.1%; p-value: 0.05). The reverse of this pattern is observed in the 'older adult' age category, where females have more affected teeth in comparison with those of males of the same age-at-death (F=24.9%, M=16.6%, p-value: 0.009). However, this result may be biased by errors in the age-at-death

	Females	*Tooth category (n, %)*				*Males*	*Tooth category (n, %)*				*N/D*	*tooth category (n, %)*			
	N	*Incisors*	*Canines*	*Premolars*	*Molars*	*N*	*Incisors*	*Canines*	*Premolars*	*Molars*	*N*	*Incisors*	*Canines*	*Premolars*	*Molars*
Early Period															
1	2	0 (0.0)	0 (0.0)	1 (50.0)	1 (50.0)	2	0 (0.0)	0 (0.0)	1 (50.0)	1 (50.0)	0	0 (0.0)	0 (0.0)	0 (0.0)	0 (0.0)
2	0	0 (0.0)	0 (0.0)	0 (0.0)	0 (0.0)	1	0 (0.0)	0 (0.0)	0 (0.0)	1 (100.0)	0	0 (0.0)	0 (0.0)	0 (0.0)	0 (0.0)
3	0	0 (0.0)	0 (0.0)	0 (0.0)	0 (0.0)	0	0 (0.0)	0 (0.0)	0 (0.0)	0 (0.0)	0	0 (0.0)	0 (0.0)	0 (0.0)	0 (0.0)
4	11	1 (9.1)	0 (0.0)	2 (18.2)	8 (72.7)	16	0 (0.0)	0 (0.0)	3 (18.8)	13 (81.2)	0	0 (0.0)	0 (0.0)	0 (0.0)	0 (0.0)
5	9	0 (0.0)	0 (0.0)	3 (33.3)	6 (66.7)	3	0 (0.0)	0 (0.0)	0 (0.0)	3 (100.0)	0	0 (0.0)	0 (0.0)	0 (0.0)	0 (0.0)
Total	**22**	**1 (4.5)**	**0 (0.0)**	**6 (27.3)**	**15 (68.2)**	**22**	**0 (0.0)**	**0 (0.0)**	**4 (18.2)**	**18 (81.8)**	**0**	**0 (0.0)**	**0 (0.0)**	**0 (0.0)**	**0 (0.0)**
Middle Period															
1	18	2 (11.1)	1 (5.5)	1 (5.5)	14 (77.9)	17	1 (5.9)	0 (0.0)	0 (0.0)	16 (94.1)	12	0 (0.0)	1 (8.3)	0 (0.0)	11 (91.7)
2	6	0 (0.0)	2 (33.3)	0 (0.0)	4 (66.6)	7	0 (0.0)	0 (0.0)	1 (14.3)	6 (85.7)	4	0 (0.0)	0 (0.0)	0 (0.0)	4 (100.0)
3	8	1 (12.5)	1 (12.5)	2 (25.0)	4 (50.0)	1	1 (100.0)	0 (0.0)	0 (0.0)	0 (0.0)	0	0 (0.0)	0 (0.0)	0 (0.0)	0 (0.0)
4	67	5 (7.5)	4 (6.0)	12 (17.9)	46 (68.6)	40	3 (7.5)	1 (2.5)	11 (27.5)	25 (62.5)	6	0 (0.0)	0 (0.0)	4 (66.7)	2 (33.3)
5	51	1 (1.9)	0 (0.0)	13 (25.5)	37 (72.6)	21	1 (4.8)	1 (4.8)	2 (9.5)	17 (80.9)	4	0 (0.0)	0 (0.0)	1 (25.0)	3 (75.0)
Total	**150**	**9 (6.0)**	**8 (5.3)**	**28 (18.7)**	**105 (70.0)**	**86**	**6 (7.0)**	**2 (2.3)**	**14 (16.3)**	**64 (74.4)**	**26**	**0 (0.0)**	**1 (3.8)**	**5 (19.2)**	**20 (76.9)**
Late/Final Period															
1	6	0 (0.0)	0 (0.0)	0 (0.0)	6 (100.0)	9	0 (0.0)	0 (0.0)	0 (0.0)	9 (100.0)	1	0 (0.0)	0 (0.0)	0 (0.0)	1 (100.0)
2	2	0 (0.0)	0 (0.0)	0 (0.0)	2 (100.0)	0	0 (0.0)	0 (0.0)	0 (0.0)	0 (0.0)	0	0 (0.0)	0 (0.0)	0 (0.0)	0 (0.0)
3	0	0 (0.0)	0 (0.0)	0 (0.0)	0 (0.0)	2	0 (0.0)	0 (0.0)	0 (0.0)	2 (100.0)	0	0 (0.0)	0 (0.0)	0 (0.0)	0 (0.0)
4	20	0 (0.0)	0 (0.0)	9 (45.0)	11 (55.0)	25	0 (0.0)	3 (12.0)	4 (16.0)	18 (72.0)	2	0 (0.0)	0 (0.0)	0 (0.0)	2 (100.0)
5	11	4 (36.4)	0 (0.0)	1 (9.1)	6 (54.5)	20	1 (5.0)	4 (20.0)	5 (25.0)	10 (50.0)	2	0 (0.0)	0 (0.0)	0 (0.0)	2 (100.0)
Total	**39**	**4 (10.3)**	**0 (0.0)**	**10 (25.6)**	**25 (64.1)**	**56**	**1 (1.8)**	**7 (12.5)**	**9 (16.1)**	**39 (69.6)**	**5**	**0 (0.0)**	**0 (0.0)**	**0 (0.0)**	**5 (100.0)**

Table 16.10. Frequencies of lesion types (from 1 to 5), between the sexes and period (by tooth); N=total teeth affected by caries; n=carious lesions divided by tooth category.

	Females				*Males*				*p (chi-squared test F vs M by occupation period)*	
	Individual		*Tooth*		*Individual*		*Tooth*		*Individual*	*Tooth*
	n/N	*%*	*n/N*	*%*	*n/N*	*%*	*n/N*	*%*		
Early Period										
Root caries (4) & gross caries/multiple caries (5)	6/6	100.0	20/22	90.9	5/7	71.4	19/22	86.4	-	0.6
Middle Period										
Root caries (4) & gross caries/multiple caries (5)	37/44	84.1	118/150	78.7	22/28	78.6	61/86	70.9	0.5	0.2
Late/Final Period										
Root caries (4) & gross caries/multiple caries (5)	10/13	76.9	31/39	79.5	17/19	89.5	45/56	80.3	0.3	0.9
Total	**53/63**	**84.1**	**169/211**	**80.1**	**44/54**	**81.5**	**125/164**	**76.2**	**0.7**	**0.4**
p (chi-squared test between occupation periods)	0.5		0.4		0.5		0.2			

Table 16.11. Frequencies of presence/absence of caries by individual and by tooth in both sexes and in individuals of indeterminate sex. The sample is divided by age groups and period, n=individual/tooth with caries; N=individual/tooth analysed. Significant p values are highlighted in bold.

estimation for individuals who have reached skeletal maturity. Table 16.12 also reports the results by period: no consistent differences between periods could be observed for individuals or for teeth. Only during the Middle period is there a slightly higher frequency of caries per tooth in females compared to males (F=12.9, M=10.1; p-value: 0.05).

Table 16.10 shows the frequencies by lesion type (from 1 to 5) in incisors, canines, premolars and molars between the sexes and by period (see also figs 16.10a and 16.10b). In all occupation periods and in both sexes, molars are the most often affected teeth, followed by premolars. In females, there is less difference between premolars and molars for the root surface (4) and gross lesions (5) in comparison with males. Table 16.12 reports the differences in lesion types between males and females by individual and affected tooth by occupation period. In none of these categories are there consistent differences and there is no clear trend in lesion types.

The results of this study show that the prevalence of carious lesions by individual and by tooth significantly increases with age at death. This is expected due to the fact that caries is a degenerative infectious disease and increments in their number reflect the greater exposure time to cariogenic factors between adults and younger individuals. In the pooled Neolithic sample, the pattern of caries frequencies appears similar between the sexes, which may suggest a similar diet in adult men and women. However, it is interesting to note that females have a greater prevalence of dental caries than males in the Middle period, which is the period of peak fertility, site complexity and presumably greater demographic pressure and population crowding. This result could hint at subtle sex-based differences in diet, health and well-being (e.g., Lukacs 2008; 2011) and, in general, access to resources, which are not expressed in other biological and cultural indicators of social differentiation, such as those based on stable isotope analysis (Pearson 2013; Pearson, Meskell 2014; Larsen et al. 2015). This result highlights the usefulness of different analytical approaches and types of data when discussing the complex and subtly nuanced biological correlates of social distinction in past populations.

Stable isotope analysis

Summary of previous stable isotope work

Previous work on human diet of those interred at Çatalhöyük has shown that it was based on C_3 plants, despite C_4 plants being present in the diet of both wild and domestic animals. Social differences in diet were limited, with males and females having diets that were identical isotopically (Pearson 2013). Later work subsequently showed that age-at-death was more important for determining not just diet (Pearson, Meskell 2013; Pearson et al. 2015b), but was used in the representation of the figurine corpus and related

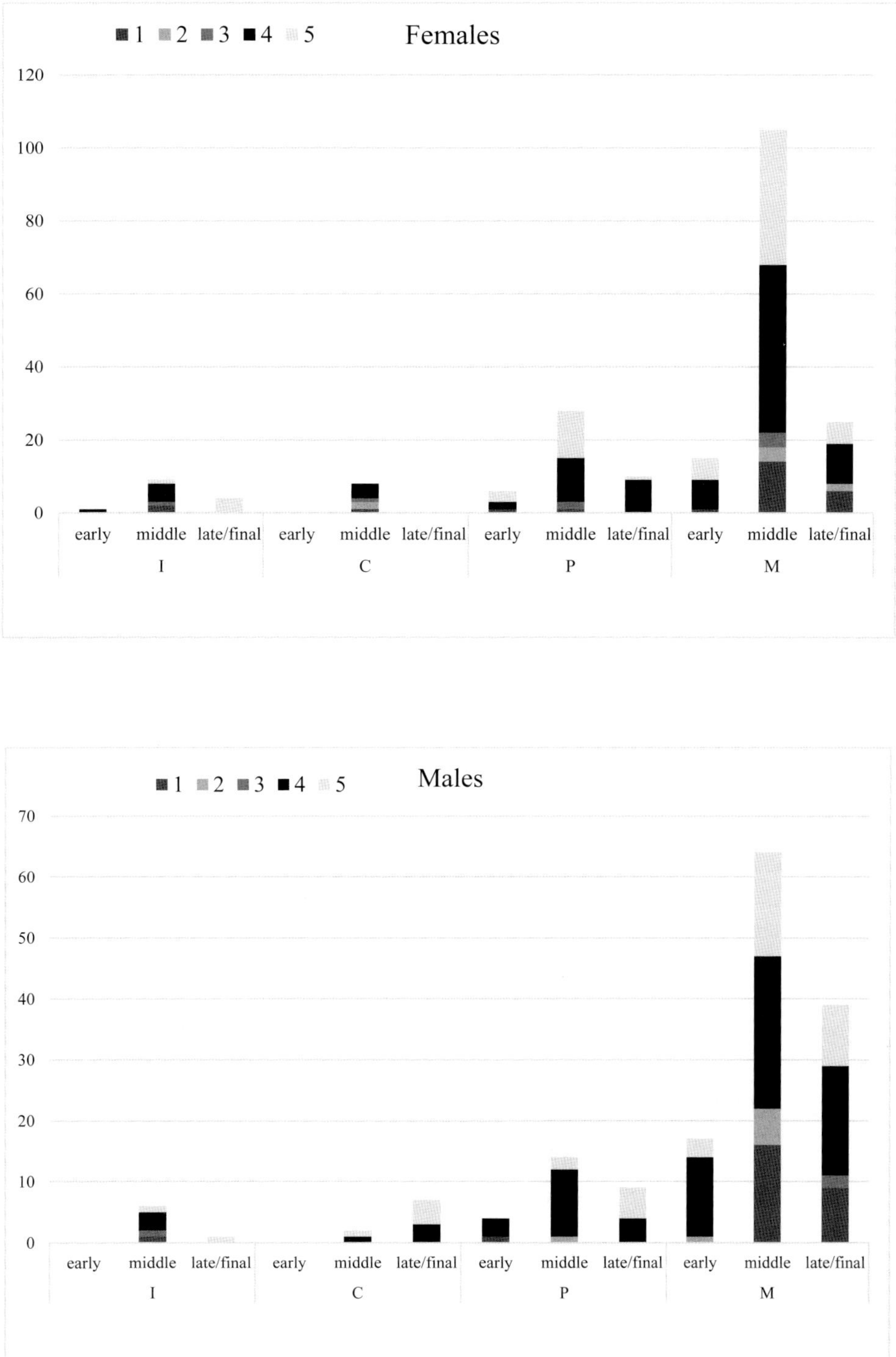

Figure 16.10. Frequencies of the severity of carious lesions scored from 1 to 5 in incisors (I), canines (C), premolars (P) and molars (M) between the sexes [(a): females, (b): males] over the course of site occupation. Upper and lower teeth and left and right sides are combined.

to numbers of grave goods as well (Nakamura, Meskell 2009; 2013a; 2013b). Attempts were made to use these isotope data to understand the nature of the use of houses, whether or not they served as kinship units insofar as diets being similar among those interred within them. However, where middens could be associated with confidence to a particular house, the faunal remains there showed little relationship with the values of the individuals buried in the associated house, perhaps suggesting a missing dietary element, such as aquatic species. Additional data are now available from these resources to determine how important they may have been.

	Females				*Males*				*Indeterminate*				*p (chi-squared test F vs M by age category)*		*Total*		
Age categories	*Individual*		*Tooth*		*Individual*		*Tooth*		*Individual*		*Tooth*		*Individual*	*Tooth*	*Individual*		*Tooth*
	n/N	*%*	*n/N*	*%*	*n/N*	*%*	*n/N*	*%*	*n/N*	*%*	*n/N*	*%*			*n/N*	*%*	*n/N*
Child (6-12 yr)	-	-	-	-	-	-	-	-	6/54	11.1	7/855	0.8	-	-	6/54	11.1	7/855
Adolescent (12-20 yr)	0/5	0.0	0/107	0.0	2/5	40.0	4/102	3.9	6/28	21.4	9/551	1.6	-	-	8/38	21.0	13/760
Young adult (20-35 yr)	11/25	44.0	31/514	6.0	10/19	52.6	26/464	5.6	1/4	25.0	1/79	1.3	0.6	0.8	22/48	45.8	58/1057
Mature adult (35-50 yr)	23/31	74.2	66/609	10.8	27/29	93.1	69/583	11.8	2/4	50.0	3/51	5.9	0.05	0.6	52/64	81.2	138/1243
Older adult (50 + yr)	23/26	88.5	91/366	24.9	12/15	80.0	49/295	16.6	3/4	75.0	7/44	15.9	0.5	0.009	38/45	84.4	147/705
Total	**57/87**	**65.5**	**188/1596**	**11.8**	**51/68**	**75.0**	**148/1444**	**10.2**	**18/94**	**19.1**	**27/1580**	**1.7**	**0.2**	**0.2**	**126/249**	**50.6**	**363/4620**
p (chi-squared test between age categories)	0.002		<0.001		0.003		<0.001		0.01		<0.001				<0.001		<0.001

	Females				*Males*				*Indeterminate*				*p (chi-squared test F vs M by period of occupation)*		*TOTAL*		
Periods	*Individual*		*Tooth*		*Individual*		*Tooth*		*Individual*		*Tooth*		*Individual*	*Tooth*	*Individual*		*Tooth*
	n/N	*%*	*n/N*	*%*	*n/N*	*%*	*n/N*	*%*	*n/N*	*%*	*n/N*	*%*			*n/N*	*%*	*n/N*
Early Period	6/8	75.0	22/210	10.5	7/9	77.8	22/227	9.7	0/6	0.0	0/168	0.0	0.9	0.8	13/23	56.5	44/605
Middle Period	44/60	73.3	150/1161	12.9	28/40	70.0	86/856	10.1	16/69	23.2	23/1037	2.5	0.7	0.05	88/169	52.1	259/3054
Late/Final Period	13/26	50.0	39/369	10.6	19/28	67.8	56/458	12.2	4/19	21.1	5/373	1.3	0.2	0.4	36/73	49.1	100/1200
Total	**63/94**	**67.0**	**211/1740**	**12.12**	**54/77**	**70.1**	**164/1541**	**10.6**	**20/94**	**21.3**	**28/1578**	**1.8**	**0.7**	**0.2**	**137/265**	**51.7**	**403/4859**
p (chi-squared test between periods of occupation)	0.09		0.3		0.8		0.4		0.8		0.3				0.8		0.6

Table 16.12. Differences in lesion types between sexes and by period. All analyses are by individuals and affected teeth; n=individual/tooth with root caries (4) and gross caries/multiple caries (5); N=total individual/tooth with caries; significant p values are highlighted in bold.

Within the houses themselves, individuals frequently had different diets to one another, with at least some of the differences likely being the result of age-related food sharing. In order to provide a clearer picture of the variation between individuals buried within different houses, the complete dataset of stable carbon and nitrogen isotope values from human bone collagen is discussed below, and has been augmented with strontium isotope analysis to establish whether some variation in diet might be the result of outsiders arriving at the site as adults.

Current work

The complete dataset of humans and animals is discussed in full by Pearson (Chapter 11) and adds a further 494 animal remains providing an overall total of 987 faunal specimen measurements used to reconstruct the human diet discussed here. A further 155 humans are also discussed below, including another 75 individuals to corroborate previous evidence of weaning practices, employing a larger dataset. The new faunal dataset has shown that the faunal remains at the site are incredibly variable isotopically. This is partly the result of fluctuating C_4 plant contribution to the diet of both domesticates (cattle, sheep and goat) and wild species (aurochsen, equids and birds), most notably in the earlier and later periods, with a hiatus in the Middle period. This suggests either some environmental fluctuation had occurred, or that the inhabitants varied their use of the landscape for hunting and herding activities. The question of what individual inhabitants ate, especially the nature of the importance of cattle/aurochs meat, has been a central question, considering the symbolic importance of these animals in the elaboration of houses. The now substantial datasets of carbon and nitrogen isotopic data for the most important meat-bearing animals indicate that these species are virtually indistinguishable from one another, preventing any assessment of how important cattle/aurochs meat was to the community. Analysis of the human remains show population mean values of $\delta^{13}C$ −18.8‰ and $\delta^{15}N$ 12.3‰. Using the trophic level shifts of 4‰ for nitrogen isotope values and 1‰ for carbon isotope values (see Chapter 11), the mean diet would be around $\delta^{13}C$ −19.8‰ and $\delta^{15}N$ 8.3‰. Since both sheep and cattle have higher carbon and nitrogen isotope values, this indicates that foods with lower values, such as plants, would have made up a significant part of the diet. Fish are currently ruled out as having made a major contribution as the data from two measurements (an amalgamation of multiple – in the 100s – of specimens which were combined to obtain enough collagen) indicate that this food group was substantially lower (by 3‰ in carbon) than would be expected if this resource had been consumed, but aquatic and terrestrial birds such as ducks, geese and bustards likely made a small contribution.

Comparing the new isotope data across age and sex categories indicates more specific use of food to reinforce age-related social structure than previously reported in Pearson and Meskell (2013) and Pearson et al. (2015b). Young and older women had different diets, as did young women compared to older men ($p=0.004$), which suggests age as the main driver, with some evidence that sex-based differentiation played a linked but more minor role. The key dietary difference was the species rather than volume of meat consumed. The nitrogen isotope values for young women and older men were similar, but the carbon isotope values were significantly lower for women (−19.3‰ in women versus −18.6‰ in men), suggesting they consumed a greater proportion of isotopically lower foods such as equids, suids and birds, in addition to plant foods with relatively higher nitrogen isotope values, whereas the older men consumed foods most similar to the rest of the population, since their overall values resemble the site mean. Secondary burial was also considered from measurement of 29 isolated crania, some with articulating mandibles. The mean values differed from the overall population values by 0.1‰ for carbon and 0.4‰ in nitrogen, suggesting the diets of individuals afforded this practice had similar diets to those that were not.

Variation in human diet according to occupational period and buildings. There is some variation in human diet over time that can be attributed to either small-scale environmental change or changes in how the landscape was used for food acquisition. There is a switch in diet in terms of the relative proportions of C_4 plants in the diets of animals in the earlier and later periods, compared to the Middle period, which is accompanied by a downwards shift in nitrogen isotope values. This downwards shift is also apparent in the human nitrogen isotope values, and is likely an environmental response, rather than a change in diet, or the reflection in human tissues of animal responses to environmental change (since the change is also observed in wild animals such as boar). When the adult male versus female carbon and nitrogen isotope data are plotted over time, the carbon isotope values show a possible hiatus in values during the Middle and Late periods (fig. 16.11), while the Final period is difficult to interpret as only one male and one female represent this period. The nitrogen isotope values (fig. 16.12) show generally increasing nitrogen isotope values over time, with a small hiatus in the Middle period in males. These data are difficult to interpret as they may relate to a number of scenarios: increasing amounts of meat in the diet over time, increasing dependence on manured as opposed to wild plants, increasing aridity in the local environment, increasing aridity in grazing areas used by herders for sheep and cattle, or generally increasing stress. Again, the Final period should be viewed with caution, with only one male and one female represented.

In terms of building-specific diets, some buildings show evidence that food was shared, but more often it seems the diets were different. However, the overall question was if stable isotope values were more similar within houses than between houses. If so, this would suggest evidence of house-based diets. PERMANOVA and PERMDISP tests were used to examine this issue (see Chapter 11 for more details). When comparing houses with at least four adults of mixed ages buried beneath them PERMANOVA, which examines the permutation of values within a house, was significant ($p=0.001$) but PERMDISP, a measure of dispersal, was not. When the effects of chronological change and the relative contribution of C_4 plants are also taken into consideration, only the Middle ($p=0.033$) and Late ($p=0.013$) periods reach significance, suggesting that house-based diets became more predominant through time.

Weaning. Previous interpretations of weaning strategies stated that the first onset of weaning began around 18 months of age and it was complete by 3 years, giving an overall weaning duration of 18 months based solely on nitrogen isotope data (Pearson 2013). Adding a further 75 individuals shows that this may have been an inaccurate assessment, based on a small sample size. The new expanded dataset instead suggests weaning started before 12 months of age (indicated by both isotopes) and continued (as total breastfeeding) until three years of age. This is based, however, on a small-sample dataset at the crucial transitions between exclusive breastfeeding (where breast milk is the only substance consumed) and total breastfeeding (where both breast milk and weaning foods are consumed) (see Pearson 2018 for a review).

Strontium isotope evidence of mobility. Understanding the origins of the population at Çatalhöyük and how many outsiders might have come to the site is based on the analysis of 80 adults. These were divided more or less equally into male/male? (n=23) and female/female? (n=33), with a further 24 of unknown sex, in order to address questions concerning whether the community practised partner-exchange kinship according to patrilocal, matrilocal or some other social mechanism. Further details on the materials and methods, including the baseline data for determining local and outsider values, can be found in Chapter 11.

The $^{87}Sr/^{86}Sr$ data reveal that most individuals buried at the site spent the majority of their childhood on the Konya Plain; only eight individuals spent any time at all away from the site at this time in their lives. One individual, skeleton (14753), an adult female from the Middle period, may have spent her childhood on the limestone terraces, and seven individuals spent part of their childhood in a more geologically radiogenic zone, towards the Taurus Mountains, or spent some time travelling between the Konya Plain and these more radiogenic zones, rather than originating from elsewhere. These individuals, both males and females covering the full range of adult ages, had no specific grave goods and were not buried in specific houses. All individuals were buried in primary or primary disturbed inhumations, except for one individual found in a secondary burial context, skeleton (20830), a probable adult female. One important observation is that these outsiders did not occur in one period or house, but strik-

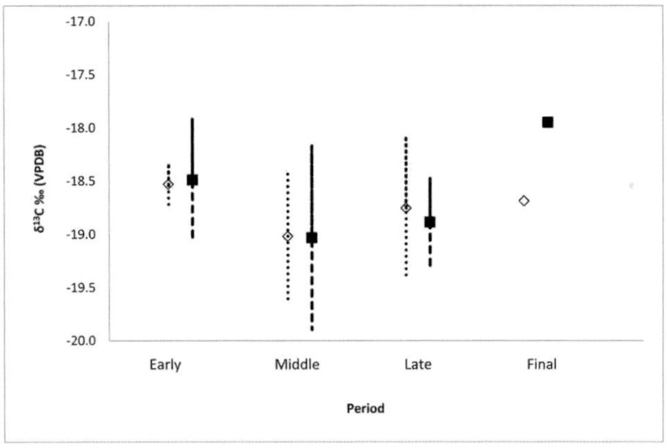

Figure 16.11. Carbon isotope values for males versus females plotted across the broad occupational periods (open diamonds: females; filled squares: males).

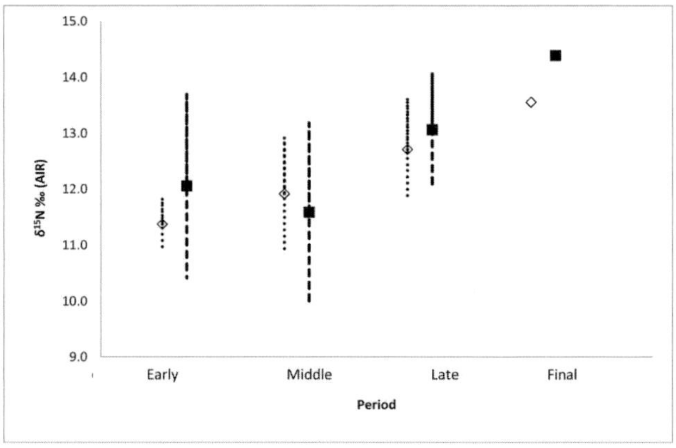

Figure 16.12. Nitrogen isotope values for males versus females plotted across the broad occupational periods (open diamonds: females; filled squares: males).

ingly only occurred once in each of the following buildings and spaces in both areas of the site: Buildings 12, 53, 65, 75, 92, 96 and Space 620 (see also Chapter 11: table 11.6). For the major occupation periods in which these activities occurred, in the North Area these burials took place in Levels North G and North H (Middle and Late periods), and similarly, in the South Area, they occurred from Levels South M-South Q, which broadly correlates with the whole of the Middle period and the beginning of the Late period of occupation.

Dental variation at Çatalhöyük

Strontium isotope analyses form a powerful complement to dental morphological and dental metrical analyses that have the potential to inform hypotheses on mobility, social structure, residence patterns and biological relationships between individuals and populations. These phenotypic data complement and augment palaeogenomic findings, such as from ancient DNA (aDNA), which may be difficult to accurately capture and require the destruction of human remains that may not always be permissible (Pilloud et al. 2016b). Dental data, on the other hand, can be collected on a large sample rather quickly and do not require the destruction of human remains. In addition to preserving well, extensive work documents the strong heritability of such traits (cf. Hughes et al. 2000; Townsend et al. 2009; Townsend et al. 2012; Pilloud et al. 2016). Moreover, there has been work relating dental variation to genetic variation in prehistoric and living populations (cf. Sofaer et al. 1972; Brewer-Carias et al. 1976; Hubbard et al. 2015) and at the individual level (Shinoda et al. 1998; Adachi et al. 2003; Corruccini et al. 2002; Shimada et al. 2004; Stojanowski, Hubbard 2017).

Variation in the permanent dentition was explored to address hypotheses surrounding social structure, post-marital residence practices and mobility. Dental morphology was recorded following the Arizona State University dental anthropology system (Turner et al. 1991; Scott, Irish 2017). Dental measurements include maximum mesio-distal and bucco-lingual dimensions of the crown (Hillson 1996; Moorrees, Reed 1964), as well as dimensions of the cervix (Hillson et al. 2005).

Previous work showed that individuals interred together in the houses at Çatalhöyük were not phenotypically similar, likely indicating that they were not close biological relatives. These results indicate that social structure was not based on biological kin relationships influencing place of burial, and instead relied on practical relationships related to the new system of food production (Pilloud, Larsen 2011). In addition, it seems that post-marital residence practices were patrilocal based on the dental metric differences between males and females (Hillson et al. 2013). Here, that finding is again tested with a slightly larger dataset. In an evaluation of all adult males and females, Levene's test of equal variances was run on all available dental metrics in SPSS version 24 (SPSS Inc. 2016). In this test, only four measurements were found to have unequal variances between males and females: the bucco-lingual cervical dimension of the lower first incisor, the bucco-lingual crown dimension of the lower fourth premolar, the bucco-lingual crown dimension of the lower second molar, and the mesio-distal crown dimension of the lower third molar. The variance for each of these measurements is greater among females, except in the lower first incisor, where males have a greater variance (table 16.13). While still based on small sample sizes, and with only a few measurements showing differences between the sexes, there is still greater variance among females, which continues to suggest they were the more biologically diverse and thus more mobile sex.

Using dental morphology frequency data for 22 traits, two diversity indices were calculated using the 'vegan' package in the R computing environment (R Core Team 2013) to compare males and females at Çatalhöyük (table 16.14). The Shannon *H* and Simpson indices were calculated; both are measures of entropy, which provides the uncertainty of the variables in a sample (Jost 2006). When these variables are high, they indicate increased diversity in the sample dataset. Both the Shannon *H* and Simpson indices indicate that females are slightly more diverse than males (table 16.15). The morphological and metric data are in agreement: females appear to have been more biologically diverse, and thus more mobile, again verifying a pattern consistent with patrilocality.

	Males				*Females*			
	n	*Mean*	*St. Dev*	*Var*	*n*	*Mean*	*St. Dev*	*Var*
LI1 BL cervix	13	5.7	0.545	0.297	20	5.65	0.365	0.133
LP4 BL crown	20	7.98	0.382	0.146	25	8.06	0.500	0.250
LM2 BL crown	17	9.98	0.388	0.151	24	9.97	0.558	0.311
LM3 MD cervix	12	8.7	0.411	0.169	20	8.17	0.837	0.701

Table 16.13. Descriptive statistics for measurements found to have different variances between males and females. L=lower, I=incisor, P=premolar, M=molar, BL=bucco-lingual, MD=mesio-distal.

Sex	*UI1 shovel*	*UI2 int groove*	*UI2 tub dent*	*UC mesial ridge*	*UC DAR*	*UM1 hypocone*	*UM1 metaconule*	*UM3 parastyle*	*UM1 en ext*	*UM3 peg*	*UM3 cong abs*
F	0.154	0.273	0.154	0.071	0.000	0.833	0.667	0.000	0.000	0.051	0.089
M	0.182	0.344	0.185	0.000	0.0067	0.917	0.550	0.000	0.000	0.049	0.083

Sex	*UI2 cong abs*	*LC DAR*	*LP4 ling cusp*	*LM1 ant fov*	*LM1 prtstyld*	*LM2 cusp 5*	*LM1 cusp 5*	*LM1 cusp 6*	*LM1 cusp 7*	*LM3 cong abs*	*LP3 cong abs*
F	0.023	0.429	0.333	0.357	0.207	0.345	0.600	0.231	0.050	0.075	0.000
M	0.023	0.545	0.591	0.250	0.107	0.304	0.632	0.111	0.077	0.080	0.000

Table 16.14. Dental morphological trait frequencies used in diversity analyses. Traits include: shovelling (shovel), interruption groove (int groove), tuberculum dentale (tub dent), mesial ridge, distal accessory ridge (DAR), hypocone, metaconule, parastyle, enamel extensions (en ext), peg-shape (peg), congenital absence (cong abs), lingual cusp of lower P4 (ling cusp), protostylid (prtstyld), cusp 5, cusp 6, and cusp 7.

	Shannon	*Simpson*
Female	2.564149	0.906564
Male	2.519752	0.900689

Table 16.15. Shannon and Simpson diversity indices between males and females from Çatalhöyük.

Measures of mobility were assessed through an analysis of dental metric data from Çatalhöyük, as well as two other Neolithic Anatolian sites: Aşıklı Höyük and Musular. The F_{st}, a measure of genetic homogeneity (Wright 1951), of these sites was calculated using Rmet (made available by John Relethford online at http://employees.oneonta.edu/relethjh/programs/) (Relethford, Blangero 1990). The F_{st} was found to be 0.0288 (Pilloud 2013). This low F_{st} value implies common ancestry and/or gene flow among groups. Based on the dating of the sites, this result could mean that populations derived from Aşıklı Höyük and Musular contributed to the population of Çatalhöyük, as it has the latest occupation sequence. Furthermore, increased variation in dental data was observed in the Çatalhöyük population sample at approximately the same time that Musular was abandoned (ca 6800 BC) (Pilloud 2009).

The results of analyses of these biological data indicate that social structure, as inferred via inclusion for burial within the same house, was likely not based on biological kin at Çatalhöyük. Additionally, there appears to have been migration between sites in Neolithic Anatolia, as indicated by similarities in dental metrics between the sample populations from two archaeological sites, Aşıklı Höyük and Musular, and those from Çatalhöyük. Finally, dental metrical and morphological evidence used to address post-marital residence practices continues to indicate that females are slightly more diverse than males. These dental data are an important line of evidence for understanding how the Neolithic population of Çatalhöyük organised their social groups, and how they moved across the landscape. These data can be further compared to other Neolithic Anatolian sites and with data from aDNA studies that show evidence for two population movements generated from Central Anatolia with connections to southern European populations and to populations in the Fertile Crescent and Levant in the Late Pre-Pottery Neolithic B (ninth to seventh millennia BC) (Kılınç et al. 2016) to better understand the complex process of Neolithisation in Anatolia.

Cross-sectional geometry

In addition to dental morphological and metrical analyses, cross-sectional limb bone geometry provides a complementary means to assess mobility in past populations that permitted them to exploit the landscape. Previous studies show that agricultural, herding and resource acquisition practices changed through time at Çatalhöyük (Ayala et al. 2017; Bogaard et al. 2017; Roffet-Salque et al. 2018; Ayala et al. this volume, Chapter 2), suggesting greater diversity of landscape use and increasing mobility over the course of the occupation of the site (Larsen et al. 2015). Behaviour and mobility can be explored using cross-sectional geometric properties to estimate bone strength, which is dependent on the distribution of cortical bone within the long bone diaphyseal shaft (cf. Ruff 2008). Here, cross-sectional geometric analyses were employed to evaluate temporal patterns in growth-related, cross-sectional long bone strength in adults and subadults spanning the Early, Middle and Late Neolithic occupation periods, with the Final period collapsed into the Late period due to its small sample size.

Cross-sectional geometric sections were reconstructed for adult and subadult femora and humeri (table 16.7 for adult sample provenience). Subadult remains excavated since 1993 were measured. These analyses require preserved periosteal surfaces and sufficient skeletal remains to estimate bone length, body size, and sex. Cross-sectional properties were reconstructed using biplanar radiographs and latex moulds to describe the cortical dimensions and external silhouettes of long bone shafts (O'Neill, Ruff 2004). Cross-sectional properties were reconstructed at femoral mid-shafts (adults at 50% of length and unfused subadults at 45.5% of length) and more distally on the humeral shaft (adults at 35% of length and unfused subadults at 41% of length) (Ruff 2002). Strength properties were calculated from the reconstructed cross-sections using an automated programme (Sylvester et al. 2010).

Estimates of bending strength were calculated as section moduli for antero-posterior (AP: Z_x) and medio-lateral (ML: Z_y) planes. A ratio of AP (antero-posterior) and ML (medio-lateral) cross-sectional bending strength (Z_x/Z_y) is an indication of mobility attributed to mechanical loading that promotes bone deposition along the antero-posterior axis of femora and tibiae (Ruff et al. 2015). Overall bending strength was estimated with the polar section modulus (Z_p), which reflects torsional and (twice) average bending strength. Estimates were standardised by body mass*bone length for comparisons of section moduli (Z_x, Z_y and Z_p) (Ruff 2000) using body mass estimates from femoral head supero-inferior breadth (Ruff et al. 2018) and maximum long bone length. Maximum bone lengths were estimated from diaphyseal lengths for unfused elements using published ratios (Ruff 2007). Given unequal sample sizes, temporal analyses use non-parametric Kruskal-Wallis ANOVA tests with Bonferroni adjusted post-hoc multiple comparison tests of mean ranks between groups.

Temporal trends within Çatalhöyük: adults

Throughout the occupation sequence at Çatalhöyük, adult AP/ML femoral bending strength increases (fig. 16.13; Kruskal-Wallis: $p=0.0008$; pairwise comparisons: Late/Final period > Middle period: $p<0.02$; Late/Final period > Early period: $p=0.001$). The pattern is statistically significant in females in the Late/Final period compared to the Early period ($p=0.007$); other pairwise comparisons with sex are non-significant, probably due in part to reduced sample sizes.

Standardised femoral bending strength (Z_x, Z_y, and Z_p) also increases through time in males (particularly, Z_x; Kruskal-Wallis: $p<0.04$). However, females show a flat or even decreasing temporal trend in body-size standardised properties that is nearly significant for Z_y (Kruskal-Wallis: $p=0.054$) between the Middle and Late/Final periods. This pattern is demonstrated in overall bone strength (Z_p), increasing through time for males from the Middle to Late/Final periods ($p<0.05$) but not in females (fig. 16.14a).

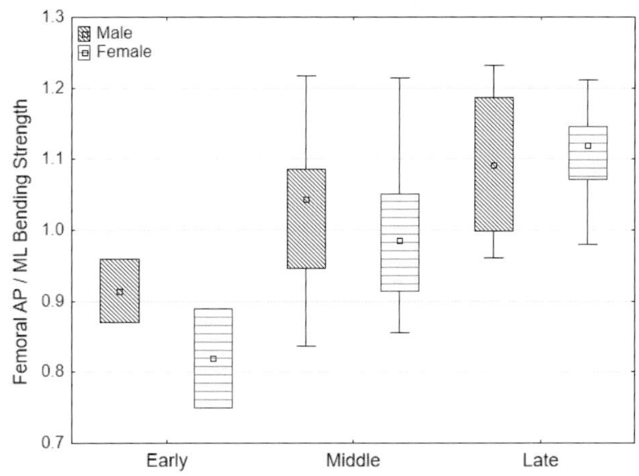

Figure 16.13. Box plot of adult femoral AP / ML bending strength (Zx / Zy) by temporal period and sex.

Temporal patterns in body-size standardised humeral strength are similar to those of femoral strength but without statistical significance. While Late/Final period male humeral strength increases (standardised Z_x, Z_y, and Z_p) compared to that of the Middle period, female humeral overall strength (Zp) decreases from the Middle to the Late/Final period (fig. 16.14b).

Broader comparisons: adults

Figure 16.15a shows Çatalhöyük femoral midshaft AP/ML bending strength in a broader temporal context. Following the Upper Palaeolithic strongly AP-oriented femora, there is a sharp decline in AP/ML bending strength in both sexes, but more markedly in males. Çatalhöyük values are generally similar to those of the Levantine Pre-Pottery Neolithic (Hershkovitz, Gopher 2008; May, Ruff 2016), while European Early Neolithic (Ruff 2018) values are comparatively lower. After the Upper Palaeolithic, sexual dimorphism diminishes, but AP/ML bending strength remains consistently greater in males.

Subdivision of the Çatalhöyük sample population temporally by period (fig. 16.15b) reveals that the Early period values are lowest, indicating more ML-oriented bending strength. The Middle period values increase and are similar to those of near contemporary Levantine Pre-Pottery Neolithic groups, while the Çatalhöyük Late period bending strength ratio is higher than that of European Mesolithic populations (Ruff 2018) and Natufian femora (Hershkovitz, Gopher 2008). Considering mobility patterns, the Early period appears relatively sedentary, the Middle period bending strength ratio is as expected for an agricultural sample, and the Late/Final period is more similar to less sedentary European Mesolithic and Natufian samples.

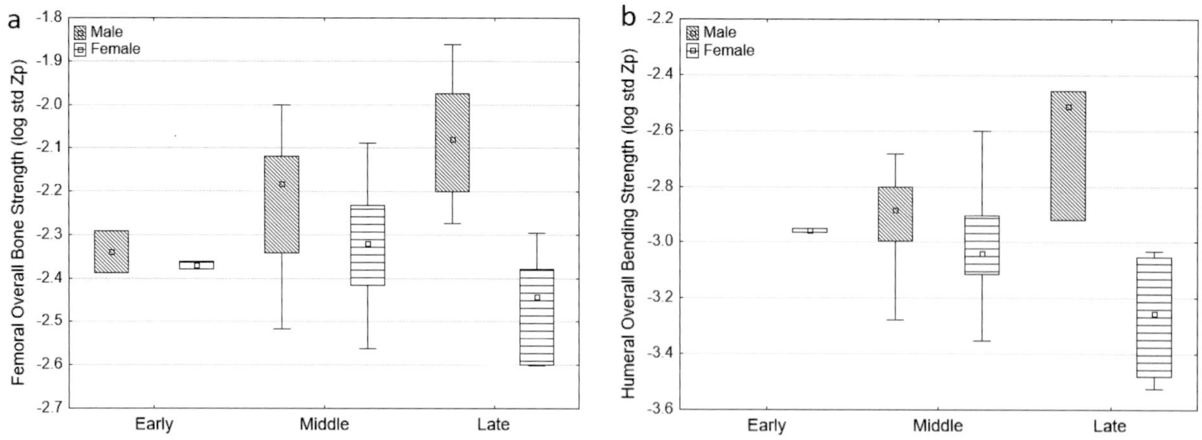

Figure 16.14. Box plot of adult (a) femoral and (b) overall bending strength (log standardised Zp) by temporal period and sex (Early period male humeri n=0).

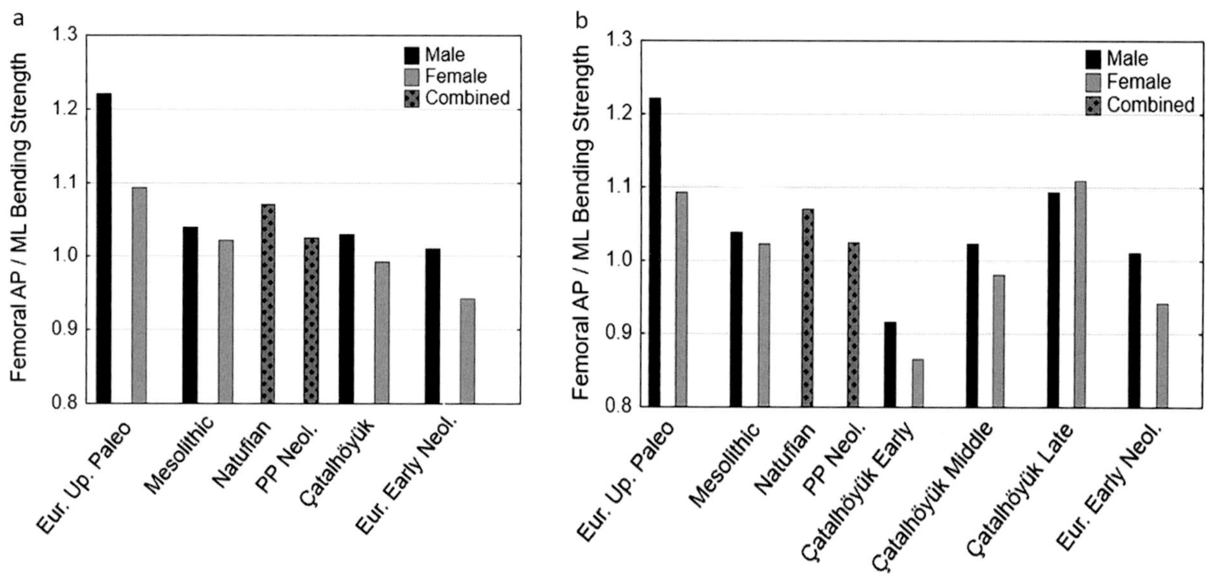

Figure 16.15. Mean femoral AP / ML bending strength (Zx / Zy) of comparative European and Mediterranean samples with (a) pooled Çatalhöyük and (b) temporally subdivided Çatalhöyük adult sample.

Thus, the early inhabitants of Çatalhöyük appear to have been a sedentary group compared to broadly contemporaneous food-producing populations that increased in daily travel through time until they were at least as mobile as some pre-agricultural populations. These results support hypotheses of increasing travel, changes in workload or type of activity, and resource exploitation through time at Çatalhöyük (Sadvari et al. 2015a) in an increasingly arid landscape possibly associated with Holocene climate change (Roffet-Salque et al. 2018, but see Ayala et al. this volume: Chapter 2; Ayala et al. 2017; Wainwright, Ayala 2019 for an alternative viewpoint).

Subadults: ontogenetic changes

Children participate in subsistence activities (Bradley 1993) – often beginning around 6 years of age. Skeletal evidence of these behaviours offers a broader understanding of behavioural adaptation through time and provides a context for the place of children in societies. The growth profile of femoral AP/ML strength ratio (Z_x/Z_y) shows bone alignment that is increasingly AP-oriented in response to the developing adult locomotor pattern and body shape (cf. Cowgill et al. 2010) (fig. 16.16a). Comparing femoral midshaft ratio residuals among the settlement periods for the entire pooled subadult sample, there is little change through time.

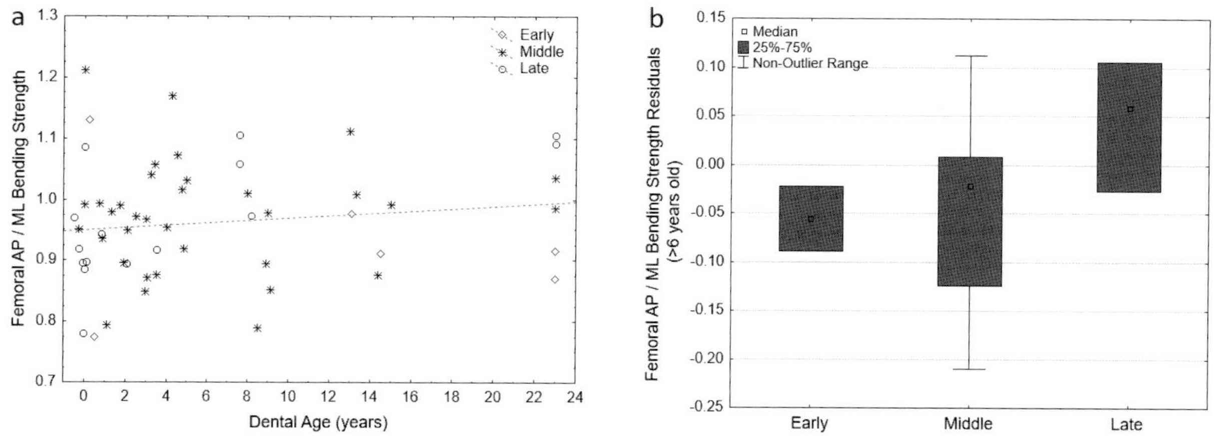

*Figure 16.16. Bending strength ratio ontogeny (a) Scatterplot of femoral AP / ML bending strength growth by dental age (=0.9495+0.002*x), categorised by period. Adult (male and female) mean values for each period are plotted at 23 years old; (b) femoral AP / ML bending strength residual for subadults ≥6 years old.*

However, for subadults ≥6 years old, there is an increase in Z_x/Z_y ratios between the Middle and Late/Final periods, albeit non-significant (fig. 16.16b).

The percentage of total area in a cross-section composed of compact bone (per cent cortical area, %CA) may reflect metabolic or protein-calorie stress during growth (markedly during early growth), where lower cortical bone may indicate increased bone apposition or absorption at the endosteal surface and poorer health (Garn et al. 1969; Mays et al. 2009; Eleazer, Jankauskas 2016). The %CA profile has a consistent age-related trend for all periods at Çatalhöyük, including a dip in infancy and early childhood. Although %CA typically declines during these periods of growth, possibly reflecting normal appositional growth processes (Frisancho et al. 1980; Ruff et al. 1994), it could also suggest nutritional (weaning?) stress (Hummert, van Gerven 1983) (fig. 16.17).

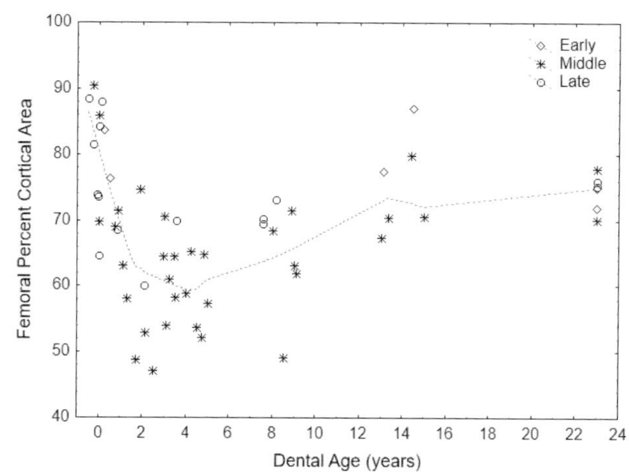

Figure 16.17. Femoral per cent cortical area (%CA) ontogeny scatterplot by dental age categorised by period with LOWESS fit (50% stiffness). Adult (male and female) mean values for each period are plotted at 23 years of age at death.

While the relationship between behaviour, mid-shaft geometry and growth is complex (Shaw, Stock 2011; Cowgill 2014), these results support the interpretation that, like adults, immature individuals may have travelled greater distances in the Late/Final periods of occupation at Çatalhöyük. These behaviours are apparent in later childhood (≥6 years old), further supporting an environmental explanation and indicating childhood involvement in subsistence work in conjunction with the adults of the community. This analysis enriches the complex story of behavioural change and juvenile activity at Çatalhöyük and in broader cultural contexts.

Tensions associated with dense living conditions

The final topics of this chapter summarise results of studies that serve to reveal the social effects of living in a densely populated community. Two of the greatest problems in densely populated agglomerations concern the management of increasing waste and avoidance of conflict leading to violence.

Parasitological study of Çatalhöyük coprolites and pelvic soil

Archaeological samples previously found to contain human faecal material (Shillito et al. 2011b) were studied to look for evidence of parasite infection within the community. This is one of the earliest sites in Europe or Asia to undergo palaeoparasitological analysis. As an early farming community, it permits insight into how parasite transmission may have changed with major changes in community structure, food sources and population size that coincided with the transition from a mobile hunter-gatherer way of life to one of settled farmers. We might expect that herding animals would have exposed people to the risk of contracting those parasites normally found in sheep, goats and cattle, or in dogs when used to manage animals. From other Neolithic sites in the Mediterranean region there is evidence for soil-transmitted helminths, including roundworm and whipworm, as well as other faecal-oral parasites such as pinworm, and zoonotic parasites such as *Taenia spp.* (tapeworms), *Fasciola sp.* (liver fluke), *Diphyllobothrium spp.* (fish tapeworm) and *Capillaria spp.* (nematodes) (Harter-Lailheugue et al. 2005; Maicher et al. 2017). Though skeletal remains from Çatalhöyük have been studied to understand health and disease in the community, it is not possible to directly assess many gastrointestinal infections from skeletal remains. For this type of study, samples that are likely to contain human faeces may also contain preserved eggs from intestinal worms that infected inhabitants.

In order to look for evidence of intestinal disease at Çatalhöyük, four human coprolites from refuse areas were analysed, and soil was collected from the pelvic areas of six human skeletons (Ledger et al. 2019). The four coprolites came from Sp.329, Level South P, dating to 6410–6150 BC. These coprolites were identified as human according to methods described in Bull et al. (2002). The skeletons sampled were excavated in buildings from the South Area of the site dating between 7100 and 6300 BC. Subsamples of each coprolite and pelvic soil sample were processed according to published standards (Anastasiou, Mitchell 2013). Digital light microscopy was used to analyse each subsample to look for the preserved eggs of intestinal helminths.

Two of the coprolites contained the eggs of whipworm, *Trichuris trichiura*, while the pelvic soil samples were negative (Ledger et al. 2019). Twelve whipworm eggs were recovered in total, at concentrations of four eggs per gram from one coprolite and eight per gram from the other. Whipworm is a soil-transmitted helminth that is transmitted by the faecal-oral route. As such, it is closely linked to sanitation and hygiene practices. The adult worms live in the large intestine where they grow up to 5cm long. The adult worm releases eggs in the faeces of the infected individual and, after a 1 to 2-week period of maturation in the soil, the eggs become infective and upon ingestion can establish an infection in a new host (Garcia 2016). Infection is often acquired through the ingestion of contaminated food or water. Whipworm has been found in many archaeological sites from the Mesolithic period onward and is thought to have evolved in hominin ancestors in Africa (Dark 2004; Mitchell 2013; Bergman, 2018).

The concentration of whipworm eggs in the coprolites is not high at 4–8 eggs/gram. This may indicate that some of the eggs did not survive to be recovered, or that the individuals infected truly had a low number of worms in their intestines. Whipworm infection is typically asymptomatic in individuals with low worm burdens and adequate nutrition. In individuals who have a higher worm burden, particularly children, whipworm can cause stunting of growth, weight loss, anaemia, abdominal pain and diarrhoea (Bethony et al. 2016; Jourdan et al. 2018). However, palaeopathological evidence from the site has suggested that children had adequate nutrition for growth and development (Hillson et al. 2013 and above), and while *cribra orbitalia* has been found, it is hard to link this to parasite infection.

Comparisons of parasite diversity found at other Neolithic sites and in later time periods in the Mediterranean reveal that the parasite profile at Çatalhöyük is much more similar to that of the Roman and Medieval periods, where parasites transmitted by the faecal-oral route are the dominant taxa found (Mitchell 2015; 2017; Williams et al. 2017; Ledger et al. 2018). The lack of zoonotic parasites (typical of hunter-gatherers) in the Çatalhöyük samples coupled with the presence of a parasite spread by poor sanitation (whipworm) would be compatible with the hypothesis that the transition from hunting and gathering to relying on farmed crops and domesticated animals may have affected the taxa of parasites present in this community. Similarly, a larger and denser population may have contributed to maintaining faecal-oral parasite transmission, as in whipworm. The results of this research demonstrate that the early agglomerated settlement inhabitants of Çatalhöyük already suffered from intestinal parasite infection from problems associated with waste disposal, but at this early date infection rates may not have approached the severity associated with some later densely inhabited urban settlements in historic periods.

Cranial injuries at Neolithic Çatalhöyük

Discrete patterns in the location, type, shape and timing of cranial injuries indicate that different mechanisms are responsible for their occurrence. Social circumstances, such as motives, socio-cultural understandings of the body, perceived threat, socio-political hierarchies and notions of transgression, as well as economic and material factors, also influence their type, number and location. Cranial injuries are therefore highly socially informative. Cranial trauma can result from interpersonal violence, accident, or be self-inflicted, but evidence for fractures of the face and head are consistent with violent assaults in medico-legal (Shermis 1984; Hussain et al. 1994; Brink et al. 1998) and bioarchaeological studies

(Walker 1997; Fibiger 2012; Glencross, Boz 2014; Knüsel 2014b). While the extent, severity and impact may vary, violent injuries significantly influence the physical, mental, economic and social well-being of individuals and communities. This section addresses the physical evidence for cranial injury observed at Neolithic Çatalhöyük, a site not previously known for evidence of violence. The results of this investigation provide strong evidence for interpersonal violence, with implications for broader insights into social interactions within this Neolithic community.

The study sample consists of skeletons from the Neolithic occupation of the East Mound. Most skeletons come from houses located in the North or South Areas of the site, with one notable exception of an individual located in a midden (Milella et al. 2016). They represent all major occupation periods from Early through Final. Of these, a total of 147 complete or partial adult crania were preserved well enough for analysis. Although not all excavated crania could be studied in the time allotted, the sample is a random one from across the site. Biological sex and age-at-death were determined where possible using standard methods applied to the entire skeleton. Of the analysed crania, sex could be assigned to 125 individuals (n=71 females and 54 males), leaving 22 adult skeletons as of indeterminate sex. Skeletons were allocated to one of seven age cohorts (that is, neonate ~ birth, infant 0–3 years, child 3–12 years, adolescent 12–20 years, young adult 20–35 years, mature adult 35–50 years, old adult 50+, and adult 20+). Assigned age-at-death for each individual represents the mid-point of that age cohort. Adults with and without injury range in age-at-death from 20 to 50+ years (tables 16.16–16.20). The different ages-at-death assigned to individual skeletons were summed to reflect the length of exposure to potential trauma and used as a denominator in comparisons of prevalence between samples. Cranial injury data were recorded following a medico-legal model (after Hussain et al. 1994) that divides the cranium and mandible into 44 separate zones. Each cranium and mandible from the sample was scored for completeness by zone, and only those portions constituting 75% or more of any given zone were included in the analysis. Cranial injuries, identified through gross examination, were recorded according to location, form (that is, fracture type, such as blunt force or penetrating trauma), shape, size (mm), and if antemortem or perimortem in nature.

The majority of injuries (n=52/55) are healed, with three mandibular fractures in a single individual, skeleton (32608), indicative of perimortem injury. All of the healed injuries show considerable evidence of remodelling. Forty-seven of the healed lesions are depressed fractures described mainly as shallow round, elliptical or ovoid impressions of the external table of the cranial vault (fig. 16.18). Other healed lesions include two penetrating wounds and three diastatic

All	*Observed*		
Age	*No.*	*No. affected*	*%*
>15 to <35	59	12	20.3
>35	66	16	24.2
Ratio			0.84

Table 16.16. Summary counts of all individuals and affected individuals by age cohort.

Females	*Observed*		
Age	*No.*	*No. affected*	*%*
>15 to <35	32	5	15.6
>35	39	11	28.2
Ratio			0.55

Table 16.17. Summary counts of all females and affected females by age cohort.

Males	*Observed*		
Age	*No.*	*No. affected*	*%*
>15 to <35	27	7	25.9
>35	27	5	18.5
Ratio			1.4

Table 16.18. Summary counts of all males and affected males by age cohort.

	Affected males	*Unaffected males*	*Total*
>15 to <35	7	20	27
>35	5	22	27
Total	12	42	54

Table 16.19. Summary counts of affected and unaffected males by age cohort.

	No. Lesions	*Unaffected bones*	*Total*
>15 to <35	14	297	311
>35	9	343	352
Total	23	640	663

Table 16.20. Summary count of male lesions and male complete bones by age cohort.

fractures affecting the cranial sutures. Locational patterning of injuries demonstrates a large number of fractures affecting the superior (35%) and posterior (31%) aspects of the cranial vault (fig. 16.19). The number increases dramatically to 87% when considered in relation to the hat-brim line (HBL), a demarcation used in medical and forensic investigations to indicate the location of injuries most likely due to a fall from standing height (below the line) or a blow (above the line) (Kremer et al. 2008). While the vast majority of injuries are located above the HBL, the HBL alone as a criterion to infer cause of cranial injury is insufficient. The absence of evidence of linear fractures suggests, however, that these injuries are unlikely to have occurred due to falls from a height, and when co-occurring with overwhelming evidence for depressed cranial fractures, are recognised as having a higher correlation with interpersonal violence (cf. Lovell 1997; Walker 2001; Schulting 2012).

The total number of cranial injuries presents a strong case for the presence of interpersonal violence at Çatalhöyük. The position of a depressed fracture on the left side of the cranium is a common pattern consistent with violent assault by a right-handed aggressor in a face-to-face encounter (Aufderheide, Rodríguez-Martín 1998: 23; Novak 2000: 99). Amongst the examined Çatalhöyük crania, a large number, 41% (7/17) of injuries are located on the posterior right side when compared to the 29% found on the posterior left side, perhaps indicating a blow coming from behind the victim. The argument for interpersonal violent confrontation is further strengthened when considered with evidence for multiple cranial injuries as opposed to single injuries, another pattern frequently associated with violent events (Kranioti 2015). The Çatalhöyük crania provide evidence of 13 individuals showing a total of 35 injuries, with some crania showing anywhere from two to five lesions. It is important to bear in mind, though, that the timing of events leading to fracture is difficult to ascertain when healed, and fractures are just as likely to be from recurrent events as simultaneous events. Penetrating and sharp force injuries and helical fractures of the mandible are also predominately associated with interpersonal violence (Kranioti 2015; Morris et al. 2015).

In a more detailed exploration of gendered and age-related patterns of cranial injury a difference was found between the young (>15 years to <35 years of age) and old (>35 years of age) male cohorts. The number of cranial injuries in young males is significantly greater than the number in older males with injuries when measured against the sum years of exposure for each cohort ($Z_{0.05,2} = 2.837$, $p = 0.0046$, 95% CI [1.46, 7.77]). The odds of cranial injury are three times more likely in young males. This counters expectations of a greater number of injuries in old males based on the reasoning that as individuals live longer, their exposure and the probability of sustaining injury increases. It is interesting to note that while the expected increase in cranial injuries in the old female cohort is evident, there remains a significant difference in the number of cranial injuries in the young male cohort when compared to the old female cohort ($Z_{0.05,2} = 2.323$, $p = 0.0202$, 95% CI [1.14, 4.80]). The odds of cranial injury are twice as likely in young males as they are in older females. The result of more cranial injuries in young males could only come about through the young male cohort being more frequently exposed to violence, a finding which accords with more recent experience in medico-legal (Hussain et al. 1994) and bioarchaeological research (see discussion in Knüsel, Smith 2014).

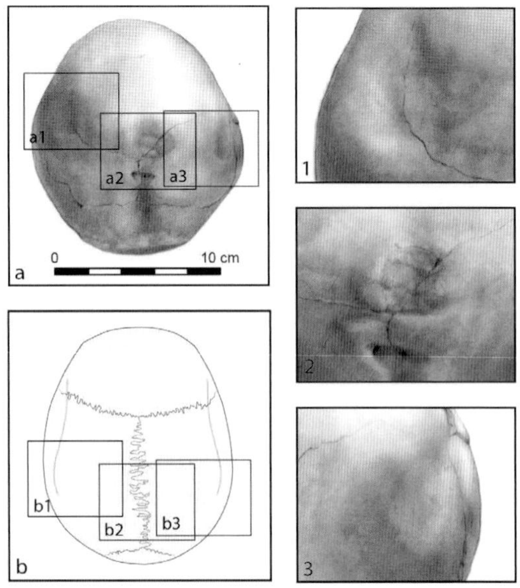

Figure 16.18. An example of depressed cranial fractures resulting from blunt force trauma (illustration: Katy Killackey).

There are also changing patterns of injury through time during the occupation of the site and differences in the spatial distribution across the site. Initial observations show the number of crania bearing injuries increase some 61% in the Middle period, which coincides with maximum population and density at the site (see above). When prevalence of lesions is compared based on length of exposure it would appear that prevalence is significantly greater in the Early period compared to the Middle period ($Z_{0.05,2} = 3.112$, $p = 0.0019$, 95% CI [1.51, 6.20]), with the odds of

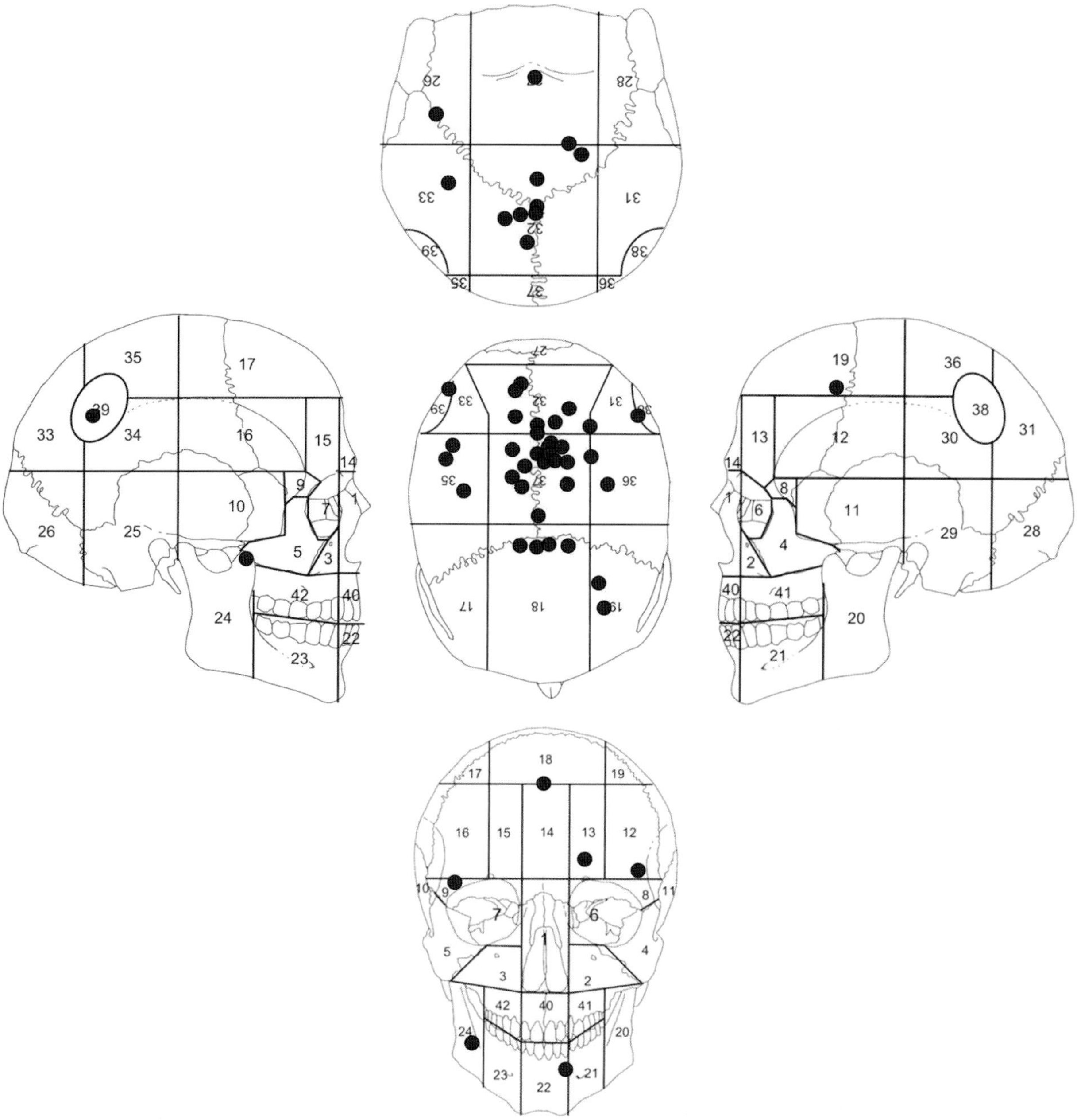

Figure 16.19. Composite representation of all affected zones and locations of cranial injuries at Neolithic Çatalhöyük.

fracture in the Early period being three times more likely, with 18 individuals from the Middle period (Levels North F and G) bearing injuries and 50 without injuries, so 18 of 68 individuals with injuries in total. However, this result must be tempered by the observation that the Early period sample contains five affected individuals from only seven examined in total. Two exhibit multiple injuries, one cranium having four lesions and a second cranium possessing three lesions, that act to inflate prevalence. When prevalence of injury in the Middle period is compared to that of the Late period, the greater number of injuries in the Middle period approaches significance ($Z_{0.05, 2} = 1.884$, $p = 0.0595$, 95% CI [0.96, 10.04]), with the odds of injury being three times more likely. Throughout the three periods of occupation, Early, Middle, and Late/Final, the site was divided into North and South Areas, with the North Area being abandoned earlier, ca 6300 BC, than the South Area (Orton et al. 2018). A comparison of cranial injury prevalence between these two areas shows significantly more injuries ($Z_{0.05, 2} = 2.115$, $p = 0.0344$, 95% CI [1.04, 3.06]) amongst inhabitants of the South Area, with the odds of cranial injury being almost two times more likely there than in the North Area.

Discussion

The two promontories that comprise the East Mound of the Neolithic site of Çatalhöyük, in addition to being well known for their internally elaborated and closely packed houses, are burial grounds. The multiple periods of funerary use include a roughly millennium-long Neolithic sequence, but also extensive human burials that date to the Roman/Byzantine period at the site, and to two Islamic phases of occupation, one of medieval date and the other of likely early post-medieval date. The most recent funerary use of the site occurred for the burial of a woman interred in the 1930s, complete with headstone (Haddow et al. 2019). The unequal distribution that characterises the Neolithic periods demands a broadly chronological approach for bioarchaeological studies to achieve samples that are preserved well enough and are numerous enough for meaningful statistical analysis. This means that many analyses comparing sub-groups/sample populations, for example by burial platform or by house, are not possible when taking a population approach – no matter how desirable they may be – due to their small sample sizes. Analyses are thus by level groupings, Early, Middle and Late, with the Final period remains, which are small in number and encountered only in the TPC/TP Area, combined with those of the Late period to bolster sample size.

The framework for the overall interpretations presented in this chapter comes from testing previously established methods to reveal the biological effects of social changes in the past. Here, though, the procedures used to arrive at these insights – using the archaeological context to determine analytical samples – are reversed, due to a lack of clarity and consensus on what type of social organisation – and changes in it – characterised the over one-millennium-long occupation of Çatalhöyük. In this endeavour human biology is used as much to inform archaeological contextual insights as the contexts are used to delimit the samples in order to monitor well-being (psycho-social and physiological stress as proposed by Hans Selye, see above), which is influenced by social organisation, broadly defined as the social interactions among individuals in a group/community.

Human groups recognise and accommodate considerable levels of social difference (see Human Development Report 1998) based on innate aspects of personal make-up (social identity), linked to age and sex (cf. Gowland, Thompson 2013), as well as social distinctions – attributes beyond those that are innate that distinguish an individual from others of a similar age and sex, the two basic structuring principles of human social organisation (cf. Brown 1981). Within and between communities these social distinctions often lead to unequal distribution of resources, information, socio-political stability and salubrious living conditions that, if maintained, lead to inter-generational (inherited or ascribed) social inequality and sometimes even complete exclusion from access to even the most basic requirements to ensure health (free from disease and disorder) and well-being (free from psycho-social and physiological stress affecting growth and development). Individuals and groups left in these unequal circumstances are lacking in well-being; some community members are exposed to poor health (morbidity) even before birth, from conception, owing to poor maternal health, such that they are unable to thrive and succumb (mortality), while others who are born bear the stigmata of having survived these insults in the development of their bodies that are severe enough to affect growth of their bones and teeth. Due to arrested development, these individuals lack the requirements necessary to achieve their full genetic potential for growth and development. Because the very young require stable access to good nutrition and stable living conditions, they and their mothers are especially prone to disruptions and reflect this absence in their biological phenotypes. Put in another way, in the absence of conditions from the bio-cultural environment that affect growth, growth is a response to bio-geographical factors (such as altitude and clinal latitudinal changes associated with warmer or colder climates, for example) in recent (Eveleth, Tanner 1990; Schell 1995; Bogin 1999; Bogin, Loucky 2007) and past populations (Schweich, Knüsel 2003; Steckel 2004).

As documented above from strontium isotopic and dental non-metric trait analysis, the population sample seems to be generally from the region of Central Anatolia, so the bio-geographical effects are likely reduced and those differences that do exist amongst community members are more likely due to bio-cultural effects. Because disruptions of growth and development are non-specific, meaning that they are linked to a variety of causes, and, moreover, due to underlying innate bodily homeostatic resilience to such disruptions, these effects, no matter how harsh, affect individuals and groups differently (Mummert et al. 2011). This chapter considered the results of studies using multiple indicators to develop an appreciation for the well-being (psycho-social and physiological stress) of community members for the roughly millennium-long occupation of the East Mound at Neolithic Çatalhöyük.

Although the adoption of agriculture is today associated with a decline in health and well-being, the picture from a society that had made that transition and adopted living in a proto-urban environment at Çatalhöyük provides a mixed view of the consequences, with some benefits as well as certain deficits. The Middle period (6700–6500 BC) at the site has previously been identified as a time during which the inhabitants experienced considerable change, marked by increased size of the settlement, with greater numbers and greater elaboration and expansion of houses, as well as burning of buildings in both the North and South Areas of the site. The Middle period has also been associated with a shift in animal diet to a greater reliance on C_3 plants, perhaps due to environmental fluctuation or, more likely, changing (and intensified) landscape use. In addition, Middle period burials display the highest levels of variability in grave inclusions/goods associations and also evince the introduction of new items, such as obsidian mirrors (Chapter 17) that indicate an elaboration of representations of the social identities of the deceased at this time. In order to deposit these items the community required the infrastructure both to obtain the raw materials and to fashion items before they became unavailable to the living through what was in many cases final deposition in burials, as well as for the construction, maintenance and elaboration of houses, burial platforms and sub-floor graves. The Middle period (6700–6500 BC) seems to be one of fundamental social change, the causes and effects of which draw interest, as much as do the founding and eventual abandonment of the site. These transitions, explored from a bioarchaeological perspective by comparing the Middle period to the preceding Early period and succeeding Late/Final period occupation of the site, are discussed here.

Demography

The juvenility index, the ratio between individuals aged between 3 and 19 years of age-at-death and the total number of individuals above 3 years of age-at-death, is used as a measure of fertility. This index changed over the course of time, increasing from the Early (9/29 [.31]) to the Middle period (96/218 [.44]), but decreasing again in the Late/Final period (20/77 [.26]) prior to the abandonment of the East Mound. This result confirms previous findings (Hillson et al. 2013; Larsen et al. 2015), but with a considerably larger sample. The Middle period population is larger in comparison to the preceding and succeeding periods, and this increase reflects increased fertility (larger numbers of births) during this period.

The results of the latest round of carbon and nitrogen stable isotope analyses attest to weaning having started earlier than previously considered, before 12 months of age, but continuing under a total breastfeeding regime, when both breast milk and weaning foods are consumed in combination (see Pearson 2018 for a review), until three years of age. While the relative proportion of cortical area as measured as a percentage in long bones (%CA) produces a consistent age-related trend for all periods at Çatalhöyük, it includes a drop in infancy and early childhood. This reduction could reflect normal appositional growth processes, or it could also be a sign of nutritional stress associated with earlier weaning. Whatever may have been the case for the well-being experienced by the very young, an earlier commencement of the weaning process shortened the period of infertility following birth, known as lactational amenorrhea, and an earlier return to fertility. This earlier return to fertility precipitated reduced birth-spacing. With the time between births thus being shortened, more than a single child could be accommodated by reproductive dyads at the same time, resulting in an increased number of children overall within the population. Moreover, prenatal mortality decreased markedly between the Early and Middle occupation periods and despite evidence for increased population density in the Middle period, far more infants were reaching full gestational age. Thus increased fecundity (that is, more children), coupled with increased fertility (that is, more conceptions) and a shortened birth-spacing interval, swelled population numbers and increased the density of inhabitants during the Middle period. In this period, middle adults were especially numerous and show a marked increase in the mortality sample from the Early to the Middle period, with no further change afterwards.

The rise in the number of middle adults, thus those in the prime of life, can be explained largely by the increased births of the Early period, and to a lesser but perceptible extent to community in-migration. Strontium isotope data reveal that most individuals buried at the site spent the majority of their childhoods on the Konya Plain, with only eight individuals of the 230 analysed showing evidence of having originated or having spent appreciable time away from the site. As indicated by similarities in isotopic analysis between sample populations across a series of buildings in the North and South Areas in the whole of the Middle and beginning of the Late/Final periods, these eight individuals, including both males and females, appear to have spent part of their lives in more distant locations or travelled to them and were finally interred at the site.

It seems that females were more likely to move to the site than males, as their dental metrical and morphological evidence indicates that they are slightly more diverse genetically than their male counterparts. The male:female ratio shows that the origins of females were not only slightly more diverse, but also that their numbers increased from the Early period, where males predominate, to the Middle period with its marked majority of females, a trend which continued into the Late/Final period. Females at Çatalhöyük were both more biologically varied (Chyleński et al. 2019) and more numerous, but this was always a plurality, rather than a significant majority in any one occupation period, or area of the settlement (North Area versus South Area).

This evidence of provenance suggests that population growth was largely, although not exclusively, generated by an increased number of births at the site. It seems that one of the defining features of this community was its capacity to generate offspring in greater numbers and more frequently over several generations, and especially between 6700 and 6500 BC, the Middle period of site occupation.

Although a proportion of these infants survived the rigours associated with the birth process, childhood mortality doubled from the Early to Middle occupation periods and then decreased by three-quarters in the Late/Final period, a marked increase in the Middle period and equally marked decrease in fertility in the Late/Final period but also of considerable infant mortality. Based on stature prior to six years of age, these children grew at a rate no different from modern middle-class Denver children, but the growth curves diverge during adolescence and adulthood (see figs 16.7a and 16.7b). Later childhood and adolescent growth was not optimum in many of these children after six years of age. The Çatalhöyük and Arikara sample populations are both shorter and of lighter body mass than those in the Denver sample. After six years of age, the Çatalhöyük children appear to be stunted in their growth for both stature and body mass in comparison to the Denver children, as indeed were the contact-period Arikara from the early protohistoric Sully site (1650–1733, South Dakota) (Owsley 1991), a group dependent on horticulture and living in densely packed villages comprised of earth lodges accommodating 10 to 15 people each (Johansson, Owsley 2002), a situation like that encountered at Çatalhöyük perhaps. The Arikara of this period were already being affected by displaced neighbouring groups, who were themselves displaced by encroaching Euro-Americans, and the presence of Euro-American traders (Jantz, Owsley 1994), bringing unfamiliar diseases such as smallpox that would decimate their numbers in the late 18th and early 19th centuries (Debo 1970: 103; Johansson, Owsley 2002). Although, as noted above, body size differences appearing during later childhood and adolescence may imply emerging genetic body size differences, rather than health disturbances, this reasoning does not really hold for groups separated by nearly 10,000 years in time, from different hemispheres and thus genetically distantly related, one of which, the Arikara, was under the social disruption of the early part of what would shortly become a full-blown contact situation (Owsley 1991). This would seem to indicate that this stunting is more likely to be due to extrinsic socio-economic factors, perhaps linked to dense living arrangements and stresses associated with settled life, than to some underlying genetic similarity. This type of growth impairment has also been found in other incipient farmers in Central and Southeast Europe, especially affecting females (Macintosh et al. 2016), a subject addressed below.

As noted in figure 16.6, however, the mortality profile of the Çatalhöyük sample does not seem to indicate that these deaths were due to a disease of mass mortality such as an epidemic. Rather, from the extent and distribution of dental fluctuating asymmetry across the site, it appears that they were linked to non-specific stressors, as greater dental asymmetry among sub-adults is significantly associated with premature mortality. This pattern not only supports an interpretation of dental fluctuating asymmetry as a faithful marker of developmental stress leading to increased mortality, but also helps to explain the appearance of the mortality profile, in line with the postulated weakened immuno-competence of juveniles in this population (Larsen et al. 2015). Further confirmation of this assessment comes from analysis of the prevalence of linear enamel defects (LEH) in the canine teeth of 122 of 123 (99%) individuals having at least one defect. While there are no statistically significant differences between LEH prevalence when different occupation periods are compared, this result is due to their uniformly high prevalence throughout the Neolithic occupation of the site from the Early to the Late/Final periods, with percentages ranging from 93.5% to 100% of individuals having two or more defects (fig. 16.9).

Cross-sectional analysis of limb bones indicates that immature individuals began to travel greater distances and worked more strenuously in the Middle and more so in the Late/Final periods of occupation at Çatalhöyük. That these behaviours emerged in later childhood (≥6 years old) indicates later childhood involvement in subsistence work, in conjunction with the adults of the community. Although there were children that survived this transition to adult activities, many did not and are represented in the higher number of children in the mortality profile. This evidence suggests that children were seen as part of, and active participants in, the workforce and were exposed to increased physiological stress reflected in an increased number of deaths in later childhood.

The increased number of middle adult deaths, those in the prime of life and who would have occupied important social positions in the community, and the concurrent peak in childhood mortality during the Middle period seems to indicate that adult well-being as well as that of their children decreased during the Middle period. Although the population as a whole was both more numerous and more diverse in its origins, these indicators of successful adaptation are offset by a greater childhood mortality and increased prime adult deaths, and thus decreased longevity in the Middle period.

Sex differences and distinctions

Sex plays an important role in physiological responses to environmental perturbations due to the buffering proffered by the more resilient female immune system (Malina et al. 1985; Stinson 1985). Males are thus more susceptible to physiological disturbance, especially during developmental years, and more readily express these disruptions in their phenotypes. The growth and development of males is thus more sensitive to fluctuations in population well-being; under changing, more demanding conditions, the growth and development of males is more likely to be impaired. Femoral bending strength, a measure of mobility, increases significantly in males through time, but females show a flat or even decreasing temporal trend between the Middle and Late/Final periods. This pattern is demonstrated in overall bone strength for both the humerus and femur that increases through time for males from the Middle to Late/Final periods, but not in females (see fig. 16.13a, b). Males seem to have been experiencing greater physiological demands through time, while those of females were not as great and even reduced over the same period. Perhaps they were not travelling as much or as far as the males, or were not exerting themselves as much as the demonstrably less well-buffered males.

Conversely, if females show physiological changes, this indicates that the stressor responsible is more virulent and/or acute (more fatal, resulting in death, rather than a long-lasting and therefore chronic condition). The birth process is one of these acute conditions. Females of child-bearing age are at increased risk of morbidity (chronic poor health) and mortality (sudden death) due to the hazards associated with childbirth. There are clear examples of this danger at Çatalhöyük in the burials of pregnant women. One such is skeleton (13162), the remains of a woman found with those of a foetus (Boz, Hager 2013: 235, fig. 19.32a); although engaged in the pelvis, the pelvic outlet had been reduced by bilateral ischial ramus (pelvic) fractures (personal observation C.J. Knüsel). In this state this young woman would have had difficulty in delivering a full-term infant naturally without obstetrical intervention; dying in this tragic, liminal state may be implicated in the later post-mortem removal of her cranium and mandible (Patton, Hager 2014).

Given sex-linked characteristics, the expectation is that women would outnumber men over the life-course as men succumbed over the years in increasing numbers, but that there would be a concomitant increase in mortality among younger women of child-bearing age due to the dangers and sequelae of childbirth. The first of these observations appears to be borne out. There are more middle adult males dying than there are similarly aged females, with reduced numbers of old adult males represented in the mortality profiles and more older adult females, except in the Early period, where a disproportionate number of males has already been noted (fig. 16.5). Moreover, based on cumulative dental disease as measured by the number of carious teeth, females appear to have had greater longevity than their male counterparts at the site. The only evidence to suggest deaths due to obstetrical complications is found in individual cases, but cannot be perceived in the proportions of females dying in the young adult category compared to that of males. This highlights the importance of individuals that stand out from the greater whole, both biologically and by their funerary treatment. Without them, these obstetrical calamities would not be perceptible.

The distribution of carious lesions by individual and by tooth significantly increases with age at death, an expected pattern reflective of longer exposure to cariogenic foods. Coupled with no difference in expression in dental caries between the sexes through time, this suggests a similar diet for adult men and women through the duration of site occupation. This pattern of similarity is interrupted by caries prevalence in Middle period females, however, who have a greater prevalence of dental caries than males during this period of site intensification. There were no consistent differences between periods for individuals or for teeth, so there is no obvious peak of poor dental health, and dental health insults occurred generally in members of the community. Within some age groups, however, there is a difference in prevalence between the sexes. Middle adult males have a slightly higher frequency of caries per individual when compared to females in the same age category. The reverse of this pattern is observed in the older adult age category, where females have more affected teeth in comparison with those of males of the same age-at-death. It thus seems that males were predilected during the prime middle adult years, but in old age there were a greater number of females with poorer dental health after many years of accumulation of such lesions.

This sex-linked difference may be due to differences in the constituents of diet – the kinds of foods and their preparations, as well as differing dental hygiene regimes – since men and women shared isotopically identical diets. The difference in dental caries experience, however, identifies a cumulative health deficit for women through the life course, and perhaps a drop in well-being due to altered access to certain foods and increased consumption of cariogenic foods that is not reflected in the results of isotope analyses. Younger and older women had different diets, as did young women compared to older men. Sex-related dietary differences based on carbon isotope values show that these were significantly lower for women (−19.3‰ in women versus −18.6‰ in men) and, in turn, this suggests that women consumed a greater proportion of certain species, equids, suids and birds, in addition to plant foods with relatively higher nitrogen isotope values, whereas the older men consumed foods most similar to the rest of the population. Young women, it seems, may have consumed a different diet that formed part of their gendered social identity. This dietary difference is not reflected, however, in those individuals who had undergone manipulation of the cephalic extremity (see Haddow, Knüsel 2017); their dietary signature resembles that of the overall population (see above). This is not the case, though, for the young adult male (32608) found in B.161, who has values suggesting a completely different diet to the other older males buried there. In general, though, individuals interred in the same buildings often did not have the same diets, but overall statistical analyses suggest that diets within houses were relatively more similar than diets between houses, and there is no correlation with elaboration index for houses, which suggests that although those buried in any particular house showed more similarity in their diets when compared with those interred in other houses, these were not exclusive differences. There appears to be no apparent social distinction linked to dietary access differences between burial locations/houses. This difference in this young man stands out because he was not accorded the more commonplace intramural funerary location beneath a platform and possesses the only example of perimortem mandibular (those showing no signs of healing) and dental fractures found to date at the site (see Haddow et al. 2016). These injuries and his unusual mortuary treatment indicate a social distinctiveness for this individual.

Thus while there are distinctions based on sex and age, especially for diet among young women, these distinctions, as with biological relatedness, do not correlate with burial in particular houses. Moreover, when middens could be associated with a particular house, the faunal remains found within them showed little relationship with the individuals buried in the associated house based on isotopic analysis. These associations may thus be more symbolic than subsistence or socio-economic ones, or the events that led to deposition were not based on a particular house. This evidence supports the interpretation that those buried in particular houses do not show any exclusive connection that would support the hypothesised house-based residential groups, a result supported by non-metric dental trait analysis that, as noted above, does not identify closely related biological groups among those buried in the houses (see also Pilloud, Larsen 2011).

Violence

While sex remains a constant throughout the life course, gender is more malleable and can be said to be age-related, because it changes for both males and females throughout the life course (Gilchrist 2004; Gowland 2006). Trauma has been demonstrated to be especially indicative of gender distinctions (that is, among men and women in human societies) in human groups, because they act to modulate the circumstances of violent encounters, the type of injuries received, their number and their extent (cf. Kjellström 2014; Tung 2014; Novak 2017).

Signs of interpersonal violence are already evident in the Early period. However, characterisation of the extent of violence in the Early period in relation to the Middle period is yet to be confirmed, given the small sample size for the Early period. When the prevalence of injury of the Middle period is compared to that of the Late period, however, the odds of injury are three times more likely in the Middle period. This correspondence adds further evidence demonstrating that the Middle period was one of social change that affected human relations in the community. It is suggested that although violent confrontations producing injuries occurred from the time of the founding of the site, they grew in prevalence as community numbers increased and continued into the Late period of occupation, being two times more common in the South Area when compared with the North Area at this time. During the Late period the South Area accommodated a greater number of inhabitants as the North Area was abandoned; so, as site population shifted from the North Area of the site to the South Area, so did the instances of violence, testimony to the correspondence between density of inhabitation and the occurrence of violence within the community.

As noted at the outset of this discussion, violence is engendered, and so it is at Neolithic Çatalhöyük. Young males are most often affected and are three times more likely to have sustained cranial trauma than older males and two times more likely than older females, who demonstrate the increase in cranial injuries that comes with the expectation of greater numbers of injuries being sustained as a consequence of the greater number of accumulated years of life. That this is not the case for young males, with fewer accumulated years of life, demonstrates that they were more frequently exposed to circumstances that predisposed them to trauma and were the social age group most at risk of sustaining injuries. Wounds, overwhelming depressed fractures from blunt force trauma, are more frequently found above the hat-brim line, a location more often found in victims of assault – that is, inter-personal violence – than of accidents. This pattern of males, and especially younger males, being at increased risk of assault violence is a density-dependent one encountered through time in a variety of societies and circumstances (see Kjellström 2013; Tung 2013; Knüsel, Smith 2013: 668–77; Walker 2001; Brink 1998; Hussain 1994). This association seems to have already been set in place by the seventh millennium in Anatolia. Moreover, that these males died young suggests that their head injuries, even if survived, may have played a part in their earlier ages at death (Boldsen et al. 2015) and, whatever the case, reflect an aspect of the lifestyle they assumed. In short, it appears that violence formed a socially distinctive part of young male identity.

The position and concentration of fractures affecting the superior (35%) and posterior (31%) aspects of the cranial vault (fig. 16.18) suggests that these were sustained when the victim was not facing the assailant and that many came from above. This location is consistent with injuries caused by projectiles and, given the ovoid shape of these lesions, thrown stones and/or other similarly shaped objects may be suspected. Some individuals were repeatedly struck, either at the same time or in succession (similar states of healing of lesions cannot distinguish between these at this time), which demonstrates that certain individuals had been targeted more than once. The occurrence of these injuries attests to the effects of social stress present within the community, which likely reached a peak during the most densely populated Middle period. That these injuries were not fatal in the majority of cases intimates that the intent may not have been to kill, but rather to punish or control individual behaviour in an attempt to curb or motivate compliance to contemporary precepts. That females also suffered similar injuries but in a lesser frequency suggests that both sexes were meted the same treatment, but females were proportionately less often exposed to trauma leaving such evidence on the skeleton.

Abandonment

Prior to the abandonment of the Neolithic East Mound and the establishment of a Chalcolithic settlement on the West Mound there is a peak in the number of neonatal deaths in the Late/Final periods that indicates relatively high mortality before two months of age. Considering the stability of fertility estimates throughout the major occupation periods, this result suggests that living conditions during the Late period of the settlement were probably not improved, even in the context of a decrease in population size and density. There is evidence for the general improvement of maternal health and well-being into the Middle period, with a marked decline in the Late/Final periods reflected in the increase of pre-term and neonatal mortality coupled with declining fertility. This Late/Final population was one in decline, with a drop in fertility, poorer maternal health and a consequent greater number of foetal/infant deaths, thus prior to and around the time of birth. Essentially, this population was no longer realising the success of the preceding periods and appears to have resulted from a decline in population vitality that co-occurs with the gradual abandonment of the East Mound. Perhaps one sign of this period of abandonment is the presence of coprolites in an open refuse area dating to 6410–6150 BC, during the Late/Final periods of the site, perhaps suggesting that at this time inhabited areas had become interspersed with larger open areas and waste ground, but population numbers and proximity to human waste were still sufficient to support orally transmitted intestinal parasite infections. Moreover, by this time, some community members had become less sedentary than their predecessors at the site based on changes in femoral bending strength ratios, a measure of mobility. While the inhabitants of the site already resemble in the Early period a relatively sedentary group when compared with broadly contemporaneous food-producing populations and had developed a pattern associated with intensive agricultural groups by the Middle period, by the Late/Final period, prior to the abandonment of the East Mound, perhaps in response to increasing aridity, they had become more mobile, similar to less sedentary, pre-agricultural European Mesolithic and Natufian populations.

Possible social correlates of developmental stress: dental fluctuating asymmetry, enamel hypoplasia, stature and mortality

A relevant research question relates to the possible development among the inhabitants of Çatalhöyük of some form of social differentiation resulting from unequal access to resources and/or differential well-being. The possible presence (or absence thereof) of social inequality among early Neolithic communities is a classic topic of discussion, characterised by diverging opinions and ambiguous results (e.g., Rollefson 2002; Hodder, Cessford 2004; Düring 2007; Kuijt 2009; Pearson et al. 2013; Wright 2014). At Çatalhöyük the long-term occupation of the site, its large size, and estimated changes in population size and density raise the question of the presence and type of social norms responsible for structuring communal life (see also section above on interpersonal violence). The appearance of marked vertical social differentiation appears clearly only during the later stages of prehistory, after 6000 BC in Mesopotamia, with the rise of regionally differentiated societies showing evidence for internally differentiated settlements, large-scale pottery and eventually bronze production, as well as larger-scale exchange of materials over greater distances. These phenomena are often equated with the rise of 'civilisation' or socially complex and stratified societies associated with the city-states that came to dominate the Middle East in the Bronze Age, after 3000 BC (Nissen 1988; Düring 2011a). These developments have been used as proxies for changes in social organisation, relations among those making these transitions, without fully integrating the evidence from human remains in their burial contexts that reflect changes in the nature of social relations. It is, in any case, worth questioning if at least the seeds of these later phenomena were already present during earlier times. While current interpretations of Çatalhöyük gravitate toward a basically egalitarian social model (Hodder, Cessford 2004; Hodder, Pels 2010), the aim of this section is to revisit this important research question by incorporating new bioarchaeological data, explicitly focusing on possible patterns of developmental stress, and to compare these with levels of architectural complexity at the site.

Briefly, if at Çatalhöyük no social differentiation were present, one might assume a lack of differences in living conditions among the members of the community. Following this reasoning, the null hypothesis would be an absence of correlation between archaeological and biological (skeletal) signatures of 'well-being'.

It should be noted that the aim here is not to test the presence at Çatalhöyük of an overtly hierarchical social system, a hypothesis which is not supported on the basis of current archaeological knowledge of the site itself and more generally of the Neolithic Near East. Rather, it is the possible presence of nuanced differences in bioarchaeological patterns which may suggest the beginning of some form of social differentiation in the context of an egalitarian system (see also Wright 2013).

In any case, when addressing such questions, the choice of variables to be tested is paramount. In terms of bioanthropological data, preference should be given to those features expressing differences in living conditions (that is, environmental differences) and not excessively influenced by additional factors of no primary relevance to the question at hand. Based on these criteria, the present discussion is based on four variables: dental fluctuating asymmetry (DFA), adult stature, linear enamel hypoplasia (LEH), and differential mortality. For the definition of DFA, please see the earlier section of this chapter, above. LEH is represented by discontinuities in dental enamel apposition (Hillson 1996; Larsen 2015) and can be (cautiously) used as a proxy of the relative exposure of an individual to stressful episodes throughout their growth. Overall, the choice of these variables was suggested by their sensitivity to the effects of environmental stressors. Briefly, the higher the prevalence of DFA and LEH, and the shorter the estimated stature, the higher the estimated level of environmental-induced deviation from the genetically targeted phenotype.

With respect to mortality, attritional (non-catastrophic) mortality models for stable populations predict higher mortality among younger subadults and older adults (Chamberlain 2009). In order to address this question, the mortality distribution of two 'intermediate' age categories: adolescents and young adults were targeted in order to check for potential deviations in age frequencies potentially associated with contrasts in mortality schedules and, by proxy, living conditions. This approach was especially chosen due to the well-known risk of confounding changes in fertility with subadult mortality (Larsen 2015).

With regard to archaeological features, pairwise comparisons between buildings incorporate the disadvantages of reduced sample sizes. An alternative strategy would have been to score each building along a continuous scale according to some proxy of complexity in order to test the possible convergence (or lack thereof) between building complexity and general indicators of well-being among the individuals buried within them. The elaboration index ('EI', Hodder, Pels 2010) is such a tool, having the advantage of being an elegant proxy for the relative architectural complexity of a given building. The EI is based on the number of floor segments, basins, benches, installations, pillars and wall paintings in the main room of the building (Hodder, Pels 2010).

A second approach would be to separate the individuals buried in history houses (houses continually rebuilt in the same place and including large numbers of burials – Hodder, Pels 2010; Hodder 2016) and compare them with the rest of the sample. This approach tests for the presence of specific living conditions among those buried in the buildings, possibly related to the postulated social role of history houses within the community at Çatalhöyük (Hodder, Pels 2010; Hodder 2016). The following analyses focus on three main hypotheses, namely:

- H1: if variability in EI reflects social differentiation at the site, then a negative correlation between DFA and EI, and a positive correlation between stature and EI is postulated.
- H2: if burial in history houses indicates a privileged social standing, then lower DFA and LEH, and increased stature of these individuals compared with the rest of the population of Çatalhöyük is postulated.
- H3: if variability in EI reflects some form of social differentiation at the site, then a relatively higher mortality of adolescents and young adults in buildings characterised by comparatively lower EI scores is postulated.

Material and methods. Analyses were performed on all stratified individuals for whom data on DFA, LEH or stature were available, and who were buried in buildings for which the EI was previously calculated (that is, individuals from contexts not associated with a specific value for the EI were excluded). Given the relatively small number of individuals, separate analyses were performed for each variable considered in the study. For the methodological details about the calculation of DFA and estimation of stature see above in this chapter. With regard to LEH, this was observed on the upper and lower first and second incisors and canines, and scored on a three-stage ordinal scale (from absence [1] to severe hypoplasia [3]). For the sake of simplicity, in this contribution each individual is scored according to the maximum expression of LEH on the upper and lower dentition. In order to minimise the bias of age and sex on the results, analysis of DFA was performed on adults and subadults separately (no differences between sexes were noted – see above), and analyses of statures were performed separately by sex.

The association between DFA, stature, age category frequencies and the EI was tested by means of a Spearman test. The effect of EI on LEH was tested by means of an ordinal logistic model. Furthermore, DFA, LEH and stature estimates were compared between individuals from history houses and non-history houses in order to further check for possible differences in relative exposure to environmental stressors between these two samples (note that given the small number of individuals, all individuals from history houses were grouped in a single sample). Specifically, the building sequences considered as part of the 'history house sample' are the following: (B.1/B.5), (B.6/B.17/B.24), (B.8/B.21/B.22/B.23), (B.10/B.44/B.56/B.65), (B.20/B.7 /B.16/B.18) and (B.59/B.60). DFA and stature medians between history house and non-history house individuals were compared by means of a Wilcoxon test, whereas the possible association between the development of LEH and type of building was tested by means of a Chi-squared test. In all tests alpha was set at 0.05.

Results. Analysis of DFA, LEH and stature estimates highlight a lack of correlation between these variables and EI (table 16.21, figs 16.20 and 16.21). With regard to age distribution, the relative frequency of adolescents and young adults shows no significant correlation with the EI (table 16.21). It is interesting to note, however, the different trajectories of these age categories: both show an increase in relative frequency in parallel with the EI until about EI=20. Adolescent mortality then slightly decreases, whereas young adult mortality tends to stay relatively constant (fig. 16.22).

a)	*n*	*Spearman ρ*	*p*
Stature (both sexes)	32	-0.1526	0.4042
Stature (Females)	18	-0.1307	0.6051
Stature (Males)	14	-0.023	0.9379
% Adolescents		0.2495	0.3342
% Young Adults		0.2631	0.3077
DFA			
UM2_MD	60	-0.11	0.3986
UM2_BL	59	-0.09	0.4808
LM2_BL	35	0.08	0.6278
LM1_MD	50	0.25	0.0855
LC_MD	41	0.10	0.5316

b)	*p (EI effect)*	
Upper LEH (n=57)	0.0231	(increase)
Lower LEH (n=55)	0.4272	
Overall LEH (n=67)	0.1387	

Table 16.21. Correlation between skeletal and demographic data and elaboration index. a) Results of Spearman test between elaboration index and adult stature, adolescents and young adult frequencies, and fluctuating asymmetry; b) significance of elaboration index (EI) as predictor of linear enamel hypoplasia (LEH). UM2=upper second molar; LM2=lower second molar; LM1=lower first molar; LC=lower canine; MD=mesio-distal; BL=bucco-lingual.

When comparing individuals from history houses with the rest of the sample, 'non-history house individuals' show a significantly higher DFA for the LM2 bucco-lingual diameter. No differences were found in LEH frequencies, whereas males from history houses present a significantly reduced stature when compared with the rest of the sample (table 16.22).

Discussion. Most of the starting hypotheses can be rejected on the basis of these results, which overall show: 1) a lack of correlation between DFA, LEH and stature and building elaboration, 2) an overall lack of differences in these variables between history houses and non-history houses (with the sole exception of a higher DFA of LM2 bucco-lingual diameter in the latter, and a shorter stature among males from history houses), and 3) no patterning in age profiles between buildings featuring contrasting EI scores.

It must be stressed that the relatively small number of analysed individuals requires some caution when interpreting these results (especially for stature data). In any case, the overall pattern appears relatively clear: the extent of building elaboration (as well as its inclusion in a history house sequence) is unrelated to the living conditions of the individuals buried therein. Interestingly, this result agrees with the results of a similar test based on nitrogen and carbon stable isotopes, which show no differences in diet according to the elaboration of a building (see Chapter 11) and with previous analyses highlighting homogenous diets and similar exposure to cariogenic food in individuals buried in history houses and the rest of the sample (Hillson et al. 2013).

The lower DFA for LM2 bucco-lingual diameter and the shorter stature of males buried in history houses are intriguing, since, among the variables considered in this study, they are the only ones suggesting physiological trends possibly linked to social differences at the site. This result accords with previous research showing increased mobility and greater prevalence of degenerative joint disease in adult males from history houses, and of infections in subadults from the same buildings (Larsen et al. 2015). It seems therefore reasonable to postulate that, although average well-being at the site does not indicate marked social differences, individuals buried in history houses (especially males) were exposed to different environmental stimuli compared with the rest of the population (in terms of stress during growth, and relative mobility and prevalence of degenerative joint disease during adulthood). Given the ambiguity of these patterns, the underlying processes are, at the moment, largely unknown and difficult to disentangle. They strongly suggest, however, that if some form of social differentiation was present in the community at Çatalhöyük, these differences had complex and variable biological consequences for community members during their lifetimes. An important point to consider, moreover, is that the hypotheses tested here are admittedly (and deliberately) simplistic, since they link social differentiation to differences in building elaboration. Given the likely presence of non-material variables defining social relationships at Çatalhöyük (Hodder, Cessford 2004; Hodder 2016), this narrows the breadth of the results of this analysis.

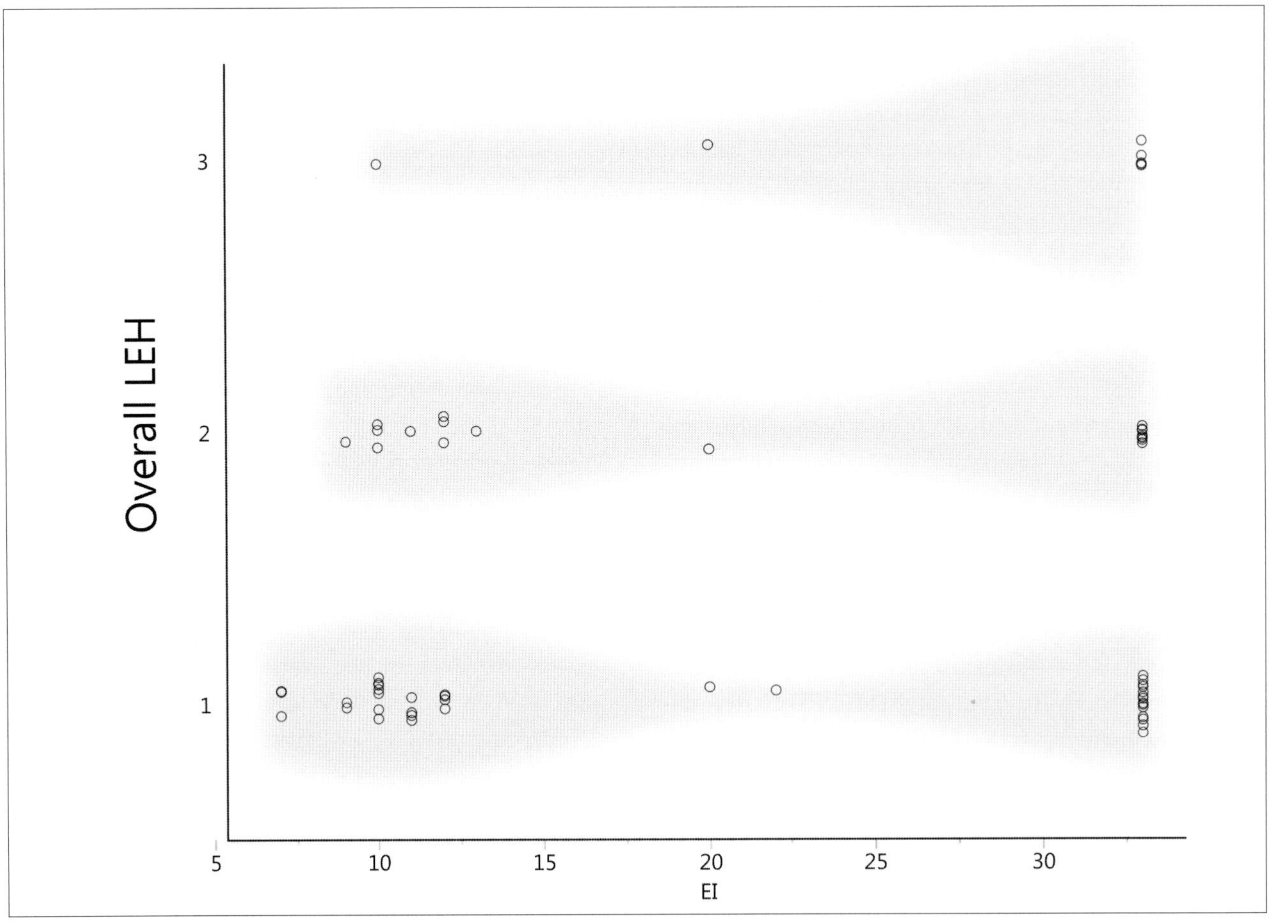

Figure 16.20. Linear enamel hypoplasia (LEH) versus elaboration index (EI). For simplicity, individual LEH values are obtained from the maximum scores for the upper and lower dentition.

17. Funerary practices II: burial associations

Milena Vasić, Christopher J. Knüsel and Scott D. Haddow

> Mortuary practices represent the complex interplay of emotions, material culture and social memories of the mourners and the deceased in the past, testimonied by the material remains of these ceremonies, namely grave goods, skeletal remains and funerary structures.
> (Chesson 2001: 1)

Burials provide invaluable information about the various aspects of the lifeways of people that were buried, as well the people who buried them. Funerary practices are not a mere passive reflection of the social structure, but are the active product of social strategies and the dynamics of social organisation (Hodder 1982; Parker Pearson 1982; 1999; Thomas 1991; Carr 1995). These practices not only illustrate how societies dealt with death and dying, but they also mediate and renegotiate the relationship between the dead and the living.

With its large skeletal and material assemblage spanning over 1,000 years of the existence of the settlement, Çatalhöyük offers an exceptional dataset for studying funerary practices in the Neolithic Middle East. In the 1960s James Mellaart excavated a significant number of intramural burials (around 480 skeletons in total, see Mellaart 1964; 1966). Mellaart wrote extensively about the funerary practices and burial goods in his preliminary reports and a site monograph and tried to find correlations in the locations of burials, age-at-death and sex of the interred individuals, body treatment and objects that were placed in those burials (Mellaart 1962; 1963; 1964; 1966; 1967). The absence of biological anthropologists during the excavations resulted in concerns regarding the interpretation of funerary practices at Çatalhöyük (Hamilton 1996; Düring 2003). The methods of the Çatalhöyük Research Project (1993–2017) involved a more systematic excavation and recording of both skeletal remains and burial goods, thus providing an opportunity for a re-assessment of Mellaart's interpretations.

Although this volume is primarily focused on the research conducted in the last segment of the project (2009–2017), this chapter includes the entire burial assemblage and the information obtained during 25 years of research at Çatalhöyük. The studied skeletal assemblage consists of 620 individuals from the stratified burial features in the North, South, GDN and TPC Areas (table 17.1). As they would bias the results and deflate the number of associations, tertiary burials were excluded from the analyses, and the focus is on primary and secondary depositions (see Chapter 15 for the definition of the deposition categories).

Over the last nine seasons of excavations 321 individuals were added to the human remains assemblage, significantly increasing the sample since the last series of publications and filling in the gaps in the assemblage, especially for the earliest levels of occupation at the site, thus providing an opportunity to develop a better diachronic overview. Although our understanding of funerary practices at Çatalhöyük has not been drastically changed, and patterns observed in the previous publications still stand true (Hamilton 1996; 1997; 1998a; 1998b; 1999; 2000; 2005; Boz, Hager 2013; Nakamura, Meskell 2013a), newly added individuals significantly improve our knowledge about funerary practices, perhaps particularly those relating to the burial of adolescents and young adults, age categories which were not well represented in the previous assemblage. In addition, recent excavations and research by the human remains team provided new insight into different ways of burying the dead that have not been considered before, such as delayed burial (see Chapter 15).

Due to the complex nature of Çatalhöyük burials, the condition and completeness of skeletal remains, their disturbance and the practice of consecutive (collective) interments at the same location (for example, under the same platform), separating objects that had been originally deposited with the interred individuals but were incorporated in the fill from the material that originates from the soil that was used for the burial infill was not always possible. Burial fills tend to have a substantial amount of material, and the density is much higher than in other types of primary contexts (for example, activity deposits and floors) (see Volume 14, Chapter 3). This material is usually considered background noise and is assumed to reflect tertiary deposition. Whilst a lot of material found in the burial fills does come from the surrounding matrix, it is important to have in mind that due to disturbance of burials, some items that were originally part of the burial and were in closer association with the interred bodies ended up in the fill.

		Early Period	*Middle Period*	*Late Period*	*Total*
Age-at-death	prenate (pre-full term)	8	8	7	23
	neonate (birth–2 months)	9	32	56	97
	infant (2 months–3 years)	12	58	19	89
	child (3–12 years)	9	93	21	123
	adolescent (12–20 years)	4	27	10	41
	adult young (20–35 years)	6	30	15	51
	adult middle (35–50 years)	4	41	26	71
	adult old (50+ years)	8	26	11	45
	adult (20+ years)	2	45	33	80
	Total	62	360	197	620
Sex	female	4	37	22	63
	female?	4	30	10	44
	male	7	31	19	57
	male?	4	20	7	31
	too young to determine	39	209	107	355
	unknown	4	33	33	70
	Total	62	360	197	620

Table 17.1. The numbers of burial features studied.

Objects were found in direct association (that is, directly on the body, or in the immediate vicinity) with a significantly higher proportion of primary undisturbed inhumations than primary disturbed individuals, which clearly demonstrates that Neolithic disturbances of inhumations have resulted in objects being displaced (fig. 17.1). Overall, half of individuals whose burials had not been disturbed had items that were found near or on the body, whilst only just over a quarter of primary disturbed interments had direct associations (excluding various organic materials used to contain the body (see below) and treatment with pigments, as the evidence is highly influenced by differences in preservation). Therefore, disturbance of burials directly affects the associations that are observed archaeologically. Furthermore, the vast majority of burial features contain single inhumations; however, over 10% of the studied burial features contained more than one individual, but these inhumations were consecutive (and therefore collective in nature), and burials with multiple (simultaneous) inhumations are quite rare. It is difficult to discern how much time had passed between each interment in these burials, but it is evident that there is a negative correlation between the occurrence of direct associations and the number of buried individuals. That is, it appears that the more individuals there are in a burial feature, the less likely it is that burial goods are in direct association with the buried individuals (fig. 17.2). Almost half of burial features with single inhumations have items that can be directly related to the buried individuals, whilst only about 20% of burials with multiple consecutive inhumations contain direct associations. However, fills of features with multiple inhumations can be quite rich in material, and it is still to be determined what portion of this material had originally been burial goods, and what represents an accidental inclusion.

For example, beads are the most frequent type of artefacts found in burial features, and they occur relatively frequently in other deposits as well, albeit in much smaller numbers. With the exception of a few placed deposits and very large midden and infill units, non-burial contexts almost never contain more than ten beads (Vasić 2018). Furthermore, soil around interred individuals that did not have beads in close association with their bodies rarely contain larger groups of beads. Out of 176 occurrences of beads in burial features, 97 were smaller groups of beads (less than ten) and, given that there were no direct associations, it is possible that some of these, if not most, represent accidental inclusions. In total, 222 beads were retrieved from these 97 features, which amounts to only 1.56% of the total bead assemblage in burials, whilst the remaining beads represent burial inclusions for 78 individuals in total. Therefore, it is clear that the vast majority of beads retrieved from the burial fills were deliberately placed in burials.

Whilst it is relatively easy to distinguish between deliberate and accidental deposition for some types of objects, such as beads, determining the nature of deposition for most materials is less straightforward. It is clear that burial fills do contain material that was originally deliberately placed with the interred individuals,

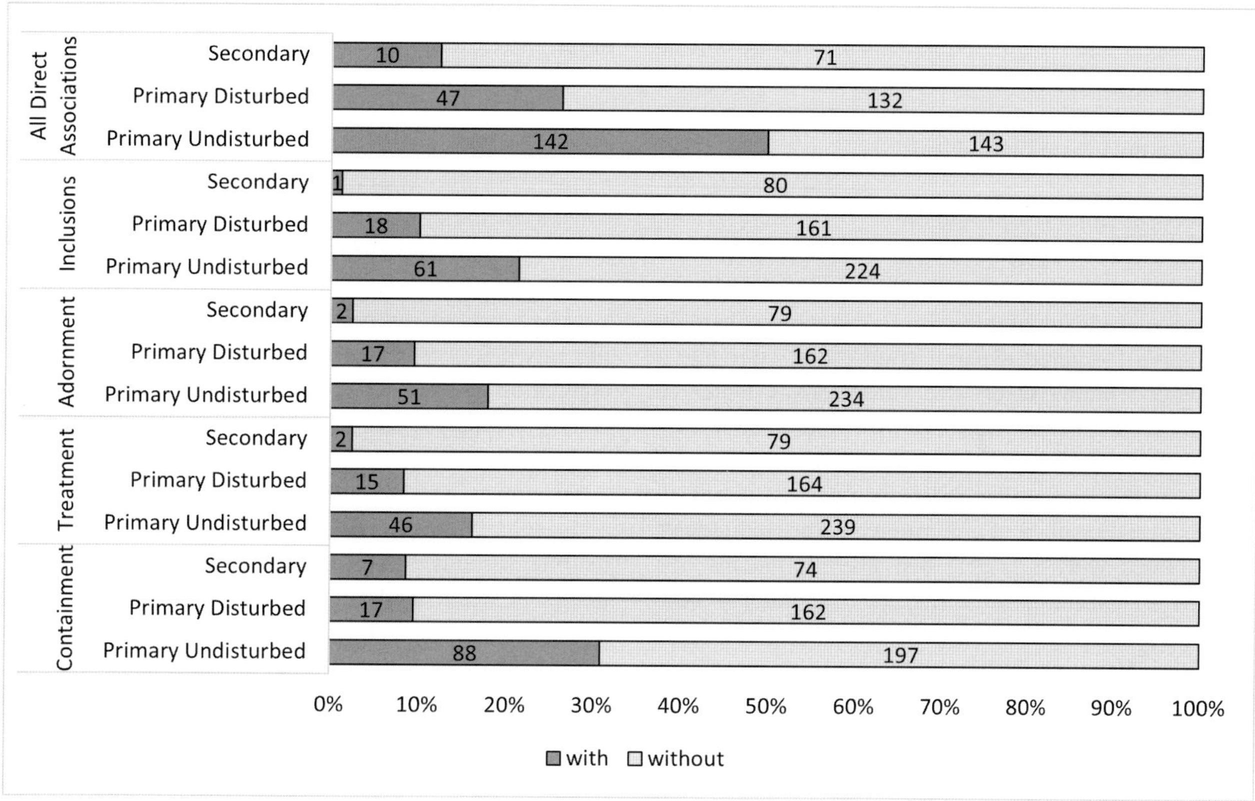

Figure 17.1. Deposition categories and burial associations.

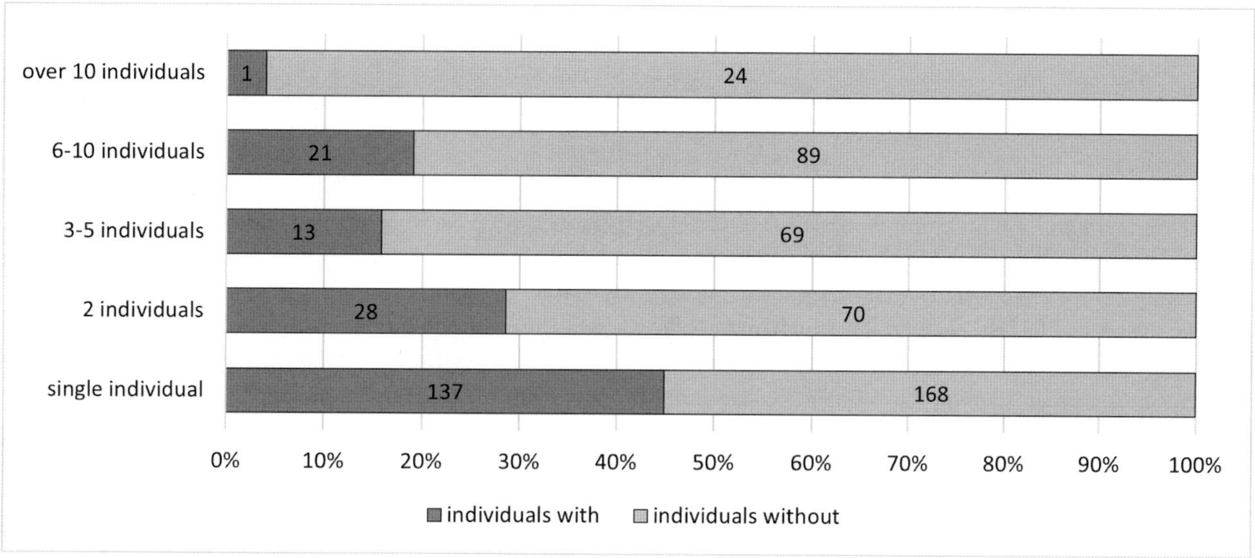

Figure 17.2. Proportion of individuals with direct associations in features with single and multiple (consecutive) inhumations.

and additional investigations need to be conducted in order to further interpret the material from burial fills. This is far from a straightforward process, due to the risk of assigning intentionality to the deposition of artefacts (especially the complete objects) and interpreting unworked material or small fragments of objects as accidental inclusions. As Nakamura and Meskell (2013a: 442–43) noted, the presence of animal scat in a few burials is a good illustration of the complexity of funerary contexts at Çatalhöyük and a reminder that meanings of materials and the connotation they carry cannot be assumed.

For that reason, the focus of this chapter is on direct associations. The material assemblage was split into four main categories: material related to the containment of the body, treatment with pigments and plaster and associated finds (pigment containers, applicators and lumps), body adornment, and inclusions; that is, objects that accompanied the deceased and were either placed directly on the body or next to it, but still in close association with the interred individual.

Overview of funerary practices

Containment of the body

Preparation for interment involved binding the bodies in order to make them tightly flexed, wrapping them in some sort of organic material (rather like later burial shrouds or winding sheets, see Volume 15, Chapter 16 for details), and placing them on mats and/or in baskets (fig. 17.3). It is also possible that some of the organic remains represent clothing, rather than material in which bodies were wrapped. Furthermore, some individuals were treated with red, blue and green pigments and adorned with different artefacts, most commonly bead strings, but also fasteners, perforated boar tusk plates (collars), bangles and finger rings. Yellow ochre was also reported to have been used in funerary practices (e.g., Hamilton 2005; Cessford 2007b). Over 20 occurrences of yellow residue were found with subadults and adult males and females. However, as the geochemical and microscopic analysis showed that four out of five analysed samples of yellow residue are actually plant remains (Shillito et al. 2013b), yellow residue was not counted in the pigment treatment category. In addition, the interred individuals were often accompanied by a variety of objects that were placed on or next to the body.

These actions did not represent a compulsory part of funerary practices, and a significant portion of the buried population showed no evidence of body treatment and had no direct associations whatsoever. Out of 620 studied individuals, 418 did not have any direct associations, and no treatment was observed on their bodies, whilst an additional 52 individuals had remains of baskets, matting, shrouds, or evidence of binding, but no pigment treatment, items of adornment, or objects in direct association. The quantity of burial associations is far from high; the majority of individuals that did have direct associations had only one or two items accompanying them,

Figure 17.3. Reconstruction of an adult burial with typical grave accompaniments: (a) bast cordage and hide straps for binding the body and wrappings; (b) common grave goods such as beads, worked bone, chipped stone objects and shells with pigment; (c) containers made of wood, hide and basketry; (d) woven matting; (e) linen textile and hide for wrapping the bodies.

whilst inhumations that contained five or more burial associations are quite rare (fig. 17.4; in order not to skew the results, fragments of animal bone, residue, phytoliths, pigments and other unquantifiable material were counted as one (per occurrence). In addition, as they would significantly inflate the numbers, beads were counted in groups (for example, one necklace with 100 beads was recorded as one, not as 100 separate objects).

Phytoliths and remains interpreted as remnants of baskets and organic material(s) which were wrapped around bodies (identified thus far as textiles and hides, see Volume 14, Chapter 16) are the most frequent form of body containment, but they have been observed with just over 5% of the interred individuals. Bands of phytoliths indicating binding occur with only 2.7% individuals; however, given the hyper-contracted and tight flexion witnessed in a much larger number of burials (see Chapter 15), it is likely that binding was more common but not always preserved. Burials from burnt buildings tend to have a lot of organic material, due to a higher level of preservation. The primary focus of the phytolith studies at Çatalhöyük has been on diet, land use and farming technologies; consequently, storage bins, middens, fire spots and fire installations have been the main contexts that were sampled (see Rosen 2005; Ryan 2013; Shillito et al. 2013b). Although basketry and matting were analysed (Rosen 2005; Wendrich 2005; Ryan 2013), there have not been any systematic phytoliths studies of the soil from burial features with the purpose of investigating hides, textiles and other organic materials that were used to cover the body of the interred individuals or to explore the remains of materials used for binding. Therefore, the data that we have is mostly based on observations in the field and is thus limited to cases in which bands of phytoliths were clearly visible. Burials in burnt buildings tend to have a higher level of preservation of organic material, which could indicate that such materials were likely ubiquitous but not preserved in other buildings. As they could be seen as means of positioning the body, wooden planks that were found under or above bodies, and one thus far identified case of a concentration of leaves being placed under the head of the buried individual, can also be grouped with other kinds of preparations that served the purpose of body containment.

Treatment with pigments and plaster

As shown in table 17.2, pigments exist in several forms in burials. Apart from the pigment residue observed on skeletons, there are also a few instances of the pigment residue being recorded in the soil surrounding the skeleton, but not directly on it. A few burials also indicate the possibility of matting or wrapping material being coloured rather than the individuals themselves, which was also noted by Mellaart (1963). Pigment residue also exists in containers (most commonly *Unio* shells) and on animal bone artefacts which were used as pigment applicators. In addition, lumps of pigment were found in burials, either directly associated with individuals, or in the fill of burial features.

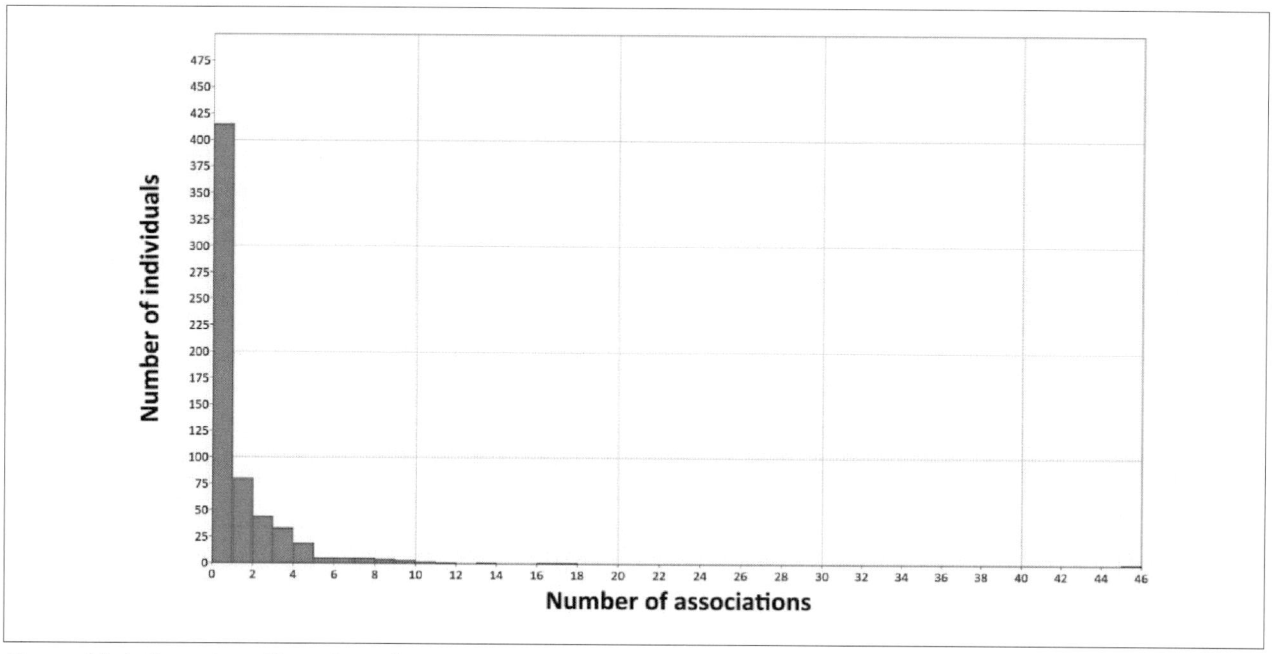

Figure 17.4. Quantity of burial goods.

Number of occurrences	*Inclusions*
66	Beads
36	Basket
35	Pigment application
34	Wrapping
23	Organic residue
18	Matting
17	Binding
16	Pouch, *Unio* shell
12	Animal bone
11	Pigment container
10	Pigment lump
8	Pigment layer, Blade, Flake
7	Wooden bowl, Bone point
5	Wooden plank, Pigment applicator, Object basket/Container
4	Ring, Plaster application, Plaster lump, Wood residue
3	Clay ball, Animal scat
2	Plaster layer, Belt hook, Belt eye, Collar, Knife, Mirror, Pressure flaking tool, Grinding tool, Macehead, Pottery, Perforator, Mineral, Projectile point, Blade/Flake, Figurine, Stone
1	Animal skeleton, Awl, Dagger, Flute, Harpoon, Stone bowl, Fish hook, Scraper, Axe, Polisher, Debitage, Obsidian core, Chalk lump, Leaves, Copper tube Necklace, Fastener, Spoon, Pigment Palette?

Table 17.2. Number of occurrences and their types (direct associations).

Pigments seem most frequently to be associated with crania of buried individuals, although there are also examples of stained elements comprising the torso, limbs and pelvic region. Due to the condition of skeletal remains and preservation of pigments, little can be said about potential patterns. It is not clear whether the colour was simply applied on individuals and whether or not the aim was always to produce specific patterns or designs. Some individuals exhibit clear patterns, as seen in the band of red pigment applied to the cranial vault of a young adult male (23972) buried in B.150 in the TPC Area (fig. 17.5), and more detailed designs could have been accomplished with pigment applicators. Peculiarly, almost all shell containers have only red residue in them, whilst pigment applicators are almost exclusively stained with blue- or green-coloured residue.

The use of pigments is one of the most common features of funerary practices at Çatalhöyük. Direct pigment traces have been observed on 35 individuals (5.6%), whilst if all recorded occurrences of pigments (pigment traces on bones, residue surrounding the interred individuals, and objects associated with pigment treatment) are considered, the proportion of individuals treated and/or associated with pigments increases to 9.4%. It appears that some individuals were treated with both cinnabar and red ochre, whereas four individuals were found with both green/blue and red pigments. An infant (8184) buried in B.3 had an applicator with green pigment and a container with red pigment. Similarly, another infant (17457) was buried in the contemporary B.49 with an applicator, blue pigment and an *Unio* container with red pigment. Both blue and red residues were found on an obsidian mirror associated with the primary burial of a child (19460) in B.129. A lump of blue pigment and an *Unio* shell containing red pigment were found in the burial fill of a feature with a prenate (4215), buried in Sp.168 (Mellaart`s house E.VII.6). Although these instances are rare, they clearly demonstrate that individuals were sometimes treated with more than one colour.

A small number of individuals also exhibit evidence for plaster being used for treatment of the deceased. Plastered skulls were found at a number of PPNB sites across the Middle East (Jericho, Tell Ramad, Nahal Hemar, ‘Ain Ghazal, Kfar HaHoresh, Beisamoun, Yiftahel and Köşk Höyük, see Bonogofsky 2003; 2005; Özbek 2009; Slon et al. 2014). A single example of a plastered skull has thus far been found at Çatalhöyük; an old adult female was buried in B.42 holding a plastered and painted skull of another adult, possibly female (Hodder 2006c; Boz, Hager 2013; Nakamura, Meskell 2013a). In addition, a plastered and painted mandible (19829) was recovered from one of the post retrieval pits in B.89 (Haddow, Knüsel 2017). There is also evidence that suggests treatment of limbs with plaster, witnessed on two individuals ((14441) and (16601)) buried in B.49, whilst the whole body (including the cranium) of a young adult (30040) buried in B.131 was treated with plaster (Haddow et al. 2016b). In addition, similarly to the occurrence of lumps of pigments in burials, lumps of plaster were found directly associated with a small number of buried individuals.

The presence of organic materials, evidence of matting, binding, wrapping, baskets and even pigments, is highly dependent on preservation factors; therefore, it is likely that they were more frequent than what is

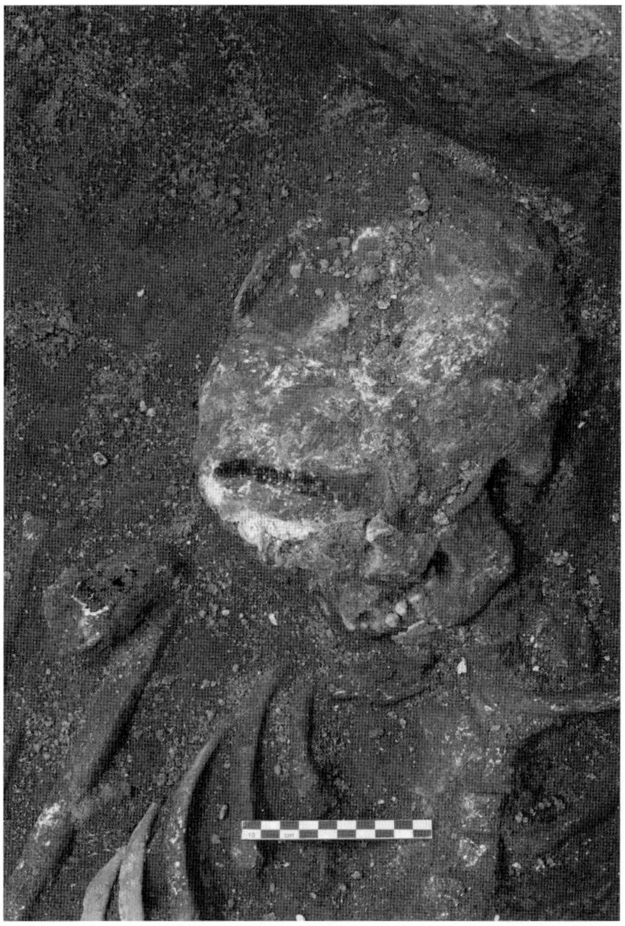

Figure 17.5. Band of red pigment on young adult male (23972) (photo: Jason Quinlan).

possible to observe today. Nevertheless, it is clear that not all individuals were buried in the same way. Quite the opposite, the burial assemblage is highly diverse; there are over 60 different object and treatment types found directly associated with buried individuals, whilst ten additional types were found in burial fills (tables 17.2 and 17.3). As Nakamura and Meskell (2013a) previously noted, the majority occur only once or twice.

Body adornment

Beads and bead strings are the most frequent burial associations and, at the same time, the most common type of adornment found outside burial contexts. Over 43,000 beads have thus far been found, making the Çatalhöyük bead assemblage one of the largest in the Neolithic Middle East (this number includes beads retrieved during Mellaart's excavations as well). The only Neolithic settlement in the region that is known to have a larger assemblage is Körtik Tepe (Özkaya Coçkun 2011). Although it is clear that beads were used in both daily life and ritualised activities (Vasić 2018), the large majority of them were retrieved from burial features (75.6% of the bead assemblage: total N=23,488). Despite their large quantity, beads were found directly associated with only 10.7% of the buried population. In addition, larger groups of beads were found in the fills of 31 individuals, thus increasing the portion of individuals with beads to 15.7%. Mellaart's material is problematic as his records are not complete; nonetheless, judging by the evidence recovered during the recent excavations, it is likely that the vast majority of beads from the 1960s were also found in burial features, especially given the excavation methodologies employed at that time.

The dead were adorned with necklaces, armlets, bracelets and anklets, with the ornaments being most common in the neck/torso region (see Volume 14, Chapter 15). Some of the buried individuals had more than one string of beads. When not in pairs, anklets are almost always found around the left ankle (the only exception is seen with an adult female (10829) that had an anklet on her right ankle). On the other hand, bracelets tend to be worn on the right wrist. It is likely that beads were used in life in a similar way (that is, as single and multi-stranded necklaces, bracelets and anklets), especially given the relatively frequent occurrence of similar beads (albeit in smaller numbers) in non-burial contexts. Nevertheless, there is evidence of beads being used in funerary practices in different ways as well; they were also found wrapped around the bodies of buried individuals, or simply placed next to them.

An adult female (15924) was buried with a necklace consisting of wooden and fluorapatite beads around the neck, whilst a different string, containing animal bone and marine shell beads, was placed on her chest (Eddisford et al. 2007). The whole body of an infant (2105) buried in B.1 was wrapped in a string or strings of over 1,700 tiny black disc beads. These beads were concentrated around the lower limbs, right knee, and chest, and at least three strands ran from the neck to the right humerus, and two strands were found below the ribs (Cessford 2007b). Strikingly, an adolescent (23126) buried in the neighbouring building B.131 also had hundreds of beads associated with different parts of the body (fig. 17.6). The burial of this adolescent yielded over 3,000 beads, some of which were used to adorn the neck and one wrist. However, the large majority of beads (which incidentally resemble the beads in B.1) were found concentrated mostly around the limbs, in multiple rows and different directions, which indicates that they were not used as standard bracelets or anklets. Therefore, it is likely that the body of this individual was also wrapped in multiple strands of beads. These examples point to a different sort of ornamentation – ornamentation that was done as part of a funerary practice, rather than self-decoration in everyday life. It is also plausible that the beads might have been sewn/embroidered onto a material (either used for wrapping the body, or an item of clothing),

Type	*Direct*	*Indirect*	*Unknown*	*Grand total*
Abrader		2		2
Animal bone/tooth	17	25	5	47
Animal scat	3		1	3
Animal skeleton	1			1
Awl	1			1
Axe	1	2	1	4
Bangle		1		1
Basket	30	1	1	32
Basket/Binding	1			1
Basket/Container for objects	5			5
Basket/Matting	8			8
Basket/Matting/ Wrapping	2			2
Basket/Matting/ Wrapping/Binding	7	9	4	20
Basket?	3		1	4
Beads	11,066	4,566	544	16,176
Belt eye	2	1		3
Belt hook	2	1		3
Binding	14			14
Binding?	4			4
Blade	11	8	2	21
Blade/Burin	1			1
Blade/Flake	1	48	4	53
Blade/Scraper		1		1
Bone Pin		2		2
Bone tool unknown	3	3		6
Cereals		1		1
Chalk lump	1	2		3
Chert tool unknown		2	1	3
Clay ball	2	154	2	158
Clay object unknown		4	1	5
Clay potstand		3		3
Clay stamp		1		1
Collar	9	3		12
Copper tube	1			1
Coprolite		2		2
Dagger	1			1
Debitage	1	5		6
Fastener	2			2
Figurine	2	81		83
Fish hook	1	1		2
Flake	8	42	5	55
Flute	1			1
Geometric clay objects	1	53		54
Grinding tool	2	25		27
Harpoon	1			1
Knife	3			3
Knife/Spearhead	1			1
Leaves pillow	1			1
Macehead	2	2		4
Matting	14			14
Matting/Wrapping	11			11
Matting/Wrapping/ Binding	2			2
Matting?	4			4
Mineral	3	3		6
Mirror	1		2	3
Needle		8		8
Obsidian core	1	9		10
Obsidian tool unknown		3	1	4
Organic residue	23	4	1	28
Painted plaster		2		2
Perforator	3	2		5
Perforator/Burin	2			2
Perforator/Cutting tool		1		1
Perforator/Scraper			1	1
Perforator?	1	2		3
Pigment application	37			37
Pigment applicator	6			6
Pigment applicator?		1		1
Pigment container	13	3		16
Pigment layer	8	3		11
Pigment lump	10	11		21
Pigment lump?	5	2		7
Pigment on object	5	1	1	6
Pigment palette?	1			1

Table 17.3. Material assemblage.

although this hypothesis has not been verified by use-wear analysis (C. Tsoraki, pers. comm.). Nevertheless, the possibility of beads being used as an attached decoration should not be completely excluded.

A large number of beads came from fills of burial features, some of which contained a series of consecutive single inhumations. Due to the disturbance of burials, it is quite difficult to establish to which of these buried

neonates and infants, although they do not appear as often as they do with prenates. Items of containment tend to change for individuals that were three years or older, and mats and shrouds become more frequent than baskets, which is likely to be due to body size. In addition, the shift from baskets to mats and shrouds coincides with the end of the period of weaning (see Chapters 11 and 16). Although this difference has been noted, it is clear that 'protection' of the body, or its separation from the soil around it by means of wrapping, matting and baskets, played an important role in funerary practices.

Both subadults and adults were buried with pigments and items related to pigment application (table 17.6). Pigment residue and/or artefacts related to pigments appear with all age categories, with the exception of neonates. Amongst the subadults, infants tend to be the most commonly found with pigments, whereas out of 23 prenates, only three were associated with pigments. Whilst young and middle adults seem to have received similar pigment treatment relatively equally, pigment residue and/or related artefacts were found with over a quarter of individuals older than 50 years.

Similarly, beads were almost never placed in burials of the youngest members of Çatalhöyük society (fig. 17.7). With the exception of a large number of stone disc beads found with one prenate (4215) and Galena and *Unio* beads with three neonates ((10498), (10450) and (10400)) buried within two successive buildings (B.53 and B.42), bodies of individuals younger than two months were not adorned. On the other hand, beads are regularly found with infants, children and older individuals, and they are especially common with individuals older than 35 years. The practices of treatment with pigments and adorning the dead with beads can roughly be correlated to age progression; the older the buried individual was, the more likely he or she was to be associated with pigments and decorated with beads (fig. 17.7 and table 17.7).

Although this correlation has been noted, the evidence for a clear-cut differentiation of adornment based on the age of individuals does not exist, and similar items are found with different age categories. Beads of the same type and colour were found with individuals belonging to different age categories (Vasić 2018, and see Volume 14, Chapter 15). Exceptions do exist, and single occurrences of certain bead types (T.6, T.13, T.14 and T.19) have been recorded only with subadults, whilst others (T.7, T.24, T.26, T.28 and T.30) have thus far been found only with adults. Similarly, both subadults and adults were found with all types of bead ornaments. Anklets are less common with adults than necklaces and bracelets are, whereas anklets and bracelets seem equally frequent with subadults. Four out of five infants with bracelets had a bracelet on each wrist, whilst one skeleton (4406) also had two anklets, and another skeleton (23805) had one anklet and a necklace. Necklaces seem to be the most frequent type of adornment with subadults, in general, and adolescents have thus far been found only with bead ornaments in the neck region.

	Matting (%)	*Wrapping (%)*	*Binding (%)*	*Basket (%)*	*Wooden container (%)*	*Plank (%)*	*Leaves (%)*	*Unknown (%)*	*Total (%)*
Prenate		4.3		34.8				13.0	52.2
Neonate	2.1	1.0	1.0	8.2	1.0			1.0	12.4
Infant	3.4	7.9	3.4	11.2				4.5	20.2
Child	0.8	4.1	3.3	4.1		0.8	0.8	6.5	17.9
Adolescent	9.8	9.8		2.4				7.3	24.4
Young adult	3.9	9.8	3.9			2.0		2.0	15.7
Middle adult	2.8	8.5	4.2	1.4				5.6	19.7
Old adult	2.2	11.1	8.9	2.2		6.7		6.7	26.7
Adult unknown	3.8	1.3		1.3				5.0	11.3
Female	3.2	6.3	6.3	1.6				7.9	22.2
Female?		11.4	2.3	6.8				2.3	18.2
Male	7.0	7.0	10.5			3.5		3.5	22.8
Male?	6.5	6.5				3.2		3.2	16.1
Female all	1.9	8.4	3.7	3.7				5.6	20.6
Male all	6.8	6.8				3.4		3.4	20.5

Table 17.5. Percentage of individuals with different types of body containment.

Age-at-death and sex		*Pigment treatment (%)*	*Associated finds (%)*	*All associations (%)*
Age-at-death	prenate	13.0	0.0	13.0
	neonate	0.0	0.0	0.0
	infant	7.9	6.7	14.6
	child	4.9	4.1	8.1
	adolescent	4.9	4.9	7.3
	young adult	9.8	5.9	15.7
	middle adult	5.6	9.9	12.7
	old adult	13.3	17.8	26.7
	adult unknown	2.5	0.0	2.5
Sex	female	1.6	12.7	14.3
	female?	4.5	6.8	9.1
	male	15.8	8.8	19.3
	male?	12.9	6.5	16.1
	female all	3.7	10.3	12.1
	male all	14.8	8.0	18.2
Total		5.6	5.0	9.7

Table 17.6. Pigments and associated finds with age-at-death and sex categories.

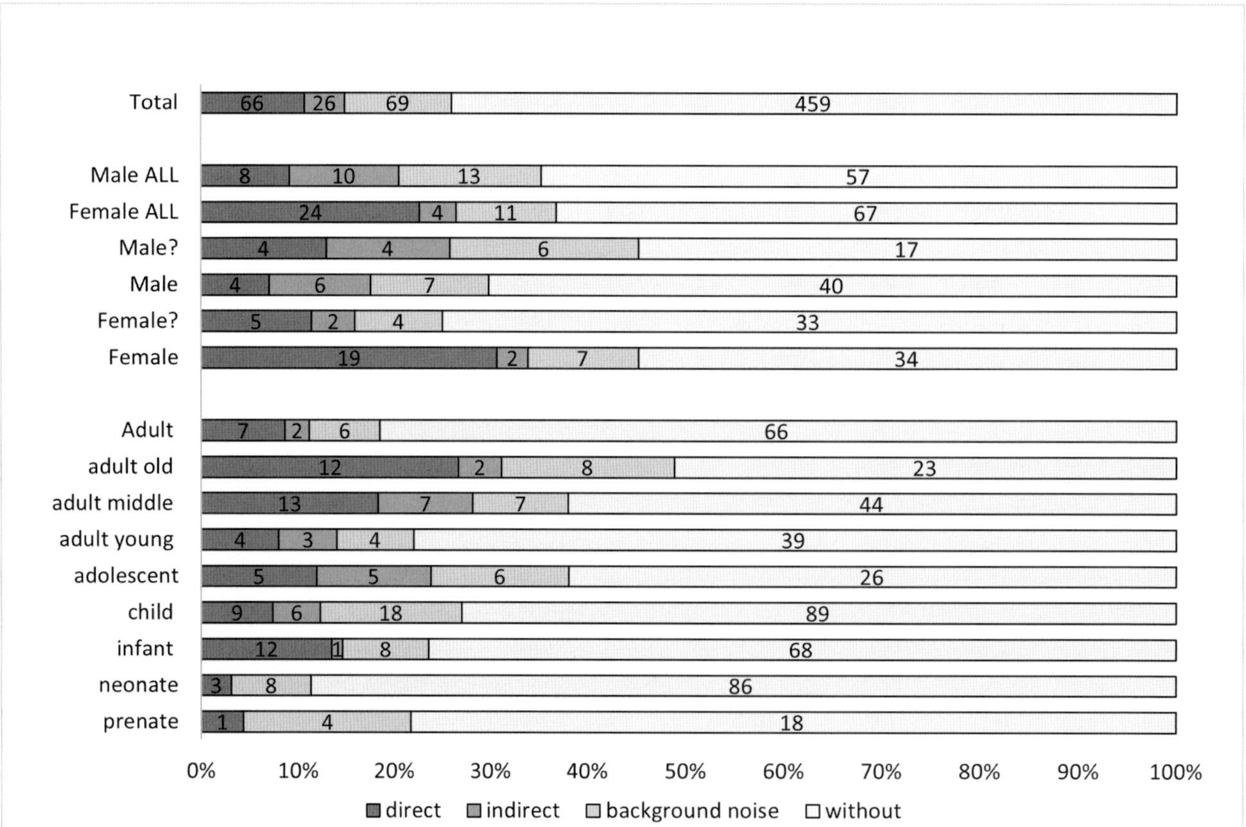

Figure 17.7. Bead associations with different age-at-death and sex groups.

Other items of adornment are rare; rings adorned the hands of three subadults and one young adult; collars occur with three or four adults (all of which also have direct bead associations); belt hooks were found with two adult males; belt eyes with an infant and an adult male, and a single example of a bangle that was retrieved from a burial context was found in the fill of a feature containing one adult and six adults. Given the small sample, it is difficult to come to firm conclusions; however, it can be tentatively inferred that apart from rings and beads, there is a tendency to place items of adornment with adults.

The importance of age-at-death in funerary practices at Çatalhöyük can also be illustrated by the occurrence of tools in burials. Subadults rarely have tools in direct association; in total, three individuals of age-at-death up to three years were found accompanied by an obsidian flake, whilst other direct associations do not exist. Similarly, only three children were found in direct association with tools (a bone tool, possible point, an obsidian flake, and an obsidian scraper). Adolescents had no direct associations, and amongst the adults, tools are by far more common (in both direct and indirect association) with individuals older than 50 years than they are with the younger adults. When both direct and indirect associations are considered together, infants are least commonly found with tools, individuals aged 3 to 35 years (children to young adults) are quite similar (the range going from 14 to 24% of the skeletal assemblage), whilst a significant difference has been observed with old adults, as 42.2% of the individuals older than 50 years were found with tools. Despite old adults having tools more often, they do not seem to be buried with specific types of tools. On the contrary, they have tools that other age categories have as well. On the other hand, young adults were buried with tools that other age categories do not have. Maceheads have thus far been retrieved from two burials only, both of them belonging to young adults. Also, the only example of a fish hook and a harpoon in direct association were also found in a burial of a young adult.

Funerary practices at Çatalhöyük appear to change for the individuals older than three months, and the older they were, the more likely they were to be buried with a higher quantity and wider range of associations. Treatment of the dead could have been related to certain life stages, but given the broad biological categories, it is difficult to correlate the material assemblage in funerary context and connotations the placed objects carried with the potential rites of passage, or the socially defined categories of life course transitions. For example, adolescents are generally treated as subadults in archaeology, due to the stages of skeletal development, but also due to our preconceived western notion of adulthood (Sofaer 2011). However, we cannot be certain how they were perceived at Çatalhöyük. As Haddow et al. show, their burial locations within houses do not differ from the location of adult inhumations (this volume Chapter 15). Similarly, the practices related to the body treatment in death and the accompanying objects are not very different from the burial preparations carried out for the young adults, although it should be noted that there are no major differences between burial of children and adolescents either (table 17.7). For this reason, burial inclusions are unfortunately inconclusive in regard to determining the position of adolescents in the society, as they were buried in similar ways as both subadults and adults. Nevertheless, the correlation of frequency of burial associations and age progression is quite remarkable, especially given that it is consistent regardless of which type of association is considered. Although burial disturbances are an important factor and should not be neglected, it is clear that the age-at-death of buried individuals did play an important role in the burial preparations, and funerary practices in general. (The highest percentage of undisturbed burials is seen with prenates (over 80%) and middle and old adults (over 60%). Therefore, to some extent, the fact that the burials of adults (especially middle and old adults) seem to have more associations could be a consequence of the fact that a majority of them were intact, whilst the disturbance rate of inhumations of adolescents and young adults could partly be a reason for these individuals not having so many associations).

On the other hand, there is no clear differentiation in burial associations based on the sex of the buried individuals. Tendencies do exist, but so does a notable variability. Material remnants of containers/wrapping occur equally with both sexes, items of adornment and other burial inclusions are more common with females, whilst the pigment treatment and associated finds are more frequent with males (fig. 17.8). However, it is important to note that similar items were found with both sexes. For example, apart from two subadults in B.49, all pouches were found in burials of adults. Pouches occur with both males and females; their contents do not differ based on either the age-at-death or sex of individuals with which they were associated.

Both direct and indirect associations suggest that sex cannot be directly correlated with the occurrence of tools in burials, as an equal portion of males and females were buried with tools (14.8% and 12.3%, respectively). However, discrete tendencies do exist based on material. Whilst artefacts made of animal bone occur equally with both sexes, objects made of obsidian and chert tend to be more commonly found in male burials. For example, 8% of males were buried with obsidian tools, whereas only 4.7% of females had similar direct associations. Males and

	Prenate	*Neonate*	*Infant*	*Child*	*Adolescent*	*Adult young*	*Adult middle*	*Adult old*	*Adult*
Animal bone		1		2.5		2.0	1.4	4.4	1.2
Animal scat					2.4			2.2	
Animal skeleton							1.4		
Animal tooth					2.4				
Beads	4.4	1.0	15.7	6.6	11.9	8.0	19.7	26.7	8.6
Chalk lump								2.2	
Chert blade						4.0	1.4	2.2	
Chert blade/burin								2.2	
Chert dagger						2.0			
Chert flake		1.0	1.1						
Chert knife						2.0	1.4		
Chert perforator								4.4	
Claw						2.0			
Clay ball		1.0						4.4	
Clay object					2.4			2.2	
Collar								2.2	1.2
Container or Clothing/Pouch						2.0			
Copper tube			1.1						
Fastener				0.8		4.0		4.4	
Figurine anthropomorphic					2.4				
Figurine zoomorphic quadruped		1.0							
Fish hook						2.0			
Grinding tool							1.4		
Groundstone polisher							1.4		
Harpoon						2.0			
Macehead						4.0			
Microfauna						2.0			
Mineral								2.2	
Mirror				0.8	2.4				
Object container								2.2	
Obsidian blade						2.0	2.8	4.4	

Table 17.7. Percentage of individuals with different direct associations (excluding containment items and pigment residue).

females were found with about 40 different types of associations (excluding unidentified objects and residues). Some items, such as belt hooks and eyes, awls, flutes, daggers, knives, fishhooks, harpoons, maceheads, rings and wooden planks exist with males, but not with females. Conversely, items that were found with females but not with males include debitage, obsidian cores, fasteners other than belt hooks and eyes, grinding tools, minerals, pigment applicators, plaster residue on bones and spoons. Nevertheless, these types of inclusions are very rare, and occur only in one or two burials. Perhaps the most notable difference is in the occurrence of pigment applicators, which have thus far been found only in three female and two infant burials. On the other hand, the only example of a potential pigment palette was found in a male burial in B.77.

Although pigment associations appear more common with males, recent excavations did not yield any examples of males associated with green and/or blue pigments. Nevertheless, an adult male cranium with green staining from Mellaart's assemblage (S.D. Haddow, pers. obs.) demonstrates that these colours were not exclusively related to females. This can also be confirmed by the equal occurrence of green and blue beads with both sexes.

	Prenate	*Neonate*	*Infant*	*Child*	*Adolescent*	*Adult young*	*Adult middle*	*Adult old*	*Adult*
Obsidian blade/Flake				0.8				2.2	
Obsidian core						2.0			
Obsidian flake		1.0						8.9	1.2
Obsidian microdebitage/flake								2.2	
Obsidian perforator/burin								2.2	
Obsidian perforator/scraper								2.2	
Obsidian projectile point						2.0		2.2	
Obsidian scraper				0.8			1.4		
Pebble								2.2	
Pigment applicator			2.2			2.0		4.4	
Pigment container			5.6	0.8		2.0	4.2	2.2	
Pigment lump			1.1	1.6			4.2	8.9	
Pigment palette?								2.2	
Plaster lump		1.0		0.8			1.4	2.2	
Plastered object								2.2	
Plastered/painted skull								2.2	
Pottery					2.4		1.4		
Pouch			1.1			6.0	7.0	11.1	
Quartz							1.4		
Ring		1.0	1.1	0.8		2.0			
Spoon									1.2
Stalactite							1.4		
Stone axe			1.1				1.4		
Stone bowl			1.1						
Stone tool?		1.0							
Unio shell		1.0	1.1	3.3		4.0	7.0	2.2	2.5
Unknown wood residue				1.6	2.4	2.0	1.4		
Wooden bowl				0.8	4.8	4.0	2.8		
Wooden object			1.1	0.8	2.4				
Wooden plate/bowl						2.0			
Yellow pigment lump?		1.0		0.8			1.4	2.2	

Table 17.7 (continued). Percentage of individuals with different direct associations (excluding containment items and pigment residue).

Beads tend to exist more often in female burials, although if both direct and indirect associations are considered (but not the clearly accidental inclusions), the difference between the portion of males and females with beads is much smaller (fig. 17.7). In addition, females have more bead types and, to date, the direct associations of a few bead types have been observed only with females. On the other hand, *Unio* pendants, which are usually found associated with subadults, were found with one male (32770), but no females had similar associations.

The patterns presented above indicate that age-at-death rather than sex was an important factor in funerary practices. However, this could be partly due to the fact that sex identification was only possible for less than a third of the skeletal assemblage. Given the rare occurrence of certain object types in burials, firm conclusions regarding the exclusive association of some artefacts with a particular sex are not possible at this point. The lack of patterning and the high variability observed in burials suggests that there was not a strict gendered division of burial inclusions; rather, the burial assem-

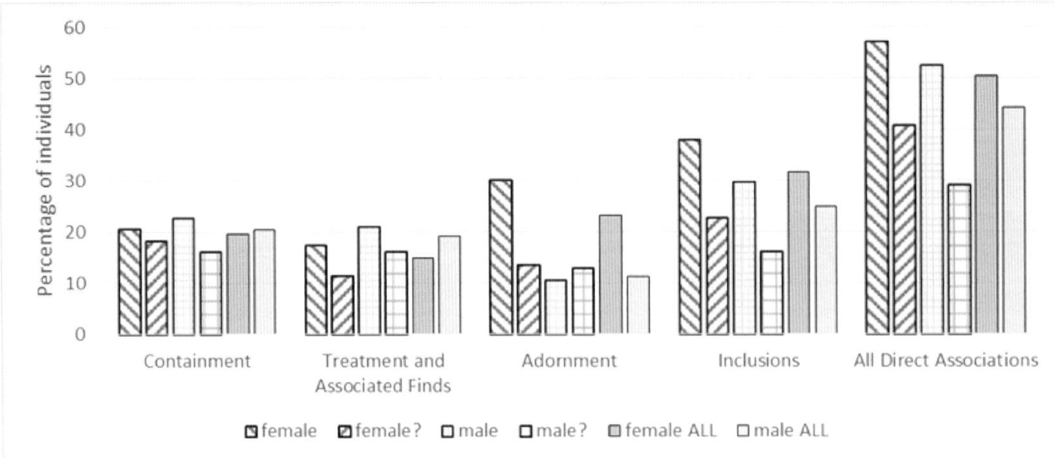

Figure 17.8. Burial associations with males and females.

blage is more indicative of subtle differences and discrete tendencies in treatments of males and females. Other data also do not endorse differentiation (for example, the diet, see Pearson 2013; or osteoperiostitis and trauma, see Agarwal et al. 2015; see also this volume, Chapter 16). Therefore, it can tentatively be concluded that if social distinctions based on sex existed in the society, they were not a focus of funerary practices.

The setting of burial

Burial location

The vast majority of burials are intramural; the people of Çatalhöyük were burying their dead in houses – under platforms and floors, and in association with other building features (for example, walls, foundation deposits, etc.). The usual internal arrangement of Çatalhöyük houses divided the space into the main and side (storage) room. Not only did interments in the main room form almost 90% of the inhumations (see Chapter 15: fig. 15.27), but also individuals buried in the side rooms had notably fewer associations, and none of them were adorned or treated with pigments (fig. 17.9).

Arbitrary horizontal division of the space into side room and west, central and east sections of the main room demonstrates that, although the eastern section of the main rooms has almost as many inhumations as the other sections combined, individuals interred in the western part of the main rooms have slightly more direct associations, although the proportions are quite similar (fig. 17.9). Evidence of body containment occurs somewhat more often in burials in the central and western sections, and the interred bodies in the central section were slightly less frequently associated with pigments. Items of adornment and other accompanying inclusions are the most common with individuals buried in the western and eastern sections (most commonly northwest and northeast corners).

Most buildings demonstrate that the northeast and east platforms were preferred burial locations. Exceptions do exist, as seen, for example, in B.88, where all individuals were buried in the side room, B.114, where the vast majority of individuals were interred under the southeast platform, or B.161, in which all inhumations were located in the west-central floor area. However, most buildings either contain burials in the northeast or east sections, or do not contain any burials at all. This does not mean that people buried their dead in other parts of buildings after the space was filled up in the north and northeast areas. Some buildings have evidence of burials in other sections that are earlier than other inhumations that were located under north and northeast platforms. For example, an individual was buried in the west-central area of B.160 in the early phase of its occupation, and two more individuals were buried in the same section whilst another individual was buried under the east platform in the late occupation phase.

Use-life varied from building to building; some were occupied for a long time, for example, some of the buildings belonging to the Early period, whilst others were short-lived, such as the Late period B.65, B.56 and B.44. There is no evidence that would suggest the existence of visible burial markers, but it is clear that people did remember where the dead were buried (for example, B.49, see Boz, Hager 2013), and given the examples of continuity in burial treatment in some buildings (see below), they also remembered with what they were buried.

The majority of inhumations took place during the occupation of buildings (see Chapter 15), whilst only 40 individuals in the studied assemblage were interred during the building foundation, and interment of ten individuals was allocated to the abandonment processes.

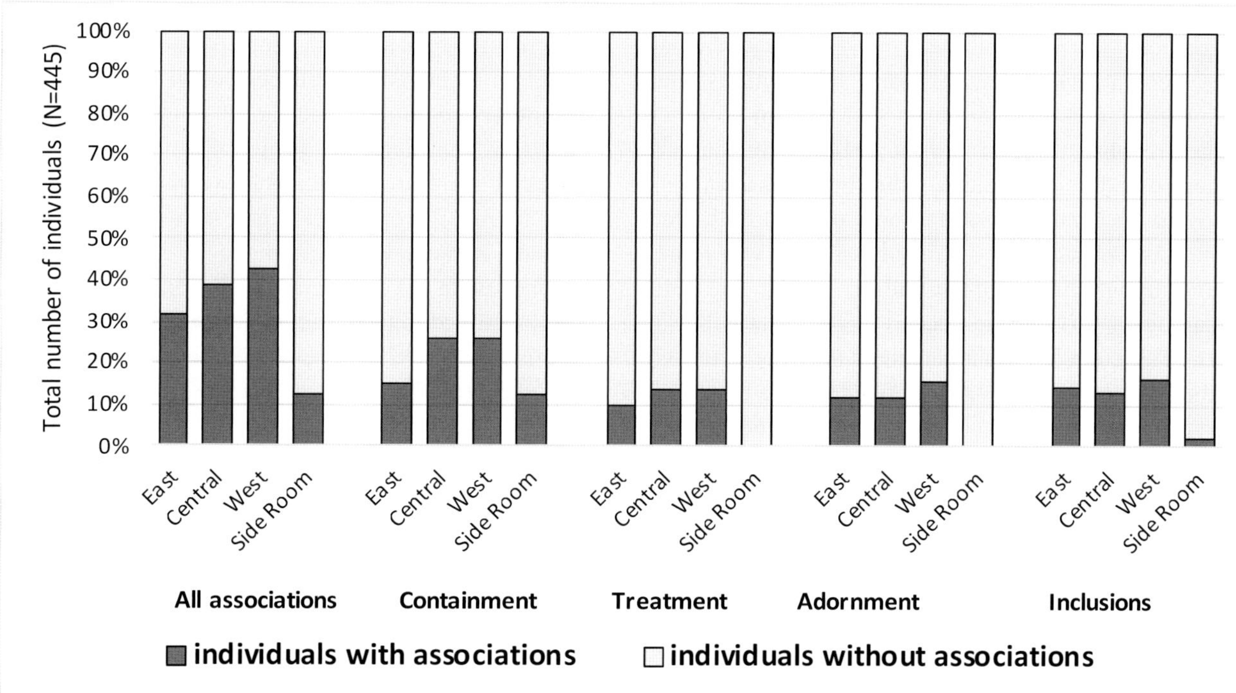

Figure 17.9. Associations according to burial location.

Foundation burials have the highest percentage of individuals with direct associations, but there is no significant differentiation in burials in foundation and occupation phases (fig. 17.10). The significant difference, however, is seen in the burials related to the abandonment phase. Individuals that were interred during this phase do not typically have any associations. The only exception is a young adult male (32608) that was not buried in a typical way, but rather, his body was placed, or even thrown, in the building during the infill process. Organic remains suggest that the body was wrapped in material, and associated with matting and possibly animal scat (whilst scat seems like a deliberate inclusion for the other individuals (see below), in this case it is not clear). Feider and Jenkins (Chapter 10) suggest that it might be accidental, especially because the building fill also contained a high density of micro-fauna. The other individuals had no evidence of treatment, and no objects were found directly associated with the body.

Secondary burials

Circulation and re-deposition of crania and other body parts are distinctive aspects of bodily manipulation in Neolithic funerary practices (Talalay 2004; Kuijt 2008). Although the majority of inhumations at Çatalhöyük represent primary depositions, there is also evidence of individuals being moved from their original location and reinterred elsewhere (Andrews et al. 2005; Boz, Hager 2013). Burial associations are quite rare in these burials; it is plausible, however, that these individuals might have originally been buried with more items, but the objects were simply not moved with them to their last burial location. In addition, more than half of burial features with secondary inhumations contained a wide range of artefacts in the burial fills. However, given that secondary deposition occurs largely in burial features that contain primary inhumations as well, it is not certain whether the material from the fill had once been placed deliberately in burials, let alone to whom this material might have originally belonged.

There are 31 subadults within the secondary burial deposition category, out of which 30 were buried without any associations. The only exception is a child (12876), buried together with a middle adult, possibly female (12863). These individuals were wrapped together and placed under a platform in B.56. Similarly, no burial associations have been retrieved from secondary burials of nine adolescents.

Peculiarly, a large concentration of secondary inhumations (51 out of 81) was clustered in three neighbouring buildings B.1, B.131 and B.77, as well as B.129, which was built on top of B.131. Evidence from these four buildings suggests that individuals in primary depositions accompanied by secondary inhumations do not necessarily have direct associations. The ones that do have directly associated objects have different types, although repetition has been noted in the placement of beads, wooden bowls, mirrors, pigment treatment and wrapping (table 17.8). The direct associations observed both with the secondary inhumations accompanying the primary inhumations and secondary burials of individuals that contain no other human remains, consist almost entirely of remains related to body containment. Other burial associations have thus far been witnessed in only three burials. One of them is a plastered and painted skull that was found in the upper limbs of an adult buried in B.42 (Sadarangani 2015). The second example is a burial feature F.3620 in B.77 that contained long bones of two adults that were carefully placed together in a pile. This bundle of bones was found in direct association with a concentration of perforated *Nassarius* shell and a few stone beads. Together with other examples of heavily disturbed burials containing direct associations, this burial demonstrates the importance of commemoration, remembrance and ancestry.

One of the most notable secondary burials at Çatalhöyük belongs to a young adult male (22196) F.3808, buried in in B.5, located directly below B.1 (fig. 17.11). The cranium and mandible with red pigment coating was placed in an organic container together with two chert blades. Plant micro remains suggest that an organic material (see Chapter 7) was wrapped around the skull before it was placed in a container, whilst the container was resting on a wooden plate or a shallow bowl (Lancelotti et al. 2016). Immediately next to this bundle were two *Unio* shells (one of which contained red pigment), four obsidian projectile points with reeds that might be suggestive of these objects being placed in another container, and an elaborate macehead. Peculiarly, neither the projectile points nor the macehead showed any evidence of use-wear (Doyle 2016; Tsoraki 2016), which could mean that they were all made immediately before the burial. Above this container were the infra-cranial elements, which were tightly wrapped together and placed on a mat. The fact that this young adult's bones appear to have been buried in two separate bundles (see Haddow et al. 2016b for further discussion) with elaborate items and a higher quantity of burial goods than most of the buried individuals on site makes it one of the most exceptional burials at Çatalhöyük.

Burials in external areas

A small portion of burials were not located intramurally within houses (4.7% of the studied skeletal assemblage). Evidence of extramural burial exists in 12 external spaces, and it appears that at least two of these spaces that were used as burial ground were chosen as the location for building construction (see Regan 2015; Haddow et al. this volume, Chapter 15 for further discussion). Out of 29 individuals that were buried in these external spaces, only six had direct and indirect associations, whilst the other 23 had no associations

Figure 17.10. Burial associations and different building occupation phases.

Building	Feature	Primary inhumation	Age	Sex	Primary associations	Number of secondary inhumations	Secondary associations
B.1	F.30	1425	adolescent	unknown	none	4	none
		1424	OA	female	none		none
		1450	infant	unknown	none		none
	F.31	1481	YA	female?	none	4	none
B.131	F.7961	23126	adolescent	female?	mirror, bowl, pottery, necklace, bracelet, beads across body, wrapping and pigment treatment	9	none
	F.7962	31705	adolescent	female	wooden bowl, beads, and wrapping	8	none
	F.7963	30044	MA	female	chert blade, binding and wrapping	8	one individual (MA male) with wrapping
B.77	F.3601	19022	adolescent	unknown	matting	1	none
	F.3620	none				2	two adult males? with a string of *Nassarius* beads
	F.3697	20683	MA	female	wooden bowl, necklace, pouch with speleothem and quartz and beads	1	wrapping/binding on mandible
	F.7853	21657	OA	male?	beads	1	adult female with basket
		21656	child	unknown	*Unio* shell and wrapping		
B.129	F.3630	19460	child	unknown	mirror, necklace, anklets	5	none
	F.3639	none				3	none
	F.3684	20601	MA	male?	none	5	none

Table 17.8. Occurrence of secondary burials and their associations in the northwestern part of the North Area.

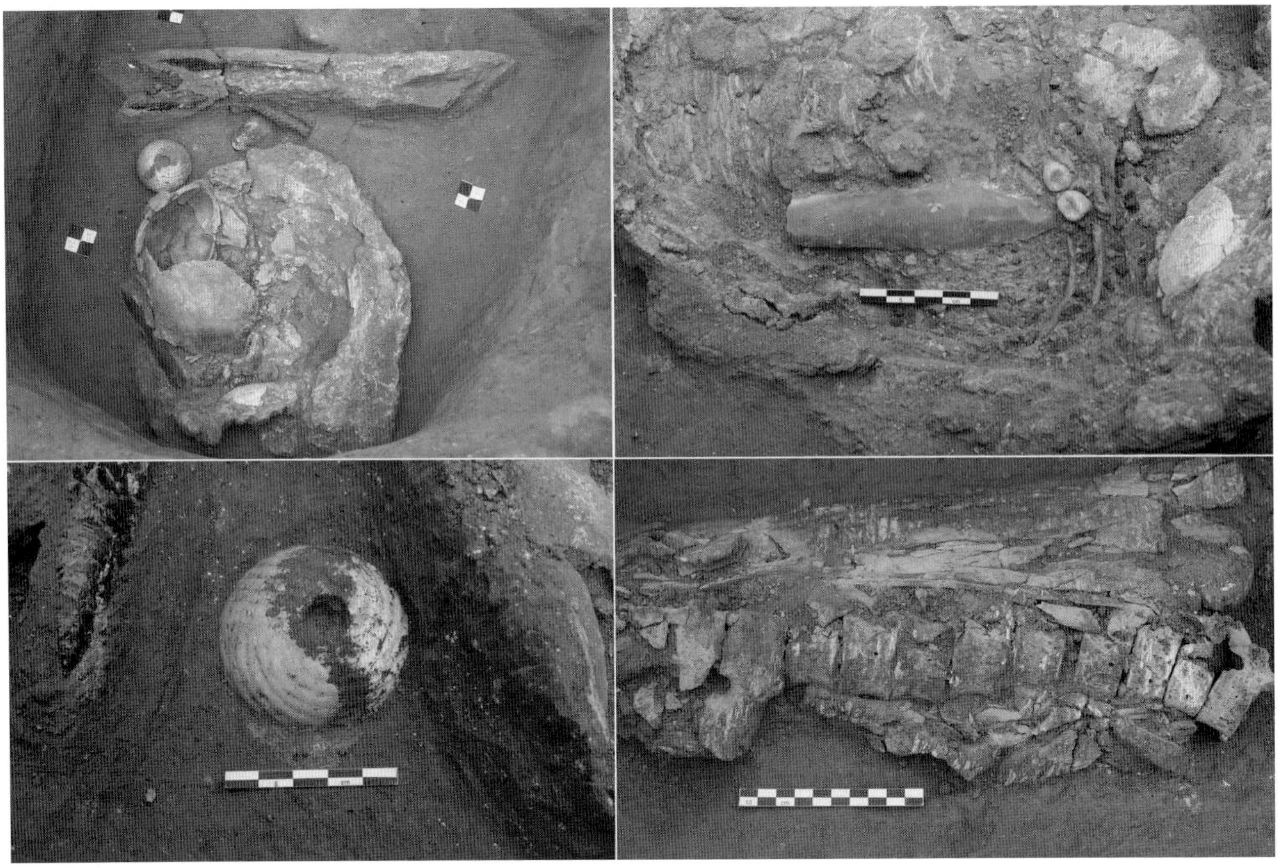

Figure 17.11. Secondary burial F.3808 (22196) in B.5.

whatsoever. In the last years of excavations, a few quite extraordinary extramural burials were found. Perhaps the most outstanding examples are burials that were found in Sp.602. This was an open space that was formed within the footprint of B.132 after its abandonment and was in use until building B.77 was erected on top of it (see Volume 12, Chapter 22). One infant, one adolescent and three adults were buried here, and they were buried with a variety of unique associations of objects and treatments of the body. Their bodies were tightly flexed and some individuals seem to have been partially skeletonised at the time of interment, with only the ligaments holding joints together (Haddow et al. 2015).

An old adult, possible female (21685) was wrapped in an organic material, and remnants of a basket were found near her (it is not clear if they represented a container for her body or for objects, although it should be noted that no objects were found within the basket). She had a string of white, black and green stone beads and a single *Antalis* bead around her neck. An *Unio* shell with pigment was lying beneath her torso, whilst another *Unio* shell with no visible pigment traces and burnt and friable animal bone fragments were also found in direct association with her body.

The skeleton of another old adult female (21672) buried here was wrapped in a material, more likely representing a shroud than actual clothing. Resting between her thorax and left femur was a pouch containing a number of objects: a plaster lump, an *Unio* shell, eight chert scrapers, a possible chert perforator, three obsidian flakes, a projectile point, two pigment applicators, greenish residue, an obsidian perforator/burin, fish vertebrae, an imitation red deer canine bead, a piece of carnelian (used on site only for bead making) and microdebitage (obsidian flakes). Close to the cranium of this individual was a basket with over a dozen fragments of bone points, two chert scrapers and an obsidian perforator/burin. The burial fill of this feature also yielded a number of objects: three mini clay balls, nine chert scrapers, a projectile point, three obsidian flakes, a chert blade, a chert perforator and two ring fragments.

A young adult male (21636) represents another remarkable burial in this space. Given the condition of the skeleton and its tight flexion, as well as phytoliths that suggest wrapping and binding, it has been suggested that this individual was kept bound for a period of time before he was buried in this space (Haddow et al. 2015). The remains of a potential pouch were found just below the region of the neck and contained a number of objects: a harpoon, a fish hook, another unidentified bone tool, a potential dagger, and a bone bead imitating red deer canine beads; in addition, an *Unio* shell and four stone beads as well as a fragment of a quern were recovered from the fill.

Given that individuals interred in external areas were not typically buried with accompanying objects, these burials are quite extraordinary, especially due to the large quantity of artefacts with which they were buried. Although repetition has been noted (imitation red deer canine beads, pouches and green pigment), tools buried with the adult female (21672) seem to be linked to pigment application and bead manufacture, whilst the pouch of the adult male (21636) contained items related to fishing. The contents of the pouches and basket are quite idiosyncratic and could be biographical (see Nakamura, Meskell 2013a for other examples of biographical burials).

Building variability and spatial distribution of burial associations

Although they follow a similar pattern with their repetitive layouts and division into clean and dirty areas, Çatalhöyük buildings were by no means uniform. Walls of some buildings were decorated with a variety of paintings and installations, others had hoards, caches and structured deposits. Burial associations exhibit a similar diversity to that observed in material culture across different media. As discussed above, associations do not exist in every burial, and the vast majority of individuals were interred without any discernible accompanying objects. That is the case with all individuals that were interred in three Late period buildings in the North Area: B.58, B.88 and B.54. However, the vast majority of inhumations in these buildings were of subadults, which is likely to be the reason why they had no direct associations (see above). In addition, the majority of individuals in B.54 and B.58 were buried in the side room, a location that does not usually have many direct associations in burials.

Amongst the buildings with ten or more interments, pigment associations do not exist in burials in buildings B.44, B.65, B.80, B.122 and B.128, and no remnants of body containment were found in B.128 and B.150. Bone tools (excluding pigment applicators) were not found (not even in the fills) in burials in buildings B.44, B.49, B.77 and B.129 in the North Area, and B.6, B.44, B.89 and B.97 in the South Area. Chert tools are completely absent from burials in B.52, B.60, B.77, B.80, B.89, B.122, B.128, B.129 and B.132; obsidian tools do not exist at all in burials in B.6, B.52, B.54, B.60, B.80, B.89 and B.128; individuals were not buried with stone tools in B.6, B.17, B.54, B.60, B.80, B.97, B.102, B.122, B.128, B.129 and B.132; bodies of individuals in B.97, B.65, B.80, B.114 and B.150 were not adorned.

The highest portion of individuals with direct associations has been observed in the subsequent buildings, B.17 and B.6, in the South Area, and in B.132 and B.77 in the North Area, as well as B.49, which is contemporary with B.77. At the same time, individuals in these buildings also have the highest quantity of burial goods (calculated as number of burial goods divided by total number of buried individuals). Pigments and associated finds are the most common with individuals in B.6, B.17 and B.132 (40% or more), whilst the evidence of matting, wrapping, binding and baskets most frequently occurs with individuals buried in B.6, B.49, B.17, B.52 and B.77, in which a third or more individuals were found with remains that had been contained within some sort of material. Bodies of 20% of individuals (or more) were adorned in B.6, B.17, B.132, B.49, B.52 and B.77, whilst objects were directly placed with 15% or more of individuals buried in buildings B.17, B.52, B.102 and B.77.

Furthermore, building size and elaboration could not be correlated with the presence of different aspects of burial treatment, and the percentage of buried individuals with pigment associations, adornment, or objects did not depend on the size of a building nor its elaboration (figs 17.12 and 17.13).

Similarly, no exceptional treatment has been observed in the burial assemblage in buildings defined as 'history houses' (sequences of buildings built on the same spot and with similar layout, see Hodder and Pels 2010 for further discussion). These buildings are characterised by an outstanding continuity, not just in their internal arrangements, but also in various repetitive activities that were noted across different media (for example, wall paintings with vultures and humans in E.VIII.8 and E.VII.8 or placement of pottery in the southeast corner in B.65 and B.56). Strikingly, the burial assemblage does not seem to follow this pattern. Not only were individuals within one building buried with different objects, if they had any, but similar items are almost never found in the successive buildings. There are, however, quite a few exceptions, such as placement of imitation red deer canine beads in two burials in B.6 and B.17, occurrence of pigments and associated items in B.22, B.21 and B.8, belt hooks and eyes in E.VIB.20 and E.VIA.20 (belt eyes were also found in the two successive buildings B.17 and E.VI.10, but it should be noted that there is a gap between these two buildings; B.17 was allocated to Level South K and E.VI.10 belongs to either South N or South O), rings in E.VII.14 and E.VI.14 (Mellaart 1964: 95) and wooden bowls in B.1 and B.5. The most notable example of continuity is found in buildings B.131 and B.129. Both of these buildings contained a number of inhumations, but two subadults ((23126) and (19460)) were buried almost

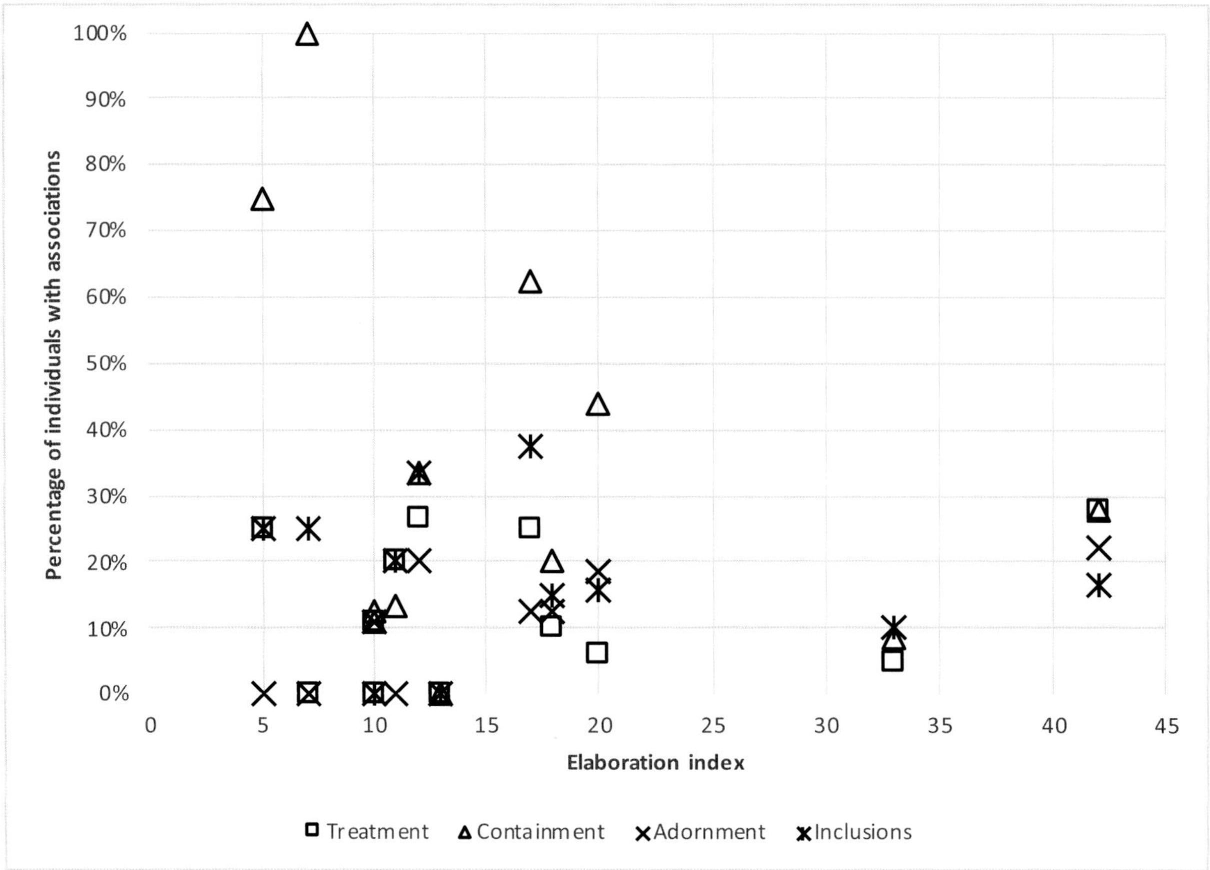

Figure 17.12. Burial associations and building elaboration index.

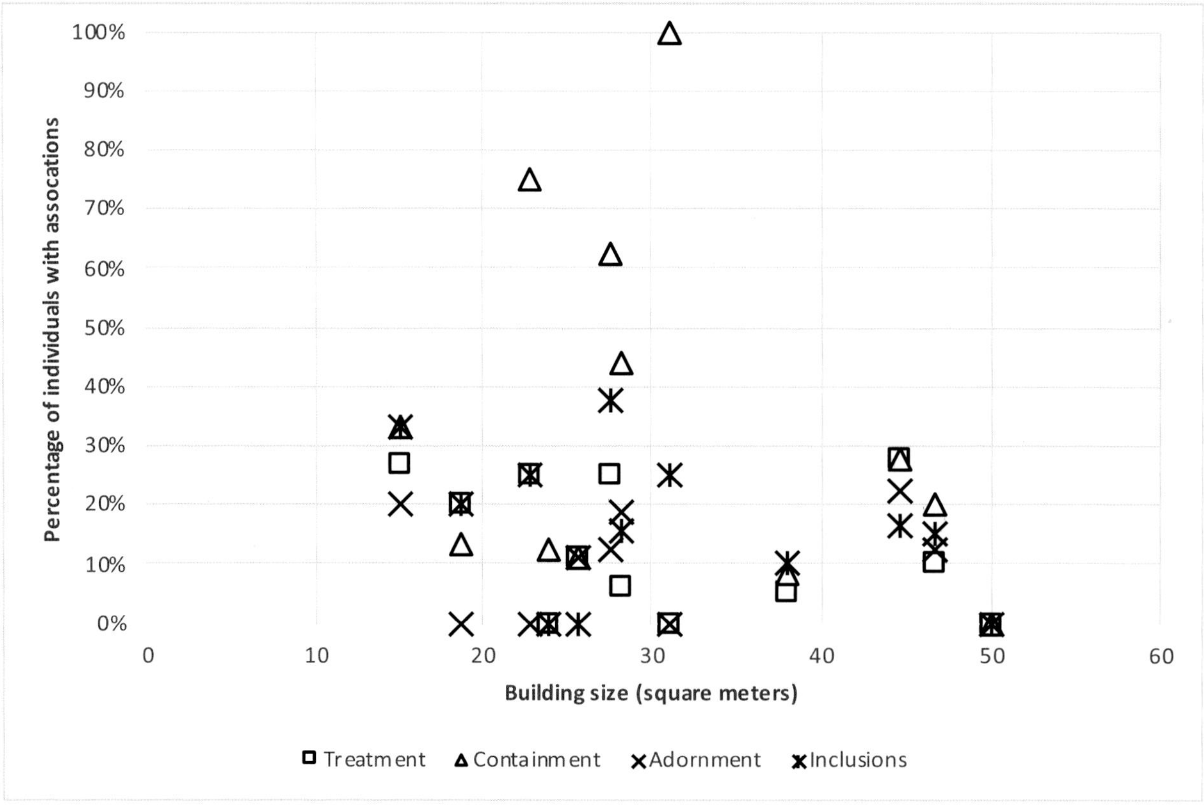

Figure 17.13. Burial associations and building size.

exactly beneath the northeast platform of the succeeding house. Both subadults were accompanied by remains of several other individuals, and both of them were buried with obsidian mirrors, pigments and hundreds of beads.

Despite the continuity observed in other aspects in 'history houses', continuity of burial associations does not seem to have been an imperative. Quite the opposite, people interred in these buildings were buried with similar items as exist in other burials, and their bodies were treated and adorned in similar ways as were others (Vasić 2018). The burial assemblage does not reflect any differentiation that could have existed between the members of a group using and/or being buried in these houses and groups of people buried elsewhere. If certain items of adornment were markers of their group or a sodality, it is not apparent in burials. However, continuity that does exist in a few houses cannot be a coincidence, and is suggestive of the reiteration of symbolism that encompassed life and death at Çatalhöyük.

Burial inclusions vary greatly across the site, and even individuals buried in the same building did not receive the same treatment. Although some houses display evidence of repetition of placement of similar objects in two or more burial features, the combination of objects and treatment is almost never the same. Similarly, neighbouring buildings seldom have similar objects in burials.

In order to explore the burial associations in neighbouring buildings, three blocks of buildings were examined: Early period buildings B.17 and B.6 (Mellaart's 'shrine' sequence E.VI-E.IX.10) and buildings B.161-B.160-B.43 (Mellaart's houses E.IX-E.VI.27) in the South Area, and Middle period buildings in the South (B.80, B.76, B.96, B.97) and North (B.1, B.131, B.77) Areas.

Building 17 had 25 burials, out of which nine individuals had no direct associations. Eight individuals exhibited evidence of treatment with pigments, eight had remnants of baskets, wrapping or matting, and a variety of inclusions were found with ten individuals. Associated items included a pouch with preforms for imitation red deer canine beads (1x), a clay ball fragment (2x), a belt eye (1x), various (finished) beads on the pelvis (2x) and neck (1x), an obsidian blade (1x), an obsidian flake (2x), an *Unio* shell (2x), animal bone fragments (2x), a lump of pigment (2x), a wooden plank (1x), bone points (1x), and an obsidian perforator/scraper (1x).

Two successive buildings, B.160 and B.161, located east of B.17, had four and eight inhumations, respectively. Three individuals buried in B.160 had direct associations. The body of an old adult male (32437) was bound and wrapped, pigment residue was found on his bones, and a wooden plank was placed above his body. The body of another adult, a middle adult male (32618), was also wrapped, but he had no other associations. Finally, the figurine of a quadruped was found next to the body of a neonate (32623) that was placed in a basket.

Building 161 had inhumations of four adult males and four subadults. An old adult male (32645) was wrapped, treated with red pigment, and placed on a wooden plank. The body of the second old adult male (32646) was also wrapped (or alternatively placed on a mat). Another old adult, possible male, (32674) had an obsidian flake placed posterior to the mandible. Perhaps the most intriguing of the four adult burials belongs to a young adult male (32608) that was found in association with a large quantity of carnivore scat (see Chapter 10), and remains of wrapping and matting. Amongst the subadults, a child (32648) was found in association with animal bone.

After these buildings were abandoned, B.6 and B.43 were built on top of them, largely respecting the layout of the former buildings. Building 6 had a smaller number of inhumations (note, however, that this building was also partially excavated by Mellaart and so the total number of inhumations is unknown) than its predecessor B.17, but eight out of ten inhumations had direct associations: baskets (4x), matting (3x), pigment (5x), scat and pellets (2x), wooden plank (1x), wooden residue on chest (1x), bracelets (1x), anklets (1x) and a necklace (1x).

Building 43 had at least nine inhumations (this building was also partially excavated by Mellaart as house E.VIII.27), out of which five individuals had direct associations. Phytoliths in association with a prenate (10527) are suggestive of a basket, pigment traces were found on bones of a prenate (22309) and an infant (22335), a child (10529) was buried with a lump of pigment, one and possibly two necklaces and a bracelet, and an adolescent (21802) was buried with an animal tooth placed on the pelvis, and two large shell pendants resting on the neck area. These pendants are thus far the only known examples of *Patella* and *Spondylus* being used on site for bead manufacture.

According to Mellaart's reports, bodies of two individuals (an adult and a subadult) buried in building E.VIII.31, adjacent to building B.43, were treated with red pigment and adorned with strings of beads, whilst Mellaart noted that the adult individual was buried with a large quantity of microfauna bones (Mellaart 1966). Whilst we cannot be certain in regard to the nature of the deposition in this case, recent excavations provided evidence of larger quantities of digested microfauna (that is, scat) indeed being used in funerary practices. The observed digestion, gnawing and puncture marks, as well as the high density of scats, suggests that they were deliberately deposited with the interred individuals (Jenkins

2012a; Jenkins, Yeomans 2013; see also this volume, Chapter 10 for further discussion).Whilst the variability in these buildings is evident, it is remarkable that all six buildings had treatment with red pigments, and beads, wooden planks and animal scat were found in burials in some of these houses, albeit not with every individual (fig. 17.14). As discussed above, treatment with red pigments was relatively common and was observed in a number of burials across the settlement. Imitation red deer canine beads, although present in different parts of the site, are not that frequent in burials, especially not in direct association with interred individuals. The most outstanding feature of burials in these buildings is the repeated placement of wooden planks, carnivore scat and carnivore pellets, something that almost never happened elsewhere (apart from these buildings, wooden planks were reported only in one or possibly two burials in the North Area). These inclusions are remarkable by themselves, but the fact that they occur in four burials located in three neighbouring (albeit not entirely contemporary) buildings makes them even more exceptional.

The variety of funerary practices can also be illustrated by the block of contemporary buildings that belong to a later level – South O: B.80, B.76, B.96 and B.97. In total, 44 individuals were buried in these four buildings. Starting from the easternmost building, out of 16 inhumations, only two burials in B.80 had objects directly associated with them. The individuals in question were interred together ((19151) and (19152)), and both of them had a bone point in association with their pelves. In addition, skeleton (19151) had a pottery sherd beneath the cervical vertebrae, whilst individual

Building	Red Pigment	Imitation Red Deer Canine Beads	Other Beads	Wooden Plank	Animal Scat and Pellets
B.6 (E.VIII.10)					
B.17 (E.IX.10)					
B.43 (E.VIII.27)					
B.160					
B.161					
E.VIII.31					

Figure 17.14. Noted repetition of certain types of associations in neighbouring buildings.

(19152) had an obsidian core, also in the pelvic area, and an *Unio* shell placed next to her left ankle. The only other direct associations are remains of matting, wrapping and baskets found with three individuals in the buildings.

The neighbouring B.76 contained only five individuals buried within it, out of which three were directly associated with items. A young adult male (18496) had a chert blade lying next to his feet, whilst a figurine depicting a seated headless female was found in the fill, by the left shoulder of an adolescent (18464), although it is not certain if this figurine represents a deliberate placement or if its deposition is tertiary. A middle adult male (18701) had a group of galena beads by his left shoulder and a group of imitation red deer canine beads in his waist area. This burial was disturbed by another inhumation, and therefore it is possible that these beads had been displaced from their original location. The galena beads might have originally been a necklace, and the bone beads were either adorning his waist, or alternatively, might have been placed in a pouch.

Building 96, located west of B.76, contained nine burials, with five having direct associations. Evidence of binding, wrapping and/or matting has been found in burials of an adult, possible female (20830), an adult female (20824), an old adult, possible male (20810) and a child (19727). In addition, an *Unio* shell was placed at the feet of (19727), (20810) had an obsidian flake on his forehead, and another old adult, possible female (20832) had a plastered object behind her lower back and by her left knee, whilst an isolated cranium of another individual was placed next to her head.

Out of 14 individuals buried in the neighbouring B.97, evidence of wrapping, binding, matting and baskets was found with four individuals, whereas one additional individual had direct inclusions. Animal bone was placed under the cranium of a neonate (20391), possibly representing a head support. This building lies immediately south of Mellaart's shrines 8 and 10. According to Mellaart, building E.VI.10 was built in level VIB (South N), but it remained in use in the following level (South O) (Mellaart 1964); therefore, it was at least partially contemporaneous with the buildings discussed above. Unfortunately, the information we have about burials in this building is limited to a single find: a belt eye that was found with an adult (Mellaart 1964: fig. 43.20).

Moving to the North Area of the settlement, three large buildings located in the northwest region had a large number of burials in them; together, their inhumations comprise over a fifth of the entire skeletal assemblage. The northernmost building, B.1, contains the highest number of burials recovered thus far. Out of 60 individuals, 16 were found with direct associations: beads (4x), rings (2x), spoon (1x), wooden bowl (1x), chert flake (1x), bone tool (1x), stone tool (1x), shell (2x), chalk lump (1x), plaster lump (2x), pigment treatment (1), yellow organic residue (8x), binding (1x), matting (1x), wood residue (1x) and phytoliths indicating matting, wrapping, binding or baskets (7x).

Contemporary B.131, lying to the south of B.1, had 41 interments, out of which ten individuals had direct associations: beads (5x), mirror (1x), chert blade (1x), pottery (1x), shell (1x), wooden bowls (4x), organic residue (1x), pigment treatment (2x), plaster treatment (1x), binding (2x), wooden container (1x), wrapping (8x), potential matting (1x) and phytoliths (1x). Peculiarly, all three occurrences of pigments were noted in burials of adolescents, whilst the wooden bowls were placed with two females and two possible females. Apart from the abovementioned adolescent (23126), buried with hundreds of beads and a mirror, perhaps the most interesting burials in this building belong to two adults that were buried under the northern platform. These two individuals were buried with similar items in a mirrored position (fig. 17.15). The young adult female (22661) was buried in flexed position on her left side, facing north with the cephalic extremity to the west and the feet to the east. A large bead was found at her neck, whilst a wooden bowl was placed between chin, right knee and right forearm. The other burial, located immediately to the east, belonged to a middle adult possible female (32324). This individual was buried on her right side, and she was also facing north. She was adorned with a necklace and two bracelets (both on her right upper limb, forearm and wrist), and a wooden bowl was placed by her left knee.

Immediately to the south of this building was B.77 with 39 inhumations, of which 22, and potentially 23, had direct associations: beads (5x, potentially 6x), collar (1x), obsidian blade/flake (1x), wooden bowl (1x), pouches (2x), shells (2x), minerals (1x), organic residue (1x), pigment container (1x), pigment application (1x), binding (3x), baskets (4x), matting (2x), wrapping (6x), indeterminate items of body containment (basket/matting or wrapping/binding) (3x).

Beads from these three buildings form over 40% of the total bead assemblage from burial features. That is due to two burials in which over 5,000 beads were found. Strings of very small black disc beads were wrapped around the body of an infant (2105) buried in B.1 and large concentrations of beads were also found associated with different parts of the body of an adolescent (23126) buried in the neighbouring B.131, likely indicating that the strings were wrapped around the body (see above). The importance of these large bead assemblages coming

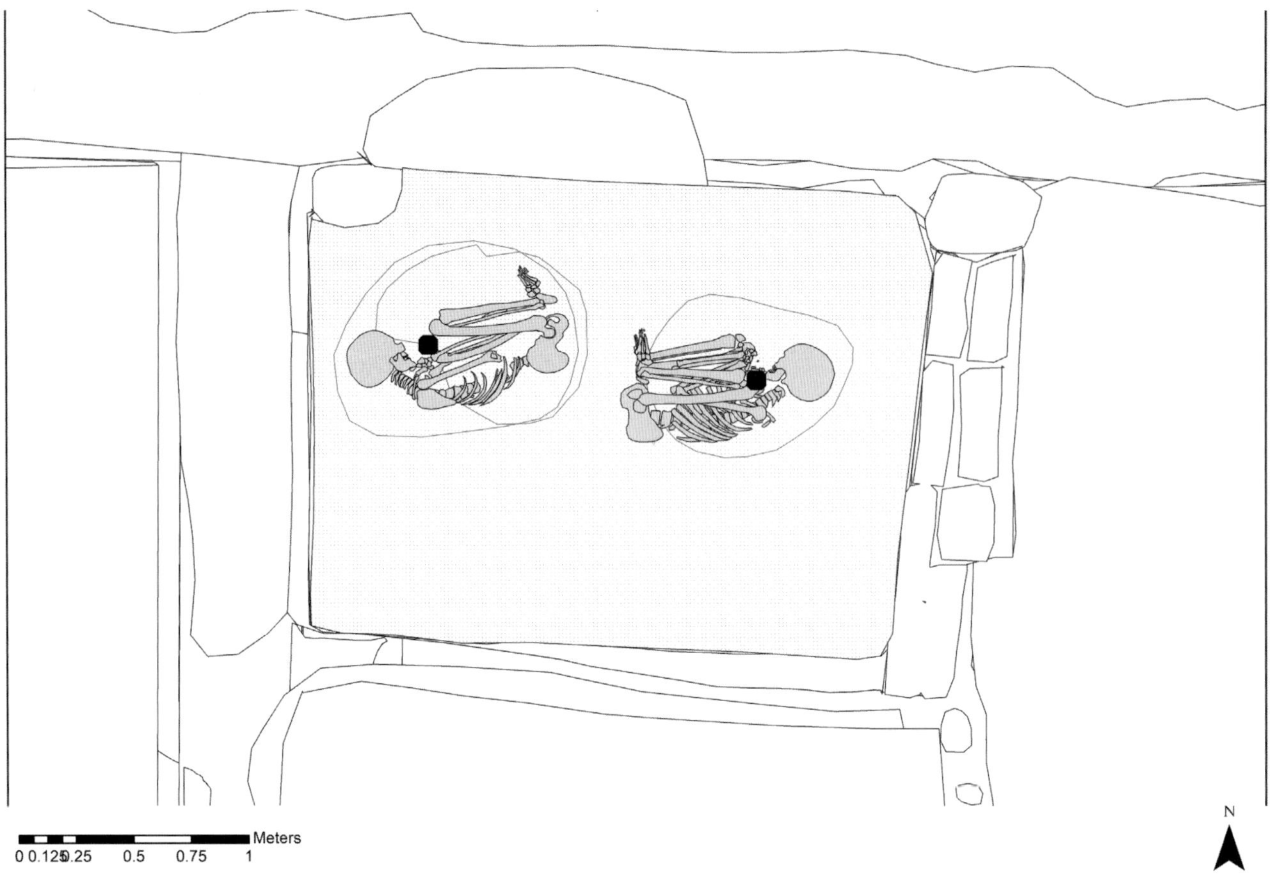

Figure 17.15. Beads and wooden bowls with two individuals in B.131 ((22661) on the left and (32324) on the right).

from two burials in neighbouring buildings is further increased by the fact that no other buildings thus far have produced similar bead associations. Furthermore, a repetition of placement of wooden bowls in burials in these three buildings is also of importance, although this might be partially due to preservation factors related to the burning of these buildings. Wooden bowls were also found in one burial in B.52 and possibly one burial in B.5, located immediately under B.1. Recent excavations have not produced any examples in the burials in the South Area, although Mellaart did mention their occurrence, especially in level VI, which is thought to be contemporary with these buildings (Mellaart 1963; 1964).

Another potential pattern is seen in the placement of maceheads in burials. These artefacts are not common on site, and they are largely absent from burials. Although Mellaart notes that several individuals were found with one or two maceheads, he only provides one context: the burial of an adult with a brachycephalic cranium in E.VIII.31 (Mellaart 1966). Only four examples have been recovered from burial features in 25 years of excavation; complete maceheads were found in burials in B.5 and B.114, whilst two fragments were recovered from the infill of two burial features in B.131. Although B.5 is earlier than the other two buildings, the proximity of these buildings makes their placement even more remarkable.

Given the variability of the burial assemblages within each house, as well as its diversity in neighbouring buildings, it is no surprise that the material assemblages do not display any significant differentiation between settlement Areas. The Middle period burials demonstrate that people were buried in very similar ways in the North and South Areas. Some tendencies do exist, as observed in a higher frequency of adornment and treatment in the North Area (fig. 17.16), although the latter is highly dependent on the extent of preservation. More specifically, items of adornment are more common with females in the North Area burials than they are with females buried in the South Area, whilst no pigment or plaster residue, nor associated finds such as applicators, lumps or containers were found directly associated with the female burials in the South Area (the noted difference is likely to be a result of an incomplete dataset and Mellaart's records; given that items of adornment exist in the 1960s assemblage, and Mellaart does mention treatment with pigments in these levels in his reports, it is likely that at least some of it comes from female burials). Another difference can be observed in the occurrence of belt hooks and eyes. According to his reports, Mellaart found at least 13 belt hooks and eyes in burials of this period. It is not clear how many burials he excavated from these levels, but out of 276 inhumations excavated in the North Area over the 25 years of Çatalhöyük Research Project, only two burials contained belt hooks and eyes. This could potentially mean that these objects were more commonly used by the people living and/or being buried in the South Area.

Burials from the Late period also do not exhibit great differences, although direct associations are slightly more common with individuals buried in the southern part of the mound (including South, GDN and TPC Areas) (fig. 17.17). The lack of strong distinctions between different areas of the settlement has also been noted in the skeletal assemblage, which does not show any significant distinction between the burials in the North and South Areas (Hillson et al. 2013; Larsen et al. 2013; Pearson 2013).

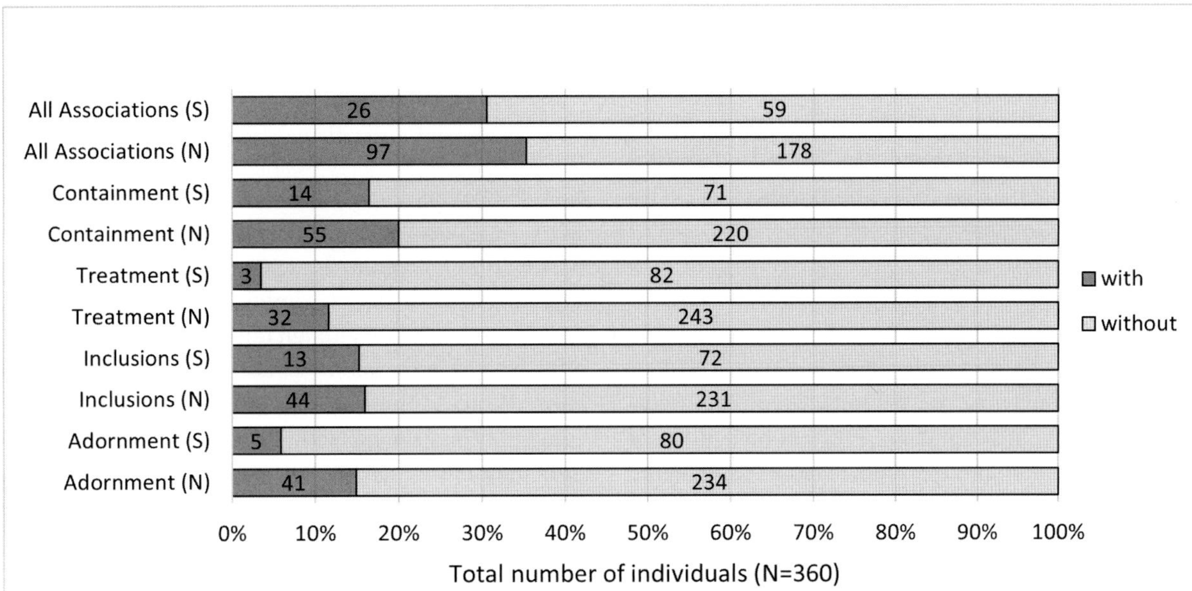

Figure 17.16. Comparison of associations in the Middle Period burials in the North and South Areas.

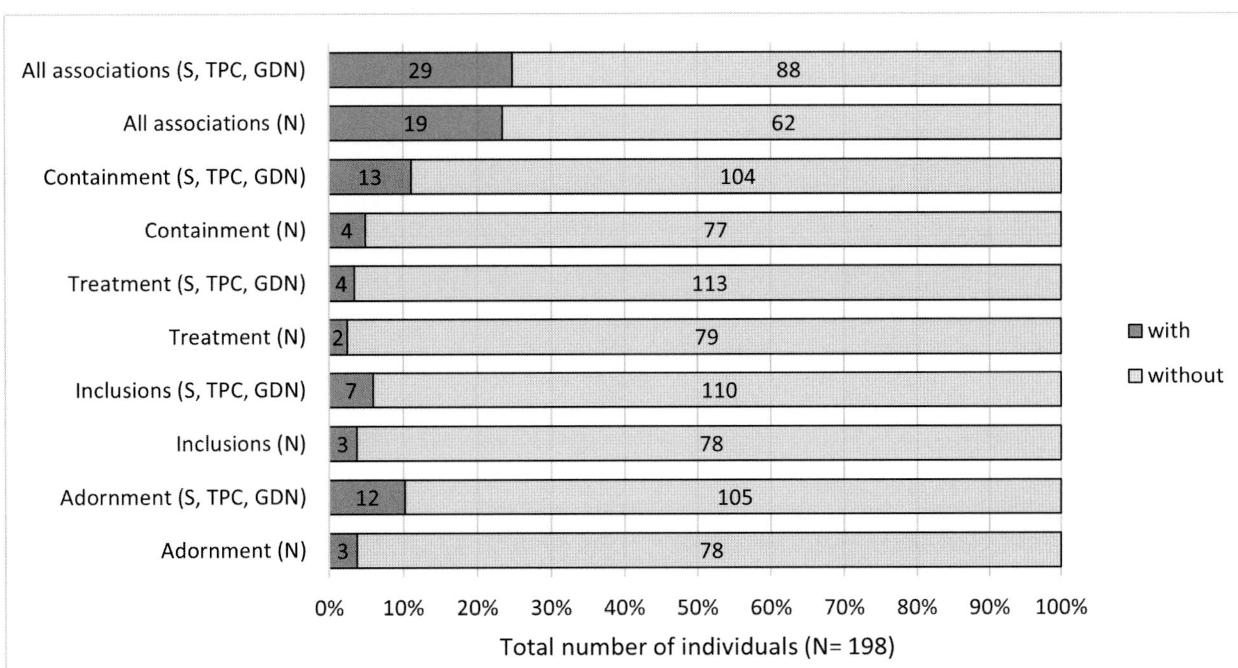

Figure 17.17. Comparison of associations in the Late Period burials in the North, South, TPC and GDN Areas.

As presented above, each building's burial assemblage is highly variable, and consequently, neighbouring buildings also have diverse associations in burials. The seemingly unpatterned distribution of burial goods goes well with other lines of evidence that suggest that social organisation at Çatalhöyük functioned through a multifarious mosaic of social networks that criss-crossed the settlement, rather than being based solely on the physical proximity of buildings (Hodder 2014e). Nevertheless, there is still evidence that would suggest a certain level of association between the neighbouring buildings. The above-discussed repeated occurrence of animal scat, wooden planks, pigments and imitation red deer canine beads in the block of buildings in the South Area, as well as maceheads in the North Area, cannot be interpreted as a mere coincidence and does suggest that a certain degree of the notion of 'neighbourhoodness' did indeed exist at Çatalhöyük, but the social connections were not exclusively determined by it.

Burial associations over time

Continuity and gradual change have been determined to be the main temporal trends at Çatalhöyük (Hodder 2014d). The arrangement of the settlement, the architecture and the layout of buildings, together with different aspects of material culture, demonstrate remarkable stability through time. Similarly, burial associations also demonstrate continuity through time, which indicates that funerary practices were to some extent conservative. The dead were treated in similar ways and accompanied by the same types of objects from the earliest levels throughout the occupation of the settlement.

Each category of burial associations exhibits a decrease in frequency over time (fig. 17.18). Even primary undisturbed burials have fewer associations (fig. 17.19); though there is a slight increase in the proportion of individuals with inclusions and items of adornment in the Middle period, over time individuals seem to have been buried with fewer objects. Despite the evident overall decrease in burial associations through time, there is variability amongst the age categories and types of associations (fig. 17.20). Items related to body containment seem to occur more or less equally among both subadults and adults, except in the Late period, when they are more common with adults. A similar proportion of infants was found with items related to body containment in the Early and Middle periods, whilst proportions of children, adolescents and young adults with this association do not change from the Middle to Late period. Old adults were buried with items of adornment much more frequently in the Middle period. Items of adornment (that is, beads) show the smallest decrease through time, whilst the difference in other types of associations is much higher. Pigment associations decrease through time, but young adults are more commonly associated with pigments in the Late period than they are in the Middle period, whilst a significant decrease in the proportion of old adults with pigments has been noted in the Late period. Burial inclusions are much more common with children, adolescents and old adults in the Early period than they are in the later levels.

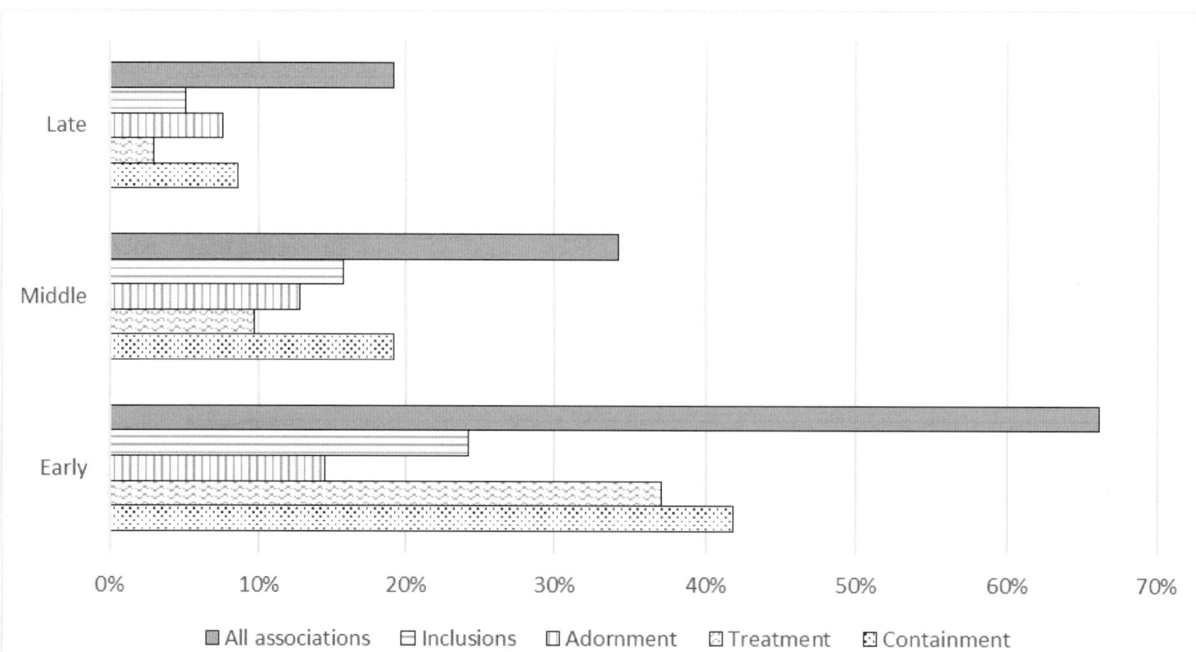

Figure 17.18. Burial associations with all individuals through time.

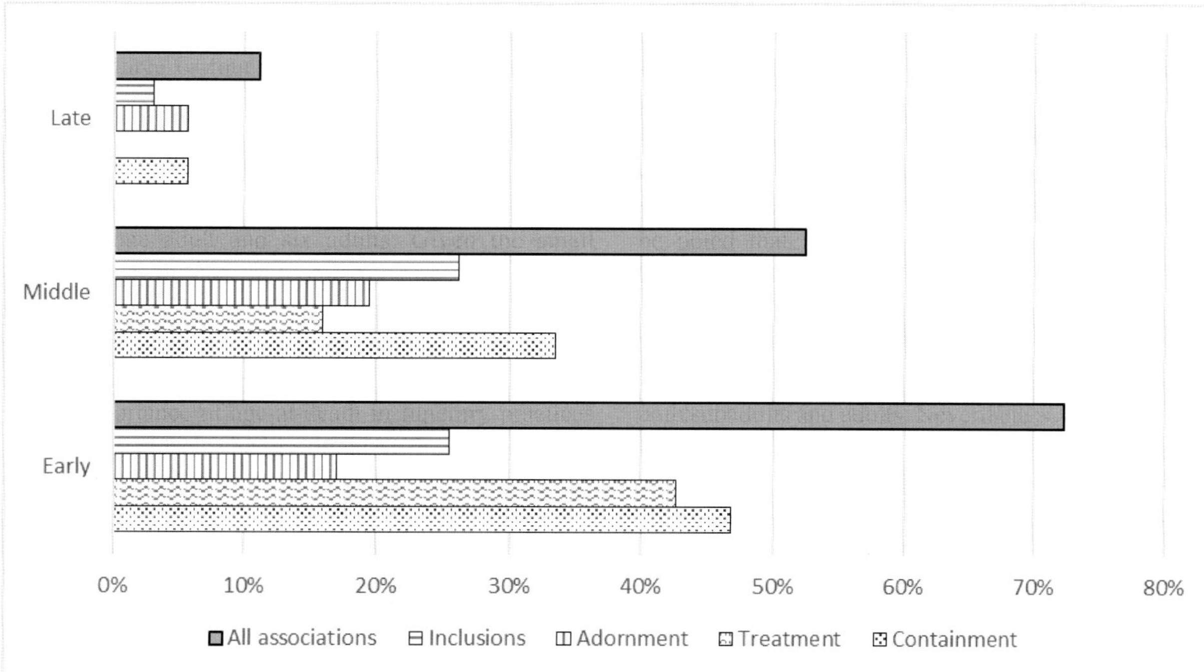

Figure 17.19. Burial associations with primary undisturbed individuals through time.

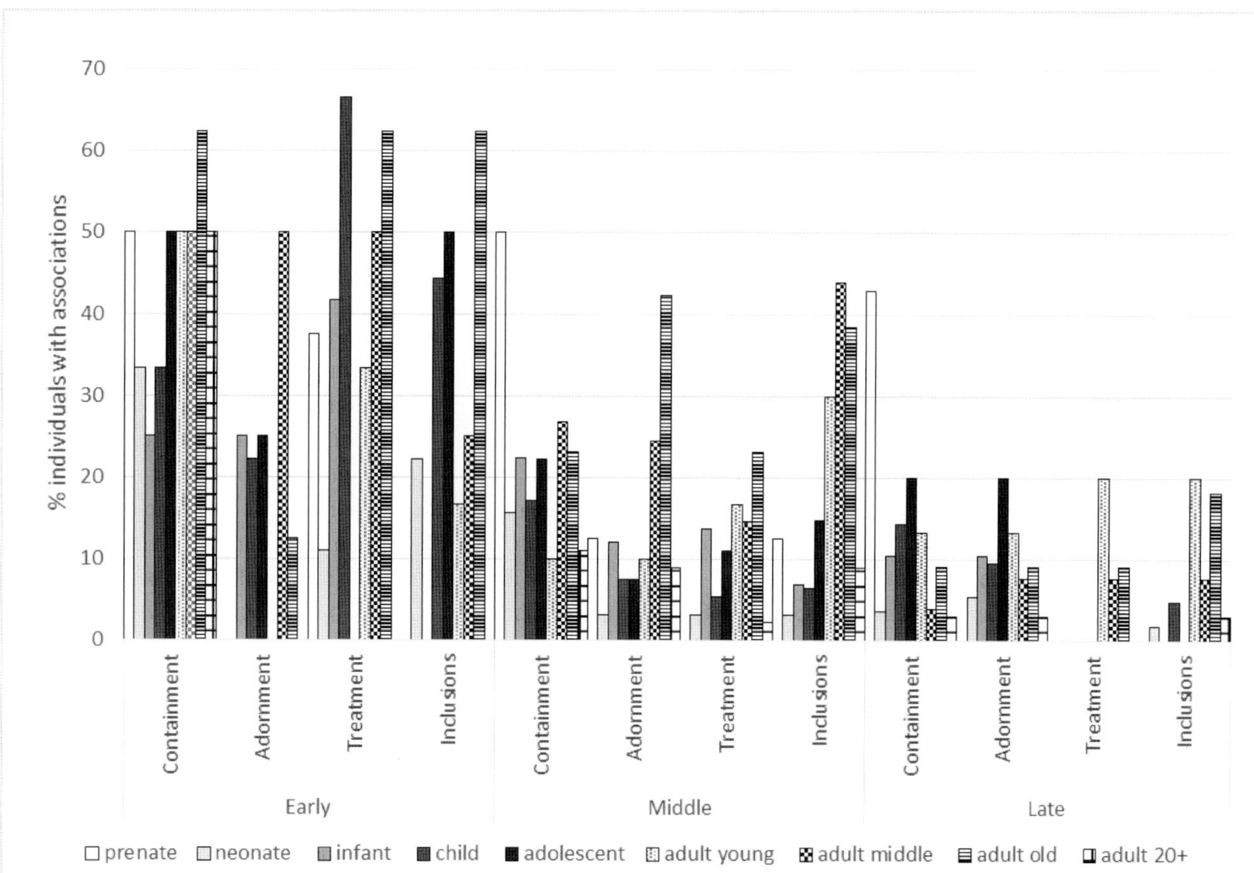

Figure 17.20. Burial associations with different age-at-death categories though time.

The average number of burial goods per person also decreases in the Late period, and amongst the individuals with direct associations, the ones that were buried during the Middle period were buried with more things than individuals interred in other periods (table 17.9). Peculiarly, Late period burials of children and adolescents have the highest number of burial goods per person, which is not the case in earlier levels. This could be related to the changes in socially ascribed status that children and adolescents had in the Late period. On the other hand, subadults do not seem to have been treated with pigments in burials of this period. The only burial of a subadult from this period that contained pigments was a burial of a child in B.129 which contained red and blue pigment residue on mirrors, but none on the skeletal remains. Slight shifts can also be observed in the frequencies of burial associations with males and females. Females seem to be more commonly found with all types of burial associations in each period, with the exceptions of containment, which is more common with males in the Early period, treatment with pigments and associated finds in the Middle and Late periods, and inclusions in the Late period (fig. 17.21).

	Early Period	*Middle Period*	*Late Period*
Prenate	1.4	1.3	1
Neonate	1.4	2	1.3
Infant	2.6	3.3	1.2
Child	2.8	2	3.5
Adolescent	2.7	4.9	3
Young adult	2.7	6	2.2
Middle adult	3.7	3.4	2.3
Old adult	3.9	6.2	1.3
Adult (20+)	1	2	1.8
Sub-adults all	2.2	2.7	1.9
Adults all	3.4	4.2	2
Female	3	4.7	1.8
Male	3.7	4.2	2.6
All individuals	2.7	4.7	1.8

Table 17.9. Average number of burial goods per buried individual.

Artefacts made of chert and ground stone (excluding beads) exist in direct association only in the Middle period burials, although it should be noted that they do occur in burial fills in Early and Late periods (table 17.10). In contrast, artefacts made of clay and obsidian seem to be more common in the Early period burials. Placement of *Unio* shell in burials has been noted in burial features of the Early and Middle periods, but not in the Late period, whilst the use of these shells as pigment containers has been observed in burials in the Middle period and a single burial from the TPC Area. On the other hand, similar observations cannot be made in the bead assemblage. Certain tendencies do exist (for example, bone beads are more common with individuals in the Early period), but apart from the

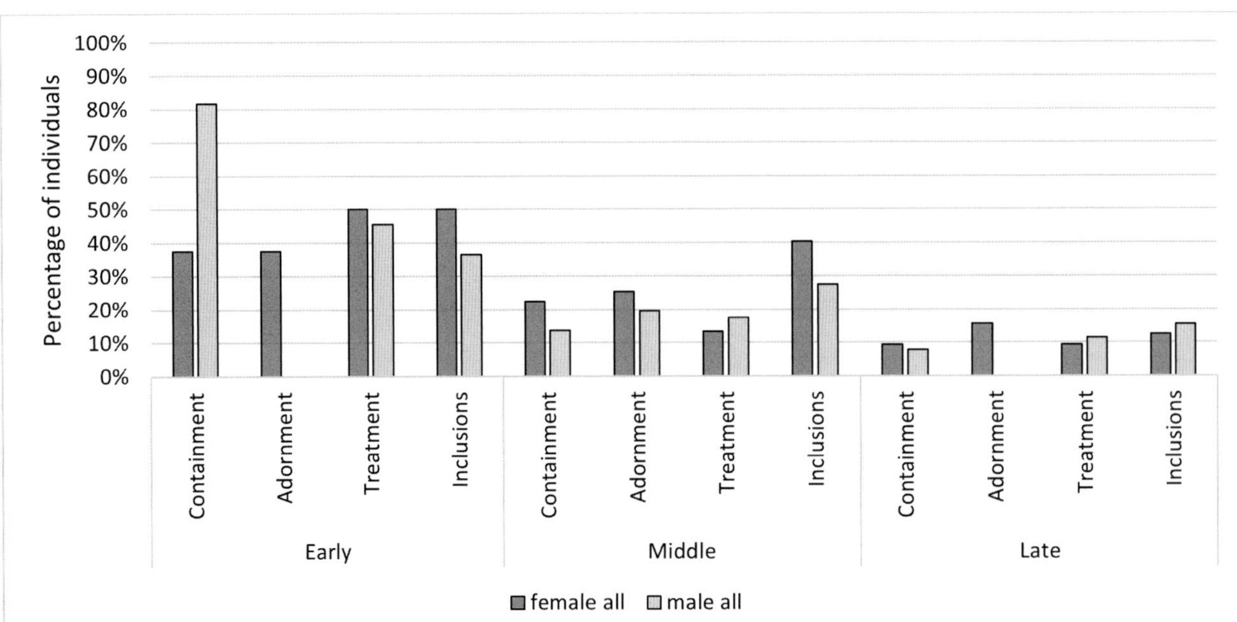

Figure 17.21. Proportion of males and females with different associations though time.

noted decrease of occurrence of beads in burials, portions of individuals that were buried with beads of a certain type of material do not exhibit visible differences through time (table 17.11). Similarly, most bead types occur equally through time, with some types being introduced in the Middle period (table 17.12). Some types are rare and occur in one or two burials at most, but amongst the types that occur with ten or more individuals (T.1, T.2, T.5, T.9), it is striking that imitation red deer canine beads (T.9) have not been found in direct associations with individuals buried in the Late period.

Whilst burials of the Middle period display the highest levels of variability in associations and also evince the introduction of new items, such as mirrors (which also reflect the beginning of the use of these objects in the settlement in general), to the funerary repertoire, the Late period exhibits no novelties. Quite the opposite, the Late period assemblage contains the same types of artefacts that were also placed with individuals in the earlier periods, and the array of types of objects is substantially reduced (from at least 59 different types of occurrences noted in the Middle period to 17 types in the Late period). The decreased frequency and diversity of occurrence of artefacts in burials most likely reflects changing circumstances in the settlement that have also been noted in other material assemblages (Hodder 2014d). Perhaps the only type of object that increases in variability in the Late period is bead strings. A notable intensification of bead diversity has been noted in the upper levels, with some unique types occurring, and the strings exhibit a higher diversity of combinations of types (Vasić 2018).

As discussed above, the quantity, diversity and frequency of burial associations decrease through time. Burial features with single inhumations represent the vast majority of burials in each period; however, the occurrence of features with multiple inhumations increases over time, and so do the primary disturbed and secondary burials. At the same time, fragmentation of buried bodies also intensifies, with the proportion of complete individuals dropping from almost 50% in the Early period to below 15% in the Late period assemblage. As previously noted, burial disturbance and completeness does affect burial associations, but burials of primary undisturbed and complete individuals also demonstrate a decline in burial associations through time.

Despite a remarkable degree of variation, the funerary practices at Çatalhöyük were quite conservative. The continuity observed in various aspects of funerary practices demonstrates the same responses to the dead and death persisting throughout the centuries of settlement occupation.

Material (%)	*Early Period*	*Middle Period*	*Late Period*
Obsidian	8.1	2.2	2
Bone	3.2	4.4	0.5
Stone		3.1	
Clay	9.0	0.3	
Chert		2.5	
Shell	3.2	5.8	0.5
Total number of individuals	62	361	198

Table 17.10. Percentage of individuals buried in direct association with artefacts of different material (excluding beads).

Discussion

> The individual human being is unpredictable, but the reactions of human mobs, Seldon found, could be treated statistically. The larger the mob, the greater the accuracy that could be achieved.
>
> (Asimov 1954)

The burial assemblage at Çatalhöyük suggests a particular focus on bodily manipulation, as attested in binding and wrapping of it, treatment with pigments and plaster, adornment, and circulation of body parts, and at the same time, it demonstrates a tendency towards placing certain types of objects with the dead.

Burials seem to conform to a certain level of standardised practices, though they also display a remarkable level of variability. In spite of the enormous observed variation, the choices made for individual burials were far from random. The complexity of funerary practices and the bewildering multifariousness of burial associations are indicative of different relationships of the living to the dead, rather than strict conventions that had to be followed. The variability witnessed in the burial assemblage is also reflective of individual decisions that were made during the burial processes. These decisions might have been based on the relationships of the person or people responsible for the deposition, but also on the individual who died, his or her role and status in society, and the circumstances of death.

Burial goods were once traditionally regarded as personal possessions and, consequently, a sign of social status and wealth. Some of the items found in burials could have indeed belonged to the interred individuals, but burial goods can also represent donations to the dead, and it is important to remember that it was the living who made the choice as to what was to be included and to accompany the dead (Thomas 1991; Parker Pearson 1999). The identity of the deceased must have played an important role, and rather than a mere indication of the accumulation of wealth, the fact that older individuals have a higher diversity, frequency and quantity of artefacts accompanying them could be indicative of their longevity and accrued status and position in the community, as well as various networks and relationships that these individuals formed during their lifetimes. Thus the notable variability in burial assemblage could be interpreted as a result of different social relationships and expression of nuanced identities that were mediated through the careful and probably personalised selection of body treatment and accompanying objects.

Percentage of individuals	*Period*	*Stone (%)*	*Bone (%)*	*Shell (%)*	*Clay (%)*	*Copper (%)*	*Wood (%)*	*Plaster (%)*
Individuals with beads	Early	55.6	33.3	44.4	11.1			
	Middle	79.5	27.3	50.0	9.1	2.3	6.8	2.3
	Late	85.7	21.4	57.1	7.1	7.1		
All individuals	Early	8.1	4.8	6.5	1.6	0.0	0.0	0.0
	Middle	9.7	3.3	6.1	1.1	0.3	0.8	0.3
	Late	6.1	1.5	4.0	0.5	0.5		0.0

Table 17.11. Percentage of individuals with direct bead associations.

% of individuals with beads	*Early (%)*	*Middle (%)*	*Late (%)*	*% of all individuals*	*Early (%)*	*Middle (%)*	*Late (%)*
T.1	44.4	63.6	64.3	T.1	6.5	7.8	4.6
T.2	22.2	22.7	21.4	T.2	3.2	2.8	1.5
T.3	11.1	13.6	7.1	T.3	1.6	1.7	0.5
T.4	11.1	9.1	14.3	T.4	1.6	1.1	1.0
T.5		15.9	28.6	T.5		1.9	2.0
T.6			7.1	T.6			0.5
T.7		4.6		T.7		0.6	
T.8		11.4	14.3	T.8		1.4	1.0
T.9	33.3	20.5		T.9		2.5	
T.10		2.3	7.1	T.10		0.3	0.5
T.11		6.8		T.11		0.8	
T.13	11.1			T.13	1.6		
T.14	11.1			T.14	1.6		
T.15		9.1	14.3	T.15		1.1	1.0
T.17	11.1			T.17	1.6		
T.19		4.6		T.19		0.6	
T.21		4.6	21.4	T.21		0.6	1.5
T.22	11.1	6.8	7.1	T.22	1.6	0.8	0.5
T.23		6.8	7.1	T.23		0.8	0.5
T.24		4.6		T.24		0.6	
T.25	33.3	11.4	7.1	T.25	4.8	1.4	0.5
T.26		4.6	7.1	T.26		0.6	0.5
T.27	11.1	9.1	14.3	T.27	1.6	1.1	1.0
T.28		6.8	14.3	T.28		0.8	1.0
T.29		9.1	14.3	T.29		1.1	1.0
T.30			14.3	T.30			1.0

Table 17.12. Portion of skeletal assemblage with directly associated bead types through time.

As Nakamura and Meskell (2013a) rightly point out, repetitive placement of tools and other objects in burials of the youngest members of the Çatalhöyük community through time does not endorse the idea of burial associations representing personal belongings. Funerary practices do not necessarily reflect the status of the dead, but can sometimes idealise or invert it (Hodder 1982: 200). Therefore, it is plausible that placement of some items might have been symbolic and, for example, objects in burials of children might have symbolised the adulthood that they never reached. Furthermore, some items display evidence of extensive use, witnessed in repair and use-wear, such as collars buried with an adult female in B.50 (10829). Whilst it could be argued that such items could have been personal belongings of the deceased, it is highly unlikely that the items that have no sign of use (for example, the above-mentioned tools buried with the adult male in B.5) belonged to the individuals they were buried with.

Donations may have been made on an individual and communal level. Placement of a smaller number of objects (typically one or two) is not suggestive of group donations. In contrast, bead strings that were buried with the dead could have been created by a group of people. Strings of beads displaying different use-wear not only demonstrate that these beads had been used separately prior to the burial, but could also point to a possibility of different people donating their beads, thus creating a newly formed ornament that was placed in the burial and leaving a permanent mark of their participation in the funeral as well as their relationship to the deceased (Vasić 2018).

The most common types of associations such as pigments, beads, mats and material wrapped around the body probably carried collective meanings through time. That is not to say that there were no differences between different types of beads, or bead strings, or different meanings that treatment with green or red pigments, or treatment of the cephalic extremity or pelvic region carried. Rather, the overall intentions lying behind these associations might have been similar. Through the repetitive placement of objects in burials, these objects became an integral aspect of burial, and through funerary practices, their values and the meanings they carried were reinforced and reformulated. On the other hand, the placement of other inclusions might have had a more nuanced connotation and could have been reflective of both the identity of the deceased and the personalised relationship between the dead and the living who put those objects in the burial.

As previously discussed, age-at-death seems to have been a more dominant structuring principle in burial associations than sex. The youngest members of the Çatalhöyük community were rarely buried with accompanying objects, and the older they were at the time of their death, the more likely they were to be buried with a higher quantity and a wider range of associations. Despite the increased frequency, diversity and quantity of burial associations noted with older individuals, similar objects were found associated with different age categories. Similarly, both males and females were buried with an array of objects. Although some of them were found only with males or females (for example, maceheads and belt hooks and eyes with males, or pigment applicators and spoons with females), due to their rare occurrence it is not possible to discern any differences that would be based on how men and women were perceived in the society. This goes well with the other lines of evidence, such as diet and stress markers, as well as the lack of emphasis on depiction of primary sex traits on anthropomorphic figurines (Hillson et al. 2013; Nakamura, Meskell 2013b; Pearson 2013; Sadvari et al. 2015b), which indicates that sex distinctions were not really emphasised.

Given the size of buildings, the act of interment could not have been witnessed by a large number of people (unless they were coming in and out to participate in the act of burying). Together with the fact that burials are usually located not by the entrance but on the opposite side of the building, it suggests that funerals, at least the final segment (that is, positioning of the body in the pit and burial closure) were quite intimate. However, parts of the funerary rite might have been carried out in another location, and even outdoors. Whilst placement of inclusions must have been done on the very spot, during the interment itself, it is plausible that treatment with pigments, adornment, and even wrapping and binding of bodies could have been done on building roofs, or in external areas, thus enabling a wider audience or more participants. This goes well with the idea of delayed burial discussed in Chapter 15. In addition, some of the foundation burials likely occurred in the open, meaning that in some cases, they may have been witnessed by a larger group of people. Strikingly, individuals that were buried in external areas, and whose funeral could have been attended by larger groups of people, typically do not have any inclusions. On the other hand, exceptional burials of adults in the external Sp.602, which are amongst the richest burials at Çatalhöyük, demonstrate the same level of elaboration as for the individuals buried inside buildings.

Funerary practices do not display strong evidence of social differentiation. Although the fact that only a small portion of the buried population displays evidence of treatment could be used to argue for social hierarchy, one would expect to find correlation between the burial association and burial location, or age and sex of the buried individuals. Prominent individuals would have been buried with a higher quantity of objects, and/or rare items in large and elaborate houses, but this was not always the case. People buried in the same building would have similar items and material culture would likely display different treatment of that group of individuals from individuals buried elsewhere. A hierarchical segregation would also entail some sort of spatial clustering of buildings with more privileges and power. However, various aspects of the material culture point to a different form of social organisation, and there is no substantial evidence that would suggest that Çatalhöyük was not an egalitarian society. Although some neighbouring buildings or sequences of buildings built on the same spot do contain similar items in their burials, the same types of objects are found in different parts of the settlement. For example, boar tusk collars were found in two neighbouring buildings in the South Area (B.50 and E.VII.12), but they were also found in burials in B.77 in the North Area. Similarly, belt hooks and eyes were found in the adjacent buildings E.VI.1 and E.VI.5, but also in E.VI.29, 10 and 20 in the South Area, as well as in buildings in the North Area. This goes well with the other lines of evidence that suggest no major differentiation between the buried populations in the different areas (for example, no indication of differentiation in the diet, Pearson 2013). Although the burial assemblage does not exhibit strong evidence of a hierarchical organisation, it does demonstrate a great level of sociality and social complexity.

Whilst the burial assemblage is not reflective of a hierarchical social structure, social distinctions are evident in unique placements such as the plastered skull in B.42 and the lamb skeleton in B.50 (Nakamura, Meskell 2013a: 462) elaborate burials in the external Sp.602, the exceptional secondary burial in B.5, or rare occurrences of elaborate objects such as maceheads and daggers. Some of these conspicuous objects might have been indicative of the status or roles individuals using them held in the society. They could have formed an important part of the external display, as wearing objects such as belt hooks and eyes or boar tusk plates could not have gone unnoticed. These objects could have represented certain roles or affiliation with particular social groups, and their use must have played an important part in construction and negotiation of identities and social organisation, as social status must have been defined and reinforced through the use of these items.

It is clear that social distinctions based on individual level existed at Çatalhöyük. As different social groups must have existed, it is likely that their members were distinguished in some way. Given the spatial distribution noted in various aspects of material culture reflecting complex social networks, it is more difficult to pinpoint the actual groups and discern the distinctions that would have existed based on their membership. Further investigations need to be conducted, but it is clear that these groups were not based primarily on biological kinship (Pilloud, Larsen 2011; Pearson 2013). Whilst some signals do exist, as witnessed in occasional repetition of objects in burials in neighbouring buildings, the high variability of burial associations in individual buildings and the fact that similar objects occur in various parts of the settlement do not demonstrate restricted access to raw material or the exclusive use of specific objects based on the group membership, which would support the idea of inequality.

The continuity of funerary practices throughout the length of occupation of the settlement, as seen in the placement of the same types of objects, similar positioning of the body, body treatment and adornment reveal a relatively permanent set of values and are indicative of an enduring symbolic system. This went beyond the individual life cycle and was transmitted from generation to generation. Inclinations observed in the selection of objects and treatment may have shifted occasionally, as witnessed, for example, in the fact that bodies of males were adorned only in the Middle period, in the treatment of the dead with scat and pellets in the Early period, or in the tendency of burying the dead under floors in the Early period as opposed to platforms in the upper levels and chambers in the final phase of occupation (thus far excavated only in the TP Area), but the general focus on the body and an overall consistency of types of objects which were to accompany it remained the same. Despite the overall continuity, burials do exhibit gradual change, which seems to culminate in the Late period. The intensification of consecutive inhumations at the same location over time might be reflective of an increased need for the reinforcement of collectivism. At the same time, an increased occurrence of skeletal remains in tertiary contexts over time (see Chapter 15), and repeated placements of singular bones or disarticulated skeletal elements in burials with primary individuals, raises the question of whether these should be interpreted as burial inclusions or as burial goods rather than inhumations. The changes observed in funerary practices coincide with the new socio-economic circumstances in the settlement of the Late period that are reflected in the intensified sheep herding, mobility, reduced density of the settlement, house dispersal, cattle domestication, and intensification of the use of stamp seals and other clay objects (Hodder 2014d). Bodily manipulation as the centre of sociality in both life and death would have been amongst the fundamental apparatuses in the changing elements of social circumstances.

Acknowledgements
We are greatly indebted to Carolyn Nakamura who created an excellent database with meticulously recorded finds. This chapter would not have been possible without her help and advice. We are also grateful to Ian Hodder and all members of the Çatalhöyük Research Project whose hard work actively contributed to this amazing dataset.

18. Ancient genomics in Neolithic central Anatolia and Çatalhöyük

Reyhan Yaka, Ayça Doğu, Damla Kaptan, Nihan Dilşad Dağtaş, Maciej Chyleński, Kıvılcım Başak Vural, N. Ezgi Altınışık, Igor Mapelli, Dilek Koptekin, Cansu Karamurat, Hasan Can Gemici, Sevgi Yorulmaz, Vendela Kempe Lagerholm, Evrim Fer, Ulaş Işıldak, Ayshin Ghalichi, Gülşah Merve Kılınç, Camilla Mazzucato, Anna Juras, Arkadiusz Marciniak, Clark S. Larsen, Marin Pilloud, Scott D. Haddow, Christopher J. Knüsel, İnci Togan, Anders Götherström, Yılmaz Selim Erdal, Elif Sürer, Füsun Özer, Çiğdem Atakuman and Mehmet Somel

Over multiple millennia, from the earliest traces of long-term occupation of camp sites (ca 20,000 BC) to the development of full-scale farming (ca 8000–6000 BC), the Neolithic transition in southwest Asia gradually shaped human societies in dramatic ways (Nadel 2002; Maher et al. 2012; Asouti, Fuller 2013). Here, we present recent insights from ancient genomics studies into these societies while focusing on two questions: the population processes driving cultural change in Neolithic central Anatolia and genetic kinship among Çatalhöyük co-burials.

Ancient genomics has most frequently been employed in inferring demographic history, which, in turn, can shed light on questions surrounding cultural change. During the Neolithic Transition, many new material cultural traits, raw materials and technologies, as well as new patterns of social organisation and belief systems were created and shared among geographically distinct communities of the Levant, Zagros Mountains, Mesopotamia, Anatolia, Cyprus and the Aegean (Ibáñez et al. 2018; Reingruber 2011). While these exchanges can be partly traced from material culture, the type of human interactions that drove these transitions remains little understood. Did new traits spread purely through cultural interaction among societies, via acculturation, emulation or material exchanges? Or did the new traits spread via human movement between regions? If such human mobility did have a role, was it a steady dispersal process involving conjugal exchange, or did it occur in pulses, as mass migration events? Was human movement sex-biased, or did it involve males and females equally? Today, we can discern among different modes of human mobility and material culture change by studying ancient genome data integrated with archaeological data.

Ancient genomics can further help investigate social traditions of Neolithic societies. Were the earliest settled societies organised as nuclear or extended families, or were they not organised around biological ties at all? Studying the subject solely based on material culture is notoriously difficult (Kuijt, Goring-Morris 2002). To date, a number of studies have estimated biological relatedness among co-buried individuals in southwest Asian Neolithic societies using either dental features (Pilloud, Larsen 2011; Alt et al. 2015) or maternal genetic markers (Chyleński et al. 2019). However, these approaches provide only low-resolution perspectives, as we will discuss later. Kinship has thus remained a largely unwritten chapter for the Neolithisation of southwest Asia.

The potential of ancient genomics in studying demography and kinship

Ancient DNA (aDNA) analysis refers to the study of degraded DNA molecules from ancient remains that contain organic matter, such as bones, teeth, coprolites, dental calculus, or sediments. In certain ways aDNA work today involves the same steps as standard modern DNA analysis: DNA purification (extraction), amplification, sequencing and bioinformatic analysis, aimed at inferring kinship, demographic history, etc. However, unique characteristics of aDNA render its analysis additionally challenging – compared to that of modern DNA. Most important is the fact that DNA molecules break down over time, eventually to the point of non-existence.

The first generation of aDNA

From the 1980s to the early 2010s, human aDNA research was an exceptionally cumbersome and much more precarious endeavour than it is today. The DNA technologies of the late 20th century permitted manipulation and sequencing of only specific short DNA regions (ca 100 base pairs), in a targeted fashion, one by one, rather than the whole set of chromosomes, that is, the genome (ca 3 billion base pairs). Most aDNA work was focused on mitochondrial DNA (mtDNA), which is the small DNA molecule of the cellular organelle passed from mother to offspring. Mitochondrial DNA was an obvious target because it is present at 100–1,000 times higher copy numbers than nuclear DNA (the main source of DNA in a human cell). Therefore, a random fragment of mtDNA is more likely to have survived in a bone sample than a random fragment of nuclear DNA. It is also a particularly variable part of the genome and even short sequences of mtDNA can contain sufficient information to distinguish between closely related lineages.

However, mtDNA provides information only about the maternal lineage and this constitutes a severe limitation when inferring genetic kinship or population history. For example, an individual will appear as unrelated to her/his father as to any random individual from the perspective of mtDNA analysis. In contrast, nuclear DNA (or whole genome data) is derived from all ancestors, and the information it provides is much more extensive. Demographic history inferred about only maternal ancestors (from mtDNA) and about all ancestors (from genome data) can be strikingly different. For instance, mtDNA analysis on the first Denisovan remains had suggested that humans and Neanderthals were sister-species while Denisovans were an outgroup; in contrast, whole genome analysis showed that for most of their ancestry Denisovans were actually sister-species to Neanderthals (Reich et al. 2010). Being mainly restricted to mtDNA was a serious drawback for aDNA research in early periods.

A second major obstacle hindering the early aDNA studies was posed by the risk of modern-day DNA contamination and the seeming inability to distinguish between authentic ancient DNA and modern human DNA, which is ubiquitous (Hofreiter et al. 2001; Malmström et al. 2005). It can be difficult, if not impossible, to fully determine whether a *single* short DNA molecule extracted from an archaeological sample is authentic (endogenous aDNA) or represents external human DNA contamination (that could have happened in the field or in the lab). Hence, despite the adoption of many precautions against contamination by laboratories, human aDNA analysis was for a long time fraught with scepticism.

The technical challenges, the limited information available per sample and the spectre of modern DNA contamination collectively impeded aDNA research on human archaeological material for about three decades. The field remained confined to a few specialised laboratories until the 2010s. During this period, most aDNA work focused on species other than *Homo sapiens* and particularly on extinct species, such as Neanderthals and woolly mammoths, the DNA sequences of which would not be easily confused with those from their present-day relatives.

Second generation aDNA

The late 2000s and early 2010s saw major shifts in aDNA analysis approaches and the emergence of a new field, that of ancient genomics (Stoneking, Krause 2011; Pickrell, Reich 2014; Skoglund, Mathieson 2018). In this approach, aDNA molecules extracted from organic material are processed and sequenced in parallel, on the order of millions to billions of DNA fragments. This yields sequence information across the whole genome of an organism within a single, relatively simple experiment. The information thus retrieved is orders of magnitude more extensive than that derived from sequencing mtDNA. Indeed, genome-wide data from even a single individual will yield information about all the individual's recent ancestors. For instance, for a non-inbred person this would translate into 128 ancestors seven generations ago, while studying mtDNA would inform about a single ancestral lineage only.

The first Neanderthal genome study was a prime example demonstrating the power of this approach. Comparisons of Neanderthal and modern human mitochondrial DNA sequences had initially revealed no evidence for Neanderthal introgression into the modern human gene pool (Serre et al. 2004). Once the Neanderthal genome was sequenced, however, this new data permitted geneticists to identify traces of recently introduced Neanderthal ancestry in modern-day non-African genomes, strongly indicating the occurrence of low-level introgression of Neanderthals in the ancestors of non-African humans (Green et al. 2010). Since that time, ancient genome analyses have been growing at rapid speed. All this was made possible by innovations in four areas, which we summarise below.

Next generation sequencing

The most fundamental development was the advent of next generation (or second generation) sequencing (NGS) technologies. This permits shotgun sequencing of all available genomic DNA from a sample – in parallel – producing much more extensive information at much lower costs than previously available. Here 'shotgun' refers to the fact that any molecule obtained from the material is processed and sequenced in parallel with others, irrespective of its origin, in contrast to earlier targeted approaches that analysed only specific parts of DNA (such as mtDNA loci). Importantly, the experiments involved in preparing the aDNA extracts for sequencing are technically not more challenging than those of the earlier 'targeted' approaches.

Novel laboratory protocols

Another line of development involved laboratory methods that boosted aDNA retrieval. These were critical for aDNA analyses of material from temperate regions, such as Neolithic bones from Anatolia because DNA decays faster at higher environmental temperatures and therefore little endogenous DNA remains in bones older than a few millennia in warm climates (Allentoft et al. 2012). Advances included the identification of the petrous temporal bone as a rich source of DNA (Gamba et al. 2014), the development of efficient aDNA isolation and processing protocols tailored for retrieving short degraded molecules (Dabney et al. 2013; Gansauge, Meyer 2013) and removing damage from ancient molecules (Briggs et al. 2010; Rohland et al. 2015). Another technique that strongly influenced ancient genomics was the hybridisation capture procedure for enriching human DNA molecules (Carpenter et al. 2013; Fu et al. 2013). Within a DNA extract from an ordinary Anatolian Neolithic bone, the vast majority (usually >99%) of molecules will not be human but rather environmental DNA, mainly from bacteria. Hybridisation capture, a type of molecular fishing, alleviates this situation by enriching human DNA molecules in total DNA extracts. Enrichment, in turn, reduces the total amount of next generation sequencing needed to achieve the same amount of information. Enrichment has been employed in two ways: targeting the whole genome (Allentoft et al. 2015; Kılınç et al. 2016), or targeting specific variants (single nucleotide polymorphisms, or SNPs) (Haak et al. 2015). The latter method appears particularly efficient in enrichment but is limited to only previously ascertained genetic variants; this kind of data may complicate downstream demographic inferences, such as population diversity measures.

In silico *identification of DNA contamination*

Another major development involved statistical methods for estimating modern DNA contamination from shotgun sequencing experiments. This development was a growing understanding of the types of modifications and post-mortem damage (PMD) that accumulate within DNA through time – specifically, cytosine deamination causing C to T mismatches at the ends of molecules (Hofreiter et al. 2001; Briggs et al. 2007). A number of *in silico* methods were thus developed to estimate the amount of human contamination in an ancient genome dataset using information on PMD or on abnormal patterns of polymorphism at haploid loci (for example, more than one type of mtDNA in the same DNA sample, which does not biologically occur) (Green et al. 2010; Fu et al. 2013). These methods not only permit identification of modern contamination in aDNA datasets, but they can also be used to select molecules with PMD signatures, thus filtering out potential contaminant molecules and saving at least part of the authentic data (Skoglund et al. 2014; Fu et al. 2015). Contamination has thereby become less of a peril and more of an economic issue for human aDNA work, although aDNA laboratories still strive to avoid it as much as possible due to the cost it incurs.

Novel computational protocols

Finally, the last decade has witnessed the development of computational protocols and statistical methods for analysing ancient genomes. This is important as most population genomic methods are tailored for high- or medium-quality modern genomes (for example, covering the full human genome 30 times, or 30X coverage). Conversely, ancient genome datasets are generally of both much lower and also more variable quality compared to modern DNA datasets. Most cover only a small fraction of the whole genome, are comprised of only short sequence reads, and contain post-mortem damage. They thus require special treatment. This added treatment frequently involves (a) considering only variants that have been identified elsewhere, (b) considering only variants that cannot be confounded with post-mortem damage, and (c) removing some of the data to equalise quality among samples (a process called 'pseudohaploidisation') (Skoglund et al. 2014; Mathieson et al. 2015).

Meanwhile, population genomic methods have been developed that can be efficiently applied to such sub-optimal genomic data. These include the outgroup f_3 statistic that measures genome-wide genetic affinity between two groups and the *D*-test (or f_4 statistic) used to test whether a 'Test' population has higher affinity to one of two alternative populations (Green et al. 2010; Patterson et al. 2012). Here, higher genetic affinity is measured as the proportion of shared variants between two populations and it indicates a relatively higher amount of shared ancestry, such as with a sister versus with a cousin. Importantly, such tests can be performed with populations represented by even single individual genomes and can result in significant signals, given the wealth of information in a single genome.

The laboratory routine

We now describe a routine second generation aDNA experiment on a human archaeological sample (fig. 18.1). The initial part is called 'pre-screening' – an initial attempt to gauge the amount of useful DNA in a sample. The first step of pre-screening involves molecular work: well-preserved parts of the human skeleton (for example, the petrous temporal bone, teeth, auditory ossicles or finger bones) are brought into the clean room, powderised, the DNA is extracted, a sequencing library (a modified collection of DNA molecules) prepared and shotgun sequenced on an NGS (Next Generation Sequencing) machine, the latter task frequently performed at specialised centres. Sequencing at this stage is in small volumes; that is, 'low coverage' or 'low depth', restricted to millions of DNA molecules.

The second step of pre-screening is computational: the resulting raw data are cleaned *in silico* and mapped to the human reference genome. This yields two important statistics: (a) the ratio of human sequences mapping (aligning) to the reference human genome over the whole set of sequences (note that the non-human portion here may be of microbial origin or be DNA artefacts); (b) the PMD signal across the mapped human sequences, that is, the proportion of C to T type of mismatches at the ends of molecules (fig. 18.1).

If the pre-screening results from a sample reveal that the human-specific proportion of DNA is too low (for example, <0.5% of molecules being human) and/or the PMD signal is too weak to be compatible with authenticity, the researcher will normally choose to discard that sample, since it is impossible to gain useful information from such material using current methods. Otherwise, the sample may be further processed. This may involve (a) deep sequencing, that is, sequencing large volumes of molecules, from the same library or newly built libraries from the same sample, with hundreds to thousands of millions of molecules produced; (b) enriching for human molecules via hybridisation capture (whole genome capture or SNP-capture) on the libraries and then performing new sequencing, to increase the efficiency and thus reduce sequencing costs.

The resulting data are then used in computationally estimating the genetic sex and the genotype of that individual. Genetic sex determination is based on the proportion of DNA sequences deriving from the X chromosome to other chromosomes. Calling genotypes refers to estimating the genotype of an individual for variants found across the genome, up to many millions. For instance, if a SNP has two alleles, G or T, individuals in a population can carry three possible genotypes: GG, TT, or GT (one allele from paternal and the other from maternal origin). Theoretically, DNA sequencing data can be used to determine which of these three genotypes an individual carries. However, aDNA data are frequently sparse, and genotype calling involves only determining if the individual carries at least one G or at least one T variant.

The genotype data can then be used to address various questions: to infer demographic history (based on tests of affinity between the genotype of an ancient individual and those of other ancient or present-day individuals), to estimate the level of genetic kinship to other individuals, to estimate certain phenotypic traits such as eye colour, or to determine how inbred the individual was. If the aDNA dataset has been produced

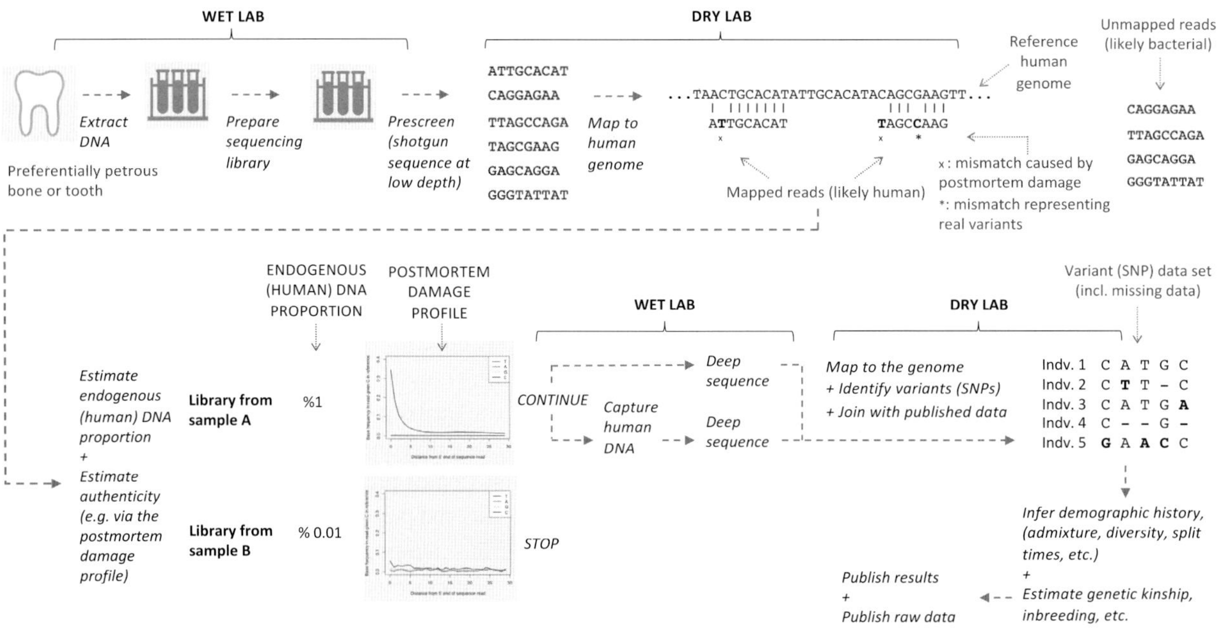

Figure 18.1. The second generation aDNA analysis routine. The first phase of an experiment – pre-screening – aims to estimate the endogenous human DNA proportion and post-mortem damage in the human molecules, as a rough estimate of authenticity. Selected libraries continue to be analysed by direct deep (large volume) sequencing, or hybridisation capture followed by deep sequencing. The resulting data are then used in population genomic analyses for estimating demographic history, genetic kinship levels among individuals, and level of inbreeding, etc. The post-mortem damage figures in the lower panel indicate the proportion of sequenced human DNA molecules that contain a T base at their ends when a C is expected in the reference genome at that same position. Observing a T instead of a C is the most prominent indication of post-mortem damage that accumulates in DNA over time.

from teeth, the DNA molecules that do not map to the human genome can be screened for matches to blood-borne pathogen genomes (Rascovan et al. 2018). Age-at-death of individuals may also be roughly estimated from DNA methylation patterns using specifically processed high quality aDNA datasets (Pedersen et al. 2014).

An important point is the conventional practice of sharing raw and/or processed datasets upon publication among population geneticists, usually by depositing the data in public databases. This enables other researchers to download and compare their new data with previously published datasets and augment knowledge accumulation. As of May 2020, there have already been more than 3,500 ancient human genomes published by various groups, available for analysis.

Sequencing technologies are still evolving, but the experimental procedures described here may still prevail in aDNA laboratories for some time yet because of constraints imposed by the degraded and low-quality nature of aDNA. Instead, we may expect statistical methods to develop even further in the coming years to make even more effective use of the currently available data. The rapidly expanding collection of ancient genomes is expanding the potential uses of the data itself.

Genomic insights into European Neolithisation: demic processes

Since the late 20th century, geneticists in Europe and North America have been tackling the demic versus cultural diffusion question in the context of European Neolithisation. Early studies used genetic data from modern populations (Menozzi et al. 1978; Cavalli-Sforza et al. 1988; Richards 2003), suggesting a role for migration of southeastern origin in the spread of farming across Europe. Later, mitochondrial ancient DNA studies (Haak et al. 2008; Malmström et al. 2009) reported discrepancies between the mtDNA profiles of early Neolithic farmers and contemporaneous European Mesolithic hunter-gatherers, also supporting an ancient migration event. Results from ancient mtDNA analyses from the Neolithic and the Iron Age Southwest Asian groups (Fernández et al. 2014; Yaka et al. 2018) have further been in line with this idea.

However, given the uncertainties pertaining to mtDNA analyses (discussed earlier), it was ancient genomics research using whole genome sequence data that provided the final verdict. These studies consistently showed that European Neolithic farmer communities, across Central Europe, Iberia and Scandinavia, were all genetically highly distinct from European Mesolithic groups from the same localities (Skoglund et al. 2012; Gamba et al. 2014; Lazaridis et al. 2014; Skoglund et al. 2014). It was further shown that early European farmers had genetic profiles highly similar to those of Neolithic and Chalcolithic west Anatolian and Aegean populations (Günther et al. 2015; Mathieson et al. 2015; Cassidy et al. 2016; Hofmanová et al. 2016; Kılınç et al. 2016; Omrak et al. 2016), as would be predicted from the archaeological record. Demic diffusion from the Aegean/Anatolia likely had a major role in driving the Neolithisation of Europe. Genomic data has further confirmed that this human movement process followed two main routes: the land route that reached central Europe and a maritime route that reached Iberia and from there, spread along the Atlantic coast (Brace et al. 2019).

That said, genomic sampling of early farmer populations is currently sparse, and the results of genomic analyses do not suggest that cultural interaction was non-existent or that local foragers were replaced by migrant populations. Genetic profiling of middle Neolithic European populations shows that migrant early farmers in Europe and local hunter-gatherer populations eventually did admix, such that modern Europeans have ancestry from both sources (Lazaridis et al. 2014). Moreover, this admixture may have started quite early, as implied by the finding of a 'genetically Mesolithic' individual buried in an early LBK site in Hungary (Gamba et al. 2014). Finally, genetic data on the Neolithisation of the Baltic region indicates that sedentism and farming there were adopted by locals instead of being brought by migrating farmers (Jones et al. 2017). In fact, as we discuss below, acculturation may be the predominant process in many instances of rapid cultural change.

Genomic insights into Neolithic development in the primary zone: acculturation and exchange

It might appear awkward that genetic studies on European Neolithisation have been more intensive than those on the initial Neolithisation process of Southwest Asia, *inside* the primary zone of incipient Neolithic developments; that is, the Levant, north Mesopotamia, the Zagros Mountains and Central Anatolia. The population dynamics behind the initial evolution of sedentism, farming and herding, and the social re-organisation associated with this transition, have been long discussed (Kuijt 2000b; Price, Bar-Yosef 2011). This discrepancy may partly be attributed to Eurocentric tendencies and the fact that many major aDNA research centres are in Europe. Tighter regulations in southwest Asian countries on archaeological sample export may also have slowed down the influx of archaeological material to European centres. At least equally important is the fact that working with aDNA from southwest Asia is far more formidable than working with European samples, due to relatively high temperatures that promote DNA degradation in the former region.

Nevertheless, multiple groups have recently succeeded in producing genomic data from different regions of the Neolithic primary zone through intensive efforts usually involving processing large numbers of individual remains. The year 2016 saw the publication of the first genome studies on the Neolithic primary zone, including individuals from Central Anatolia (Kılınç et al. 2016), the Zagros (Broushaki et al. 2016; Gallego-Llorente et al. 2016) and both the Levant and Zagros (Lazaridis et al. 2016). This was later followed by re-analyses of these data sets (Kılınç et al. 2017) and recently a new publication on Central Anatolia (Feldman et al. 2019). Overall, this work revealed a number of notable points.

1) At the start of the Holocene, western Eurasia hosted genetically differentiated populations. Before the advent of the full Neolithic package (pre-7000 BCE) we have evidence for three relatively isolated human gene pools in the Levant, in the Caucasus and Zagros regions (fig. 18.2) and in Europe, including mainland and steppe (Kılınç et al. 2017). Of particular note here is the genetic isolation between the Zagros region and the Levant, despite their geographic proximity and evidence for cultural contacts, such as the widespread circulation of Anatolian obsidian in these regions (Goring-Morris, Belfer-Cohen 2011; Ibáñez et al. 2018). It is likely that the gene pools were shaped by the shrinkage of available habitats during the Last Glacial Maximum, driving isolation and differentiation through genetic drift (Brewster et al. 2014).

Meanwhile, Epipalaeolithic and Aceramic Central Anatolians (represented by the populations of Epipalaeolithic Pınarbaşı and Aceramic Boncuklu) shared affinities with all three gene pools (Kılınç et al. 2017; Feldman et al. 2019). In a sense, Central Anatolians were 'in the middle' with connections to all groups. How this Central Anatolian population came into being before the Holocene is an open question that will require studying even earlier genetic material from the region. The genetic profiles of north Mesopotamian and Aegean populations of the pre-7000 BCE period are also as yet unknown.

The studies mentioned here were conducted using genetic data from a small number of individuals from a few settlements, sometimes even a single individual representing a settlement. As mentioned earlier, owing to the fact that the genome of a single individual contains information from that individual's many ancestors, it is possible to make wide inferences with even this limited data, assuming one was not unlucky enough to sample a single fresh migrant in a mainly local population. That said, increasing sample size in the future would facilitate making bolder statements.

2) The Pre-Pottery/Aceramic villagers of the Levant, of the Zagros and of Central Anatolia were, in their genomic profiles, not only distinct from each other, but also closely related to the Epipalaeolithic populations of those same regions (Lazaridis et al. 2016; Feldman et al. 2019). The same pattern is also reflected in the mtDNA profiles of these populations (Chyleński et al. 2019). These results strongly imply that the Neolithic transition within Southwest Asia was largely enacted through cultural interactions between local populations, as previously predicted by material culture studies (Baird 2012). In other words, the inter-regional spread of technology, rituals and domestic species mainly occurred through cultural exchange but not large-scale population dispersal.

Still, this does not rule out any population movement among regions. On the contrary, *D*-tests suggest that the Aceramic Boncuklu population, relative to a single Epipalaeolithic Pınarbaşı individual sequenced, received gene flow from eastern (Zagros and/or Caucasus) regions (Feldman et al. 2019). Feldman and colleagues estimate that 10 per cent of Boncuklu's ancestors may have arrived from the east, although these estimates may vary widely. Of course, it should also be noted that this human movement signal may actually represent gene flow from southeast Anatolia/Mesopotamia, which is not yet genetically profiled, instead of gene flow from the Zagros or the Caucasus.

3) From the Pre-Pottery/Aceramic period to the Pottery/Ceramic Neolithic period, human dispersal between Anatolia and neighbouring regions appears to have gained momentum. Populations of the PPNB Levant (for example, 'Ain Ghazal) show higher genetic affinity to Neolithic Anatolians relative to Natufian groups who lived in the same region 2,000 years earlier (Kılınç et al. 2017). This can be explained by a simple scenario: after the Natufian, certain Anatolian-related groups moved in and mixed with the Levantine locals and therefore PPNB Levantine groups carry more Anatolian variants than did the Natufian groups.

The Pottery Neolithic Central Anatolian population of Tepecik-Çiftlik likewise had closer affinity to Levantine and Zagros populations than did Aceramic Boncuklu people who lived in the same broad region 2,000 years earlier. These changing genetic affinities can also be explained by gene flow into Anatolia from its southern and eastern peripheries. A pattern of accelerating inter-regional gene flow is captured by increasing within-population genetic diversity levels during the Neolithic transition (Kılınç et al. 2016; Lazaridis et al. 2016).

That said, alternative complex scenarios, such as complex population structure (for example, spatial heterogeneity in the population) and variation in admixture tendencies among sub-populations could also explain the observed genomic data without the occurrence of regional admixture events. These more complicated scenarios may be tested in the future when denser data are available.

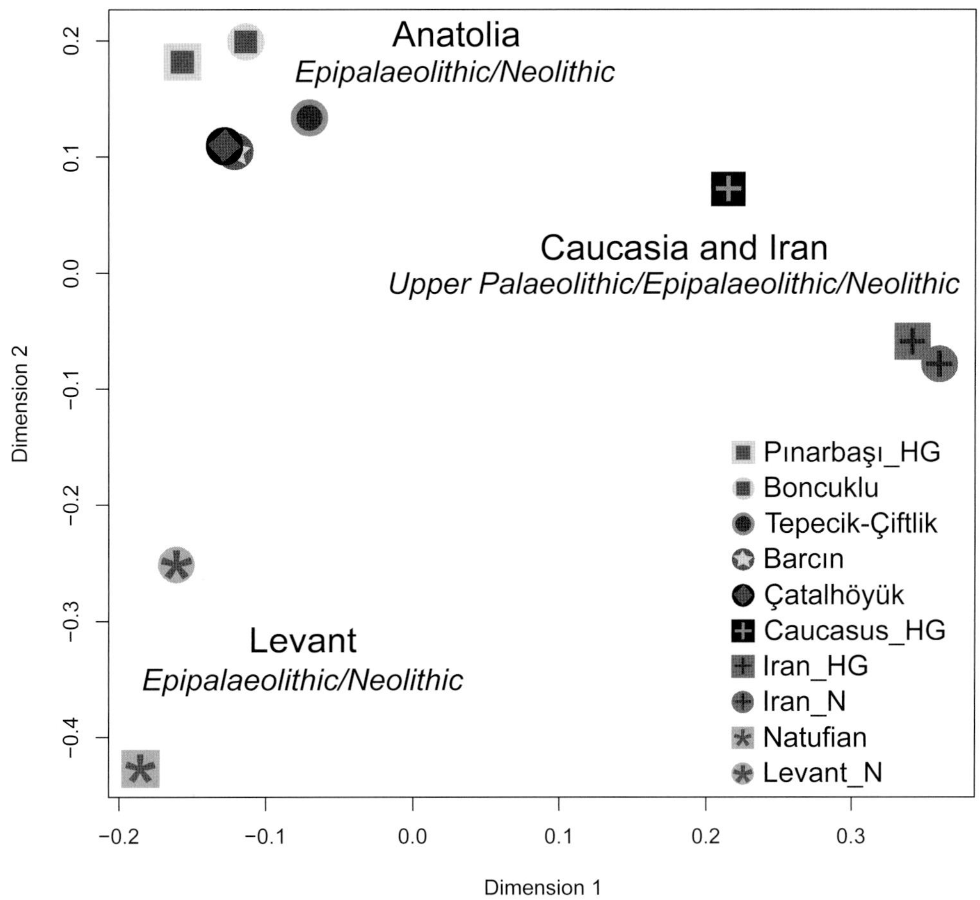

Figure 18.2. Genetic similarities between pre-Neolithic and Neolithic populations of Southwest Asia. The multi-dimensional scaling plot summarises f_3-statistics that measure genome-wide similarity between pairs of populations (Patterson et al. 2012). The statistics were calculated among regional populations with published genomic data: Pınarbaşı_HG: Epipalaeolithic from Pınarbaşı (Feldman et al. 2019); Boncuklu: Aceramic Boncuklu Höyük (Kılınç et al. 2016; Feldman et al. 2019); Tepecik-Çiftlik: Pottery Neolithic Tepecik-Çiftlik (Kılınç et al. 2016); Barcın: Pottery Neolithic Barcın Höyük (Mathieson et al. 2015; Hofmanová et al. 2016); Caucasus_HG: Upper Palaeolithic Caucasus hunter-gatherer (Jones et al. 2015); Iran_HG: Iran Epipalaeolithic from Hotu Cave (Lazaridis et al. 2016); Iran_N: Iran Pre-Pottery Neolithic from Ganj Dareh, Tepe Abdul Hosein and Wezmeh Cave (Broushaki et al. 2016; Lazaridis et al. 2016); Natufian: Natufian from Raqefet Cave (Lazaridis et al. 2016); Levant_N: Levant Pre-Pottery Neolithic from Ain Ghazal, Mota, Ba'ja and Kfar HaHoresh (Lazaridis et al. 2016; Feldman et al. 2019). Çatalhöyük data are unpublished (Yaka et al. in preparation).

4) A little-understood point is the source of the Neolithic North Aegean communities that seem to have emerged post-7000 BCE. Neolithic European farmers were genetically closest to these Neolithic North Aegean communities (Barcın in northwest Anatolia and Revenia in northeast Greece) than to any other primary zone population hitherto studied (Kılınç et al. 2017).

So, who were these Neolithic Aegean people? Were they seafaring colonists from the Levant, a composite of Mediterranean and inland Anatolian migrants who merged on the Aegean coast, or local Mesolithic populations acculturated by their maritime or inland connections? The archaeological evidence has remained equivocal (Özdoğan 2011b; Perlès et al. 2011; Reingruber 2011; Baird 2012; Çilingiroğlu, Çakırlar 2013; Horejs et al. 2015), while genetic data now provides interesting clues.

First, the north Aegean population was genetically closer to that of Central Anatolia than to any other potential source population, including the Neolithic Levantine people (Kılınç et al. 2017; Chyleński et al. 2019). This rules out a Levantine-derived colonist origin. Surprisingly, relative to Central Anatolians, the north Aegean Neolithic populations also carried genetic affinities to distant populations, from Europe and the Caucasus to the Levant (fig. 18.2) (Kılınç et al. 2017). The Aegeans thus appeared perhaps 'too diverse' for a migrant group of recent Central Anatolian origin.

We may thus consider an alternative hypothesis to the idea of migration/colonisation of the Aegean by farmers from the Neolithic primary zone. Perhaps Aegean Neolithic societies were largely local forager populations who underwent rapid acculturation via connections with the Neolithic primary zone (Kılınç et al. 2017). This conclusion resonates well with archaeological evidence indicating a heterogeneous cultural background in the Aegean; for example, the intensive use of obsidian from the Aegean island of Melos or the non-existence of Near Eastern traditions of lithic tool production, such as pressure-flaking, before the seventh millennium BCE in the region (Düring 2011b; Perlès et al. 2011; Reingruber 2011; Yakar 2016). If true, acculturation could have had a more prominent role, at least for the early Neolithisation of the Aegean sphere, than currently thought. If so, it becomes plausible to suggest that population movements drove the Neolithisation of Europe from a wider geographical area encompassing the Aegean and Anatolia, rather than merely Anatolia. Full investigation of these scenarios awaits further data, specifically from pre-Neolithic populations in the region.

First genomic insights into Neolithic Çatalhöyük humans

At the METU Ancient DNA Laboratory, genomic analyses of Çatalhöyük human material started in 2015, but it was more than three years before useable amounts of genetic data were obtained. The work was focused on East Mound sub-floor burials. By early 2020, the METU aDNA Laboratory had prepared over 200 Illumina sequencing libraries from 157 individuals' teeth and petrous bones and subjected these to pre-screening (fig. 18.1) in collaboration with Stockholm University; selected libraries were further deep-sequenced after hybridisation capture. In collaboration with METU, the aDNA group at the Adam Mickiewicz University in Poznań (AMUP) also initiated a study on Çatalhöyük genetics focused on mtDNA. This work led to the first DNA-based study of Çatalhöyük humans, published in 2019 (Chyleński et al. 2019).

DNA preservation and the question of adult burial practices

Between 2015 and 2018 the METU team had prepared sequencing libraries based on >20 individuals' remains, and none had provided useful amounts of endogenous DNA (>0.5%). This precluded us from including Çatalhöyük aDNA data in a 2016 publication on Neolithic Anatolian populations, which had included a Boncuklu Höyük (Aceramic Neolithic Central Anatolia) individual with >6% endogenous DNA, and four out of six Tepecik-Çiftlik (Pottery Neolithic Central Anatolia) individuals with endogenous DNA proportions close to 1% (Kılınç et al. 2016). Chyleński et al. also reported low DNA preservation in Çatalhöyük (Chyleński et al. 2019).

This had raised the question of whether Çatalhöyük human bones may have been overall more poorly preserved than those of other Neolithic Central Anatolian sites. However, recent sampling and sequencing efforts showed this not to be the case: the current data from recent pre-screening results suggest no systematic difference among Tepecik-Çiftlik, Boncuklu Höyük and Çatalhöyük in the endogenous DNA proportions (fig. 18.3) (Kruskal-Wallis test $p>0.10$).

Strikingly, analysis of this pre-screening dataset revealed that endogenous DNA proportions clearly depend on the age-at-death of Çatalhöyük individuals, with adults showing significantly lower DNA preservation than sub-adults, including newborns, infants and children (fig. 18.3) (Mann-Whitney U test $p<0.001$). This pattern was wholly unexpected, as it is generally believed that adult bones preserve DNA better due to their compactness. In fact, early sampling efforts in Çatalhöyük had for this reason tried to avoid subadult material, leading to the earlier observation on low DNA preservation at this site.

The reason for lower adult vs subadult DNA preservation at Çatalhöyük is as yet unknown, as well as whether it applies to other Neolithic Anatolian settlements. It is tempting to speculate that unique burial traditions were practised mainly for adults, such as collective burials, secondary burials and interventions, that intervened in the decomposition process in ways such as defleshing (Pilloud et al. 2016a; Haddow, Knüsel 2017) and could underlie DNA preservation differences. Nevertheless, the as yet small sample size and high variation in DNA preservation rates, even among individuals buried in the same location, preclude reaching an absolute conclusion at this point.

Çatalhöyük genetic and genomic data

In the first DNA-based study of Çatalhöyük people published, the aDNA laboratory at AMUP generated whole mitochondrial genomes using pre-screening followed by mitochondrial DNA capture from ten individuals (Chyleński et al. 2019). All were burials within four neighbouring buildings (B.80, B.89, B.96, B.97) belonging to Levels South O-North G, 6700–6500 BCE. Seven were subadults.

As of March 2020, the METU aDNA Laboratory, in collaboration with the AMUP and Stockholm University (SU) laboratories, produced low-coverage genomic data from 12 individuals from the levels South K, M and N and North G (B.17, B.50, B.89, B.91, B.114) and two from Late-Neolithic phases of Çatalhöyük (TP and TPC). Seven of these individuals' remains were radiocarbon dated and all dates were consistent with the archaeological contexts (ca 6700–6500 BCE for Middle phase and ca 6400–6100 BCE for Late phase individuals).

The genome data were produced using both shotgun sequencing from the pre-screening experiments and whole genome capture experiments followed by deep sequencing. Across the studied individuals the genome coverages range from 0.01x to 0.27x; in other words, one can observe only 1% to 27% of the genome-wide variants of each individual. Despite these low proportions, given the presence of millions of common variants across the human genome, it is still possible to extract information on tens of thousands of variants per individual. This amount of information can be sufficient for simple demographic analyses (for example, comparing affinities of a Çatalhöyük individual to one from the Levant or from Boncuklu in a *D*-test). It also permits estimation of close biological kinship.

All except one of the 14 individuals with sufficient genome data were subadults; the exception was one Late Neolithic individual. Ten were found to be genetically female. Because the sex of subadults cannot be assessed reliably from study of their skeletal remains, we cannot assess the overall match with the anthropological sex assignment.

DNA-based kinship patterns among Çatalhöyük co-burials

The most striking result from the Chyleński et al. study was that, among the ten individuals whose mitochondrial genomes had been sequenced, all carried distinct mtDNA lineages (Chyleński et al. 2019). That is, all ten individuals had different mothers, maternal grandmothers, etc., despite being interred within four neighbouring buildings.

This intriguing result could be explained in two ways: Çatalhöyük burial traditions might have been strictly patrilocal, a possibility that has previously been raised based on the study of similarities in dental markers across larger Çatalhöyük samples (Pilloud, Larsen 2011). Alternatively, individuals buried in the same buildings may not have been biologically closely related kin.

The genomic data currently available (Yaka et al. in preparation), although sparse, supports the latter notion. Out of the 14 individuals we generated genomic data for, ten were buried in B.17, B.50 and B.114. We studied genetic kinship among a total of 14 such co-buried pairs,

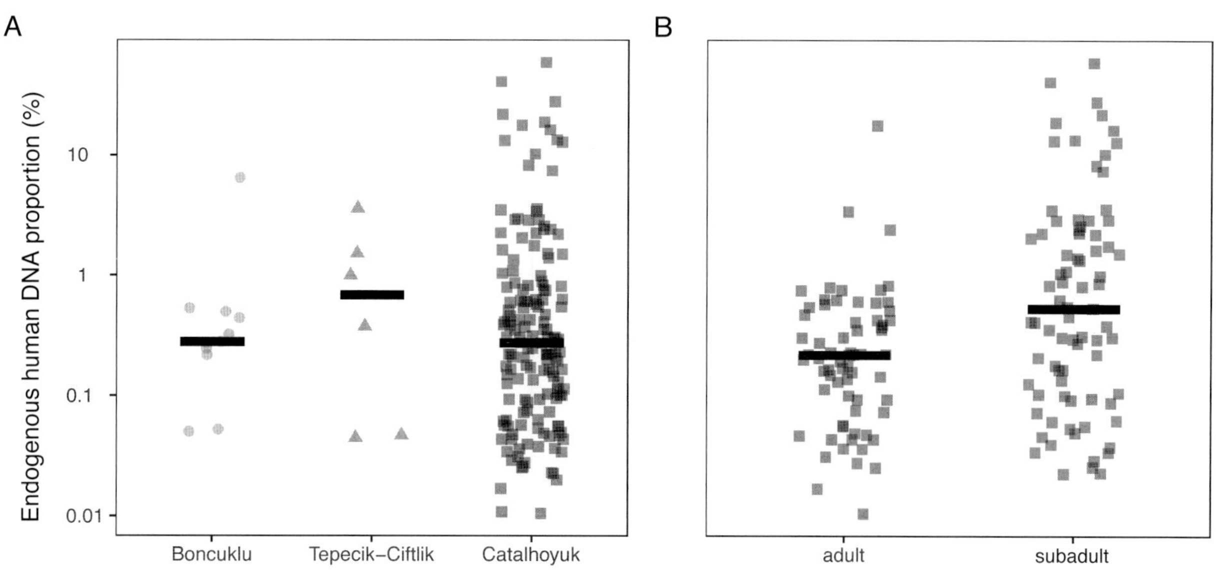

Figure 18.3. Endogenous DNA proportions in Çatalhöyük. (A) Proportions calculated from Boncuklu (n=9), Tepecik-Çiftlik (n=6) and Çatalhöyük (n=157) individuals' bone and teeth samples. (B) Proportions calculated for adult (n=67) and subadult (n=79) Çatalhöyük individuals. Adolescents were not included. Endogenous DNA proportions were measured as the percentage of shotgun sequenced DNA molecules from pre-screening experiments mapping to the reference human genome. Each individual is represented by a single observation. The highest proportion was used for a minority of cases where multiple libraries had been produced per individual. The horizontal black lines indicate the median per group.

where the polymorphism data would allow us to detect up to the third degree of relatedness (that is, cousins or avuncular relations).

Out of the 17 pairs scanned, only one pair of female subadults were found to be siblings (Yaka et al. in preparation). None of the other co-buried subadults were closely related. Although the data are as yet too limited to reach a clear conclusion, they point to the exciting prospect that non-biological social kinship structures influenced co-burial at Çatalhöyük (cf. Pilloud, Larsen 2011) and possibly other contemporaneous Neolithic societies in southwest Asia. We discuss the implications of this finding in the last section.

Population-level similarities between Çatalhöyük and other southwest Asian Neolithic groups

Preliminary population genetics and kinship analyses have revealed a number of patterns that we summarise here (Yaka et al. in prep.):

1) Treated together as a population, the 14 Çatalhöyük individuals with genomic data belong to the Early Holocene central/west Anatolian gene pool. Specifically, the Çatalhöyük population is genetically close to individuals from other Epipalaeolithic and Neolithic Central Anatolian populations (Pınarbaşı, Boncuklu Höyük, Tepecik-Çiftlik) and to those from the northwest Anatolian Neolithic Barcın Höyük, but distinct from Levantine or Zagros populations (fig. 18.2). This overall pattern is supported by *D*-tests (Yaka et al. in prep.). It is also consistent with the report by Chyleński et al., who found that the overall haplogroup composition of this sample was similar to that of other Central Anatolian and northwest Anatolian groups (Chyleński et al. 2019).
2) There is limited genetic differentiation among these Çatalhöyük individuals and no conspicuous difference between Middle- and Late-period Çatalhöyük individuals. That said, we cannot yet rule out possible subtle differences in the genetic profiles between Middle- and Late-period Neolithic Çatalhöyük populations (which could be caused by gene flow into Çatalhöyük from distant regions). More genomes will be needed to address this.
3) In *D*-tests, the Çatalhöyük population appears equally related to Aceramic Boncuklu from the Konya Plain and the Pottery Neolithic Tepecik-Çiftlik from the Cappadocia region. Meanwhile, it appears to carry more Levantine variants compared to the Boncuklu sample, similar to that reported for Tepecik-Çiftlik (Kılınç et al. 2017). This result is compatible with some level of Levantine gene flow into Central Anatolia between 7500 and 6500 BCE.

Conclusion and future directions

Ancient genomics analyses are shedding light on various aspects of the Neolithic transition in Southwest Asia. We can already see that, during the initial development of sedentism and later farming, populations on the east, west and northwest borders of the Neolithic primary zone largely remained genetically distinct, with limited gene flow among them. However, this is only a first glimpse and the picture remains full of gaps. From genetic data we know nothing, or close to nothing, about human movement to or from Neolithic Mesopotamia. The Aegean question – if local Neolithisation occurred in this region – is open. The colonisation of Cyprus also remains an enigma. Still, we can underline some tentative outcomes. First, the Neolithisation across different regions of the Fertile Crescent, including Central Anatolia, involved limited gene flow across regions (Kılınç et al. 2016; Lazaridis et al. 2016). Also, the Neolithisation of West Anatolia and the general Aegean sphere do not seem to be related to migrations from the Levant (*contra* Horejs et al. 2015), whereas during the PPNB some human movement between Central Anatolia and the Levant seems to have occurred. These broad human movement patterns also involve the ancestry of the Çatalhöyük population, which was genetically highly similar to the Aceramic Boncuklu Höyük population, albeit slightly more diverse due to past admixture (Yaka et al. in preparation).

We further remain largely ignorant regarding the role of biological kinship in these early Neolithic societies. What role did biological relatedness have in the organisation of these communities? Social kinship can take on diverse forms and genetic relatedness is not a prerequisite (Sahlins 2013; Johnson, Paul 2016). Despite the widespread assumption that forager groups are genetically kin-based, recent research has shown that within-group genetic relatedness is conspicuously low in many modern-day foragers (Hill et al. 2011) and post-marital residence is not necessarily sex-biased (Dyble et al. 2015). In contrast, in many farmer communities, genetic kinship is central to social organisation and patrilocality is predominant (Stone 2014). Recent ancient genomics work on Upper Palaeolithic Siberian hunter-gatherers (Sikora et al. 2017) and LBK and Bronze Age farmers from Europe (Mittnik et al. 2019; Sánchez-Quinto et al. 2019) also point in the same direction: a limited role of genetic kinship in pre-farming societies and predominant patrilineal patterns once farming had been established. If social organisation of pre-Neolithic societies was indeed akin to those of modern-day hunter-gatherers, then genetic kin-based and patrilocal/patrilineal traditions could have emerged either early or late and variably from one region or community to another during the Neolithic transition.

At least for the case of Çatalhöyük, our preliminary results suggest that co-buried individuals in the same building may have been frequently not closely related. This, along with earlier observations based on morphological data (Pilloud, Larsen 2011) and mitochondrial DNA analyses (Chyleński et al. 2019), raises doubts regarding a fully kin-centred social organisation at Çatalhöyük. Indeed, they contradict the hypothesis that extended kinship-based families lived and were buried within the Neolithic houses (Kuijt et al. 2011). Instead, the results appear more consistent with the 'history house' concept proposed by Ian Hodder and Peter Pels (Hodder and Pels 2010), where biologically non-kin can become related, that is, can become 'practical kin', through their identification with the physical structure of a 'house', which in fact is the primary agent of making and naturalising relations between people and place. Hodder suggests that those people who were buried in a particular building need not even have lived within that architectural structure and may involve 'adoptive, foster or fictive kin held together by memory and history making' (Hodder 2016). At the same time, based on Jessica Pearson's studies (Pearson 2013), it is possible to say that a social 'house' may incorporate more than one building, linked through co-eating and co-burying. Now, the goal before us is to generate a full-scale map of population dynamics and traditions during the Neolithic, through comprehensive and systematic sampling for ancient genomics and integrated analysis of this data with archaeological and bioarchaeological datasets.

Acknowledgements

The authors acknowledge support from TÜBİTAK (grant no. 117Z229), the European Research Council (ERC) (grant no. ERC-772390 NEOGENE), the National Genomics Infrastructure in Stockholm funded by Science for Life Laboratory, the Knut and Alice Wallenberg Foundation and the Swedish Research Council, and SNIC/Uppsala Multidisciplinary Center for Advanced Computational Science for assistance with massively parallel sequencing and access to the UPPMAX computational infrastructure. This research received financial support from the French State via the 'Investments for the Future' framework programme, Initiative d'Excellence de l'Université de Bordeaux (IdEx), (Award No. ANR-10-IDEX-03-02 to CJK).

Bibliography

Aas, J.A., Griffen, A.L., Dardis, S.R., Lee, A.M., Olsen, I., Dewhirst, F.E., Leys, E.J., Paster, B.J. 2008: 'Bacteria of dental caries in primary and permanent teeth in children and young adults' *Journal of Clinical Microbiology* 46: 1407–17

Adachi, N., Dodo, Y., Oshshima, N., Doi, N., Yoneda, M., Matsumura, H. 2003: 'Morphologic and genetic evidence for the kinship of juvenile skeletal specimens from a 2,000 year-old double burial of the Usu-Moshiri site, Hokkaido, Japan' *Anthropological Science* 111: 347–63

Agarwal, S.C., Beauchesne, P., Glencross, B., Larsen, C.D., Meskell, L., Nakamura, C., Pearson, J.A., Sadvari, J.W. 2015: 'Roles for sexes' in I. Hodder, A. Marciniak (eds), *Assembling Çatalhöyük*. Leeds, Maney Publishing: 87–96

Akkermans, P.M.M.G. 1987: 'A late Neolithic and early Halaf village at Sabi Abyad, Northern Syria' *Paléorient* 13.1: 23–40

— 2008: 'Burying the dead in Late Neolithic Syria' in J.M. Cordoba, M. Molist, C. Perez, I. Rubio, S. Martinez (eds), *Proceedings of the 5th International Congress on the Archaeology of the Ancient Near East*. Madrid, Universidad Autónoma of Madrid: 621–45

Albarella, U. 1995: 'Depressions on sheep horncores' *Journal of Archaeological Science* 22: 699–704

— 2002: '"Size matters": how and why biometry is still important in zooarchaeology' in K. Dobney, T. O'Connor (eds), *Bones and the Man: Studies in Honour of Don Brothwell*. Walnut Creek, Oxbow Books: 51–62

Albert, A., Greene, D. 1999: 'Bilateral asymmetry in skeletal growth and maturation as an indicator of environmental stress' *American Journal of Physical Anthropology* 110: 341–49

Albert, R.M., Weiner, S. 2001: 'Study of phytoliths in prehistoric ash layers using a quantitative approach' in J.D. Meunier, F. Colin (eds), *Phytoliths, Applications in Earth Sciences and Human History*. Lisse, A. A. Balkema Publishers: 251–66

Albert, R.M., Shahack-Gross, R., Cabanes, D., Gilboa, A., Lev-Yadun, S., Portillo, M., Sharon, I., Boaretto, E., Weiner, S. 2008: 'Phytolith-rich layers from the Late Bronze and Iron Ages at Tel Dor (Israel): mode of formation and archaeological significance' *Journal of Archaeological Science* 35: 57–75

Albert, R.M., Ruiz, J.A., Sans, A. 2016: 'PhytCore ODB: a new tool to improve efficiency in the management and exchange of information on phytoliths' *Journal of Archaeological Science* 68: 98–105

Aldridge, D.C. 1999: 'The morphology, growth and reproduction of Unionidae (Bivalvia) in a fenland waterway' *Journal of Molluscan Studies* 65: 47–60

Alexandrowicz, S.W. 1999: '*Bithynia tentaculata* (Linnaeus, 1758) as an indicator of age and deposition environment of quaternary sediments' *Folia Malacologica* 7.2: 79–88

Allaby, R.G., Stevens, C., Lucas, L., Maeda, O., Fuller, D.Q. 2017: 'Geographic mosaics and changing rates of cereal domestication' *Philosophical Transactions of the Royal Society B: Biological Sciences* 372. https://doi.org/10.1098/rstb.2016.0429

Allentoft, M.E., Collins, M., Harker, D., Haile, J., Oskam, C.L., Hale, M.L., Campos, P.F., Samaniego, J.A., Gilbert, M.T.P., Willerslev, E., Zhang, G., Scofield, R.P., Holdaway, R.N., Bunce, M. 2012: 'The half-life of DNA in bone: measuring decay kinetics in 158 dated fossils' *Proceedings of the Royal Society B: Biological Sciences* 279: 4724–33. https://doi.org/10.1098/rspb.2012.1745

Allentoft, M.E., Sikora, M., Sjögren, K.-G., Rasmussen, S., Rasmussen, M., Stenderup, J., Damgaard, P.B., Schroeder, H., Ahlström, T., Vinner, L., Malaspinas, A.-S., Margaryan, A., Higham, T., Chivall, D., Lynnerup, N., Harvig, L., Baron, J., Della Casa, P., Dąbrowski, P., Duffy, P.R., Ebel, A.V., Epimakhov, A., Frei, K., Furmanek, M., Gralak, T., Gromov, A., Gronkiewicz, S., Grupe, G., Hajdu, T., Jarysz, R., Khartanovich, V., Khokhlov, A., Kiss, V., Kolář, J., Kriiska, A., Lasak, I., Longhi, C., McGlynn, G., Merkevicius, A., Merkyte, I., Metspalu, M., Mkrtchyan, R., Moiseyev, V., Paja, L., Pálfi, G., Pokutta, D., Pospieszny, Ł., Price, T.D., Saag, L., Sablin, M., Shishlina, N., Smrčka, V., Soenov, V.I., Szeverényi, V., Tóth, G., Trifanova, S.V., Varul, L., Vicze, M., Yepiskoposyan, L., Zhitenev, V., Orlando, L., Sicheritz-Pontén, T., Brunak, S., Nielsen, R., Kristiansen, K., Willerslev, E. 2015: 'Population genomics of Bronze Age Eurasia' *Nature* 522.7555: 167–72

Alley, R.B., Águstsdóttir, A.M. 2005: 'The 8k event: cause and consequence of a major Holocene abrupt climate change' *Quaternary Science Reviews* 24: 1123–49

AlQahtani, S.J., Hector, M.P., Liversidge, H.M. 2010. 'Brief communication: the London atlas of human tooth development and eruption' *American Journal of Physical Anthropology* 142.3: 481–90

Alt, K.W., Benz, M., Vach, W., Simmons, T.L., Goring-Morris, A.N. 2015: 'Insights into the social structure of the PPNB site of Kfar HaHoresh, Israel, based on dental remains' *PLoS ONE* 10.9: e0134528

Altınsaçlı, S., Altınsaçlı, S., Pacal, F.P. 2014: 'Macrophytic flora of the lakes Balıklı and Kayı in Gülağac, Aksaray, Turkey' *Asian Journal of Plant Science and Research* 4.2: 13–21

Altman, J., Hedl, R., Szabó, P., Mazůrek, P., Riedl, V., Müllerová, J., Kopecký, M., Doležal, J. 2013: 'Tree-rings mirror management legacy: dramatic response of standard oaks to past coppicing in central Europe' *PLoS ONE* 8.2: 1–11

Ambrose, S.H. 1990: 'Preparation and characterization of bone and tooth collagen for isotopic analysis' *Journal of Archaeological Science* 17: 431–51

— 1991: 'Effects of diet, climate and physiology on nitrogen isotope abundances in terrestrial food webs' *Journal of Archaeological Science* 18: 293–317

— 1993: 'Isotopic analysis of paleodiets: methodological and interpretative considerations' in M.K. Sandford (ed.), *Investigations of Ancient Human Tissue: Chemical Analyses in Anthropology*. Pennsylvania, Gordon and Breach Science Publishers: 59–130

— 2000: 'Controlled diet and climate experiments on nitrogen isotope ratios of rats' in S.H. Ambrose, M.A. Katzenberg (eds), *Biogeochemical Approaches to Paleodietary Analysis* (Advances in Archaeological and Museum Science 5). New York, Kluwer Academic/Plenum Publishers: 243–63

Ambrose, S.H., Norr, L. 1993: 'Experimental evidence for the relationship of the carbon isotope ratios of whole diet and dietary protein to those of bone collagen and carbonate' in J. Lambert, G. Grupe (eds), *Prehistoric Human Bone: Archaeology at the Molecular Level*. Berlin, Springer-Verlag: 1–37

Anastasiou, E., Mitchell, P.D. 2013: 'Simplifying the process of extracting intestinal parasite eggs from archaeological sediment samples: a comparative study of the efficacy of widely-used disaggregation techniques' *International Journal of Paleopathology* 3: 204–07

Ancient Human DNA - uMap. n.d. http://umap.openstreetmap.fr/en/map/ancient-human-dna_41837#2/-0.7/56.6

Anderson, D.G., Lovers, J.P.L., Schoer, S.A., Wishart, R.P. 2017: 'Architectures of domestication: on emplacing human-animal relations in the North' *Journal of the Royal Anthropological Institute* 23: 398–418

Anderson, E., Almond, M.J., Matthews, W. 2014: 'Analysis of wall plasters and natural sediments from the Neolithic town of Çatalhöyük (Turkey) by a range of analytical techniques' *Spectrochimica Acta Part A: Molecular and Biomolecular Spectroscopy* 133: 326–34

Anderson, S., Ertug-Yaras, F. 1998: 'Fuel fodder and faeces: an ethnographic and botanical study of dung fuel use in Central Anatolia' *Environmental Archaeology* 1: 99–109

Andrews, P. 1990: *Owls, Caves and Fossils*. London, Natural History Museum Publications

Andrews, P., Bello, S. 2006: 'Pattern in human burial practice' in C. Knüsel, R. Gowland (eds), *The Social Archaeology of Funerary Remains*. Oxford, Oxbow Books: 14–29

Andrews, P., Molleson, T., Boz, B. 2005: 'The human burials at Çatalhöyük' in I. Hodder (ed.), *Inhabiting Çatalhöyük: Reports from the 1995–99 Seasons*. London, British Institute at Ankara; Cambridge, McDonald Institute for Archaeological Research: 261–78

AnimalBase Project Group. 2005–2019: AnimalBase. Early Zoological Literature Online. www.animalbase.uni-goettingen.de

Anspach, M.R. 2019: 'Paired leopards and encircled prey' in I. Hodder (ed.), *Violence and the Sacred in the Ancient Near East: Girardian Conversations at Çatalhöyük*. Cambridge, Cambridge University Press: 129–50

Anvari, J., Brady, J., Franz, I., Naumov, G., Orton, D., Ostaptchouk, S., Stroud, E., Willett, P.T., Rosenstock, E., Biehl, P.F. 2017: 'Continuous change: venturing into the early Chalcolithic at Çatalhöyük' in S.R. Steadman, G. McMahon (eds), *The Archaeology of Anatolia Volume II: Recent Discoveries (2015–2016)*. Newcastle upon Tyne, Cambridge Scholars Publishing: 6–39

Aranguren, B., Becattini, R., Mariotti Lippi, M., Revedin, A. 2007: 'Grinding flour in Upper Palaeolithic Europe (25000 years bp)' *Antiquity* 81: 845–55

Araujo, R., Gomez, I., Machordom, A. 2005: 'The identity and biology of *Unio mancus* Lamarck, 1819 (= *U. elongatulus*) (Bivalvia: Unionidae) in the Iberian Peninsula' *Journal of Molluscan Studies* 71: 25–31

Araus, J.L., Febrero, A., Buxó, R., Rodríguez-Ariza, M.O., Molina, F., Camalich, M.D., Martín, D., Voltas, J. 1997: 'Identification of ancient irrigation practices based on the carbon isotope discrimination of plant seeds: a case study from the South-East Iberian Peninsula' *Journal of Archaeological Science* 24: 729–40

Araus, J.L., Slafer, G.A., Romagosa, I., Molist, M. 2001: 'FOCUS: estimated wheat yields during the emergence of agriculture based on the carbon isotope discrimination of grains: evidence from a 10th millennium BP site on the Euphrates' *Journal of Archaeological Science* 28: 341–50

Arbuckle, B.S. 2013a: 'The late adoption of cattle and pig husbandry in Neolithic Central Turkey' *Journal of Archaeological Science* 40: 1805–15

— 2013b: 'Köşk Höyük Faunal Data' Released: 2013-05-17. Open Context. <http://opencontext.org/projects/05F5B702-2967-49B1-FEAA-9B2AA0184513> https://doi.org/10.6078/M74Q7RW8. ARK (Archive): https://n2t.net/ark:/28722/k2nc5sf2h

— 2013c: 'Erbaba Höyük and Suberde Zooarchaeology' Released: 2013-05-17. Open Context. <http://opencontext.org/projects/CDD40C27-62ED-4966-AF3D-E781DD0D4846> https://doi.org/10.6078/M70Z715B. ARK (Archive): https://n2t.net/ark:/28722/k21n7xp5p

Arbuckle, B.S., Atıcı, L. 2013: 'Initial diversity in sheep and goat management in Neolithic south-western Asia' *Levant* 45: 219–35

Arbuckle, B.S., Makarewicz, C.A. 2009: 'The early management of cattle (*Bos taurus*) in Neolithic central Anatolia' *Antiquity* 83: 669–86

Arbuckle, B.S., Kansa, S.W., Kansa, E., Orton, D., Çakırlar, C., Gourichon, L., Atıcı, L., Galik, A., Marciniak, A., Mulville, J., Buitenhuis, H., Carruthers, D., de Cupere, B., Demirergi, A., Frame, S., Helmer, D., Martin, L., Peters, J., Pöllath, N., Pawłowska, K., Russell, N., Twiss, K., Würtenberger, D. 2014: 'Data sharing reveals complexity in the westward spread of domestic animals across Neolithic Turkey' *PLoS ONE* 9: e99845

Arbuckle, B.S., Price, M.D., Hongo, H., Öksüz, B. 2016: 'Documenting the initial appearance of domestic cattle in the Eastern Fertile Crescent (northern Iraq and western Iran)' *Journal of Archaeological Science* 72: 1–9

Armelagos, G.J., Goodman, A.H. 1991: 'The concept of stress and its relevance to studies of adaptation in prehistoric populations' *Collegium Anthropologicum* 15: 45–58

Armelagos, G.J., Brown, P.J., Turner, B. 2005: 'Evolutionary, historical and political economic perspectives on health and disease' *Social Science and Medicine* 61: 755–65

Arslan, N., Kara, D., Odabaşı, D.A. 2013: 'Twelve new records (Clitellata, Chironomidae and Gastropoda) from Lake Gölbaşı (Hatay-Turkey)' *Turkish Journal of Fisheries and Aquatic Sciences* 13: 869–73

Asimov, I. 1954: *Second Foundation*. New York: Gnome Press

Asouti, E. 2003: 'Woodland vegetation and fuel exploitation at the prehistoric campsite of Pınarbaşı, south-central Anatolia, Turkey: the evidence from the wood charcoal macro-remains' *Journal of Archaeological Science* 30.9: 1185–201

— 2005: 'Woodland vegetation and the exploitation of fuel and timber at Neolithic Çatalhöyük: report on the wood charcoal macro-remains' in I. Hodder (ed.), *Inhabiting Çatalhöyük: Reports from the 1995–99 Seasons*. London, British Institute at Ankara; Cambridge, McDonald Institute for Archaeological Research: 213–58

— 2013: 'Woodland vegetation, firewood management and woodcrafts at Neolithic Çatalhöyük' in I. Hodder (ed.), *Humans and Landscapes of Çatalhöyük: Reports from the 2000–2008 Seasons*. London, British Institute at Ankara; Los Angeles, Cotsen Institute of Archaeology Press: 129–61

Asouti, E., Austin, P. 2005: 'Reconstructing woodland vegetation and its exploitation by past societies, based on the analysis and interpretation of archaeological wood charcoal macro-remains' *Environmental Archaeology* 10.1: 1–18

Asouti, E., Fuller, D.Q. 2013: 'A contextual approach to the emergence of agriculture in Southwest Asia: reconstructing early Neolithic plant-food production' *Current Anthropology* 54: 299–345

Asouti, E., Kabukcu, C. 2014: 'Holocene semi-arid oak woodlands in the Irano-Anatolian region of Southwest Asia: natural or anthropogenic?' *Quaternary Science Reviews* 90: 158–82

Aufderheide, A.C. 2003: *The Scientific Study of Mummies*. Cambridge, Cambridge University Press

Aufderheide, A.C., Rodríguez-Martín C. 1998: *The Cambridge Encyclopaedia of Human Palaeopathology*. Cambridge, Cambridge University Press

Ayala, G., Wainwright, J., Walker, J., Hodara, R., Lloyd, J.M., Leng, M., Doherty, C. 2017: 'Palaeoenvironmental reconstruction of the alluvial landscape of Neolithic Çatalhöyük, central southern Turkey: the implications for early agriculture and responses to environmental change' *Journal of Archaeological Science* 87: 30–43

Bailit, H.L., Workman, P.L., Niswander, J.D., Maclean, C.J. 1970: 'Dental asymmetry as an indicator of genetic and environmental conditions in human populations' *Human Biology* 42: 626–38

Baird, D. 2002: 'Early Holocene settlement in Central Anatolia: problems and prospects as seen from the Konya Plain' in F. Gérard, L. Thissen (eds), *The Neolithic of Central Anatolia: Internal Developments and External Relations During the 9th–6th Millennia cal BC, Proceedings of the International CANeW Round Table, Istanbul, 23–24 November 2001*. Istanbul, Ege Yayınları: 139–59

— 2005: 'The history of settlement and social landscapes in the Early Holocene in the Çatalhöyük area' in I. Hodder (ed.), *Çatalhöyük Perspectives: Themes from the 1995–99 Seasons*. London, British Institute at Ankara; Cambridge, McDonald Institute for Archaeological Research: 55–74

— 2007a: 'The Boncuklu Project: the origins of sedentism, cultivation and herding in Central Anatolia' *Anatolian Archaeology* 13: 14–18

— 2007b: 'Pınarbaşı: from Epipalaeolithic camp site to sedentarising village in central Anatolia' in M. Özdoğan and N. Başgelen (eds), *The Neolithic in Turkey: New Excavations and New Discoveries*. Istanbul: Arkeoloji ve Sanat Yayinlari: 285–311

— 2012: 'The Late Epipaleolithic, Neolithic and Chalcolithic of the Anatolian Plateau, 13,000–4,000 BC' in D.T. Potts (ed.), *A Companion to the Archaeology of the Ancient Near East* I. Hoboken, Wiley-Blackwell: 431–65

Baird, D., Carruthers, D., Fairbairn, A., Pearson, J. 2011: 'Ritual in the landscape: evidence from Pınarbaşı in the seventh-millennium cal BC Konya Plain' *Antiquity* 85: 380–94

Baird, D., Fairbairn, A., Martin, L., Middleton, C. 2012: 'The Boncuklu Project; the origins of sedentism, cultivation and herding in Central Anatolia' in M. Özdoğan, N. Başgelen, P. Kuniholm (eds), *The Neolithic in Turkey: New Excavations and New Research* 3. *Central Turkey*. Istanbul, Ege Yayınları: 219–44

Baird, D., Fairbairn, A., Jenkins, E., Martin, L., Middleton, C., Pearson, J., Asouti, E., Edwards, Y., Kabukcu, C., Mustafaoğlu, G., Russell, N., Bar-Yosef, O., Jacobsen, G., Wu, X., Baker, A., Elliott, S. 2018: 'Agricultural origins on the Anatolian plateau' *Proceedings of the National Academy of Sciences* 115: E3077–86

Ball, T., Chandler-Ezell, K., Duncan, N., Dickau, R., Hart, T.C., Iriarte, J., Lentfer, C., Logan, A., Lu, H., Madella, M., Pearsall, D.M., Piperno, D., Rosen, A.M.,Vrydaghs, L., Weisskopf, A., Zhang, J. 2015: 'Phytoliths as a tool for investigations of agricultural origins and dispersals around the world' *Journal of Archaeological Science* 68: 32–45

Banning, E.B. 1998: 'The Neolithic Period: triumphs of architecture, agriculture and art' *Near Eastern Archaeology* 61.4: 188–237

Barański, M. 2016: 'GDN Research on the Late Neolithic architecture' *Çatalhöyük Archive Report 2016*. http://www.catalhoyuk.com/archive_reports/2016

Barański, M.Z., García-Suárez, A., Klimowicz, A., Love, S., Pawłowska, K. 2015: 'The architecture of Neolithic Çatalhöyük as a process: complexity in apparent simplicity' in I. Hodder, A. Marciniak (eds), *Assembling Çatalhöyük*. Leeds, Maney Publishing: 111–126

Barboza, P.S., Farley, S.D., Robbins, C.T. 1997: 'Whole-body urea cycling and protein turnover during hyperphagia and dormancy in growing bears (*Ursus americanus* and *U-arctos*)' *Canadian Journal of Zoology* 75: 2129–36

Barker, D.J.P. 2004: 'The developmental origins of adult disease' *Journal of the American College of Nutrition* 23: 588–95

Barrett, C.K., Guatelli-Steinberg, D., Sciulli, P.W. 2012: 'Revisiting dental fluctuating asymmetry in neandertals and modern humans' *American Journal of Physical Anthropology* 149: 193–204

Bartel, B. 1982: 'A historical review of ethnological and archaeological analyses of mortuary practice' *Journal of Anthropological Archaeology* 1: 32–58

Bar-Yosef Mayer, D.E. 2013: 'Mollusc exploitation at Çatalhöyük' in I. Hodder (ed.), *Humans and Landscapes of Çatalhöyük: Reports from the 2000–2008 Seasons*. London, British Institute at Ankara; Los Angeles, Cotsen Institute of Archaeology Press: 329–38

Bar-Yosef Mayer, D.E., Leng, M.J., Aldridge, D.C., Arrowsmith, C., Gümüs¸ B.A., Sloane, H.J. 2012: 'Modern and early-middle Holocene shells of the freshwater mollusc *Unio*, from Çatalhöyük in the Konya Basin, Turkey: preliminary palaeoclimatic implications from molluscan isotope data' *Journal of Archaeological Science* 39: 76–83

Bar-Yosef Mayer, D.E., Leng, M.J., Aldridge, D.C., Arrowsmith, C., Gümüş, B.A., Sloane, H.J. 2013: '*Unio* shells from Çatalhöyük: preliminary palaeoclimatic data from isotopic analyses' in I. Hodder (ed.), *Humans and Landscapes of Çatalhöyük: Reports from the 2000–2008 Seasons*. London, British Institute at Ankara; Los Angeles, Cotsen Institute of Archaeology Press: 87–91

Bar-Yosef, O. 2002: 'The Upper Palaeolithic revolution' *Annual Review of Anthropology* 31: 363–93

Bayliss, A., Brock, F., Farid, S., Hodder, I., Southon, J., Taylor, R.E. 2015: 'Getting to the bottom of it all: a Bayesian approach to dating the start of Çatalhöyük' *Journal of World Prehistory* 28: 1–26

Beckett, R.G., Nelson, A.J. 2015: 'Mummy restoration project among the Anga of Papua New Guinea' *The Anatomical Record* 298: 1013–25

Beckett, R.G., Lohmann U., Bernstein, J. 2011: 'A field report on the mummification practices of the Anga of Koke Village, Central Highlands, Papua New Guinea' *Yearbook of Mummy Studies* 1: 11–17

Bell, L.S. (ed.) 2012: *Forensic Microscopy for Skeletal Tissues*. New York, Springer

Bentley, R.A. 2006: 'Strontium isotopes from the earth to the archaeological skeleton: a review' *Journal of Archaeological Method and Theory* 13: 135–87

Bentley, R.A., Price, T.D., Stephan, E. 2004: 'Determining the "local" $^{87}Sr/^{86}Sr$ range for archaeological skeletons: a case study from Neolithic Europe' *Journal of Archaeological Science* 31: 365–75

Berggren, Å., Dell'Unto, N., Forte, M., Haddow, S., Hodder, I., Issavi, J., Lercari, N., Mazzucato, C., Mickel, A. and Taylor, J.S. 2015: 'Revisiting reflexive archaeology at Çatalhöyük: integrating digital and 3D technologies at the trowel's edge' *Antiquity* 89: 433–48

Bergman, J. 2018: 'Stone age disease in the north – human intestinal parasites from a Mesolithic burial in Motala, Sweden' *Journal of Archaeological Science* 96: 26–32

Best, J., Presslee, S., Demarchi, B. 2015: 'The avian eggshell from Çatalhöyük' *Çatalhöyük 2015 Archive Report*. http://www.catalhoyuk.com/archive_reports/2015

Bethony, J., Brooker, S., Albonico, M., Geiger, S.M., Loukas, A., Diemert, D., Hotez, P.J. 2006: 'Soil-transmitted helminth infections: ascariasis, trichuriasis, and hookworm' *Lancet* 367: 1521–32

Bıçakçı, E., Godon, M., Çakan, Y.G. 2012: 'Tepecik-Çiftlik' in M. Özdoğan, N. Başgelen, P. Kuniholm (eds), *The Neolithic in Turkey: New Excavations and New Research* 3. *Central Turkey*. Istanbul, Arkeoloji ve Sanat Yayınları: 89–134

Biehl, P. 2012: 'The transition of the megasite Çatalhöyük in the Late Neolithic and Early Chalcolithic' in R. Matthews, J. Curtis (eds), *Proceedings of the 7th International Congress on the Archaeology of the Ancient Near East* 1. *Mega-Cities and Mega-Sites. The Archaeology of Consumption and Disposal, Landscape, Transport and Communication*. Wiesbaden, Harrassowitz: 17–34

Bienert, H.-D. 1991: 'Skull cult in the prehistoric Near East' *Journal of Prehistoric Religion* 5: 9–23

Bienert, H.-D., Bonogofsky, M., Gebel, H.-G.K., Kuijt, I., Rollefson, G.O. 2004: 'Where are the dead?' in H.-D. Bienert, H.-G.K. Gebel, R. Neef (eds), *Central Settlements in Neolithic Jordan* (Studies in Early Near Eastern Production, Subsistence and Environment 5). Berlin, ex-oriente: 157–76

Bigalke, A. 1973: 'The exploitation of shellfish by coastal tribesmen of the Transkei' *Annals of the Cape Provincial Museums (Natural History)* 9: 159–75

Binski, P. 1996: *Medieval Death: Ritual and Representation*. London, British Museum Press

Birck, J.L. 1986: 'Precision K–Rb–Sr isotopic analysis—application to Rb–Sr chronology' *Chemical Geology* 56: 73–83

Bleicher, N. 2014: 'Four levels of patterns in tree-rings: an archaeological approach to dendroecology' *Vegetation History and Archaeobotany* 23: 615–27

Bloch, M. 2010: 'Is there religion at Çatalhöyük...or are there just houses?' in I. Hodder (ed.), *Religion and the Emergence of Civilization: Çatalhöyük as a Case Study*. Cambridge, Cambridge University Press: 146–62

Bocaege, E. 2015: *Childhood Growth in the Neolithic: A Detailed Case Study of Çatalhöyük*. Ph.D. thesis, University College London, London

Bocquentin, F., Kodaş, E., Ortiz, A. 2016: 'Headless but still eloquent! Acephalous skeletons as witnesses of Pre-Pottery Neolithic North-South Levant connections and disconnections' *Paléorient* 42.2: 35–55

Bocquet-Appel, J.-P. 2002: 'Paleoanthropological traces of a Neolithic demographic transition' *Current Anthropology* 43: 637–50

— 2008: 'Explaining the Neolithic demographic transition' in J.-P. Bocquet-Appel, O. Bar-Yosef (eds), *The Neolithic Demographic Transition and Its Consequences*. New York, Springer: 35–55

— 2011: 'When the world's population took off: the springboard of the Neolithic demographic transition' *Science* 333: 560–61

Boessneck, J. 1969: 'Osteological differences between sheep (*Ovis aries Linné*) and goat (*Capra hircus Linné*)' in D.R. Brothwell, E.S. Higgs (eds), *Science in Archaeology: A Survey of Progress and Research* (2nd edition). London, Thames & Hudson: 311–58

Bogaard, A. 2005: '"Garden agriculture" and the nature of early farming in Europe and the Near East' *World Archaeology* 37: 177–96

— 2011: *Plant Use and Crop Husbandry in an Early Neolithic Village: Vaihingen an der Enz, Baden-Württemberg* (Frankfurter Archäologische Schriften). Bonn, Habelt-Verlag

— 2017: 'Neolithic "cooperatives": assessing supra-household cooperation in crop production at Çatalhöyük and beyond' in M. Benz, H.-G.K. Gebel, T. Watkins (eds), *The Construction of Neolithic Corporate Identities (ICAANE Workshop Basel, June 9–10th, 2014)*. Berlin, ex oriente: 119–33

Bogaard, A., Isaakidou, V. 2010: 'From megasites to farmsteads: community size and the nature of early farming in the Near East and Europe' in B. Finlayson, G. Warren (eds), *Landscapes in Transition*. Oxford, Oxbow: 192–207

Bogaard, A., Heaton, T.H.E., Poulton, P., Merbach, I. 2007: 'The impact of manuring on nitrogen isotope ratios in cereals: archaeological implications for reconstruction of diet and crop management practices' *Journal of Archaeological Science* 34: 335–43

Bogaard, A., Charles, M., Twiss, K.C., Fairbairn, A., Yalman, N., Filipović, D., Demirergi, A., Ertuğ, F., Russell, N., Henecke, J. 2009: 'Private pantries and celebrated surplus: storing and sharing food at Neolithic Çatalhöyük, Central Anatolia' *Antiquity* 83: 649–68

Bogaard, A., Charles, M., Livarda, A., Ergun, M., Filipović, D., Jones, G. 2013: 'The archaeobotany of mid-later occupation levels at Neolithic Çatalhöyük' in I. Hodder (ed.), *Humans and Landscapes of Çatalhöyük: Reports from the 2000–2008 Seasons*. London, British Institute at Ankara; Los Angeles, Cotsen Institute of Archaeology Press: 93–128

Bogaard, A., Henton, E., Evans, J.A., Twiss, K.C., Charles, M.P., Vaiglova, P., Russell, N. 2014a: 'Locating land use at Neolithic Çatalhöyük, Turkey: the implications of ^{87}Sr/^{86}Sr signatures in plants and sheep tooth sequences' *Archaeometry* 56.5: 860–77

Bogaard, A., Ryan, P., Yalman, N., Asouti, E., Twiss, K.C., Mazzucato, C., Farid, S. 2014b: 'Assessing outdoor activities and their social implications at Çatalhöyük' in I. Hodder (ed.), *Integrating Çatalhöyük: Themes from the 2000–2008 Seasons*. London, British Institute at Ankara; Los Angeles, Cotsen Institute of Archaeology Press: 123–48

Bogaard, A., Fuller, D.Q., Filipović, D., Charles, M. 2015: 'Macro- and micro-botanical remains' *Çatalhöyük Archive Report 2015*. http://www.catalhoyuk.com/archive_reports/2015

Bogaard, A., Hodgson, J., Nitsch, E., Jones, G., Styring, A., Diffey, C., Pouncett, J., Herbig, C., Charles, M., Ertuğ, F., Tugay, O., Filipović, D., Fraser, R. 2016: 'Combining functional weed ecology and crop stable isotope ratios to identify cultivation intensity: a comparison of cereal production regimes in Haute Provence, France and Asturias, Spain' *Vegetation History and Archaeobotany* 25: 57–73

Bogaard, A., Filipović, D., Fairbairn, A., Green, L., Stroud, E., Fuller, D., Charles, M. 2017: 'Agricultural innovation and resilience in a long-lived early farming community: the 1,500-year sequence at Neolithic to early Chalcolithic Çatalhöyük, Central Anatolia' *Anatolian Studies* 67: 1–28

Bogaard, A., Styring, A., Ater, M., Hmimsa, Y., Green, L., Stroud, E., Whitlam, J., Diffey, C., Nitsch, E., Charles, M., Jones, G., Hodgson, J. 2018: 'From traditional farming in Morocco to early urban agroecology in northern Mesopotamia: combining present-day arable weed surveys and crop isotope analysis to reconstruct past agrosystems in (semi-)arid regions' *Environmental Archaeology* 23: 303–22

Bogin, B. 1999: *Patterns of Human Growth* (2nd edition). Cambridge, Cambridge University Press

Bogin, B., Loucky, J. 1997: 'Plasticity, political economy, and physical growth status of Guatemala Maya children living in the United States' *American Journal of Physical Anthropology* 102: 17–32

Bogutskaya, N.G. 1992: 'A revision of species of the genus *Pseudophoxinus* (Leuciscinae, Cyprinidae) from Asia Minor' *Mitteilungen aus dem Hamburgischen Zoologischen Museum und Institut* 89: 61–290

Boivin, N. 2000: 'Life rhythms and floor sequences: excavating time in rural Rajasthan and Neolithic Çatalhöyük' *World Archaeology* 31: 367–88

Boldsen, J.L, Milner, G.R., Weise, S. 2015: 'Cranial vault trauma and selective mortality in medieval to early modern Denmark' *Proceedings of the National Academy of Sciences of the USA* 12.6: 1721–26

Bonogofsky, M. 2003: 'Neolithic plastered skulls and railroading epistemologies' *Bulletin of the American Schools of Oriental Research* 331: 1–10

— 2004: 'Including women and children: Neolithic modeled skulls from Jordan, Israel, Syria and Turkey' *Near Eastern Archaeology* 67.2: 118–19

— 2005: 'A bioarchaeological study of plastered skulls from Anatolia: new discoveries and interpretations' *International Journal of Osteoarchaeology* 15: 124–35

— 2006: 'Complexity in context: plain, painted and modeled skulls from the Neolithic Middle East' in M. Bonogofsky (ed.), *Skull Collection, Modification and Decoration* (BAR International Series 1539). Oxford, Archaeopress: 15–28

Booth, T.J. 2016: 'An investigation into the relationship between funerary treatment and bacterial bioerosion in European archaeological human bone' *Archaeometry* 58.3: 484–99

Booth, T.J., Chamberlain, A.T., Parker Pearson, M. 2015: 'Mummification in Bronze Age Britain' *Antiquity* 89.347: 1155–73

Borgerhoff Mulder, M., Fazzio, I., Irons, W., McElreath, R.L., Bowles, S., Bell, A., Hertz, T., Hazzah, L. 2010: 'Pastoralism and wealth inequality' *Current Anthropology* 51: 35–48

Bourdieu, P. 1977: *Outline of a Theory of Practice*. Cambridge, Cambridge University Press

Boyer, P. 1999: *A Geoarchaeological Approach to Late Quaternary Environmental Change in South Central Turkey*. PhD thesis, Loughborough University, Loughborough

Boyer, P., Roberts, N., Baird, D. 2006: 'Holocene environment and settlement on the Çarşamba alluvial fan, south-central Turkey: integrating geoarchaeology and archaeological field survey' *Geoarchaeology* 21.7: 675–98

Boyer, P., Roberts, N., Merrick, J. 2007: 'KOPAL excavations at Çatalhöyük 1996–2000' in I. Hodder (ed.), *Excavating Çatalhöyük: South, North and KOPAL Area Reports from the 1995–99 Seasons*. London, British Institute at Ankara; Cambridge, McDonald Institute for Archaeological Research: 551–70

Boz, B., Hager, L.D. 2004: 'Human remains' *Çatalhöyük 2004 Archive Report*. http://www.catalhoyuk.com/archive_reports/2004

— 2013: 'Living above the dead: intramural burial practices at Çatalhöyük' in I. Hodder (ed.), *Humans and Landscapes of Çatalhöyük: Reports from the 2000–2008 Seasons*. London, British Institute at Ankara; Los Angeles, Cotsen Institute of Archaeology Press: 413–40

— 2014: 'Making sense of social behaviour from disturbed and commingled skeletons: a case study from Çatalhöyük, Turkey' in A.J. Osterholtz, K.M. Baustian, D.L. Martin (eds), *Commingled and Disarticulated Human Remains: Working toward Improved Theory, Method and Data*. New York, Springer: 17–33

Brace, S., Diekmann, Y., Booth, T.J., van Dorp, L., Faltyskova, Z., Rohland, N., Mallick, S., Olalde, I., Ferry, M., Michel, M., Oppenheimer, J., Broomandkhoshbacht, N., Stewardson, K., Martiniano, R., Walsh, S., Kayser, M., Charlton, S., Hellenthal, G., Armit, I., Schulting, R., Craig, O.E., Sheridan, A., Parker Pearson, M., Stringer, C., Reich, D., Thomas, M.G., Barnes, I. 2019: 'Ancient genomes indicate population replacement in Early Neolithic Britain' *Nature Ecology & Evolution* 3: 765–71

Bradley, C. 1993: 'Women's power, children's labor' *Cross-Cultural Research* 27: 70–96

Bréhard, S., Beeching, A., Vigne, J.-D. 2010: 'Shepherds, cowherds and site function on middle Neolithic sites of the Rhône Valley: an archaeozoological approach to the organization of territories and societies' *Journal of Anthropological Archaeology* 29: 179–88

Brewer-Carias, C.A., LeBlanc, S.A., Neel, J.V. 1976: 'Genetic structure of a tribal population, the Yanomama Indians. XIII. Dental microdifferentiation' *American Journal of Physical Anthropology* 44.1: 5–14

Brewster, C., Meiklejohn, C., von Cramon-Taubadel, N., Pinhasi, R. 2014: 'Craniometric analysis of European Upper Palaeolithic and Mesolithic samples supports discontinuity at the Last Glacial Maximum' *Nature Communications* 5: 4094

Briggs, A.W., Stenzel, U., Johnson, P.L.F., Green, R.E., Kelso, J., Prü, K., Meyer, M., Krause, J., Ronan, M.T., Lachmann, M., Pä, S. 2007: 'Patterns of damage in genomic DNA sequences from a Neandertal' *Proceedings of the National Academy of Science of the United States of America* 104: 14616–21

Briggs, A.W., Stenzel, U., Meyer, M., Krause, J., Kircher, M., Pääbo, S. 2010: 'Removal of deaminated cytosines and detection of in vivo methylation in ancient DNA' *Nucleic Acids Research* 38: e87

Brink, O., Vesterby, A., Jense, J. 1998: 'Pattern of injuries due to interpersonal violence' *Injury* 29.9: 705–09

Brönmark, C., Hansson, L.-A. 2005: *The Biology of Lakes and Ponds*. Oxford, Oxford University Press

Brönnimann, D., Portmann, C., Pichler, S.L., Booth, T.J., Röder, B., Vach, W., Schibler, J., Rentzel, P. 2018: 'Contextualising the dead – combining geoarchaeology and osteo-anthropology in a new multi-focus approach in bone histotaphonomy' *Journal of Archaeological Science* 98: 45–58

Brooks, S., Suchey, J.M. 1990: 'Skeletal age determination based on the *os pubis*: a comparison of the Acsádi-Nemeskéri and Suchey-Brooks methods' *Human Evolution* 5.3: 227–38

Brothwell, D. 1981: 'The Pleistocene and Holocene archaeology of the house mouse and related species' *Symposia of the Zoological Society of London* 47: 1–13

Broushaki, F., Thomas, M.G., Link, V., López, S., van Dorp, L., Kirsanow, K., Hofmanová, Z., Diekmann, Y., Cassidy, L.M., Díez-del-Molino, D., Kousathanas, A., Sell, C., Robson, H.K., Martiniano, R., Blöcher, J., Scheu, A., Kreutzer, S., Bollongino, R., Bobo, D., Davoudi, H., Munoz, O., Currat, M., Abdi, K., Biglari, F., Craig, O.E., Bradley, D.G., Shennan, S., Veeramah, K.R., Mashkour, M., Wegmann, D., Hellenthal, G., Burger, J. 2016: 'Early Neolithic genomes from the eastern Fertile Crescent' *Science* 353: 499–503

Brown, D.A. 1984: 'Prospects and limits of a phytolith key for grasses in the central United States' *Journal of Archaeological Science* 11: 345–68

Brown, J.A. 1981: 'The search for rank in prehistoric burials' in R. Chapman, I. Kinnes, K. Randsborg (eds), *The Archaeology of Death*. London, Carlisle Press: 25–37

Brown, T.A., Nelson, D.E., Vogel, J.S., Sothon, J.R. 1988: 'Improved collagen extraction by modified Longin Method' *Radiocarbon* 30: 171–77

Buikstra J.E., Ubelaker, D.H. 1994: *Standards for Data Collection from Human Skeletal Remains*. Fayetteville, Arkansas Archaeological Survey

Bull, I.D., Lockheart, M., Elhmmali, M., Roberts, D., Evershed, R. 2002: 'The origin of faeces by means of biomarker detection' *Environment International* 27.8: 647–54

Bull, I.D., Elhmmali, M.M., Perret, V., Matthews, W., Roberts, D.J., Evershed, R.P. 2005: 'Biomarker evidence of faecal deposition in archaeological sediments at Çatalhöyük' in I. Hodder (ed.), *Inhabiting Çatalhöyük: Reports from the 1995–99 Seasons*. London, British Institute at Ankara; Cambridge, McDonald Institute for Archaeological Research: 415–20

Çakırlar, C. 2012: 'The evolution of animal husbandry in Neolithic Central-West Anatolia: the zooarchaeological record from Ulucak Höyük (c. 7040–5660 cal. BC, Izmir, Turkey)' *Anatolian Studies* 62: 1–33

— 2013a: 'Rethinking Neolithic subsistence at the gateway to Europe with new archaeozoological evidence from Istanbul' in M. Groot, D. Lenties, J. Zeiler (eds), *Barely Surviving or More than Enough? The Environmental Archaeology of Subsistence, Specialisation and Surplus Food Production*. Leiden, Sidestone Press: 59–79

— 2013b: 'Zooarchaeology of Neolithic Ulucak' Released: 2013-08-11. Open Context. <http://opencontext.org/projects/99BDB878-6411-44F8-2D7B-A99384A6CA21> https://doi.org/10.6078/M7KS6PHV ARK (Archive): https://n2t.net/ark:/28722/k2jw88k6d

Çamurcuoğlu, D. 2015: *The Wall Paintings of Çatalhöyük (Turkey): Materials, Technologies and Artists*. PhD thesis, University of London, London

Can, M.F., Şereflişan, H. 2018: 'Age, growth and mortality of *Unio tigridis* (Bourguignat, 1852)' *Turkish Journal of Agriculture – Food Science and Technology* 6.5: 613–17

Canti, M.G. 1999: 'The production and preservation of faecal spherulites: animals, environment and taphonomy' *Journal of Archaeological Science* 26: 251–58

Carpenter, M.L., Buenrostro, J.D., Valdiosera, C., Schroeder, H., Allentoft, M.E., Sikora, M., Rasmussen, M., Gravel, S., Guillén, S., Nekhrizov, G., Leshtakov, K., Dimitrova, D., Theodossiev, N., Pettener, D., Luiselli, D., Sandoval, K., Moreno-Estrada, A., Li, Y., Wang, J., Gilbert, M.T.P., Willerslev, E., Greenleaf, W.J., Bustamante, C.D. 2013: 'Pulling out the 1%: whole-genome capture for the targeted enrichment of ancient DNA sequencing libraries' *American Journal of Human Genetics* 93: 852–64

Carr, C. 1995: 'Mortuary practices: their social, philosophical-religious, circumstantial and physical determinants' *Journal of Archaeological Method and Theory* 2.2: 105–200

Carruthers, D. 2006: 'Pınarbaşı 1994: animal bones' Released: 2006-03-25. Open Context. <http://opencontext.org/projects/TESTPRJ0000000004> ARK (Archive): https://n2t.net/ark:/28722/k2zs2vk0z At link: https://opencontext.org/projects/TESTPRJ0000000004

Carter, T., Milić, M. 2014: 'The chipped stone: a visual characterisation of the Çatalhöyük obsidian' in I. Hodder (ed.), *Substantive Technologies at Çatalhöyük: Reports from the 2000–08 Seasons*. London, British Institute at Ankara; Los Angeles, Cotsen Institute of Archaeology Press: 417–78

Cassidy, L.M., Martiniano, R., Murphy, E.M., Teasdale, M.D., Mallory, J., Hartwell, B., Bradley, D.G. 2016: 'Neolithic and Bronze Age migration to Ireland and establishment of the insular Atlantic genome' *Proceedings of the National Academy of Sciences of the United States of America* 113.2: 368–73

Castex, D., Kacki, S. 2016: 'Demographic patterns distinctive of epidemic cemeteries in archaeological samples' *Microbiology Spectrum* 4.4: PoH-0015-2015

Cauvin, J. 2000: *The Birth of the Gods and the Origins of Agriculture*. Cambridge, Cambridge University Press

Cavalli-Sforza, L.L., Piazza, A., Menozzi, P., Mountain, J. 1988: 'Reconstruction of human evolution: bringing together genetic, archaeological, and linguistic data' *Proceedings of the National Academy of Sciences of the United States of America* 85: 6002–06

Ceci, L. 1984: 'Shell midden deposits as coastal resources' *World Archaeology* 16.1: 62–74

Çek, S., Şereflişan, H. 2006: 'Certain reproductive characteristics of the freshwater mussel *Unio terminalis delicates* (Lea, 1863) in Gölbaşi Lake, Turkey' *Aquaculture Research* 37: 1305–15

Cessford, C. 2005: 'Estimating the Neolithic population of Çatalhöyük' in I. Hodder (ed.), *Inhabiting Çatalhöyük: Reports From the 1995–99 Seasons*. London, British Institute at Ankara; Cambridge, McDonald Institute for Archaeological Research: 323–28

— 2007a: 'Overall discussion of Buildings 1 and 5' in I. Hodder (ed.), *Excavating Çatalhöyük: South, North and KOPAL Area Reports from the 1995–99 Seasons*. London, British Institute at Ankara; Cambridge, McDonald Institute for Archaeological Research: 531–50

— 2007b: 'Neolithic excavations in the North Area, East Mound, Çatalhöyük 1995–99' in I. Hodder (ed.), *Excavating Çatalhöyük: South, North and KOPAL Area: Reports from the 1995–99 Seasons*. London, British Institute at Ankara; Cambridge, McDonald Institute for Archaeological Research: 345–552

Çetik, A.R. 1985: *Türkiye Vejetasyonu I: İç Anadolunun Vejetasyon ve Ekolojisi*. Konya, Selçuk Üniversitesi Yayınları

Chamberlain, A. 2009: 'Archaeological demography' *Human Biology* 81.2–3: 275–87

Chapman, J. 2010: '"Deviant" burials in the Neolithic and Chalcolithic of central and south eastern Europe' in K. Rebay-Salisbury, M.L.S. Sørensen, J. Hughes (eds), *Body Parts and Bodies Whole: Changing Relations and Meanings*. Oxford, Oxbow Books: 30–45

Chapman, R., Kinnes, I., Randsborg, K. (eds) 1981: *The Archaeology of Death*. Cambridge, Cambridge University Press

Charles, M. 1998: 'Fodder from dung: the recognition and interpretation of dung-derived plant material from archaeological sites' *Environmental Archaeology* 1: 111–22

Charles, M., Bogaard, A. 2010: 'Charred plant macro-remains from Jeitun: implications for early cultivation and herding practices in western Central Asia' in D.R. Harris (ed.), *Excavations at Jeitun*. Philadelphia, University of Pennsylvania Museum Press: 150–65

Charles, M., Jones, G., Hodgson, J.G. 1997: 'FIBS in archaeobotany: functional interpretation of weed floras in relation to husbandry practices' *Journal of Archaeological Science* 24: 1151–61

Charles, M., Bogaard, A., Jones, G., Hodgson, J.G., Halstead, P. 2002: 'Towards the archaeobotanical identification of intensive cereal cultivation: present-day ecological investigation in the mountains of Asturias, northwest Spain' *Vegetation History and Archaeobotany* 11: 133–42

Charles, M., Doherty, C., Asouti, E., Bogaard, A., Henton, E., Larsen, C.S., Ruff, C.B., Ryan, P., Sadvari, J.W., Twiss, K.C. 2014: 'Landscape and taskscape at Çatalhöyük: an integrated perspective' in I. Hodder (ed.), *Integrating Çatalhöyük: Themes from the 2000–2008 Seasons*. London, British Institute at Ankara; Los Angeles, Cotsen Institute of Archaeology Press: 71–89

Chávez, A., Martínez, C., Soberanes, B. 2000: 'The effect of malnutrition on human development: a 24-year study of well-nourished and malnourished children living in a poor Mexican village' in A.H. Goodman, D.L. Dufour, G.H. Pelto (eds), *Nutritional Anthropology: Biocultural Perspectives on Food and Nutrition*. Mountain View, Mayfield Publishing: 234–52

Chow, V.T. 1959: *Open-Channel Hydraulics*. New York, McGraw-Hill

Christian, P., Mullany, L.C., Hurley, K.M., Katz, J., Black, R.E. 2015: 'Nutrition and maternal, neonatal, and child health' *Seminars in Perinatology* 39: 361–72

Chyleński, M., Ehler, E., Somel, M., Yaka, R., Krzewińska, M., Dabert, M., Juras, A., Marciniak, A. 2019: 'Ancient mitochondrial genomes reveal the absence of maternal kinship in the burials of Çatalhöyük people and their genetic affinities' *Genes* 10: E207

Çilingiroğlu, Ç., Çakırlar, C. 2013: 'Towards configuring the Neolithisation of Aegean Turkey' *Documenta Praehistorica* 40: 21–29

Claassen, C. 1998: *Shells*. Cambridge, Cambridge University Press

Cohen, A., Serjeantson, D. 1996: *A Manual for the Identification of Bird Bones from Archaeological Sites*. London: Archetype Publications

Cohen, M.N., Armelagos, G.J. (eds) 1984: *Paleopathology at the Origins of Agriculture*. Orlando, Academic Press

Collins C., Asouti E., Grove M., Kabukcu C., Bradley L., Chiverrell, R. 2018: 'Understanding resource choice at the transition from foraging to farming: an application of palaeodistribution modelling to the Neolithic of the Konya Plain, south-central Anatolia, Turkey' *Journal of Archaeological Science* 96: 57–72

Corcuera, L., Camarero, J.J., Gil-Pelegrín, E. 2004: 'Effects of a severe drought on growth and wood anatomical properties of *Quercus faginea*' *IAWA Journal* 25: 185–204

Cormie, A.B., Schwarcz, H.P. 1994: 'Stable isotopes of nitrogen and carbon of North American white-tailed deer and implications for paleodietary and other food web studies' *Palaeogeography, Palaeoclimatology, Palaeoecology* 107: 227–41

— 1996: 'Effects of climate on deer bone $\delta^{15}N$ and $\delta^{13}C$: lack of precipitation effects of $\delta^{15}N$ for animals consuming low amounts of C_4 plants' *Geochimica et Cosmochimica Acta* 60: 4161–66

Corruccini, R.S., Shimada, I., Shinoda, K.-I. 2002: 'Dental and mtDNA relatedness among thousand-year-old remains from Huaca Loro, Peru' *Dental Anthropology* 16: 9–14

Costa, R.L. Jr. 1986: 'Asymmetry of the mandibular condyle in Haida Indians' *American Journal of Physical Anthropology* 70: 119–23

Cowgill, L.W. 2014: 'Femoral diaphyseal shape and mobility: an ontogenetic perspective' in K.J. Carlson, D. Marchi (eds), *Reconstructing Mobility: Environmental, Behavioral, and Morphological Determinants*. New York, Springer: 188–94

Cowgill, L.W., Warrener, A., Pontzer, H., Ocobock, C. 2010: 'Waddling and toddling: the biomechanical effects of an immature gait' *American Journal of Physical Anthropology* 143: 52–61

Crawford, S. 1991: 'When do Anglo-Saxon children count?' *Journal of Theoretical Archaeology* 2: 17–24

— 1993: 'Children, death and the afterlife in Anglo-Saxon England' *Anglo-Saxon Studies in Archaeology and History* 6: 83–91

Crawford, S., Shepherd, G. 2007: *Children, Childhood and Society* (BAR International Series 1696). Oxford: Archaeopress

Crevecoeur, I., Schmitt, A., Schoep, I. 2015: 'An archaeothanatological approach to the study of Minoan funerary practices: case-studies from the Early and Middle Minoan cemetery at Sissi, Crete' *Journal of Field Archaeology* 40: 283–99

Crittenden, A.N., Zes, D.A. 2015: 'Food sharing among Hadza hunter-gatherer children' *PloS ONE* 10.7: e0131996

Croucher, K. 2006: 'Getting ahead: exploring meanings of skulls in the Neolithic Near East' in M. Bonogofsky (ed.), *Skull Collection, Modification and Decoration* (BAR International Series 1539). Oxford, Archaeopress: 29–44

— 2010: 'Bodies in pieces in the Neolithic Near East' in K. Rebay-Salisbury, M.L.S. Sørensen, J. Hughes (eds), *Body Parts and Bodies Whole: Changing Relations and Meanings*. Oxford, Oxbow Books: 6–19

Czajkowska, B. I., Bogaard, A., Charles, M., Jones, G., Kohler-Schneider, M., Mueller-Bieniek, A., Brown, T. A. 2020. 'Ancient DNA typing indicates that the "new" glume wheat of early Eurasian agriculture is a cultivated member of the *Triticum timopheevii* group' *Journal of Archaeological Science* 123: 105258.

Czeszewska, A. 2014: 'Wall paintings at Çatalhöyük' in I. Hodder (ed.), *Integrating Çatalhöyük: Themes from the 2000–2008 seasons*. London, British Institute at Ankara; Los Angeles, Cotsen Institute of Archaeology Press: 185–96

D'Ortenzio, L., Brickley, M., Schwarcz, H., Prowse, T. 2015: 'You are not what you eat during physiological stress: isotopic evaluation of human hair' *American Journal of Physical Anthropology* 157: 374–88

Dabney, J., Meyer, M., Pääbo, S. 2013: 'Ancient DNA damage' *Cold Spring Harbor Perspectives in Biology* 5.7: a012567

Dahl, G., Hjort, A. 1976: *Having Herds: Pastoral Herd Growth and Household Economy*. Stockholm, University of Stockholm

Dal Corso, M., Nicosia, C., Balista, C., Cupitò, M., Dalla Longa, E., Leonardi, G., Kirleis, W. 2017: 'Bronze Age crop processing evidence in the phytolith assemblages from the ditch and fen around Fondo Paviani, northern Italy' *Vegetation History and Archaeobotany* 26: 5–24

Damm, C. 1991: 'From burials to gender roles: problems and potentials in postprocessual archaeology' in I.D Walde, N.D. Willows (eds), *The Archaeology of Gender*. Calgary, The University of Calgary: 130–35

Dark, P. 2004: 'New evidence for the antiquity of the intestinal parasite *Trichuris* (whipworm) in Europe' *Antiquity* 78: 676–81

Darton, Y., Rodet-Belarbi, I. 2018: 'Damage caused by permanent fetters in present-day sheep on the island of Delos (Greece)' in L. Bartosiewicz, E. Gál (eds), *Care or Neglect? Evidence of Animal Disease in Archaeology*. Oxford, Oxbow Books: 240–46

Davies S.J., Lamb, H.F., Roberts, S.J. 2015: 'Micro-XRF core scanning in palaeolimnology: recent developments' in R.G. Croudace, I.W. Rothwell (eds), *Micro-XRF Studies of Sediment Cores*. Dordrecht, Springer: 189–226

Davis, S.J.M. 1981: 'The effects of temperature change and domestication on the body size of Late Pleistocene to Holocene mammals of Israel' *Paleobiology* 7: 101–14

— 2000: 'The effect of castration and age on the development of the Shetland sheep skeleton and a metric comparison between bones of males, females and *castrates*' *Journal of Archaeological Science* 27: 373–90

de Cupere, B., Duru, R., Umurtak, G. 2008: 'Animal husbandry at the Early Neolithic to Early Bronze Age site of Bademağacı (Antalya Province, SW Turkey): evidence from the faunal remains' in E. Vila, L. Gourichon, A.M. Choyke, H. Buitenhuis (eds), *Archaeozoology of the Near East VIII: Proceedings of the Eighth International Symposium on the Archaeozoology of Southwestern Asia and Adjacent Areas*. Lyons, Maison de l'Orient et de la Méditerranée: 367–405

de Meester, T. (ed.) 1970: *Soils of the Great Konya Basin, Turkey*. Wageningen, Centre for Agricultural Publishing and Documentation

de Moulins, D. 1997: *Agricultural Changes at Euphrates and Steppe Sites in the Mid-8th to the 6th Millennium B.C.* (BAR International Series 683). Oxford, Archaeopress

Debo, A. 1970: *A History of the Indians of the United States*. London, Pimlico

de Ridder, N.A. 1965: 'Sediments of Konya Basin, Central Anatolia, Turkey' *Paleogeography, Paleoclimatology, Palaeoecology* 1: 225–54

Degerbøl, M. 1970: 'The Urus (*Bos primigenius bojanus*) and Neolithic domesticated cattle (*Bos taurus domesticus Linné*) in Denmark' *Det Kongelige Danske Videnskabernes Selskab Biologiske Skrifter* 17: 5–177

DeLeon, V.B. 2007: 'Fluctuating asymmetry and stress in a medieval Nubian population' *American Journal of Physical Anthropology* 132: 520–34

Demarchi, B., Presslee, S., Sakalauskaite, J., Fischer, R., Best, J. 2020: 'The role of birds at Çatalhöyük revealed by the analysis of eggshell' *Quaternary International* 543: 50–60

Demirergi, G.A. 2015: *Food Sharing during the Transition to Agriculture at Neolithic Çatalhöyük, Central Anatolia*. PhD thesis, Stony Brook University, Stony Brook

Demirergi, G.A., Twiss, K.C., Bogaard, A., Green, L., Ryan, P., Farid, S. 2014: 'Of bins, basins and banquets: storing, handling and sharing at Neolithic Çatalhöyük' in I. Hodder (ed.), *Integrating Çatalhöyük: Themes from the 2000–2008 Seasons*. London, British Institute at Ankara; Los Angeles, Cotsen Institute of Archaeology Press: 91–108

DeNiro, M.J. 1985: 'Post-mortem preservation and alteration of in-vivo bone collagen isotope ratios in relation to paleodietary reconstruction' *Nature* 317: 806–09

DeNiro, M.J., Epstein, S. 1978: 'Influence of diet on the distribution of carbon isotopes in animals' *Geochimica et Cosmochimica Acta* 42: 495–506

— 1981: 'Influence of diet on the distribution of nitrogen isotopes in animals' *Geochimica et Cosmochimica Acta* 45: 341–51

Denys, C., Stoetzel, E., Andrews, P., Bailon, S., Rihane, A., Huchet, J.B., Fernandez-Jalvo, Y., Loroulandie, V. 2017: 'Taphonomy of small predators multi-taxa accumulations: palaeoecological implications' *Historical Biology* 30.6: 868–81

Der, L., Issavi, J. 2017: 'The urban quandary and the "mega-site" from the Çatalhöyük perspective' *Journal of World Prehistory* 30: 189–206

Dewbury, A.G., Russell, N. 2007: 'Relative frequency of butchering cutmarks produced by obsidian and flint: an experimental approach' *Journal of Archaeological Science* 34: 354–57

Diffey, C. 2018: *Feeding the City: A Comparative Study of Agricultural Production in Bronze Age Systems of Western Asia*. DPhil thesis, University of Oxford, Oxford

Diffey, C., Neef, R., Bogaard, A. 2017: 'The archaeobotany of large-scale hermetic cereal storage at the Hittite capital of Hattusha' in A. Schachner (ed.), *Innovation versus Beharrung: Was macht den Unterschied des hethitischen Reichs im Anatolien des 2. Jahrtausends v. Chr? Internationaler Workshop zu Ehren von Jürgen Seeher, Istanbul 23-24. Mai 2014*. Istanbul, Ege Yayinlari: 185–201

Diffey, C., Neef, R., Seeher, J., Bogaard, A. 2020. 'The agroecology of an early state: new results from Hattusha' *Antiquity* 94: 1204–23

Dobney, K., Goodman, A.H. 1991: 'Epidemiological studies of the dental enamel hypoplasias in Mexico and Bradford: their relevance to archaeological skeletal studies' in H. Bush, M. Zvelebil (eds), *Health in Past Societies: Biocultural Interpretations of Human Skeletal Remains in Archaeological Contexts*. Oxford, British Archaeological Reports: 81–100

Dodson, P. Wexlar. D. 1979: 'Taphonomic investigations of owl pellets' *Paleobiology* 5: 279–84

Doherty, C. 2017: *Living with Clay: Materials, Technology, Resources and Landscape at Çatalhöyük*. PhD thesis, University of Leicester, Leicester

Domínguez-Solera, S., Domínguez-Rodrigo, M. 2011: 'A taphonomic study of a carcass consumed by griffon vultures (*Gyps fulvus*) and its relevance for the interpretation of bone surface modifications' *Archaeological and Anthropological Sciences* 3.4: 385–92

Douché, C., Willcox, G. 2018: 'New archaeobotanical data from the Early Neolithic sites of Dj'ade el-Mughara and Tell Aswad (Syria): a comparison between the Northern and Southern Levant' *Paléorient* 44.2: 45–57

Dowle, M., Srinivasan, A. 2017: 'data.table: Extension of data.frame'. R package version 1.10.4-3. https://CRAN.R-project.org/package=data.table.

Doyle, S. 2016: 'Chipped stone from the North and South Areas' *Çatalhöyük 2016 Archive Report*. http://www.catalhoyuk.com/archive_reports/2016

Doyle, W.J., Johnston, O. 1977: 'On the meaning of increased fluctuating dental asymmetry: a cross populational study' *American Journal of Physical Anthropology* 46: 127–34

Ducos, P. 1988: *Archaeozoologie Quantitative: Les Valeurs Numériques Immediates à Çatal Hüyük* (Les Cahiers du Quaternaire 12). Paris, Éditions du Centre National de la Recherche Scientifique

Duday, H. 2008 'Archaeological proof of an abrupt mortality crisis: simultaneous deposit of cadavers, simultaneous deaths?' in D. Raoult. M. Drancourt (eds), *Palaeomicrobiology: Past Human Infections*. Berlin and Heidelberg, Springer: 49–54

— 2009: *The Archaeology of the Dead: Lectures in Archaeothanatology*. Oxford, Oxbow Books

Duday, H., Courtaud, P., Crubézy, E., Sellier, P., Tillier, A.-M. 1990: 'L'Anthropologie "de terrain": reconnaissance et interprétation des gestes funéraires' *Bulletins et Mémoirs de la Société d'Anthropologie de Paris* 2.3–4: 29–50

Düring, B.S. 2001: 'Social dimensions in the architecture of Neolithic Çatalhöyük' *Anatolian Studies* 51: 1–18

— 2003: 'Burials in context: the 1960s inhumations of Çatalhöyük East' *Anatolian Studies* 53: 1–15

— 2006: *Constructing Communities: Clustered Neighbourhood Settlements of the Central Anatolian Neolithic, ca. 8500–5500 cal BC*. PhD thesis, University of Leiden, Leiden

— 2007: 'Reconsidering the Çatalhöyük community: from households to settlement systems' *Journal of Mediterranean Archaeology* 20.2: 155–82

— 2011a: *The Prehistory of Asia Minor: From Complex Hunter-Gatherers to Early Urban Societies*. Cambridge, Cambridge University Press

— 2011b: 'Fortifications and fabrications: reassessing the emergence of fortifications in prehistoric Asia Minor' in B.S. Düring, A. Wossink, P.M.M.G. Akkermans (eds), *Correlates of Complexity: Essays in Archaeology and Assyriology Dedicated to Diederik J.W. Meijer in Honour of his 65th Birthday*. Leiden, Nederlands Instituut voor het Nabije Oosten: 69–85

Düring, B.S., Marciniak, A. 2006: 'Households and communities in the central Anatolian Neolithic' *Archaeological Dialogues* 12: 165–87

Duru, R., De Cupere, B. 2003: 'Faunal remains from Neolithic Höyücek (SW-Turkey) and the presence of early domestic cattle in Anatolia' *Paléorient* 29.1: 107–20

Dury, G.H. 1973: 'Magnitude-frequency analysis and channel morphology' in M. Morisawa (ed.), *Fluvial Geomorphology*. London, Allen and Unwin: 91–121

Dyble, M., Salali, G.D., Chaudhary, N., Page, A., Smith, D., Thompson, J., Vinicius, L., Mace, R., Migliano, A.B. 2015: 'Sex equality can explain the unique social structure of hunter-gatherer bands' *Science* 348: 796–98

Eastwood, W.J., Fairbairn, A., Stroud, E., Roberts, N., Lamb, H., Yiğitbaşıoğlu, H., Çetin Şenkul, G., Moss, A., Turner, R., Boyer, P. 2018: 'Comparing pollen and archaeobotanical data for Chalcolithic cereal agriculture at Çatalhöyük, Turkey' *Quaternary Science Reviews* 202: 4–18

Eddisford, D.F., Hassan, S.A., Lizarlde, M., McCann, S., Sadarangani, F., Taylor, J., Yeomans, L. 2007: 'Foundation trenches' *Çatalhöyük 2007 Archive Report*. http://www.catalhoyuk.com/archive_reports/2007

Edwards, Y.H., Martin, L. 2007: 'Fauna from the Natufian and PPNA cave site of Iraq ed-Dubb in Highland Jordan' *Paléorient* 33.1: 143–74

Eleazer, C.D., Jankauskas, R. 2016: 'Mechanical and metabolic interactions in cortical bone development' *American Journal of Physical Anthropology* 160: 317–33

El-Najjar, M.Y., DeSanti, M.V., Ozebek, L. 1978: 'Prevalence and possible etiology of dental enamel hypoplasia' *American Journal of Physical Anthropology* 48: 185–92

Emberling, G., Robb, J., Speth, J.D., Wright, H.T. 2002: 'Kunji Cave: Early Bronze Age burials in Luristan' *Iranica Antiqua* 37: 47–104

Emslie, S. 1982: 'Osteological identification of long-eared and short-eared owls' *American Antiquity* 47: 155–57

Entwistle, N., Heritage, G., Milan, D. 2019: 'Ecohydraulic modelling of anabranching rivers' *River Research and Applications* 35: 353–64

Ercan, E., Gaygusuz, Ö., Tarkan, A.S., Reichard, M., Smith, C. 2013: 'The ecology of freshwater bivalves in the Lake Sapanca basin, Turkey' *Turkish Journal of Zoology* 37: 730–38

Erdal, Y.S. 2015: 'Bone or flesh: defleshing and post-depositional treatments at Körtik Tepe (Southeastern Anatolia, PPNA Period)' *European Journal of Archaeology* 18.1: 4–32

Erdoğu B. 2013: 'Uğurlu, a Neolithic settlement on the Aegean island of Gökçeada' in M. Özdoğan, N. Başgelen, P. Kuniholm (eds), *Neolithic in Turkey: New Excavations and New Research. Northwestern Turkey and Istanbul.* Istanbul: Arkeoloji ve Sanat Yayınları: 1–33

Ergun, M., 2018: 'Where the wild things are. Contextual insights into wild plant exploitation at Aceramic Neolithic Aşıklı Höyük, Turkey' *Paléorient* 44.2: 9–28

Ergun, M., Bogaard, A., Charles, M. 2013: 'The archaeobotany of the IST Area' in I. Hodder (ed.), *Çatalhöyük Excavations: The 2000–2008 Seasons*. London, British Institute at Ankara; Los Angeles, Monographs of the Cotsen Institute of Archaeology Press: 655–57

Ergun M., Tengberg, M., Willcox. G., Douché, C. 2018: 'Plants of Aşıklı Höyük and changes through time: first archaeobotanical results from the 2010–14 excavation seasons' in M. Özbaşaran, G. Duru, M. Stiner (eds), *The Early Settlement at Aşıklı Höyük: Essays in Honour of Ufuk Esin*. Istanbul, Ege Yayınları: 191–217

Ertug, F. 2004: 'Recipes of old tastes with einkorn and emmer wheat' *Tüba-AR* 7: 177–88

Esin, U., Harmanakaya, S. 1999: 'Aşıklı in the frame of central Anatolian Neolithic' in M. Özdoğan, N. Başgelen (eds), *Neolithic in Turkey: The Cradle of Civilization*. Ankara: Arkeoloji ve Sanat Yayinlari: 115–32

Esper, J., Großjean, J., Camarero, J.J., García-Cervigón, A., Olano, J.M., González-Rouco, F., Domínguez-Castro, F., Büntgen, U. 2015: 'Atlantic and Mediterranean synoptic drivers of central Spanish juniper growth' *Theoretical and Applied Climatology* 121: 571–79

Evans, J., Stoodley, N., Chenery, C. 2006: 'A strontium and oxygen assessment of a possible fourth century immigrant population in a Hampshire cemetery, southern England' *Journal of Archaeological Science* 33: 265–72

Eveleth, P.B., Tanner, J.M. 1990: *Worldwide Variation in Human Growth*. Cambridge, Cambridge University Press

Faber, P., Johnstone, A.M., Gibnet, E.R., Elia, M., Stubbs, R.J., Roger, P.L., Milne, E., Buchan, W., Lobley, G.E. 2003: 'The effect of rate and extent of weight loss on urea salvaging in obese male subjects' *British Journal of Nutrition* 90: 221–31

Fairbairn, A., Near, J., Martinoli, D. 2005a: 'Macrobotanical investigation of the North, South and KOPAL Area excavations at Çatalhöyük East' in I. Hodder (ed.), *Inhabiting Çatalhöyük: Reports from the 1995–99 Seasons*. London, British Institute at Ankara; Cambridge, McDonald Institute for Archaeological Research: 137–202

Fairbairn, A., Asouti, E., Russell, N., Swogger, J.G. 2005b: 'Seasonality' in I. Hodder (ed.), *Çatalhöyük Perspectives: Themes from the 1995–99 Seasons*. London, British Institute at Ankara; Cambridge, McDonald Institute for Archaeological Research: 93–108

Fairbairn, A., Martinoli, D., Butler, A., Hillman, G. 2007; 'Wild plant seed storage at Neolithic Çatalhöyük East, Turkey' *Vegetation History and Archaeobotany* 16.6: 467–79

Farid, S., 2007a: 'Neolithic excavations in the South Area, East Mound, Çatalhöyük 1995–99' in I. Hodder (ed.), *Excavating Çatalhöyük: South, North and KOPAL Area Reports from the 1995–99 Seasons*. London, British Institute at Ankara; Cambridge, McDonald Institute Monographs: 41–342

— 2007b: 'Level IX relative heights, Building 2, Buildings 22 & 16 and Building 17' in I. Hodder (ed.), *Excavating Çatalhöyük: South, North and KOPAL Area Reports from the 1995–99 Seasons*. London, British Institute at Ankara; Cambridge, McDonald Institute for Archaeological Research: 139–226

— 2007c: 'Level VIII: Space 161, Space 162, Building 4, Space 115, Buildings 21 & 7, Building 6 and relative heights of Level VIII' in I. Hodder (ed.), *Excavating Çatalhöyük: South, North and KOPAL Area Reports from the 1995–99 Seasons*. London, British Institute at Ankara; Cambridge, McDonald Institute for Archaeological Research: 227–82

— 2008: 'A review of the Mellaart Level system and the introduction of a new phasing system at Çatalhöyük 2008' *Çatalhöyük 2008 Archive Report*. http://www.catalhoyuk.com/archive_reports/2008

— 2013a: 'Buildings 51/52' in I. Hodder (ed.), *Çatalhöyük Excavations: The 2000–2008 Seasons*. London, British Institute at Ankara; Los Angeles, Cotsen Institute of Archaeology Press: 359–98

— 2013b: 'Timelines: phasing Neolithic Çatalhöyük' in I. Hodder (ed.), *Çatalhöyük Excavations: The 2000–2008 Seasons*. London, British Institute at Ankara; Los Angeles, Cotsen Institute for Archaeology Press: 91–130

— 2014: 'Timelines: phasing Neolithic Çatalhöyük' in I. Hodder (ed.), *Çatalhöyük Excavations: The 2000–2008 Seasons*. London, British Institute at Ankara; Los Angeles, Cotsen Institute of Archaeology Press: 91–129

Faure, G., Mensing, T.M. 2005: *Isotopes: Principles and Applications*. Hoboken, Wiley

Feldman, M., Fernández-Domínguez, E., Reynolds, L., Baird, D., Pearson, J., Hershkovitz, I., May, H., Goring-Morris, N., Benz, M., Gresky, J., Bianco, R.A., Fairbairn, A., Mustafaoğlu, G., Stockhammer, P.W., Posth, C., Haak, W., Jeong, C., Krause, J. 2019: 'Late Pleistocene human genome suggests a local origin for the first farmers of central Anatolia' *Nature Communications* 10: 1218

Fernández, E., Pérez-Pérez, A., Gamba, C., Prats, E., Cuesta, P., Anfruns, J., Molist, M., Arroyo-Pardo, E., Turbón, D. 2014: 'Ancient DNA analysis of 8000 B.C. Near Eastern farmers supports an Early Neolithic pioneer maritime colonization of mainland Europe through Cyprus and the Aegean Islands' *PLoS Genetics* 10.6: e1004401

Fernández, F.J., Montalvo, C.I., Fernández-Jalvo, Y., Andrews, P., Lopez, J.M. 2017: 'A re-evaluation of the taphonomic methodology for the study of small mammal fossil assemblages of South America' *Quaternary Science Reviews* 155: 37–49

Fernández-Jalvo, Y., Andrews, P. 1992: 'Small mammal taphonomy of Gran Dolina, Atepuerca (Burgus), Spain' *Journal of Archaeological Sciences* 19: 407–28

Fernandez-Jalvo, Y., Andrews, P. 2003: 'Experimental effects of water on bone fragments' *Journal of Taphonomy* 1.3: 147–63

Fernández-Jalvo, Y., Peter, A. 2016: *Atlas of Taphonomic Identifications*. Dordrecht, Springer

Fibiger, L. 2012: 'Investigating trauma in the German Wartburg Culture' in R.J. Schulting, L. Fibiger (eds), *Sticks, Stones, and Broken Bones: Neolithic Violence in a European Perspective*. Oxford, Oxford University Press: 175–206

— 2013: 'Misplaced childhood: interpersonal violence and children in Neolithic Europe' in C.J. Knüsel, M.J. Smith (eds), *The Routledge Handbook of the Bioarchaeology of Human Conflict*. London, Routledge: 127–45

Filipović, D., 2012: *An Archaeobotanical Investigation of Plant Use, Crop Husbandry and Animal Diet at Early-Mid Neolithic Çatalhöyük, Central Anatolia*. DPhil thesis, University of Oxford, Oxford

— 2014: *Early Farming in Central Anatolia: An Archaeobotanical Study of Crop Husbandry, Animal Diet and Land Use at Neolithic Çatalhöyük* (BAR International Series 2667). Oxford, Archaeopress

Flohr, P., Fleitmann, D., Matthews, R., Matthews, W., Black, S. 2016: 'Evidence of resilience to past climate change in Southwest Asia: early farming communities and the 9.2 and 8.2 ka events' *Quaternary Science Reviews* 136: 23–39

Fraser, R., Bogaard, A., Heaton, T., Charles, M., Jones, G., Christensen, B.T., Halstead, P., Merbach, I., Poulton, P.R., Sparkes, D., Styring, A.K. 2011: 'Manuring and stable nitrogen isotope ratios in cereals and pulses: towards a new archaeobotanical approach to the inference of land use and dietary practices' *Journal of Archaeological Science* 38: 2790–2804

French, D.H., Hillman, G.C., Payne, S., Payne, R.J. 1972: 'Excavations at Can Hasan III 1969–1970' in E.S. Higgs (ed.), *Papers in Economic Prehistory*. Cambridge, Cambridge University Press: 181–90

Frisancho, A., Guire, K., Babler, W., Borkan, G., Way, A. 1980: 'Nutritional influence on childhood development and genetic control of adolescent growth of Quechuas and Mestizos from the Peruvian lowlands' *American Journal of Physical Anthropology* 52: 367–75

Fu, Q., Meyer, M., Gao, X., Stenzel, U., Burbano, H.A., Kelso, J., Pääbo, S. 2013: 'DNA analysis of an early modern human from Tianyuan Cave, China' *Proceedings of the National Academy of Sciences of the United States of America* 110.6: 2223–27

Fu, Q., Hajdinjak, M., Moldovan, O.T., Constantin, S., Mallick, S., Skoglund, P., Patterson, N., Rohland, N., Lazaridis, I., Nickel, B., Viola, B., Prüfer, K., Meyer, M., Kelso, J., Reich, D., Pääbo, S. 2015: 'An early modern human from Romania with a recent Neanderthal ancestor' *Nature* 524: 216–19

Fuller, D.Q., Stevens, C.J., 2017: 'Open for competition: domesticates, parasitic domesticoids and the agricultural niche' *Archaeology International* 20: 112–23

Fuller, D.Q., Weisskopf, A. 2014: 'Barley: origins and development' in C. Smith (ed.), *Encyclopedia of Global Archaeology*. New York, Springer 763–66

Fuller, B.T., Fuller, J.L., Sage, N.E. Harris, D.A, O'Connell, T.C., Hedges, R.E.M. 2004: 'Nitrogen balance and δ^{15}N: why you're not what you eat during nutritional stress' *Rapid Communications in Mass Spectrometry* 19: 2497–2506

Harris, E.F., Nweeia, M.T. 1980: 'Dental asymmetry as a measure of environmental-stress in the Ticuna Indians of Colombia' *American Journal of Physical Anthropology* 53: 133–42

Harris, I., Jones, P., Osborn, T., Lister, D. 2014: 'Updated high-resolution grids of monthly climatic observations - the CRU TS3.10 dataset' *International Journal of Climatology* 34: 623–42

Harrison, D.L., Bates, P.J.J. 1991: *The Mammals of Arabia*. Sevenoaks, Harrison Zoological Museum

Harrison, K., Martin, V., Webster, B. 2013: 'Structural fires at Çatalhöyük' in I. Hodder (ed.), *Substantive Technologies at Çatalhöyük: Reports from the 2000–2008 Seasons*. London, British Institute at Ankara; Los Angeles, Cotsen Institute of Archaeology Press: 137–46

Harter-Lailheugue, S., Le Mort, F., Vigne, J.-D., Guilaine, J., Le Brun, A., Bouchet, F. 2005: 'Premiéres données parasitologiques sur les populations humaines précéramiques chypriotes (VIIIe et VIIe millénaires av. J.-C.)' *Paléorient* 31.2: 43–54

Harvey, E.L., Fuller, D.Q. 2005: 'Investigating crop processing using phytolith analysis: the example of rice and millets' *Journal of Archaeological Science* 32.5: 739–52

Hastorf, C.A. 2001: 'Making the invisible visible: the hidden jewels of archaeology' in P.M. Drooker (ed.), *Fleeting Identities: Perishable Material Culture in Archaeological Research*. Carbondale, Southern Illinois University Press: 27–42

— 2005: 'Macrobotanical investigation: field methods and laboratory analysis procedures' in I. Hodder (ed.), *Inhabiting Çatalhöyük: Reports from the 1995–99 Seasons*. Research; London, British Institute at Ankara; Cambridge, McDonald Institute for Archaeological Research: 129–35

— 2016: 'The archaeological study of food activities' in C.A. Hastorf (ed.), *The Social Archaeology of Food: Thinking about Eating from Prehistory to the Present*. Cambridge, Cambridge University Press: 83–141

Hastorf, C.A., Near, J. 1998: 'Archaeobotanical archive report' *Çatalhöyük 1998 Archive Report*. http://www.catalhoyuk.com/archive_reports/1998

Heaton, T.H.E. 1987: 'The $^{15}N/^{14}N$ ratios of plants in South Africa and Namibia: relationship to climate and coastal/saline environments' *Oecologia* 74: 236–46

— 1999: 'Spatial, species, and temporal variations in the $^{13}C/^{12}C$ ratios of C_3 plants: implications for palaeodiet studies' *Journal of Archaeological Science* 26: 637–49

Heckathorn, S.A., McNaughton, S.J., Coleman, J.S. 1999: 'C_4 plants and herbivory' in R.F. Sage, R.K. Monson (eds), *C_4 Plant Biology*. New York, Academic Press: 285–312

Hecker, H.M. 1982: 'Domestication revisited: its implication for faunal analysis' *Journal of Field Archaeology* 9: 217–36

Hedges, R.E.M., Millard, A.R., Pike, A.W.G. 1995: 'Measurements and relationships of diagenetic alteration of bone from three archaeological sites' *Journal of Archaeological Science* 22: 201–09

Hedges, R.E.M., Reynard, L.M. 2007: 'Nitrogen isotopes and the trophic level in humans in archaeology' *Journal of Archaeological Science* 34: 1240–51

Helmer, D., Gourichon, L., Monchot, H., Peters, J., Saña Segui, M. 2005: 'Identifying early domestic cattle from Pre-Pottery Neolithic sites on the Middle Euphrates using sexual dimorphism' in J.-D. Vigne, J. Peters, D. Helmer (eds), *First Steps of Animal Domestication: New Archaeozoological Approaches*. Oxford: Oxbow Books: 86–95

Hendy, J., Colonese, A.C., Franz, I., Fernandes, R., Fischer, R., Orton, D., Lucquin, A., Spindler, L., Anvari, J., Stroud, E., Biehl, P.F., Speller, C., Boivin, N., Mackie, M., Jersie-Christensen, R.R., Olsen, J.V., Collins, M.J., Craig, O.E., Rosenstock, E. 2018: 'Ancient proteins from ceramic vessels at Çatalhöyük West reveal the hidden cuisine of early farmers' *Nature Communications* 9: 4064

Henry, A.G. 2012: 'Recovering dietary information from extant and extinct primates using plant microremains' *International Journal of Primatology* 33: 702–15

Henry, A.G., Hudson, H.F., Piperno, D.R. 2009: 'Changes in starch grain morphologies from cooking' *Journal of Archaeological Science* 36: 915–22

Henry, A.G., Brooks, A.S., Piperno, D.R. 2011: 'Microfossils in calculus demonstrate consumption of plants and cooked foods in Neanderthal diets (Shanidar III, Iraq; Spy I and II, Belgium)' *Proceedings of the National Academy of Sciences* 108: 486–91

Henton, E. 2012: 'The combined use of oxygen isotopes and microwear in sheep teeth to elucidate seasonal management of domestic herds: the case study of Çatalhöyük, central Anatolia' *Journal of Archaeological Science* 39: 3264–76

— 2013: 'Oxygen stable isotope and dental microwear evidence of herding practices at Çatalhöyük' in I. Hodder (ed.), *Humans and Landscapes of Çatalhöyük: Reports from the 2000–2008 Seasons*. London, British Institute at Ankara; Los Angeles: Cotsen Institute of Archaeology Press: 299–316

Herrero, J., Porta, J. 2000: 'The terminology and the concepts of gypsum-rich soils' *Geoderma* 96: 47–61
Hershkovitz, I., Gopher, A. 2008: 'Demographic, biological and cultural aspects of the Neolithic revolution: a view from the Southern Levant' in J.-P. Bocquet-Appel, O. Bar-Yosef (eds), *The Neolithic Demographic Transition and its Consequences*. New York, Springer: 441–79
Hershkovitz, I., Livshits, G., Moskona, D., Arensburg, B., Kobyliansky, E. 1993: 'Variables affecting dental fluctuating asymmetry in human isolates' *American Journal of Physical Anthropology* 91: 349–65
Hesse, B. 1979: 'Rodent remains and sedentism in the Neolithic: evidence from Tepe Ganj Dareh, Western Iran' *Journal of Mammalology* 60: 856–57
— 1984: 'These are our goats: the origin of herding in West Central Iran' in B.J. Clutton, C. Grigson (eds), *Animals and Archaeology 3: Early Herders and their Flocks* (BAR International Series 202). Oxford, Archaeopress: 242–44
Hill, K.R., Walker, R.S., Bozicevic, M., Eder, J., Headland, T., Hewlett, B., Hurtado, A.M., Marlowe, F., Wiessner, P., Wood, B. 2011: 'Co-residence patterns in hunter-gatherer societies show unique human social structure' *Science* 331: 1286–89
Hillman, G.C. 2000: 'Plant food economy of Abu Hureyra' in A.M.T. Moore, G.C. Hillman, A.J. Legge (eds), *Village on the Euphrates*. Oxford, Oxford University Press: 372–92
Hillman, G.C., Davies, M.S. 1990: 'Measured domestication rates in wild wheats and barley under primitive cultivation, and their archaeological implications' *Journal of World Prehistory* 4: 157–222
Hillman, G.C., Davies, M.S. 1992: 'Domestication rates in wild wheats and barley under primitive cultivation: preliminary results and archaeological implications of field measurements of selection coefficient' in P.C. Anderson (ed.), *Préhistoire de l'Agriculture: Nouvelles Approaches Expérimentales et Ethnographiques*. Paris, Editions du Centre National de la Recherche Scientifique: 113–58
Hillman, G., Hedges, R., Moore, A., Colledge, S., Pettitt, P. 2001: 'New evidence of Lateglacial cereal cultivation at Abu Hureyra on the Euphrates' *The Holocene* 11: 383–93
Hillson, S. 1996: *Dental Anthropology*. Cambridge, Cambridge University Press
— 2001: 'Recording dental caries in archaeological human remains' *International Journal of Osteoarchaeology* 11: 249–89
— 2014: *Tooth Development in Human Evolution and Bioarchaeology*. Cambridge, Cambridge University Press
Hillson, S., FitzGerald, C., Flinn, H. 2005: 'Alternative dental measurements: proposals and relationships with other measurements' *American Journal of Physical Anthropology* 126: 413–26
Hillson, S.W., Larsen, C.S., Boz, B., Pilloud, M.A., Sadvari, J.W., Agarwal, S.C., Glencross, B., Beauchesne, P., Pearson, J.A., Ruff, C.B., Garofalo, E.M., Hager, L.D., Haddow, S.D. 2013: 'The human remains I: interpreting community structure, health, and diet in Neolithic Çatalhöyük' in I. Hodder (ed.), *Humans and Landscapes of Çatalhöyük: Reports from the 2000–2008 Seasons*. London, British Institute at Ankara; Los Angeles, Cotsen Institute of Archaeology Press: 339–96
Hodder, I. 1982: *Symbols in Action: Ethnoarchaeological Studies of Material Culture*. Cambridge, Cambridge University Press
— 1984: 'Burials, houses, women and men in the European Neolithic' in D. Miller (ed.), *Ideology, Power and Prehistory*. Cambridge, Cambridge University Press: 51–68
— (ed.) 1996: *On the Surface: Çatalhöyük 1993–95* (Çatalhöyük Research Project Series 1). London, British Institute at Ankara; Cambridge, McDonald Institute for Archaeological Research.
— (ed.) 2000a: *Towards Reflexive Method in Archaeology: The Example at Çatalhöyük* (Çatalhöyük Research Project Series 2). London, British Institute at Ankara; Cambridge, McDonald Institute for Archaeological Research
— 2000b: 'Developing a reflexive method in archaeology' in I. Hodder (ed.), *Towards Reflexive Method in Archaeology: The Example at Çatalhöyük*. Cambridge, McDonald Institute for Archaeological Research; London, British Institute at Ankara: 3–15
— (ed.) 2005: *Inhabiting Çatalhöyük: Reports from the 1995–99 Seasons* (Çatalhöyük Research Project Series 4). London, British Institute at Ankara; Cambridge, McDonald Institute for Archaeological Research
— (ed.) 2006a: *Changing Materialities at Çatalhöyük: Reports from the 1995–99 Seasons* (Çatalhöyük Research Project Series 5). London, British Institute at Ankara; Cambridge, McDonald Institute for Archaeological Research
— (ed.) 2006b: *Çatalhöyük Perspectives: Themes from the 1995–99 Seasons* (Çatalhöyük Research Project Series 6). London, British Institute at Ankara; Cambridge, McDonald Institute for Archaeological Research
— 2006c: *The Leopard's Tale: Revealing the Mysteries of Çatalhöyük*. Thames and Hudson, London

— (ed.) 2007: *Excavating Çatalhöyük: South, North and KOPAL Area Reports from the 1995–99 Seasons* (Çatalhöyük Research Project Series 3). London, British Institute at Ankara; Cambridge, McDonald Institute for Archaeological Research

— 2010: *Religion in the Emergence of Civilization: Çatalhöyük as a Case Study*. Cambridge, Cambridge University Press

— 2011: 'Human-thing entanglement: towards an integrated archaeological perspective' *Journal of the Royal Anthropological Institute (N.S.)* 17: 154–77

— 2012: *Entangled. An Archaeology of the Relationships between Humans and Things*. Oxford: Wiley Blackwell

— (ed.) 2013a: *Humans and Landscapes of Çatalhöyük: Reports from the 2000–2008 Seasons* (Çatalhöyük Research Project Series 8). London, British Institute at Ankara; Los Angeles, Cotsen Institute of Archaeology Press

— 2013b: *Substantive Technologies at Çatalhöyük: Reports from the 2000–2008 Seasons* (Çatalhöyük Research Project Series 9). London, British Institute at Ankara; Los Angeles, Cotsen Institute of Archaeology Press

— 2013c: 'Dwelling at Çatalhöyük' in I. Hodder (ed.), *Humans and Landscapes of Çatalhöyük: Reports from the 2000–2008 Seasons*. London, British Institute at Ankara; Los Angeles, Cotsen Institute of Archaeology Press: 1–29

— (ed.) 2014a. *Çatalhöyük Excavations: The 2000–2008 Seasons* (Çatalhöyük Research Project Series 7). London, British Institute at Ankara; Los Angeles: Cotsen Institute of Archaeology Press

— (ed.) 2014b: *Integrating Çatalhöyük: Themes from the 2000–2008 Seasons* (Çatalhöyük Research Project Series 10). London, British Institute at Ankara; Los Angeles, Cotsen Institute of Archaeology Press

— (ed.) 2014c: *Religion at Work in a Neolithic Society: Vital Matters*. Cambridge, Cambridge University Press

— 2014d: 'Temporal trends: the shapes and narratives of cultural change at Çatalhöyük' in I. Hodder (ed.), *Integrating Çatalhöyük: Themes from the 2000–2008 Seasons*. London, British Institute at Ankara; Los Angeles, Cotsen Institute of Archaeology Press: 169–84

— 2014e: 'Mosaics and networks: the social geography of Çatalhöyük' in I. Hodder (ed.), *Integrating Çatalhöyük: Themes from the 2000–2008 Seasons*. London, British Institute at Ankara; Los Angeles, Cotsen Institute of Archaeology Press: 149–67

— 2016: 'More on history houses at Çatalhöyük: a response to Carleton et al.' *Journal of Archaeological Science* 67: 1–6

— (ed.) 2018a: *Religion, History and Place in the Origin of Settled Life*. Boulder, University of Colorado Press

— 2018b: *Where are We Heading?: The Evolution of Humans and Things*. New Haven, Yale University Press

— (ed.) 2019: *Violence and the Sacred in the Ancient Near East: Girardian Conversations at Çatalhöyük*. Cambridge, Cambridge University Press

— (ed.) 2020: *Consciousness, Creativity and Self at the Dawn of Settled Life*. Cambridge, Cambridge University Press

Hodder, I., Cessford, C. 2004: 'Daily practice and social memory at Çatalhöyük' *American Antiquity* 69.1: 17–40

Hodder, I., Doherty, C. 2014: 'Temporal trends: the shapes and narratives of cultural change at Çatalhöyük' in I. Hodder (ed.), *Integrating Çatalhöyük: Themes from the 2000–2008 Seasons*. London, British Institute at Ankara; Los Angeles, Cotsen Institute of Archaeology Press: 169–83

Hodder, I. and Marciniak, A. (eds) 2015: *Assembling Çatalhöyük* (European Association of Archaeologists, Themes in Contemporary Archaeology 1). Leeds, Maney

Hodder, I., Meskell, L. 2011: 'A "curious and sometimes a trifle macabre artistry": some aspects of symbolism in Neolithic Turkey' *Current Anthropology* 52.2: 1–29

Hodder, I., Pels, P. 2010: 'History houses: a new interpretation of architectural elaboration at Çatalhöyük' in I. Hodder (ed.), *Religion in the Emergence of Civilization: Çatalhöyük as a Case Study*. Cambridge, Cambridge University Press: 163–86

Hofmanová, Z., Kreutzer, S., Hellenthal, G., Sell, C., Diekmann, Y., Díez-Del-Molino, D., van Dorp, L., López, S., Kousathanas, A., Link, V., Kirsanow, K., Cassidy, L.M., Martiniano, R., Strobel, M., Scheu, A., Kotsakis, K., Halstead, P., Triantaphyllou, S., Kyparissi-Apostolika, N., Urem-Kotsou, D., Ziota, C., Adaktylou, F., Gopalan, S., Bobo, D.M., Winkelbach, L., Blöcher, J., Unterländer, M., Leuenberger, C., Çilingiroğlu, Ç., Horejs, B., Gerritsen, F., Shennan, S.J., Bradley, D.G., Currat, M., Veeramah, K.R., Wegmann, D., Thomas, M.G., Papageorgopoulou, C., Burger, J. 2016: 'Early farmers from across Europe directly descended from Neolithic Aegeans' *Proceedings of the National Academy of Sciences of the United States of America* 113: 6886–91

Hofreiter, M., Jaenicke, V., Serre, D., von Haeseler, A., Pääbo, S. 2001: 'DNA sequences from multiple amplifications reveal artifacts induced by cytosine deamination in ancient DNA' *Nucleic Acids Research* 29: 4793–99

Hollund, H.I., Jans, M.M.E., Collins, M.J., Kars, H., Joosten, I., Kars, S.M. 2012: 'What happened here? Bone histology as a tool in decoding the postmortem histories of archaeological bone from Castricum, the Netherlands' *International Journal of Osteoarchaeology* 22.5: 537–48

Hongo, H., Meadow, R.H. 2000: 'Faunal remains from Prepottery Neolithic Levels at Çayönü, southeastern Turkey: a preliminary report focusing on pigs (*Sus sp.*)' in M. Mashkour, A.M. Choyke, H. Buitenhuis, F. Poplin (eds), *Archaeozoology of the Near East IV: Proceedings of the Fourth International Symposium on the Archaeozoology of Southwestern Asia and Adjacent Areas* (ARC Publicatie 32). Groningen, Centre for Archeological Research and Consultancy: 121–41

Hongo, H., Pearson, J., Öksüz, B., Ilgezdi, G. 2009: 'The process of ungulate domestication at Çayönü, southeastern Turkey: a multidisciplinary approach focusing on *Bos* sp. and *Cervus elaphus*' *Anthropozoologica* 44: 63–78

Hoover, K.C., Matsumura, H. 2008: 'Temporal variation and interaction between nutritional and developmental instability in prehistoric Japanese populations' *American Journal of Physical Anthropology* 137: 469–78

Hoover, K.C., Corruccini, R.S., Bondioli, L., Macchiarelli, R. 2005: 'Exploring the relationship between hypoplasia and odontometric asymmetry in Isola Sacra, an imperial Roman necropolis' *American Journal of Human Biology* 17: 752–64

Horden, P., Purcell, N. 2000: *The Corrupting Sea. A Study of Mediterranean History*. Oxford, Blackwell Publishing

Horejs, B., Milić, B., Ostmann, F., Thanheiser, U., Weninger, B., Galik, A. 2015: 'The Aegean in the early 7th millennium BC: maritime networks and colonization' *Journal of World Prehistory* 28.4: 289–330

Houben, H., Guillaud, H. 1994: *Earth Construction: A Comprehensive Guide*. London, Intermediate Technology Publications

Hubbard, A.R., Guatelli-Steinberg, D., Irish, J.D. 2015: 'Do nuclear DNA and dental nonmetric data produce similar reconstructions of regional population history? An example from modern coastal Kenya' *American Journal of Physical Anthropology* 157.2: 295–304

Hughes, T., Dempsey, P., Richards, L., Townsend, G.C. 2000: 'Genetic analysis of deciduous tooth size in Australian twins' *Archives of Oral Biology* 45.11: 997–1004

Hummert, J., van Gerven, D. 1983: 'Skeletal growth in a medieval population from Sudanese Nubia' *American Journal of Physical Anthropology* 60.4: 471–78

Humphrey, L. 2014: 'Isotopic and trace element evidence of dietary transitions in early life' *Annals of Human Biology* 41: 348–57

Hunter, R.W. 1961: 'Life cycles of four freshwater snails in limited populations in Loch Lomond, with a discussion of infraspecific variation' *Proceedings of the Zoological Society of London* 137: 135–71

Hussain, K., Wijetunge, D.B., Grubnic, S., Jackson, I.T. 1994: 'A comprehensive analysis of craniofacial trauma' *The Journal of Trauma* 36.1: 34–47

Ibáñez, J.J., González-Urquijo, J., Teira-Mayolini, L.C., Lazuén, T. 2018: 'The emergence of the Neolithic in the Near East: a protracted and multi-regional model' *Quaternary International* 470: 226–52

Ingold, T. 2000: *The Perception of the Environment. Essays on Livelihood, Dwelling and Skill*. London, Routledge

Isendahl, C., Stump, D. 2019: *The Oxford Handbook of Historical Ecology and Applied Archaeology*. Oxford, Oxford University Press

Issavi, J. 2017: 'External Spaces 636, 631 and 610' *Çatalhöyük 2017 Archive Report*. http://www.catalhoyuk.com/archive_reports/2017

Jackson, M.P.C., Moore, S.V. 2018: 'Taphonomies of landscape: investigating the immediate environs of Çatalhöyük from prehistory to the present' *Anatolian Studies* 68: 177–207

Jans, M.M.E., Nielsen-Marsh, C.M., Smith, C.I., Collins, M.J., Kars, H. 2004: 'Characterisation of microbial attack on archaeological bone' *Journal of Archaeological Science* 31: 87–95

Jantz, R.L., Owsley, D. 1984: 'Long bone growth variation among Arikara skeletal populations' *American Journal of Physical Anthropology* 63: 13–20

— 1994: 'White traders in the Upper Missouri: evidence from the Swan Creek site' in D.W. Owsley, R.L. Jantz (eds), *Skeletal Biology in the Great Plains: Migration, Health, Warfare, and Subsistence*. Washington, D.C., Smithsonian Institution Press: 189–201

Jenkins, E.L. 2005: 'Çatalhöyük microfauna' in I. Hodder (ed.), *Inhabiting Çatalhöyük: Reports from the 1995–99 Seasons*. London, British Institute at Ankara; Cambridge, McDonald Institute for Archaeological Research: 111–16

— 2009: *Unwanted Inhabitants? The Microfauna from Çatalhöyük and Pınarbaşı*. Saarbrücken, VDM-Verlag

— 2012a: 'Mice, scats and burials: unusual concentrations of microfauna found in human burials at the Neolithic site of Çatalhöyük, Central Anatolia' *Journal of Social Archaeology* 12: 380–403

— 2012b: 'The microfauna of the BACH Area' in R. Tringham, M. Stevanović (eds), *Last House on the Hill: BACH Area Reports from Çatalhöyük, Turkey*. Los Angeles, Cotsen institute of Archaeology Press: 253–60

Jenkins, E.L., Yeomans, L. 2013: 'The Çatalhöyük microfauna' in I. Hodder (ed.), *Humans and Landscapes of Çatalhöyük: Reports from the 2000–2008 Seasons*. London, British Institute at Ankara; Los Angeles, Cotsen Institute of Archaeology Press: 259–69

Jenkins, E.L., Allcock, S.L., Elliott, S., Palmer, C., Grattan, J. 2017: 'Ethno-geochemical and phytolith studies of activity related patterns: a case study from Al Ma'tan, Jordan' *Environmental Archaeology: The Journal of Human Palaeoecology* 22.4: 412–33

Jenkins, R. 2004: *Social Identity*. London, Routledge

Johansson, S.R., Owsley, D.W. 2002: 'Welfare history on the Great Plains: mortality and skeletal health, 1650–1900' in R.H. Steckel, J.C. Rose (eds), *The Backbone of History: Health and Nutrition in the Western Hemisphere*. Cambridge, Cambridge University Press: 524–60

Johnsen, W.A. 2019: 'Mimetic theory, the wall paintings, and the domestication, de-domestication, and sacrifice of cattle at Çatalhöyük' in I. Hodder (ed.), *Violence and the Sacred in the Ancient Near East*. Cambridge, Cambridge University Press: 153–64

Johnson, K.M., Paul, K.S. 2016: 'Bioarchaeology and kinship: integrating theory, social relatedness, and biology in ancient family research' *Journal of Archaeological Research* 24.1: 75–123

Johnston, F., Wainer, H., Thissen, D., MacVean, R. 1976: 'Hereditary and environmental determinants of growth in height in a longitudinal sample of children and youth of Guatemalan and European ancestry' *American Journal of Physical Anthropology* 44: 469–75

Jones, E.R., Gonzalez-Fortes, G., Connell, S., Siska, V., Eriksson, A., Martiniano, R., McLaughlin, R.L., Gallego Llorente, M., Cassidy, L.M., Gamba, C., Meshveliani, T., Bar-Yosef, O., Müller, W., Belfer-Cohen, A., Matskevich, Z., Jakeli, N., Higham, T.F.G., Currat, M., Lordkipanidze, D., Hofreiter, M., Manica, A., Pinhasi, R., Bradley, D.G. 2015: 'Upper Palaeolithic genomes reveal deep roots of modern Eurasians' *Nature Communications* 6.1: 8912

Jones, E.R., Zarina, G., Moiseyev, V., Lightfoot, E., Nigst, P.R., Manica, A., Pinhasi, R., Bradley, D.G. 2017: 'The Neolithic transition in the Baltic was not driven by admixture with early European farmers' *Current Biology* 27.4: 576–82

— 1991: 'Numerical analysis in archaeobotany' in W. van Zeist, K. Wasylikowa, K.-E. Behre (eds), *Progress in Old World Palaeoethnobotany*. Rotterdam, A.A. Balkema: 63–80

— 2002: 'Weed ecology as a method for the archaeobotanical recognition of crop husbandry practices' *Acta Palaeobotanica* 42: 185–93

Jones, G., Bogaard, A., Charles, M., Hodgson, J.G. 2000: 'Distinguishing the effects of agricultural practices relating to fertility and disturbance: a functional ecological approach in archaeobotany' *Journal of Archaeological Science* 27: 1073–84

Jones, G., Bogaard, A., Halstead, P., Charles, M., Smith, H. 1999: 'Identifying the intensity of crop husbandry practices on the basis of weed floras' *Annual of the British School at Athens* 94: 167–89

Jones, G., Valamoti, S., Charles, M. 2000: 'Early crop diversity: a "new" glume wheat from northern Greece' *Vegetation History and Archaeobotany* 9: 133–46

Jones, G.G., Sadler, P. 2012: 'Age at death in cattle: methods, older cattle and known-age reference material' *Environmental Archaeology* 17: 11–28

Jones, M.D., Leng, M.J., Roberts, C.N., Türkes, M., Moyeed, R. 2005: 'A coupled calibration and modelling approach to the understanding of dry-land lake oxygen isotope records' *Journal of Paleolimnology* 34: 391–411

Jost, L. 2006: 'Entropy and diversity' *Oikos* 113.2: 363–75

Jourdan, P.M., Lamberton, P.H.L., Fenwick, A., Addiss, D.G. 2018: 'Soil-transmitted helminth infections' *Lancet* 391: 252–65

Juggins, S. 2019: 'Rioja: analysis of quaternary science data' https://cran.r-project.org/web/packages/rioja/index.html accessed 26/4/2019

Kabukcu, C. 2013: 'Fuel experiments with reeds (*Phragmites australis*) at Neolithic Boncuklu, Konya Plain, Central Anatolia'. Paper presented at the 16th Conference of the International Workgroup for Paleoethnobotany, 17 June 2013 (Aristotle University, Thessaloniki, Greece)

— 2015: *Prehistoric Vegetation Change and Woodland Management in Central Anatolia: A Study of Late Pleistocene-Mid Holocene Anthracological Remains from the Konya Plain*. PhD thesis, University of Liverpool, Liverpool

— 2017: 'Woodland vegetation history and human impacts in south-central Anatolia 15,000–7000 cal BP: anthracological results from five prehistoric sites in the Konya Plain' *Quaternary Science Reviews* 176: 85–100

— 2018a: 'Identification of woodland management practices and tree growth conditions in archaeological fuel waste remains: a case study from the site of Çatalhöyük in central Anatolia, Turkey' *Quaternary International* 463: 282–97

— 2018b: 'Wood charcoal analysis in archaeology' in E. Pişkin, A. Marciniak, M. Bartkowiak (eds), *Environmental Archaeology: Current Theoretical and Methodological Approaches*. Cham, Springer: 133–54

Kalyoncu, H., Barlas, M., Yildirim, M.Z., Yorulmaz, B. 2008: 'Gastropods of two important streams of Gökova Bay (Muğla, Turkey) and their relationships with water quality' *International Journal of Science & Technology* 3.1: 27–36

Kamp, K.A. 1998: 'Social hierarchy and burial treatments: a comparative assessment' *Cross-Cultural Research* 32: 79–115

Karkanas, P., Efstratiou, N. 2009: 'Floor sequences in Neolithic Makri, Greece: micromorphology reveals cycles of renovation' *Antiquity* 83: 955–67

Karul, N., Avcı, M.B. 2011: 'Neolithic communities in the eastern Marmara region: Aktopraklık C' *Anatolica* 37: 1–15

Katz, O., Cabanes, D., Weiner, S., Maeir, A.M., Boaretto, E., Shahack-Gross, R. 2010: 'Rapid phytolith extraction for analysis of phytolith concentrations and assemblages during an excavation: an application at Tell es-Safi/Gath, Israel' *Journal of Archaeological Science* 37: 1557–63

Keane, W. 2010: 'Marked, absent, habitual: approaches to Neolithic religion at Çatalhöyük' in I. Hodder (ed.), *Religion in the Emergence of Civilization: Çatalhöyük as a Case Study*. Cambridge, Cambridge University Press: 187–219

Keepax, C.A. 1981: 'Avian eggshell from archaeological sites' *Journal of Archaeological Science* 8: 315–35

Kenyon, K.M. 1956: 'Jericho and its setting in Near Eastern history' *Antiquity* 30.120: 184–97

Kieser, J.A., Groeneveld, H.T., Preston, C.B. 1986: 'Fluctuating dental asymmetry as a measure of odontogenic canalization in man' *American Journal of Physical Anthropology* 71: 437–44

Kirwan, G.M., Boyla, K.A., Castell, P., Demirci, B., Özen, M., Welch, H., Marlow, T. 2008: *The Birds of Turkey: The Distribution, Taxonomy and Breeding of Turkish Birds*. London, Christopher Helm

Kılınç, G.M., Omrak, A., Özer, F., Günther, T., Büyükkarakaya, A.M., Bıçakçı, E., Baird, D., Dönertaş, H.M., Ghalichi, A., Yaka, R., Koptekin, D., Açan, S.C., Parvizi, P., Krzewińska, M., Daskalaki, E.A., Yüncü, E., Dağtaş, N.D., Fairbairn, A., Pearson, J., Mustafaoğlu, G., Erdal, Y.S., Çakan, Y.G., Togan, İ., Somel, M., Storå, J., Jakobsson, M., Götherström, A. 2016: 'The demographic development of the first farmers in Anatolia' *Current Biology* 26.19: 2659–66

Kılınç, G.M., Koptekin, D., Atakuman, Ç., Sümer, A.P., Dönertaş, H.M., Yaka, R., Bilgin, C.C., Büyükkarakaya, A.M., Baird, D., Altınışık, E., Flegontov, P., Götherström, A., Togan, İ., Somel, M. 2017: 'Archaeogenomic analysis of the first steps of Neolithization in Anatolia and the Aegean' *Proceedings of the Royal Society B: Biological Sciences* 284.1867: 20172064

Kiviat, E., Hamilton, E. 2001: '*Phragmites* use by Native North Americans' *Aquatic Botany* 69: 341–57

Kjellström, A. 2013: 'Interpreting violence: a bioarchaeological perspective from medieval central Sweden' in C.J. Knüsel, M.J. Smith (eds), *The Routledge Handbook of the Bioarchaeology of Human Conflict*. London, Routledge: 237–50

Klaus, H.D., Tam, M.E. 2009: 'Contact in the Andes: bioarchaeology of systemic stress in colonial Mórrope, Peru' *American Journal of Physical Anthropology* 138.3: 356–68

— 2015: 'Requiem aeternam?: archaeothanatology of mortuary ritual in colonial Mórrope, north coast of Peru' in I. Shimada, J.L. Fitzsimmons (eds), *Living with the Dead in the Andes*. Tucson, University of Arizona Press: 267–303

Klimowicz, A. 2017: 'Building 132, Spaces 633 and 531 and underlying external area Space 630' *Çatalhöyük 2017 Archive Report*. http://www.catalhoyuk.com/archive_reports/2017

Klingenberg, C.P. 2003: 'A developmental perspective on developmental instability: theory, models and mechanisms' in M. Polak (ed.), *Developmental Instability: Causes and Consequences*. Oxford, Oxford University Press: 14–34

Kluyver, T.A., Jones, G., Pujol, B., Bennett, C., Mockford, E.J., Charles, M., Rees, M., Osborne, C.P. 2017: 'Unconscious selection drove seed enlargement in vegetable crops' *Evolution Letters* 1.2: 64–72

Knipper, C., Mittnik A., Massy K., Kociumaka, C., Kucukkalipci, I., Maus, M., Wittenborn, F., Metz, S.E., Staskiewicz, A., Krause, J. 2017: 'Female exogamy and gene pool diversification at the transition from the Final Neolithic to the Early Bronze Age in central Europe' *Proceedings of the National Academy of Sciences of the USA* 114.38: 10083–88

Knüsel, C.J. 2014a: 'Crouching in fear: terms of engagement for funerary remains' *Journal of Social Archaeology* 14.1: 26–58

— 2014b: 'Courteous knights and cruel avengers: a consideration of the changing social context of medieval warfare from the perspective of human remains' in C.J. Knüsel, M.J. Smith (eds), *The Routledge Handbook of the Bioarchaeology of Human Conflict*. London, Routledge: 263–81

Knüsel, C.J., Glencross, B. 2017a: 'Çatalhöyük, archaeology, violence' *Contagion: Journal of Violence, Mimesis, and Culture* 23: 23–36

— 2017b: 'Çatalhöyük, archaeology, violence' in J. Allison, W. Palaver (eds), *Handbook of Mimetic Theory and Religion*. New York: Palgrave MacMillan: 69–75

Knüsel, C.J., Robb, J.E., 2016: 'Funerary taphonomy: an overview of goals and methods' *Journal of Archaeological Science: Reports* 10: 655–73

Knüsel, C.J., Smith, M.J. 2014: 'The osteology of conflict: what does it all mean?' in C.J. Knüsel, M.J. Smith (eds), *The Routledge Handbook of the Bioarchaeology of Human Conflict*. London, Routledge: 656–94

Knüsel, C.J., Batt, C.M., Cook, G., Montgomery, J., Müldner, G., Ogden, A.R., Palmer, C., Stern, B., Todd, J., Wilson, A.S. 2010: 'The identity of the St. Bees Lady, Cumbria: an osteobiographical approach' *Medieval Archaeology* 54: 275–317

Knüsel, C.J., Haddow, S.D., Sadvari, J.W., Byrnes, J. 2012: 'Çatalhöyük human remains team archive report 2012' *Çatalhöyük 2012 Archive Report*. http://www.catalhoyuk.com/archive_reports/2012

Knüsel, C.J., Haddow, S.D., Sadvari, J.W., Betz, B. 2013: 'Çatalhöyük human remains team archive Report 2013' *Çatalhöyük 2013 Archive Report*. http://www.catalhoyuk.com/archive_reports/2013

Knüsel, C.J., Glencross, B., Milella, M. 2019: 'A Girardian framework for violent injuries at Neolithic Çatalhöyük in their Western Asian context' in I. Hodder (ed.), *Violence and the Sacred in the Neolithic: Girardian Conversations at Çatalhöyük*. Cambridge, Cambridge University Press: 60–95

Köbbing, J.F., Thevs, N., Zerbe, S. 2013: 'The utilization of reed (*Phragmites australis*): a review' *Mires and Peat* 13: 1–14

Kohler-Schneider, M. 2001: *Verkohlte Kultur- und Wildpflanzen aus Stillfried an der March als Spiegel spätbronzezeitlicher Landwirtschaft im Weinviertel*. Vienna, Verlag der Österreichischen Akademie der Wissenschaften

Koşal Şahin, S., Yildirim, M.Z. 2007: 'The mollusk fauna of Lake Sapanca (Turkey: Marmara) and some physicochemical parameters of their abundance' *Turkish Journal of Zoology* 31: 47–52

Koşal Şahin, S., Zeybek. M. 2016: 'Distribution of mollusca fauna in the streams of Tunceli Province (East Anatolia, Turkey) and its relationship with some physicochemical parameters' *Turkish Journal of Fisheries and Aquatic Sciences* 16: 187–95

Kranioti, E. 2015: 'Forensic investigation of cranial injuries due to blunt force trauma: current best practice' *Research and Reports in Forensic Medical Science* 5: 25–37

Kuijt, I. 1996: 'Negotiating equality through ritual: a consideration of Late Natufian and Pre-Pottery Neolithic A period mortuary practices' *Journal of Anthropological Archaeology* 15: 313–36

— 2000a: 'Keeping the peace: ritual, skull caching, and community integration in the Levantine Neolithic' in I. Kuijt (ed.), *Life in Neolithic Farming Communities: Social Organization, Identity, and Differentiation*. New York, Kluwer Academic/Plenum Publishers: 137–64

— 2000b: 'Life in Neolithic farming communities: an introduction' in I. Kuijt (ed.), *Life in Neolithic Farming Communities: Social Organization, Identity, and Differentiation*. New York, Kluwer Academic/Plenum Publishers: 3–13

— 2008: 'The regeneration of life: Neolithic structures of symbolic remembering and forgetting' *Current Anthropology* 49.2: 171–97

— 2009: 'What do we really know about food storage, surplus, and feasting in preagricultural communities?' *Current Anthropology* 50.5: 641–44

— 2018: 'Material geographies of house societies: reconsidering Neolithic Çatalhöyük, Turkey' *Cambridge Archaeological Journal* 28.4: 565–90

Kuijt, I., Goring-Morris, N. 2002: 'Foraging, farming, and social complexity in the Pre-Pottery Neolithic of the Southern Levant: a review and synthesis' *Journal of World Prehistory* 16.4: 361–440

Kuijt, I., Guerrero, E., Molist, M., Anfruns, J. 2011: 'The changing Neolithic household: household autonomy and social segmentation, Tell Halula, Syria' *Journal of Anthropological Archaeology* 30.4: 502–22

Kujanova, M., Bigoni, L., Veleminska, J., Veleminsky, P. 2008: 'Limb bone asymmetry and stress in medieval and recent populations of Central Europe' *International Journal of Osteoarchaeology* 18: 476–91

Kumari, M., Kangur, K., Haldna, M. 2007: 'Variation of macrozoobenthos communities in the reed *Phragmites australis* belt of two large shallow lakes' *Proceedings of the Estonian Academy of Sciences: Biology/Ecology* 56: 141–53

Kuzucuoğlu, C. 2002: 'Environmental setting and evolution from the 9th to the 5th millennium cal BC in Central Anatolia: an introduction to the study of relations between environmental conditions and the development of human societies' in L. Thissen, F. Gérard (eds), *The Neolithic of Central Anatolia, Internal Developments and External Relations During the 9th–6th Millennia cal BC, Proceedings of the International CANeW Round Table, Istanbul 23–24 November 2001*. Istanbul, Ege Yayınları: 33–58

Kuzucuoğlu, C., Roberts, N. 1997: 'Évolution de l'environment en Anatolie de 20,000 á 6,000 BP' *Paléorient* 23.2: 7–24

Kuzucuoğlu, C., Bertaux, J., Black, S., Denefle, M., Fortugne, M., Karabiyikoğlu, M., Kashima, K., Limondin-Lozouet, N., Mouralis, D., Orth, P. 1999: 'Reconstruction of climatic changes during the Late Pleistocene, based on sediment records from the Konya Basin (Central Anatolia, Turkey)' *Geological Journal* 34: 175–98

Laland, K.N., O'Brien, M.J. 2011: 'Cultural niche construction: an introduction' *Biological Theory* 6: 191–202

Lancelotti, C., Madella, M. 2012: 'The "invisible" product: developing markers for identifying dung in archaeological contexts' *Journal of Archaeological Science* 39: 953–63

Lancelotti, C., Madella. M., Barastegui, J. 2016: 'Plant microremains (phytoliths and starch grains)' *Çatalhöyük Archive Report 2016*. http://www.catalhoyuk.com/archive_reports/2016

Larsen, C.S. 1995: 'Biological changes in human populations with agriculture' *Annual Review of Anthropology* 24: 185–213

— 2006: 'The agricultural revolution as environmental catastrophe: implications for health and lifestyle in the Holocene' *Quaternary International* 150.1: 12–20

— 2015: *Bioarchaeology: Interpreting Behavior from the Human Skeleton* (2nd edition). Cambridge, Cambridge University Press

— 2018: 'The bioarchaeology of health crisis: infectious disease in the past' *Annual Review of Anthropology* 47: 295–313

Larsen, C.S., Hillson, S.W., Ruff, C.B., Sadvari, J.W., Garofalo, E.M. 2013: 'The human remains II: interpreting lifestyle and activity in Neolithic Çatalhöyük' in I. Hodder (ed.), *Humans and Landscapes of Çatalhöyük: Reports from the 2000–2008 Seasons*. London, British Institute at Ankara; Los Angeles, Cotsen Institute of Archaeology Press: 397–412

Larsen, C.S., Hillson, S.W., Boz, B., Pilloud, M.A., Sadvari, J.W., Agarwal, S.C., Glencross, B., Beauchesne, P., Pearson, J., Ruff, C.B., Garofalo, E.M., Hager, L.D., Haddow, S.D., Knüsel, C.J. 2015: 'Bioarchaeology of Neolithic Çatalhöyük: lives and lifestyles of an early farming society in transition' *Journal of World Prehistory* 28: 27–68

Larsen, C.S., Knüsel, C.J., Haddow, S.D., Pilloud, M.A., Milella, M., Sadvari, J.W., Pearson, J., Ruff, C.B., Garofalo, E.M., Bocaege, E., Betz, B.J., Dori, I., Glencross, B. 2019. 'Bioarchaeology of Neolithic Çatalhöyük reveals fundamental transitions in health, mobility, and lifestyle in early farmers' *Proceedings of the National Academy of Sciences* 116.26: 12615–23

Lazaridis, I., Patterson, N., Mittnik, A., Renaud, G., Mallick, S., Kirsanow, K., Sudmant, P.H., Schraiber, J.G., Castellano, S., Lipson, M., Berger, B., Economou, C., Bollongino, R., Fu, Q., Bos, K.I., Nordenfelt, S., Li, H., de Filippo, C., Prüfer, K., Sawyer, S., Posth, C., Haak, W., Hallgren, F., Fornander, E., Rohland, N., Delsate, D., Francken, M., Guinet, J.-M., Wahl, J., Ayodo, G., Babiker, H.A., Bailliet, G., Balanovska, E., Balanovsky, O., Barrantes, R., Bedoya, G., Ben-Ami, H., Bene, J., Berrada, F., Bravi, C.M., Brisighelli, F., Busby, G.B.J., Cali, F., Churnosov, M., Cole, D.E.C., Corach, D., Damba, L., van Driem, G., Dryomov, S., Dugoujon, J.-M., Fedorova, S.A., Gallego Romero, I., Gubina, M., Hammer, M., Henn, B.M., Hervig, T., Hodoglugil, U., Jha, A.R., Karachanak-Yankova, S., Khusainova, R., Khusnutdinova, E., Kittles, R., Kivisild, T., Klitz, W., Kučinskas, V., Kushniarevich, A., Laredj, L., Litvinov, S., Loukidis, T., Mahley, R.W., Melegh, B., Metspalu, E., Molina, J.,

Mountain, J., Näkkäläjärvi, K., Nesheva, D., Nyambo, T., Osipova, L., Parik, J., Platonov, F., Posukh, O., Romano, V., Rothhammer, F., Rudan, I., Ruizbakiev, R., Sahakyan, H., Sajantila, A., Salas, A., Starikovskaya, E.B., Tarekegn, A., Toncheva, D., Turdikulova, S., Uktveryte, I., Utevska, O., Vasquez, R., Villena, M., Voevoda, M., Winkler, C.A., Yepiskoposyan, L., Zalloua, P., Zemunik, T., Cooper, A., Capelli, C., Thomas, M.G., Ruiz-Linares, A., Tishkoff, S.A., Singh, L., Thangaraj, K., Villems, R., Comas, D., Sukernik, R., Metspalu, M., Meyer, M., Eichler, E.E., Burger, J., Slatkin, M., Pääbo, S., Kelso, J., Reich, D., Krause, J. 2014: 'Ancient human genomes suggest three ancestral populations for present-day Europeans' *Nature* 513.7518: 409–13

Lazaridis, I., Nadel, D., Rollefson, G., Merrett, D.C., Rohland, N., Mallick, S., Fernandes, D., Novak, M., Gamarra, B., Sirak, K., Connell, S., Stewardson, K., Harney, E., Fu, Q., Gonzalez-Fortes, G., Jones, E.R., Roodenberg, S.A., Lengyel, G., Bocquentin, F., Gasparian, B., Monge, J.M., Gregg, M., Eshed, V., Mizrahi, A.-S., Meiklejohn, C., Gerritsen, F., Bejenaru, L., Blüher, M., Campbell, A., Cavalleri, G., Comas, D., Froguel, P., Gilbert, E., Kerr, S.M., Kovacs, P., Krause, J., McGettigan, D., Merrigan, M., Merriwether, D.A., O'Reilly, S., Richards, M.B., Semino, O., Shamoon-Pour, M., Stefanescu, G., Stumvoll, M., Tönjes, A., Torroni, A., Wilson, J.F., Yengo, L., Hovhannisyan, N.A., Patterson, N., Pinhasi, R., Reich, D. 2016: 'Genomic insights into the origin of farming in the ancient Near East' *Nature* 536.7617: 419–24

Ledermann, S. 1969: *Nouvelles Tables-types de Mortalité* (Travaux et Documents [Institut national d'études démographiques] 53). Paris, Presses unniversitaire de France

Ledger, M.L., Stock, F., Schwaiger, H., Knipping, M., Brückner, H., Ladstätter, S., Mitchell, P.D. 2018: 'Intestinal parasites from public and private latrines and the harbour canal in Roman Period Ephesus, Turkey (1st c. BCE to 6th c. CE)' *Journal of Archaeological Science: Reports* 21: 289–97

Ledger, M.L., Anastasiou, E., Shillito, L.-M., Mackay, H., Bull, I.D., Haddow, S.D., Knüsel, C.J., Mitchell, P.D. 2019: 'Parasite infection at the early farming community of Çatalhöyük, Turkey (7100–6150 BC)' *Antiquity* 93: 573–87

Lewis J.P., Leng, M.J., Dean, J.R., Marciniak, A., Bar-Yosef Mayer, D.E., Wu, X. 2016: 'Early Holocene palaeoseasonality inferred from the stable isotope composition of *Unio* shells from Çatalhöyük, Turkey' *Environmental Archaeology* 22.1: 79–95

Liang, E., Lu, X., Ren, P., Li, X., Zhu, L., Eckstein, D. 2011: 'Annual increments of juniper dwarf shrubs above the tree line on the central Tibetan Plateau: a useful climatic proxy' *Annals of Botany* 109.4: 1–8

Liversidge, H. 1994: 'Accuracy of age estimation from developing teeth of a population' *International Journal of Osteoarchaeology* 4: 37–46

— 2008: 'Timing of human mandibular third molar formation' *Annals of Human Biology* 35: 294–321

Liversidge, H., Molleson, T. 2004: 'Variation in crown and root formation and eruption of human deciduous teeth' *American Journal of Physical Anthropology* 123: 172–80

Liversidge, H., Herdeg, B., Rosing, F.W. 1998: 'Dental age estimation of non-adults: a review of methods and principles' in K.W. Alt, R.W. Rosing, M. Teschler-Nicola (eds), *Dental Anthropology, Fundamentals, Limits and Prospects*. Vienna, Springer: 419–42

Lockie, J.D., Stephen, D. 1959: 'Eagles, lambs and land management on Lewis' *Journal of Animal Ecology* 28: 43–50

Lodge, D.M. 1991: 'Herbivory on freshwater macrophytes' *Aquatic Botany* 41: 195–224

Lodge, D.M., Kelly, P. 1985: 'Habitat disturbance and the stability of freshwater gastropod populations' *Oecologia* 68: 111–17

Long, S.P. 1999: 'Environmental responses' in R.F. Sage, R.K. Monson (eds), *C_4 Plant Biology*. New York, Academic Press: 215–49

Longin, R. 1971: 'New method of collagen extraction for radiocarbon dating' *Nature* 230: 241–42

Lopes-Lima, M., Seddon, M.B. 2014: Unio mancus. *The IUCN Red List of Threatened Species 2014*: e.T22737A42466471. http://dx.doi.org/10.2305/IUCN.UK.2014-3.RLTS.T22737A42466471.en

López, P.R. 2012: *The Built Environment and Public Health*. San Francisco, Jossey-Bass

Lorentz, K.O. 2010: 'Parts to a whole: manipulations of the body in prehistoric Eastern Mediterranean' in K. Rebay-Salisbury, M.L.S. Sørensen, J. Hughes (eds), *Body Parts and Bodies Whole: Changing Relations and Meanings*. Oxford, Oxbow Books: 20–29

Love, S. 2012: 'The geoarchaeology of mudbricks in architecture: a methodological study from Çatalhöyük, Turkey' *Geoarchaeology* 27.2: 140–56

Lovejoy, C.O., Meindl, R.S., Pryzbeck, T.R., Mensforth, R.P. 1985: 'Chronological metamorphosis of the auricular surface of the ilium - a new method for the determination of adult skeletal age at death' *American Journal of Physical Anthropology* 68.1: 15–28

Lovell, N.C. 1997: 'Trauma analysis in paleopathology' *Yearbook of Physical Anthropology* 40: 139–70

Lukacs, J.R. 2008: 'Fertility and agriculture accentuate sex differences in dental caries rates' *Current Anthropology* 49.5: 901–14

— 2011: 'Sex differences in dental caries experience: clinical evidence, complex etiology' *Clinical Oral Investigations* 15.5: 649–56

Lyman, R.L. 2008: *Quantitative Paleozoology*. Cambridge, Cambridge University Press

Lyman, R.L., Vanpool, T.L. 2009: 'Metric data in archaeology: a study of intra-analyst and inter-analyst variation' *American Antiquity* 74: 485–504

Mace, R. 1993: 'Nomadic pastoralists adopt subsistence strategies that maximise long-term household survival' *Behavioral Ecology and Sociobiology* 33: 329–34

Machiafava, V., Bonucci, L., Ascenzi, A. 1974: 'Fungal osteocalasia: a model of dead bone resorption' *Calcified Tissue Research* 14: 195–210

Macintosh, A.A., Pinhasi, R., Stock, J.T. 2016: 'Early life conditions and physiological stress following the transition to farming in Central/ Southeast Europe: skeletal growth impairment and 6000 years of gradual recovery' *PLoS ONE* 11.2: e0148468

Macko, S.A., Estep, M.L.F., Hare, P.L., Hoering, T.C. 1982: 'Stable nitrogen and carbon isotopic composition of individual amino acids isolated from cultured microorganisms' *Annual Report of the Director, Geophysical Laboratory, Carnegie Institution, Washington, 1981–1982*: 404–10

Macphail, R.I., Courty, M.A., Hartner, J., Wattez, J. 1997: 'The soil micromorphological evidence of domestic occupation and stabling activities' in R. Maggi (ed.), *Arene Candide: A Functional and Environmental Assessment of the Holocene Sequences Excavated by Bernardo Brea (1940–1950)*. Rome, Istituto Italiano di Paleontologia Umana: 53–88

Madella, M., García-Granero, J. 2014. 'Phytoliths and starch grains' *Çatalhöyük 2014 Archive Report*. http://www.catalhoyuk.com/archive_reports/2014

Madella, M., Lancelotti, C., Ruiz-Giralt, A., Barastegui, A., García-Granero, J. 2015: 'Phytoliths' *Çatalhöyük 2015 Archive Report*. http://www.catalhoyuk.com/archive_reports/2015

Maeda, O., Lucas, L., Silva, F., Tanno, K.-I., Fuller, D.Q. 2016: 'Narrowing the harvest: increasing sickle investment and the rise of domesticated cereal agriculture in the Fertile Crescent' *Quaternary Science Reviews* 145: 226–37

Maher, L.A., Richter, T., Macdonald, D., Jones, M.D., Martin, L., Stock, J.T. 2012: 'Twenty thousand-year-old huts at a hunter-gatherer settlement in Eastern Jordan' *PLoS ONE* 7.2: e31447

Maicher, C., Hoffmann, A., Côté, N.M.L., Palomo Pérez, A., Saña Segui, M., Le Bailly, M. 2017: 'Paleoparasitological investigations on the Neolithic lakeside settlement of La Draga (Lake Banyoles, Spain)' *Holocene* 27: 1659–68

Malina, R., Little, B.B., Bushang, P.H., DeMoss, J., Selby, H.A. 1985: 'Socioeconomic variation in the growth status of children in a subsistence agricultural community' *American Journal of Physical Anthropology* 68: 385–91

Malmström, H., Storå, J., Dalén, L., Holmlund, G., Götherström, A. 2005: 'Extensive human DNA contamination in extracts from ancient dog bones and teeth' *Molecular Biology and Evolution* 22.10: 2040–47

Malmström, H., Gilbert, M.T.P., Thomas, M.G., Brandström, M., Storå, J., Molnar, P., Andersen, P.K., Bendixen, C., Holmlund, G., Götherström, A., Willerslev, E. 2009: 'Ancient DNA reveals lack of continuity between Neolithic hunter-gatherers and contemporary Scandinavians' *Current Biology* 19.20: 1758–62

Manning, K., Timpson, A., Shennan, S., Crema, E. 2015: 'Size reduction in early European domestic cattle relates to intensification of Neolithic herding strategies' *PLoS One* 10: e0141873

Marciniak, A. 2013: 'Çatalhöyük Area TP zooarchaeology'. Released: 2013-03-02. Open Context. <http://opencontext.org/projects/02594C48-7497-40D7-11AE-AB942DC513B8> https://doi.org/10.6078/M7B8562H ARK (Archive): https://n2t.net/ark:/28722/k2tb10t14

Marciniak, A., Czerniak, L. 2007: 'Social transformations in the Late Neolithic and the Early Chalcolithic periods in Central Anatolia' *Anatolian Studies* 57: 115–30

Marciniak, A., Barański, M.Z., Bayliss, A., Czerniak, L., Goslar, T., Southon, J., Taylor, R.E. 2015a: 'Fragmenting times: interpreting a Bayesian chronology for the Late Neolithic occupation of Çatalhöyük East, Turkey' *Antiquity* 89: 154–76

Marciniak, A., Asouti, E., Doherty, C., Henton, E. 2015b: 'The nature of household in the upper levels at Çatalhöyük' in I. Hodder, A. Marciniak (eds), *Assembling Çatalhöyük*. Leeds, Maney Publishing: 151–65

Marguerie, D., Hunot, J.-Y. 2007: 'Charcoal analysis and dendrology: data from archaeological sites in northwestern France' *Journal of Archaeological Science* 34: 1417–33

Marom, N., Bar-Oz, G. 2009: 'Culling profiles: the indeterminacy of archaeozoological data to survivorship curve modelling of sheep and goat herd maintenance strategies' *Journal of Archaeological Science* 36: 1184–87

— 2013: 'The prey pathway: a regional history of cattle (*Bos taurus*) and pig (*Sus scrofa*) domestication in the Northern Jordan Valley, Israel' *PLoS One* 8: e55958

Martin, L., Meskell, L. 2012: 'Animal figurines from Neolithic Çatalhöyük: figural and faunal perspectives' *Cambridge Archaeological Journal* 22: 401–19

Marwick, B. 2016: 'Computational reproducibility in archaeological research: basic principles and a case study of their implementation' *Journal of Archaeological Method and Theory* 24: 424–50

Mathieson, I., Lazaridis, I., Rohland, N., Mallick, S., Patterson, N., Roodenberg, S.A., Harney, E., Stewardson, K., Fernandes, D., Novak, M., Sirak, K., Gamba, C., Jones, E.R., Llamas, B., Dryomov, S., Pickrell, J., Arsuaga, J.L., de Castro, J.M.B., Carbonell, E., Gerritsen, F., Khokhlov, A., Kuznetsov, P., Lozano, M., Meller, H., Mochalov, O., Moiseyev, V., Guerra, M.A.R., Roodenberg, J., Vergès, J.M., Krause, J., Cooper, A., Alt, K.W., Brown, D., Anthony, D., Lalueza-Fox, C., Haak, W., Pinhasi, R., Reich, D. 2015: 'Genome-wide patterns of selection in 230 ancient Eurasians' *Nature* 528.7583: 499–503

Matthews, W. 2005a: 'Micromorphological and microstratigraphic traces of uses and concepts of space' in I. Hodder (ed.), *Inhabiting Çatalhöyük: Reports from the 1995–99 Seasons*. London, British Institute at Ankara; Cambridge, McDonald Institute for Archaeological Research: 355–98, 553–72

— 2005b: 'Life-cycles and life-courses of buildings' in I. Hodder (ed.), *Çatalhöyük Perspectives: Themes from the 1995–99 Seasons*. London, British Institute at Ankara; Cambridge, McDonald Institute for Archaeological Research: 125–49

— 2012: 'Household life-histories and boundaries: microstratigraphy and micromorphology of architectural surfaces in Building 3 of the BACH Area' in R. Tringham, M. Stevanovic (eds), *Last House on the Hill: BACH Area Reports from Çatalhöyük, Turkey*. Los Angeles, Cotsen Institute of Archaeology Press: 205–24

— 2016: 'Humans and fire: changing relations in early agricultural and built environments in the Zagros, Iran, Iraq' *The Anthropocene Review* 3: 107–139

— 2018: 'Creating settled life: micro-histories of community, ritual and place - the Central Zagros and Çatalhöyük' in I. Hodder (ed.), *Religion History and Place: The Origins of Settled Life*. Denver, University of Colorado Press: 64–98

Matthews, W., Portillo, M. 2017: 'Built environment and livestock dung: integrated micromorphology, phytolith and chemical analyses' *Çatalhöyük 2017 Archive Report*. http://www.catalhoyuk.com/archive_reports/2017

Matthews, W., French, C.A.I., Lawrence, T., Cutler, D.F. 1996: 'Multiple surfaces: the micromorphology' in I. Hodder (ed.), *On the Surface: Çatalhöyük 1993–1995*. London, British Institute at Ankara; Cambridge, McDonald Institute for Archaeological Research: 301–42

Matthews, W., Hastorf, C., Andrews, P., Molleson, T. 2000: 'Integrating archaeological science' in I. Hodder (ed.), *Towards Reflexive Method in Archaeology: The Example at Çatalhöyük*. London, British Institute at Ankara; Cambridge, McDonald Institute for Archaeological Research: 37–50

Matthews, W., Shillito, L.-M., Almond, M.J., Walker, G., Bowen, J., Anderson, E., Koromila, G. 2010: 'Integrated micro-contextual approaches: micromorphology and sub-surface terahertz imaging of architectural materials and paintings, and organic residue, phytolith and geochemical analysis of middens' *Çatalhöyük 2010 Archive Report*. http://www.catalhoyuk.com/archive_reports/2010

Matthews, W., Almond, M.J., Anderson, E., Wiles, J., Williams, H., Rowe, J. 2013: 'Biographies of architectural materials and buildings: integrating high-resolution micro-analysis and geochemistry' in I. Hodder (ed.), *Substantive Technologies at Çatalhöyük: Reports from the 2000–2008 Seasons*. London, British Institute at Ankara; Los Angeles, Cotsen Institute of Archaeology Press: 115–36

Matthews, W., Shillito, L.-M., Elliott, S., Bull, I.D., Williams, J. 2014: 'Neolithic lifeways: microstratigraphic traces within houses, animal pens and settlements' in A. Whittle, P. Bickle (eds), *Early Farmers: The View from Archaeology and Science*. London, British Academy: 251–79

May, H., Ruff, C. 2016: 'Physical burden and lower limb bone structure at the origin of agriculture in the Levant' *American Journal of Physical Anthropology* 161: 26–36

Mayhew, D.F. 1977: 'Avian predators as accumulators of fossil mammal material' *Boreas* 6: 25–31

Mays, S., Ives, R., Brickley, M. 2009: 'The effects of socioeconomic status on endochondral and appositional bone growth, and acquisition of cortical bone in children from 19th century Birmingham, England' *American Journal of Physical Anthropology* 140: 410–16

Mazzucato, C. 2019: 'Socio-material archaeological networks at Çatalhöyük: a community detection approach' *Frontiers in Digital Humanities* 6: 8
McCammon, R. 1970: *Growth and Development*. Springfield, Charles C. Thomas
McCarthy, M.A., Masters, P.I.P. 2005: 'Profiting from prior information in Bayesian analyses of ecological data' *Journal of Applied Ecology* 42: 1012–19
McElreath, R.L. 2016a: *Rethinking: Statistical Rethinking Book Package*. R package version 1.59
— 2016b: *Statistical Rethinking: A Bayesian Course with Examples in R and Stan*. London, Chapman & Hall/CRC
Meadow, R. 1984: 'Animal domestication in the Middle East: a view from the eastern margin' in J. Clutton-Brock, C. Grigson (eds), *Animals and Archaeology 3: Early Herders and their Flocks* (BAR International Series 202). Oxford, British Archaeological Reports: 309–37
— 1999: 'The use of size index scaling techniques for research on archaeozoological collections from the Middle East' in C. Becker, H. Manhart, J. Peters, J. Schibler (eds), *Historia Animalium ex Ossibus: Beiträge zur Paläoanatomie, Archäologie, Ägyptologie, Ethnologie und Geschichte der Tiermedizin*. Leidorf, Verlag Marie: 285–300
Meehan, B. 1982: *Shell Bed to Shell Midden*. Canberra, Australian Institute for Aboriginal Studies
Mellaart, J. 1962: 'Excavations at Çatal Hüyük: first preliminary report, 1961' *Anatolian Studies* 12: 41–65
— 1963: 'Excavations at Çatal Hüyük, 1962: second preliminary report' *Anatolian Studies* 13: 43–103
— 1964: 'Excavations at Çatal Hüyük, 1963, third preliminary report' *Anatolian Studies* 14: 39–119
— 1965: 'Çatal Hüyük: a Neolithic city in Anatolia' *Proceedings of the British Academy* 51: 201–13
— 1966: 'Excavations at Catal Hüyük, 1965: fourth preliminary report' *Anatolian Studies* 16: 165–91
— 1967: *Çatal Hüyük: A Neolithic Town in Anatolia*. London: Thames and Hudson
Mellars, P., Stringer, C. 1989: *The Human Revolution: Behavioural and Biological Perspectives on the Origins of Modern Humans*. Edinburgh, Edinburgh University Press
Mellett, J.S. 1974: 'Scatological origin of microvertebrate fossil accumulations' *Science* 5: 350
Melville, M.D., Atkinson, G. 1985: 'Soil colour: its measurement and its designation in models of uniform colour space' *Journal of Soil Science* 36.4: 495–512
Menozzi, P., Piazza, A., Cavalli-Sforza, L. 1978: 'Synthetic maps of human gene frequencies in Europeans' *Science* 201.4358: 786–92
Merrett, D.C., Meiklejohn, C. 2007: 'Is House 12 at Bouqras a charnel house?' in M. Faerman, L.K. Horwitz, T. Kahana, U. Zilberman (eds), *Faces from the Past. Skeletal Biology of Human Populations from the Eastern Mediterranean* (BAR International Series 1603). Oxford, Archaeopress: 127–39
Meskell, L.M. 1999: *Archaeologies of Social Life: Age, Sex, Class et cetera in Ancient Egypt*. Oxford, Blackwell
— 2001: 'Archaeologies of identity' in I. Hodder (ed.), *Archaeological Theory Today*. Cambridge, Polity Press: 187–213
Metcalf, P. 1981: 'Meaning and materialism. The ritual economy of death' *Man* (New Series) 16.4: 563–78
Mickleburgh, H.L., Wescott, D.J. 2018: 'Controlled experimental observations on joint disarticulation and bone displacement of a human body in an open pit: implications for funerary archaeology' *Journal of Archaeological Science: Reports* 20: 158–67
Middleton, C. 2014: *Stable Isotope Evidence for the Development of Herding in Central Anatolia, 13,000–5,000 BC*. PhD thesis, University of Liverpool, Liverpool
— 2018: 'The beginning of herding and animal management: the early development of caprine herding on the Konya Plain, Central Anatolia' *Anatolian Studies* 68: 1–31
Milano, S., Prendergast, A.L., Schöne B.R. 2016: 'Effects of cooking on mollusk shell structure and chemistry: implications for archaeology and paleoenvironmental reconstruction' *Journal of Archaeological Science: Reports* 7: 14–26
Milella, M., Knüsel, C.J., Haddow, S.D. 2016: 'A Neolithic case of fibrous dysplasia from Çatalhöyük (Turkey)' *International Journal of Paleopathology* 15: 10–18
Milella, M., Betz, B.J., Knüsel, C.J., Larsen, C.S., Dori, I. 2018: 'Population density and developmental stress in the Neolithic: A diachronic study of dental fluctuating asymmetry at Çatalhöyük (Turkey, 7,100–5,950 BC)' *American Journal of Physical Anthropology* 167.4: 737–49
Millard, A.R. 2001: 'The deterioration of bone' in D.R. Brothwell, A.M. Pollard (eds), *Handbook of Archaeological Sciences*. Chichester, John Wiley and Sons: 637–47
Miller, C.E., Conard N.J., Goldberg, P., Berna, F. 2010: 'Dumping, sweeping and trampling: experimental micromorphological analysis of anthropogenically modified combustion features' in I. Théry-Parisot, L. Chabal, S. Costamagno (eds), *The Taphonomy of Burned Organic Residues and Combustion Features in Archaeological Contexts: Proceedings of the Round Table, Valbonne, May 27–29, 2008. P@lethnologie* 2: 25–37

Mills, B. 2014: 'Relational networks and religious sodalities at Çatalhöyük' in I. Hodder (ed.), *Religion at Work in a Neolithic Society: Vital Matters*. Cambridge, Cambridge University Press: 159–86

Mitchell, P.D. 2013: 'The origins of human parasites: exploring the evidence for endoparasitism throughout human evolution' *International Journal of Paleopathology* 3: 191–98

— 2015: 'Human parasites in Medieval Europe: lifestyle, sanitation and medical treatment' *Advances in Parasitology* 90: 389–420

— 2017: 'Human parasites in the Roman World: health consequences of conquering an empire' *Parasitology* 144: 48–58

Mitrović, S., Vasić, M. 2013: 'An integrated perspective on the uses of materials at Çatalhöyük based on the analysis of heavy residues' in I. Hodder (ed.), *Substantive Technologies at Çatalhöyük: Reports from the 2000–2008 Seasons*. London, British Institute at Ankara; Los Angeles, Cotsen Institute of Archaeology Press: 27–49

Mittnik, A., Massy, K., Knipper, C., Wittenborn, F., Pfrengle, S., Carlichi-Witjes, N., Deeg, H., Furtwängler, A., Harbeck, M., Heyking, K., Kociumaka, C., Kucukkalipci, I., Lindauer, S., Metz, S., Staskiewicz, A., Thiel, A.,Wahl, J., Haak, W., Pernicka, E., Schiffels, S., Stockhammer, P.W., Krause, J. 2019: 'Kinship-based social inequality in Bronze Age Europe' *Science* 366: 731–34

Molleson, T., Andrews, P., Boz, B., 2005: 'Reconstruction of the Neolithic people of Çatalhöyük' in I. Hodder (ed.), *Inhabiting Çatalhöyük: Reports from the 1995–99 Seasons*. London, British Institute at Ankara; Cambridge, McDonald Institute for Archaeological Research: 279–300

Molnar, S. 1971: 'Human tooth wear, tooth function and cultural variability' *American Journal of Physical Anthropology* 34: 175–89

— 1972: 'Tooth wear and culture: a survey of tooth functions among some prehistoric populations' *Current Anthropology* 13: 511–26

Monchot, H., Léchelle, J. 2002: 'Statistical nonparametric methods for the study of fossil populations' *Paleobiology* 28: 55–69

Monchot, H., Mashkour, M., Vigne, J.-D. 2005: 'Kernel smoothing and mixture analyses for the determination of the sex ratios at death, at the beginning of the domestication of ungulates' in J.-D. Vigne, J. Peters, D. Helmer (eds), *First Steps of Animal Domestication: New Archaeozoological Approaches*. Oxford, Oxbow Books: 55–60

Moore, A.M.T., Molleson, T.I. 2000: 'Disposal of the dead' in A.M.T. Moore, G.C. Hillman, A.J. Legge (eds), *Village on the Euphrates: From Foraging to Farming at Abu Hureyra*. Oxford, Oxford University Press: 277–300

Moore, A.M.T., Hillman, G.C., Legge, A.J. 2000: *Village on the Euphrates: From Foraging to Farming at Abu Hureyra*. Oxford, Oxford University Press

Moore, J., Scott, E. (eds) 1997: *Invisible People and Processes: Writing Gender and Childhood into European Archaeology*. London, Leicester University Press

Moorrees, C., Fanning, E., Hunt, E. 1963: 'Age variation of formation stages for ten permanent teeth' *Dental Research* 42.6: 1490–1502

Moorrees, C.F.A., Reed, R.B. 1964; 'Correlations among crown diameters of human teeth' *Archives of Oral Biology* 9: 685–97

Morel, P. 1990: 'Quelques remarques à propos de coquilles d'oeufs découvertes dans une tombe de l'époque augustéenne à Sion Petit-Chasseur' in J. Schibler, J. Sedlmeier (eds), *Festschrift für Hans R. Stampfli. Beiträge zur Archäozoologie, Archäologie, Anthropologie, Geologie und Paläontologie*. Basel, Helbing & Lichtenhahn: 141–46

Morales, M.M., Giannini, N.P. 2013: 'Ecomorphology of the African felid ensemble: the role of the skull and postcranium in determining species segregation and assembling history' *Journal of Evolutionary Biology* 26: 980–92

Morris, C., Bebeau, N.P., Brockhoff, H., Tandon, R., Tiwana, P. 2015: 'Mandibular fractures: an analysis of the epidemiology and patterns of injury in 4,143 fractures' *Journal of Oral Maxillofacial Surgery* 73.5: 951.e1-951.e12

Moses, S. 2012: 'Sociopolitical implications of Neolithic foundation deposits and the possibility of child sacrifice: a case study at Çatalhöyük, Turkey' in A. Porter and G.M. Schwartz (eds), *Sacred Killing: The Archaeology of Sacrifice in the Ancient Near East*. Winona Lake, Eisenbrauns: 57–88

Moss, M.L. 1993: 'Shellfish, gender, and status on the Northwest Coast: reconciling archaeological, ethnographic and ethnohistorical records of the Tlingit' *American Anthropologist* 95.3: 631–52

Mouthon, J, Daufresne, M. 2008: 'Population dynamics and life cycle of *Pisidium amnicum* (Müller) (Bivalvia: Sphaeriidae) and *Valvata piscinalis* (Müller) (Gastropoda: Prosobranchia) in the Saône river, a nine-year study' *International Journal of Limnology* 44.4: 241–51

Mulholland, S.C., Rapp, G.J. 1992: 'A morphological classification of grass silica-bodies' in G.J. Rapp, S.C. Mulholland (eds), *Phytolith Systematics: Emerging Issues, Advances in Archaeological and Museum Science*. New York, Plenum Press: 65–89

Müller, E.N., Wainwright, J., Parsons, A.J., Turnbull, L. (eds) 2013: *Patterns of Land-Degradation in Drylands: Understanding Self-Organized Ecogeomorphic Systems*. Dordrecht, Springer

Müller, P., Staudigel, P.T., Murray, S.T., Vernet, R., Barusseau, J.P., Westphal, H., Swart, P.K. 2017: 'Prehistoric cooking versus accurate palaeotemperature records in shell midden constituents' *Scientific Reports* 7.1: 3555 https://doi.org/10.1038/s41598-017-03715-8

Mulville, J., Twiss, K.C., Wolfhagen, J., Demirergi, G.A., Orton, D., Foster, H., Daujat, J., Madgwick, R., Jones, J. R., Knüsel, C. 2014: 'Faunal remains' *Çatalhöyük Archive Report*. http://www.catalhoyuk.com/archive_reports/2014

Mummert, A., Esche, E., Robinson, J., Armelagos, G.J. 2011: 'Stature and robusticity during the agricultural transition: evidence from the bioarchaeological record' *Economics and Human Biology* 9: 284–301

Munsell Colour Company 1994: *Munsell Soil Color Charts*. New Windsor, Macbeth Division of Kollmorgen Instruments Corporation

Myzyk, S. 2007: 'Life cycle of *Valvata piscinalis* (O.F. Müller, 1774) (Gastropoda: Heterobranchia) in the laboratory' *Folia Malacologica* 15.4: 145–74

Nadel, D. 2002: *Ohalo II. A 23,000-Year-Old Fisher-Hunter-Gatherer's Camp on the Shore of the Sea of Galilee*. Haifa, Reuben and Edith Hecth Museum, University of Haifa

Nadel, D., Piperno, D.R., Holst, I., Snir, A., Weiss, E. 2012: 'New evidence for the processing of wild cereal grains at Ohalo II, a 23,000 year-old campsite on the shore of the Sea of Galilee, Israel' *Antiquity* 86: 990–1003

Nagel, K.O, Castagnolo, L. 1991: 'Fish hosts for the glochidium of *Unio mancus*' *Rivista di Idrobiologia* 30.2–3: 339–46

Nakamura, C., Meskell, L.M. 2009: 'Articulate bodies: forms and figures at Çatalhöyük' *Journal of Archaeological Method and Theory* 16: 205–30

— 2013a: 'Figurine worlds at Çatalhöyük' in I. Hodder (ed.), *Substantive Technologies at Çatalhöyük: Reports from the 2000–2008 Seasons*. London, British Institute at Ankara; Cambridge, McDonald Institute for Archaeological Research: 201–34

— 2013b: 'The Çatalhöyük burial assemblage' in I. Hodder (ed.), *Humans and Landscapes of Çatalhöyük: Reports from the 2000–2008 Seasons*. London, British Institute at Ankara; Cambridge, McDonald Institute for Archaeological Research: 441–66

Nanson, G.C., Knighton, A.D. 1996: 'Anabranching rivers: their cause, character and classification' *Earth Surface Processes and Landforms* 21.3: 217–39

Neumann, K., Strömberg, A.E.C., Ball, T., Albert, R.M., Vrydaghs, L., Scott-Cummings, L. (International Committee for Phytolith Taxonomy ICPT). 2019: 'International Code for Phytolith Nomenclature (ICPN) 2.0' *Annals of Botany*

Newton, M.W. 1996: *Dendrochronology at Çatalhöyük: A 576-year Tree-Ring Chronology for the Early Neolithic of Anatolia*. M.A. thesis, Cornell University, Ithaca

NGRIP Dating Group 2006: *Greenland Ice Core Chronology 2005. (GICC05). IGBP PAGES/World Data Center for Paleoclimatology Data Contribution Series #2006-118*. Boulder, NOAA/NCDC Paleoclimatology Program

Nissen, H. 1988: *The Early History of the Ancient Near East, 9000–2000 B.C.* Chicago, University of Chicago Press

Novak, S.A. 2000: 'Battle-related trauma' in V. Fiorato, A.E. Boylston, C.J. Knüsel (eds), *Blood Red Roses: The Archaeology of a Mass Grave from Towton, AD 1461*. Oxford, Oxbow Books: 90–102

— 2017: 'On the stories of men and the substance of women: interrogating gender through violence' in S.C. Agarwal, J.K. Wesp (eds), *Exploring Sex and Gender in Bioarchaeology*. Albuquerque, University of New Mexico Press: 128–64

O'Brien, M.J., Laland, K.N. 2012: 'Genes, culture, and agriculture: an example of human niche construction' *Current Anthropology* 53: 434–70

O'Connor, T.P. 1988: *Bones from the General Accident Site, Tanner Row* (Archaeology of York 15/2). London: Council for British Archaeology

— 2000: *The Archaeology of Animal Bones*. Gloucestershire, Sutton Publishing Limited

O'Connor, T.P. 2013: *Animals as Neighbours: The Past and Present of Commensal Animals*. East Lansing, Michigan State University Press

O'Leary, M.H. 1981: 'Carbon isotope fractionation in plants' *Phytochemistry* 20: 553–67

O'Neill, M.C., Ruff, C. 2004: 'Estimating human long bone cross-sectional geometric properties: a comparison of noninvasive methods' *Journal of Human Evolution* 47: 221–35

O'Shea, J.M. 1984: *Mortuary Variability: An Archaeological Investigation*. New York, Academic Press

Odabaşı, D.A., Glöer, P., Yildirim, M.Z. 2015: 'The *Valvata* species of Turkey with a description of *Valvata kebapcii* n. sp. (Mollusca: Valvatidae)' *Ecologica Montenegrina* 2.2: 135–42

Okumura, M., Eggers, S. 2014: 'Cultural formation processes of the bioarchaeological record of a Brazilian shell mound' in M. Roksandic, S.M. de Souza, S. Eggers, M. Burchell, D. Klokler (eds), *The Cultural Dynamics of Shell-Matrix Sites*. Albuquerque, University of New Mexico Press: 103–12

Omrak, A., Günther, T., Valdiosera, C., Svensson, E.M., Malmström, H., Kiesewetter, H., Aylward, W., Storå, J., Jakobsson, M., Götherström, A. 2016: 'Genomic evidence establishes Anatolia as the source of the European Neolithic gene pool' *Current Biology* 26.2: 270–75

Orton, D. 2010: 'Animal bones 2010: West Mound Trench 5' *Çatalhöyük 2010 Archive Report*. http://www.catalhoyuk.com/archive_reports/2010

Orton, D., Russell, N., Twiss, K., Martin, L., Frame, S. 2013: 'Çatalhöyük zooarchaeology'. Released: 2013-02-13. Open Context. <http://opencontext.org/projects/1B426F7C-99EC-4322-4069-E8DBD927CCF1> ARK (Archive): https://n2t.net/ark:/28722/k2f47js2k (link: https://opencontext.org/projects/1B426F7C-99EC-4322-4069-E8DBD927CCF1)

Orton, D., Anvari, J, Gibson, C., Last, J., Bogaard, A., Rosenstock, E., Biehl, P. 2018: 'A tale of two tells: dating the Çatalhöyük West Mound' *Antiquity* 92: 620–39

Osborn, D.J., Helmy, I. 1980: *The Contemporary Land Mammals of Egypt Including Sinai* (Fieldiana Zoology new series 5). Chicago, Field Museum of Natural History

Österling, M., Schneider, L.D. 2012: The thick-shelled river mussel (*Unio crassus*) – host fish suitability and rearing of juvenile mussels' in *The Thick-Shelled River Mussel Brings Back Life to Rivers*, LIFE10 NAT/SE/000046, http://www.ucforlife.se/wp-content/uploads/2012/12/7.2.3_TECHNICAL-REPORT_C1b.pdf

Out, W., Madella, M. 2017: 'Towards improved detection and identification of crop by-products: morphometric analysis of bilobate leaf phytoliths of *Pennisetum glaucum* and *Sorghum bicolor*' *Quaternary International* 434: 1–14

Owsley, D.W. 1991: 'Temporal variations in femoral cortical thickness of North American Plains Indians' in D. Ortner, A. Aufderheide (eds), *Human Paleopathology. Current Syntheses and Future Options*. Washington, D.C., Smithsonian Institution Press: 105–09

Özbaşaran, M. 2011: 'The Neolithic on the plateau' in S.R. Steadman, G. McMahon (eds), *The Oxford Handbook of Ancient Anatolia*. Oxford, Oxford University Press: 99–124

Özbek, M. 2009: 'Remodeled human skulls in Köşk Höyük (Neolithic Age, Anatolia): a new appraisal in view of recent discoveries' *Journal of Archaeological Science* 36: 379–86

Özdoğan, M. 2010: 'Westward expansion of the Neolithic way of life: sorting the Neolithic package into distinct packages' in P. Matthiae, F. Pinnock, L. Nigro, N. Marchetti (eds), *Proceedings of the 6th International Congress on the Archaeology of the Ancient Near East*. Wiesbaden, Harrassowitz Verlag: 883–98

— 2011a: 'Eastern Thrace: the contact zone between Anatolia and the Balkans' in S.R. Steadman, G. McMahon (eds), *The Oxford Handbook of Ancient Anatolia*. Oxford, Oxford University Press: 657–82

— 2011b: 'Archaeological evidence on the westward expansion of farming communities from Eastern Anatolia to the Aegean and the Balkans' *Current Anthropology* 52: S415–S430

Özgül, A., Tuljapurkar, S., Benton, T.G., Pemberton, J.M., Clutton-Brock, T.H., Coulson, T. 2009: 'The dynamics of phenotypic change and the shrinking sheep of St. Kilda' *Science* 325: 464–67

Özkaya, V., Coçkun, A. 2011: 'Körtik Tepe' in M. Özdogan, N. Baçgelen, P. Kuniholm (eds), *The Neolithic in Turkey: New Excavations and New Research* 1. *The Tigris Basin*. Istanbul, Archaeology and Art Publications: 89–127

Pagán-Jiménez, J.R. 2015: *Almidones. Guía de Material Comparativo Moderno del Ecuador para los Estudios Paleoetnobotánicos en el Neotrópico*. Buenos Aires, Aspha Ediciones

Pagán-Jiménez, J.R., Guachamín-Tello, A.M., Romero-Bastidas, M.E., Constantine-Castro, A.R. 2016: 'Late ninth millennium B.P. use of *Zea mays L.* at Cubilán area, highland Ecuador, revealed by ancient starches' *Quaternary International* 404: 137–55

Palaver, W. 2019: 'Death at Çatalhöyük' in I. Hodder (ed.), *Violence and the Sacred in the Ancient Near East: Girardian Conversations at Çatalhöyük*. Cambridge, Cambridge University Press: 41–59

Palmer, A.R., Strobeck, C. 2003: 'Fluctuating asymmetry analyses revisited' in M. Polak (ed.), *Developmental Instability: Causes and Consequences*. Oxford, Oxford University Press: 279–319

Panagiotakopulu, E., Buckland, P.C., Day, P.M. 1995: 'Natural insecticides and insect repellents in antiquity: a review of the evidence' *Journal of Archaeological Science* 22: 705–10
Papageorgopoulou, C., Shved, N., Wanek, J., Rühli, F.J. 2015: 'Modeling Ancient Egyptian mummification on fresh human tissue: macroscopic and histological aspects' *The Anatomical Record* 298: 974–87
Paradis, S., Dufraisse, A., Allée, P. 2013: 'Radius of curvature measurements and wood diameter: a comparison of different image analysis techniques' in F. Damblon (ed.), *Proceedings of the 4th International Meeting of Anthracology. Brussels, 8–13 September 2008. Royal Belgian Institute of Natural Sciences* (BAR International Series 2486). Archaeopress, Oxford: 173–82
Parker Pearson, M. 1982: 'Mortuary practices, society and ideology: an ethnoarchaeological study' in I. Hodder (ed.), *Symbolic and Structural Archaeology*. Cambridge, Cambridge University Press: 99–113
— 1993: 'The powerful dead: archaeological relationships between the living and the dead' *Cambridge Archaeological Journal* 3.2: 203–29
— 1999: *The Archaeology of Death and Burial*. Stroud, Sutton Publishing Limited
Parker Pearson, M., Chamberlain, A., Craig, O., Marshall, P., Mulville, J., Smith, H., Chenery, C., Collins, M., Cook, G., Craig, G., Evans, J., Hiller, J., Montgomery, J., Schwenninger, J.-L., Taylor, G., Wess, T. 2005: 'Evidence for mummification in Bronze Age Britain' *Antiquity* 79.305: 529–46
Parsons, A.J., Abrahams, A.D. (eds) 2009: *Geomorphology of Desert Environments* (2nd edition). Berlin, Springer
Parsons, A.J., Wainwright, J., Stone, P.M., Abrahams, A.D. 1999: 'Transmission losses in rills on dryland hillslopes' *Hydrological Processes* 13: 2897–2905
Patterson, N., Moorjani, P., Luo, Y., Mallick, S., Rohland, N., Zhan, Y., Genschoreck, T., Webster, T., Reich, D. 2012: 'Ancient admixture in human history' *Genetics* 192.3: 1065–93
Patton, K.C., Hager, L.D. 2014: '"Motherbaby": A death in childbirth at Çatalhöyük' in I. Hodder (ed.), *Religion at Work in a Neolithic Society: Vital Matters*. Cambridge, Cambridge University Press: 225–58
Pawłowska, K. in prep.: 'Animals at Late Neolithic Çatalhöyük: subsistence, food processing and depositional activities' in A. Marciniak (ed.), *Çatalhöyük Research Project monograph* (TP volume 2)
Payne, S. 1973: 'Kill-off patterns in sheep and goats: the mandibles from Aşvan Kale' *Anatolian Studies* 23: 281–303
Pearson, J.A. 2004: *Ancient Diet in Neolithic Anatolia: Isotope Analysis of Biological Remains and their Archaeological Implications*. D. Phil. thesis, Keble College, Oxford
— 2013: 'Human and animal diet as evidenced by stable carbon and nitrogen isotope analysis' in I. Hodder (ed.), *Humans and Landscapes of Çatalhöyük: Reports from the 2000–2008 Seasons*. London, British Institute at Ankara; Los Angeles, Cotsen Institute of Archaeology Press: 271–98
— 2018: 'Biocultural influences of total versus exclusive breastfeeding: stable isotope evidence of European and Asian trends for the last 10,000 years' in S.C. Agarwal, P. Beauchesne (eds), *Children and Childhood in the Past*. Gainesville, University Press of Florida: 61–97
Pearson, J.A., Grove, M.J. 2013: 'Counting sheep: sample size and statistical inference in stable isotope analysis' *World Archaeology* 47: 373–87
Pearson, J.A., Meskell, L.M. 2013: 'Isotopes and images: fleshing out bodies at Çatalhöyük' *Journal of Archaeological Method and Theory* 22: 461–82
— 2014: 'Biographical bodies: flesh and food at Çatalhöyük' in A. Whittle, P. Bickle (eds), *Early Farmers: The View from Archaeology and Science* (Proceedings of the British Academy 198). Oxford, Oxford University Press: 215–32
Pearson, J.A., Buitenhuis, H., Hedges, R., Martin, L., Russell, N., Twiss, K. 2007: 'New light on early caprine herding strategies from isotope analysis: a case study from Neolithic Anatolia' *Journal of Archaeological Science* 34: 2170–79
Pearson, J.A., Grove, M., Özbek, M., Hongo, H. 2013: 'Food and social complexity at Çayönü Tepesi, southeastern Anatolia: stable isotope evidence of differentiation in diet according to burial practice and sex in the early Neolithic' *Journal of Anthropolical Archaeology* 32.2: 180–89
Pearson, J.A., Bogaard, A., Charles, M., Hillson, S., Larsen, C.S., Russell, N., Twiss, K. 2015a: 'Stable carbon and nitrogen isotope analysis at Neolithic Çatalhöyük: evidence for human and animal diet and their relationship to households' *Journal of Archaeological Science* 57: 69–79
Pearson, J.A., Haddow, S.D., Hillson, S.W., Knüsel, C.J., Larsen, C.S., Sadvari J. 2015b: 'Stable carbon and nitrogen isotope analysis and dietary reconstruction through the lifecourse at Neolithic Çatalhöyük, Turkey' *Journal of Social Archaeology* 15: 210–32

Pedersen, J.S., Valen, E., Velazquez, A.M.V., Parker, B.J., Rasmussen, M., Lindgreen, S., Lilje, B., Tobin, D.J., Kelly, T.K., Vang, S., Andersson, R., Jones, P.A., Hoover, C.A., Tikhonov, A., Prokhortchouk, E., Rubin, E.M., Sandelin, A., Gilbert, M.T.P., Krogh, A., Willerslev, E., Orlando, L. 2014: 'Genome-wide nucleosome map and cytosine methylation levels of an ancient human genome' *Genome Research* 24.3: 454–66

Peres, T.M. 2017: 'Foodways archaeology: a decade of research from the southeastern United States' *Journal of Archaeological Research* 25: 421–60

Perkins, D. 1969: 'Fauna of Çatal Hüyük: evidence for early cattle domestication in Anatolia' *Science* 164: 177–79

Perlès, C., Takaoğlu, T., Gratuze, B. 2011: 'Melian obsidian in NW Turkey: evidence for early Neolithic trade' *Journal of Field Archaeology* 36.1: 42–49

Peršič, M. 1998: 'Dormouse hunting as part of Slovene national identity' *Natura Croatica* 7.3: 199–211

Perzigian, A.J. 1977: 'Fluctuating dental asymmetry – variation among skeletal populations' *American Journal of Physical Anthropology* 47: 81–88

Peters, J., Arbuckle, B., Pollath, N. 2015: 'Subsistence and beyond: animals in Neolithic Anatolia' in M. Özdoğan, N. Başgelen, P. Kuniholm (eds), *The Neolithic In Turkey* 6. Istanbul, Archaeology and Art Publications: 1–65

Pickrell, J.K., Reich, D. 2014: 'Toward a new history and geography of human genes informed by ancient DNA' *Trends in Genetics* 30.9: 377–89

Pilloud, M.A. 2009: *Community Structure at Neolithic Çatalhöyük: Biological Distance Analysis of Household, Neighborhood, and Settlement*. PhD thesis, Ohio State University, Columbus

— 2013: 'Postmarital residence in Neolithic Anatolia' *American Journal of Physical Anthropology* 150.S56: 221–22

Pilloud, M.A., Larsen, C.S. 2011: '"Official" and "practical" kin: inferring social and community structure from dental phenotype at Neolithic Çatalhöyük, Turkey' *American Journal of Physical Anthropology* 145.4: 519–30

Pilloud, M.A., Haddow, S.D., Knüsel, C.J., Larsen, C.S. 2016a: 'A bioarchaeological and forensic re-assessment of vulture defleshing and mortuary practices at Neolithic Çatalhöyük' *Journal of Archaeological Science: Reports* 10: 735–43

Pilloud, M.A., Edgar, H.J.H., George, R., Scott, G.R. 2016b: 'Dental morphology in biodistance analysis' in M.A. Pilloud, J.T. Hefner (eds), *Biological Distance Analysis: Forensic and Bioarchaeological Perspectives*. San Diego, Academic Press: 109–33

Pinto Llona, A.C.P., Andrews, P. 1999: 'Amphibian taphonomy and its application to the fossil record of Dolina (middle Pleistocene, Atapuerca, Spain)' *Palaeogeography, Palaeoclimatology, Palaeoecology* 149: 411–29

Piperno, D.R. 2006: *Phytoliths: A Comprehensive Guide for Archaeologists and Paleoecologists*. Lanham, AltaMira Press

Piperno, D., Weiss, E., Holst, I., Nadel, D. 2004: 'Processing of wild cereal grains in the Upper Palaeolithic revealed by starch grain analysis' *Nature* 430: 670–73

Poch, R.M., Artieda, O., Herrero, J., Lebedeva-Verba, M. 2010: 'Gypsic features' in G. Stoops, V. Marcelino, F. Mees (eds), *Interpretation of Micromorphological Features of Soils and Regoliths*. Amsterdam, Elsevier: 195–216

Poland, J.G. 2018: *A Methodological Approach to the Identification of Duck and Goose Remains from Archaeological Sites with an Application to Roman Britain*. PhD thesis, University of Sheffield, Sheffield

Popkin, P.R.W., Baker, P., Worley, F., Payne, S., Hammon, A. 2012: 'The Sheep Project (1): determining skeletal growth, timing of epiphyseal fusion and morphometric variation in unimproved Shetland sheep of known age, sex, castration status and nutrition' *Journal of Archaeological Science* 39: 1775–92

Portillo, M., Matthews, W. 2020: 'Investigating use of space and human-animal interactions in agricultural built environments: the geo-ethnoarchaeology of livestock dung' in A. Otto, M. Herles, K. Kaniuth (eds), *Proceedings of the 11th International Congress on the Archaeology of the Ancient Near East*. Wiesbaden, Harrassowitz Verlag: 497–508

Portillo, M., Valenzuela, S., Albert, R.M. 2012: 'Domestic patterns in the Numidian site of Althiburos (northern Tunisia): the results from a combined study of animal bones, dung and plant remains' *Quaternary International* 275: 84–96

Portillo, M., Kadowaki, S., Nishiaki, Y., Albert, R.M. 2014: 'Early Neolithic household behavior at Tell Seker al-Aheimar (Upper Khabur, Syria): a comparison to ethnoarchaeological study of phytoliths and dung spherulites' *Journal of Archaeological Science* 42: 107–18

Portillo, M., Belarte, M.C., Ramon, J., Kallala, N., Sanmartí, J., Albert, R.M. 2017: 'An ethnoarchaeological study of livestock dung fuels from cooking installations in northern Tunisia' *Quaternary International* 431: 133–44

Portillo, M., García-Suárez, A., Klimowicz, A., Barański, M.Z., Matthews, W. 2019: 'Animal penning and open area activity at Neolithic Çatalhöyük, Turkey' *Journal of Anthropological Archaeology* 56: 101106

Portillo, M., García-Suárez, A., Matthews W. 2020: 'Livestock faecal indicators for animal management, penning, foddering and dung use in early agricultural built environments in the Konya Plain, Central Anatolia' *Archaeological and Anthropological Sciences* 12: 40

Powell, M.L. 1985: 'The analysis of dental wear and caries for dietary reconstruction' in R.I. Gilbert, J.H. Mielke (eds), *The Analysis of Prehistoric Diets*. Orlando, Academic Press: 307–38

Preece, C., Livarda, A., Christin, P.-A., Wallace, M., Martin, G., Charles, M., Jones, G., Rees, M., Osborne, C.P. 2017: 'How did the domestication of Fertile Crescent grain crops increase their yields?' *Functional Ecology* 31: 387–97

Presslee, S., Wilson, J., Woolley, J., Best, J., Russell, D., Radini, A., Fischer, R., Kessler, B., Boano, R., Collins, M., Demarchi, B. 2017: 'The identification of archaeological eggshell using peptide markers' *STAR: Science & Technology of Archaeological Research* 3: 89–99

Price, T.D., Bar-Yosef, O. 2011: 'The origins of agriculture: new data, new ideas' *Current Anthropology* 52: S163–S174

Qazar, I.A.A. 2016: 'The relationship between aquatic macrophytes and some gastropoda (snails) in the lower reaches of Hammar marsh' *Mesopotamia Environmental Journal* 2.4: 23–32

R Core Team 2013: *R: A Language and Environment for Statistical Computing*. Vienna, Austria: R Foundation for Statistical Computing. ISBN 3-900051-07-0, URL http://www.R-project.org

— 2017: *R: A Language and Environment for Statistical Computing*. Vienna, R Foundation for Statistical Computing: https://www.R-project.org/

Rascovan, N., Sjögren, K.G., Kristiansen, K., Nielsen, R., Willerslev, E., Desnues, C., Rasmussen, S. 2019: 'Emergence and spread of basal lineages of *Yersinia pestis* during the Neolithic decline' *Cell* 176.1-2: 295–305

Rasmussen, P. 1993: 'Analysis of sheep/goat faeces from Egolzwil 3, Switzerland: evidence for branch and twig foddering of livestock in the Neolithic' *Journal of Archaeological Science* 20: 479–502

Rasmussen, S.O., Vinther, B.M., Clausen, H.B., Andersen, K.K. 2007: 'Early Holocene Greenland ice core chronology 2005 (GICC05) and 10 year means of oxygen isotope data from ice core DYE/3. *PANGAEA*' 10.1594/PANGAEA.586863

Redding, R.W. 1981: *Decision-Making in Subsistence Herding of Sheep and Goats in the Middle East*. PhD thesis, University of Michigan, Ann Arbor

— 1984: 'Theoretical determinants of a herder's decisions: modeling variation in the sheep/goat ratio' in J. Clutton-Brock, C. Grigson (eds), *Animals and Archaeology 3: Early Herders and their Flocks* (BAR International Series 202). Oxford, British Archaeological Reports: 223–41

Reese, D. 2005: 'The Çatalhöyük shells' in Hodder I. (ed.), *Inhabiting Çatalhöyük: Reports from the 1995–99 Seasons*. London, British Institute at Ankara; Cambridge, McDonald Institute for Archaeological Research: 123–28

Regan, R. 2014: 'The sequence of Buildings 75, 65, 56, 69, 44 and 10 and external Spaces 119, 129, 130, 144, 299, 314, 319, 329, 333, 339, 367, 371 and 427' in I. Hodder (ed.), *Çatalhöyük Excavations: The 2000–2008 Seasons*. London, British Institute at Ankara; Los Angeles, Cotsen Institute of Archaeology Press: 131–89

Reich, D., Green, R.E., Kircher, M., Krause, J., Patterson, N., Durand, E.Y., Viola, B., Briggs, A.W., Stenzel, U., Johnson, P.L.F., Maricic, T., Good, J.M., Marques-Bonet, T., Alkan, C., Fu, Q., Mallick, S., Li, H., Meyer, M., Eichler, E.E., Stoneking, M., Richards, M., Talamo, S., Shunkov, M.V., Derevianko, A.P., Hublin, J.-J., Kelso, J., Slatkin, M., Pääbo, S. 2010: 'Genetic history of an archaic hominin group from Denisova Cave in Siberia' *Nature* 468.7327: 1053–60

Reingruber, A. 2011: 'Early Neolithic settlement patterns and exchange networks in the Aegean' *Documenta Praehistorica* 38: 291

Relethford, J.H., Blangero, J. 1990: 'Detection of differential gene flow from patterns of quantitative variation' *Human Biology* 62.1: 5–25

Rennie, M.J., Jackson, L.J. 2005: 'The influence of habitat complexity on littoral invertebrate distributions: patterns differ in shallow prairie lakes with and without fish' *Canadian Journal of Fisheries and Aquatic Sciences* 62: 2088–99

Richards, M. 2003: 'The neolithic invasion of Europe' *Annual Review of Anthropology* 32: 135–62

Richards, M.P., Pearson, J.A. 2005: 'Stable-isotope evidence of diet at Çatalhöyük' in I. Hodder (ed.), *Inhabiting Çatalhöyük: Reports from the 1995–99 Seasons*. London, British Institute at Ankara; Cambridge, McDonald Institute for Archaeological Research: 313–22

Richards, M.P., Pearson, J.A., Russell, N., Molleson, T.I., Martin, L. 2003: 'Stable isotope evidence of diet at Neolithic Çatalhöyük, Turkey' *Journal of Archaeological Science* 30: 67–76

Riehl, S., Benz, M., Conard, N.J., Darabi, H., Deckers, K., Fazeli Nashli, H., Zeidi-Kulehparcheh, M. 2012: 'Plant use in three Pre-Pottery Neolithic sites of the northern and eastern Fertile Crescent: a preliminary report' *Vegetation History and Archaeobotany* 21: 95–106

Riehl, S., Zeidi, M., Conard, N.J. 2013: 'Emergence of agriculture in the foothills of the Zagros mountains of Iran' *Science* 341.6141: 65

Rindos, D. 1984: *The Origins of Agriculture*. London, Academic Press

Robb, J.E. 2013: 'Creating death: an archaeology of dying' in L. Nilsson Stutz, S. Tarlow (eds), *The Oxford Handbook of the Archaeology of Death and Burial*. Oxford, Oxford University Press: 441–57

Robb, J.E., Elster, E., Isetti, E., Knüsel, C.J., Tafuri, M.A., Traverso, A. 2015: 'Cleaning the dead: Neolithic ritual processing of human bone at Scaloria Cave, Italy' *Antiquity* 89.343: 39–54

Roberts, N. 1991: 'Late Quaternary geomorphological change and the origins of agriculture in south central Turkey' *Geoarchaeology* 6.1: 1–26

— 1995: 'Climatic forcing of alluvial-fan regimes during the Late Quaternary in the Konya Basin, south central Turkey' in J. Woodward, J. Lewin, M. Macklin (eds), *Mediterranean Quaternary River Environments*. Rotterdam, A.A. Balkema: 207–17

Roberts, N., Rosen, A. 2009: 'Diversity and complexity in early farming communities of Southwest Asia: new insights into the economic and environmental basis of Neolithic Çatalhöyük' *Current Anthropology* 50: 393–402

Roberts, N., Boyer, P., Parish, R. 1996: 'Preliminary results of geoarchaeological investigations at Çatalhöyük' in I. Hodder (ed.), *On the Surface: Çatalhöyük 1993–1995*. London, British Institute at Ankara; Cambridge, McDonald Institute for Archaeological Research: 19–40

Roberts, N., Black, S., Boyer, P., Eastwood, W.J., Griffiths, H.I., Lamb, H.F., Leng, M.J., Parish, R., Reed, J.M., Twigg, D., Yiğitbaşioğlu, H. 1999: 'Chronology and stratigraphy of Late Quaternary sediments in the Konya Basin, Turkey: results from the KOPAL Project' *Quaternary Science Reviews* 18: 611–30

Roberts, N., Reed, J., Leng, M.J., Kuzucuoğlu, C., Fontugne, M., Bertaux, J., Woldring, H., Bottema, S., Black, S., Hunt, E., Karabıyıkoğlu, M. 2001: 'The tempo of Holocene climatic change in the eastern Mediterranean region: new high-resolution crater-lake sediment data from Central Turkey' *Holocene* 11: 719–34

Roberts, N., Boyer, P., Merrick, J. 2007: 'The KOPAL on-site and off-site excavations and sampling' in I. Hodder (ed.), *Excavating Çatalhöyük: South, North and KOPAL Area Reports from the 1995–99 Seasons*. London, British Institute at Ankara; Cambridge, McDonald Institute for Archaeological Research: 553–72

Roberts, N., Jones, M.D., Benkaddour, A., Eastwood, W.J., Filippi, M.I., Frogley, M.R., Lamb, H.F., Leng, M., Reed, J.M., Stein, M., Stevens, L., Valero-Garces, B., Zanchetta, G. 2008: 'Stable isotope records of Late Quaternary climate and hydrology from Mediterranean lakes: the ISOMED synthesis' *Quaternary Science Reviews* 27: 2426–41

Roberts, N., Allcock, S.L., Arnaud, F., Dean, J.R., Eastwood, W.J., Jones, M.D., Leng, M.J., Metcalfe, S.E., Malet, E., Woodbridge, J., Yiğitbaşıoğlu, H. 2016: 'A tale of two lakes: a multi-proxy comparison of Late Glacial and Holocene environmental change in Cappadocia, Turkey' *Journal of Quaternary Science* 31.4: 348–62

Robins, R.P., Stock, E.C. 1990: 'The burning question: a study of molluscan remains from a midden on Moreton Island' *Tempus* 2: 80–100

Robinson, D., Handley, L.L., Scrimgeour, C.M., Gordon, D.C., Forster, B.P., Ellis, R.P. 2000: 'Using stable isotope natural abundances ($\delta^{15}N$ and $\delta^{13}C$) to integrate the stress responses of wild barley (*Hordeum spontaneum C. Koch*) genotypes' *Journal of Experimental Botany* 51: 41–50

Roffet-Salque, M., Marciniak, A., Valdes, P.L, Pawłowska, K., Pyzel, J., Czerniak, L., Krüger, M., Roberts, C.N., Pitter, S., Evershed, R.P. 2018: 'Evidence for the impact of the 8.2-kyBP climate event on Near Eastern early farmers' *Proceedings of the National Academy of Sciences* 115: 8705–09

Rohland, N., Harney, E., Mallick, S., Nordenfelt, S., Reich, D. 2015: 'Partial uracil – DNA – glycosylase treatment for screening of ancient DNA' *Philosophical Transactions of the Royal Society B: Biological Sciences* 370.1660: 20130624

Rollefson, G.O. 2001: 'The Neolithic period' in B. McDonald, R. Adams, P. Bienkowski (eds), *The Archaeology of Jordan*. Sheffield, Sheffield Academic Press: 67–105

— 2002: 'Ritual and social structure and Neolithic 'Ain Ghazal' in I. Kuijt (ed.), *Life in Neolithic Farming Communities: Social Organization, Identity, and Differentiation*. New York, Kluwer Academic: 165–90

Roodenberg, J., von As, A., Jacobs, L., Winjen, M.H. 2003: 'Early settlement in the plain of Yenişehir (NW Anatolia). The basal occupation layers at Menteşe' *Anatolica* 29: 17–59

Roseboom, T.J., van der Meulen, J.H.P., Ravelli, A.C.J. 2001: 'Effects of prenatal exposure to the Dutch Famine on adult disease in later life: an overview' *Molecular and Cellular Endocrinology* 185: 93-98

Rosen, A. 1992: 'Preliminary identification of silica skeletons from Near Eastern archaeological sites: an anatomical approach' in G.J. Rapp, S.C. Mulholland (eds), *Phytolith Systematics: Emerging Issues, Advances in Archaeological and Museum Science*. New York, Plenum Press: 129–47

— 1998: 'Archive report on phytolith studies at Çatalhöyük' *Çatalhöyük 1998 Archive Report*. http://www.catalhoyuk.com/archive_reports/1998

— 1999: 'Phytolith archive report for 1999 season at Çatalhöyük' *Çatalhöyük 1999 Archive Report*. http://www.catalhoyuk.com/archive_reports/1999

— 2000: 'Phytolith archive report for 2000 season at Çatalhöyük' *Çatalhöyük 2000 Archive Report*. http://www.catalhoyuk.com/archive_reports/2000

— 2005: 'Phytolith indicators of plant and land use at Çatalhöyük' in I. Hodder (ed.), *Inhabiting Çatalhöyük: Reports from the 1995–99 Seasons*. London: British Institute at Ankara; Cambridge, McDonald Institute for Archaeological Research: 203–12

Rosen, A.M., Roberts, N. 2005: 'The nature of Çatalhöyük: people and their changing environments on the Konya Plain' in I. Hodder (ed.), *Çatalhöyük Perspectives: Themes from the 1995–99 Seasons*. London: British Institute at Ankara; Cambridge, McDonald Institute for Archaeological Research: 39–53

Rössner, C., Deckers, K., Benz, M., Özkaya, V., Riehl, S. 2018: 'Subsistence strategies and vegetation development at Aceramic Neolithic Körtik Tepe, southeastern Anatolia, Turkey' *Vegetation History and Archaeobotany* 27: 15–29

Rothwell, I.W., Croudace, R.G, 2015: 'Micro-XRF studies of sediment cores: a perspective on capability and application in the environmental sciences' in R.G. Croudace, I.W. Rothwell (eds), *Micro-XRF Studies of Sediment Cores*. Dordrecht, Springer: 1–21

Rowe, N.H. 1982: 'Dental caries' in P.F. Steel (ed.), *Dimensions of Dental Hygiene* (3rd edition). Philadelphia, Lea & Febiger: 209–37

Rowley-Conwy, P., Albarella, U., Dobney, K. 2012: 'Distinguishing wild boar from domestic pigs in prehistory: a review of approaches and recent results' *Journal of World Prehistory* 25: 1–44

Rozas, V., 2003: 'Regeneration patterns, dendroecology, and forest-use history in an old-growth beech–oak lowland forest in Northern Spain' *Forest Ecology and Management* 182: 175–94

Ruff, C. 2000: 'Body size, body shape, and long bone strength in modern humans' *Journal of Human Evolution* 38: 269–90

— 2002: 'Variation in human body size and shape' *Annual Review of Anthropology* 31: 211–32

— 2007: 'Body size prediction from juvenile skeletal remains' *American Journal of Physical Anthropology* 133: 698–716

— 2008: 'Biomechanical analyses of archaeological human skeletons' in M.A. Katzenberg, S.R. Saunders (eds), *Biological Anthropology of the Human Skeleton*. New York, Alan R. Liss: 183–206

— (ed.) 2018: *Skeletal Variaton and Adaptaton in Europeans: Upper Paleolithic to the Twentieth Century*. Hoboken, Wiley-Blackwell

Ruff, C., Walker, A., Trinkaus, E. 1994: 'Postcranial robusticity in Homo. III: Ontogeny' *American Journal of Physical Anthropology* 93.1: 35–54

Ruff, C.B., Holt, B.M., Niskanen, M., Sladek, V., Berner, M., Garofalo, E., Garvin, H.M., Hora, M., Maijanen, H., Niinimaki, S., Salo, K., Schuplerova, E., Tompkins, D. 2012: 'Stature and body mass estimation from skeletal remains in the European Holocene' *American Journal of Physical Anthropology* 148: 601–17

Ruff, C.B., Burgess, M.L., Squyres, N., Junno J.-A., Trinkaus, E. 2018: 'Lower limb articular scaling and body mass estimation in Pliocene and Pleistocene hominins' *Journal of Human Evolution* 115: 85–111

Russell, N. 1998: 'Cattle as wealth in Neolithic Europe: where's the beef?' in D. Bailey, S. Mills (eds), *The Archaeology of Value: Essays on Prestige and the Processes of Valuation* (BAR International Series 730). Oxford, Archaeopress: 42–54

— 2005: 'The Çatalhöyük worked bone' in I. Hodder (ed.), *Changing Materialities at Çatalhöyük: Reports from the 1995–99 Seasons*. London, British Institute at Ankara; Cambridge, McDonald Institute for Archaeological Research: 339–68

— 2012: 'Mammals from the BACH Area at Çatalhöyük' in R.E. Tringham, M. Stevanović (eds), *Last House on the Hill: BACH Area Reports from Çatalhöyük, Turkey*. Los Angeles, Cotsen Institute of Archaeology Press: 225–42

— 2019a: 'Feathers and talons: birds at Neolithic Çatalhöyük, Turkey' *Archaeological and Anthropological Sciences*. doi.org/10.1007/s12520-018-0681-z

— 2019b: 'Spirit birds at Neolithic Çatalhöyük' *Environmental Archaeology*. doi.org/10.1080/14614103.2017.1422685

Russell, N., Griffitts, J.L. 2013: 'Çatalhöyük worked bone: South and 4040 Areas' in I. Hodder (ed.) *Substantive Technologies at Çatalhöyük: Reports from the 2000–2008 Seasons*. London, British Institute at Ankara; Los Angeles, Cotsen Institute of Archaeology Press: 277–306

Russell, N., Martin, L. 2005: 'The Çatalhöyük mammal remains' in I. Hodder (ed.), *Inhabiting Çatalhöyük: Reports from the 1995–99 Seasons*. London, British Institute at Ankara; Cambridge, McDonald Institute for Archaeological Research: 33–98

Russell, N., McGowan, K.J. 2003: 'Dance of the cranes: crane symbolism at Çatalhöyük and beyond' *Antiquity* 77: 445–55

— 2005: 'Çatalhöyük bird bones' in I. Hodder (ed.), *Inhabiting Çatalhöyük: Reports from the 1995–99 Seasons*. London, British Institute at Ankara; Cambridge, McDonald Institute for Archaeological Research: 99–110

Russell, N., Meece, S. 2006: 'Animal representations and animal remains at Çatalhöyük' in I. Hodder (ed.), *Çatalhöyük Perspectives: Reports from the 1995–99 Seasons*. London, British Institute at Ankara; Cambridge, McDonald Institute for Archaeological Research: 209–30

Russell, N., Twiss, K.C. 2017: 'Digesting the data: dogs as taphonomic agents at Neolithic Çatalhöyük, Turkey' in M. Mashkour, M. Beech (eds), *Archaeozoology of the Near East IX: Proceedings of the Ninth International Symposium on the Archaeozoology of Southwestern Asia and Adjacent Areas*. Oxford, Oxbow: 59–73

Russell, N., Martin, L., Buitenhuis, H. 2005: 'Cattle domestication at Çatalhöyük revisited' *Current Anthropology* 46: S101–S108

Russell, N., Martin, L., Twiss, K.C. 2009: 'Building memories: commemorative deposits at Çatalhöyük' *Anthropozoologica* 44: 103–28

Russell, N., Twiss, K.C., Orton, D.C., Demirergi, G.A. 2013: 'More on the Çatalhöyük mammal remains' in I. Hodder (ed.), *Humans and Landscapes of Çatalhöyük: Reports from the 2000–2008 Seasons*. London, British Institute at Ankara; Los Angeles, Cotsen Institute of Archaeology Press: 213–58

Russell, N., Twiss, K.C., Orton, D.C., Demirergi, G.A. 2014a: 'Changing animal use at Neolithic Çatalhöyük, Turkey' in B. De Cupere, V. Linseele, S. Hamilton-Dyer (eds), *Archaeozoology of the Near East X: Proceedings of the Ninth International Symposium on the Archaeozoology of Southwestern Asia and Adjacent Areas*. Leuven, Peeters: 45–68

Russell, N., Wright, K.I., Carter, T., Ketchum, S., Ryan, P., Yalman, E.N., Regan, R., Stevanović, M., Milić, M. 2014b: 'Bringing down the house: house-closing deposits at Neolithic Çatalhöyük, Turkey' in I. Hodder (ed.), *Integrating Çatalhöyük: Themes from the 2000–2008 Seasons*. London, British Institute at Ankara; Los Angeles, Cotsen Institute of Archaeology Press: 109–21

Ryan, P. 2008: 'The phytoliths' *Çatalhöyük 2008 Archive Report*. http://www.catalhoyuk.com/archive_reports/2008

— 2009: *Diversity of Plant and Land Use During the Near Eastern Neolithic: Phytolith Perspectives from Çatalhöyük*. PhD thesis, University College London, London

— 2011 'Plants as material culture in the Near Eastern Neolithic: perspectives from the silica skeleton artifactual remains at Çatalhöyük' *Journal of Anthropological Archaeology* 30.3: 292–305

— 2013: 'Plant exploitation from household and landscape perspectives: the phytolith evidence' in I. Hodder (ed.), *Humans and Landscapes of Çatalhöyük: Reports from the 2000–2008 Seasons*. London, British Institute at Ankara; Los Angeles, Cotsen Institute of Archaeology Press: 163–90

Ryan, P., Rosen, A.M. 2006: 'Phytolith archive report for 2006 season at Çatalhöyük' *Çatalhöyük 2006 Archive Report*. http://www.catalhoyuk.com/archive_reports/2006

Sadarangani, F. 2014: 'The sequence of Buildings 53 and 42 and external spaces 259, 260 and 261' in I. Hodder (ed.), *Çatalhöyük Excavations: The 2000–2008 Seasons*. London, British Institute at Ankara; Los Angeles, Cotsen Institute of Archaeology Press: 191–219

Sadras, V.O. 2007: 'Evolutionary aspects of the trade-off between seed size and number in crops' *Field Crops Research* 100: 125–38

Sadvari, J.W., Charles, M., Ruff, C.B., Carter, T., Vasić, M., Larsen, C.S., Bar-Yosef Mayer, D. E., Doherty, C. 2015a: 'The people and their landscape(s): changing mobility patterns at Neolithic Çatalhöyük' in I. Hodder, A. Marciniak (eds), *Assembling Çatalhöyük*. Leeds, Maney Publishing: 167–77

Sadvari, J., Tsoraki, C., Dogiama, L., Knüsel, C.J. 2015b: 'Reading the bones, reading the stones' in I. Hodder, A. Marciniak (eds), *Assembling Çatalhöyük*. Leeds: Maney Publishing: 59–74

Sahlins, M. 2013: 'On the culture of material value and the cosmography of riches' *HAU: Journal of Ethnographic Theory* 3.2: 161–95

Salvagno, L., Albarella, U. 2017: 'A morphometric system to distinguish sheep and goat postcranial bones' *PLoS One* 12: 37

Sánchez-Quinto, F., Malmström, H., Fraser, M., Girdland-Flink, L., Svensson, E.M., Simões, L.G., George, R., Hollfelder, N., Burenhult, G., Noble, G., Britton, K., Talamo, S., Curtis, N., Brzobohata, H., Sumberova, R., Götherström, A., Storå, J., Jakobsson, M. 2019: 'Megalithic tombs in western and northern Neolithic Europe were linked to a kindred society' *Proceedings of the National Academy of Sciences of the United States of America* 116.19: 9469–74

Santana, J., Velasco, J, Ibáñez, J.J., Braemer, F. 2012: 'Crania with mutilated facial skeletons: a new ritual treatment in an Early Pre-Pottery Neolithic B cranial cache at Tell Qarassa North (South Syria)' *American Journal of Physical Anthropology* 149: 205–16

Sass-Klaassen, U., Leuschner, H.H., Buerkert, A., Helle, G. 2008: 'Tree-ring analysis of *Juniperus excelsa* from the northern Oman Mountains' in D. Elferts, G. Brumelis, H. Gärtner, G. Helle, G. Schleser (eds), *TRACE - Tree Rings in Archaeology, Climatology and Ecology 6: Proceedings of the DENDROSYMPOSIUM 2007*. Potsdam, GeoForschungsZentrum Potsdam: 83–90

Savage, S.H. 2000: 'The status of women in Predynastic Egypt as revealed through mortuary analysis' in A.E. Rautman (ed.), *Reading the Body: Representations and Remains in the Archaeological Record*. Philadelphia, University of Pennsylvania Press: 77–94

Savard, M., Nesbitt, M., Jones, M.K. 2006: 'The role of wild grasses in subsistence and sedentism: new evidence from the northern Fertile Crescent' *World Archaeology* 38: 179–96

Saxe, A.A. 1971: *Social Dimensions of Mortuary Practices in a Mesolithic Population from Wadi Haifa, Sudan*. PhD thesis, University of Michigan, Ann Arbor

Schell, L.M. 1995: 'Human biological adaptability with special emphasis on plasticity: history, development and problems for future research' in C.G.N. Mascie-Taylor, B. Bogin (eds), *Human Variability and Plasticity*. Cambridge, Cambridge University Press: 213–37

Schmandt-Besserat, D. 2013: 'The plastered skulls' in D. Schmandt-Besserat (ed.), *Symbols at 'Ain Ghazal* ('Ain Ghazal Excavation Reports 3). Berlin, ex-oriente: 213–43

Schotsmans, E.M.J., Haddow, S.D., Pilloud, M.A., Milella, M., Glencross, B., Betz, B.J., Knüsel, C.J. 2017: 'Manipulation of the dead: exploring delayed burial practices at Neolithic Çatalhöyük' *American Journal of Physical Anthropology* 162.s64: 249–50

Schour, I., Massler, M. 1941: 'The development of the human dentition' *Journal of the American Dental Association* 28: 1153–60

Schulting, R.J. 2012: 'Skeletal evidence for interpersonal violence: beyond mortuary monuments in southern Britain' in R.J. Schulting, L. Fibiger (eds), *Sticks, Stones, and Broken Bones: Neolithic Violence in a European Perspective*. Oxford, Oxford University Press: 223–48

Schwarcz, H.P., Dupras, T.L., Fairgrieve, S.I. 1999: '^{15}N Enrichment in the Sahara: in search of a global relationship' *Journal of Archaeological Science* 26: 629–36

Schweich, M., Knüsel, C.J. 2003: 'Bio-cultural effects in medieval populations' *Economics and Human Biology* 1.3: 367–77

Schweingruber, F.H. 2007: *Wood Structure and Environment*. Berlin and Heidelberg, Springer-Verlag

Schweingruber, F., Eckstein, D., Serre-Bachet, F., Bräker, O. 1990: 'Identification, presentation and interpretation of event years and pointer years in dendrochronology' *Dendrochronologia* 8: 9–37

Scott, G.R., Irish, J.D. 2017: *Human Crown and Root Morphology: The Arizona State University Dental Anthropology System*. Cambridge, Cambridge University Press

Sealy, J., van der Merwe, N.J., Lee-Thorp, J.A., Lanham, J.L. 1987: 'Nitrogen isotope ecology in southern Africa: implications for environmental and dietary tracing' *Geochimica et Cosmochimica Acta* 51: 2707–17

Sellier, P. 2016: 'Différent types de Sépulture ou différentes étapes d'une même séquence funéraire? Un exemple démonstrative de châine opératoire mortuaire chez les anciens Marquisiens' *Bulletins et Mémoires de la Société d'Anthropologie de Paris* 28: 45–52

Selye, H. 1936: *The Stress of Life*. New York, McGraw-Hill

Şereflişan H., Gökçe, M.A. 2016: 'Growth performance of the freshwater mussel, *Unio terminalis delicates* (Lea, 1863) (Mollusca: Bivalvia: Unionidae) in the Gölbaşı Lake, Turkey' *Pakistan Journal of Zoology* 48.4: 1109–15

Şereflişan H., Yildirim M.Z., Şereflişan M. 2009: 'The gastropod fauna and their abundance, and some physico-chemical parameters of Lake Gölbaşı (Hatay, Turkey)' *Turkish Journal of Zoology* 33.3: 287–96

Serre, D., Langaney, A., Chech, M., Teschler-Nicola, M., Paunovic, M., Mennecier, P., Hofreiter, M., Possnert, G., Pääbo, S. 2004: 'No evidence of Neandertal mtDNA contribution to Early Modern Humans' *PLoS Biology* 2.3: e57

Shahack-Gross, R. 2011: 'Herbivorous livestock dung: formation, taphonomy, methods for identification, and archaeological significance' *Journal of Archaeological Science* 38: 205–18

Shaikh, B., Kanakaeaj, P., Karthika, K. 2016: 'Scope and effect of retting treatment on tensile characteristics of natural fiber extracted from *Phragmites australis* (Naanal)' *Journal of Textile and Apparel, Technology and Management* 10.2: 1–7

Shapland, F., Lewis, M. 2013: 'Brief communication: a proposed osteological method for the estimation of pubertal stage in human skeletal remains' *American Journal of Physical Anthropology* 151: 302–10

Shaw, C.N., Stock, J.T. 2011: 'The influence of body proportions on femoral and tibial midshaft shape in hunter-gatherers' *American Journal of Physical Anthropology* 144: 22–29

Shennan, S. 2018: *The First Farmers of Europe: An Evolutionary Perspective*. Cambridge, Cambridge University Press

Shillito, L.-M. 2011: 'Taphonomic observations of archaeological wheat phytoliths from Neolithic Turkey and the use of conjoined phytolith size as an indicator of water availability' *Archaeometry* 53.3: 631–41

— 2017a: 'Multivocality and multiproxy approaches to the use of space: lessons from 25 years of research at Çatalhöyük' *World Archaeology* 49.2: 237–59

— 2017b: 'Biofuels and respiratory health: the potentials of the archaeological record at Çatalhöyük' *Çatalhöyük 2017 Archive Report*. http://www.catalhoyuk.com/archive_reports/2017

Shillito, L.-M., Matthews, W. 2013: 'Geoarchaeological investigations of midden-formation processes in the Early to Late Ceramic Neolithic levels at Çatalhöyük, Turkey ca. 8550–8370 cal BP' *Geoarchaeology: An International Journal* 28: 25–49

Shillito, L.-M., Ryan, P. 2013: 'Surfaces and streets: phytoliths, micromorphology and changing use of space at Neolithic Çatalhöyük (Turkey)' *Antiquity* 87: 684–700

Shillito, L.-M., Almond, M.J., Nicholson, J., Pantos, M., Matthews, W. 2009: 'Rapid characterisation of archaeological midden components using FT-IR spectroscopy, SEM-EDX and micro-XRD' *Spectrochimica Acta Part A: Molecular and Biomolecular Spectroscopy* 73.1: 133–39

Shillito, L.-M., Matthews, W., Almond, M.J., Bull, I.D. 2011a: 'The microstratigraphy of middens: capturing daily routine in rubbish at Neolithic Çatalhöyük, Turkey' *Antiquity* 85: 1024–38

Shillito, L-M., Bull, I.D., Matthews, W., Almond, M.J., Williams, J.M., Evershed, R.P. 2011b: 'Biomolecular and micromorphological analysis of suspected faecal deposits at Neolithic Çatalhöyük, Turkey' *Journal of Archaeological Science* 38: 1869–77

Shillito, L.-M., Matthews, W., Almond, M. 2013a: 'Ecology, diet and discard practices: new interdisciplinary approaches to the study of middens – integrating micromorphological, phytolith and geochemical analyses' in I. Hodder (ed.), *Humans and Landscapes of Çatalhöyük: Reports from the 2000–2008 Seasons*. London, British Institute at Ankara; Los Angeles: Cotsen Institute of Archaeology Press: 65–76

Shillito, L.-M., Matthews, W., Bull, I.D., Almond, M.J., Williams, J.M., Evershed, R.P. 2013b: 'Integrated geochemical and microscopic analysis of human coprolites, animal dung and organic remains in burials' in I. Hodder (ed.), *Humans and Landscapes of Çatalhöyük: Reports from the 2000–2008 Seasons*. London, British Institute at Ankara; Los Angeles, Cotsen Institute of Archaeology Press: 75–78

Shillito, L.-M., Haddow, S., Issavi, J., García-Suárez, A., Mackay, H., Namdeo, A. Forthcoming: 'Waste disposal, pollution and public health' in I. Hodder (ed.), *Synthesis of Work on the East Mound 2009–2017*. London, British Institute at Ankara

Shimada, I., Shinoda, K.-I., Farnum, J., Corruccini, R.S., Watanabe, H. 2004: 'An integrated analysis of pre-Hispanic mortuary practices: a Middle Sican case study' *Current Anthropology* 45.3: 369–402

Shinoda, K., Matsumura, H., Nishimoto, T. 1998: 'Genetical and morphological analysis on kinship of the Nakazuma Jomon people using mitochondrial DNA and tooth crown measurements' *Zoo-archaeology* 11: 1–21

Sichert, B., Rentzel, P., Demarchi, B., Best, J., Negri, A., Deschler-Erb, S. 2019: 'Incubated eggs in a Roman burial? A preliminary investigation on how to distinguish between the effects of incubation and taphonomy on avian eggshell from archaeological sites' *Journal of Archaeological Science: Reports* 26: 101845

Sidell, E.J., Scudder C. 2005: 'The eggshell from Çatalhöyük: a pilot study' in I. Hodder (ed.), *Inhabiting Çatalhöyük: Reports from the 1995–99 Seasons*. London, British Institute at Ankara; Cambridge, McDonald Institute for Archaeological Research: 117–21

Sidell, J. 1993: *A Methodology for the Identification of Archaeological Eggshell*. Philadelphia, The University Museum of Archaeology and Anthropology

Sikora, M., Seguin-Orlando, A., Sousa, V.C., Albrechtsen, A., Korneliussen, T., Ko, A., Rasmussen, S., Dupanloup, I., Nigst, P.R., Bosch, M.D., Renaud, G., Allentoft, M.E., Margaryan, A., Vasilyev, S.V., Veselovskaya, E.V., Borutskaya, S.B., Deviese, T., Comeskey, D., Higham, T., Manica, A., Foley, R., Meltzer, D.J., Nielsen, R., Excoffier, L., Mirazon Lahr, M., Orlando, L., Willerslev, E. 2017: 'Ancient genomes show social and reproductive behavior of early Upper Palaeolithic foragers' *Science* 358.6363: 659–62

Silver, I.A. 1969: 'The ageing of domestic animals' in D.R. Brothwell, E.S. Higgs (eds), *Science in Archaeology: A Survey of Progress and Research*. New York, Praeger Publishers: 283–302

Skinner, M.F., Goodman, A.H. 1992: 'Anthropological uses of developmental defects of enamel' in S. Saunders, M.A. Katzenberg (eds), *Skeletal Biology of Past Peoples: Research Methods*. New York, Wiley-Liss: 153–74

Skoglund, P., Mathieson, I. 2018: 'Ancient genomics of modern humans: the first decade' *Annual Review of Genomics and Human Genetics* 19.1: 381–404

Skoglund, P., Malmstrom, H., Raghavan, M., Stora, J., Hall, P., Willerslev, E., Gilbert, M.T.P., Götherström, A., Jakobsson, M. 2012: 'Origins and genetic legacy of Neolithic farmers and hunter-gatherers in Europe' *Science* 336.6080: 466–69

Skoglund, P., Northoff, B.H., Shunkov, M.V, Derevianko, A.P., Pääbo, S., Krause, J., Jakobsson, M. 2014: 'Separating endogenous ancient DNA from modern day contamination in a Siberian Neandertal' *Proceedings of the National Academy of Sciences of the United States of America* 111.6: 2229–34

Slon, V., Sarig, R., Hershkovitz, I., Khalaily, H., Milevski, I. 2014: 'The plastered skulls from the Pre-Pottery Neolithic B site of Yiftahel (Israel) – a computed tomography-based analysis' *PLoS One* 9.2: e89242

Smith, B.D. 2007: 'Niche construction and the behavioral context of plant and animal domestication' *Evolutionary Anthropology* 16: 188–99

Smith, B.H. 1984: 'Patterns of molar wear in hunter-gatherers and agriculturalists' *American Journal of Physical Anthropology* 63.1: 39–56

— 1991: 'Standards of human tooth formation and dental age assessment' in M.A. Kelly, C.S. Larsen (eds), *Advances in Dental Anthropology*. New York, Wiley-Liss: 143–68

Smith, R.N. 1969: 'Fusion of ossification centres in the cat' *Journal of Small Animal Practice* 10: 523–30

Snow, D.W., Perrins, C.M. 1998: *The Birds of the Western Palearctic. Concise Edition*. Oxford, Oxford University Press

Sofaer Derevenski, J. 1997: 'Engendering children, engendering archaeology' in J. Moore, E. Scott (eds), *Invisible People and Processes: Writing Gender and Childhood into European Archaeology*. London and New York, Leicester University Press, 192–200

Sofaer, J. 2011: 'Towards a social bioarchaeology of age' in S.C. Agarwal, B.A. Glencross (eds), *Social Bioarchaeology*. Chichester, Wiley-Blackwell: 285–311

Sofaer, J.A., Bailit, H.L., MacLean, C.J. 1971: 'A developmental basis for differential tooth reduction during hominid evolution' *Evolution: International Journal of Organic Evolution* 25: 509–17

Sofaer, J.A., Niswander, J., Maclean, C., Workman, P. 1972: 'Population studies on southwestern Indian tribes. V. tooth morphology as an indicator of biological distance' *American Journal of Physical Anthropology* 37: 357–66

Sørensen, M.L.S. 1997: 'Reading dress: the construction of social categories and identities in Bronze Age Europe' *Journal of European Archaeology* 5.1: 93–114

— 2000: *Gender Archaeology*. Cambridge, Polity Press

Soylu, E. 2014: 'Metazoan parasites of fish species from Lake Gala (Edirne, Turkey)' *Ege Journal of Fish Aquaculture* 31.4: 187–93

Spradley, M.K, Hamilton, M., Giordano, A. 2012: 'Spatial patterning of vulture scavenged human remains' *Forensic Science International* 219: 57–63

SPSS Inc. 2016: *IBM SPSS Statistics for Windows, Version 24.0*. Armonk, NY, IBM Corporation

Stan Development Team 2017: *Stan Modeling Language: User's Guide and Reference Manual*. Stan Version 2.17.0

— 2018: *RStan: The R Interface to Stan*. R package version 2.17.3: http://mc-stan.org

Steckel, R.H. 2004: 'New light on the "Dark Ages": the remarkably tall stature of northern European men during the Medieval Era' *Social Science History* 28.2: 211–29

Steckel, R.H., Larsen, C.S., Sciulli, P.W., Walker, P.L. 2006: 'Data collection codebook' *The Global History of Health Project* https://www.uv.es/paleolab/Codebook-08-25-051%5B1%5D.pdf

Stewart, G.R., Turnbull, M.H., Schmidt, S., Erskine, P.D. 1995: '^{13}C Natural abundance in plant communities along a rainfall gradient: a biological integrator of water availability' *Australian Journal of Plant Physiology* 22: 51–55

Stewart, G.S., Smith, C.P. 2005: 'Urea nitrogen salvage mechanisms and their relevance to ruminants, non-ruminants and man' *Nutrition Research Reviews* 18: 49–62

Stiner, M.C., Buitenhuis, H., Duru, G., Kuhn, S.L., Mentzer, S.M., Munro, N.D., Pöllath, N., Quade, J., Tsartsidou, G., Özbaşaran, M. 2014: 'A forager-herder trade-off, from broad-spectrum hunting to sheep management at Aşıklı Höyük, Turkey' *Proceedings of the National Academy of Sciences* 111.23: 8404–09

Stinson, S. 1985: 'Sex differences in environmental sensitivity during growth and development' *Yearbook of Physical Anthropology* 28: 123–47

Stojanowski, C.M., Hubbard, A.R. 2017: 'Sensitivity of dental phenotypic data for the identification of biological relatives' *International Journal of Osteoarchaeology* 27: 813–27

Stone, L. 2014: *Kinship and Gender: An Introduction* (5th edition). New York, Westview Press

Stoneking, M., Krause, J. 2011: 'Learning about human population history from ancient and modern genomes' *Nature Reviews Genetics* 12.9: 603–14

Stoops, G. 2003: *Guidelines for Analysis and Description of Soil and Regolith Thin Sections*. Madison, Soil Science Society of America

Stroud, E. 2016. *An Archaeobotanical Investigation into the Chalcolithic Economy and Social Organisation of Central Anatolia*. PhD thesis, University of Oxford, Oxford.

Stroud, E. submitted. 'Macrobotanical remains' in P. Biehl, J. Anvari, E. Rosenstock (eds), *The End of Çatalhöyük: The West Mound Excavations*. Los Angeles, Cotsen Institute Press.

Stroud, E., Bogaard, A., Charles, M. submitted. 'A stable isotope and functional weed ecology investigation into Chalcolithic cultivation practices in Central Anatolia: Çatalhöyük, Çamlıbel Tarlası and Kuruçay' *Journal of Archaeological Science: Reports*

Stuiver, M., Grootes, P.M., Braziunas, T.F. 1995: 'The GISP2 $\delta^{18}O$ climate record of the past 16,500 years and the role of the sun, ocean and volcanoes' *Quaternary Research* 44: 341–54

Styring, A.K., Ater, M., Hmimsa, Y., Fraser, R., Miller, H., Neef, R., Pearson, J.A., Bogaard, A. 2016: 'Disentangling the effect of farming practice from aridity on crop stable isotope values: a present-day model from Morocco and its application to early farming sites in the eastern Mediterranean' *The Anthropocene Review* 3: 2–22

Styring, A.K., Charles, M., Fantone, F., Hald, M.M., McMahon, A., Meadow, R.H., Nicholls, G.K., Patel, A.K., Pitre, M.C., Smith, A., Sołtysiak, A., Stein, G., Weber, J.A., Weiss, H., Bogaard, A. 2017: 'Isotope evidence for agricultural extensification reveals how the world's first cities were fed' *Nature Plants* 3: 17076

Swadling, P. 1976: 'Changes induced by human exploitation in prehistoric shellfish populations' *Mankind* 10.3: 156–62

— 1977: 'Central province shellfish resources and their utilisation in the prehistoric past of Papua New Guinea' *The Veliger* 19.3: 293–302

Sykes, N. 2014: *Beastly Questions: Animal Answers to Archaeological Issues*. London, Bloomsbury

Sylvester, A., Garofalo, E., Ruff, C. 2010: 'Technical note: an R program for automating bone cross-section reconstruction' *American Journal of Physical Anthropology* 142: 665–69

Tainter, J.A. 1978: 'Mortuary practices and the study of prehistoric social systems' *Advances in Archaeological Method and Theory* 1: 105–41

Talalay, L.E. 2004: 'Heady business: skulls, heads and decapitation in Neolithic Anatolia and Greece' *Journal of Mediterranean Archaeology* 17.2: 139–63

Tanno, K.-I., Willcox, G. 2006: 'How fast was wild wheat domesticated?' *Science* 311.5769: 1886

— 2012: 'Distinguishing wild and domestic wheat and barley spikelets from early Holocene sites in the Near East' *Vegetation History and Archaeobotany* 21: 107–15

Tchernov, E. 1991: 'Of mice and men: biological markers for long-term sedentism; a reply' *Paléorient* 17.1: 153–60

Terry, R.C. 2007: 'Inferring predator identity from skeletal breakage of small-mammal prey remains' *Evolutionary Ecology Research* 9: 199–219

Testart, A. 2008: 'Des crânes et des vautour ou la guerre oubliée' *Paléorient* 34.1: 35–58

Thiébault, S., 2006: 'Wood-anatomical evidence of pollarding in ring porous species: a study to develop?' in A. Dufraisse (ed.), *Charcoal Analysis: New Analytical Tools and Methods for Archaeology* (BAR International Series 1483). Oxford, Archaeopress: 95–102

Thomas, J. 1991: *Rethinking the Neolithic*. Cambridge, Cambridge University Press

Tibbetts, B. 2017: 'Perinatal death and cultural buffering in a Neolithic community' in E. Murphy, M. Le Roy (eds), *Children, Death and Burial: Archaeological Discourses*. Oxford, Oxbow Books: 35–42

Tieszen, L.L. 1991: 'Natural variations in the carbon isotope values of plants: implications for archaeology, ecology, and palaeoecology' *Journal of Archaeological Science* 18: 227–48

Tieszen, L.L., Fagre, T. 1993: 'Carbon isotopic variability in modern and archaeological maize' *Journal of Archaeological Science* 20: 25–40

Tirlea, D., Beaudoin, A.B., Vinebrooke, R.D. 2014: 'Freeze-dried is as good as frozen: evaluation of differential preservation of pollen grains in stored lake sediments' *Review of Palaeobotany and Palynology* 215: 46–56

Tomek, T., Bocheński, T. 2000: *The Comparative Osteology of European Corvids (Aves: Corvidae), With a Key to the Identification of Their Skeletal Elements*. Krakow, Institute of Systematics and Evolution of Animals

Townsend, G.C., Hughes, T., Luciano, M., Bockmann, M. and Brook, A. 2009: 'Genetic and environmental influences on human dental variation: a critical evaluation of studies involving twins' *Archives of Oral Biology* 54 (Supplement 1): S45–S51

Townsend, G.C., Bockmann, M., Hughes, T., Brook, A. 2012: 'Genetic, environmental and epigenetic influences on variation in human tooth number, size and shape' *Odontology* 100.1: 1–9

Treherne, P. 1995: 'The warrior's beauty: the masculine body and self-identity in Bronze-Age Europe' *Journal of European Archaeology* 3.1: 105–44

Tringham, R., Stevanović, M. (eds) 2012: *House Lives: Building, Inhabiting, Excavating a House at Çatalhöyük, Turkey. Reports from the BACH Area, Çatalhöyük, 1997–2003*. Los Angeles, Cotsen Institute of Archaeology Press

Tsartsidou, G., Lev-Yadun, S., Albert, R.-M., Miller-Rosen, A., Efstratiou, N., Weiner, S. 2007: 'The phytolith archaeological record: strengths and weaknesses evaluated based on a quantitative modern reference collection from Greece' *Journal of Archaeological Science* 34: 1262–75

Tsartsidou, G., Lev-Yadun, S., Efstratiou, N., Weiner, S. 2008: 'Ethnoarchaeological study of phytolith assemblages from an agro-pastoral village in Northern Greece (Sarakini): development and application of a Phytolith Difference Index' *Journal of Archaeological Science* 35: 600–13

Tsartsidou G., Lev-Yadun., S, Albert., R.M., Rosen., A.M., Efstratiou, N., Weiner, S. 2011: 'The phytolith archaeological record: strengths and weaknesses evaluated based on a quantitative modern reference collection from Greece' *International Journal of Innovations in Biological and Chemical Sciences* 1: 22–27

Tsoraki, C. 2016: 'Ground stone technologies' *Çatalhöyük 2016 Archive Report*. http://www.catalhoyuk.com/archive_reports/2016

Tung, B. 2013: 'Building with mud: an analysis of architectural materials at Çatalhöyük' in I. Hodder (ed.) *Substantive Technologies at Çatalhöyük: Reports from the 2000–2008 Seasons*. London, British Institute at Ankara; Los Angeles, Cotsen Institute of Archaeology Press: 67–80

Tung, T.A. 2013: 'Gender-based violence in the Wari and Post-Wari era of the Andes' in C.J. Knüsel, M.J. Smith (eds), *The Routledge Handbook of the Bioarchaeology of Human Conflict*. London, Routledge: 333–54

Turner, C.G., Nichol, C.R., Scott, G.R. 1991. 'Scoring procedures for key morphological traits of the permanent dentition: the Arizona State University dental anthropology system' in M.A Kelley, C.S. Larsen (eds), *Advances in Dental Anthropology*. New York: Wiley-Liss: 13–31

Turner-Walker, G. 2019: 'Light at the end of the tunnels? The origins of microbial bioerosion in mineralised collagen' *Palaeogeography, Palaeoclimatology, Palaeoecology* 529: 24–38

Twiss, K.C. 2006: 'A modified boar skull from Çatalhöyük' *Bulletin of the American Schools of Oriental Research*: 1–12

— 2012: 'The complexities of home cooking: public feasts and private meals inside the Çatalhöyük house' in S. Pollock (ed.), *Between Feasts and Daily Meals: Toward an Archaeology of Commensal Spaces* (eTopoi: Journal for Ancient Studies Special Volume 2): 53–73

Twiss, K.C., Russell, N. 2010: 'Taking the bull by the horns: ideology, masculinity, and cattle horns at Çatalhöyük' *Paléorient* 35.2: 17–29

Twiss, K.C., Bogaard, A., Bogdan, D., Carter, T., Charles, M.P., Farid, S., Russell, N., Stevanović, M., Yalman, E.N., Yeomans, L. 2008: 'Arson or accident? The burning of a Neolithic house at Çatalhöyük' *Journal of Field Archaeology* 33: 41–57

Twiss, K.C., Bogaard, A., Charles, M., Henecke, J., Russell, N., Martin, L., Jones, G. 2009: 'Plants and animals together: interpreting organic remains from Building 52 at Çatalhöyük' *Current Anthropology* 50.6: 885–95